Aerosols

Science, Technology, and Industrial Applications
of Airborne Particles

Aerosols

Science, Technology, and Industrial Applications of Airborne Particles

Proceedings of the First International Aerosol Conference held
September 17–21, 1984, Minneapolis, Minnesota, U.S.A.

Editors:

Benjamin Y.H. Liu
Particle Technology Laboratory, University of Minnesota,
Minneapolis, Minnesota, U.S.A.

David Y.H. Pui
Particle Technology Laboratory, University of Minnesota,
Minneapolis, Minnesota, U.S.A.

Heinz J. Fissan
University of Duisburg, Duisburg, West Germany

Elsevier
New York • Amsterdam • Oxford

This book has been registered with the Copyright Clearance Center, Inc. For further information, please contact the Copyright Clearance Center, Salem, Massachusetts.

Published by:

Elsevier Science Publishing Company, Inc.
52 Vanderbilt Avenue, New York, New York 10017

Sole distributors outside the United States and Canada:

Elsevier Science Publishers B. V.
P. O. Box 211, 1000 AE Amsterdam, The Netherlands

Library of Congress Cataloging in Publication Data

Main entry under title:

Aerosols: science technology and industrial applications
    of airborne particles.

    Includes index.
    1. Aerosols. I. Liu, Benjamin Y. H., 1934–
    II. Pui, David Y. H.   III. Fissan, Heinz J.
TP244.A3A346   1984   660.2′94515   84-13626
ISBN 0-444-00947-7

Manufactured in the United States of America

Preface

This volume contains the extended abstracts of papers presented at the First International Aerosol Conference, held at the Amfac Hotel in Minneapolis, Minnesota, September 17–21, 1984. It constitutes the proceedings of the conference.

The First International Aerosol Conference is the first such international aerosol conference jointly sponsored by the American Association for Aerosol Research (AAAR) and its European counterpart, Gesellschaft fuer Aerosolforschung (GAeF) (Association for Aerosol Research). It is the plan of the associations to hold such joint international conferences in the future every two years, alternately in the U.S. and in Europe. The Second International Aerosol Conference will be held in West Berlin, West Germany, in 1986. The two associations will hold their respective national meetings separately in 1985.

Interests in aerosols have continued to grow in recent years, as evidenced by the large number—approximately 300—of papers presented at this conference. Interests in the more classical subject areas of aerosol research such as basic properties and behavior, instrumentation and measurement techniques, particulate control technology, health effects, etc., have not abated. New areas of research are also beginning to emerge—for instance, particulate contamination control in electronic and computer manufacturing, and the production of fine particle materials through aerosol processes. It is the hope of the conference organizers to stimulate such new research interests and facilitate the exchange of information in the field.

The timely publication of scientific and technical information is the great challenge confronting the scientific community today. We have chosen this particular form of publication for the following reasons. The publication of a conference proceeding with full length manuscripts is a tedious and time-consuming process and the publication of a short one-page abstract would not make the publication very useful or substantive. As a compromise, we have chosen to publish a book of extended abstracts, each abstract being limited to four pages. We hope that by this means the authors can communicate their major research findings to the scientific and technical communities in a timely and yet substantive fashion.

This book of extended abstracts is divided into four parts covering the subject areas of Instrumentation and Measurement (Part I), Aerosol Characterization (Part II), Gas Cleaning, Clean Rooms, Particle Production and Combustion Aerosol (Part

III), and Basic Properties and Behavior and Health Effects (Part IV). The four parts are in turn divided into twenty-two chapters to allow a more coherent organization of the subject matter. The book also provides an author and a subject index to facilitate access of the information.

B.Y. H. Liu

D.Y. H. Pui

H. J. Fissan

Acknowledgment

A large undertaking such as this one cannot succeed without the help of many individuals. We wish to thank the two sponsoring associations, whose directors and officers are listed below, for entrusting us with the responsibility for organizing the conference. We wish to thank the members of the Technical Program Committee and the Special Session Organizers for helping us to put together the technical program for the conference and for the Local Arrangement Committee for making the conference itself a success. For the many others who put in endless hours in connection with the conference and the proceedings, we wish to mention in particular Joseph M. Kroll, Moira A. Keane, and Beverly Ringsak of the Department of Conferences, University of Minnesota, who have helped us on this and many other occasions; and our secretary, Ms. Linda Stoltz, whose energy, enthusiasm, dedication, loyalty, and devotion are a source of inspiration for us all. Last, but not least, we wish to thank all the authors, whose interest and contributions are what have made such an undertaking meaningful, possible, and worthwhile.

Officers and Directors of AAAR

David T. Shaw (president), State University of New York, Buffalo, NY
Simon L. Goren (vice-president), University of California, Berkely, CA
David S. Ensor (treasurer/secretary), Research Triangle Institute,
    Research Triangle Park, NC
Peter W. Dietz, General Electric Company, Schenectady, NY
Sheldon K. Friedlander, University of California, Los Angeles, CA
Milton Kerker, Clarkson College of Technology, Potsdam, NY
Alvin Lieberman, Particle Measurement System, Boulder, CO
Benjamin Y. H. Liu, University of Minnesota, Minneapolis, MN
Roger McClellan, Inhalation Toxicology Research Institute, Albuquerque, NM
Morris S. Ojalvo, National Science Foundation, Washington, DC
William R. Pierson, Ford Scientific Laboratory, Dearborn, MI
Othmar Preining, University of Vienna, Vienna, Austria
Adel F. Sarofim, Massachusetts Institute of Technology, Cambridge, MA
John H. Seinfeld, California Institute of Technology, Pasadena, CA
William E. Wilson, U.S. Environmental Protection Agency,
    Research Triangle Park, NC

viii

Officers and Directors of GAeF
Michael Benarie, Ircha, Bretigny, France
Axel Berner, Institut für Experimentalphysik, Vienna, Austria
Heinz J. Fissan, University of Duisburg, Duisburg, West Germany
James Gentry, University of Maryland, College Park, MD
D. Hochrainer, Fraunhofer Institut, Schmallenber, West Germany
Vittorio Prodi, University of Bologna, Bologna, Italy
Wolfgang Schikarski, Kernforschungszentrum Karlsruhe, Karlsruhe,
    West Germany
K. Spurny, Institut für Toxikologie u. Aerosolforschung, Schmallenberg,
    West Germany
Willi Stahlhofen, Ges. f. Strahlen– und Umweltforschung, Frankfurt,
    W. Germany
Werner Stoeber, Fraunhofer Institut, Munster-Roxel, West Germany

Technical Program Committee and Special Session Organizers
Thomas G. Dzubay, Environmental Protection Agency,
    Reseach Triangle Park, NC
David Ensor, Research Triangle Institute, Research Triangle Park, NC
Heinz J. Fissan, University of Duisburg, Duisburg, West Germany
Sheldon K. Friedlander, University of California, Los Angeles, CA
David B. Kittelson, University of Minnesota, Minneapolis, MN
Benjamn Y. H. Liu, University of Minnesota, Mineapolis, MN
Ralph Lutwack, Jet Propulsion Laboratory, Pasadena, CA
C. V. Mathai, AeroVironment, Inc., Pasadena, CA
Virgil A. Marple, University of Minnesota, Minneapolis, MN
Peter H. McMurry, University of Minnesota, Minneapolis, MN
William R. Pierson, Ford Motor Company, Dearborn, MI
David Y. H. Pui, University of Minnesota, Minneapolis, MN
David T. Shaw, State University of New York, Buffalo, NY

Local Arrangement Committee
Dan Japuntich, 3M Company
Benjamin Y. H. Liu, University of Minnesota
Virgil A. Marple, University of Minnesota
Peter H. McMurry, University of Minnesota
David Y. H. Pui, University of Minnesota
Kenneth R. Rubow, University of Minnesota
Robin E. Schaller, Donaldson Company
Gilmore Sem, TSI, Inc.

Liu   **Pui**   **Fissan**

Benjamin Y. H. Liu is a Professor of Mechanical Engineering and the Director of the Particle Technology Laboratory, University of Minnesota. He has been involved in aerosol research for some twenty years and to a lesser extent also in solar energy research. He has served as a visiting professor at the University of Paris VI, a visiting scientist at the Oak Ridge National Laboratory, a member of the Particulate Control Technology Committee of the National Academy of Sciences, a member of the editorial advisory boards of the Journal of Aerosol Science and the Journal of Colloid and Interface Science, and as a technical consultant to many industrial firms. He is best known for his basic research on electrical charging and precipitation, aerosol sampling and transport, filtration, and inertial impaction. He is the developer or co-developer of several widely used aerosol generating, measuring, and sampling instruments, including the widely used vibrating orifice monodisperse aerosol generator, the pulse precipitating electrostatic aerosol sampler, the electrical mobility analyzer and classifier, and the constant output atomizer. He is the editor of the book *Fine Particles,* a co-editor of the books *Application of Solar Energy for Heating and Cooling of Buildings* and *Aerosols in the Mining and Industrial Work Environment,* and an editor-in-chief of the journal *Aerosol Science and Technology.* He is a Fellow of the American Society of Mechanical Engineers, a Fellow of the American Association for the Advancement of Science, a Guggenheim Fellow, and the recipient of a Senior U. S. Scientist Award from the Alexander Von Humboldt Foundation of West Germany.

David Y. H. Pui is a Research Associate and Manager, Particle Technology Laboratory, University of Minnesota. He received his Ph.D. degree from the University of Minnesota and has been involved in basic and applied aerosol research at Minnesota for some ten years. The major emphasis of his research has been in aerosol instrumentation, basic charge transfer processes in aerosol systems, studies on ultrafine aerosol particles, and coal gasification. He was involved in the design and development of the electrical aerosol analyzer, the electrostatic aerosol classifier, the electrical aerosol detector, and the PM-10 size selective aerosol sampling inlet. He has published some 40 papers and reports in the field and has served on a number of governmental and industrial committees relating to air quality and contamination control.

Heinz J. Fissan is a Professor of Electrical Engineering, University of Duisburg, West Germany. He received his Dr.-Ing. degree and "Habilitation" from the Rheinische Westfaelische Technische Hochschule in Aachen, West Germany, and spent two years in the United States (University of California, La Jolla, and University of Minnesota) for postdoctoral research. His major research areas include kinetics, heat transfer, and soot formation in flames, flame spectroscopy, particulate emissions from spark ignition engines and other combustion sources, smoke detection, and aerosol instrumentation and measurement techniques. He has published some 50 papers and reports. He is a member of Verein Deutscher Ingeniure (VDI), Gesellschaft fuer Aerosolforschung (GAeF), Normemausschuss Siebboeden und Kornmessung (NASK) and the American Association for Aerosol Research (AAAR). He is currently the Secretary General of GAeF, the co-sponsor of this Conference.

PART I.
INSTRUMENTATION AND MEASUREMENT

CHAPTER 1.
OPTICAL MEASUREMENT

CHAPTER 2.
CONDENSATION DETECTION

CHAPTER 3.
ELECTRICAL MEASUREMENT

CHAPTER 4.
SAMPLING METHODS AND DEVICES

CHAPTER 5
SAMPLING, TRANSPORT AND DEPOSITION

† Denotes a plenary lecture given at the conference

CHAPTER 6
MEASUREMENT AND DATA ANALYSES

PART II.
AEROSOL CHARACTERIZATION

CHAPTER 7.
ATMOSPHERIC AEROSOLS

† Denotes a plenary lecture given at the conference

CHAPTER 8.
RECEPTOR MODEL

CHAPTER 12.
WORK ENVIRONMENT AND INDOOR AEROSOLS

PART III.
GAS CLEANING; CLEAN ROOM; PARTICLE PRODUCTION; COMBUSTION AEROSOLS

CHAPTER 13.
FILTRATION

CHAPTER 14.
CONTROL TECHNOLOGY

† Denotes a plenary lecture given at the conference

CHAPTER 15.
CLEAN ROOM

CHAPTER 16.
INDUSTRIAL FINE PARTICLE PRODUCTION

CHAPTER 17.
AEROSOL GENERATION

CHAPTER 18.
COMBUSTION AEROSOLS

PART IV.
BASIC PROPERTIES AND BEHAVIOR; HEALTH EFFECTS

Chapter 19.
BASIC PROPERTIES AND BEHAVIOR

† Denotes a plenary lecture given at the conference

CHAPTER 20.
COAGULATION, CONDENSATION AND NUCLEATION

CHAPTER 21.
GAS-PARTICLE INTERACTION

† Denotes a plenary lecture given at the conference

PART I.

INSTRUMENTATION AND MEASUREMENT

CHAPTER 1.

OPTICAL MEASUREMENT

Published 1984 by Elsevier Science Publishing Co., Inc.

Aerosols, Liu, Pui, and Fissan, editors

COHERENT EXTINCTION

by

Donald R. Pettit
Los Alamos National Laboratory
Los Alamos, NM 87545
USA

Extended Abstract

Consider a single particle submerged in the beam waist region provided by a pair of lenses arranged so the emerging beam of light remains colli- mated. A detector which receives the emerging beam will witness a decrease in intensity due to the presence of the particle in the beam waist, thus recording an extinction signal. If the light beam comes from a coherent source, and the particle-scattering event is contained within a region bounded by the resolution-limited blur spot diameter of the downstream lens, the Van Cittert-Zernike theorem will show that the gathered forward scattered light will be coherent with the source. In assessing the signal from such a system, the individual scattered wave-field amplitudes are first summed and then squared to determine the final intensity, thus defining the concept of *coherent extinction.*

Most of the energy in the beam will be unaffected by the presence of the particle and will be designated as noncarrier. The wave-field amplitude associated with this noncarrier will act as a local oscillator beam and will amplify the weak forward scattered signal via the principles of optical hom- odyne detection. The noncarrier, acting as a local oscillator, does not have the capability of relative phase adjustment with the signal as in clas- sic two beam homodyning. This single beam system, then, defines the concept of *self-homodyning* , where the beam contains its own internal local oscil- lator without any possible phase adjustment between it and the signal.

The forward scattered wave-field amplitude in this coherent extinction scheme is found from the cumulative sum of the individual scattered wave- field amplitudes over the collection angle of the downstream lens. The non- carrier wave-field amplitude is the difference between the total wave-field amplitude in the beam and the maximum amount of wave-field amplitude inter- acted upon by the particle. In this manner self-homodyning, when applied to coherent extinction, can detect the light scattered out of the beam through its effects on the noncarrier.

The maximum wave-field amplitude interacted upon by the particle would intuitively be found from a cumulative sum of the individual scattered wave- field amplitudes over all spherical angles, much like what is done in regular incoherent extinction, where a cumulative sum of individual scattered inten- sities is made over all spherical angles. This approach is incorrect; a cumulative sum of wave-field amplitudes over all spherical angles does not necessarily give a larger value than a sum over a smaller piece of the spherical pie. The maximum wave-field amplitude needed in calculating the noncarrier is found from a cumulative sum of the individual wave field amplitudes up to some maximum angle.

A plot of maximum cumulative angle vs particle diameter for polystyrene latex spheres and HeNe laser illumination shows a maximum angle less than 25 degrees for particle diameters greater than 2 micrometers. For particle diameters less than 0.08 micrometers, the maximum angle limits to 90 degrees, which is not surprising for particles in the Rayleigh regime. Only at particle diameters of 0.4 and 1.2 micrometers does the maximum angle reach 180 degrees, thus including all spherical angles.

In comparing coherent extinction to classic incoherent extinction, the resulting signals show little difference for particle diameters greater than 5 micrometers. The coherent extinction signal for submicrometer particles is significantly enhanced over incoherent extinction, due to the self-homodyne amplification. Based on a common noise level, predicted detection limits for incoherent extinction approach 3 micrometers, while coherent extinction exhibits a limit of 0.3 micrometers. The multivaluedness associated with classic extinction is predicted to be much worse for coherent extinction, especially in the submicrometer range.

With the use of a laser and a pair of microscope objectives for simple extinction measurements, coherent extinction could be automatically realized and possibly transparent to the experimenter. A significant discrepancy in the submicrometer range may be apparent if the concepts of coherent extinction and self-homodyning are not incorporated into the analysis.

Published 1984 by Elsevier Science Publishing Co., Inc.
Aerosols, Liu, Pui, and Fissan, editors

COUNTING EFFICIENCY AND SIZING CHARACTERISTICS OF OPTICAL PARTICLE COUNTERS

by

J. Gebhart[o], P. Blankenberg[*] and C. Roth[o].

[o]Gesellschaft für Strahlen- und Umweltforschung, 6000 Ffm.
[*]Trox GmbH, Filtertechnik, 4133 Neukirchen-Vluyn; FR-Germany.

Several optical particle counters were investigated experimentally with respect to their counting efficiency and classification properties. Tests were carried out with monodisperse aerosols of polystyrene spheres (refractive index: 1.59) and of oil droplets consisting of di-2-ethylhexyl-sebacate (refractive index: 1.45). The polystyrene latex aerosol additionally passed an aerosol neutralizer (Modell 3054 TSI) at the outlet of the generator. No discharge was required for the oil droplets produced by condensation of vapour. For the characterization of the test aerosols independent methods like electron microscopy and sedimentation cell analysis were applied. During an experimental run several instruments simultaneously sampled aerosol from a 300 l-tank which was continously supplied with a constant stream of aerosol and diluting air. Two aerosol size spectrometers of the laboratory (Gebhart et al. 1976; Roth et al. 1978) which were checked independently with an electrostatic aerosol sampler (Model 3100 TSI), served as reference instruments for the actual number concentration in the tank.

The instruments investigated are listed in the table. They differ in the angular range of scattered light, the kind of illumination, the sensitivity and the sample flow rate. For the evaluation of the sizing characteristics the cumulative size distributions of the test aerosols were compared to those ones measured with the instruments. In five instruments using white light illumination all measured diameters of the polystyrene particles were shifted systematically to smaller sizes by more than 20%. All other instruments indicated the diameters of the polystyrene spheres within an accuracy of 10% with the exception of the 0.95 μm-particles. Due to an ambiguous part in the calibration curve all instruments with near forward scattering (10-30°) classified those particles by about a factor 1.5 too small.

Number concentration c of the aerosol, sample flow rate $\dot{V}$ and counting rate $\dot{N}$ of a particle counter are connected by the relation:

$$\dot{N} = c\,\dot{V} \qquad\qquad (1)$$

Equation (1) assumes that each particle contained in the sample

flow produces a single count. In praxis, however, counting losses
due to coincidences occur which can be estimated by the equation
(van der Meulen et al. 1980):

$$\frac{c}{c_k} = e^{-c_k \dot{V} t_r} \qquad (2)$$

where c is the measured and c_k the corrected number concentration
and t_r the recovery time necessary between two successive counting
events. For an estimation of the coincidence losses an experimental pulse time τ was determined for each instrument by connecting
the output of the counter with an oscilloscope. This measured time
τ may not be identical with the recovery time t_r but can be considered as a useful approximation.

For an evaluation of the counting efficiency η the coincidence corrected concentration c_k was related to the concentration
c_o obtained by the reference methods according to:

$$\eta = \frac{c_k}{c_o} \qquad (3)$$

By this definition counting efficiency is not affected by coincidence losses and thus reflects a specific property of an instrument. Replacing the time t_r by the pulse time τ and substituting
equation (3) into equation (2) finally leads to:

$$\frac{c}{\eta c_o} = e^{-\eta c_o \dot{V} \tau} \qquad (4)$$

Equation (4) was applied to calculate the counting efficiency
from the measured quantities c, c_o and τ at a given flow rate $\dot{V}$.
Results are presented in the table. In the submicron range the
counting efficiency of most instruments decreases with decreasing
particle diameter so that the lower detection limit is rather a
gradual transition than a sharp step. In the hypermicron size
range satisfactory results are obtained. However for particle
diameters above 2 µm losses in the sampling system of the counter
may again reduce the counting efficiency especially if instruments with low sampling rates are used in connection with plastic
tubes and charged particles (see Royco 236). For all ten instruments using white light the counting efficiency of 0.327 µm particles remained below 0.2 at flow rates of 2800 cm^3/min. For the
Laser—instruments on the other hand a counting efficiency of
about 1 was achieved in the whole submicron range.

In the particle counter HC 15 the sensing volume is formed
by optical means only. As concequency the sample flow rate cannot
be defined exactly. Whereas small particles are only counted when
they pass the central region of the stop bigger particles can
also give rise to countable pulses at the periphery of the imaging

system even if these peripheral light flashes are partially shadowed. Thus the effective sample flow rate depends on particle size. The bigger the particles the more peripheral light flashes constribute to a measured cumulative number distribution (Helsper et al. 1980). A method to overcome this problem has been discussed in a recent paper (Umhauer 1982).

Variations of the counting efficiency in the micron range of the instruments may also be caused by incorrect sample flow rates. This was the case in about half of the instruments investigated. Only with 3 instruments the sample flow rate could be observed and adjusted on the panel plate. Instruments which could be operated at different flow rates changed their counting efficiency by more than a factor 2 when they were switched over. Even after an adjustment of the sample flow rates with an external flow meter the counting efficiency at 280 cm^3/min remained in the average about 50% higher than at 2800 cm^3/min flow rate (see Kartel Partoskop A, A/R; Royco 227; Royco 4120). Observations on an oscilloscope showed that at 280 cm^3/min flow rate a lot of stray particles having pulse times up to 250 µs passed the sensing volume. Since pulse times of this magnitude exceed the recovery time of the electronics more than one count can originate from a single stray particle.

References:

Gebhart, J., Heyder, J., Roth, C. and Stahlhofen, W. (1976) Fine Particles (Edited by Liu B.Y.H.) p. 793 Acad. Press, New York, San Francisco, London.

Helsper, C. and Fißan, H.J. (1980) 8.th Annual Meeting of the Gesellschaft für Aerosolforschung, Schmallenberg, Germany, F.R. 22.-24. Oktober.

van der Meulen, A., Plomp, A., Oeseburg, F., Buringh, E., von Aalst, R.M. and Hoevers, W. (1980) Atmosph. Environm. 14, 495.

Roth, C. and Gebhart, J. (1978) Microscopica Acta 81, 119.

Umhauer, H. (1982) 10.th Annual Meeting of the Gesellschaft für Aerosolforschung, Bologna, Italy, 14.-17. September.

type of instrument	scattering angle	illumi-nation	d_{min} μm	τ μs	$\dot{V}$ cm³ min⁻¹	counting efficiency for particle dia. d/μm				
						0.163	0.327	0.47	0.95	2.02
HC 15	90°-range	white	0.3	~10	5.6	-	0.46	1.06	1.51	-
Climet CI-208 C	15 -105°	white	0.3	4.5	28000	-	0.17	0.64	0.82	0.76
DAP-Test 2000	90°-range	white	0.5	~15	500	-	-	0.085	0.89	1.06
Kratel Part. A/R 32 channels	10 - 30°	white	0.3 (0.5)	20	28000	-	-	-	0.37	-
				25	2800	-	0.02	0.21	0.64	1.01
				*		-	0.03	0.76	0.82	0.89
				30	280	-	0.065	1.07	1.63	1.78
				*		-	0.052	1.09	1.21	1.20
Kratel Part. A/R 5 channels	10 - 30°	white	0.3 (0.5)	20	28000	-	-	-	0.44	0.65
				25	2800	-	0.033	0.25	0.72	1.20
				30	280		0.15	1.40	2.87	-
Kratel Part. A 32 channels	10 - 30°	white	0.3 (0.5)	25	2800	-	0.03	0.21	0.76	0.90
				*		-	0.04	0.72	0.71	0.78
				35	280	-	0.15	1.06	3.17	-
				*		-	0.10	1.6	1.3	1.58
Kratel Part. A 5 channels	10 - 30°	white	0.3 (0.5)	25	2800	-	0.18	0.76	0.71	0.99
				25	280	-	0.68	2.29	2.63	-
Royco 247	10 - 30°	white	0.5	25	28000	-	-	-	1.10	1.39
Royco 227	10 - 30°	white	0.3	25	2800	-	0.12	0.66	0.62	0.89
				30	280	-	0.16	0.92	1.05	-
Royco 4102	10 - 30°	white	0.3	25	2800	-	0.178	1.01	0.98	-
				~40	280	-	0.62	1.30	1.47	-
Royco 226	35 - 120°	He Ne Laser	0.12	9	300	1.03	0.99	1.02	1.03	-
Royco 236	35 - 120°		0.12	9	300	1.04	0.98	0.91	1.05	0.71△
Royco 1000	30 - 170°	Laser	0.3	8	28000	-	0.93	1.06	0.95	-

General specifications and measured counting efficiencies of the optical particle counters involved in the program.
*results obtained after servicing; △ valid for charged particles sampled through a 3 mm plastic tube of about 40 cm length.

LASER SPECTROMETER: THEORY AND PRACTICE*

S. C. Soderholm and G. C. Salzman**
Industrial Hygiene Group and Experimental Pathology Group**
Los Alamos National Laboratory
Los Alamos, NM 87545
U. S. A.

EXTENDED ABSTRACT

Introduction

Laser spectrometers determine the size distribution of an aerosol of spherical particles of known refractive index by measuring the amount of light scattered into a detector by individual particles and accumulating the counts in a multi-channel analyzer. They offer high resolution, speed, and convenience in measuring aerosols having particle diameters between 0.1 and several microns. This paper describes theoretical calculations of the size response of one laser spectrometer (Model LAS-X manufactured by Particle Measuring Systems, Inc.) and relates practical experience gained through its use in Los Alamos research programs.

Theory

The Model LAS-X determines the diameter of an individual spherical particle from the amount of light scattered into the detector as the particle passes through the intra-cavity beam of a He-Ne laser. The manufacturer has calibrated the instrument with polystyrene latex spheres, so particle diameter corrections must be calculated to size spheres of other materials. In this work, calculations were performed to predict the dependence of the scattered light intensity on a) the particle diameter, b) the particle refractive index, and c) the position of the particle relative to the nodes and anti-nodes of the standing wave(s) inside the laser cavity.

The relevant features of the intra-cavity laser beam are:
1) the laser beam width is approximately 600 μm, according to the manufacturer, with a radially symmetric gaussian distribution of light intensity,
2) the light is plane polarized due to the Brewster window at one end of the laser gas tube,
3) the light is nearly monochromatic with a wavelength $\lambda = 0.6328$ μm,
4) a distribution of wavelengths is possible due to the Doppler effect with the intensity distribution having a full-width at half maximum of only 2×10^{-6} μm (Sinclair and Bell, 1969), and
5) a series of discrete wavelengths are found within the Doppler intensity envelope due to the requirement of having a standing wave

*Work performed at Los Alamos National Laboratory under the auspices of the U. S. Department of Energy, Airborne Waste Management Program Office, Contract No. W-7405-ENG-36.

between the laser end mirrors (For the Model LAS-X which has a laser mirror separation of roughly 20 cm, the discrete wavelengths differ by 1×10^{-6} μm, so there are approximately 4 discrete wavelengths of appreciable intensity within the Doppler intensity envelope of this laser.).

The scheme for calculating the amount of light scattered into the detector of the Model LAS-X for spherical particles utilizes Mie calculations contained in a subroutine from Dave (1968). Mie calculations give the complex amplitude, which contains amplitude and relative phase information, of light scattered from a plane wave as a function of particle diameter, refractive index, scattering angle, incident light polarization, and incident light intensity. The relative phase of two incident plane waves of the same wavelength is retained in the two scattered waves. The detailed calculations consist of the following elements:

1) The standing wave for each discrete wavelength in the cavity is composed of two plane waves traveling in opposite directions.
2) The relative phase of the two incident plane waves (for each discrete wavelength) depends on the distance Δ between the center of the particle and the nearest node of the standing wave.
3) Because the incident light is plane polarized, each plane wave is composed of two components, one parallel to the scattering plane and one perpendicular.
4) Mie calculations give the complex amplitudes of the scattered waves arising from the 4 incident plane waves (two polarizations and two directions of travel) for each discrete wavelength in the laser cavity. The intensity of light scattered into each element of solid angle of the detector is calculated from a coherent addition of the scattered waves.
5) Integration over all solid angles subtended by the detector (all azimuthal angles and polar angles of 35° to 120°) gives a measure of the light signal which would be detected by the Model LAS-X as a function of the distance Δ.

Some interesting results of this theoretical study are:
1) For each discrete wavelength in the laser cavity, the relative intensity of light scattered into the detector $I(d,n,\lambda,\Delta)$ can be written as:

$$I(d,n,\lambda,\Delta) = I_1(d,n,\lambda) - I_2(d,n,\lambda) \cos\left(\frac{4\pi\Delta}{\lambda}\right) - I_3(d,n,\lambda) \sin\left(\frac{4\pi\Delta}{\lambda}\right)$$

where d and n are the particle diameter and (real) refractive index, respectively, λ is the wavelength of incident light, and Δ is the axial distance (positive or negative) from the particle center to the nearest node of the standing wave.
2) The intensity of scattered light typically varies by ±40 per cent from the mean as Δ varies from 0 (at a node) to ±λ/4 (at an anti-node).
3) Averaging over all possible values of Δ ($-\lambda$/4 to $+\lambda$/4), the mean intensity is $I_1(d,n,\lambda)$. This would be the mean scattered light intensity from particles of diameter d in a monodisperse aerosol.

4) Comparison of this work with previous calculations by Pinnick and Auvermann (1979) for several laser spectrometer models other than the Model LAS-X and by Salzman, et al. (1983) for the Model LAS-X reveals that they calculated only I_1+I_2, the light scattered from particles passing through an anti-node in the standing wave.
5) There is no significant effect on the response of the Model LAS-X due to the existence of several discrete wavelengths because:
   a) The absolute differences in wavelength are so small that they result in only undetectable differences in the results of the basic Mie calculations.
   b) In some spectrometers, there could be a significant effect due to the differences in phase of the standing waves at the point in the laser cavity where the particles intersect the beam. However, the differences in phase between two standing waves are very small near the ends of the cavity (where the particles intersect the light beam in the Model LAS-X) so the discrete wavelengths have little effect in the Model LAS-X.
6) The calculated scattered light intensity averaged over Δ, $I_1(d,n,\lambda)$, may be a very good approximation to the light scattered from each particle of diameter d. Calculated excursions from the mean due to the center of a particle located at a node or anti-node, for example, would be expected to occur only if the particle trajectory were exactly parallel to the wave fronts so the distance Δ would be constant throughout the whole transit of the 600 μm wide beam. However, it is much more likely that each particle will pass along several standing waves during its transit through the beam (averaging over Δ, effectively), because the laser and particle beams are probably not perfectly perpendicular and because the laser wave fronts are probably somewhat non-planar over the beam width because of the curved mirror at one end of the laser cavity.
7) The quantity $I_1(d,n,\lambda)$ can be calculated from the incoherent addition of the 4 scattered waves for each discrete wavelength. It is a smoother function of particle diameter than the results of Salzman, et al., so it is easier to correct the bin diameters for changes in particle refractive index.

Practice

We have gained practical experience with the Model LAS-X in a laboratory setting and in the Department of Energy Filter Test Facilities (FTFs). The Model LAS-X is generally stable in operation and is operable by FTF technicians who are not trained in aerosol science. They check its calibration daily with a single size of polystyrene latex spheres and acquire data with a computerized data analysis system developed at Los Alamos.

Our maintenance experience leads us to expect to send each unit to the manufacturer annually to replace the laser tube. Some units have had a power supply replaced. Generally, we have not needed to clean the mirrors, since the manufacturer cleans, tunes, and re-calibrates each unit annually when it is sent for laser tube replacement.

The computerized data acquisition and analysis system has several features of note:

1) It utilizes a Hewlett-Packard Model 85 micro-computer with 32K of programmable memory.

2) The computer's printer provides convenient tabular and graphical presentations of the raw data and the lognormal distribution having the same number concentration, geometric mean diameter, and geometric standard deviation as the raw data.

3) Provision has been made to allow the user to enter quantities from the keyboard which are associated with the data collection, such as dilution ratio, temperature, barometric pressure, and relative humidity. These "keyboard quantities" are easily reprogrammed to fit the user's needs.

4) Up to 99 sets of raw data and the associated "keyboard quantities" can be conveniently stored on the built-in cassette tape for later analysis with the same software. After computation, summary data including all calculated parameters can be stored on the same cassette tape.

5) The software is "menu-driven" and has been used by technicians at the FTFs to reliably acquire and analyze data from the Model LAS-X.

Summary

The theoretical analysis described gives us reasonable confidence in correcting the bin diameters of the Model LAS-X for the refractive index of spherical particles. The computerized system developed for acquiring and analyzing data from the Model LAS-X is convenient and works reliably.

References

Dave, J. V. (1968) "Subroutines for Computing the Parameters of the Electromagnetic Radiation Scattered by a Sphere", IBM Scientific Center Report No. 320-3237.

Pinnick, R. G. and Auvermann, H. J. (1979) "Response Characteristics of Knollenberg Light-scattering Aerosol Counters", J. Aerosol Sci. 10:55.

Salzman, G. C., Ettinger, H. J., Tillery, M. I., Wheat, L. D. and Grace, W. K. (1983) "Potential Application of a Single Particle Aerosol Spectrometer for Monitoring Aerosol Size at the DOE Filter Test Facilities", Proceedings of the 17th DOE Nuclear Air Cleaning Conference Held in Denver, Colorado 2-5 August 1982 (Edited by M. W. First), National Technical Information Service, CONF-820833, Vol. 2, p. 801.

Sinclair, D. C. and Bell, W. E. (1969) Gas Laser Technology, Holt, Rinehart and Winston, Inc., New York.

Published 1984 by Elsevier Science Publishing Co., Inc.

Aerosols, Liu, Pui, and Fissan, editors

METHOD FOR CORRECTING COINCIDENCE ERRORS OF OPTICAL PARTICLE COUNTERS:
CONCENTRATION AND SIZE DISTRIBUTION

K. Janka, V. Kulmala and J. Keskinen
Tampere University of Technology, Department of Electrical Engineering,
Physics,
P.O. Box 527,
SF-33101 Tampere 10, Finland

In optical particle counters the upper limit of the measurable concentration range is determined by a coincidence error. Coincidence influences errors on the measured concentration and on the size distribution as well. Correcting only the concentration error for monodispersive particles is quite simple /e.g. 1, 2/.

If also the size-distribution error should be corrected, the form of detected pulses should be known, and in addition, quite complicated mathematics is needed. However, optical particle counters may produce pulses with varied and unknown duration and with irregular shape.

In this study a new method for performing the coincidence correction for both the concentration and the size distribution is presented. The method is based on the measurement of, not only the number of triggered pulses, but also durations of the detected pulse levels. Due to the additional measurement information any knowledge about the form of pulses is not needed, and pulses with varied durations and irregular shape are allowed. Moreover, the algorithm performing the correction is simple.

1. K. Janka and V. Kulmala, Rev. Sci, Instrum. 54 (1983)

2. K. Janka and V. Kulmala, 11th Annual Meeting of GAF, September 14-16, 1983, Munich.

Published 1984 by Elsevier Science Publishing Co., Inc.
Aerosols, Liu, Pui, and Fissan, editors

16

SEDIMENTATION METHOD IN CALIBRATING OPTICAL PARTICLE COUNTERS

M. Lehtimäki[1], J. Keskinen[2], M. Rajala[2], K. Janka[2]

1 Occupational Safety Engineering Laboratory Technical Research Centre
  of Finland, P.O. Box 656, Tampere 10, Finland

2 Department of Physics, Tampere University of Technology
  P. O. Box 527, 33101 Tampere 10, Finland

Optical particle counters are widely used in aerosol studies, mainly
due to their convenience in direct measurement of particle number concentration and the size distribution. Unfortunately these devices are sensitive
to the optical properties of the particles, which means that instead of
geometric or aerodynamic particle size, only an equivalent optical
diameter can be measured. The possibility to measure the aerodynamic
particle size of different test aerosols (e.g. Arizona road dust, NaCl,
methylene blue) would be of great importance especially in testing air
cleaning devices and in calibrating size selective particle samplers
cyclones, impactors).

The problem caused by the optical properties of the particles can be
avoided using an aerodynamic particle size analyzer (1). These instruments are, however, relatively expensive and are so far available in some
advanced laboratories only. The aerodynamic particle size calibration of an
optical particle counter can be carried out with the aid of impactors.
This method used by Marple (2) is relatively accurate but somewhat tedious.
The present work presents an alternative method which is based on the
sedimentation of particles in an elutriator. The penetration properties of particles in an elutriator are less sharply size dependent but
the penetration properties can be easily varied in a relatively wide
range simply by varying the air flow velocity in the elutriator.

The experimental work has been carried out using a ceramic elutriator
which contains parallel square channels separated by thin walls. The
monodisperse DOP particles generated by vibrating orifice generator have
been used to determine the penetration characteristics for different flow
velocities. The results of these measurements, which are in agreement
with the theory, have been then used in studying the properties of a
particle counter (PMS, LasX) in the case of Arizona road dust particles.

The penetration properties of particles registered to different size
channels of the particle size analyzer have been compared to the ones
measured for DOP particles and the corresponding aerodynamic particle sizes
have been determined. The results indicate that the simple elutriator
can be conveniently used in an approximate calibration of particle counters
for unknown test aerosols.

1. Remiarz, R.J. et al., Aerosols In the Mining and Industrial Work
    Volume 3, (Editors Marple, V.A. and Liu, B.Y.H.), Ann Arbor Science
    Publishers, 1983.

2. Marple, V.A., Aerosol Measurement (Editors Lundgren, D.A. et al.)
    University of Florida Press, 1979.

Published 1984 by Elsevier Science Publishing Co., Inc.
Aerosols, Liu, Pui, and Fissan, editors

18

VIRTUAL IMPACTOR AS AN ACCESSORY TO OPTICAL PARTICLE COUNTERS

J. Keskinen[1], K. Janka[1], M. Lehtimäki[2]

1 Department of Physics, Tampere University of Technology
  P.O. Box 527, 33101 Tampere 10, Finland

2 Occupational Safety Engineering Laboratory
  Technical Research Centre of Finland
  P.O. Box 656, 33101, Tampere 10, Finland

One of the basic methods to measure particle concentrations is the optical particle counter. In this method there is, however, a problem which is caused by the polydispersivity of the particles to measured. In order to be able to measure small particles and high concentrations without significant coincidence errors, a low sample air flow rate must be used. The low flow rate leads, however, to a poor sensitivity when large particles are to be measured and unpractically long counting periods are unavoidable.

In the present work the possibility to increase the sensitivity of optical particle counters for large particles is studied. The method used is the connection of a virtual impactor to the inlet of an optical particle counter. The minor air flow of the virtual impactor which contains an increased concentration of large particles is used as the sample air flow. The properties of a virtual impactor connected to a particle counter with low sample air flow rate have been studied experimentally in the size range 1-10 microns. The results indicate that in the size range $D_p > 2$ microns the number concentration of particles can be increased by a factor 10...50.

A practical application of the present method is the study of air cleaning devices using a polydisperse test aerosol and a particle size analyzer. The increase in the large particle sensitivity makes it possible to measure more reliably the filtration efficiencies for large particles. The differences in the size distributions measured with and without the virtual impactor can be used in estimating the relationship between the aerodynamic and optical sizes of different test aerosols.

Published 1984 by Elsevier Science Publishing Co., Inc.

Aerosols, Liu, Pui, and Fissan, editors

AEROSOL MEASUREMENT BY PHOTON CORRELATION SPECTROSCOPY (II)

M. Itoh and K. Takahashi
Institute of Atomic Energy
Kyoto University
Uji, Kyoto 611
JAPAN

EXTENDED ABSTRACT

Introduction

The photon correlation spectroscopy (PCS) has been utilized in polymer science and recently in aerosol science for the measeurement of single or multicomponent diffusion coeffisients of aerosols (Itoh and Takahashi, 1981; Dahneke, 1983). PCS is advantageous comparing with usual measurement of aerosols for its high sensitivity and real in-situ characteristics.

PCS is especially applicable to measuring dynamic behavior and physical properties of sub-micron aerosol particles, however, it is necessary to modify the simple photon count autocorrelation function. Points of modification are in the applicability of quasi-elastic light scattering theory to the Mie scattering region and in the treatment of the theory so as to fit to non-Gaussian nature of aerocolloidal system due to small number concentration of particles. One of the case of modified intensity autocorrelation function is

$$g^{(2)}(t) = 1 + \underbrace{C_1 \exp[-2Dq^2\{t-(t_c-t_c \exp(-t/t_c))\}]}_{(i)} + \underbrace{C_2 \exp(-Gt)}_{(ii)}, \qquad (1)$$

where C_1, C_2, D, q, G, t_c and t is the coherence factor, the particle concentration factor, the diffusion coefficient, the scattering vector, the relaxation time of concentration fluctuation, relaxation time of particle and the delay time of correlation, respectively. The term (i) of Eq.(1) is the usual Gaussian term which involves non-Malkovian correction, and the term (ii) is the non-Gaussian term (Schaefer and Berne, 1972) which depends on particle concentration fluctuation. It is the purpose of this study to determin the limitation of unmodified autocorrelation function for aerosol measurment, because of fundamental importance are to understand the effect of above problems and to make clear the applicability of PCS to in-situ measurement of aerosol particles.

Experimental

The experiments on the reliability of PCS were made for aerosol and hydrosol PSL particles. Both experiments were made using the same PCS system shown in Fig.1. Correlation data processor is the single clipped autocorrelator (Malvern K7023) with scaling unit (RR95). The polarization of collimated Argon laser beam was adjusted by polarization rotator to get linear polarization of arbitraly polarization angle. A glass cell (50mm dia. x 80mm height) was used for hydrosol experiment, and glass ring cell (180mm dia. x 20mm height) was for aerosol experiment. The latter is a thermal diffusion type chamber designed to stabilize measuring condition

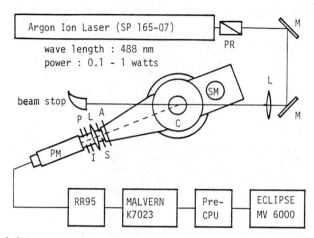

Fig.1 Schematic diagram of photon correlation spectroscopy system.
PR : Polarization Rotator, M : Mirror, L : Lens, A : Analyser,
S : Slit (0.1 – 0.5 mm), I : Iris, P : Pin-hole (0.1 mm),
SM : Stepping Motor, C : Thermal Diffusion Chamber or Cell.

from various disturbances such as convection. Nominal diameter of PSL
particles are 0.091, 0.197, 0.257, 0.497 and 0.913 μm . He, N_2 and Ar are
used for medium gas of aerosols. The particle concentration of hydrosol was
determined from turbidity of liquid with a photometric spectroscopy.

Experimental conditions and parameters are shown in Table.1. The
scaling was tested for the non-Gaussian situation where clipped correlation
reveals an eventual departure from full correlation. A variety of medium
gas and temperature are made use of checking the kinetic effect of gas for
Cunningham slip correction factor. The Mie effect is related to diameter of
particle and to scattering angle. The non-Gaussian effect depends on the

Table 1. Set up conditions and experimental parameters for Hydrosol
and Aerosol PSL particles. (* : Hydrosol only, + : Aerosol only,
No mark : Both, SV : Light Scattering Volume in PCS)

--|--	Conditions & Parameters	__
+ *	Correlator Operation Mode Polarization Angle Medium Gas Medium Liquid	Single Clipping, and/or Scaling V/V, H/H, V/H (see, note 1) Helium, Nitrogen, Argon Isopropyl Alcohol
*	Nominal Dia. of PSL (μm) PSL Particle Conc. (#/SV) Temperature (°C) Scattering Angle (degree)	.091, .197, .257, (.330*), .497, .913 0.1, 1, 10, 100, 1000, 10000, 100000 20, 55 (+) 20, 30, 45, 60, (80*), 90, (100*), 120

Note 1) V/H : Vertical for incident/Horizontal for scattered light.

---

number concentration of PSL particle in which the central limit
theorm cannot be assured. Ratio of
depolarized scattered light to
polarized scattered light
indicates the degree of
agglomeration of aerosol
particles.

Results and Discussion

Typical effect of concentration are shown in Fig.2 for
0.497μm PSL hydrosol particles.
The highest concentration is
limitted by the degree of multiple
scattering. The lowest limit is
determined by SN ratio and the
photon arrival interval which can
reconstruct the spectrum of
Doppler shift by Brownian motion.
The accuracy of PCS is shown in
Fig.3 for the same 0.497μm PSL
aerosol particles in different
media gases. Table.2 shows a part
of experimental results for
various parameters.

The measured diameter of hydrosols agrees with the nominal
diameter for whole range. Diameter
of aerosol particles is also in
agreement with the nominal diameter in such a limitted case that
the particle is larger than
0.257μm and the photon count rate
is higher than 0.05 counts per
sample time. Within the limit of
condition, kinds of medium gas and
difference of temperature have no
significant effect on particle
sizing. The departure from nominal
diameter, however, becomes clear
in smaller aerosol particles. It
is thought that thin film of emulsifier of PSL solvent prevents
clean up of aerosol surface,
because very small amount of existence of agglomeration was
observed by electron-microscopy
for dispersed PSL aerosol particles. It is also infered by the
weak depolarized light scattering
originated in rotational diffusion
of irregular particles.

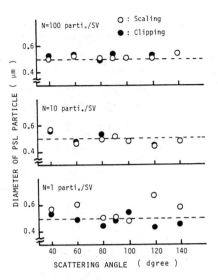

Fig.2 Influence of correlation mode
and number concentration on particle
sizing of 0.497μm PSL hydrosols.

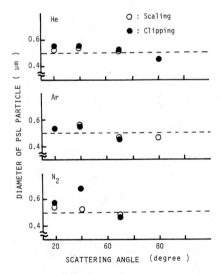

Fig.3 Influence of correlation mode
and medium gas on particle sizing
of 0.497μm PSL aerosols.

Table 2. Summary of experimental results for particle sizing
(* : Result by scaled correlation, SV : Scattering volume.).

Nominal Diameter (um)	Hydrosol PSL Particle Dia. (Particle Conc)	Medium Gas	Measured Dia. (averaged)[+]	
			Aerosol	Particle
0.091	0.93 0.93 * (1000 part./SV)	He Ar N$_2$	0.27 0.28 0.28	0.31 0.24 0.27
0.197	-----	He Ar N$_2$	0.25 0.26 0.27	0.28 0.25 0.25
0.257–Aero 0.330–Hydro	0.34 0.36 * (1000 part./SV)	He Ar N$_2$	0.31 0.30 0.30	0.26 0.28 0.30
0.497	0.50 0.49 * (10 part./SV)	He Ar N$_2$	0.52(0.53*) 0.51(0.50*) 0.57(0.52*)	---- ---- ----
0.913	0.99 1.03 * (10 part./SV)	He Ar N$_2$	0.78(0.89*) 0.95(0.88*) 0.75(0.72*)	1.03(1.00*) 0.55(0.82*) 0.98(0.90*)
	20	Temperature (C)	20	55

+) Averaging for the data that photon count rate per sample time is
higher than 0.05.

    As a conclusion, the accuracy of PCS is affected by scattered light
intensity, and forward scattering is preferable at normal condition using
clipped correlator. In this situation, PCS is applicable to in–situ measurement of sub–micron aerosol sizing (<10% error) and to study of dynamics
of aerosols. However, there was no significant difference in the applicable
range of particle concentration between scaled and clipped correlation in
this experiment. The result is thought that the clipping correlator is not
able to reproduce the spectrum of Doppler shift by Brownian motion of
aerosols because of the low data rate by scaling of photon count. It is
recommended to utilize a real time fast multibit correlator for getting the
higher data rate, and then for expanding the applicable range and improving the accuracy of PCS in aerosol study.

References

Dahneke,B.E. (1983) Measurment of Suspended Particles by Quasi–Elastic
Light Scattering (Edited by B.E. Dahneke), Wiely, New York.
Itoh,   M.   and Takahashi, K. (1981) "Aerosol Measurement by Photon
Correlation Spectroscopy (I)", Aerosols in Science, Medicine and
Technology, Proc. 9th Conf. GAeF, 273.
Schaefer, D. B. and Berne, B. J. (1972) "Light Scattering from Non–Gaussian
Concentration Fluctuations", Phys. Rev. Letts., 28:475.

Published 1984 by Elsevier Science Publishing Co., Inc.

Aerosols, Liu, Pui, and Fissan, editors

OPTICAL DIAMETERS OF AGGREGATE AEROSOLS*

B. T. Chen and Y. S. Cheng
Lovelace Inhalation Toxicology Research Institute
P. O. Box 5890
Albuquerque, NM 87185
U. S. A.

EXTENDED ABSTRACT

Introduction

The behavior of irregularly-shaped particles has been investigated in many systems. The equivalent diameters of aggregate aerosols have been derived from measurements of aerodynamic diameter, electrical mobility, and diffusion coefficient. A few experiments have been performed to measure the response of simple, close-packed cluster aggregates in optical particle counters (OPC). Gebhart et al. (1976) reported the light-scattering equivalent diameter of cluster aggregates composed of polystyrene latex (PSL) spheres in terms of geometric diameter. These experiments are difficult to repeat because the instruments are not commercially available. In the study, we intended to make similar measurements to understand more about the optical diameters of PSL cluster aggregates produced by nebulization. Three commercial optical counts--a Royco 226, Royco 236, and Climet 208A--were studied, and results were then compared with previous data.

Experimental

Primary PSL spheres between 0.3 and 2.4 μm in diameter were used in the study. The emulsifier used to prevent the later spheres from aggregating in the suspension was removed by centrifugation to increase the fraction of aggregates of uniform PSL spheres in the latex suspension. As a result, a highly concentrated suspension of latex spheres was prepared for the Lovelace nebulizer. After nebulization, aerosol was passed through a diffusion dryer, a Kr-85 bipolar-ion source, and a mixing chamber before entering the OPC. An aerosol sample for scanning electron microscopy (SEM) was taken from the chamber with an electrostatic precipitator. A multi-channel analyzer was used to classify the output signals from the OPC into discrete channels.

Results and Discussion

The SEM micrographs of PSL particles deposited on the precipitator indicate that the aggregate PSL particles from the nebulization always tend toward a compact configuration (cluster aggregate). The pulse height spectra displayed in the multi-channel analyzer contain many distinct peaks,

---

*Research performed under U. S. Department of Energy Contract Number
DE-AC04-76EV01013.

each representing a cluster aggregate having a specific number of primary spheres. Each peak can easily be corresponded to a certain aggregate by comparing the peak height in the spectra with the relative number of aggregates in the SEM micrographs.

The light-scattering equivalent diameter (D_{sc}, n) of an aggregate formed by n uniform spheres with primary diameter, D_1, and refractive index, m, is defined as the diameter of a spherical particle with refractive index, m, that has the same partial scattering cross-section as that of the aggregate. The D_{sc}, n of the PSL aggregate can be determined from the output signal of the OPC according to its size-response curve (Chen et al., 1984).

Figure 1 shows the relationship between measured diameter (D_{sc}, n) of doublet aggregates (n = 2) and D_1 with Royco instruments. It is clear that D_{sc}, n ~ n 1/3 D_1 for D_1 < 0.76 μm. This is because small particle behavior conforms more closely to the situation for Rayleigh scattering in which the scattering cross-section is independent of particle shape and proportional to the square of the volume of the dielectric particle (Hodkinson, 1966). As a result, D_{sc}, n is identical with the

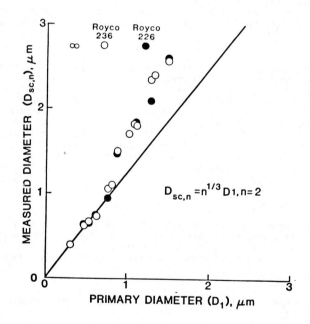

Figure 1. The relationship between optical equivalent diameters (D_{sc}, n) measured by the Royco counters and the primary particle diameter (D_1) for doublet aggregates (n = 2).

volume equivalent diameter, $n^{1/3} D_1$. Figure 2 shows the relationship between D_1 and D_{sc}, n measured from the Climet 208A. Good agreement is shown among data points and the straight line, D_{sc}, $n = n^{1/2} D_1$ ($n = 2$) for $D_1 > 0.76$ µm ($n^{1/2} D_1$ is the area equivalent diameter). These data are consistent with the statement made by Hodkinson (1966) and Gebhart et al. (1976) that for particles larger than the wavelength, the light scattering becomes either a surface effect or a projected area effect, depending on the optical configuration of the instrument.

References

Chen, B. T., Y. S. Cheng, and H. C. Yeh (1984) J. Aerosol Sci. 15(4), August 1984.

Gebhart, J., J. Heyder, C. Roth, and W. Stahlohofen (1976) in Fine Particles (Edited by B. Y. H. Liu), Academic Press, New York, p. 794.

Hodkinson, J. R. (1966) in Aerosol Science (Edited by C. N. Davies), Academic Press, New York, p. 287.

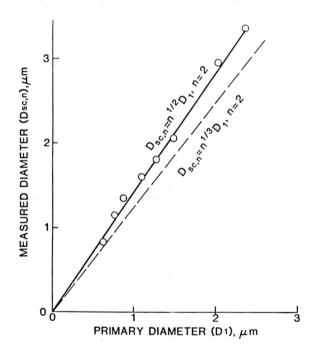

Figure 2. The relationship between optical equivalent diameters (D_{sc}, n) measured by the Climet 208A counter and the primary particle diameter (D_1 for doublet aggregates (n = 2).

Published 1984 by Elsevier Science Publishing Co., Inc.

Aerosols, Liu, Pui, and Fissan, editors

ANALYSIS OF AEROSOLS BY POLARIZATION MEASUREMENTS

R. H. Zerull
Bereich Extraterrestrische Physik
Ruhr-Universität Bochum
4630 Bochum 1
F.R.G.

Introduction

Light scattering measurements are a basic technique for the analysis of aerosol size and -concentration. It may be applied to single particles and mixtures of particles, as well, and can be optimized concerning the scattering angle(s) selected for particular applications. Further improvement can be achieved by additional polarization measurements. Whereas the inclusion of circular polarization is profitable only for particles of very specific shape, measurements of linear polarization and cross-polarization are generally useful. In particular, distinction between different types of aerosols within an mixture are possible, provided that the components exhibit different scattering features. Therefore, they must be distinctly different concerning at least one of their physical properties, i.e. size, shape, structure, or material.

Formalism

The scattering problems can be described proceeding from a linear transformation (see van de Hulst, 1957)

$$\begin{pmatrix} I_{s1} \\ I_{s2} \\ U_s \\ V_s \end{pmatrix} = \frac{\lambda^2}{4\pi^2 r^2} \begin{vmatrix} A_{ij} \end{vmatrix} \begin{pmatrix} I_{o1} \\ I_{o2} \\ U_o \\ V_o \end{pmatrix}$$

Index o characterizes the incident, index s the scattered radiation; λ is the wavelength; r is the distance between the scattering particle and the observer. For the case where the incident light is polarized perpendicular or parallel to the scattering plane and only the light polarized in the same direction is measured, the scattering event is described by the elements A_{11} and A_{22} of the scattering matrix alone.

Based on these coefficients, the total measured intensity is proportional to $A_{11} + A_{22}$, and the degree of linear polarization can be defined as

$$P = \frac{A_{11} - A_{22}}{A_{11} + A_{22}}$$

If, in addition, the scattered light component polarized perpendicular to the incident beam is measured, the cross-polarizing elements A_{12} and A_{21}, respectively, become relevant for nonspherical particles.

Measuring Concept

All elements of the scattering matrix mentioned in the previous section depend on the material, size, shape, and structure of the particles. To make use of this close relationship for particle discrimination, scattering angles must be selected where especially strong effects can be observed. Based on extensive microwave and laser experiments (Zerull and Giese 1974; Zerull et al. 1977, 1980; Weiss-Wrana, 1983) the following concept was developed and tested with an appropriate laser-setup. It consists of three sensors:

Sensor 1   for measurement of enhancement of the near forward
            scattered intensity
Sensor 2   for measurment of intensity, degree of linear polarization
            and cross-polarization at 90⁰
Sensor 3   for measurement of enhancement of backscattering.

The following table illustrates the power of polarization and cross-polarization measurements. They allow conclusions concerning size, material, shape, and structure, whereas pure intensity measurements just provide information concerning the particle size.

Measured Quantity	Scattering Angle	Distinction Criterion	Particle Properties addressed
Intensity	near 0⁰ ≈ 90⁰ near 180⁰	relative enhancement of forward and backward scattering	Size
Linear Polarization	≈ 90⁰	zero weakly positive strongly positive weakly negative strongly negative	Material, Shape, Structure, Size
Cross-Polarization	≈ 90⁰	zero weak strong	Material, Shape, Size

Detailed results will be presented at the conference.

There are mainly two applications for this concept in aerosol science: One can get information on the physical properties (size, shape, structure, material) of an unknown one-component aerosol by simple optical in situ measurements. Furthermore, concentrations of different components within a mixture of aerosols can be measured independently. For this purpose it has to be provided, that the aerosol components are distinctly different concerning size, shape, or material. The measuring principle may be optimized for each special problem by choice of light source and appropriate arrangement of sensors. Best results are obtained, if the instrument is calibrated with all components in question separately.

References

Hulst, van de, H.C. (1957) Light Scattering by Small Particles. Wiley, N.Y.

Zerull, R., Giese, R.H. (1974) Planets, Stars and Nebulae Studies with Photopolarimetry (T. Gehrels, Editor) University of Arizona Press.

Zerull, R.H., Giese, R.H., Weiss, K. (1977) Appl. Optics, 16, 777.

Zerull, R.H., Giese, R.H., Weiss, K. (1977) "Optical Polarimetry", SPIE, Vol. 112, 191-199.

Zerull, R.H., Giese, R.H., Schwill, S., Weiß, K. (1980) Light Scattering by Irregularly Shaped Particles (Ed. D.W. Schuerman) Plenum Press, New York.

Weiss-Wrana, K. (1983) "Optical properties of interplanetary dust: comparison with light scattering by larger meteoritic and terrestrial grains", Astron. Astrophys., 126, 240-250.

Published 1984 by Elsevier Science Publishing Co., Inc.

Aerosols, Liu, Pui, and Fissan, editors

APPLICATION OF AN ELECTRO-OPTICAL CONTINUOUS AEROSOL MONITOR TO ON-LINE PARTICLE SIZE ANALYSIS IN INDUSTRIAL PROCESSES

G. Kreikebaum, F. M. Shofner, and A. C. Miller, Jr.
ppm, Inc.
Knoxville, TN 37922
U. S. A.

EXTENDED ABSTRACT

Introduction

The importance of particle size measurement to many industrial processes has been well established. Abrasives, cement, explosives, foods, glass sand, ores and toners are just a few examples of products that require particle size analysis to ensure product specifications. The application of an automatic on-line particle size analysis system for these processes offers such advantages as speed of response, operator independence and convenient data presentation. The speed of response allows this method to ultimately close the loop for control of particle size. This paper describes the use of the Continuous Aerosol Monitor (CAM) in such an application.

Theoretical

The principles of operation of the Continuous Aerosol Monitors and its application to industrial hygiene have been previously reported (Kreikebaum, et al, 1983; Smith, 1984). The unique combination of a nephelometer (or photometer) and a single particle counter allows light-scattering measurements of aerosol concentration and particle size from $0.2\mu m$ up with solid state components such as an infrared LED as a light source and a pin photodiode as a detector. These rugged, low energy components can satisfy the requirements of plant environment and plant safety. The presentation will include the discussions of near-forward light scattering as it applies to the particle size analysis of various materials.

There are a number of mechanisms that can lead to errors in particle size analysis (Novak, 1984). They range from obtaining a non-representative sample from the process to material degradation or agglomeration in the material analysis preparation and to attrition in the material transport. Discussion of the system components will concentrate on how the system design minimizes these effects.

Application Data

Data will be presented on precision vs. analysis time and material feed rate.

30

Included are comparisons of the on-line system data to various
off-line analyses by a number of methods from sieving to laser
diffraction pattern analysis. The discussion will also focus on
possible differences in particle size analysis results, especially
for non-spherical material.

References

Kreikebaum, G., Miller, A. C., and Shofner, F. M. (1983) "Electro-
Optical Mass Concentration and Approximate Size Measurement",
Aerosols in the Mining and Industrial Work Environments, (Edited
by V. A. Marple and B. Y. H. Liu). Vol. 3, p. 719.

Smith, J. P. (1984) "Response Characteristics of Scattered Light
Aerosol Monitors Used for Control Monitoring", American Industrial
Hygiene Conference, Detroit, Michigan.

Novak, J. W. and Thompson, J. R. (1984) "Progress in On-Line
Particle Size Monitoring Systems for Process Control", Proceedings, 9th Powder and Bulk Solids Conference, Chicago, Illinois.

Published 1984 by Elsevier Science Publishing Co., Inc.

Aerosols, Liu, Pui, and Fissan, editors

PERFORMANCE OF PMS MODEL LAS-X OPTICAL PARTICLE COUNTER

W. C. Hinds and G. Kraske
Division of Environmental and Occupational Health Sciences
UCLA School of Public Health
Los Angeles, CA   90024
U. S. A.

EXTENDED ABSTRACT

Introduction

This paper describes our evaluation of the performance of the PMS optical particle counter model LAS-X in three areas: the effect of particle refractive index, the effect of coincidence error on measured size, and inlet losses as a function of particle size.

The performance of PMS (Particle Measuring Systems,Inc.; Boulder, CO) optical particle counters have been characterized by several investigators [Pinnick and Auvermann(1979), Pinnick et al (1981), and Garvey and Pinnick (1983)]. Garvey and Pinnick (1983) evaluated the performance of PMS model ASASP-X and gave response curves which were confirmed with experimental measurements.

The model LAS-X and the model ASASP-X have the same optics and both come in two versions. One version has four ovelapping size ranges, each with 15 equal interval channels, covering the range of 0.09 to 3.0 µm. The other has three ranges covering the size range 0.12 to 7.5 µm and a fourth ranges that covers the entire range from 0.12 to 7.5 µm with unequal width channels. Each range has a 16th channel that gives the count of oversize particles. Both versions use He-Ne laser illumination and collect light over scattering angles of 35° to 120° by means of a parabolic mirror so that the full 360° range of polarization is collected. A focused stream of particles passes through the focal point of the parabolic mirror. This illuminated region is within the laser cavity so illumination is equal in the forward and backward directions. Earlier work by Garvey and Pinnick (1983) evaluated the response characteristics of the 0.09 to 3.0 µm version; we present here our evaluation of the 0.12 to 7.5 µm version.

Effect of Refractive Index

The effect of refractive index was evaluated theoretically by calculating the relative signal produced for 31 particle sizes for each of 34 refractive indexes. The refractive indexes used were 1.3, 1.333, 1.35, 1.4, 1.448, 1.45, 1.471, 1.485, 1.5, 1.544, 1.55, 1.5905, 1.6, 1.65, 1.7, 1.8, 1.9, 2.0, 1.2-(.1, .3,.5,.7,1,1.5,2.0)i, 1.5-(.1,.3,.5,.7,1,1.5,2)i, 1.59-0.66i, and 1.67-0.26i. Signals were calculated using the light scattering calculation subroutine DAMIE written by Dave(1968). Relative output signals were determined by combining the scattering amplitudes produced by forward and backward illumination and integrating over all scattering angles. As pointed out by Garvey and Pinnick (1983) it is the scattering amplitudes rather than the scattering intensities that must be combined because the scattered light from the forward and backward illumination are not independent. Thus the relative signal R has the form

$$R = K \int_{35^0}^{120^0} \left[\left| S_1(\theta) + S_1(\pi-\theta) \right|^2 + \left| S_2(\theta) + S_2(\pi-\theta) \right|^2 \right] \sin\theta \ d\theta \qquad (1)$$

where K is a constant, theta is the scattering angle and S_1 and S_2 are the scattering amplitudes in the horizontal and vertical directions and the straight brackets indicate complex conjugation. A single constant relating the relative signal calculated by (1) and the instrument calibration was determined by a least squares fit between the manufacturer's calibration data for nine polystyrene latex particle sizes and the calculated response curve for m = 1.5905 – 0.0i.

Response curves were plotted with the adjusted manufacturer's calibration curve to show departures of the scattered signal from the calibration curve for a given particle size. A more useful presentation is to project the signal back onto the diameter axis to give the indicated size for a particle having a given size and refractive index. In many situations aerosols of unknown or mixed refractive index are present and it is more useful to know the extent of the errors in size that are introduced by the range of refractive indexes present. This is shown in Figure 1. for three refractive index ranges. The curves have been smoothed to show the general contours of the error curves.

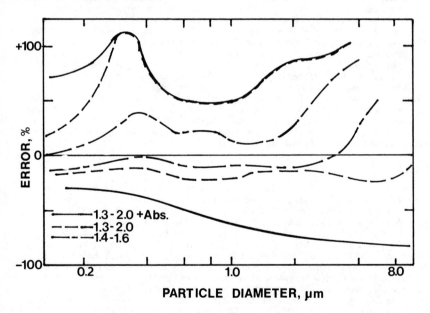

Figure 1. Maximum error in size measurement versus particle size

Data are truncated at both ends by the extreme channel limits. For nonabsorbing particles oversizing or positive errors are greater then negative errors (undersizing). When absorbing and nonabsorbing particles are present together the errors are more symetrical. All refractive index groups show significant oversizing error for particles larger than about 2.0 μm.

Effect of Coincidence on CMD

The effect of coincidence on measured size distribution was evaluated by measuring the size distribution of a submicrometer DOP aerosol at different dilutions. Computer fitting was used to determine the best fitting log normal distribution for each measured size distribution. For number concentrations less than 10^4/cm^3 there was less than a 10% error in CMD. Errors were calculated assuming that the low concentration plateau in CMD was the correct CMD. This is consistent with similar measurements made with polystyrene latex spheres that showed less than a one channel width (average channel width is 10.6% of the lower limit)shift in the peak size for concentrations up to 10^4/cm^3. These results can be compared to a theoretical loss of counts by coincidence of 30% for 10^4/cm^3 using the sensing volume given by the manufacturer of 0.04 mm^3.

Inlet Losses

Because this version of the LAS-X sizes particle up to 7.5 μm and registers particles greater than 7.5 μm in the oversize channel it is of interest to identify size selective particle losses in the inlet and associated tubing. Monodisperse methylene blue particles of 1.8, 6.3, and 12.5 μm aerodynamic diameter were generated with a vibrating orifice aerosol generator. The standard PMS inlet tube (stainless steel 3.2 mm OD, 2.1 mm ID, and 100 mm length) was positioned so that 28 mm protruded into a 115 mm diameter chamber. Simutaneous samples were taken for microscopic counting with a 13 mm open face filter holder positioned at the same level and 4 cm away from the PMS inlet tube. The aerosol passed through a neutralizer and flowed vertically downwards in the chamber at a velocity of about 1 cm/s. This velocity is sufficiently low that sampling is equivalent to still air sampling for both samplers (Hinds, 1982). Results for a sampling flow rate of 5 cm^3/s are shown in Figure 2 with the mean and +/- two standard deviations given for each point. Two measurements at 7 cm^3/s are also given for 12.5 μm diameter particles. The latter gives an indication of the effect of sampling flow rate on sampling efficiency. The sampling efficiency measured here does not differentiate between inlet losses, internal losses, and counting losses.

34

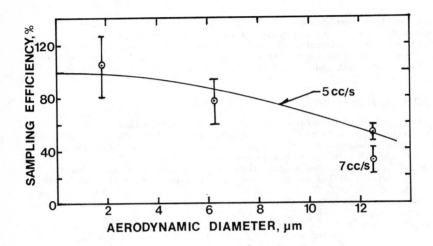

Figure 2.   Inlet losses versus particle size.

References

Dave, J. V. (1968) "Subroutines for Computing the Parameters of
Electromagnetic Radiation Scattered by a Sphere" Report No. 320-3237, IBM
Scientific Center, Palo Alto, CA.

Garvey, D. M. and Pinnick, R. G. (1983) "Response Characteristics of the
Particle Measuring Systems Active Scattering Aerosol Spectrometer Probe
(ASASP-X)" Aerosol Sci. Tech. 2:477-488.

Hinds, W. C. (1982) Aerosol Technology: Properties, Behavior, and
Measurement of Airborne Particles ,Wiley Interscience, New York.

Pinnick, R. G. and Auvermann, H. J. (1979) "Response Characteristics of
Knollenberg light-scattering aerosol counters", J Aerosol Sci. 10:55-74.

Pinnick, R. G., Garvey, D. M., and Duncan, L. D. (1981) "Calibration of
Knollenberg FSSP Light-scattering Counters for Measurement of Cloud
Droplets", J, Applied Meteorol. 20:1049-1057.

Published 1984 by Elsevier Science Publishing Co., Inc.

Aerosols, Liu, Pui, and Fissan, editors

GRADE EFFICIENCY CURVES OF A SAMPLING CYCLONE AND MEASUREMENT OF PARTICLE SIZE DISTRIBUTIONS OF AIRBORNE PARTICLES WITH AN OPTICAL PARTICLE COUNTER

H. BUETTNER
Particle technology and fluid mechanics laboratory
Mechanical Engineering Departement
Universität Kaiserslautern
6750 Kaiserslautern
F R G

EXTENDED ABSTRACT

Introduction

In order to characterize flow separation and classification processes particle size distributions must be measured in airborne state, without affecting
the state of dispersion or disturbing the flow. Light scattering devices with
optically defined measuring volume (see figure 1) are specially designed for
this purpose.

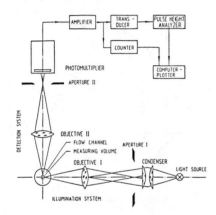

Figure 1. Schematic of a light
scattering device
with optically defi-
ned measuring volume

The light scattering device must be calibrated. Primarily measured particle
characteristic is the pulse amplitude of scattered light pulses of a single
particle.To obtain a calibration curve, normally monodisperse Latex particles
are used. If the particles are irregular shaped and differ in their optical
properties from the Latex particles this calibration is unsatisfying. In these
cases it is necessary to calibrate within the twophase flow using the generally non ideal particles of this flow. Preferably a particle size directly related to the particle motion should be measured.

Grade efficiency curve of the sampling cyclone

Calibration on the basis of equivalent settling diameters can be achieved by
measuring with the optical particle counter before and after an impactor. This
first was suggested by Marple and Rubow (1976). The result is the grade effi-

ciency curve of the impactor against the pulse amplitude of scattered light
pulses. If the grade efficiency curve of the impactor against aerodynamic diameters (or Stokes diameter) is known, these both curves can be compared to win
the calibration curve. Often only the cut sizes are compared. The grade efficiency against aerodynamic diameter of the used impactor is known experimentally and theoretically. With this method a calibration curve for a droplet
system of a mixture Glycerol/water was established. Own measurements on solid
particles showed that the impactor did not work very well because of blow up,
bounce effects and wall losses. In many cases the deposited particles caused
a change of the grade efficiency curve of the impactor. The quantities of dust
that can be examined are very small. These disadvantages are eliminated on replacing the impactor by a sampling cyclone. The disadvantage of a sampling cyclone is the fact that a theoretical calculation of its grade efficiency curve
is currently impossible.With the calibration curve of the droplet system gly-
cerol/water it is possible to measure the grade efficiency curve of the used
sampling cyclone with the light scattering device. The results together with
the cyclone are shown in figure 2.

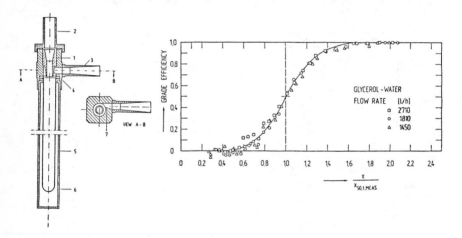

Figure 2. Sampling cyclone and dimensionless grade efficiency curve
($x_{50,t}$ MEAS. = measured cut size)

The measured grade efficiency curves correlate well in the tested range, if
the grade efficiencies are plotted against a normalized particle diameter.

Calibration curves of the light scattering device

With this result it is possible to use the sampling cyclone to obtain calibration curves pulse amplitude against Stokes diameter of the optical particle
counter for irregular shaped particles.
Figure 3 shows such calibration curves for different particulate systems.
Every material has his own curve dependent on particle shape and optical constants.

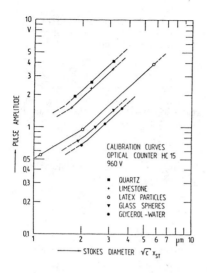

Figure 3. Calibration curves of the optical particle counter

The measured cumulative number distributions against pulse amplitude can be converted by the help of calibration curves in cumulative volume distributions against Stokes diameter. Such particle size distributions measured with an optical particle counter can be compared with measurements of the same distributions with a sedimentation analysis because in both cases the Stokes diameter is the particle characteristic. This comparison is shown in figure 4.

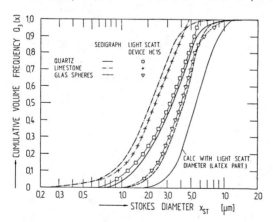

Figure 4: Comparison of particle size distributions measured with an optical counter and with a sedigraph

The agreement is good and shows the practicability of the used calibration method. The differences in the range of the coarser particles is caused by the limits in the measuring range of the light scattering device.

Conclusion

With a combination, optical particle counter sampling cyclone, exists a method for measuring particle size distributions in the airborne state without disturbing the flow or affecting the state of dispersion. The measured distributions can be plotted against the Stokes diameter.

References

Marple, V.A. and Rubow, K.L. (1976) "Aerodynamic Particle Size Calibration of Optical Particle Counters", J. Aerosol Sci. 7: 425

Published 1984 by Elsevier Science Publishing Co., Inc.

Aerosols, Liu, Pui, and Fissan, editors

DETERMINATION OF SIZE DISTRIBUTION PARAMETERS OR REFRACTIVE
INDICES OF AIRBORNE PARTICLES USING A LASER LIGHT SCATTERING
SPECTROMETER

K. Blum and H. J. Fissan
Prozess- und Aerosolmesstechnik
Universität Duisburg, FRG

EXTENDED ABSTRACT

Introduction

Optical measurement techniques permit quasi-continous measurements of aerosols without separating the particles from the
suspending gas, thereby lessening the chance of causing a change
in particle size during measurement. In most instruments the
scattered light in a fixed range of angles with respect to the
incident light beam is observed. In optical particle counters the
scattered light of single particles is detected (Fissan, Helsper
1983). In other instruments the scattered light from a particle
cloud is analyzed (Armbruster, Breuer, Gebhart, Neulinger 1984).

In this paper we report on the measurement of scattered light
intensities of a particle cloud as a function of scattering
angle. The chosen experimental set-up is intended for the analysis of aerosols from combustion processes. The aerosol contain
rather small particles ($D_p < 1 \ \mu m$) whose properties may change
if they are sampled. The information on the intensity distributions can be used for instance in the design of fire detectors
and other optical instruments for particle analysis. The measured
scattered light intensities can also be used to derive certain
properties of the particles, such as the size distribution parameters and refractive index.

Experimental Set-up

A He-Ne-Laser (λ = 633 nm) is used as the light source. The
aerodynamically focussed aerosol stream is drawn through the
intersection aerea of the expanded laser beam and the small angle
field of view of the detector. The signal is processed in a lock-
in amplifier and plotted.

Results

Intensity distributions of glass fibers:
A glass fiber was positioned in the center of the measuring
volume and the scattered light intensity distribution was measured. These experiments were mainly performed to check the behavior of the experimental set-up. Thus problems with unstable
scatterers like aerosols were avoided. The measured scattered
light intensities of a fiber with a diameter D=12 μm is shown
in fig.1. To eliminate the dependance of the intensity distribution on particle number concentration in the case of aerosols, on
incident light intensity and other changes in the optical behavior of the set-up, the ratio of intensity at an angle θ to that

40

at θ = 100° is plotted.

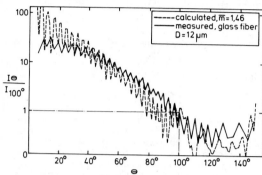

Figure 1. Measured and Calculated
Scattered Light Intensity Ratios
of a Glass Fiber

This ratio shows strong fluctuations with angle. Also shown is the calculated intensity ratio using Mie-theory for spherical particles with D_p = 12 μm and the refractive index of quartz $\overline{m}$ = 1.46.

A more detailed analysis of the scattering process of a fiber positioned perpendicular to the plane of the laser and detector reveals that only the absolute intensities of a fiber and a spherical particle are different.

Therefore the intensity ratios are very similar (fig. 1) especially with respect to the position of maxima and minima. In the range of small angles in the forward direction, the measured intensities are smaller than the calculated values. This is probably caused by the non-linear behavior of the detector at high intensities.

Theoretical calculations based on Mie-theory for monodisperse, spherical particles with different diameter show that the angle between two adjacent maxima or minima is dependent upon the particle size. In fig. 2 the diameter D is plotted as a function of the angle difference Δ θ at a scattering angle of about 40°.

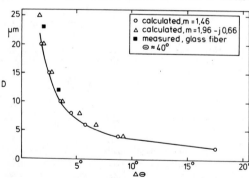

Figure 2. Objekt Diameter D against
Angle Difference between Intensity
Maxima

The calculations were performed for particles with two different refractive indices, covering the range of interest in the present study. It is interesting to note that the dependance of Δ θ on D is relatively unaffected by refractive index. The determination of the angles at which maxima or minima of intensity occurs can be used for size determination especially for monodisperse particles. The method has been checked by determining the angular difference of two glass fibers of known diameter. The results confirm the theoretical findings.

The application of this method is limited to the particle

range $2 \mu m < D_p < 25$ μm $(\theta = 40^{\circ})$, and to aerosols, which
produce intensity fluctuations, especially nearly monodisperse
and spherical particles.

The intensity distributions also contain information about
the refractive index of particle material. The intensity ratios
of particles with different refractive indices but equal size are
shifted (fig. 3), as theoretical calculations show.

The determination of particle diameter and refractive index
can be based on a comparison of measured and calculated intensity
distributions. The exact comparison of theory and experiment is
difficult to perform, as the measured intensity distributions may
contain errors. Thus the method may not be conclusive. The method
yields better results if the size or the refractive index is
known. In the case of a polydisperse aerosol one must know the
type of the size distribution in advance to reduce the number of
unknown parameters. A computer code has been developed for the
comparison of theoretical and experimental intensity distributions, from which missing parameters are derived.

In fig. 3 the measured intensity distribution of monodisperse
DOP-particles produced with a vibrating orifice generator is
shown. In this case all parameters are known. The calculated
intensity distribution is also shown. The comparison of these two
distributions shows that the measured curve is similar to the
calculated one, but it does not show the detailed fine structure
given by the theoretical calculations. This can be explained by
the finite angle of the detector aperture, the fluctuation in
aerosol concentration during the measuring period of t= 6 seconds
and the low signal-to-noise ratios. These problems will be solved
in the future by improving the experimental set-up and by adjusting the theoretical model to the experimental conditions.

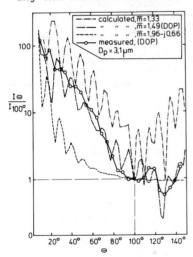

Figure 3. Scattered Light
Intensity Ratios of Dif-
ferent Particle Materials

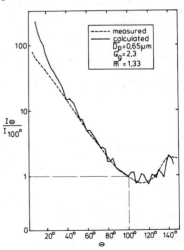

Figure 4. Measured and
Calculated Scattered Light
Intensity Ratios of an Atomizer

Nevertheless, the results show clearly that the determination of refractive index of particles of known size is possible.

In the case of polydisperse aerosols the intensity fluctuations are decreased. The informations about the aerosol parameters have to be derived from the position and the shape of the intensity distribution. In this case the comparison yields only good results, if only two parameters, for instance mean diameter and standard deviation have to be determined and the other two real and imaginary part of refractive index are known. The method has been applied to an aerosol of water droplets containing a small amount of salt produced with an atomizer.

In this case the refractive index is known (water $\overline{m}$ = 1.33). In fig. 4 the measured intensity ratios are shown. Using the program for comparison of measured and calculated intensity distributions the median diameter and geometric standard deviation have been determined. ($\overline{D_p}$ = 0.65 μm; σ_g= 2.3). The determined geometric standard deviation has been confirmed by measuring the size distribution of the residue salt particles after evaporation of the water using an electrical aerosol analyzer. The data have been used to calculate the intensity distribution. The corresponding intensity ratios are given in fig. 4. The experimental curve is smoother than the calculated one for reasons given before. The deviation at small angles again is caused by non-linearity effects of the detector.

Summary

An experimental set-up has been described which allows the determination of scattered light intensities as a function of scattering angle. The measured intensity distributions of a glass fiber, monodisperse DOP-particles and polydisperse water particles have been used to determine the unknown parameters such as the size distribution parameters or refractive index by comparison with calculated intensity distributions. The limitations of this method are discussed. It can be applied to determine size or optical properties of particles, if the particles are spherical and if some information about particle size or refractive index is available in advance.

References

Fissan, H. J. a. Helsper, C. (1983)
"Calibration of Polytec HC-15 and HC-70 Opticle Particle Counter"
Aerosols in the Mining and Industrial Work Environments
Vol. 3, Instrumentation Ed. by V. Marple, B. Liu, Ann Arbor Science, S. 825-829

Armbruster, Breuer, Gebhart a. Neulinger (1984)
"Photometric determination of respirable dust concentration without elutriation of coarse particles"
3. European Symposium on Particle Characterization, Nürnberg, May, 1984

Published 1984 by Elsevier Science Publishing Co., Inc.

Aerosols, Liu, Pui, and Fissan, editors

TRANSMISSION AND GRANULOMETRY OF AN AEROSOL CLOUD
APPLICATION TO THE DETERMINATION OF
THE COMPLEX INDEX OF AEROSOLS

By P. Adam*, P. Herve** and P. Masclet*

1-INTRODUCTION

To understand the optical behavior of particles suspended in the
atmosphere, we need to know their thermooptical properties: transmission,
reflection, emission. The experimental measurement of the complex refractive
index of the material constituting the aerosol makes it possible for us to
calculate all these quantities.

Some methods consist in depositing the aerosol to obtain a liquid or a
solid body and then to use some well-known technical measurements:
Kubelka-Munck [1,2], Frensnel coefficients [3,4], or emissivity [5,6].
However the disadvantage of such methods - sedimentation time (several months
for atmospheric dust [7]) and the impossibility to follow the chemical and
physical evolution of the product - led us to develop an "in situ" measurement
technique of the complex refrative index, i. e. on the sample in suspension.
It consists in a simultaneous measurement of the transmission and the granulometry.

2-THEORY

2.1-TRANSMISSION

Beer's law gives the transmission through a cloud of particles:
T=exp(-βL) where L is the optical thickness. β is the extinction coefficient
and depends both upon the granulometry and the complex refractive index:

$$\beta = \int_{o}^{\infty} \pi Q(n^*,\alpha) N(a) a^2 da \qquad (1)$$

where N(a) is the granulometry (number of particles per unit volume whose
radii are between a and a+da), n the complex refactive index ($n^*=n_r-j\chi$) and
the size parameter ($\alpha=2\pi a/\lambda$),

Q is the extinction efficiency factor. The problem is to solve eq. (1)
to find n^*, once β and N(a) are experimentally determined. For spherical particles, we have developed an algorithm with Mie's theory [8] using the technique of continuous fractions, on a PDP 11/23 computer. The curves on fig. 1
take about 20 min of calculation and contain 200 points.

*E.T.C.A, 16 b. Rue Prieur de la Cote d'Or, 94114 ARCEUIL CEDEX (FRANCE)
** I.U.T., 1 Chemim Desvallieres, 92410 VILLE D'AVRAY (FRANCE)

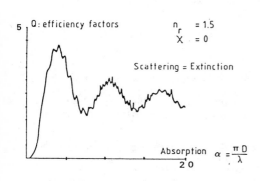

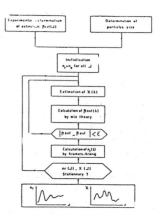

Fig.1 Efficiency factors   Fig.2 Resuming flowchart

The Kramers-Kronig relation 9 completes eq.(1) and enables us to solve it and to find n. Fig. 2 shows the general flowchart of our method. This iterative process converges in the domain $0 < n_r$ and $0 < X < 2$, which is verified for most of the atmospheric aerosols.

2.2-GRANULOMETRY

To determine the granulometry, we use a He-Ne laser beam, diffracted by the aerosol particles. Using Babinet's theorem, the elementary diffraction patterns of each particle add to each other. After a mathematical transform, we are led to a classical Fourier transform problem. The most important advantage of this method is to provide us with a continuous granulometery of the particles.

3-MEASURING APPARATUS

The measuring devices are built on a Fourier transform spectrometer, model 608X from Nicolet, which enables us to acquire up to 50 spectra of transmission per second. Two detectors, a photomultipier and a germanium bolometer, cooled by liquid helium make it possible for us to obtain a spectral domain lying from 0.3 µm to 1000 µm.

In the middle of the apparatus (fig. 3), there is an experiment chamber in which two optical beams are crossing, one for transmission measurements and the other for granulometer measurements.

The diffraction pattern of the cloud of particles, which enables us to calculate the granulometer, is obtained (fig. 4) by means of a chopped laser beam and a lens (f=300 mm). It is then analyzed in one direction by a scanning mirror and detected by the photomultiplier.

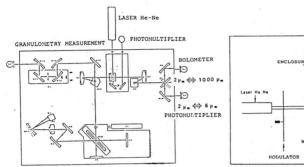
Fig.3 Measuring apparatus

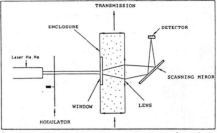
Fig4 Measuring system for granulometry

The position of the experiment chamber in the image plan makes it possible for us to modify, as we wish, the scale of the diffraction pattern detected and to choose the measuring range of granulometry (.3μm - 200 μm). All these devices together allow us to obtain more than 50 spectra per second.

4-RESULTS

Measurements have been performed on three different types of samples: a chemically non-evoluting cloud of coal ashes, an aerosol during its formation period, generated by Tetrachloride of Titanium (TiC14) and on a not very turbulent flame.

4.1-COAL ASHES

On Fig. 5 we can see the diffraction pattern of coal ashes and the corresponding granulometry in the range .2 μm - 2 μm. These coal ashes show aggregates of particles whose size is about 60 μm. These aggregates can be destroyed by ultrasounds.

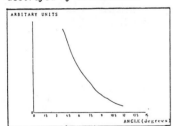
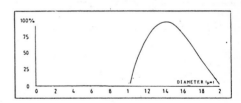

Fig.5 Diffraction pattern and granulometry of coal ashes

46

4.2-TETRACHLORIDE OF TITANIUM (TiCl4)

The corresponding granulometry has been compared with that obtained using an industrial particle sizer ASASP-X. These two results are in very good agreement (fig. 6).

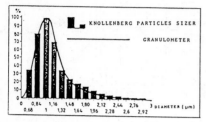

Fig.6 Granulometry of an aerosol of TiCl4

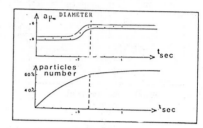

Fig.7 Evolution of the granulometry

Fig. 7 shows the evolution of the granulometry during the formation time of the aerosol. Fig. 8 represents the optical transmission of the aerosol.

The last figure (fig. 9) shows the complex refractive index of the aerosol. This result allows us to compute the thermooptical properties of the cloud.

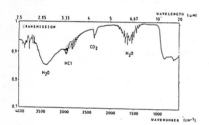

Fig.8 Optical transmission of the aerosol

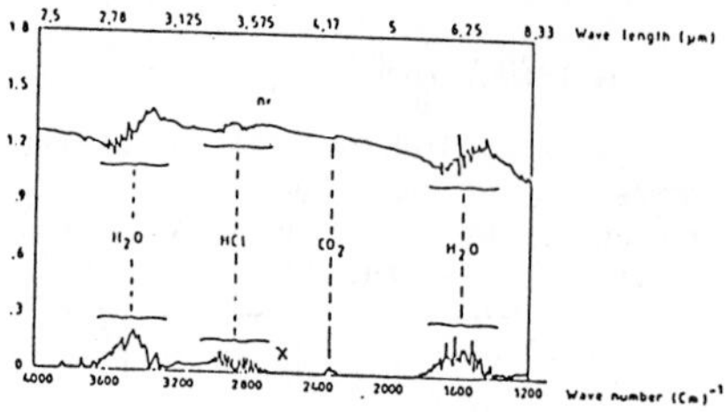

Fig.9 Complex refractive index

5-CONCLUSION

     We have achieved some techniques to make simultaneous measurements of the
infrared transmission and the granulometery of a cloud of particles. A new
advantage of our method is to obtain a continuous granulometer of the size of
the particles in a large range - .3 μm to 200 μm - with a time of analysis
smaller than 1/50 sec. The transmission measurements of the spectrum ranging
from 2 μm to 25 μm are performed in the same time interval. We can then
calculate the complex refractive index of the aerosol by using Mie's theory
and Kramers-Kronig relation. Hence, the physical and chemical evolution of
a spherical aerosol is perfectly determined by the knowledge of its granulometry and its complex refractive index known every 1/50 of a second. For
example, we can continuously follow a particle jet such as coal ashes or the
formation of an obscuring smoke. Measurements have already been performed
over not very turbulent flames and we are improving the acquistion of time-
basis of our equipments to be able to analyze phenomena even in turbulent com-
bustions.

BIBLIOGRAPHY

1.   P. Kubelka and F. Munck
     Zeitung fur Technische Physik, V12, p. 593 (1931)

2.   J. D. Lindberg and L. S. Laude
     Applied Optics, V13, p. 1923 (1974)

3.   V. Tomaselli and al.
     Applied Optics, V20, p. 3961 (1981)

4.   A. Rusk and D. Williams
     J.O.S.A. V61, p. 895 (1971)

5.   P. Herve
     These de Doctorate d'Etat.
          Influence de l'etate de surface sur l'emission thermique des
          materiaux. Paris (1977)

6.   P. Masclet
     These de Docteur-Ingenieur.
     Mesure de l'emissivite et de l'indice complexe des solides et des
     liquides dans l'infrarouge. PARIS (1982)

7.   J. D. Lindberg and J. B. Gillespis
     Applied Optics, V16, p. 2628 (1977)

8.   H. C. Van De Hulst
     Light scattering by small particles, John Wiley (1957)

9.   L. Landau and E. Lifchitz
     Electrodynamics of continuous media, MIR editions - Moscow (1969)

Published 1984 by Elsevier Science Publishing Co., Inc.

Aerosols, Liu, Pui, and Fissan, editors

Light Scattering On Iron Oxide Aerosols

by

Bidyut Bhattacharyya and D. T. Shaw

Laboratory for Power and Environmental Studies
State University of New York at Buffalo
Buffalo, New York

ABSTRACT

A technique has been developed to perform the dynamic-light-scattering experiment on iron oxide aerosols which are normally formed by combustion process. In this method iron oxide particles are made to flow in the air between two tubes separated by a distance 0.5 cm. These particles are used to scatter the He-Ne laser light in order to measure the auto-correlation function of the scattered light. This gives a quantitative information about the fluctuations of the motion of the particles, including all accurate determination of their correlation time. From the correlation time one can estimate the total diffusion coefficient (including translation as well as rotation). The application of the technique for the determination of the time-dependent size distribution function of a fast-coagulating chain-aggregate aerosol will be discussed.

CHAPTER 2.

CONDENSATION DETECTION

Published 1984 by Elsevier Science Publishing Co., Inc.
Aerosols, Liu, Pui, and Fissan, editors

SIZE-SUPERSATURATION RELATIONSHIP IN
ACTIVATION OF ULTRAFINE AEROSOL PARTICLES

Y. Kousaka, K. Okuyama and T. Niida
Department of Chemical Engineering
Sakai 591, Japan

EXTENDED ABSTRACT

Introduction

The activation of ultrafine aerosol particles in various supersaturated vapors has been studied. As the condensation nuclei, monodisperse NaCl, $ZnCl_2$, Ag and silicon oil particles having diameters up tp 15 nm were used, and water, hexanol, DBP and DEHS(di-(2ethylhexyl) sebacate) were used as the supersaturated vapors. Using these particles and vapors, number fractions of enlarged particles by condensation were observed at various supersaturation ratios.

Kelvin effect

The vapor pressure at the surface of a spherical liquid droplet, p_d, is given as the following Kelvin equation.

$$p_d/p_s = \exp 4\bar{v}\sigma/(d_p RT) \qquad (1)$$

p_s is the saturated vapor pressure on a flat surface, R is the gas constant, $\bar{v}$, σ, d_p and T are molar volume, surface tension, droplet diameter and temperature, respectively. This equation gives the critical relationship between the droplet size and the surrounding supersaturation($S=p_d/p_s$) where neither condensation nor evaporation occurs. In the case where the substance of the surrounding supersaturated vapor is the same as that of the droplet, the droplet is expected to be enlarged by condensation if the supersaturation ratio S surrounding the droplet is larger than p_d/p_s given by Eq.(1). In the case where supersaturated vapor condenses on a foreign particle whose substance is different from that of the vapor, the interactions between the vapor and the particle surface become important and the consideration of surface chemistry is required. The supersaturation ratios at which particles initiate to grow are observed in the several systems where substances of particle and vapor are different, and the deviation from Eq.(1) is discussed.

Control of supersaturation

The method to form supersaturation is the same in principle as that reported in our previous paper(Kousaka et al., 1982), where a room temperature aerosol is instantaneously and continuously mixed with the clean air flow which is heated and saturated with a certain vapor. This method has been applied to CNC(mixing type CNC). Supersaturation can be controlled in this method by changing the temperature of the heated saturated air or by changing the mixing ratio of the heated air flow rate to the total flow rate, R_h. The value of the supersaturation ratio S thus obtained is calculated by the analysis derived from the material balance of the vapor and the enthalpy balance of the system.

Experimental method

NaCl, ZnCl$_2$, Ag and silicon oil aerosols were produced by an evaporation-condensation type aerosol generator(Kousaka et al., 1982) and were
electrically classified by a differential mobility analyzer(DMA) to obtain
the monodisperse particles as the condensation nuclei. As the vapor, water
and hexanol which have low boiling point, and DBP and DEHS which have
rather high boiling point were used.

Figure 1 shows the schematic diagram of the apparatus for water and
hexanol vapor experiment. The mixing type CNC is the improved one having
the similar structure as that developed previously(Kousaka at el., 1982).
The supersaturation ratio was controlled by changing the temperature of the
saturated air leaving the mixing ratio R_h constant. The numbers of grown
droplets at various supersaturation ratios were automatically counted by
the ultramicroscope-video system.

Figure 2 shows the schematic diagram of the apparatus for DBP and DEHS
vapor experiment. Two particle size magnifiers(PSMs) which are the same in
principle as the mixing type CNC(Okuyama et al., 1984) and two DMAs were
combined in series in this case. The electrically classified particles in
the first DMA were mixed with the warm air saturated by a vapor in the
first PSM, and then they were admitted into the second DMA, which is quite
same as the first DMA, without passing through the neutralizer. The
particles which were not enlarged in the first PSM were drawn from the
second DMA since the collector rod voltage of the second DMA was set at the
same as that of the first DMA, and then they were enlarged in the second
PSM to detect the number by an optical counter. The reason of the necessity
of this kind of complicated procedure is that the increase in size by
condensation of DBP or DEHS vapor is too small to directly detect particles
by an optical counter.

Experimental result and discussion

Figure 3 shows experimental results of the number ratio of particles
being not activated to the total par-
ticles, n/n_0, at the stated supersaturation ratio S.

Figure 1. Set up of condensation
experiment for water or
hexanol vapor

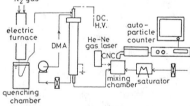

Figure 2. Set up of
condensation
experiment for
DBP or DEHS

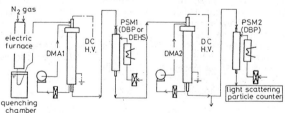

It is seen that the number ratio n/n_0 decreases with the increase in
supersaturation, or particle enlargement is enhanced with the increase in
the surrounding supersaturation S. If it can be assumed that the relationship between the critical supersaturation ratio $S(=p_d/p_s)$ and the
particle diameter d_p is expressed as Eq.(1) for all particle-vapor systems
in Fig. 3, the abscissa S of Fig. 3 can be transformed into particle
diameter d_p using Eq.(1). Figure 4 shows the result thus transformed. These
figures may be said to be "Kelvin equivalent cumulative size distribution"
in the given particle-vapor system. The solid lines in Fig. 4 shows the
cumulative size distributions of electrically classified particles in DMA.
Deviation of the experimental results from the solid lines may be caused
firstly by the interactions between the vapor and the particle surface and
second the shape of particles. Further, it should be noticed that the size
distributions of electrically classified particles are not absolute if
particles are not spherical or aggregated; they just show the electrical
mobility equivalent size distribution.

Figure 5 shows the comparison of the Kelvin equivalent diameter with
the electrical mobility equivalent diameter(electrically classified diameter) represented by geometric mean. It is seen for instance that NaCl

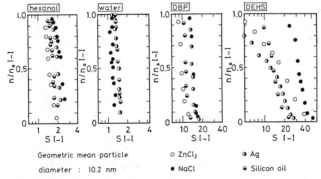

Geometric mean particle   o ZnCl$_2$ ◑ Ag

diameter : 10.2 nm   ● NaCl   ◒ Silicon oil

Figure 3. Number fraction of particles which are not activated, n/n_0,
as the fraction of supersaturation ratio S

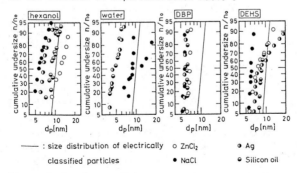

——— : size distribution of electrically   o ZnCl$_2$ ◑ Ag

classified particles   ● NaCl   ◒ Silicon oil

Figure 4. Cumulative Kelvin equivalent size distribution

54

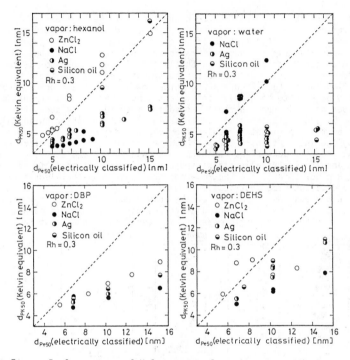

Figure 5. Comparison of Kelvin equivalent diameter with
electrically classified diameter(geometric mean diam.)

particle is easy to enlarge in water vapor but is difficult in other vapor.

Conclusion

It was found that activation of ultrafine aerosol particles in
supersaturated vapors greatly depends upon the combination of particle
with vapor. The interpretation of the results obtained in this study is, so
far, not clear, but the technique developed here will be useful to approach
the subject in question.

References

Kousaka, Y., T. Niida, K. Okuyama and H. Tanaka (1982) "Development of a
Mixing Type Condensation Nucleus Counter", J. Aerosol Sci. 13:231

Okuyama, K., Y. Kousaka and T. Motouchi (1984) "Condensational Growth of
Ultrafine Aerosol Particles in New Particle Size Magnifier", Aerosol Sci.
and Technol., in press.

Published 1984 by Elsevier Science Publishing Co., Inc.
Aerosols, Liu, Pui, and Fissan, editors

TEMPERATURE IN THE POLLAK COUNTER

David Sinclair & Earl O. Knutson
Environmental Measurements Laboratory
376 Hudson Street
New York, NY 10014
U. S. A.

EXTENDED ABSTRACT

Introduction

The temperature drop occurring during adiabatic expansion in the Pollak condensation nucleus counter has been studied by several investigators. Measurements and calculations have been made which show wide disagreement between theory and practice.

Pollak and Metnieks (1959) calculated a temperature drop of 23 °C in their photographic counter using the equations for dry adiabatic expansion, which we find are not applicable to the wet conditions in the counter. Measurements were made by Ohta (1959). He calculated the wet expansion temperature drop to be 7 °C at his initial temperature of 25 °C; and he measured 6 °C. The difference he ascribed to wall effects.

The oft-quoted measurements by Isreal and Nix (1966) require new interpretation. They measured the dry adiabatic temperature drop to be 6 °C. We are convinced their counter was not dry. When they measured the wet adiabatic expansion they got 5.5 °C drop, substantially the same temperature drop as when "dry". In interpreting the latter measurement they say: "The experiments with moist interior of the counter show that the extinction of light in the counter begins 7 milliseconds after the beginning of the expansion. At that time, the initial temperature (20 °C) is decreased by 0.3 °C, corresponding to a supersaturation of about 1.9%. The attenuation of the intensity of light 'stops to increase' as soon as the expansion is terminated." Further they say: "The experiments lead to the conclusion that the supersaturation during the whole expansion stays below a value of 2%."

However, their oscillograph shows that the extinction continues to increase for 150 milliseconds, when the temperature falls to its lowest value of 14.5 °C. This temperature corresponds to a supersaturation of 16.5%. The fact that the extinction began at a supersaturation of 1.9% simply meant that there were some particles of 14 nm diameter present, the size that serve as condensation nuclei at this supersaturation. We have discussed the above paper at considerable length because its evident misinterpretation has caused some anxiety in the condensation nucleus counter community.

This anxiety is expressed by Kassner et. al. (1968). They say: "The disconcerting results of Israel and Nix can be adequately explained in terms of actual thermocouple response."

Podzimek et. al., (1982) calculated the dry and wet adiabatic temperature drop to be 15.5 °C and 7.5 °C, respectively, when the pre-expansion temperature was 20 °C. However, they measured only about 2.75 °C temperature drop in wet expansion, primarily because the diameter of their thermocouple wire was too large (0.003 in.).

Method of Measurement

Instead of the usual thermocouple we used a resistance wire 0.0005 in. diameter composed of 70% nickel and 30% iron. This wire has the advantage that it has no bead of three times the wire diameter and high heat capacity. This wire has a high temperature coefficient of resistance: 0.40% per degree centigrade.

A 19 inch length of this wire was soldered to two silver electrodes mounted, one at each end, in insulators in the side wall of the Pollak counter. The electrodes extended into the center of the fog tube so that the resistance wire was mounted along the axis. The resistance of this wire was about 725 ohms at room temperature of about 25 °C. The transient decrease in resistance caused by the adiabatic cooling was observed on an oscilloscope and photographed with a Polaroid camera. The transient pressure drop was measured on the oscilloscope with a Statham pressure transducer. The light extinction produced by the fog during wet expansion was measured with the oscilloscope connected to a photocell.

Measurements

A variety of photographs were taken to observe the temperature, pressure and extinction during both dry and wet adiabatic expansion, as well as the effect of varying the exhaust rate with different types of valves. It was found that the final temperature and the final extinction were substantially unchanged over a range of exhaust times from 5 ms to 300 ms. The initial pressure was 160 torr above the ambient 760 torr, giving the pressure ratio 1.21. The initial temperature was 25 °C.

Figure 1 and Figure 2 show results of dry adiabatic expansion. The vertical scale for temperature was 20 mv/div. Twenty-one such photos were taken for which the average temperature drop was 16 °C. Prolonged drying of the ceramic lining (which contains 200 cm^3 water) with dry air from a tank was necessary before this temperature was observed. Figure 3 and Figure 4 show results of wet adiabatic expansion of room aerosol of 50 K particles per cm^3. The vertical temperature scale was 10 mv/div. The wavy line in the extinction curve shows changes in water drop diameter. The average of twenty-one such photos gave 8 °C temperature drop.

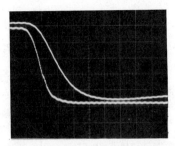

Fig. 1. Dry adiabatic expansion.
Temp., upper curve, 16 °C drop.
Pressure, lower curve. Horiz. time
scale, 50 ms/div.

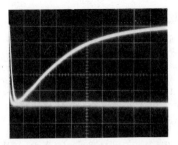

Fig. 2. Dry adiabatic expansion.
Temp., upper curve, 16 °C drop.
Pressure, lower curve. Horiz. time
scale, 500 ms/div.

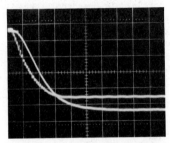

Fig. 3. Wet adiabatic expansion.
Temp., upper/lower curve, 8 °C drop.
Pressure, lower/upper curve. Horiz.
time scale, 20 ms/div.

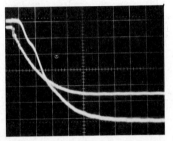

Fig. 4. Wet adiabatic expansion.
Extinction, upper/lower curve.
Pressure, lower/upper curve. Horiz.
time scale, 20 ms/div.

Calculations

The theoretical temperature drop in dry adiabatic expansion can be readily calculated from the well known equation: $T_2 = T_1 (P_2/P_1)^{(\lambda-1)/\lambda}$, where: T_2 is the final absolute temperature; T_1 is the initial absolute temperature, 298 °K; P_2 is the final pressure, 760 mm; P_1 is the initial pressure, 920 mm; and λ is the ratio of the specific heats, 1.4. This gives a final temperature of 282 °K, a temperature drop of 16 °C.

The theoretical temperature drop in wet adiabatic expansion, which applies to the Pollak counter, is not so readily calculated. One of us (EOK) has derived an equation based on the fact that the process is isentropic. Lack of space in this abstract does not permit derivation of this equation. The results showed about 8 °C temperature drop and about 4×10^{-6} grams released water per cm^3 air.

58

Discussion

We have shown that the temperature and supersaturation developed in the Pollak are in accord with theory in both dry and wet adiabatic expansion. The measurements show that there is ample time for the temperature drop to occur without interference from the walls of the chamber. The expansion time had negligible effect on the temperature until the time was delayed to 1.5 sec. by means of a partially opened needle valve. Furthermore, the temperature required 5 sec. to return to the initial value.

The question arises as to what is the minimum particle size detectable by the Pollak counter? This size depends on the supersaturation which depends on the temperature drop. This is usually assumed to be the temperature drop in dry adiabatic expansion, about 16 °C, depending on the initial temperature. Miller and Bodhaine (1982) list 16 authors who use the dry adiabatic equation to calculate the supersaturation and the corresponding minimum droplet radius from the Kelvin equation. They calculate the supersaturation 1.31 and the minimum droplet radius 1.3×10^{-7} cm corresponding to a temperature drop from 20 °C to 4.4 °C. In our experiments the temperature drop was 8 °C, from 25 °C to 17 °C, in wet adiabatic expansion. The corresponding vapor pressures give a supersaturation of 0.36 and the minimum detectable radius 3.55×10^{-7} cm.

References

Israel, H. and Nix, N. (1966), Journal de Recherches Atmospheric 2:185-187.
Kassner, J. L. Jr. (1968) et.al., Journal of Atmospheric Sciences 25:919-926.
Miller, S. W. and Bodhaine, B. A., (1982), J. Aerosol Sci. 13:481-490.
Ohta, S. (1959), Geophysical Magazine 29:229-287.
Podzimek, J. et.al. (1982), Atmospheric Environment 16:1-11.
Pollak, L.W. and Metnieks, A.L.(1959), Geofisica Pura e Applicata 43:285-301.

Published 1984 by Elsevier Science Publishing Co., Inc.

Aerosols, Liu, Pui, and Fissan, editors

A THEORETICAL MODEL FOR AN ULTRAFINE AEROSOL
CONDENSATION NUCLEUS COUNTER

M. R. Stolzenburg and P. H. McMurry
Particle Technology Laboratory
Mechanical Engineering Department
University of Minnesota
Minneapolis, MN 55455
U. S. A.

EXTENDED ABSTRACT

Introduction

Ultrafine aerosols (particle diameter less than 0.02µm) play an important role in many aerosol systems. While commercial instrumentation for detecting, counting and sizing larger particles is currently available, no generally accepted technique for routine measurements of ultrafine aerosols is available. An improved design for a continuous-flow condensation nucleus counter (CNC) is described and theoretically modeled. Modifications of a commercially produced CNC have resulted in improved activation efficiency of ultrafine particles as well as faster response time.

Design

The CNC is a modified TSI Model 3020 consisting of a sequential saturator and condenser with continuous flow and n-butyl alcohol as the working fluid (Agarwal and Sem,1980). In the original instrument, the sample aerosol flow (5 cm^3/sec) is drawn through the saturator which is maintained at a temperature of 35°C. The alcohol-laden sample then passes through the condenser tube the walls of which are held at 10°C. The resulting supersaturation causes alcohol to condense on the particles growing them to roughly 10µm in diameter. These alcohol droplets are then detected optically to determine the aerosol concentration.

In the new design (Figure 1) an aerosol sheathing system (Wilson et al.,1983) has been introduced such that the aerosol flow (0.5 cm^3/sec) is injected near the end of the saturator where it is sheathed by alcohol-laden clean air (4.5 cm^3/sec) from the saturator. A short vertical saturator section allows time for the alcohol to diffuse into the aerosol flow core before entering the condenser. Because the particles are confined to the center of the flow where the highest supersaturation occurs, the activation efficiency for ultrafine aerosols is greater than in the original instrument in which the particles are spread over the entire condenser tube cross-section.

In addition, the aerosol inlet system has been optimized for ultrafine aerosol sampling. By extracting the sample from cores of transport flows and reinjecting it into the core of the condenser flow, wall effects have been minimized. This, together with eliminating the saturator from the path of the aerosol flow, has greatly reduced particle diffusion losses and improved the response time of the instrument. Furthermore, the built-in dilution factor of ten means higher aerosol concentrations can be measured in the more reliable single-particle counting mode.

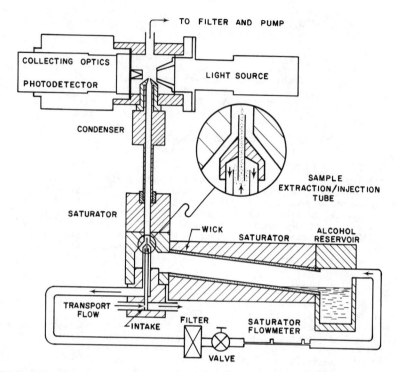

Figure 1. Flow system of modified TSI Model 3020 CNC. Saturator, condenser and aerosol inlet system shown to scale.

Theory

The activation and subsequent condensational growth of nuclei in the CNC can be theoretically modeled. Calculation of simultaneous radial and axial heat and mass transfer within the vertical saturator and the condenser (in the absence of particles) yields profiles of temperature, mass concentration and supersaturation of the working fluid. As a particle moves through the condenser, it may be lost to the wall by diffusion or it may fail to encounter a supersaturation high enough to activate condensational growth on its surface. The analysis of particle convection and diffusion in conjunction with the supersaturation profile produces the nucleus activation efficiency as a function of particle Kelvin diameter defined by

$$D_p = 4 \, \sigma \, M \, / \, \rho_l \, R \, T \, \ln S$$

where σ, M, ρ_l and S are, respectively, the surface tension, molecular weight, liquid density and supersaturation of the condensable vapor.

The subsequent condensational growth of alcohol droplets is modeled in the free-molecule, transition and continuum regimes. The growth theory considers coupled heat and mass transfer to the droplets due to ordinary diffusion and thermal conduction as well as higher order effects such as Stefan flow, thermal diffusion and the diffusion thermal effect (Wagner,1982). The model yields a measure of the mean size and spread of the alcohol droplet distribution produced by the CNC as a function of the size of the nuclei being sampled.

Results and Conclusions

The foregoing analysis has been performed for both the original TSI CNC and the modified instrument. Figure 2 shows radial supersaturation profiles at several different distances along the condenser axis of the Model 3020 CNC with standard operating parameters. R is the radius of the condenser, r and x are, respectively, radial and axial distances within the condenser and $r'=r/R$ and $x'=x/R$ are the corresponding dimensionless coordinates. The supersaturation profiles for the modified instrument are very similar but have slightly lower values due to the subsaturated aerosol core flow at the entrance to the condenser.

Because the overall peak supersaturation in the condenser occurs along the axis, the smaller the particle the nearer the centerline it must be to be activated. Hence, in the ultrafine range, when the aerosol is spread across the entire cross-section of the condenser tube as in the Model 3020, a significant fraction of the particles may fail to be activated (or be lost to the walls by diffusion) even for sizes well above the minimum detection

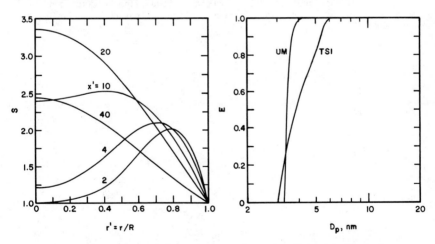

Figure 2. Supersaturation, S, versus dimensionless radial coordinate, $r'=r/R$, for various dimensionless axial distances, $x'=x/R$, down condenser of TSI Model 3020 CNC.

Figure 3. Theoretical single particle activation efficiency, E, versus particle Kelvin diameter, D_p, for condensers of TSI Model 3020 CNC and modified UM CNC.

62

limit. In the modified design, since the aerosol is introduced around the centerline of the condenser, these effects are greatly reduced.

The theoretically calculated single particle (no particle-particle interaction) activation efficiency, E, as a function of particle Kelvin diameter, D_p, is shown in Figure 3 for the two instruments. Both calculations are for the standard operating parameters of the TSI CNC, that is, saturator and condenser temperatures of 35°C and 10°C, respectively, and a total flow of 5 cm^3/sec. The modified design has a much sharper efficiency cutoff giving it a more clearly defined lower detection limit. It has a slightly higher minimum detectable size due to the lower overall peak supersaturation attained in the condenser. However, by increasing the temperature difference between the saturator and the condenser the detection limit of the modified instrument can be lowered to that of the original instrument without causing homogeneous nucleation. The final product is a CNC with significantly improved detection capabilities for ultrafine aerosols.

References

Agarwal, J. K., and G. J. Sem (1980) "Continuous Flow, Single-Particle-Counting Condensation Nucleus Counter", J. Aerosol Sci. 11:343.

Wagner, P. E. (1982) "Aerosol Growth by Condensation", Aerosol Microphysics II, Chemical Physics of Microparticles (Edited by W. H. Marlow), Springer-Verlag Berlin Heidelberg New York.

Wilson, J. C., D. H. Hyun, and E. D. Blackshear (1983) "The Function and Response of an Improved Stratospheric Condensation Nucleus Counter", J. Geophys. Res. 88(C11):6781.

Published 1984 by Elsevier Science Publishing Co., Inc.

Aerosols, Liu, Pui, and Fissan, editors

APPLICATION OF A MULTISTEP CONDENSATION NUCLEI COUNTER AS A DETECTOR FOR PARTICLE SURFACE COMPOSITION

R. Niessner[x], C. Helsper[∇] and G. Rönicke[+]

[x]Universität Dortmund, Postfach 50 05 00, D-4600 Dortmund
[∇]Universität Duisburg, Bismarckstrasse 81, D-4100 Duisburg 1
[+]Aerosolmesstelle Schauinsland, D-7081 Schallstadt
F. R. G.

EXTENDED ABSTRACT

Introduction

Information on the chemical nature and the physical properties of particle surface is necessary for the understanding of
the chemical reactions taking place between particles and the
surrounding gas. To obtain this information for submicron
particles, a method has been developed, which is based on the
influence of particle surface on the condensation process.
An electrostatic classifier is used to generate a monodisperse
aerosol of known electrical mobility diameter. The Kelvin equivalent diameter of this aerosol is then determined using a
simple Nolan-Pollak type condensation nuclei counter which has
been modified to run automatically through a series of different
water vapour supersaturations. The comparison between the Kelvin
size and the electrical mobility diameter of these monodisperse
particles yields information on the surface structure of the
particles. For polydisperse particles such an interpretation is
not possible.

Experimental Setup for the Determination of the Kelvin Equivalent Diameter

Figure 1 shows the experimental setup used for the study.
The particles generated by the electrostatic classifier (TSI,
Model 3070) are introduced into the expansion chamber of the
condensation nuclei counter (CNC). The expansion chamber is a
brass tube, 60 cm in length and 32 mm in inner diameter.

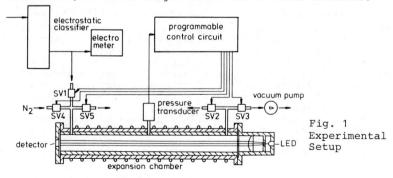

Fig. 1
Experimental
Setup

The inside surface of the tube is covered by a mixture of
sawdust and cement soaked with water to humidify the aerosol.
The whole chamber is temperature controlled at 293 $\pm$ 0.02 K by
a coolant flowing through a copper pipe coil soldered to the
outer tube wall. Measurement of the droplet concentration in the
chamber is achieved by measuring the light extinction along the
chamber axis. The aerosol is drawn through the chamber by a
vacuum pump while the solenoid valves 1 and 3 are opened. Next,
the pressure in the chamber is increased by introducing clean
compressed nitrogen into the chamber through valve 4. After the
desired pressure is reached, the control circuit closes valve 4.
After a short time interval necessary to humidify the air, valve
2 and 5 are opened for expansion. The expansion is accomplished
within a time of 10 to 15 ms. The system automatically runs
through 16 increasing pressure steps, thus increasing the saturation and decreasing the smallest detectable particle size.

Calibration Results

The CNC has been calibrated using monodisperse NaCl-Particles produced by an evaporation and recondensation technique
(Scheibel and Porstendörfer, 1983) in connection with the electrical mobility analyzer. Figure 2 shows some of the calibration
results. For these different monodisperse fractions a normalized

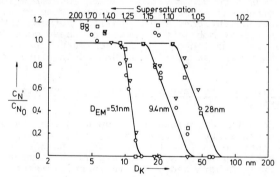

Fig. 2 Calibration Results for the Condensation Nuclei Counter

concentration readout of the CNC is plotted as a function of the
Kelvin diameter and the supersaturation, respectively. Increasing
the supersaturation leads to a steep increase of the normalized
number concentration to a certain constant level at which all
particles are activated as condensation nuclei. Further increase
of supersaturation yields a further increase of light extinction.
This means, that the already existing droplets are taking up more
water. The concentration has been normalized in a way, that the
constant level after the first steep increase has been assigned
a value of one. The different symbols refer to different particle
concentrations. The size resolution in the size range below 10 nm
is rather good. These results are in good agreement with those

obtained by Porstendörfer et al., (1984) using the Size Analyzing
Nucleus Counter (SANC) (Pohl and Wagner, 1980).

Response of the CNC to Different Particle Surface Compositions

In a first experiment sulfuric acid particles were produced
by a modified Sinclair-La Mer generator (Niessner and Klockow,
1980). By mixing NH_3 to this aerosol, the particle material
could be converted to ammonium sulfate. Figure 3 shows the
behavior of these particles in the CNC.

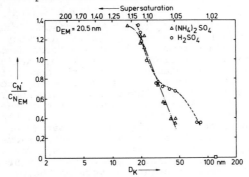

Fig. 3   Condensation Behavior for Particles of the Same Electrical
Mobility and Different Material

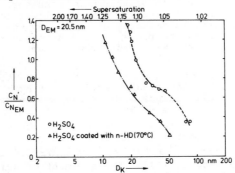

Fig. 4   Condensation Behavior of Coated and Uncoated
H_2SO_4-Particles

In contrast to figure 2, the indicated number concentrations
are normalized here to the actual concentration determined by an
electrometer. It can be seen, that the sulfuric acid particles
start to grow at much lower supersaturations than the ammonium
sulfate particles. In another experiment sulfuric acid particles
were coated with n-Hexadecanol (Niessner, 1984) to change their
surface from a hygroscopic to a non hygroscopic character. From
parallel neutralization experiments with ammonia as a trace gas

it is known that the H_2SO_4 droplet surface is completely covered
by a n-Hexadecanol layer, because no neutralization reaction is
observed. The result of this experiment is shown in figure 4.
The curve represented by circles gives the instrument's response
to uncovered sulfuric acid particles. The triangles refer to the
coated particles. It can be seen that the condensation process
reacts rather sensitively to the changed particle surface. The
coated particles need a much higher supersaturation to grow to
the same size as the uncoated ones.

Conclusions

These first results show, that a comparison of the electric
mobility diameter with the Kelvin equivalent diameter can give
insight into the surface composition of submicron particles. In
contrast to pure salt particles, e.g. $(NH_4)_2SO_4$, sulfuric acid
droplets are already activated during the residence time (20 sec)
of the aerosol in the humidifier tube and therefore condensational
growth can be expected at much lower supersaturations. In the case
of completely covered sulfuric acid aerosol the change in surface
composition retards the condensational growth. The behavior of
n-HD coated sulfuric acid droplets and salt particles is not so
different; the existence of a monolayer of water seems to be
sufficient to initiate the condensation process. For a detailed
discussion, however, more refinded experiments are necessary.
Studies on the influence of organic layers on the water attachment
to particles or droplets (e.g. in pharmaceutical industry or in
investigations of "interstitial particles" in cloud systems) are
typical fields of application of this new technique.
For the determination of the Kelvin size excellent resolution is
available below a particle diameter of 10 nm. By use of smaller
supersaturation steps in the range 1 - 5 % a better resolution
will be obtained for particles larger than 15 nm.

References

Niessner, R. and D. Klockow (1980) "Modified La Mer-Generator for
the Production of Small Amounts of Sulfuric Acid Aerosol",
Analytical Chemistry 52: 594 - 595

Niessner, R. (1984) "Coated Particles: Preliminary Results of Laboratory Studies on Interaction of Ammonia with Coated Sulfuric Acid
Droplets or Hydrogensulfate Particles", Sci.Total Environ., in press

Pohl, F.G. and P.E. Wagner (1980) "Measurement of Kelvin-Equivalent Size Distributions of Urban Atmospheres", J. Phys. Chem. 84:
1642 - 1644

Porstendörfer, J.; H.G. Scheibel; F.G. Pohl; O. Preining; G.
Reischl and P.E. Wagner (1984) "Heterogeneous Nucleation of Water
Vapor on Monodispersed Ag- and NaCl-Particles with Diameters
between 6 and 18 nm", Aerosol Sci. Technol., in press

Scheibel, H.G. and J. Porstendörfer (1983) "Generation of Monodisperse Ag- and NaCl-Aerosol with Particle Diameters between 2 nm
and 300 nm", J. Aerosol Sci. 14: 113 - 126

Published 1984 by Elsevier Science Publishing Co., Inc.

Aerosols, Liu, Pui, and Fissan, editors

HIGH FLOWRATE CNC : DESIGN AND CALIBRATION

Y. Metayer* and G. Madelaine*

*Laboratoire de Physique et de Métrologie des Aérosols
CEA/IPSN/DPT/SPIN - Centre d'Etudes Nucléaires de Fontenay-aux Roses,
B.P. n° 6 - 92260 Fontenay Aux Roses, France

The quality of clean rooms used for certain industrial operations appears as a more and more crucial question. Checking the efficiency of operating devices is currently performed by means of Optical Particles Counters in the size range $Dp > 0.3$ μm.

For smaller particles, no such counting instrument is available : expansion type CNCs are not sensitive at concentration levels ranging about 100 p. (litre)$^{-1}$. As for the commercial Continuous Flow CNC, its small flow rate (0.3 litre. mn^{-1}) is not suitable for a quick and accurate detection of a slight concentration increase above the permitted value.

Therefore we have developed a high flow rate CNC based on the conventional Continuous Flow CNC concept. It behaves as a Particle "Magnifier" consisting essentially of a saturator tube and a condenser part. This device is operated at a 28 litre. mn^{-1} flowrate, downstream a classical Optical Counter registering only the number of pulses.

A modelization describing the heat and mass transfer occurring in a cooled pipe has been used to determine the proper dimensions and working parameters of the instrument.

Different low volatile fluids have been used : Di Octyl Phtalate (DOP) and Di Ethyl Sebacate (DES). Calibration of the magnifier has been achieved with NaCl particles produced by electrostatic classification in the range 0.02 μm - 0.3 μm. The overlap of indications obtained with the High Flowrate CNC and an Optical Counter has been checked with aerosols generated by the Lamer-Sinclair method.

CHAPTER 3.

ELECTRICAL MEASUREMENT

Published 1984 by Elsevier Science Publishing Co., Inc.

Aerosols, Liu, Pui, and Fissan, editors

AUTOMATED DIFFERENTIAL MOBILITY PARTICLE SIZER

Patricia B. Keady, Frederick R. Quant and Gilmore J. Sem
TSI Incorporated, P.O. Box 64394, St. Paul, MN 55164 USA

EXTENDED ABSTRACT

The differential mobility particle sizer (DMPS) measures the size distribution of aerosol particles in the 0.01-1.0 μm diameter range. The polydispersed aerosol particles are size classified with high resolution by a differential mobility analyzer (DMA) and counted by a continuous-flow condensation nucleus counter (CNC). A microcomputer automates the system by controlling the instruments, collecting raw data, performing the data inversion to obtain concentration as a function of diameter and providing data output. Several approaches for the data reduction have been developed and refined over the past twelve years. The approach chosen for the commercial version is based on methods developed by ten Brink et al (1983) and Fissan (1982).

A photograph and schematic diagram of the DMPS system are shown in Figures 1 and 2.

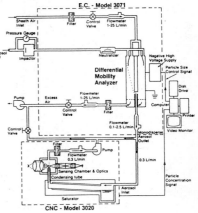

Figure 1. Differential mobility particle sizer for submicron size distribution measurement. The system includes an electrostatic classifier (TSI Model 3071) a CNC (TSI Model 3020) and an Apple II personal computer.

Figure 2. Schematic of the differential mobility particle sizer.

The incoming polydisperse aerosol is neutralized with a radioactive source and brought to a known bipolar electrical charge distribution. The electrical mobility of a particle is a function of diameter and number of electrical charges the particle possesses. The majority of the particles have either one or no charges. The DMA extracts particles within a narrow mobility bandwidth by

72

attracting positively-charged particles from the aerosol flow toward an exit slit in the negatively-charged center electrode. The mobility selected is determined by the analyzer geometry, flow, and electric field. The entire range of particle mobilities can be measured by setting the system flowrates and then adjusting the voltage potential on the center rod of the DMA in a step-wise fashion from about 10 KV to near zero. A factor of 1.222... between voltage steps results in perfectly adjoining mobility bandwidths. The probability of a particle of a given mobility being extracted by the DMA is 100% at the midpoint mobility falling to zero at the band endpoints. The triangularly-shaped transfer function means half of the available particles within the selected mobility bandwidth pass on to the CNC for concentration measurements.

The 'neutral' charge distribution of the aerosol is determined from Boltzmann's law of charge equilibrium. However, below 0.1 µm, there is still debate among scientists about the correct charge distribution. We have chosen to use Fuchs' calculation for negative ions below 0.1 µm. A comparison of the two theories is shown in Figure 3. Fuchs' theory has been experimentally verified by Kousaka (1983) and Reischl (1983).

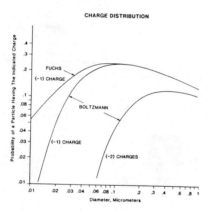

Figure 3. Charge distribution of a neutralized aerosol.

We place an impactor at the aerosol inlet to the DMPS to remove large particles. Two different impactor nozzles are provided to give 50% cut diameters just below the maximum size range of the DMPS. The data inversion to number size distribution is based strongly on no particles entering the DMA with diameters larger than the upper size limit of the DMPS. Knowing that all particles measured in the largest particle diameter channel have a single charge, one can calculate the true concentration of particles of that size and also the percentage of particles of that size with multiple charges. The multiply-charged particles will behave like smaller singly-charged particles because of their increased electrical mobility. The contribution that these multiply-charged particles will make to the smaller diameter channels can be subtracted out in the computer data reduction. Since the multiply-charged particles of a given diameter channel don't necessarily fall entirely within a single smaller diameter channel, a fractional portion of the total is subtracted from each of the affected channels. The computer program considers up to six electrical charges on each particle as it proceeds through the entire distribution.

The DMA was originally designed to set and control four interrelated flows with three valves and mass flowmeters. The incoming polydisperse aerosol flow has no valve or flowmeter. Even with as little as a 2% error in the mass flowmeters, the worst-case additive affect on the unmeasured aerosol flow could result in as much as 40% error. The impactor now provides an added feature of measuring the aerosol flow by the pressure drop across the impactor nozzle. Each impactor nozzle is calibrated for flowrate so the overall accuracy for concentration measurements with the system is greatly improved.

The computer program corrects the aerosol concentration measurement for the efficiency of the CNC as given by Agarwal and Sem (1980) and for coincidence in the CNC count mode.

The computer automates the system by controlling the voltage to the DMA and measuring the concentration for each voltage step. Features of the computer program include full graphical and tabular display of the size distributions, a background subtraction and comparison option, real-time date and time measurement and the option for using the CNC in the count-mode for more accurate sampling of low concentrations.

The size distribution of alumina particles is shown in Figure 4. The solid

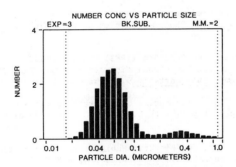

Figure 4.   DMPS size distribution measurement of an atomized alumina
             suspension. The dotted lines indicate the measuring range
             of the DMPS determined by the system flowrates.

particles were suspended in water and atomized with a Collison atomizer, dried and neutralized. The high resolution of the DMPS clearly shows the bimodal nature of the aerosol. This information could easily be lost in a lower resolution instrument such as a diffusion battery or electrical aerosol analyzer.

74

References

Agarwal, J.K. and G.J. Sem (1980) "Continuous Flow, Single-Particle-Counting Condensation Nucleus Counter", J. Aerosol Sci. 11:343-357.

Fissan, H.J., C. Helsper, and J.H. Thielen (1982) "Determination of Particle Size Distributions by Means of an Electrostatic Classifier", Proceedings of the 10th Annual Conference, Gesellschaft für Aerosolforschung, Bologna, Italy.

Kousaka, Y., M. Adachi, K. Okuyama, N. Kitada and T. Motouchi (1983) "Bipolar Charging of Ultrafine Aerosol Particles", Aerosol Sci. Tech. 2:421-427.

Reischl, G.P. (1983) "The Bipolar Charging of Aerosol: Experimental Results in the Size Range Below 20-nm Particle Diameter", J. Colloid Interface Sci. 91:1.

ten Brink, H.M., A. Plomp, H. Spoelstra and J.F. van de Vate (1983) "A High-Resolution Electrical Mobility Aerosol Spectrometer (MAS)", J. Aerosol Sci. 14:5, pp. 589-597.

Bibliography

Alofs, D.J. and P. Balakumar (1982) "Inversion to Obtain Aerosol Size Distributions from Measurements with a Differential Mobility Analyzer", J. Aerosol Sci. 13:513-527.

Haaf, W. (1980) "Accurate Measurements of Aerosol Size Distribution - II. Construction of a New Plate Condenser Electric Mobility Analyzer and First Results", J. Aerosol Sci. 11:201-212.

Hoppel, W.A. (1978) "Determination of the Aerosol Size Distribution from the Mobility Distribution of the Charged Fraction of Aerosols", J. Aerosol Sci. 9:41-54.

Keady, P.B., F.R. Quant, and G.J. Sem (1983) "Differential Mobility Particle Sizer: A New Instrument for High-Resolution Aerosol Size Distribution Measurement Below 1 μm", TSI Quarterly 9, Issue 2.

Knutson, E.O. (1972) "The Distribution of Electric Charge Among the Particles of an Artificially Charged Aerosol", Ph.D. Thesis, University of Minnesota, Minneapolis, MN.

Knutson, E.O. (1976) "Extended Electric Mobility Method for Measuring Aerosol Particle Size and Concentration", Fine Particles, B.Y.H. Liu, editor, Academic Press, New York, pp. 739-762.

Knutson, E.O. and K.T. Whitby (1975) "Accurate Measurement of Aerosol Electric Mobility Moments", J. Aerosol Sci. 6:453-460.

Reischl, G.P. (1981) "On the Equivalence of Low Pressure Impactor and Differential Mobility Analyzer Data for Submicron Particle Size Distributions", Proceedings of 9th Annual Conference, Gesellschaft für Aerosolforschung, Duisburg, W. Germany.

Published 1984 by Elsevier Science Publishing Co., Inc.
Aerosols, Liu, Pui, and Fissan, editors

MEASUREMENTS OF PARTICLE LOSSES IN A DIFFERENTIAL MOBILITY ANALYSER AND THEIR INFLUENCE ON THE EVALUATED SIZE DISTRIBUTION

A. Reineking, J. Porstendörfer
Isotopenlaboratorium der Georg-August-Universität
Burckhardtweg 2, D-3400 Göttingen, F.R.G.

EXTENDED ABSTRACT

Introduction

Electric mobility measurements are used to determine aerosol size distributions, especially in the particle size range below 200 nm. Hewitt (1957) described a differential mobility analyser (DMA), which allows mobility distributions to be obtained directly. The first practical instrument for measuring size distributions (Whitby and Clark, 1966) used unipolar charging by means of a corona discharge. Bipolar aerosol charging has the advantages of lower concentration ratios of multiply to singly charged particles and the charge equilibrium is attained. Lui and Pui (1974) described a cylindrical DMA with bipolar charging by means of the ions of a radioactive source. Knutson and Whitby (1975 a,b) published two papers on this device, concerning its application of separating out a monodisperse fraction and to measure the size distribution of an introduced polydisperse aerosol. This instrument becomes commercial available from TSI 1976 and is often used in aerosol research of small particles.
The differential mobility method to determine the size distribution of an aerosol consists of two steps: The measurement of the mobility distribution (by means of stepwise increasing voltage settings) and the calculation of the particle size distribution under consideration of the fraction of singly and multiply charged particles in the bipolar charge equilibrium and the particle losses in the analyser system.
Recently we measured the bipolar charge of monodisperse aerosols and developed a new charging theory based on the charging theory of Fuchs (Porstendörfer et al. (1984), Hussin et al. (1983)). The second important point at the evaluating of differential mobility analyser data is the knowledge of particle losses in the total mobility analyser system as a function of particle size and flow rates. Therefore we measured the loss function using monodisperse particles with diameters between 3 - 110 nm.

Experimental Set-Up

A condensation type aerosol generator (Scheibel and Porstendörfer, 1983), together with electrostatic classification, produced monodisperse Ag-aerosols in the range of diameters between 2 and 150 nm. The particle concentrations vary from 10^3 - 10^6 particles/cm^3 depending on the particle size, and the standard deviation of the number size distribution is about 12 %.
The experimental system to determine the aerosol losses in the mobility analyser (TSI, Model 3071) is shown in figure 1. The singly charged aerosols of the generator were drawn through a second electrostatic aerosol classifier (the first classifier is part of the aersol generator) and an equivalent volume (volume of the bipolar ion source) without neutralisation by ions. By measuring the current produced by the singly charged particles with an

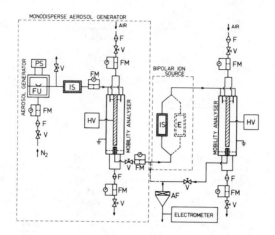

Figure 1: Arrangement for measuring the particle losses, the particle size
distribution and the total number concentration.
FM: flowmeter, F: filter, FU: furnace, IS: bipolar ion source
V: valve, HV: high voltage, PS: power supply, E: equivalent
volume, AF: aerosolfilter

aerosol filter electrometer the number concentrations of aerosol particles at
the input- and outputside were determined.

Particle Loss Function

The particle losses in the analyser, bipolar ion source and the piping due to
diffusion are depending on the flow rates and the particle diameter. Therefore
the losses were measured for different aerosol flowrates (0.3 l/min and 2.0
l/min) in the size range 2.5 to 110 nm. The results are shown in figure 2. For
small flow rates (0.3 l/min) internal losses below 60 nm are not negligible;
at 20 nm nearly 50 % of the input particles are lost. For higher flow rates
(2.0 l/min) the particle losses have to consider in the size range below 30 nm
(10% for 20 nm and 50 % for 7 nm).

Particle Size Distribution and Total Number Concentration

To calculate the particle size distribution from the data of the mobility
distribution we used an inversion technique described by Knutson (1976). In
figure 3 are shown some size distributions in the size range 8 - 40 nm to
examine the data evaluation (developed charging theory (Hussin et al.,1983),
inversion technique, particle losses). These calculated values are compared
with directly measured concentrations.
The experimental system is the same as shown in figure 1 only the flowrates

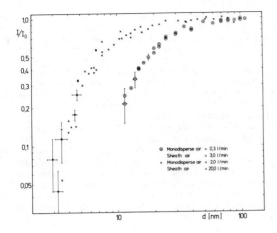

Figure 2: Particle losses of an electrostatic classifier, TSI Model 3071

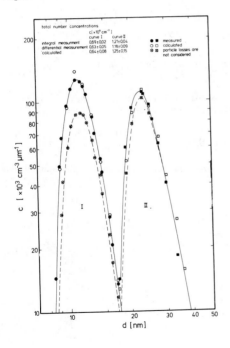

Figure 3: Size distributions and total number concentrations. The flowrates were 2.0 l/min for the aerosol air and 20 l/min for the excess air. The calculated values were obtained from the mobility distribution and the new charging function (Hussin et al.,1983)

of the first classifier were changed to have polydisperse single charged particles at the inlet of the second classifier. The flowrates of the second classifier were 2.0 and 20 l/min for the aerosol air and the excess air. In the first step the total number concentration of the charged particles were measured and in a second and third step the extracted mobility fractions were determined without and with using a bipolar ion source, respectively.

The measured and calculated size distributions are in good agreement. The area under the curves in each case is the total number concentration of the aerosol. These concentrations are also in accordance with the integral measured concentration, the deviations are smaller than 10 %.

In figure 3 are also shown the not corrected measured size distribution to demonstrate the influence of the particle losses. The mean and the slopes of the distributions are found to be nearly correct, but the total number concentrations of the size distributions are too low. If particle losses are not considered, one calculates from the distribution in the small size range (8-17 nm, curve I) $0.45*10^6$ particles/cm^3 . The correct value ($0.69*10^6$ particles/cm^3) is greater by a factor of 1.53. In the size range 17-40 nm (curve II) the integral measured concentration is only 12 % higher than the concentration without considering the particle losses .

References

Hussin, A.; Scheibel. H.G.; Becker, K.H. and Porstendörfer, J. (1983):
     J. Aerosol Sci. 14, 671
Hewitt, G.W. (1957): J. Electr. Engng 76, 30
Knutson, E.O. and Whitby, K.T. (1975a): J. Aerosol Sci. 6, 443
Knutson, E.O. and Whitby, K.T. (1975b): J. Aerosol Sci. 6, 453
Knutson, E.O. (1976): in Fine Particles; Lui, B.Y.H. (Ed.)
     Acadamic Press, New York
Lui, B.Y.H. and Pui, D.Y.H. (1974): J. Colloid Interf. Sci. 47, 155
Porstendörfer, J.; Hussin, A.; Scheibel, H.G. and Becker, K.H. (1984):
     J. Aerosol Sci. 15,
Scheibel, H.G. and Porstendörfer, J. (1983): J. Aerosol Sci. 14, 113
Whitby, K.T. and Clark, W. (1966): Tellus 18, 573

Published 1984 by Elsevier Science Publishing Co., Inc.

Aerosols, Liu, Pui, and Fissan, editors

DEPOSITION RATE OF UNIPOLAR CHARGED AEROSOLS
IN UNIFORM ELECTRIC FIELD AND LAMINAR FLOW
IN PARALLEL PLATES ELECTRODE

Fumio Isahaya
Hitachi Plant Engineering & Construction Co.,Ltd.
Research Laboratory
537, Kami-hongo, Matsudo-shi, Chiba-ken, Japan

EXTENDED ABSTRACT

Introduction

In general the equation for expressing the fundamental performance characteristics of the deposition rate, collection efficiency, for an electrostatic
air cleaner and precipitator where the aerosols are charged by using corona
discharge and then deposited on the collecting electrodes with an externally
applied intensive electric field, has been widely accepted in the following
forms. For the region of low Reynolds number of approximately 2,000 or less
with respect to gas flow, so-called laminar flow and also the almost uniform
particle size, we have

$$\eta_c = WL_c/P_c V_g, \tag{1}$$

where η_c is the deposition rate of unipolar charged aerosols, W the migration
velocity of aerosol particle, L_c the effective length of collecting electrode
in the direction of gas stream, P_c the spacing of parallel plates electrode and
V_g the gas velocity. Furthermore, for the region of large Reynolds number of
approximately 10,000 or more, so-called fully turbulent flow, we have

$$\eta_c = 1 - exp. - (WL_c/P_c V_g). \tag{2}$$

In practice, however, since the some uncertainty phenomena such as an aerodynamic re-entrainment, aerodynamic disturbance due to ionic wind, space charge
effect and particle agglomeration will be present, it has been widely used a
realistic effective migration velocity on the basis of empirical means, arrived
at from laboratory experiments and actual operation tests for full-scale plant,
to design a precipitator to give required efficiency, deposition rate, for certain specified gas and aerosol conditions.

Nevertheless, it has been living yet the problem awaiting solution whether
equation (1) is consistent with experimental results in the range of low Reynolds number of approximately 500 to 2,000. Hereupon, in order to answer the
question as mentioned before, this paper describes that in what form, equation
(1) and (2) shoud be modified on the basis of the experimental investigations.

Experimental Methods

Figure 1 shows the shematic presentation of cross sectional view of the
testing electrodes having a charging section using positive corona and a collecting section consisting of parallel plates electrode with applying the electric
field. The deposition rate was obtained by the number concentration of aerosol
particles at the outlet of testing electrodes being energized and de-energized

electrically. The number concentration was measured with an optical single particle counter (Shibata SP-800) using the incense stick smoke[1] as testing aerosol which shows the similar properties to the side stream of tobacco smoke with particle size of approximately 0.77 μm. Since the depositions occur more or less in the charging section, hereupon the net deposition rate η_c of the collecting section was obtained to make an exception of   that of charging section . Furthermore, the deposition rate distribution on the collecting plate electrode was measured with a radioactive tracer technique using Thorium B as testing radioactive aerosols[2] with the particle size of approximately 0.1 μm.

These experiments were carried out under within the range of Reynolds number of approximately 500 to 2,000, the gas velocity of 1.0 to 4.0 m/s, the electric field intensity of collecting section of approximately 1.25 to 6.25 kV/cm and the positive corona current density of the discharge wire in the charging section of 1.0 to 4.0 μA/cm, furthermore the product $N_0 t$ of free ion density N_0 multiplied by charging time t of approximately 0.2×10^{11} to 1.8×10^{11} sec/cm^3.

Experimental Results

Figures 2 and 3 show the representative examples of experimental data for the performance characteristics of the deposition rate, furthermore Figure 4 shows summarily the performance characteristics of deposition rate versus the comprehensive design parameters. In accordance with Figure 4, the deposition rate can be expressed as an exponential equation by

$$\eta_c = 1 - \exp.-(KE_c\sqrt{i}L_c/P_c V_g^\delta), \tag{3}$$

where K is an empirical constant related to the aerosol properties and other parameters, in this case $K = 1.645 \times 10^{-2}$, δ the coefficient depended upon aerodynamic re-entrainment of particles placed empirically δ =1.5 to 2.0. The calculating results derived from equation (1) are also shown in Figures 2 and 3 in comparison with these experimental results where the migration velocity W can be derived from a force balance between Stokes' law and Coulomb's law in the theoretical consideration. Furthermore, with regard to the particle charging by electric field and ions diffusion mechanisms, such theoretical equations have respectively been given by Pauthenier(1932), White(1951), Murphy(1959), Liu and Yeh(1968) and Smith and McDonald(1976). However, the existence of a more or less difference between each calculating results derived from these equations can be seen which was well summarized by Smith and McDonald[3]. For this reason, hereinafter the particle charge related to the theoretical migration velocity was decided on Hewitt's experiments[4] as the best source of data because Hewitt devised both a corona charger and a screen charger(so-called a tri-electrodes charger) to measure the particle charge with which is about the same as our experimental conditions.

Figure 5 shows the representative example for the deposition rate distribution on the collecting plate electrode which clearly represents the tendency having an exponential decreasing toward the down stream. Such a tendency can also support the validity for the exponential expression in equation (3).

Conclusion

The experimental investigations described in this paper have led to the

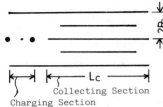

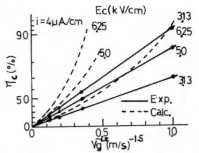

Fig.1 Shematic presentation of
testing two-stage type
precipitator.

Fig.2 Representative example for deposition
rate versus gas velocity with electric
field intensity of collecting section
as parameter.

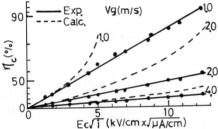

Fig.3 Representative example for deposition rate
versus product of electric field intensity of
collecting section and square root of corona
current density of charging section with gas
velocity as parameter.

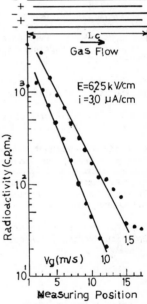

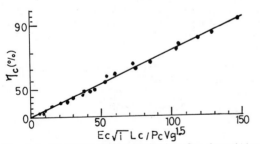

Fig.4 Performance characteristics for deposition
rate versus comprehensive design parameters.

Fig.5 Thorium B aerosols
deposition rate distribution
versus downstream location of
collecting plate electrode
with gas velocity as parameter.

following conclusion. The aerosols deposition rate even in this low Reynolds number of approximately 500 to 2,000 are not always consistent with equation (1) and the existence of the critical length beyond which the aerosols are completely deposited. Such causes can be considered as follows,

(1) aerodynamic re-entrainment[2]
(2) turbulence intensity or diffusivity[5] in the distance for developing untill fully laminar or parabolic flow
(3) aerosol particles agglomeration
(4) stochastics irregularity for particle charging[4]
(5) deposition by electric image force[6]
(6) diffusional deposition by space charge effect[7]
(7) inertial force.

However, since the influences by causes (4),(5),(6) and (7) are negligibly small in this experiment, accordingly it is a reasonable assumption to suppose that the predominant effects are due to cause (1),(2) and (3).

References

(1) Isahaya,F. A New Air Cleaning and Measuring Technology. Special Publication of the Hitachi Hyoron, p.44-48, 1964.

(2) Isahaya,F. A New Correction Method of Migration Velocity in Deutsch Efficiency Equation for Sizing of Full-scale Electrostatic Precipitators Using Pilot-scale Precipitator. The Fourth Symposium on the Transfer and Utilization of Particulate Control Technology. U.S.A.E.P.A.,Houston, Texas, 1982.

(3) Smith,W.B. and McDonald,J.R. Development of a Theory for the Charging of Particles by Unipolar Ions. J.Aerosol Sci., $\underline{7}$, p.151-166, 1976.

(4) Hewitt,G.W. The Charging of Small Particles for Electrostatic Precipitation . AIEE Trans.,$\underline{76}$, p.294-306, 1957.

(5) Leonard, G.L.,Michner,M. and Self,S.A. Particle Transport in Electrostatic Precipitators. Atmos. Environ.,$\underline{14}$,p.1289-1299, 1980.

(6) Yu,C.P. and Chanda,K. Deposition of Charged Particles from Laminar Flows in Rectangular and Cylindrical Channels by Image Force. J.Aerosol Sci.,$\underline{9}$, p.175-180, 1978.

(7) Yu,C.P. Precipitation of Unipolarly Charged Particles in Cylindrical and Spherical Vessels. J.Aerosol Sci.,$\underline{8}$,p.237-241, 1977.

Published 1984 by Elsevier Science Publishing Co., Inc.
Aerosols, Liu, Pui, and Fissan, editors

FIELD CALIBRATION OF THE ELECTRICAL AEROSOL ANALYZER
AND THE DIFFERENTIAL MOBILITY PARTICLE SIZER

Lida Xu and Dale A. Lundgren
Environmental Engineering Sciences
University of Florida
Gainesville, FL 32611

Introduction

Proper operation of an Electrical Aerosol Analyzer (EAA) or Differential Mobility Particle Sizer (DMPS) is of great concern to an aerosol researcher. Most laboratories do not possess the required equipment to perform a primary instrument calibration. Factory calibration is both expensive and time consuming and need only be done when an instrument is not working properly. A simple and inexpensive calibration check procedure is needed to indicate improper instrument operation.

A simple calibration check procedure would involve using the instrument to measure a known, reproducible, standard polydisperse aerosol. Ideally, this standard aerosol would have a geometric standard deviation of about 2.0, a controlled number mean diameter of about 0.03 to 0.1 μm, and a number concentration of about 100,000 particles per cubic centimeter, which would vary by no more than ±10% from run to run, day to day, equipment setup to equipment setup, or operator to operator. Previous experiences suggest that an aerosol generating system using the standard Collison atomizer[1] might meet these requirements.

Aerosol Generating System

In general, the Collison atomizer can produce a stable aerosol, but the concentration is too high for direct checking of an EAA or DMPS. Therefore, a wind tunnel dilution system was used to reduce the number concentration of output aerosol to about 100,000 part./cm^3 (see Figure 1). This dilution system consists of a high volume air sampler, an absolute type filter and a mixing chamber or tunnel. A sodium chloride - water solution (of 0.1% to 10%) is nebulized by a standard 3-jet Collison atomizer at a set pressure of 10 to 30 psig. This aerosol is then mixed with 15 l/min. of clean dry air before entering the wind tunnel. A disk and orifice is used to make sure that the aerosol is well mixed. A Kr-85 particle charge neutralizer and metal tubing are used to reduce particle charge and wall losses in tubing leading to measure instruments.

This generating system can easily be made in every laboratory. This equipment can be assembled or disassembled in about 10 minutes.

To generate a stable, constant aerosol concentration, the atomizer itself must be in proper operating condition. The atomizers orifices must not be plugged, either entirely or partly. A simple orifice check procedure has been developed.

Experimental Results

It was found that the aerosol size distribution generated by this system can be very stable, with the median size depending mainly on the concentration of the solution atomized. The aerosol number concentration is more sensitive to operation variables.

Number Concentration
The aerosol number concentration is influenced by solution concentration and several other parameters including atomizing pressure and wind tunnel flowrate. For a 1% sodium chloride solution concentration in water, at an atomizing pressure of 20 psig and a wind tunnel flowrate of 55 cfm, the number concentration measured by the EAA or a Condensation Nuclei Counter (CNC) is about 100,000 part./cm^3. In general:

1) The number concentration is repeatable.
2) The number concentration measured by EAA and CNC are pretty close, but the concentration measured by DMPS is lower than the others.
3) The number concentration measured by two EAA are slightly different.

It has been found that the number concentration of generated aerosol decreases slowly with time. In general, the variation is less than 10% in 1 hour. Experiments have shown the variation in 5 minutes is about ±2%. The stability of aerosol number concentration is quite good. To obtain reproducible aerosol number concentration, fresh-made solution must be used. If a used solution is atomized again, the number concentration will be changed, even though the size distribution will remain the same.

Size Distribution
Several experiments have been performed to determine the effects of solution concentration, atomizing pressure, wind tunnel flowrate and resident time on the aerosol size distribution. The dominant parameter to influence the size distribution is concentration of solution atomized.

Conclusion
Collison atomizer - wind tunnel aerosol generating system can produce very stable aerosol in both number concentration and size distribution. Aerosol generated by this system can be used as a standard aerosol to check both EAA and DMPS. The response of EAA to a standard aerosol generated by 1% sodium chloride solution, atomizing pressure 20 psig and 55 cfm flowrate of wind tunnel should be NMD = 0.035 µm ±10%, σg = 2.0 ±10% and number concentration 1 x 10^5 part./cm^3 ±15%, and for DMPS it should be NMD = 0.051 µm ±10%, σg = 2.0 ±10% and number concentration 5 x 10^4 part./cm^3 ±15%.

Reference
1) British Standards Institute, British Standard 2577, 2 Park Street, London, WI (1955).

2) Patricia B. Keady, Frederick R. Quant, and Gilmore J. Sem. Differential Mobility Partial Sizer, A New Instrument for High-Resolution Aersol Size Distribution Measurement Below 1 um. TSI Quarterly, Vol. IX, Issue 2, April/June 1983.

3) Richard J. Remiarz, Gilmore J. Sem, Jugal K. Agarwal, Charles E. McManns. An Automated Diffusion Battery/Condensation Nucleus Counter System for Sizing Submicron Aerosol. TSI Quarterly, Vol. VI, Issue 2, May/June 1980.

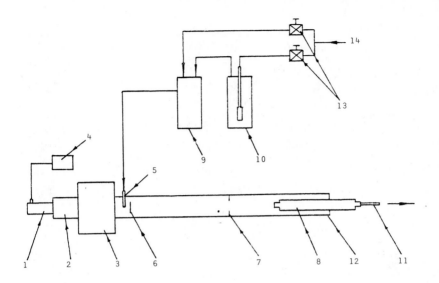

1 Standard Orifice

2 Hi-Vol Sampler's Blower

3 Filter Box

4 Manometer

5 Inlet of Aerosol

6 Disk

7 Orifice

8 Kr-85 Neutralizer

9 Mixing Bottle

10 Collison Atomizer

11 Metal Tubing (to instrument)

12 Wind Tunnel

13 Valves

14 Clean, Dry Compressed Air

FIGURE 1   Schematic Diagram of Aerosol Generating System

CHAPTER 4.

SAMPLING METHODS AND DEVICES

Published 1984 by Elsevier Science Publishing Co., Inc.

Aerosols, Liu, Pui, and Fissan, editors

RESULTS OF FRACTIONATING PARTICULATE MEASUREMENTS
IN THE UPPER PARTICLE SIZE RANGE

L. Laskus
Institute for Water, Soil and Air Hygiene
of the Federal Health Office
Corrensplatz 1, D - 1000 Berlin 33
Federal Republic of Germany

EXTENDED ABSTRACT

Introduction

In 1979 the U.S. Environmental Protection Agency (EPA) defined inhalable particles (IP) as those of less than 15 μm aerodynamic diameter. At present, the EPA is considering to fix the IP upper size limit at 10 μm. Recently, a further size limit for IP of 6 μm was discussed (Wedding and Weigand, 1983). Each of these recommendations for the different IP size limits are based on the consideration that only particles up to each of the above-mentioned size limits can mainly penetrate the larynx and be deposited in the conducting airways and the gas-exchange areas of the thorax. This thoracic deposition of fine particles is attributed a higher health risk than the extrathoric deposition of coarser particles. However, it should be taken into consideration that the proportion of suspended particulate matter (SPM) up to 6 μm diameter may differ from that proportion of SPM up to 15 μm by approx. 20 %. Moreover, inspired droplets or soluble components of inspired solid particles - present in the case of foggy smog weather conditions with a high pollution level in the coarser particle size range - may be deposited also in the upper respiratory tract and may cause damage near their deposition sites if they are corrosive or active.

Results obtained with size selective instruments

Different types of size selective sampling devices for SPM of a particle size up to 10 or 15 μm have been used in the U.S.A. for a few years. In the Federal Republic of Germany, comparative measurements using two types of 10 μm samplers, impactors and non-fractionating samplers are still carried out at sites in Berlin and Frankfurt.

The Institute for Water, Soil and Air Hygiene in Berlin conducts a comparison study of the following aerosol samplers: SSI hi-vol sampler (cut-off: 10 μm), 8-stage cascade impactor (cut-offs: st.1 - >10 μm, st.2 - 10 μm, st.3 - 6 μm, st.4 - 3,5 μm, st.5 - 2,1 μm, st.6 - 1,0 μm, st.7 - 0,6 μm, st.8 - 0,35 μm, back-up filter; flow rate: 2,5 m^3/h), standard hi-vol sampler, small filter device GS050/3 (non-fractionating device; aspiration in upward direction via circle slit; flow rate: 2,8 m^3/h). Results of non-fractionating measurements obtained with the hi-vol sampler (HVS), the GS050/3 sampler and the cascade impactor (CI) are shown in Table 1:

Table 1. Results of comparative measurements
         using the HVS, the GSO50/3 and the CI

Device		No. of values	Wind velocity (m/s)	Concentration (μg/m^3)	Stand. dev. (μg/m^3)	Corr. coeff.	Ratio of $\frac{1}{2}$ (%)
HVS	(1)	158	<4	95	41	0,96	103
GSO50/3	(2)			92	42		
CI (summed up over all stages)	(1)	35		124	64	0,95	103
GSO50/3	(2)			120	58		
HVS	(1)	74	4 - ~10	66	40	0,89	125
GSO50/3	(2)			53	34		

Conclusions:
- At low wind velocities it can be assumed that the HVS, the
  GSO50/3 and the impactor collect approximately the total SPM.
- At high wind velocities the HVS collects distinctly more of the
  SPM than the GSO50/3 and the correlation between both samplers
  decreases from the value of 0,96 (low wind velocity) to a value
  of 0,89 (high wind velocity).
These relations of the separation efficiency of the HVS and the
GSO50/3 are to be attributed to two effects:
- At high wind velocities, additional particles - mainly coarse
  particles - which cannot be related to the aspirated air volume
  are carried into the samling head (oversampling). This effect
  is also expressed by the fact that the correlation coefficient
  for the comparison HVS/GSO50-3 at the low wind velocity range
  is distinctly higher than that at the higher wind velocity range.
- The small smpling head of the GSO50/3 aspirating in upward direction collects particles only with a distinctly lower efficiency at higher wind velocities.

    For the comparison of the HVS and the SSI hi-vol sampler (SSI)
similar relations are found (see Table 2).

Table 2. Results of comparative measurements
         using the HVS and the SSI

Wind velocity (m/s)	No. of values	HVS (stand. dev.) (μg/m^3)	SSI (stand. dev.) (μg/m^3)	Corr. coeff.	Ratio of SSI/HVS (%)
<4	77	103 ± 74	74 ± 45	0,95	72
4 - 8,5	26	58 ± 38	33 ± 16	0,87	57

The results obtained at the higher wind velocity range show also
a clear decrease of the ratio SSI/HVS and of the correlation
coefficient as compared with the results obtained at the low wind
velocity range.

According to these results, direct comparative measurements using the GS050/3, the HVS, the SSI and the CI were carried out only when the average wind velocity was 1 m/s or less (see Table3). Thereby, it should be guaranteed:
- correct sampling of the total SPM by all devices used,
- performance of measurements at high concentrations of SPM which usualy decrease with increasing wind velocities.

Table 3. Results of comparative measurements using the GS050/3, the HVS, the CI and the SSI wind velocity: ≤1 m/s; No. of compared values: 27

	GS050/3	HVS	CI (summed up over all stages)	CI (summed up up to the 10 μm-stage)	SSI
Concentration ($\mu g/m^3$)	135	132	137	119	104
Ratio to GS050/3 (%)	100	98	101	88	77

Conclusions:
- The GS050/3, the HVS and the CI (summed up over all stages) are comparable with each other.
- The CI (summed up up to the 10 μm-stage) and the SSI differ from each other by approx. 13 %. These differences are not due to the "wall losses" of the impactor. According to the particle size distribution at the measuring site the wall losses up to the 10 μm-stage of the impactor amount to a maximum of 4 % of the total collected particle mass.

Studies of the Federal Environmental Agency in Frankfurt using the GS050/3 and the Sierra med-flo sampler (cut-off: 10 μm) lead to the results shown in Table 4. These studies were also carried out at low wind velocities.

Table 4. Results of comparative measurements using the GS050/3 and the med-flo sampler

	No. of values	Wind velocity (m/s)	Concentration ($\mu g/m^3$)	Ratio of med-flo/GS050-3 (%)
GS050/3	19	<3	84	64
Med-flo sampler			54	

According to the results of the both studies using size selective instruments and to the results of impactor measurements taken in parallel at both sites, the size selective devices show the following cutpoints: SSI (10 μm) - cut-off: approx. 6 μm; med-flo sampler (10 μm) - cut-off: approx. 3,5 μm.

Moreover, the following results of comparative measurements using the standard HVS and several size selective instruments are known from international literature:

92

Dichotomous VI (15 μm)/HVS - approx. 60 %; SSI (15 μm)/HVS - approx. 80 %
Wedding PM-10 (10 μm) /HVS -   "   63 %; SSI (10 μm)/HVS -   "   65 %

These relations of results obtained with several size selective instruments and the standard HVS are inconsistent. Moreover, the differing results cannot be interpreted so that the measurements have been conducted at different sites, since some of these results have been obtained in parallel measurements using different types of size selective instruments (Solomon et al., 1982).

Conclusion

In order to determine which fraction of the total atmospheric aerosol should be sampled by means of size selective devices, measurements of the total mass distribution of the atmospheric particles are now performed in U.S.A. using a sampling system proposed by Lundgren and Paulus (Wide Range Aerosol Classifier - WRAC). Previous studies (1975) of the mass distribution of the atmospheric particles using a similar sampling system described by Lundgren and Paulus yielded results showing a distinctly higher mass proportion of the coarse particle mode than the results obtained in other known studies. However, it is doubtful whether this sampling system produces correct results concerning the size distribution of the atmospheric aerosol. It cannot be excluded that this system collects also a considerable proportion of coarse dust fall particles which are largely dispersed inhomogeneously and are not related to a reference air volume like aerosol particles. The high mass proportion of the coarse mode measured by means of the sampling system used by Lundgren and Paulus can be due to an over-sampling-effect.

Instead of attempting to measure the correct particle size distribution and to orient size selective instruments according to these results, it would be more sensible to define a mass-related reference method without a rigid cutpoint for the sampled aerosol. The sampling head should aspirate the air in an upward direction in order to prevent an oversampling-effect.

References

Wedding, J.B. and Weigand, M.A. (1983) "A Thoracic Particle Inlet (d_{50} = 6 um, d_o = 10 um) for the High-Volume Sampler", _Atmos. Environ._ 17:1203.

Solomon, P., Derrick, M. and Moyers, J. (1982) "Performance Comparison of Three Samplers of Suspended Airborne Particulate Matter", _J. Air Poll. Contr. Ass._ 32:373.

Lundgren, D.A. and Paulus, H.J. (1975) "The Mass Distribution of Large Atmospheric Particles", _J. Air Poll. Contr. Ass._ 25:1227.

Published 1984 by Elsevier Science Publishing Co., Inc.

Aerosols, Liu, Pui, and Fissan, editors

THE USE OF AN ALUMINA FIBRE MAT
FOR A PARTICULATE SAMPLER FILTER IN
HIGH TEMPERATURE CORROSIVE ATMOSPHERES

R. Hesbol, W. Piispanen[*], L. Ström
Studsvik Energiteknik
*Battelle Columbus Laboratories
The Marviken Project
S-61024 Vikbolandt
Sweden

Introduction

In a test program designed to characterize vapors and aerosols produced from over heated core materials in typical water type nuclear reactors, a sampling system for aerosols and for vapor products was designed. Due to the corrosive nature of the aerosols and vapors and considering the extreme of sampling temperature of 800°C, it was necessary to investigate the availability of a suitable filter media for the collection of these aerosols.

The material in addition to a resistivity to strong basic condition and stability at high temperatures, must efficiently collect aerosol particles while allowing vapor phase species to pass through unchanged. A limited number of materials suit these qualifications and preliminary experiments looked at the following materials:

1)   nickel filters pore diameters 2 μm, 10 μm
2)   316 L stainless steel(ss) filters, 5 μm pore
3)   0.8 m and 0.12 mm ss wire packets
4)   ss wool
5)   glass fibre wool
6)   inconel felt fibrous filters
7)   ss felt fibrous filters
8)   titanium fibres
9)   alumina fibre wool

Of the above, the metal filters (numbers 1 and 2) exhibited rapid clogging in a 100° steam aerosol stream. The ss wool (number 4) was not acceptable for particle collection. The glass fibre wool (number 5) was acceptable for particle collection but was not thermally stable at 800°C. The stainless steel wire (number 3) was acceptable for both thermal stability and particle collection. Also the metalic felt fibrous filters (numbers 6,7, and 8) were good for particle collection up to 500°C. The alumina fibre wool was considered good for particle collection and is thermally stable up to 1550°C. As a result, further investigations of the alumina fibre wool were conducted.

The material selected for testing was Saffil[TM]. Saffil is the trade name for an alumina fibre material that is produced by ICI, Mond Division. The material is an Al_2O_3 (plus SiO_2) wool which is normally used for high temperature insulation. Iron and sodium oxides, and other metalic contaminants are reported as <0.1%.

Tests of Saffil Material

In considering the background contamination of Saffil, tests were
conducted with extraction of the filter material with water, 5N nitric acid
and aqua regia acid. The Saffil extraction showed less than 100 ppm of iron,
with the 5N nitric acid as the highest of the three treatments. The water and
agua regia extracts were less than 10 ppm. For the glass fibre filter the
iron background was approximately 200 ppm with aqua regia but less than 10 ppm
for water and 5N nitric acid. With glass fibre, the sodium background was
quite high for all three treatments but very low in the Saffil filter.

The hygroscopic character of the Saffil was examined by comparing weights
of a filter before and after 5 days of dessication. For a 1.8 g filter the
weight change was -0.03 per cent thus indicating that the Saffil material
should not be a problem with regard to moisture gain. On a thermo balance,
no weight change was observed at temperatures up to 1550°C in air.

Tests to demonstrate vapor retention of the test materials (CsOH, CsI and
Te) were conducted. The following results for gas retention of test species
on alumina fibre filters were produced:

Test with elemental iodine

Run	Temp	Atmosphere	Retention
1	800°C	Argon	0.015%

....
....
....

Test with cesium hydroxide vapor

Run	Temp	Atmosphere	Retention
1	800°C	Argon	1.9 %
2	800°C	Argon	1.5 %
3	800°C	Steam	1.5 %

Test with tellurium vapor

Run	Temp	Atmosphere	Retention
1	800°C	Steam	∿0.5%

. . . .
. . . .
. . . .

The results show that the retention due to chemical reaction between the gaseous phases of iodine, cesium hydroxide, and tellurium vapor was low and acceptable. Any degradation of the alumina fibre filter material was not visible in any case.

In order to examine the loading capacity of cesium hydroxide on Saffil, three tests were performed (in air). In the tests, solid cesium hydroxide of known weight was placed onto several layers of Saffil filter and heated for 1 hour. After the test the weights of unattacked Saffil were determined. Tests were performed at 20°C, 450°C and 715°C. It was concluded that cesium hydroxide partly evaporates during the heating period before it reacts with the Saffil. The presumably formed cesium aluminate thus had a low volatility and was temperature stable up to at least 715°C. The freshly prepared cesium aluminate was observed to be hard and brittle. Like cesium hydroxide, the cesium aluminate was hygroscopic.

As a result of the reactivity tests a capacity limit of Saffil for CsOH (the major component of the test aerosol) was determined to be 3.6 g of CsOH per g of Saffil. This capacity limit was applied to subsequent tests to determine filter size to sample volume.

As final considerations in the choice of filter materials were 1) the filter material strength and 2) weight of the filter material relative to the collected material. Ideally the material should be relatively light in weight compared to the weight of material collected. For Saffil as used in the first test, the filters were approximately 7g. Up to 2.2 g of aerosol material had been collected during these tests. The material should also be easy to handle since weighing, transfer and storage of samples are required in the analytical procedures. Saffil is slightly brittle at room temperature but does cut easily with a die. The material as a 0.5 cm thick mat can be easily handled with forceps without damage. The material is somewhat flexible and also compressible thus allowing proper placement in the filter assembly.

Test Performance

In actual tests using a 168 mm x 5 mm Saffil filter mat, the filters performed reasonably well. There were occasional ruptures but these were due to physical damage or else to corrosion from overloading of CsOH. As a result the filter mass was increased from ∿ 7 g to 14 g (10 mm thick). Due to the stickiness of the aerosol some of the Saffil often was stuck to the stainless steel filter housing. As a result, quantitative recoveries of the Saffil mat were poor. Also there is a tendency for Saffil to pulverize if it is caught in between the filter housing faces. This results in a loss of Saffil from the mat but this is recoverable in the housing wash. Overall the Saffil is still the perferred filter material due to (1) thermal stability (up to 1500°C), (2) handling properties, (3) low density and, (4) low chemical reactivity with vapors.

Published 1984 by Elsevier Science Publishing Co., Inc.

Aerosols, Liu, Pui, and Fissan, editors

A SAMPLING SYSTEM FOR HIGH TEMPERATURE HIGH PRESSURE APPLICATIONS

J. L. Alvarez, V. J. Novick, A. D. Appelhans
Idhao National Engineering Laboratory
EG&G Idhao, Inc.
P. O. Box 1625
Idaho Falls, ID   83415

Many industrial operations require the transport and reaction of particles in a flowing gas stream that is at high temperature and pressure. Measurement of the aerosols and their properties is important to quantifying the processes. The same particles may produce a hazard upon loss of containment or in waste gas streams. Evaluation of this hazard requries the characterization of the particles. A system has been designed to sample particles from a high temperature high pressure gas stream and an appropriate test vessel for characterizing and calibrating the sampling system was constructed.

The sampling system comprises a dilution/cooling gas delivery system, a ceramic sampling probe, a cyclone/valve, and a virtual impactor. The dilution/cooling gas is controlled by a choking orifice and a dome pressure regulator. The dilution gas is delivered annularly immediately behind the ceramic probe. The dome pressure regulator uses a differential piston technique to regulate the flow by means of a reference pressure to the regulator dome. The reference pressure in this case is the sample pressure or the pressure downstream of the choking orifice. The supply pressure thus varies with the sample pressure and a constant volumetric flow is supplied to the sample probe. The diluted and cooled sample passes through a transport tube which is held at a constant temperature. A cyclone with a 20μm aerodynamic cut is placed before the virtual impactor. The cyclone is modified to be a valve so that system beyond the cyclone can be isolated from the pressure system. The virtual impactor is 2 stage and can collect a time integrated sample on filters or collect time dependently on a moving tape.

The mass flow equation can be written to obtain the sample volume flow as

$$Q_2 = \frac{R_2}{R_3} Q_3 - \frac{P_1}{P_3} \frac{R_2}{R_1} \frac{T_3}{T_1} Q_1$$

where the subscripts refer in order to the dilution gas, the sample, and the combined sample plus dilution gas. The temperature and pressure of the dilution gas have the subscript 3 because measurement and control are at the combintation conditions. A dilution gas to sample ratio of 10 to 1 will have a 20% change in the sample volume for a 1% change in any of the ratios of the above eequation. The change in the ratios is held to a minimum by various means. The temperatures of the choking orifices are held constant by immersing a length of tubing containing the orifices in a constant temperture bath, therefore $T_1 = T_3$. No changes will occur in the gas constants R_x if control is maintained and there is no change in the gases.

In one application of this system, steam changes to hydrogen as parts of the experiment oxidize. The gas constants are held constant, in this case, by placing a copper oxide recombiner before the choking orifice.

Published 1984 by Elsevier Science Publishing Co., Inc.
Aerosols, Liu, Pui, and Fissan, editors

The Cascade Impactor As A Particulate-Chromatograph
For Microanalysis of Aerosol Particles

Raymond L. Chuan   William W. Chiang
Brunswick Corporation   California Measurements, Inc.
Costa Mesa, California 92626   Sierra Madre, California 91024

EXTENDED ABSTRACT

Introduction

When aerosol particles of a wide range of sizes are collected
together it is often difficult to analyze them morphologically or
chemically. Most naturally occurring or non-laboratory-generated
aerosols are made up of groups of particles of distinct physical and/or
chemical characteristics; and these groups are usually distinguished
by particle size. It then becomes possible to pre-fractionate diverse
aerosol particles into distinctive chemical/physical groups by means
of inertial size separation, before subjecting them to detailed, and,
usually, single particle microanalysis by such means as scanning-electron-
microscopy (SEM), X-ray energy spectroscopy (XES or EDX, energy-dispersive
x-ray), Auger-electron-spectroscopy (AES) and other advanced analytical
techniques. In this sense the scheme is quite analogous to the combination
of gas-chromotography and mass-spectrometry (GC-MS) commonly used for
analyzing mixtures of gases spanning a wide range of molecular weights.
This technique, which may be called "particulate-chromatography-micro-
analysis" or PC-MA, is particularly useful in situations where the
quantity of aerosol particles collected is not amenable to the usual
methods of bulk analysis. Important, but sparse in number and often
very small, particles whose presence can be easily obscured in bulk
analysis by the dominance of other more numerous and usually larger
particles can thus be segregated and analyzed individually.

Experimental Demonstration

An example of an unsegregated sample is shown in the scanning-
electron-micrograph (SEM) in Fig. 1, which is the collection of all
particles larger than 0.05 μm aerodynamic diameter. At a magnification
of 650x we see the largest particles are about 7 μm while the smallest
are obscured by the larger particles. The elemental composition (obtained
with XES, which shows elements with atomic numbers ≥11) of this unsegregated sample is shown in the inset. We see a significant amount of Cl
and lesser amounts of Na, and S, and a very small amount of Mg. Other
elements, if present, are not seen against the dominance of the three
major elements: Na, S, Cl. (The dark lines in the X-ray spectrum
are from the substrates into which the particles are collected by
impaction.)

When the same sample is acquired with a cascade impactor, in this
case with a Quartz Crystal Microbalance Cascade Impactor (Chuan, 1975),
the distribution of aerosol mass by size is shown in Fig. 2, with a

major mode near stages 5 and 6 (geometric mean diameters of 2-3 and
4.5 µm) and a minor mode centered about stage 9. The sparsely populated
stage 4 (Fig. 3a) reveals at least two types of particles: deliquescent
halite (Fig. 3b) and mineral fragment magnesium olivine (Fig. 3c).
Stage 6 (Fig. 4a) reveals two major groups of particles: sodium hydroxide
crystals (Fig. 4b) and unidentified chloride precipitated from droplets
(Fig. 4c) mixed with trace amounts of Na, Mg, S and Fe. Stage 8 (Fig. 5a)
shows soot particles (Fig. 5b) and sulfuric acid droplets (Fig. 5c).

References

Chuan, R.L., <u>Fine Particles, Aerosol Generation, Measurement, Sampling,</u>
<u>and Analysis,</u> 763, B.Y.H. Liu, Ed. Academic Press, New York, 1975.

Fig. 1 UNSEGREGATED SAMPLE - 650X SEM; X-RAY SPECTRUM

Fig. 2 AEROSOL MASS DISTRIBUTION
BY SIZE OBTAINED WITH
QCM CASCADE IMPACTOR
Geo Mean Dia.

Fig. 3a TYPICAL PARTICLES

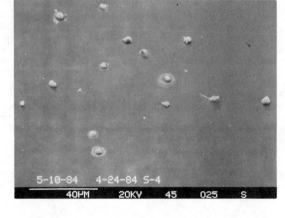

101

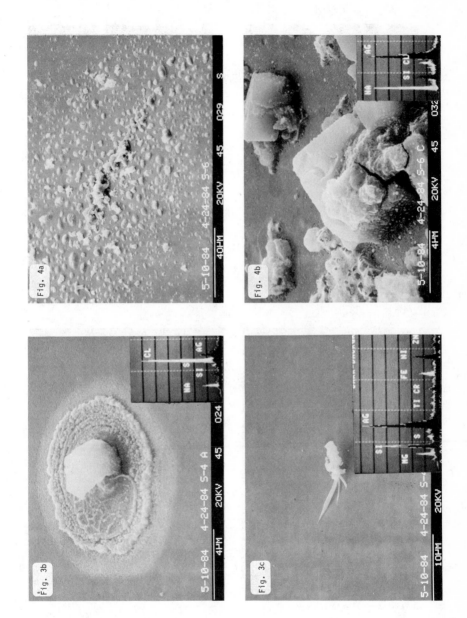

Fig. 4a

Fig. 4b

Fig. 3b

Fig. 3c

102

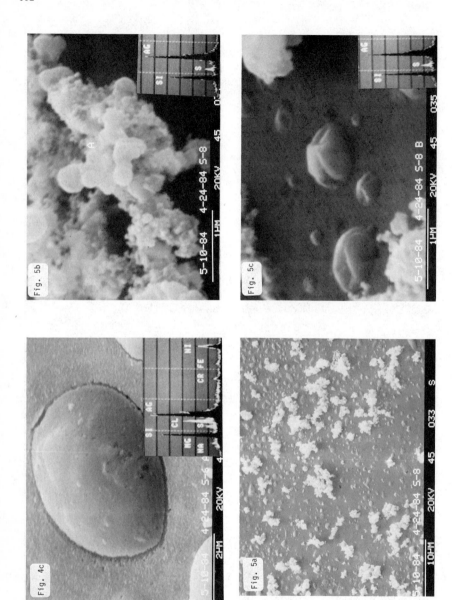

Fig. 4c

Fig. 5a

Fig. 5b

Fig. 5c

Published 1984 by Elsevier Science Publishing Co., Inc.

Aerosols, Liu, Pui, and Fissan, editors

EVALUATION OF A CALIBRATION TECHNIQUE FOR INERTIAL PARTICLE
SIZING DEVICES USING POLYDISPERSE TEST AEROSOLS

Ashley D. Williamson, William E. Farthing, and Wallace B. Smith
Southern Research Institute, Birmingham, AL

The particle collection characteristics of inertial sampling devices have
historically been characterized using monodisperse test aerosols. The
collection efficiency of the sampler for the known particle size is measured
for a range of experimental conditions. Often several runs at different
conditions are necessary to determine a characteristic cut diameter (D_{50}) for
a single stage of an inertial classifier.

In order to reduce the effort required to calibrate particle sizing
devices, a technique has been developed which employs polydisperse test
aerosols. In principle, a single run can generate an entire size-dependent
particle collection efficiency curve. For calibration studies, it is
sufficient to identify the D_{50} of a device under the experimental conditions
used; this can be done unambiguously with a single data run.

The calibration system used for these studies is shown schematically in
Figure 1. A small spray nozzle fed by a syringe pump injects a polydisperse
mist of a dye solution into a mixing and drying column. The dye aerosol
passes through a flow splitter which serves as a crude virtual impactor. Most
of the aerosol stream is exhausted into a filter and pump. The remaining
fraction, enriched in larger particles, continues through the silica gel
diffusional dryer and a heat-traced section of tubing to a branch point
mounted at the top of a convection oven. Two ball valves with teflon seats
route the gas stream through an inertial sizing device or through a bypass
line. The gas stream is cooled to near ambient temperature before passing
through a mass flow meter and optical particle counter (Climet 208). The
scattered light pulses are counted and sorted by a multichannel analyser (MCA)
to give a pulse height spectrum for both sampler outlet and bypass legs. The
two spectra can be processed either manually or by an on-line microcomputer to
give sampler collection efficiency for particles corresponding to any pulse
height region of interest. As can be seen in Figure 1, the apparatus was
designed with the capability of elevated temperature operation. This feature
has allowed experimental study of the behavior of small cyclones (Smith, et
al, 1983) and of cascade impactors (Farthing, 1983) under conditions typical
of their normal use.

In order to convert the collection efficiency vs. pulse height data to
efficiency curves related to aerodynamic particle size, it is necessary to
know the pulse height response of the optical particle counter to particles in
the size range of interest. Two independent methods were used to relate
particle counter pulse heights to particle size. First, monodisperse dye
particles were generated using a vibrating orifice aerosol generator and
measured with the optical particle counter. The pulse heights corresponding
to particles of several sizes in the range of 2-15 μm were recorded. The
particle diameters and size uniformity were verified using optical microscopy.
The second method, used by Smith, et al (1983) for small cyclone calibrations
and Farthing (1983) for impactor stages, gives an aerodynamic measure of

104

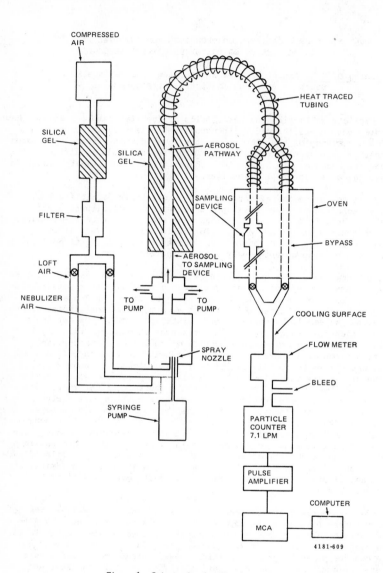

Figure 1. Schematic of calibration apparatus.

optical particle counter response. In this technique, inertial samplers previously calibrated with monodisperse test aerosols were used as aerodynamic size references. The polydisperse dye aerosol is used to measure the pulse height corresponding to the (known) D_{50} of calibrated samplers. Use of several samplers allows calibration for particle diameters in the range 0.3 to 15 μm. A response function for particles of a blue dye test aerosol is plotted in Figure 2.

In order to test the capabilities and limits of the polydisperse calibration technique, several possible sources of error were investigated. One potentially serious error source would be failure or loss of calibration of the mass flow meter, which is of necessity exposed to the dye aerosol. To avoid this, all flow readings during actual data runs are checked against measurements using an external calibrated orifice. For elevated temperature operation, this backup flow measurement is essential, since the mass flow meter used in the system is subject to significant thermal drift when the oven and gas heaters are in operation. A second potential problem is drift in the electronic components of the system—the particle counter, amplifier, and MCA—so that the measured response to a given particle size is not constant. A daily calibration protocol has been developed to detect and compensate for electronic drift. The instrument response to standard voltage pulses from a pulse generator and to aspirated monodisperse latex particles is recorded before each day's operation. If necessary, the photomultiplier voltage or amplifier gain are adjusted to reproduce earlier (standard) instrument response. A third possible problem is thermal alteration of the calibration aerosol during the test. Indeed, it is apparently possible to flash volatilize ammonium fluorescein at temperatures above 450°F. However, when all surfaces in the apparatus are kept below 375°F, no alteration in the microscopic appearances or measured pulse heights of monodisperse fluorescein dye particles was noted. Other dyes, such as the 8GLP Fast Turquoise used in earlier studies, may have higher temperature limits.

The polydisperse system has been used to calibrate several small cyclones and impactor stages at both ambient and elevated temperatures. Studies to date confirm that the technique allows considerable time savings compared to conventional techniques using monodisperse test aerosols. For example, Figure 3 shows complete efficiency curves obtained for single impactor stages with two different adhesives, one which is quite tacky at room temperature and one which is rather stiff. These data, which graphically show the effects of particle bounce, were obtained in about 15 min. total run time.

References

Farthing, W. E. (1983) "Impactor Behavior at Low Stage Re", Aerosol Sci. Technol, 2:242.

Smith, W. B. and D. L. Iozia (1983) "Performance of Small Cyclones For Aerosol Sampling". J. Aerosol Sci., 14(3):402-409.

106

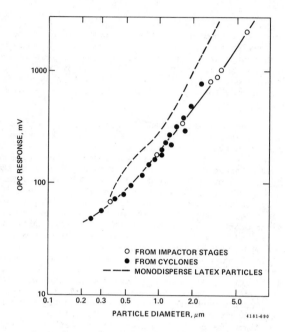

Figure 2. Response of the optical particle counter to 8GLP dye particles.

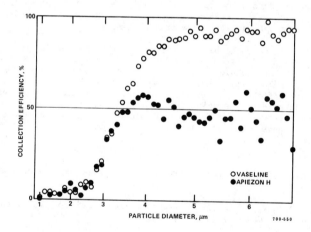

Figure 3. Measured collection efficiency of an impactor stage with two substrate greases using polydisperse calibration technique.

Published 1984 by Elsevier Science Publishing Co., Inc.
Aerosols, Liu, Pui, and Fissan, editors

CO-CALIBRATION OF FOUR, LARGE-PARTICLE IMPACTORS

R.W. Vanderpool, D.A. Lundgren, Dept. of Environmental Engineering Sciences, University of Florida, Gainesville, FL 32611, V.A. Marple, and K.L. Rubow, Particle Technology Laboratory, Mechanical Engineering Dept., University of Minnesota, St. Paul, MN 55455.

Extended Abstract

In a study to determine the size distribution of ambient aerosols, a mobile, large-particle sampling system was designed and constructed at the University of Florida. The system contains four, high-flowrate, single stage impactors for fractionation of the sampled aerosol into five distinct size fractions. The four rectangular impactors are operated in parallel in order to reduce particle losses and to allow a greater volume of ambient air to be sampled than could be achieved through a cascade arrangement of the stages.

A detailed laboratory calibration of the impactors was required for a number of reasons. First, although the impactors' design was based upon well-developed theory, calibration accounts for parameters such as high fluid Reynolds numbers, high particle Reynolds numbers, and gravity effects which have not yet been incorporated into standard impactor theory. Experimental calibration also allows quantification of particle losses and determination of substrate collection characteristics which cannot be predicted theoretically.

This paper presents the results of two independent laboratory calibrations of the four impactor stages. Calibration was performed with monodisperse test aerosols generated with the vibrating orifice aerosol generator in conjunction with greased impactor substrates. At the University of Minnesota, solid ammonium fluorescein test aerosols up to 35 um were generated for the calibration while liquid oleic acid aerosols were employed for collection efficiency tests of aerosols above 35 um. Calibration performed at the University of Florida employed solid ammonium fluorescein particles up to 70 um aerodynamic diameter for the calibration.

Results of the independent calibrations are plotted in Figure 1. As can be seen, there is strong agreement between the two calibrations not only with respect to the position of the collection efficiency curves but in their slopes as well. Agreement was also good between results of particle loss tests and efficiency tests performed with glass-fiber filter substrates.

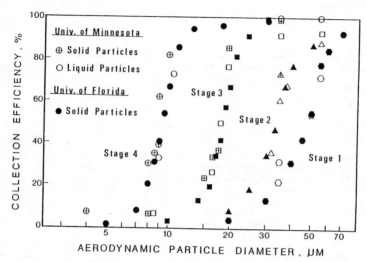

FIGURE 1. COMPARISON OF INDEPENDENT LABORATORY CALIBRATIONS

The impactor cut-points from the two calibrations are presented in Table 1. The strong agreement among the Dp_{50} cut-points illustrates the fact that similar results can be achieved from tests performed under identical conditions if careful laboratory techniques are employed. The failure of standard impactor theory to exactly predict the impactor cutpoints should not be totally unexpected. The impactors' large cutpoints and correspondingly large dimensions and operational flowrates are rather novel. Differences between the predicted and experimental results exemplify the value of laboratory calibration when exact knowledge of an impactor's performance is required.

Table 1. Summary of Impactor Calibrations

Dp_{50} (microns)

Stage	Univ. of Florida	Univ. of Minnesota	Impactor Theory
1	47	49	49
2	34	34	30
3	18.5	18.0	14.5
4	9.3	9.2	7.2

Published 1984 by Elsevier Science Publishing Co., Inc.

Aerosols, Liu, Pui, and Fissan, editors

ON THE INERTIAL IMPACTION OF HEAVY MOLECULES

J. Fernandez de la Mora , B. L. Halpern,
M. Kramer, A. Yamashita and J. Schmitt

Yale University, Departments of Mechanical Engineering and
of Chemical Engineering, New Haven, Connecticut 06520, U.S.A.

EXTENDED ABSTRACT

Having demonstrated in earlier theoretical [1,2] and experimental works [3] that heavy molecules entrained in a light carrier gas can exhibit very significant inertial effects, we now report on the use of heavy molecules as convenient laboratory models on the dynamics of ultrafine aerosols. The relevance of this work to the discipline of aerosols derives principally from the stability and ease of generation and detection of heavy molecules, which is in sharp contrast to the very great experimental difficulties one encounters when handling ultrafine aerosols. Also, due to their very considerable diffusivities, our simulated aerosols are ideally suited for fundamental studies on the simultaneous effects of inertia and diffusion on transport rates. Another experimental advantage of our model aerosols is in the possibility of inferring their speed of collision against solid obstacles from their probability of decomposition by impact, which is often a unique function of the impact energy. $W(CO)_6$, and other relatively unstable molecules that yield condensible products upon decomposition and dissociate at impact energies of only a few eV, are ideally suited for such experiments.

The present talk will be organized as follows. After a brief theoretical introduction on the relation between the flow parameters, Re=Reynolds number, M=Mach number, Sc=Schmidt number, m_p/m=molecular weight ratio (heavy-light) and the heavy molecule Stokes number,

$$ Stk = \frac{m_p \gamma}{m Sc} \frac{M^2}{Re} $$

we will present some results on the impaction of tungsten oxide molecules carried in Helium axisymmetric jets in a geometry typical of an aerosol impactor. The main experimental variable is the width of the deposits, which decreases enormously as the Stokes number increases (Fig.1). Some theoretical calculations in the region of Stk $\ll$ 1 will be compared with the data, and the relationship between inertia and pressure diffusion will be discussed. Subsequently we present results on the impaction of He-WO$_x$ jets against cylindrical probes. Then, before moving on to supersonic impaction experiments, we discuss the most important technical difficulties we have encountered in handling molecular impactors, mostly associated with the low values of the Reynolds number (Re $\leq$ 100) at which one is forced to operate.

The second part of this talk will be devoted to presenting and discussing new

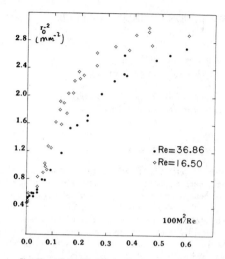

Fig.1 Characteristic width of the deposits of tungsten oxide carried in helium, as
a function of the jet mach number

experimental results on highly supersonic jets of H_2 carrying $W(CO)_6$ molecules and colliding perpendicularly against a flat plate. The carbonyl molecules decompose when impacting at speeds exceeding 1.4 Km/s [4], and their decomposition probability can be used to infer the impact velocity. These experiments provide a rather impressive proof of the very great importance of the inertia of heavy molecules, which retain much of their free stream velocity prior to colliding with a large object, even though the light gas background pressure is high enough to make it behave as a continuum fluid. But our results also provide much practical information on the performance of supersonic impactors: why the deposit sometimes has the form of a thin anulus, why it spreads more or less, depending on the prevailing conditions (Figs.2 and 3), etc. Some of these observations relate to previous papers published on aerosol beams by Schwartz and Andres [5], Dahneke and Hoover [6], Dahneke and Friedlander [7], and others. Finally, we conclude discussing the potential of heavy molecules as models for aerosols in various areas of the field.

Acknowledgments

This work has been supported by the National Science Foundation, Grant CPE-8205378, the Yale Younger Scientists Fund, and the Yale Summer Research Program.

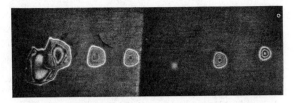

Fig.2 Interference fringes formed by the reflection of red light on the deposits resulting from the decomposition of $W(CO)_6$ by impact on a glass plate located 1 mm below the nozzle. Upstream pressures in torr are 760,462,176,130,66,113 and 95 from left to right. The third deposit is 0.6 mm horizontally across. Nozzle diameter=0.1 mm. The carrier gas was H_2. Notice the focusing of the beam at diminishing pressures.

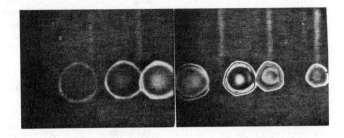

Fig.3 Deposits formed similarly as in Fig.3 at still smaller values of the Stokes number. Upstream pressures in torr from left to right are 1417(barely visible), 1263,1093,947,826,668,410 and 183.

References

1. J. Fernandez de la Mora, J.M. Mercer, D.E. Rosner and J.B. Fenn, in *Rarefied Gas Dynamics*, S.S. Fisher, editor, Prog. in Astronautics & Aeronautics vol. **74**, pp.617-626, 1981, AIAA, New York.

2. J. Fernandez de la Mora, Phys. Rev. A, **25**, 1108-1122 (1982).

3. J. Fernandez de la Mora, J.A. Wilson and B.L. Halpern, Aerosol Science & Technology **2**, 147 (1983).

4. M.S. Connolly, E.F. Greene, C. Gupta, P. Marzuk, T.H. Morton, C. Parks, and G. Staker,J. Phys. Chem. **85**, 235-260 (1981).

5. M.H. Schwartz and R.P. Andres, J. Aerosol Sci. **7**, 281-296 (1976).

6. B.F. Dahneke and J. Hoover Ref.[1], pp. 504-14.

7. B.F. Dahneke and S.K. Frielander, J. Aerosol Sci. **1**, 324-339 (1980).

Published 1984 by Elsevier Science Publishing Co., Inc.

Aerosols, Liu, Pui, and Fissan, editors

112

DEVELOPMENT OF MICROORIFICE IMPACTORS

by

V. A. Marple
University of Minnesota
Particle Technology Laboratory
130 Mechanical Engineering
111 Church Street Southeast
Minneapolis, MN 55455-0111

EXTENDED ABSTRACT

One method of designing impactors for submicron particle cut sizes is to use nozzles of very small diameters. A prototype impactor of this design, called a horizontal flow uniform deposit microorifice impactor, was built and used for several studies in our Laboratory (Marple, Liu and Kuhlmey, 1981; Kuhlmey, Liu and Marple, 1981). Recently two new versions of the microorifice impactor have been designed. For both designs the nozzle diameters range from about 45 to 124 μm in diameter for cut sizes in the range of approximately .024 to 1.1 μm aerodynamic diameter. Due to the small sizes of the nozzles, approximately 2,000 nozzles per stage are required for each 30 lpm of flow rate.

One of the new impactors using the microorifice nozzles has been designed as a eight-stage uniform deposit cascade impactor (UDI) at a flow rate of 30 l/min. The first four stages have conventionally drilled nozzles and the last four stages have microorifice nozzles. The cut size ranges from .024 μm diameter at the smallest size to 5 microns at the largest. A uniform deposit is obtained by locating each nozzle of a stage at a prescribed distance from the center of the stage such that when the nozzle plate is rotated relative to the impaction plate, a uniform deposit is obtained. The impactor is designed so that if every other stage is rotated, while the remaining stages are stationary, the impaction plate will be rotated relative to the nozzle plate of each impactor stage. The rotation is accomplished by an electric motor and a set of gears.

The second new design is a high-volume microorifice impactor (HVMOI). Again, there are four stages covering the submicron size range. The design and flow rate requirements are such that a high-volume impactor of either Andersen or Sierra design type can serve as the upper stages of the cascade impactor. Due to the high flow rate, it is not possible to have the nozzles in a spiral pattern as is the case with the uniform deposit microorifice impactor. Thus, the nozzles are laid out in rectangular arrays with approximately 17 rows of nozzles per centimeter of width of the nozzle plate. Since approximately 58,000 nozzles are required per stage, the nozzles have been divided into 8 plates of 1 cm width by 25 cm length. Each stage has eight of these nozzle plates. Since there is a possibility of the deposits being overloaded due to the small nozzles, provisions have been made to move the impaction plate and, thus, spreading out the deposit. Unlike the uniform deposit impactor however, the deposit on the impaction plate will not be uniform.

Two version of the UDI (horizontal and vertical configurations) have been designed, built and calibrated. The HVMOI, however, has only been designed and built with calibration currently underway.

The horizontal version of the UDI has been extensively calibrated with liquid and solid particles using impaction substrates of membrane filters dry aluminum foils and greased aluminum foils. The resulting collection efficiency curves are shown in Figure 1 for liquid DOP particles on foil and filter substrates. The letters on the efficiency curves refer to various nozzle plates described in Table 1. Any combination of these nozzle plates can be put into the UDI stages.

As shown in Figure 1, it was found that a foil substrate provided a sharper collection efficiency curve than a filter substrate. In addition the filter curve is shifted to smaller particle sizes than the foil curve for any stage. This indicates that even though the membrane filter is very smooth it may appear quite rough to the submicron particles collected on these stages.

V. A. Marple, B. Y. H. Liu, and G. A. Kuhlmey, "A Uniform Deposit Impactor," J. Aerosol.Sci. 12:333-337 (1981).

Kuhlmey, G. A., B. Y. H. Liu, and V. A. Marple, "A Micro-Orifice Impactor for Sub-Micron Aerosol Size Classification," presented at the 89th National Meeting of the American Institute of Chemical Engineers in Portland, Oregon (August 20, 1980). Am. Ind. Hyg.Assoc. J. 42:790-795 (1981).

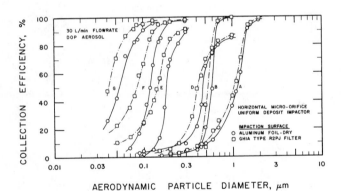

AERODYNAMIC PARTICLE DIAMETER, μm

Figure 1. Characteristic particle collection efficiency for stages A, B, D, E, F, and G using liquid particles on filter and aluminum foil impaction substrates.

114

Table 1. Operating Parameters and Experimentally Determined
Particle Cut-off Diameters For Each Stage of the
Uniform Deposit Microorifice Impactor

| Parameter | Stage - | | | | | |
	A	B	D	E	F	G
Nozzle diameter, μm:	704	467	104	73.2	61.2	46.3
Number of nozzles	85	85	2000	2000	2000	2000
S/W[1]	1.00	1.52	6.83	9.7	11.6	15.4
Impactor configuration[2]	A	AB	ABD	ABDE	ABDF	AG
Impactor pressure drop[3], psi	0.44	0.63	0.88	1.8	2.15	4.4
Pressure downstream of impactor, psia[4]	13.86	13.67	13.42	12.5	12.15	9.9
Re	670	1000	193	274	328	434
Cut-off diameter, μm:						
Ghia filter substrate	1.03	0.53	0.40	0.142	0.083	0.037
Aluminum foil substrate	1.13	0.60	0.47	0.179	0.120	0.053

[1]S/W = Jet-to-plate-distance/nozzle diameter
[2]Stages used during evaluation
[3]Based on flow rate of 30 L/min
[4]Based on ambient pressure of 14.3 psia

Published 1984 by Elsevier Science Publishing Co., Inc.
Aerosols, Liu, Pui, and Fissan, editors

A NEW AXIAL FLOW CASCADE CYCLONE
FOR SIZE CLASSIFICATION OF AIRBORNE PARTICULATE MATTER

B. Y. H. Liu and K. L. Rubow
Particle Technology Laboratory
Mechanical Engineering Department
University of Minnesota
Minneapolis, MN 55455
U. S. A

EXTENDED ABSTRACT

Introduction

A cascade cyclone has been designed, constructed and tested for the size classification of airborne particulate matter. The specific cyclone developed has five stages with cut-point diameters of 12.2, 7.9, 3.6, 2.05 and 1.05 um and a final filter stage to collect particles below 1 um in diameter at a design sampling flowrate of 30 L/min.

The cascade cyclone is based on the "axial flow," rather than the more conventional "tangential flow" geometry. In the axial flow geometry, the particle laden gas enters the cyclone through a set of helical turning vanes and flows generally along the direction of the cyclone body. The rotational motion of the gas imparted to the gas by the helical vanes causes the particles to experience a centrifugal force and precipitate toward the cyclone walls. This is in contrast to the conventional tangential flow cyclone in which the gas enters the cylone tangentially along the outer perimeter of the cyclone (see for example Smith et al, 1979). The main advantage of the axial flow cyclone is the relative ease with which the individual cyclone stages can be connected to form a multi-stage, cascade cyclone. In contrast, the conventional tangential cyclone would generally require more complicated flow passages between the stages.

Cyclone Description

Two stages of the new axial flow cascade cyclone are shown schematically in Figure 1a. Each cyclone stage is comprised of a helical turning vane, A, a thin-walled stainless steel cup, B, and an exit tube, C. The particles collected in the cup can be weighed directly by placing the cup on a balance. Three different cup sizes, ranging in mass from 22 to 49 g, were used.

Figure 1b is a simplified diagram of a cyclone stage and the dimensions of the cyclone are shown in Table 1. A double helix is used as the turning vane for the first four stages. However, because of the small size of stage 5, a single helix was used.

The principle advantage of the cyclone is its ability to collect a relative large amount of material without particle bounce and reentrainment. This is particularly important when sampling a dense particle cloud with high mass loading. However, the cyclone is relatively more complicated to design and construct than the cascade impactor. Therefore, cyclone collectors are likely to be limited to those applications where their unique ability to sample high concentration of particulate material is required.

Table 1.   Dimensions of the Cyclone Stages*

Stage	Helix	Pitch	T	D	O	L	S
1	Double	3.39	0.98	5.04	1.85	5.1	1.16
2	Double	2.54	0.65	5.04	1.85	5.1	1.16
3	Double	1.27	0.60	3.04	1.21	3.2	0.71
4	Double	1.02	0.32	2.08	1.15	3.2	0.71
5	Single	0.51	0.132	2.08	1.15	3.2	0.71

* All dimensions are in centimeters

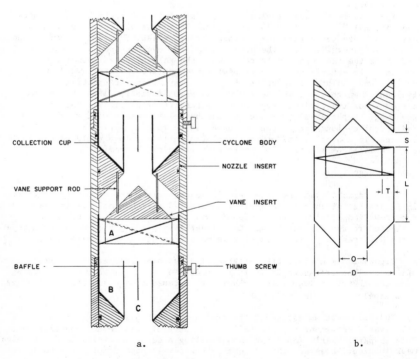

Figure 1.   Schematic diagram of the axial flow cyclone showing: a.) various
components of each stage and b.) the principle cyclone dimensions.

Cyclone Evaluation

The collection efficiency and loss characteristics of the cyclone were
determined as a function of particle size using solid, monodisperse aerosols
of ammonium fluorescein. The ammonium fluorescein aerosol was generated by a
TSI Model 3050 vibrating orifice monodisperse aerosol generator. The use of
ammonium fluorescein for aerosol generation was first described by Stober and
Flachsbart (1973). The test aerosol was sampled into the cyclone under test
for periods ranging from 5 to 20 minutes, and the cyclone was then
disassembled and the fluorescent material collected on various parts of the
system removed by rinsing the respective parts with 0.1 normal NH_4OH
solution. A fluorometer was then used to determine the concentration of the
fluorescein in the wash liquid. From these data the particle collection
efficiency and loss within the cyclone stages were determined. The
experiments were performed with particles ranging in aerodynamic diameter
form 0.8 to 23 um .

The measured efficiency of each cyclone stage is shown in Figure 2. Four
sets of collection efficiency curves are shown for each stage corresponding
to the different ways in which the collected materials in the cyclone are
combined to arrive at the efficiency. These are explained in the figure.

With regard to the particle loss in the cyclone, the loss on the inside
of the cyclone body is less than 1%, while the sum of the losses on the body
and the turning vane range from a low of 7% for 1 um particles to a high of
23% for particle of 23 um. The total loss in the system, i.e. the sum of the
losses on the cyclone body, the vane insert and inside of the exit tube of
the collection cup, ranges from 15% for 1 um particle to 33.3% for particles
of 8 um diameter. Since these last two losses are generally recoverable,
they are of minor concern in the application of the cyclone for actual
measurement purposes.

A series of test were performed using a dry dust aerosol to determine the
effect of mass loading on cyclone performance. The size distributions
obtained from six tests with Arizona road dusts are presented in Figure 4.
The total mass loadings in the tests ranged from 0.036 to 1.68 g, the
different mass loadings being obtained by varying the sample collection time
while keeping the aerosol mass concentration constant at about 4.5 g/m^3.
The material deposited in each stage ranged from 2 to 12 mg at the lowest
mass loading to 120 to 530 mg at the highest mass loading. The data show no
evidence of stage overloading, which would be indicated by an increase in the
fraction of materials collected in the smaller size ranges and a change in
the measured size distribution as the mass loading is increased.

References

Smith, W. B., R. R. Wilson and D. B. Harris (1979) "A Five-Stage Cyclone
System for in Situ Sampling", Environ. Science and Technol. 13:1387.

Stober, W. and H. Flachbart (1973) "An Evaluation of Nebulized Ammonium
Fluorescein as a Laboratory Aerosol", Atmos. Environ. 7:737.

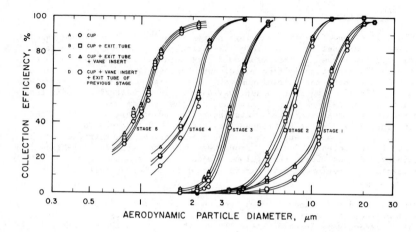

Figure 2. Collection efficiency of each cyclone stage.

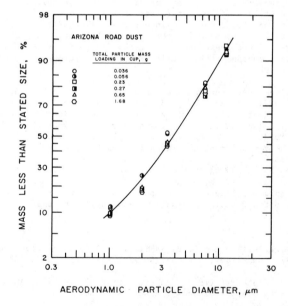

Figure 3. Measured size distribution of the Arizona road dust at various mass loadings in the sample cups.

Published 1984 by Elsevier Science Publishing Co., Inc.

Aerosols, Liu, Pui, and Fissan, editors

119

THE EFFECT OF GRAVITY ON DEPOSITION DISTANCE
OF PARTICLES IN AN INERTIAL SPECTROMETER

E.F. Aharonson, N. Dinar and R. Reichman
Israel Institute for Biological Research
P.O.Box 19 Ness-Ziona 70450 ISRAEL

EXTENDED ABSTRACT

Introduction

An inertial spectrometer has been proposed by Prodi et al. (1979) as a
powerful device for size assesment of polydisperse aerosols of the size range
1 to 7 μm aerodynamic equivalent diameter.

In the spectrometer, clean air is drawn through a rectangular cross-
section channel with a 90° bend, and a sheath of the tested aerosol is
injected via a thin nozzle which is positioned slightly above the bend
(see fig. 1). The outer wall past the bend is made of a membrane filter
through which the air is pumped out. As a result of the sudden change in the
airstream direction at the bend, suspended particles are separated in the flow
field and deposited on the filter at different distances according to their
inertia. Consequently, after appropriate calibration with homogeneous aerosols,
the aerodynamic equivalent diameters of polydisperse aerosol particles can be
evaluated from their deposition distances on the membrane filter.

The possible advantage of operating the spectrometer in other than the
upright position is considered in the present work. For that purpose, the
effect of gravity on deposition distance of monodisperse aerosols has been
examined, both experimentally and theoretically. This effect proved significant under certain operating conditions.

Figure 1. Schematic diagram of the
inertial spectrometer

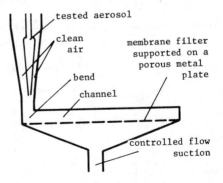

Theoretical

A two dimensional air-flow field could serve to describe the conditions pertaining in the relevant part of the channel, i.e. at and following the bend. This field was calculated as the numerical solution of the incompressible steady-state Navier-Stokes equations for a rectangle with the appropriate dimensions and boundary conditions. A primitive representation was used, involving the air-flow vector $\vec{u} = (u, \omega)$ and the pressure P. The Navier-Stokes equations were discretized, and a system of finite difference equations was defined in a staggered grid. To solve these equations, a fast multigrid algorithm, similar to that described by Brandt and Dinar (1979) was used.

The trajectory of a single spherical particle or droplet discharged in the above flow field was then obtained by solving the following general equations of motion, Brodkey (1967):

$$\frac{\pi}{6} d_p^3 \rho_f \frac{d\vec{u}_p}{dt} = \pi d_p^2 \rho_a (\vec{u} - \vec{u}_p) C_D |\vec{u} - \vec{u}_p| - \frac{\pi}{6} d_p^3 \nabla P + \frac{\pi}{6} d_p^3 \vec{G} (\rho_a - \rho_p) + $$

$$+ \frac{\pi}{12} \rho_a \frac{d}{dt}(\vec{u} - \vec{u}_p) + \frac{3}{2} \rho_a d_p^2 \sqrt{\pi \nu} \int_0^t (t-\theta)^{-\frac{1}{2}} \frac{d}{d\theta}(\vec{u} - \vec{u}_p) d\theta \quad [1]$$

$$\frac{d\vec{r}}{dt} = \vec{u}_p \quad [2]$$

where

$\vec{u} = (u, w)$ fluid velocity (a solution of the Navier-Stokes equations)

P   static pressure in the fluid (a solution of the Navier-Stokes equations)

$\vec{u}_p = (u_p, w_p)$ velocity of the particle

$\vec{r} = (x, y)$ location of the particle

$\vec{G}$ gravity acceleration

ρ_a, ρ_p density of air and particle respectively

ν kinematic viscosity of air

d_p particle diameter

C_D drag coefficient, function of Reynolds number for the particle $Re_p = d_p |\vec{u} - \vec{u}_p| / \nu$

The left hand side term of equation [1] represents the force required to accelerate the particle. The terms on the right represent, respectively: the viscous drag, the force caused by pressure gradient in the fluid, the external force of gravity, the force accelerating the apparent mass of the particle relative to the fluid, and the force derived from the effect of acceleration on the viscous force.

The initial conditions, at $t = 0$ are given by $x = x_0$ $(0 < x_0 < \ell)$, $y = 0$, $u_p = u$, and $w_p = w$ where ℓ is the length of the slit.

The gravity term was considered in either of the four different directions illustrated in fig. 2: towards the filter (a) $\vec{G} = (0, g)$, opposite to it (b) $\vec{G} = (0, -g)$, in the flow direction past the bend (c) $\vec{G} = (g, 0)$, or against it (d) $\vec{G} = (-g, 0)$.

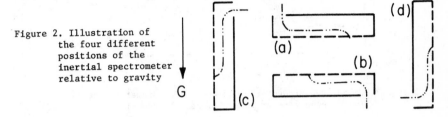

Figure 2. Illustration of the four different positions of the inertial spectrometer relative to gravity

Experimental

The effect of gravity on deposition distance in the spectrometer was demonstrated by shifts of the deposition distance caused by operation of the instrument in other than upright position. The air streamlines are obviously not affected by these inversions, but the force of gravity is acting on the particles in different relative directions.

The system used enables aerosol generation, dilution and sampling by an inertial spectrometer. Monodisperse aerosol droplets of dioctylphthalate (C.P. grade, Eastman Kodak, USA) dissolved in Isopropanol (analytical grade, Merk, W. Germany) and pigmented by Waxoline Violet A (ICI, England) were produced by a commercial vibrating orifice generator (Berglund & Liu Model 50A, RNB Assoc. Inc., USA). Diluting air supply was kept low in order to achieve the desirable concentration in the aerosol of larger droplets, aggregates of double and triple elementary droplets. After complete evaporation of the solvent, droplet sizes were in the range 3.3. to 10 μm in diameter. Droplet diameters calculated from the operation conditions of the generator agreed very well with those measured under the microscope (Model Optiphot, Nikon, Japan) after deposition on fluorochemical (FC 208, 3M Chemical Division, USA) coated microscope slides.

The spectrometer was operated consecutively, sometimes without replacing the membrane filter, in one or more of the four different positions for the same aerosol. This procedure proved very useful for demonstration of shifts in deposition distance. Two different flow-rates were used: 3 1/m and 2 1/m of the clean air, with 50 ml/m and 30 ml/m respectively of aerosol. These correspond to inlet linear velocity in the spectrometer of approximately 120 and 80 cm/s

Results

Calculated trajectories of two spheres, 4 and 10 μm in diameter, simulating aerosol droplets discharged into an airflow of 2 1/m at the spectrometer in the upright (a) and the inverted (b) positions, are presented

122

in fig. 3. Normally trajectories end when the spheres approach the filter, and deposition distance is defined to this point. In the case of the spectrometer in b position however, it is possible for a droplet to be lost against the other wall.

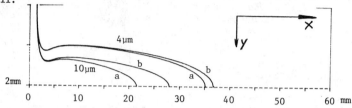

Figure 3. Calculated trajectories of 4 and 10μm diameter spheres in the spectrometer in positions (a) and (b).

A photograph of a membrane filter on which a well defined trimodal aerosol had been characterized in normally positioned (a) and inverted (b) spectrometer is presented in fig. 4. The measured deposition distances derived from several such experiments were used for comparison with the calculated values.

Figure 4. Photograph of a membrane filter on which a trimodal aerosol of 6.0, 7.5, and 8.6μm diameter spheres deposited in normally positioned (a) and inverted (b) spectrometers

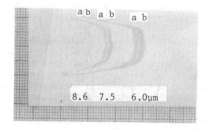

Discussion and conclusions

An algorithm for calculating deposition distances in the inertial spectrometer has been tested. It proved satisfactory for estimation of calibration curves correlating deposition distances with droplet diameters. Changes in the calibration curves as a result of an inversion or 90° rotation were predicted and confirmed experimentally. These changes may be used to improve the resolving capacity of the spectrometer in characterization of fine aerosols.

References

1. Brandt, A. and Dinar, N. (1979) "Multigrid Solution to Elliptic Flow Problem", Proc. of the Advanced Seminar on Numerical Solutions of Partial Differential Equations (Edited by S. Parter), Academic Press, New York.
2. Brodkey, R.S. (1967), pp621, The Phenomena of Fluid Motion, Adison-Wesley, Reading, Mass.
3. Prodi, V. et al. (1979) "An Inertial Spectrometer for Aerosol Particles" J. Aerosol Sci. 101:411

Published 1984 by Elsevier Science Publishing Co., Inc.

Aerosols, Liu, Pui, and Fissan, editors

EFFECT OF PARTICLE GRAVITATIONAL FORCES
ON THE CALCULATION OF IMPACTOR EFFICIENCY CURVES

D.J. Rader and V. A. Marple
Particle Technology Laboratory
Mechanical Engineering Department
University of Minnesota
Minneapolis, MN 55455
U. S .A

EXTENDED ABSTRACT

Introduction

In previous work (Marple, 1970), theoretical impactor collection efficiencies were determined by the numerical solution of the Navier-Stokes equations and subsequent numerical integration of the equation of motion of the particles. Although theoretical and experimental curves were typically found to be in good agreement (Marple and Rubow, 1976), many investigators (e.g. Ranz and Wong, 1952; Mercer and Stafford, 1969; and Lundgren, 1967) reported experimental curves which appear less steep than the sharp cuts generally predicted by theory. In an effort to explain such observations, later work (Rader and Marple, 1984) extended the theory to include an ultraStokesian drag coefficient and a facility for handling particle interception. Neither of these improvements resulted in efficiency curves displaying the "S" shape encountered experimentally.

In recent work, the numerical method is extended to include the effect of gravity on particle trajectories. Gravitational forces were originally neglected based on comparisons of particle settling speeds with typical fluid velocities. The sedimentation velocity of a 5 micron particle is approximately 0.6 cm/sec, which represents a very small fraction of the flow velocity through the nozzle. Yet, experimental work with round impactors (May, 1975) revealed a substantial shift in the collection efficiency when the calibration was performed with an ascending rather than descending jet. This effect was seen with aerosols in the 5-10 micron size range.

A possible explanation for the contribution of gravity in this size range is indicated by the flow field shown in Figure 1. For round impactors, Marple found that the flow above the impaction plate gradually approaches the plate as the stream radially expands. Thus, the particles are contained within a very thin boundary layer characterized by velocities which are much smaller than those in the jet throat. Under these conditions, even relatively low sedimentation velocities contribute to particle collection at the plate. Such considerations would also indicate that variations in the impaction surface, due to imperfections or previous deposits, might shift the collection efficiencies as well.

124

Theoretical

The influence of gravity is accomplished by including an additional force in the particle equation of motion. For the case in which Stokes drag law is deemed valid and for which the gravitational acceleration acts in the direction of flow and parallel to the centerline, the following set of non-dimensional equations apply in two dimensions:

$$\frac{dU_p'}{dt'} = \frac{2}{St}\ (U_f' - U_p')$$

$$\frac{dV_p'}{dt'} = \frac{2}{St}\ (V_f' - V_p') + \frac{1}{Fr}$$

where:

$$St = (\rho_p\ C\ V_o\ D_p^{\ 2})/(9\ \mu\ W)$$

$$Fr = V_o^{\ 2}/gW.$$

Here W is the throat diameter, V_o the average air velocity the local acceleration due to gravity, μ the viscosity of air, D_p and ρp the particle diameter and density, St the Stokes number and Fr the Froude number. The radial and axial velocities, U' and V', have been subscripted with p for the particle and f for the fluid.

The dimensionless numbers St and Fr represent the ratios of forces relevant to the present system. The dimensionless value St/2 is the ratio of the force necessary to bring a particle with initial velocity V_o to rest in a distance W/2 (inertial force) to the fluid resistance to a particle moving through it with relative velocity V_o (viscous force). Similarly, the ratio of inertial to gravitational forces is simply the Froude number. The influence of gravity cannot be safely neglected when the Froude number becomes of the same order as the Stokes number.

The particle equation of motion including the gravitational force is integrated using earlier methods (Marple, 1970). The results of a typical calculation are shown in Figure 2, where efficiency versus the square root of the Stokes number (dimensionless particle diameter) is plotted for several Froude numbers. Two effects of gravity should be noted. First, as the Froude number decreases, the collection efficiency for a particular size particle increases, shifting the curves to the left. This is expected, for gravitational settling increases particle losses in the descending-jet case. Second, the slope of the curves decreases as the Froude number decreases. For a Froude number of 10, the efficiency curve displays a distinct "S" shape. Particles in the low-efficiency tail typically originate near the centerline, and are entrained in the region of the boundary layer nearest the plate. There, low fluid velocities enhance the influence of sedimentation. The shift at the high-efficiency end of the curve is less pronounced. Here, particles originate near the outer nozzle wall, and typically follow trajectories which include them in the high-velocity regions of flow outside the impaction plate boundary layer. Thus, the influence of settling is

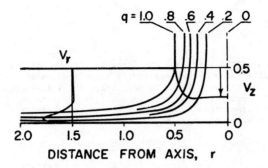

Figure 1. Flow field in a round impactor computed for Re_f=3000 and
S/W=0.5.

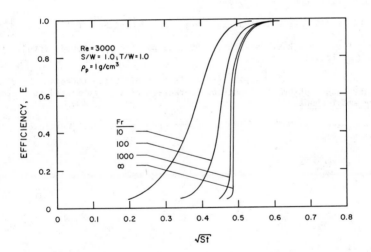

Figure 2. Characteristics for a round impactor computed for various
values of the Froude number.

126

markedly reduced.

Conclusions

The inclusion of the gravitational force in the numerical calculation of round impactor efficiency curves results in significant shifts in efficiencies when the Froude number is small.   In practice, the influence of gravity may be noticeable for particle diameters as small as 5 microns.   Sedimentation also tends to increase the "S" shaped nature of the efficiency curves, and under certain conditions may explain experimental results.

References

Lundgren, D.A. (1967) "An Aerosol Sampler for Determination of Particle Concentration of Size and Time", J. Air Poll. Control Assoc. 17:225.

Marple, V.A. (1970) "A Fundamental Study of Inertial Impactors", Ph.D. Dissertation, Univ. of Minnesota, Particle Technology Laboratory Publ. #144.

Marple, V.A. and K. Rubow (1976) "Aerodynamic Particle Size Calibration of Optical Particle Counters", J. Aerosol Sci. 7:425.

May, K.R. (1975) "Aerosol Impaction Jets", J. Aerosol Sci. 6:403.

Mercer, T.T. and R.G. Stafford (1969) "Impaction from Round Jets", Ann. Occup. Hyg. 12:41.

Rader, D.J. and V.A. Marple (1984) "Effect of UltraStokesian Drag and Particle Interception on Impaction Characteristics", submitted to Aerosol Science and Technol.

Ranz, W.E. and J.B. Wong (1952) "Impaction of Dust and Smoke Particles", Ind. Eng. Chem. 44:1371.

Published 1984 by Elsevier Science Publishing Co., Inc.

Aerosols, Liu, Pui, and Fissan, editors

THE EFFECT OF THE BOUNDARY LAYER
ON THE INERTIAL SPECTROMETER PERFORMANCE

F.Belosi, L.& A.,Via Mazzini 75
40137 Bologna
V.Prodi, Physics Dept. of the University
40126 Bologna and L.& A.,40137 Bologna
ITALY

EXTENDED ABSTRACT

Introduction

Under the assumption of inviscid air in laminar flow, in a previous work (Belosi and Prodi,1983) the behaviour of a particle along the curvature of the inertial spectrometer was found to be represented by the system of equations

$$\frac{d\omega_p}{dt} = \frac{1}{\tau} (\omega - \omega_p)$$

$$\tau r \omega_p^2 = \frac{dr}{dt}$$

(1)

where τ is the particle relaxtion time, ω and ω_p are the angular velocities of air and particle and r is the radial coordinate of the particle, with the boundary conditions r = Ri (injection coordinate of the aerosol) for t = 0 and $\omega_p = \omega_{po} = Uf(Dp)/Ri$ where Uf(Dp) is the velocity of the particle of aerodynamic diameter Dp at the exit of the convergent nozzle.A semianalytical solution of (1) was found for ω = cost., in good agreement with the experimental values up to 5 μm but with a progressively larger disagreement for bigger sizes.A boundary layer, due to the viscosity of air, can explain the discrepancy.

Convergent Nozzle

In the converging section of the throat an analytical solution of the Prandtl equation for the boundary layer exists.In this case the pressure increment is negative, and the boundary layer is thin, without separation.

In our case, the particles are injected outside of the boundary layer and continue to travel outside of it along the throat.

Curvature

Fluid flows along curved channels are unstable because of viscosity and centrifugal forces:eddies are formed, as a consequence, with axis parallel to the direction of motion, known as secondary flow (Dean 1927,Greenspan,1973).The particle motion in a curved pipe has been studied (Crane and Evans,1977), but generally the section was circular and the ratio of curvature radius to pipe radius is high.In the INSPEC the cross section is rectangular and this ratio is only 2.7.

The secondary flow has no direct influence on particle behaviour in the central part of the channel, as it can be seen from the deposition pattern, caracteristic of an indefinite gap.

This secondary flow affects only the effective velocity in this central region.

The equations of the boundary layer over a curved surface are amenable to the situation over a flat plane provided that (Jones and Watson,1966)

$$\frac{\partial H}{\partial X} < 1 \quad \text{and} \quad \frac{\partial H}{\partial Z} < 1$$

with Z and X are respectively the coordinates perpendicular to the plate and along the curved flow lines, and $H = 1 + kZ$, where k is the surface curvature radius.

Both the conditions are met and the boundary layer is approximated by the flat plate (Schlichting 1968).

$$u = U_{pot} \, f(\eta)$$

$$f(\eta) = \frac{3}{2} \eta - \frac{1}{2} \eta^3$$

$$\eta = z / 5 \sqrt{\frac{x \, \nu}{U_{pot}}}$$

with x is the distance from the leading edge of the plate and U_{pot} is the potential velocity.

In the INSPEC the inner radius is 0.15 cm. and the outer

radius increases from 0.29 to 0.33 cm. The channel cross section is therefore

$$A(\vartheta) = 2\ (R(\vartheta) - 0.15)$$

$$R(\vartheta) = a\vartheta + b$$

where a and b are costants.
   The potential velocity is

$$U_{pot} = \frac{Q}{A(\vartheta)}$$

and the angular velocity at the radius r is

$$\omega = \frac{U_{pot}}{r}\ f(\eta)$$

with

$$\eta = \frac{R(\vartheta) - r}{5\sqrt{\dfrac{R(\vartheta)\vartheta\ \nu}{U_{pot}}}}$$

where x has been replaced with $R(\vartheta)\vartheta$ and this gives the system of equations

$$\frac{d\omega_p}{d\vartheta} = \frac{1}{\tau}\ (\frac{U_{pot}\ f(\eta)}{r} - \omega_p)\ \frac{1}{\omega_p}$$

$$\frac{dr}{d\vartheta} = \tau\ r\ \omega_p$$

wich has been solved numerically with the Runge-Kutta method.
   The effective cross section $A(\vartheta)$ for the potential velocity has a depth smaller than the geometrical value of 2 cm., because of the side vortices. This depth can be evaluated from the unaffected width of the deposit: e.g. this is 1.3 cm. at 5 lpm.
   By introducing this value, a very good agreement is found between calculated and measured values, as apparent from fig. 1

130

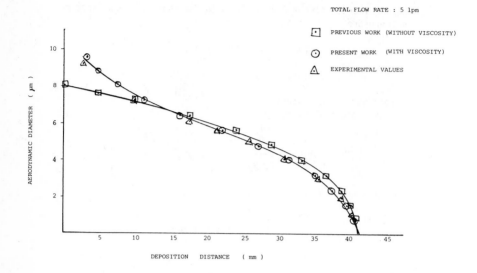

Figure 1. Comparison between experimental and theoretical values

References

Belosi F., V.Prodi (1983) "Calculation of the deposition distance in the inertial spectrometer" presented at the 11th Conference of the GAeF,Munich.

Crane R.I., R.L.Evans (1977) "The inertial deposition of particles in a bent pipe" J.Aerosol Sci. 161:170.

Dean W.R.,(1927) "Note on the motion of fluid in a curved pipe" Phil.Mag. 208:223.

Greenspan D.,(1973) "Secondary flow in a curved tube" J.Fluid Mech. 167:176.

Jones C.,E.J. Watson (1966) "Two-dimension boundary layers",Laminar Boundary Layers (Edited by Rosenhead),Clarendon Press,Oxford.

Schlichting H.,(1968) Boundary-Layer Theory Mc Graw-Hill Book Company,New York.

Published 1984 by Elsevier Science Publishing Co., Inc.
Aerosols, Liu, Pui, and Fissan, editors

A HIGH FLOW INERTIAL SPECTROMETER

V. Prodi, Physics Dept. of the University
    40126 Bologna and L.& A.,40137 Bologna
F. Belosi and A. Mularoni, L.& A.,Via Mazzini 75
    40137 Bologna ,ITALY

EXTENDED ABSTRACT

Introduction

The Inertial Spectrometer (INSPEC) has been described in the literature together with several applications (Prodi et al., 1979 - 1982 - 1983) Aerosol particles are injected into a clean air stream in a curved channel. The particle persist in their initial velocity by inertia but they are also dragged by the flow.

The particles are separated while airborne according to their aerodynamic diameter: the separation is magnified when the whole stream is drawn through a filter. Aerosol particles then deposit on it at a distance from inlet which is a unique function of the aerodynamic size at a given total flow rate.

The INSPEC is compact enough for field use and the resolution is generally higher than other field instruments, without suffering of the disadvantages frequently encountered of particle bouncing and reentrainment. The sampling flow rate though, of the order of some hundred cm^3 $min.^{-1}$ may be too low for mass evaluation in environmental sampling.

The high flow spectrometer

Therefore a spectrometer has been developed with the aim at sampling at a much higher flow rate. The arrangement has been completely changed, and built around a circular filter, with a cylindrical geometry. The throat is circular and not rectilinear as in the standard INSPEC, in which the injection is along the short side of a rectangular filter. The filter diameter is of 150 mm therefore the overall length of the throat is of factor 25 higher than in the standard unit. A cross section of the High Flow unit is shown in fig. 1: both the throat walls and the aerosol injection nozzle are fixed to the top plate.

All these are made of nikel plated hard steel. A plexiglass lid on the top plate carries the connections for the aerosol (intermediate circle) and for inner and outer clean air, which have to be separately supplied to the instrument. A consequence of this arrangement is that the stream as no lateral confinement and therefore there is no side-wall effect: the particles deposit at a radial distance from the inlet which is unique function of the aerodynamic size, i.e. on each concentric circle, the particles are monodisperse.

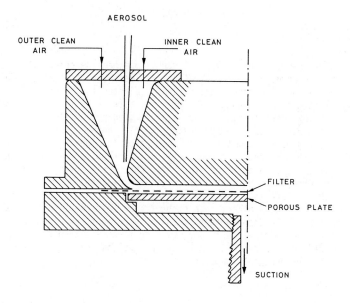

Figure 1. Schematic cross-section of the High Flow Inertial Spectrometer

The spectrometer is operated in a sealed circuit with a controlled which is replaced by the sampling flow. The air is moved by a oilless vane pump with "V" ring sealed shaft.

In fig. 2, three calibration curves are shown, for typical situations of 4000 lh^{-1} of total clean air, (3200 outer, 800 inner), of 5000 lh^{-1} (4000 outer, 1000 inner) and of 8000 lh^{-1} (6500 outer 1500 inner) and with 100 lh^{-1} lpm aerosol sampling flow rate.

Each circular strip has particles of a well defined size range: therefore if the size distribution of any element or compound is needed, the filter can be cut into concentric rings and each one analyzed. In this way, the differential size distribution is obtained and from this, the cumulative one can be derived.

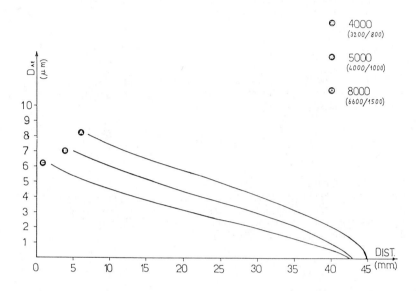

Figure 2. Calibration curves of the HIgh Flow Inertial Spectrometer

With respect to the standard INSPEC, the circular unit shows radial separation increasing from the perifery to the center, since an equal area ring is wider the closer it is to the center.

Therefore the calibration curve is stretched at the lower size end, thus extending the separation in this region.

The other advantages of the standard INSPEC are maintained, like continuous distribution, fiber alignement and separation according to the diameter.

In addition, in the absence of a lateral confinement, the whole deposit can be used for the assessment of the deposited amount, thus multiplying by an additional factor of 1.5 to 2 the useful flow rate.

134

Acknowledgements

This work has been performed under the contract CR/11/82 and CR/13/83 of the Joint Research Center, Establishment of Karlsruhe with Lavoro e Ambiente scrl, Bologna - Italy

References

Prodi V., C.Melandri, T.DeZaiacomo, M.Formignani, G.Tarroni, D.Hochrainer (1979) "An Inertial Spectrometer for Aerosol Particle" J.Aerosol Sci. 10:411.

Prodi V., T.DeZaiacomo, D.Hochrainer, K.Spurny (1982) "Fiber Collection and Measurement with an Inertial Spectrometer" J. Aerosol Sci. 13:49.

Prodi V., T.DeZaiacomo, C.Melandri, G.Tarroni, M.Formignani, P.Olvieri, L.Barilli, G.Oberdoester (1983) "Description and Applications of the Inertial Spectrometer" Aerosols in the Mining and Industrial Work andEnvironment (Edited by V.A.Marple and B.Y.H.Liu) Ann Arbor Science,Ann Arbor Mich.

Published 1984 by Elsevier Science Publishing Co., Inc.

Aerosols, Liu, Pui, and Fissan, editors

AN AUTOMATED TAPE COLLECTION SYSTEM FOR PRECISE AEROSOL ANALYSIS

T. Raunemaa*, K.-M. Pihkala, A. Hautojärvi, P. Pihkala

University of Helsinki, Department of Physics
Siltavuorenpenger 20 D, SF-00170 Helsinki 17, Finland

*University of Kuopio, PB 6, SF-70211 Kuopio 21

EXTENDED ABSTRACT

Introduction

Sampling of fine particles has been shown to be possible through the use of capillary gas flow and impaction onto a thin backing material (1). Transport through long ($\sim$ 20-40 m) capillaries has also been demostrated (2). The sampled material can be analyzed by, e.g. the X-ray emission (PIXE) technique, which determines the concentrations of 20-30 elements with good precision. This sampling method is especially suited for investigations in remote locations and for control measurements. To monitor air quality in e.g. work and industrial environments, on the other hand, a continuously operating device is preferable in most cases.

The facility

This work introduces a tape registering system, which employs a large size cassette. An aluminized 7 mm broad Mylar tape serves for aerosol collection and storage. The cassette similar to that employed in nuclear physics studies at CERN (3) is used. The cassette is enclosed in an aluminum box with dimensions 20 x 28 x 3 cm^3. In this box are situated the necessary electronics and mechanical controls.

The capillary is led into a small volume chamber, about 30 cm^3. Roughly, this chamber can be ventilated at a rate of 1-20 s^{-1} depending on the flow rate through the capillary. For the aerosols possibly not deposited on the tape this would mean a travel time of $\geq$ 1 s to the walls. For 0.001 μm aerosols movement through about 0.1 cm by diffusion, 10^{-7} cm by settling in 1 s can be assumed. For electrically charged particles ($\sim$ 0.01 μm size) $\sim$ 0.02 cm/s speed in typical field is possible. Turbulent deposition remains as the most probable deposition mechanism.

In the present model a 50 or 100 mm^2 Si(Au) alpha detector is situated inside the box. An almost immediate measurement of aerosol radioactivity is thus possible. The registration is effected through a conventional amplifying and analog-to-digital circuitry. Optionally, a 16 ch ADC with a microcomputer system can be connected. The system layout is shown in Fig. 1.

The capillary function is to collect only particles below
3-5 μm size, which results in a fairly homogeneous sample layer.
The small sample spot is well suited for analysis with a large
solid angle detector. The thickness of the sample can be
controlled and its mass concentration evaluated by using an
external alpha source. E.g. ^{229}Th source can be well applied
for the purpose as it emits several alpha radiations from
4.8 MeV to 8.4 MeV. Taken together they give a rather precise
mass attenuation, and correspondingly thickness, of the sample
layer. An average energy loss of 100 keV/μm of the backing
substrate is demonstrated in Fig. 2.

Control circuit

The operation is performed stepwise, i.e. collection of a new
sample and detection of the previous sample occur simultaneously.
The tape movement is controlled by logic circuitry.

The logics unit consists of a programmable timer/counter, an
optocontroller and a binary counter, see Fig. 3.

The time constant can be preset to n_τ values with $n = 2^m$,
$m = 0,1,...,7$ and τ varying from seconds to tens of minutes.
The micromotor (12 V, 0.8 W) driving the tape also turns a saw-
tooth plastic sheet, which determines the transfer length. For
starting, and to enable manual operation, there are switches on
the box. A tape length controller and a timer with digital
display can included.

Experimental

The system was tested by collecting aerosol alpha activity outdoors and in the laboratory. The effect of the humidity in
the sampling air was also tested. Capillaries from 1.1 m to
6.5 m length and 0.8 to 2 mm diameter were applied. The
timing was set to be optimal for ^{214}Po (E_α = 7.7 MeV) (2).

The radon in the air comes from the ground and from the
building materials. The outdoor concentration was assumed to
vary between 400-600 Bq/m at the time of the tests (4).

Typical result is sketched in Fig. 4. Experiments in other
environs were also performed. Further data collection is in progress.

This work has been financially supported by the Academy of
Finland.

References

1. T. Raunemaa, A. Hautojärvi, Aerosols in the Mining and Industrial Work Env., Vol. 3, Eds. V.A. Marple, B.Y.H. Liu, Ann Arbor Sci., Ann Arbor 1983, 971

2. T. Raunemaa, A. Hautojärvi, J. Aerosol Sci. 15, 1984, 123-132

3. Ph. Dam, E. Hagberg, B. Jonson and The Isolde Collaboration, CERN, Nucl. Instrum. Methods 161, 1979, 427-430

4. Mattsson, R., Finn. Met. Inst., Helsinki, 1984, private communication

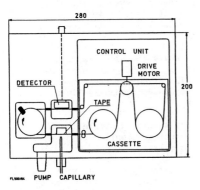

Fig. 1. The tape collection system.

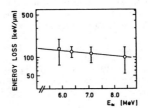

Fig. 2. Energy loss per unit thickness at various alpha energies from ^{229}Th source. 1 µm corresponds to about 100 µg/cm^2.

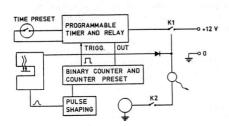

Fig. 3. The logics circuitry for tape control.

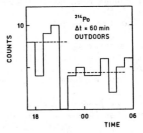

Fig. 4. ^{214}Po aerosol alpha activity variation in surface air. The average value apparently drops during night.

Published 1984 by Elsevier Science Publishing Co., Inc.
Aerosols, Liu, Pui, and Fissan, editors
138

DEVELOPMENT OF LOW COST CASCADE IMPACTORS

M. Lehtimäki[1], and J. Keskinen[2]

1 Occupational Safety Engineering Laboratory, Technical Research Centre
of Finland, P.O. Box 656, 33101 Tampere 10, Finland

2 Department of Physics, Tampere University of Technology,
P.O. Box 527, 33101 Tampere 10, Finland

The size selective particle sampling is of great importance in many
fields of the aerosol research. In the industrial hygiene the knowledge of
particle size distribution is necessary e.g. in estimating the concentration
of inhalable and respirable particles. The particle size distribution is
also important in the selection of proper air cleaning device. The size
selective particle sampling is usually realized with cascade impactors,
which are commercially available ranging from relatively massive low pressure
impactors to low weight personal cascade impactor.

In the field measurements it is often necessary to collect a great
number of samples. This is why the possibility to use several impactors
simultaneously would be of great importance. The prices of the commercially
available impactors are, however, relatively high, which usually limits the
number of impactors available.

The purpose of the present work has been to develop a simple low cost
and low weight cascade impactor to be used in particle sampling from fixed
measuring points, and as well as, in personal sampling.

The construction of simple and low cost cascade impactor has been recently
discussed by Jones et al. 1) In this case the impactor was constructed by
machining the standard plastic filter holders. In the present work the construction of a low cost cassette impactor has been further developed. The
middle part of a standard 3-section filter holder (Millipore) has been used
as a frame for an impactor stage. The aluminium plate containing the drilled
holes is fitted to the frame with the aid of cement. The design procedure by
Marple 2) has been used to determine the proper dimensions for the holes in
each impactor stage. Particle collection takes place on a ring-shaped glass-
fiber or aluminium foil plate. The impactor stages are packed to the standard
filter holder to form a small and low weight cascade impactor. The cost of
the present impactor is neglible, which means that even a great number of
impactors can be constructed without significant cost. The great number of
impactors makes it possible to collect particle samples almost in the similar
way as in the conventional filter sampling.

1) W. Jones, J. Jankovic and P. Baron, 1983
   Am. Ind. Hyg. Assoc. J. 44, 409-418.

2) V.A. Marple and K. Willeke, 1979: Aerosol Measurement
   (Lundgren et al. editors), 90-107.

Published 1984 by Elsevier Science Publishing Co., Inc.
Aerosols, Liu, Pui, and Fissan, editors

DESIGN PRINCIPLES OF THE AERAS LOW PRESSURE IMPACTOR

A.Berner
Institut für Experimentalphysik
Universität Wien
A 1090 Vienna, Austria

Extended abstract

Introduction

Low pressure impactors are usefull tools for classifying aerosols
down to sizes of 10 nm aed. (aed.: aerodynamic equivalent diameter). Because
of their higher flow rate in comparison to other instruments working in the
same size range, they are extremely usefull in collecting material for
chemical analysis. In many cases thin foils can be used for aerosol collection, and extremely high accuracies in weighing the samples are obtained. As
a consequence, the well known trimodal mass size distributions of traffic
aerosols have been measured directly by low pressure impactors.
    The design of the low pressure impactors started by operating ordinary
impactors, i.e.those who work at atmospheric pressures, in vacuum vessels,
where the expansion of the aerosol was performed by a special expansion
nozzle. Similar techniques have been used later in other impactors, where
some of the stages were operated at atmospheric pressures and the low pressure stages were coupled to the upper stages by an expansion nozzle.
    In passing the aerosol through an expansion nozzle, the particles
undergo strong shear stresses and perhaps penetrate standing shock waves, as
the expansion nozzle is working at overcritical flow conditions. These forces can be strong enough as to disintegrate the particles.
    In the low pressure impactor described here the pressure reduction
which is necessary for the deposition of finest aerosol particles is performed stepwise from stage to stage. The velocities in these stages are
fairly high, but do not exceed the velocity of sound in any part of the
impactor. The stress on the particles is therefore reduced considerably.

Design principles of the Aeras low pressure impactor.

As any cascade impactor the Aeras low pressure impactor has a certain
number of stages. The coarse particle stages are operated at atmospheric
pressure as the aerosol jet velocities are below 50 m/sec. Under these conditions the flow can be treated as incompressible. In this flow regime the
theory of particle deposition by impaction is well developed and validated
by a wealth of experimental data.
    The condition of operating impactors in the regime of incompressible
flow limits the particle size range which can be classified by these instruments. When the aerosol jet velocities are kept below 50 m/sec, approximately, and when the dimensions of the impaction nozzles are not smaller
than 0,3 mm for practical reasons, the limit below which particle can not be
classified is roughly at 0,3 μm aed..
    Certain ways have been attempted to extend the measuring range of the

cascade impactors. Microorifice impactors and low pressure impactors with internal expansion nozzles are to be mentioned here. The Aeras impactor is an other approach for separating finest aerosol particles from a gas. In some of the stages of this impactor high aerosol jet velocities are applied, and the instrument therefore becomes a more or less complex flow device with some essential differences to other impactors.

Particle separation by impaction is governed by the well known dimensionless inertia parameter

$$\Phi = (D_p \cdot \rho_p \cdot C_{p,g} \cdot u_g)/(18 \cdot \eta_g \cdot W_g) \tag{1}$$

where D_p: particle diameter; ρ_p: particle density; $C_{p,g}$: slip correction factor; η_g: viscosity of the gas; u_g: aerosol jet velocity; W_g: width of the aerosol jet.

For separating particles smaller than 0,3 µm aed. from the gas, three means can be applied:

* high velocities, u_g, of the aerosol jets ,
* higher values for the slip correction factos, $C_{p,g}$,
* smaller widths ,W_g, of the aerosol jets.

When velocities of more than 50 m/sec are applied the regime of incompressible flow is left. Instead of, the gas expands adiabatically when flowing into the nozzles, and therefore the temperature and the density of the gas decrease. In fluid dynamics the flow through nozzles is well understood. The velocity through a straight nozzle with ideal entrance configuration is given by

$$u_{g,o} = ((2k/(k-1)) \cdot (p_i^2/R.T) \cdot (1-(p_e/p_i)^{(k-1)/k}))^{1/2} \tag{2}$$

where the pressure ratio is given by the pressur, p_i, ahead of the nozzle, and by the pressure, p_e, behind the nozzle. The assumption is made that the fluid velocity ahead of the nozzle is negligibly small in comparison to $u_{g,o}$. Further, k is the isentropy coefficient, T the temperatur of the gas ahead of the nozzle and R the specific gas constant. According to equation (2) the velocity will increase when the pressure ratio falls, but for a certain critical value p_e/p_i = 0,53 for air a maximal velocity is obtained which can not be overcome by further reduction of pressure. Under this condition the velocity of sound is reached in the narrowest cross section of the flow. Here, the velocity of sound can not be exceeded.

In practice the theoretical velocity, $u_{g,o}$, is not obtained for the reason that the entrance configuration is non ideal. Making the ideal entrance is a severe problem especially for small nozzles as they are needed in impactors. The velocity has to be corrected for by certain factors, the velocity correction factors, φ , and therefore

$$u_g = \varphi \cdot u_{g,o} \tag{3}$$

where u_g is the non ideal velocity. These correction factors depend on certain parameters, among them the pressure ratio p_e/p_i and the Reynolds-number. In practice, the velocity correction factors which usually obtain values between 0,7 and 0,8 for recangular entrance configurations are determined by experiments.

An impactor stage,however, is a more complicated flow device. It consists of a nozzle plate, which usually carries more than one nozzle, a stagnation plate, which deflects the aerosol jets produced by the nozzle plate, and a spacer element for separating both plates. Investigations which have been performed with single nozzle and multiple nozzle plates demonstrate that the aerosol jet velocity,u_g, can be described by equation (3), when the correction factors are determined separately. For an impactor stage the values of the correction factors range from 0,65 to 0,85.

Nozzles with a rectangular entrance configuration are not ideal from the fluid dynamics point of view. But accuracy of the nozzle diameter and reproducibility of the entrance configuration is an other demand in making thenozzles. It has to be pointed out that the inertia parameter,and therefore the cut off point of a stage, depends on the third power of the aerosol jet width or the nozzle diameter. This is easily seen by replacing the velocity by volumetric flow rate and nozzle cross section. Nozzles, however, can be drilled to a high degree of accuracy in the order of 1μm diameter.

A remark has to made on the width of the aerosol jets. In nozzle flow the effect of the vena contracta is encountered. For non ideal entrance configurations the flow detaches from the nozzle walls and forms a cross section which is considerably smaller than the nozzle cross section and moreover depends on the dynamics of the flow. So a vena contracta is formed in nozzles with a rectangular entrance configuration, but when the nozzles are sufficiently long in comparison to their widths, the flow rebounds to the nozzle walls and leavesthe nozzle with a width identical to the width of the nozzle. This relation holds as long as α, critical pressure ratio is not superceeded.

In reading equation (2) in a different manner one can state that the pressures, p_e, in stages with high aerosol jet velocities are considerably smaller than the pressures,p_i, ahead of the stages. Consequently by a cascade of stages with high velocities the pressure is decreased stepwise from stage to stage. On the other hand the pressure reductions influence the slip correction factor,$C_{p,g}$, which is defined by

$$C_{p,g} = 1 + (2 \cdot (\ell_g/\mathcal{D}_p)) \cdot (1{,}257 + 0{,}4 \cdot \exp(-1{,}1 \cdot (\mathcal{D}_p/2 \cdot \ell_g))9) \qquad (4)$$

where l_g is the mean free path of the gas molecules. The mean free path depends on the pressure, p_e, in the stage according to the relation $l_{g,e} \cdot p_e = l_{g,o} \cdot p_o$, with $l_{g,o}$ as the mean free path at atmospheric pressure p_o. Therefore a pressure reduction leads to an increase of the mean free path and consequently to increase of the slip correction factor,$C_{p,g}$. This effect is desired as it contributes positively to the separartion of finest aerosol particles.

In operating an impactor equipped with several low pressure stages the problem arises, how to control the pressure in the stages and how to regulate the flow rate. Fortunately the design of the Aeras low pressure impactor bears the solution in itself. Here a critical nozzle is installed in the exhaust line after the final stage. At critical flow conditions, when the pressure ratio p_e/p_i of the nozzle is below its above mentioned critical value the fluid reaches the velocity of sound in the narrowest flow cross section within the nozzle. This cross section can be the wall of the nozzle

or it is defined dynamically when the vena contracta occurs. Once the velocity of sound has been obtained the discharge rate through the nozzle can
not be increased even by further reduction of the down stream pressure. As a
consequence the discharge rate is at its highest value and the pressure
ahead of the nozzle, i.e. the pressure, p_e, in the final stage is locked,
and aswell all pressures, p_e, of the preceeding stages.

Experiments demonstrate, that the volumetric flow rate, V, does not depend on atmospheric pressure, p_o, once the impactor is operated on a critical
orifice. However, the temperature ,T, is of some influence, as the velocity of sound, which governs the behaviour of the impactor, increases with temperature. The flow rate follows almost exactly the relation

$$V = V_o \cdot (T/T_o)^{1/2} \tag{5}$$

where T is the temperature of the environment the impactor is operated in
and V_o is the standard flow rate at standard temperature, T_o.

A last point has to be mentioned. In the coarse particle stages of impactors similar to the Aeras low pressure impactor in nozzle configuration
and stage design well developed deposition spots are not observed when the
Reynolds-number of the aerosol jet flow drops below values of Re = 150, approximately. A badly developed pattern of the deposits can indicate coupled
nozzle flow and severe shifts and degradation of the efficiency curves are
expected. In low pressure impactors Reynols-numbers smaller than Re = 150
are obtained, if e.g. nozzle dimensions are around 0,05 cm, velocities are
around 50 m/sec and pressures are as low as 50 mbar. The Reynolds-number of
the aerosol jet flow has a value of Re = 80.

In spite of its complex design the Aeras low pressure impactor is a useful instrument, which has found its application in many fields of aerosol
studies. One of the high lights was the detection of trimodal mass size distributions of traffic aerosols, which have been measured from a car moving
in the automotive traffic in the urban area of Vienna. Very successful were
extended studies on exhaustaerosol from automobiles. Even in the field of
atmospheric aerosols   essential studies have been performed concerning
the structure of urban and other atmospheric aerosols and the distribution
of certain chemical species.

Published 1984 by Elsevier Science Publishing Co., Inc.

Aerosols, Liu, Pui, and Fissan, editors

DESIGN OF A MULTI-STAGE VIRTUAL IMPACTOR

V. J. Novick, J. L. Alvarez, A. D. Appelhans
EG&G Idaho, Inc.
P. O. Box 1625
Idaho Falls, ID   83415

Extended Abstract

A multi-stage virtual impactor shown in Figure 1 has been designed, fabricated and tested. The impactor segregates particles at two aerodynamic sizes providing three size fractions for collection. A prototype, designed and constructed to operate at flow rates of 2 liters per minute, can be operated at flow rates as high as 10 liters per minute without an increase in internal losses. These flow rates, using ambient air through the impactor, provided a range of possible first stage cut points from 3.3 μm to 7.3 μm and a second stage range of 0.9 μm to 2.0 μm. The efficiency of each stage was measured as a function of the Stokes number (STK). Both the values of $(STK_{50})^{1/2}$ and the sharpness of the efficiency curve were found to be in agreement with Marple's (1980) theoretical predictions (Figure 2). Internal losses were found to be lower than Marple's (1980) theoretical values but not as low as Loo's (1979) dichotomous sampler. The impactor can easily handle mass loadings as high as 40 mg and can collect as high as 80 mg.

This multi-stage virtual impactor is extremely versatile due to its ability to operate at a wide range of flow rates, handle large quantities of mass and its small physical size. Planned uses include high temperature, high pressure measurements of harsh environments and ambient personal sampling.

References

Loo, B. W. et. al. (1979) "A Second Generation Dichotomous Sampler for Large Scale Monitoring of Airborne Particulate Matter", Lawrence Berkley Laboratory Report No. 8725.

Marple, V. A. and Chien, C. M. (1980) "Virtual Impactors: A Theoretical Study", Environmental Science and Technology, 14, pp. 976-984.

144

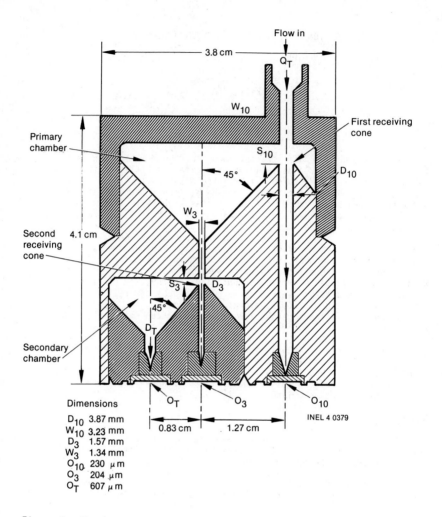

Figure 1. Two Stage Virtual Impactor.

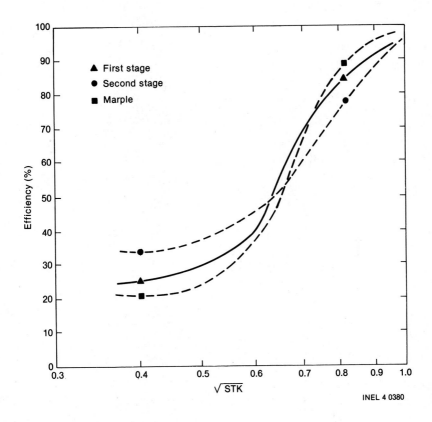

Figure 2.   Two Stage Virtual Impactor.
Efficiency Compared to Theoretical

CHAPTER 5.

SAMPLING, TRANSPORT AND DEPOSITION

Published 1984 by Elsevier Science Publishing Co., Inc.
Aerosols, Liu, Pui, and Fissan, editors

ON THE SAMPLING OF HIGH TEMPERATURE AEROSOLS

H.J. Fissan, H. Bartz
Process- and Aerosol Measurement Technology
University of Duisburg, FRG

EXTENDED ABSTRACT

Introduction

One of the major sources of man made aerosols are combustion processes. There are a lot of unsolved aerosol problems in this area. Our special interest during the last few years was on the soot formation in flames (Fissan, 1974), the phase distribution of polyaromatic hydrocarbons in the exhaust gases of internal combustion engines (Fissan and Franzen, 1980), the particle emission from fires with regard to the response of smoke detectors (Helsper et al., 1980), the emission of heavy metals from welding processes (Körber and Fissan, 1981) and the conversion of organic particle material into inorganic less harmful material in processes such as waste water combustion (Bartz and Fissan, 1983) or sintering. In all these cases and many more an analysis of the particles in the gases with respect to quantities like particle concentration, size and chemical composition has to be performed. The thermodynamic state of the aerosol differs from ambient aerosol. Here especially the analysis of particles in exhaust gases at elevated temperatures is considered.

Analysis without sampling

Combustion aerosols at elevated temperatures are very unstable. They will change their properties rather quickly because of condensation, evaporation and coagulation processes, if the thermodynamic state of the aerosol is changed. Changes are most likely to occur if the aerosol is sampled. The chances for changes are reduced, however, if remote sensing methods are used, such as light extinction or scattering methods.

Extinction and emission measurements at two wavelengths were performed scanning across a laminar diffusion flame of rotational symmetry (Fissan, 1974). Using Abel inversion techniques local extinction coefficients were determined. According to Mie-theory the ratio of these coefficients at two wavelengths is a function of the mean diameter, defined as the ratio of the third to the second moment of the particle size distribution. The extinction coefficient also allows the determination of the particle volume concentration.

Another possibility is to investigate the light scattered by the particles, especially from single particles. The optically defined scattering volume has to be rather small to avoid coincidence. One also has to take care of the light emission of the particles at very high temperatures. Since aerosols with defined properties at high temperatures are difficult to generate, the calibration of the scattering device has to rely on theory with all its limitations. Signal to noise problems limit the method to particles with diameter $Dp > 1$ μm.

In exhaust gases at temperatures below 200°C optical methods may be calibrated by comparison with other measurement techniques. These are in case of extinction measurements the sampling of particles on a high temperature resistent filter and relating the extinction signal to the gravitational analysis. In case of light scattering techniques, the calibration curve can be obtained by using a set of precollectors with known penetrations, such as an impactor (Fissan et al., 1984) or a cyclone (Büttner, 1984).

Analysis with sampling

With remote sensing techniques the kind of quantities which can be determined is limited. Especially in case of chemical analysis one has to collect the particles from the main aerosol stream or at least extract a representative sample. This is normally achieved by inserting a sampling nozzle against the flow direction through which the sample is pulled at the same velocity which occurs in the main aerosol steam (isokinetic sampling). For deviating velocities the sample concentration differs from the main stream concentration. The ratio of these two concentrations, known as the aspiration coefficient, was determined by many researches as a function of Stokes number (Belyaev and Levin, 1974). Sampling errors increase with increasing ratio of face velocity to suction velocity and Stokes number. On the other hand the Stokes number at high temperatures is smaller than at ambient temperature for particles with diameter $Dp > 0.3$ μm. Thus sampling errors are reduced at high temperatures compared to ambient conditions.

The finite dimensions of the sampling nozzles, especially if the nozzle is cooled, influence the flow field at the nozzle entrance. Special care has to be taken to design the tip of the nozzle (Fernandez et al., 1974). To over-come the problem with changing flow velocities, a nozzle applicable at high temperatures is developed based on the self adjusting concept given by Steen (Steen, 1977). Following the proposal by Liu and Pui (Liu and Pui, 1984) a second sampling nozzle is inserted into the Steen nozzle. Because the cross section of the Steen nozzle increases, the face velocity for the second nozzle is reduced. Thus, the sampling errors in case of unisokinetic sampling become smaller. This sampling method is especially of interest, if the analysis technique used requires a constant flow rate, e. g. in the case of an impactor.

To avoid changes of the aerosol one should try whenever possible to perform the analysis at the given temperature. Thus far only collection devices such as filters, cascade impactors or cyclones have been used. The reason is that other techniques such as electrical aerosol analyzers are not built to stand higher temperatures. Also the behaviour of the aerosol at higher temperatures in these instruments is not well understood.

For classification of particles according to their size at elevated temperatures, cascade impactors are most appropriate since they can stand higher temperatures and their behaviour at higher temperatures can be described theoretically. In combustion exhaust gases often two types of particles occur: small ones from condensation followed by coagulation processes, larger ones from dispersion processes. In case of a large amount of large particle material, a cup-like impactor as a precollector is used (McFarland et al., 1979). To reduce wall losses and to be able to perform a quantitative analysis of the large particle material a modified precollector was developed, which is integrated into the Andersen Stack Sampler.

Very often for the collection and/or analysis techniques it is necessary to transport and/or to reduce the temperature of the aerosol. For the transport metallic tubes are used. It should be checked, however, that these tubes do not emit particles by themselves. This was observed for corrosion resistent stainless steels. It is thus strongly recommended for high temperature applications to use high temperature stainless steels.

In a theoretical study the losses of particles in a tube at elevated temperatures have been investigated. The different effects were considered to be independant. The calculation shows that losses are reduced with increasing temperature in case of inertial effects like sedimentation, impaction or turbulent deposition. The losses increase in case of laminar diffusion and coagulation.

The temperature can be reduced by wall cooling or by dilution with cold gases. In case of wall cooling thermophoretic forces will cause losses in addition to the other already mentioned loss effects. The particles are changed by condensation and coagulation processes. In a project of the Special Research Program (SFB 209) at the University of Duisburg the changes of the physical properties of a high temperature aerosol flowing through a cooled tube or mixed with inert particle free gases, are investigated. The goal is to quantify these changes theoretically and verify the theoretical results experimentally.

Sampling Systems for High Temperature Aerosols

Since the results of particle analysis are always depending on the sampling system, guidelines for sampling aerosols have been developed all over the world. In the U.S. the EPA Method 5 is used. In the Federal Republic of Germany a sampling train is described in VDI-Guideline No. 2066.

Because both systems aim at emission control and dust collector performance, emphasis lies mainly on large particles which contribute to the emitted particle mass. These methods, however, are not suited to perform particle size measurements without any specific modifications. In general it is very difficult to develop a sampling train for high temperature aerosols, which fits the needs of all applications. We feel that each problem may ask for a specific sampling train. Therefore, the basic knowledge of the processes involved is very important.

For the investigation of the incineration of waste waters containing organic salts as well as for a special sintering process a sampling train was developed in our institute (Bartz and Fissan, 1982). It allows the size dependent collection of particles for chemical analysis. The aerosol is cooled in a well defined way so that inorganic and organic vapours except water are condensed. The size dependent chemical analysis then allows the discussion of the conversion processes within the system under consideration.

This example demonstrates clearly that the problem to be solved defines the requirements for the sampling system for aerosols at high temperatures.

References

Bartz, H.; Fissan, H. J. (1982) "Particulate Measurements in a Waste Water Combustion System", VDI-Bericht Nr. 429: 121-126

Bartz, H.; Fissan, H. J. (1983) "Sampling and Characterization of Aerosols in and from a Waste Water Combustion Process", Proceedings of the VIth World Congress on Air Quality, 1: 383-390

Belyaev, S. P.; Levin, L. M. (1974) "Techniques for Collection of Representative Aerosol Samples", J. Aerosol Sci. 5: 325-338

Büttner, H. (1984) "Messung von Partikelgrößenverteilungen an bewegten dispersen Systemen mit einem optischen Partikelzähler unter Zuhilfenahme von Impaktoren und Meßzyklonen", Preprints, 3. Eur. Symp. on Particle Characterization, Nuremberg, FRG: 287-305

Fernandez, E.; Suter, P. (1974) "Absaugsonden für Gas/Partikel Suspensionen ohne Störung der Zuströmung", Brennst. - Wärme - Kraft 26: 502-506

Fissan, H. J. (1974) "Rußkonzentrations- und Temperaturverteilungen in einer Methan-Luft-Diffusionsflamme", Habilitation, RWTH - Aachen, FRG

Fissan, H. J.; Franzen, H. (1980) "Characterization of Particle Emission of a Spark Ignition Engine", J. Aerosol Sci. 11: 249

Fissan, H. J.; Helsper, C.; Kasper, W. (1984) "Kalibrierung der Partikelgrößenbestimmung von optischen Partikelzählern", Preprints, 3. Eur. Symp. on Particle Characterization, Nuremberg, FRG: 263-274

Helsper, C.; Fissan, H. J.; Muggli, J.; Scheidweiler, A. (1980) "Particle Number Distribution of Aerosols from Test Fires", J. Aerosol Sci. 11: 439-446

Körber, D.; Fissan, H. J. (1981) "Particle Emission from Welding Painted Steel", Intern. J. Anal. Chem. 10: 13-21

Liu, B. Y. H.; Pui, D. (1984) "Private Communication"

McFarland, A.R.; Ortiz, C. A.; Bertch, R.W. "A High Capacity Preseparator for Collecting Large Particles", Atm. Envir. 13: 761-765

Steen, B. (1977) "A New Simple Isocinetic Sampler for the Determination of Particle Flux", Atm. Envir. 11: 623-627

Published 1984 by Elsevier Science Publishing Co., Inc.
Aerosols, Liu, Pui, and Fissan, editors

SAMPLING EFFICIENCY OF AEROSOLSAMPLERS FOR LARGE WINDBORNE PARTICLES

E. Vrins[1], P. Hofschreuder[1], W.M. ter Kuile[2], R. van Nieuwland[2] and F. Oeseburg[3]

1. Department of Air Pollution, Agricultural University Wageningen, P.O. Box 8129, 6700 EV WAGENINGEN, The Netherlands.

2. Institute for Environmental and Occupational Hygiene TNO, P.O. Box 214, 2600 AE DELFT, The Netherlands.

3. Prince Maurits Laboratories TNO, P.O. Box 45, 2280 AA RIJSWIJK, The Netherlands.

Introduction

Within the framework of the Dutch National Program on Coal Research guided by the "Bureau Energie Onderzoek Projecten", the sampling efficiency of thirteen aerosol samplers for large windborne particles was tested. The experiments were conducted in the open field in the period May 1983-October 1983 in joint effort by the Agricultural University Wageningen, the Institute of Environmental and Occupational Hygiene TNO Delft, Prince Maurits Laboratories TNO Rijswijk and the Netherlands Research Foundation Petten.
Aims of the project are selection of proper instruments for sampling coarse aerosols like coal dust, evaluation of the sampling efficiency of widely used commercial instruments and assessment of the inspirability curve for the human head in the open air.

The experimental set up

The experimental set up was an extension of that described in a preliminary report (1). A monodisperse test aerosol (di-ethyl-hexyl-sebacate) was emitted by four May spinning top generators with particle diameters of 10, 20 and 50 μm. The spinning tops were placed in one line perpendicular to the wind, 8 m apart from each other. The aerosol samplers were placed on one line perpendicular to the wind, 30 m downwind from the spinning tops. Most samplers were placed on an aluminium frame, that easily could be turned in the right direction. The distance between two samplers was in general 1.50 m. The sampling height was 1.5 m and distance between sampling heads and frame 0.9 m. Between these samplers ten rotorods were placed, which served as reference samplers. Each experiment lasted one hour.

The tested instruments were: LIS/P sampler according to VDI (Germany), EPA High Volume Sampler, aerosol tunnel sampler (1), TSI fugitive dust sampler, Sierra 10 μm Inlet for dichotomous sampler, Hirst spore trap, Globe of Liège, three inlet tubes with different orientation to the wind, dummy of mans head according to Vincent (U.K.), two special samplers and rotorod samplers.

While dust generation from coal stock piles will increase with higher wind-velocities and the stock piles will introduce excess turbulence, special attention is paid to these meteorological conditions in the testing program.

Over 200 experiments were performed with windvelocities ranging from 1 to 10 m/s. During some experiments windshields were placed upwind of the experimental set up to imitate coal stock piles and introduce excess turbulence. The turbulent spectrum was determined with the aid of a hot film anemometer, the frequencies ranging from 1 to 1000 Hz (3).

Results

The tested aerosol samplers greatly differ in sampling rate, ranging from 0,18 till 100 m³/hr. Several samplers could only collect a sufficient amount of aerosol, while sampling during several successive experiments.

The vast amount of meteorological data could not be totally elaborated by now (April 1984), nor all aerosol sampling data. The results presented in this paper relate to the 20 and 50 μm diameter particles and the mean collection efficiency of the samplers in three windvelocity-classes (Table I).

The rotorod sampler was used as a reference sampler. Extensive research was already devoted to the sampling characteristics of this apparatus (2). For determining the reference aerosol concentration corrections were made for collection efficiency, windvelocity and geometrical variations of the rods. Figure 1 shows some characteristic aerosol concentration profiles along the sampling line, determined by the rotorods. The smooth profiles show the high reliability of the rotorod samplers for determining relative concentration differences along the sampling line. This, however, does not guarantee an exact measurement of the absolute concentration. Problems in determining the absolute concentration occur for low windvelocities (average windspeed < 3 m/s) and 10 μm particles, when the collection efficiency of the rotorod sampler is decreasing. In evaluating the other aerosol samplers this must be taken in consideration.
As the EPA High Volume Sampler was widely used in the past, this apparatus was tested. The collection efficiency, however, is low as was concluded in previous investigations (1, 4).
The collection efficiency of the LIS/P sampler for 20 μm particles, at low windvelocities is pretty good, but decreases rapidly with high windvelocities and for 50 μm particles.
An aerosol tunnel sampler (1) turned out to be the apparatus with highest efficiency and least dependent on windvelocity. The apparently high collection efficiency for low windvelocities may be due to a decrease of the efficiency of the reference sampler.
The Globe of Liège consists of an aluminium globe of 0.12 m diameter, freely exposed to the ambient air. Considered as an impactor with windvelocity as the impaction velocity, the impaction efficiency is extremely low for 20 and 50 μm particles. Obviously the Globe of Liège is inapt for measuring the aerosolconcentration. The collection of aerosol is more due to turbulent deposition than to impaction.
The Gromoz is an aerosol sampler designed by the Institute for Environmental and Occupational Hygiene TNO. The apparatus consists of a 0.14 m diameter filter that is downward oriented and placed right in the entrance of the conical filter head. The Gromoz shows a pretty good collection efficiency, however, decreasing with high windvelocities.
The other tested instruments had a low sampling rate and/or a low sampling efficiency which asked for sampling during several hours. As a consequence the number of gathered data is lower. For this reason the instruments are not mentioned in Table I.

156

The TSI fugitive dust sampler has an extremely low sampling rate,
$D = 0.071.W$ m³/hr (W = windvelocity in m/s). Figure 2 shows the determined
collection efficiencies for 20 and 50 μm particles. Especially for low windvelocities the high collection efficiency can be due to improper functioning
of the reference sampler.
The Hirst spore trap is an impactor, that finds ample application in aero-
biological research. The sampling rate is 0.6 m³/hr. Obviously the collection
efficiency decreases rapidly for 50 μm particles (Figure 5).

At the moment no results can be presented of the Sierra 10 μm inlet, a
dummy of a manshead and a total dust helmet. The latter apparatus is
designed by the Institute for Environmental and Occupational Hygiene TNO
and is meant for personal total dust sampling.
Three inlet tubes were mounted on a windvane with different orientation
to the wind, the angle between tube-axis and winddirection amounting 0°,
45° and 90° (downward). The sampling velocity was 1.25 m/s, the sampling
rate 1.02 m³/hr, tube diameter 0.017 m and tube length 0.04 m. The
efficiency of the filter as well as the total inlet efficiency was
determined, the difference being the inner wall losses. The determined
efficiencies vary greatly in value, probably due to influences of atmospheric
turbulence. Figures 4 and 5 show the average efficiency as a function of
mean wind velocity for the three tubes for 20 and 50 μm particles. As is
expected the inlet efficiency of the winddirected tubes, exceeding one, and
wall losses increase with windvelocity. The inlet efficiency of the tube
perpendicular to the winddirection is expected to decrease with windvelocity. However, for 50 μm particles and high windspeeds, the inlet
efficiency is increasing, due to aerosol deposition on the inner wall of
the tube.

Conclusions

- Due to turbulence in outdoor air, results of aerosol sampling instruments
  for the same particle diameter and the same average windspeed, vary more
  than the standarddeviation of 10% that stems from the analytical procedure.
- The aerosol tunnel sampler is the best instrument available for sampling
  the whole aerosolspectrum. The TSI fugitive dust sampler shows more
  spreading in results but may be adequate. More data on this instrument
  should be gathered.
- An openface filterholder of large diameter like the Gromoz collects more
  aerosol than a (modified) vertical elutriator.
- A more detailed report is in preparation.

References

1. Hofschreuder, P., Vrins, E. and Boxel, J. van (1983) "Sampling efficiency
   of aerosol samplers for large windborne particles", J. Aer. Sci. 14,
   pp. 65 - 68.
2. Vrins, E. and Hofschreuder, P. (1982) "Sampling total suspended particulate
   matter", J. Aer. Sci 14, pp. 318 - 332.
3. Kuile, W.M. ter (1983) "Monsternemingstechnieken voor opwaaiend kolenstof"
   Rapport F 2117, National Program on Coal Dust, project OWS 1 (in Dutch).
4. Wedding, I.B., Mc Farland, A.R., Cermake, J.E. (1977) "Large particle
   collection characteristics of ambient aerosol samplers", Env. Sci. and
   Techn. 11 no. 4, pp. 387 - 390.

TABLE 1.

Instrument	particle size (μm)	wind velocity 0 - 3 m/s	3 - 6 m/s	> 6 m/s
EPA HVS	20	0.39	0.33	0.23
	50	0.31	0.20	0.11
LIS/P sampler	20	1.04	0.84	0.36
	50	0.34	–	0.31
Aerosol tunnel sampler	20	1.21	1.13	0.96
	50	2.05	1.22	0.90
Globe of Liège	20	0.017	0.008	0.007
	50	0.15	0.08	0.20
Gromoz	20	1.24	1.00	0.57
	50	0.88	0.65	–

158

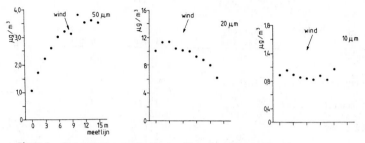

Figure 1. Some aerosol concentration profiles along the sampling line.

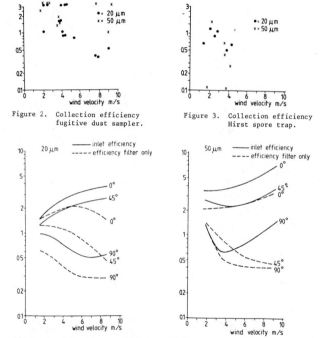

Figure 2. Collection efficiency
fugitive dust sampler.

Figure 3. Collection efficiency
Hirst spore trap.

Figures 4 and 5. Average collection efficiency of three inlet tubes at an
angle of 0°, 45° and 90° with winddirection.

Published 1984 by Elsevier Science Publishing Co., Inc.

Aerosols, Liu, Pui, and Fissan, editors

ISOKINETIC SAMPLING PROBLEM

J. Podzimek, H. D. Ayres, and J. Mundel
Department of Mechanical and Aerospace Engineering and
Graduate Center for Cloud Physics Research
University of Missouri-Rolla
Rolla, MO 65401

Semi-empirical Formulas

Principles of the isokinetic sampling are explained in many books and articles on aerosol physics (e.g. Fuchs, 1964; Cadle, 1975; Fiedlander, 1977). Our concern is to review several of the existing formulas used for isokinetic sampling of submicron and micron particles and to design a suitable system for particle sampling on board an aircraft flying horizontally at a velocity between 60 and 90 ms^{-1}.

Assuming a very thin horizontal sampling tube with the aerosol containing air velocity, V_s, and aerosol concentration, N_s, Watson (1954) deduced an equation for the aerosol true concentration, N_o, if the aircraft is moving with a steady speed, V_o:

$$\frac{N_s}{N_o} = \frac{V_o}{V_s} \{1 + f(St) \ [\left(\frac{V_s}{V_o}\right)^{1/2} - 1]\}^2 \ . \tag{1}$$

The "inertial" parameter $f(St)$, which is a function of Stokes number, St, has for St $\ll$ 1.0 the value close to 1.0 and for St = 3.0 approximately 0.25.

Badzioch (1959, 1960) deduced another formula for incompressible and steady flow with noninteracting monodisperse spherical particles of diameter, d_p, density, ρ_p, moving with a velocity, V_p:

$$\frac{N_s}{N_o} = 1 - \alpha + \alpha \frac{V_o}{V_s} \ . \tag{2}$$

If the total mass of particles, M_s and that in an undisturbed medium, M_o, are considered, then Eq. (2) will be modified to

$$\frac{M_s}{M_o} = (1 - \alpha) \frac{V_s}{V_o} + \alpha \ . \tag{3}$$

In a semi-empirical approach Badzioch related the parameter, α, to the distance, L, where the particles ahead of the sampling tube start to deviate from a straight line

$$\alpha = [1 - \exp(-L/\ell_p)]/[L/\ell_p] \ . \tag{4}$$

ℓ_p is the stop distance of the particles which in our case (for $\rho_p = 1.0 \times 10^3$ kg m^{-3}; $\rho_a = 1.23$ kg m^{-3}; μ (dynamic viscosity) = 1.78×10^{-5} kg m^{-1} s^{-1} and V_o

160

$= 60 \text{ m s}^{-1}$) had the values 6.0686×10^{-6}, 2.1847×10^{-4} and 5.4617×10^{-3} m for particle diameters 0.2, 1.2 and 6.0 µm.

Davies (1968), using Badzioch's measurements performed at $7.50 < V_o < 24.4$ m s^{-1} and $0.20 < V_o/V_s < 4.62$, suggests for the range of sampling tube diameters $0.6 < d_s < 1.8$ cm an approximate formula for $\alpha(St)$.

$$\alpha(St) = 1 - \{1 - 0.3 \text{ St } \ln [1 - (d_s)/2]\}^{-1} . \qquad (5)$$

The Davies formula is supported by Sehmel (1967) measurements. Considerable simplification of the values of α can be made for extreme values of L/ℓ_p (Badzioch):

For $\quad L/\ell_p < 0.333 \quad \ldots\ldots \quad \alpha \cong 1 - L/(2 \ell_p$,

for $\quad L/\ell_p > 4.00 \quad \ldots\ldots \quad \alpha \cong \ell_p/L$.

In our case we assumed $d_s = 1.031$ cm, $L \cong 4.2$ cm and compared the ratios of the total amount of particulate matter collected in a sampling tube, C_p, and the true particulate matter concentration, C_o, for Watson, Badzioch (simple form) and Davies formulas. Figures 1, 2 and 3 support the idea that the differences while using different (simplified) formulas are considerable and that Davies formula possibly exagerates the value of correction, especially if applied to outside air (aircraft) velocity overriding many times that of the aerosol sampling.

In addition to losses from non-isokinetic flow, there are particle losses by aerosol diffusion towards the tube wall and gravitational effects in the tubes leading the aerosol from the sampler to the instrument. These effects can be reasonably well estimated from the available semi-analytical or numerical models (e.g. Gormley and Kennedy, 1949, Thomas, 1967). Other mechanisms of particle deposition in tubes such as electrostatic charges are neglected.

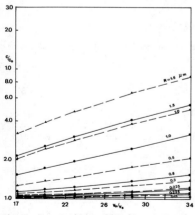

Figure 1. $C_p/C_o = f(V_o/V_s, r_p, \rho_p)$ according to Watson.

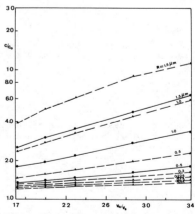

Figure 2. $C_p/C_o = f(V_o/V_s, r_p, \rho_p)$ according to Badzioch.

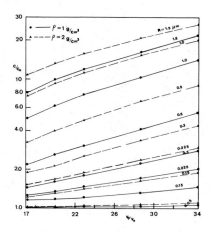

Figure 3. Ratio of the observed and true particle concentration $C_p/C_o = f(V_o/V_s, r_p, \rho_p)$ calculated for aircraft anisokinetic sampling according to Davies.

Design of the Equipment

The equipment for aircraft sampling of fine atmospheric aerosol is composed of a nozzle (Fig. 4) which is attached outside to the aircraft, the metallic tubes and attachment pieces leading the aerosol from the nozzle to the cascade impactor or ASAS 300A (Particle Measuring Systems Incorporated, Boulder) light scattering aerosol spectrometer installed inside the aircraft. The typical range of particles to be sampled is 0.1 to 5.0 µm radius. A sampling tube of inside diameter 1.031 cm was used, which gives a flowrate of 17.5 L min^{-1} (for the impactor) and average velocity of 3.942 m s^{-1}. Because of the limited accuracy of correction formulas, an adjustable nozzle (or automatically operated nozzle) will be used for meeting the conditions for isokinetic sampling.

The nozzle designed for an aircraft operation at 65 to 80 m s^{-1} is conical in shape enabling the air velocity to slow down in the aerosol sampling tube. To prevent vortices from forming, a 7° angle for the front nozzle and a 20° angle for the rear nozzle were used (Fig. 4). In this version the nozzle should be adjusted prior to the sampling to the position corresponding to the aircraft's sampling speed. A calibration of the sampling nozzle in an aerodynamic wind tunnel is anticipated. The sampling tube is made of chromium plated brass, all other parts of the sampling nozzle are made of duraluminum. All metal tubing with large curvature radii is used before the counting instrument to avoid the static charge build up.

Conclusion

Large discrepancies have been found if different formulas for isokinetic sampling conditions, especially for particles with radii larger than 1.0 µm, are used. In addition, trajectories of the heavier particles, when in the nozzle, will probably (due to their inertia) cross the pathlines of smaller particles. This might result in a particle coagulation and in a larger particle concentration in the center of the nozzle (higher counts of large

162

particles).

The adjustable nozzle which can slide on a track and correct the counts
for different aircraft sampling speeds is probably the best solution of the
isokinetic sampling problem.

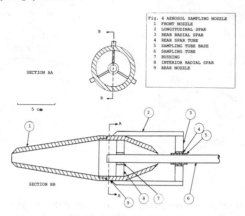

Fig. 4 AEROSOL SAMPLING NOZZLE
1   FRONT NOZZLE
2   LONGITUDINAL SPAR
3   REAR RADIAL SPAR
4   REAR SPAR TUBE
5   SAMPLING TUBE BASE
6   SAMPLING TUBE
7   BUSHING
8   INTERIOR RADIAL SPAR
9   REAR NOZZLE

References

Badzioch, S. (1959) "Collection of Gas-Borne Dust Particles by Means of An
Aspirated Sampling Nozzle", Brit. J. Appl. Phys. 10:26.

Badzioch, S. (1960) "Correction for Anisokinetic Sampling of Gas-Borne Dust
Particles", Brit. J. Inst. Fuel., March, 106.

Cadle, R. D. (1975) The Measurement of Airborne Particles, John Wiley, New
York.

Davies, C. N. (1968) "The Sampling of Aerosols", Staub-Reinhalt. Luft, 28:219.

Friedlander, S. K. (1977) Smoke, Dust and Haze, John Wiley, New York.

Fuchs, N. A. (1964) The Mechanics of Aerosols, Pergamon Press, Oxford.

Gormley, P. and M. Kennedy (1949) "Diffusion from a Stream Flowing Through a
Cylindrical Tube", Roy. Irish Acad. Proc. 52A:163.

Sehmel, G. A. (1967) Ann. Occup. Hyg. 10:73.

Thomas, J. W. (1967) "Particle Loss in Sampling Conduits", Symp. Proc.
Assessment of Airborne Radioact., IAEA, Vienna, 701.

Watson, H.H. (1954) "Errors Due to Anisokinetic Sampling of Aerosols", Am. Ind.
Hyg. Assoc. Quart. 15:21.

Published 1984 by Elsevier Science Publishing Co., Inc.
Aerosols, Liu, Pui, and Fissan, editors

THE EFFECTS OF TURBULENCE ON THE ENTRY OF AIRBORNE
PARTICLES INTO A BLUNT DUST SAMPLER

J.H. Vincent, P.C. Emmett and D. Mark
Physics Branch, Institute of Occupational Medicine,
8 Roxburgh Place, Edinburgh, U.K.

EXTENDED ABSTRACT

Introduction

Dust samplers are widely used in workplace and ambient atmospheres in
order that dust-laden air may be aspirated, collected onto a filter, and
assessed for the determination of the concentration of the airborne dust.
All such devices are 'blunt' samplers since their very presence must impose
some obstruction to the movement of the surrounding air. The sharp-edged
isokinetic probe is merely a limiting case as 'bluntness' tends towards its
minimum.

Since it is well-known that turbulence can have profound effects on the
transport of airborne particles, it is perhaps surprising that so little
attention appears to have been given to the problem in relation to dust
sampling. This is particularly so in view of the fact that practical dust
sampling frequently takes place in turbulent air.

Theory

The theory describing the entry of airborne particles into a blunt dust
sampler operating in a smooth freestream has been given in a number of
previous papers from this laboratory (Vincent and Mark 1982, Vincent,
Hutson and Mark 1982 and Vincent 1984) and elsewhere (Ingham 1981). For
a turbulent freestream, a model has now been developed which takes account
of two superimposed but competing particle transport processes. In the
first we have the reorganisation of the spatial distribution of the particles
near the sampler as a result of inertial effects associated with their movement in the distorted mean flow. Its primary effect is to establish
particle concentration gradients in the region in question. In the second
process we have turbulent diffusion which tries to counteract the establishment of such gradients. In the first instance, this picture has important
practical consequences with regard to isokinetic sampling where, since there
is by definition no flow distortion, there should be no induced particle
concentration gradients and therefore no effects of freestream turbulence
on aspiration efficiency.

For the more general case of a blunt sampler of arbitrary fixed shape,
size, sampling flow rate and facing directly into a wind whose speed also
is fixed, it may be argued from quantitative considerations of the pre-

164

ceding physical model that

$$\frac{A - A_o}{\Lambda R} = f(k) \tag{1}$$

should hold, in which A and A_o are aspiration efficiencies in the
turbulent and smooth freestreams respectively, and f is a unique function
of the turbulence inertial parameter k.   In this equation Λ is a
dimensionless parameter representing turbulent diffusion and R another which
compares the magnitude of the distortion associated with the turbulent
fluctuations with that of the distortion associated with the mean flow.
Both are functions of the freestream turbulence intensity and integral
length scale respectively.

Results and Discussion

The physical models we have outlined have been tested in a large wind
tunnel in which the turbulence conditions may be controlled by the use of
square-mesh, lattice-type grids placed across the entrance to the working
section and measured by thermal anemometry.   Experiments were carried out
to investigate the performances, firstly, of sharp-edged samplers of
diameters 10 and 22 mm respectively operating isokinetically and, secondly,
of a single idealised blunt sampler consisting of a flat disc of diameter
40 mm having a central sampling orifice of 4 mm and sampling at 9.5 l/min
in a wind speed of 2 m/s.   Test aerosols of human serum albumen, wax and
aloxite were employed.

The results for the isokinetic samplers indicated that, over a wide
range of turbulence conditions (intensity up to 15% and length scale up to
10 cm), there is no significant systematic effect due to turbulence.   This
is consistent with the physical model.

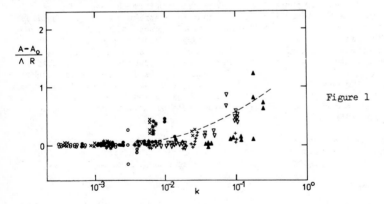

Figure 1

For the idealised blunt sampler, the results plotted in the form suggested by Equation (1) are given in Figure 1, in which A_0, having not been obtainable experimentally, was determined from blunt sampler theory (Vincent 1984). Despite the inevitable scatter in the data points and the fact that Equation (1) does not lend itself to sensitive verification at the larger turbulence length scales, these results do suggest that Equation (1) is a fair reflection of the underlying physical basis. Further, the upward trend in the function f for increasing k is consistent with the decrease in particle turbulent diffusivity for increasing particle aerodynamic size (as described by Soo 1967 and others).

In conclusion, this work suggests in general that freestream turbulence can influence the performances of dust samplers, the effect being the greatest the more marked the distortion of the mean flow near the sampler and for the largest particles. However, preliminary experiments with some 'practical' samplers intended for the collection of inspirable dust have shown, for particle aerodynamic diameters up to 40 µm and for turbulence conditions broadly consistent with those found in coal mines, that such effects are small.

References

Ingham, D.B. (1981). "The entrance of airborne particles into a blunt sampling head", J. Aerosol Sci. 12: 541-549.

Soo, S.L. (1967). Fluid Dynamics of Multiphase Systems. Blaisdell, Waltham (Mass.).

Vincent, J.H. and Mark, D. (1982). "Application of blunt sampler theory to the definition and measurement of inhalable dust", Inhaled Particles V (Edited by W.H. Walton), Pergamon Press, Oxford.

Vincent, J.H., Hutson, D. and Mark, D. (1982). "The nature of air flow near the inlets of blunt dust sampling probes", Atmos.Environ. 18: (in press).

Published 1984 by Elsevier Science Publishing Co., Inc.
Aerosols, Liu, Pui, and Fissan, editors
166

RECENT DEVELOPMENTS IN DUST SAMPLING IN
RELATION TO HEALTH EFFECTS

J.H. Vincent and D. Mark
Physics Branch, Institute of Occupational Medicine,
8 Roxburgh Place, Edinburgh, U.K.

EXTENDED ABSTRACT

Introduction

In recent years, much research at the Institute of Occupational Medicine
and elsewhere has been directed at the question of how to sample airborne
dust in a manner which reflects the dose received by the respiratory tracts
of exposed subjects.   This involves taking into account the physical transport processes by which airborne particles are conveyed from the ambient air
outside the body to the biological surfaces inside it.   The first criterion
for such a dust sampler, aimed at collecting dust relevant to health effects,
is that its entry characteristics should match those of the human subject
during breathing.   To provide a basis for such a criterion, extensive wind
tunnel studies have been carried out to investigate the sampling characteristics of a man (using inert mannequins) over a wide range of relevant conditions.   From all this work, a quantitative definition of _inspirability_
(originally referred to as 'inhalability') has been constructed (Vincent and
Armbruster 1981), taking the form of a single curve which relates the efficiency of inspiration (I) to particle aerodynamic diameter (d_{ae}).   Inspirability takes values close to unity for small d_{ae}, falling steadily to level
off at 0.5 for all d_{ae} > 30 μm (with no suggestion of a cut-off, even for
particles with d_{ae} as large as 100 μm).   More recently, this definition
has been recommended by the ACGIH for describing _inspirable particulate_
matter (IPM).   It is now widely held that this criterion should be used
to describe 'total' dust, the definition of which has, until relatively
recently, tended to be defined simply and rather arbitrarily in terms of the
instrument chosen to measure it.

A new static total/inspirable dust sampler

The new sampler consists of a sampling head which is mounted on a combined pump and drive system.   The purpose of the drive system is to
provide slow rotation of the head about a vertical axis, leak-free
connection of the head to the pump being achieved by means of a spring-
loaded rotating PTFE seal.   The head itself has a short cylindrical body
50 mm in diameter and 52 mm long, containing a single transverse slot 3 mm
wide and 15 mm long.   This in turn has thin 'lips' protruding 2 mm outwards
from the sampler body (in order to prevent sampling errors due to particle
blow-off).   In practice, the lips and entry slot form integral parts of a
dust-collecting capsule which is located inside the head.   This capsule
contains the filter, and it is weighed complete before and after each
sampling run so that dust evaluation is based on all that enters through
the entry plane of the sampling slot.   This means that performance is not
complicated by internal wall losses.   Overall the sampler has a diameter
of 150 mm and height 250 mm, it weighs 2.5 kg, and samples at 3 l/min.

Its performance in relation to the inspirability curve is shown in Figure 1, where it is seen that, for a representative range of wind speeds and a wide range of particle sizes, the comparison is good.

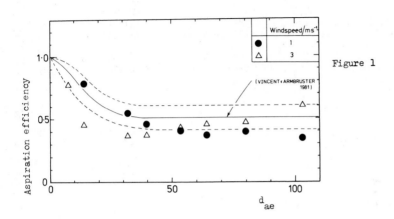

Figure 1

Work is now in progress to develop a personal sampler which similarly matches the inspirability definition.

Inspirable dust spectrometers

If the inspirable fraction of airborne dust (as sampled, for example, by an entry like the one which is described above) is classified so that its aerodynamic size distribution is obtained, then in principle all the information is available to enable determination of the mass concentration and size distribution of any biologically-relevant subfraction which may be defined numerically (Vincent and Mark 1984). An instrument which can achieve this — an inspirable dust spectrometer — may therefore have particularly useful applications in a wide range of research into dust-related disease, where the effects of correlating various deposition subfractions, breathing parameters, etc. with the incidence of various respiratory diseases, can be investigated. With this in mind, static and personal sampling devices, based on the cascade impactor principle, are being developed in our laboratory (Mark et al 1984) and elsewhere in Europe specifically for research applications relating to respiratory disease in the mining industry.

References

Mark, D., Vincent, J.H., Gibson, H., Aitken, R.J. and Lynch, G. (1984). "The development of an inhalable dust spectrometer", Ann.occup.Hyg. 28: 125-143.

Vincent, J.H. and Armbruster, L. (1981). "On the quantitative definition of the inhalability of airborne dust", Ann.occup.Hyg. 24: 245-248.

Vincent, J.H. and Mark, D. (1984). "Inhalable dust spectrometers as versatile samplers for studying dust-related health effects", Ann.occup.Hyg. 28: 117-124.

Published 1984 by Elsevier Science Publishing Co., Inc.
Aerosols, Liu, Pui, and Fissan, editors

CHARACTERIZATION OF SIMPLE INLETS IN INDUSTRIAL AND AMBIENT AIR CURRENT CONDITIONS

R. W. Wiener, K. Willeke, and L. Silverman
Aerosol Research Laboratory
Dept. of Environmental Health
University of Cincinnati
Cincinnati, Ohio 45267

EXTENDED ABSTRACT

Introduction

During the past decade American Industry has developed a great need for ways to assess the presence of particles in workplace air. This is necessary for both reasons of health, as in testing for coal dust, asbestos, or crystalline silica, and for reasons of productivity as in clean room operations, and paint spraying operations. Unfortunately, the measurement of small particles suspended in the air, especially those of respirable size is a difficult process requiring sensitive and expensive instruments that produce particle counts and size measurements biased by the method of detection and the characteristics of the collection. This means that a particular counter may only count that which it sees, so how it sees and how it obtains what it sees introduce systematic errors.

It is hoped that the data assembled by investigation of the sampling efficiency of simple inlets may be used to produce predictive models and aid in the design of complex inlets. This will help in the assessment of worker exposure and the collection of more nearly representative samples.

Experimental

In the Aerosol Research Laboratory at the University of Cincinnati Medical Center we have been sampling aerosol particles with diameters from 5 to 50 micrometers under ambient wind speed conditions in a specially designed low speed (125-1000 cps) wind tunnel. Initial studies sought to characterize the performance of a single simple inlet under a variety of conditions. The inlet was a straight-brass tube with 1/4 inch outer diameter and 7/32 inch inner diameter by 20 cm long.

In this study liquid or solid particles have been generated using a vibrating orifice aerosol generator then introduced into a mixing chamber and carried by the wind through the test section of the wind tunnel where the inlet and optical particle counter probe reside. The Royco 245 Optical Particle Counter used in these experiments has been modified to handle much larger flow rates than for which it had been designed. It has also been adapted to serve as the platform for a variety of different simple inlets.

The experimental methodology considered the true particle concentration to be determined by sampling isokinetically and isoaxially with a thin walled inlet, assuming that aspiration efficiency and entry efficiency are 100%. The transmission efficiency is found by counting the particles lost to the inner wall of the inlet. This is done by using uranine tagged particles, while performing the experiment, then removing the inlet section and washing the inside of the tube. The uranine concentration is proportional to the flourescence level of the wash. From this the number of particles that impacted the walls may be calculated. This number of particles is added to the number of particles detected by the optical particle counter to give the total number of particles present in the sampled air. The transmission efficiency becomes equal to the number of particles registered by the optical particle counter divided by the total number of particles. Therefore the sampling efficiency for the isokinetic-isoaxial experiment is equal to the transmission efficiency.

The overall isokinetic sampling efficiency is related to other sampling conditions by using a relative sampling efficiency. The relative sampling efficiency may be measured in real time by the optical particle counter and multichannel analyzer. It is defined as the ratio of the particle concentration penetrating the inlet and being detected by the optical particle counter for a given sampling condition to that being detected during isoaxial-isokinetic sampling. The overall sampling efficiency is the product of the isoaxial-isokinetic transmission efficiency and the relative sampling efficiency for a given set of conditions.

Preliminary testing using the 1/4 inch outer diameter inlet have shown that the following are true:

1. For particles above 10 micrometers in diameter there is a difference in sampling efficiency for upward and downward sampling at +/-15 degrees and at +/- 30 degrees. This is suspected to be due to the interplay of inertial and gravitational forces. For upward sampling the two are synergistic and for downward they are antagonistic.

2. Figure 1 shows the effect of sampling efficiency as a function of particle size for various angles to the mean wind at a velocity of 500 cm/sec. The data are plotted for different sampling ratios, R, defined as the ratio of wind velocity to the average air velocity of the inlet. At isoaxial sampling the curves for different R values cross over each other at a particle size of about 30 micrometers due to the dominating effect of gravitational settling on large particles at low inlet velocities. At an angle of 60 degrees, a different cross-over occurs at a much smaller particle size. It appears that impaction onto the inner wall increases as the angle between inside wall and wind direction increases. When impaction dominates, losses are greatest for conditions of inlet velocity less than wind velocity for which the particle trajectory has the least opportunity for turning into the inlet. More than

half the losses due to impaction in all experiments occurred within the first one centimeter of the tube.

3. Figure 2 compares solid and liquid particles for a fixed inlet
    geometry. At the angle of 30 degrees it is obvious that solid
    particles greater than 10 micrometers are much more efficiently
    sampled than liquid particles. This is due to particle bounce.
    Liquid particles adhere to the inlet walls while the solid parti-
    cles are transmitted through.

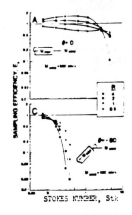

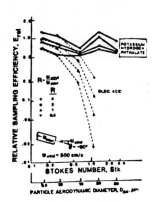

Figure 1. Change in the balance   Figure 2. Change in relative
          of gravitational forces   sampling efficiency
          and inertial forces with   for solid and liquid
          sampling angle.   particles.

     Currently, the wind tunnel facility is being used to study inlets of
different diameters. A series of interchangeable inlets has been built
with interior diameters of 1/8, 7/32, 13/32, and 20/32 inch (3.2, 5.6,
10.3, and 15.8 mm). Similar tests will be made with these inlets as were
made in the preliminary study inlet. Relative and transmission efficiences
will be obtained for the sampling angles and velocities previously studied.
Two types of inlet faces will be examined for each inlet I.D. sampler.
One will be a classical thin walled inlet with an apex angle of 15 degrees
and an inlet face ratio of the outer diameter to the inner diameter of 1.1
or less. The other will have a face ratio of 2.0. By using solid and
liquid particles at the same fixed sampling positions entry efficiency will
be studied.

A hot-film anemometry system has also been added to the system to better characterize wind conditions in the tunnel. Velocity and turbulence measurements will be made to test the influence of turbulence levels on sampling efficiency. Preliminary testing with turbulence levels of 3 to 8% have shown no discernible effects. Screens have been introduced before the mixing chamber to increase turbulence levels to 15% of the mean wind velocity. Further testing will be performed with this enhanced capability.

References:

Willeke, K and Tufto, P., "Sampling Efficiency Determination of Aerosol Sampling Inlets." Aerosols in the Mining and Industrial Work Environments, V. I, pp.321-346, Marple, V.A., and Liu, B.Y.H. eds., Ann Arbor Science, Ann Arbor (1983).

Tufto, P. and Willeke, K. "Dynamic Evaluation of Aerosol Sampling Inlets." Environ. Sci. Technol. 16:607-609 (1982).

Published 1984 by Elsevier Science Publishing Co., Inc.
Aerosols, Liu, Pui, and Fissan, editors

DEPOSITION OF AEROSOL PARTICLES IN A LIQUID LINED TUBE

C. S. Kim and M. A. Eldridge
Aerosol Research Laboratory
Division of Pulmonary Disease
Mount Sinai Medical Center
Miami Beach, FL 33140
U.S.A.

EXTENDED ABSTRACT

Deposition of aerosol particles increases in turbulent flow and the deposition further increases with incrasing surface roughness. When a liquid
layer is formed on the inside surface of a tube and undertakes wavy and
vibratory motion, this creates flow turbulence as well as surface roughness.
Enhanced aerosol deposition in the tube is thus expected. This situation
is related to a number of important engineering problems in two-phase gas-
liquid flow as well as deposition of inhaled medications or other aerosols
in patients with excessive mucus lining in the lung.

We measured aerosol deposition in a tube with a uniform lining of liquid
layer moving in the direction of aerosol flow. The liquid layer was made
with several different liquid solutions with varying degrees of viscoelasti-
city which were prepared by dissolving polyethylene oxide powder in filtered
and deionized water. The deposition tube was made by connecting two identical
glass tubes (1.0 cm I.D. and 20 cm length) in series with a 1 mm gap between
the tubes so that a liquid could be supplied into the tube through the circum-
ferential gap. The tube assembly was positioned vertically with the lower
tube connected to the top of the aerosol chamber. Initially, a liquid was
supplied into the tube at a constant rate by a syringe pump while a clean
air of 20 - 70 L/min was drawn through the tube by a vacuum pump. When the
liquid moved upward and covered the entire surface of the upper tube, the
air supply was stopped and an oleic acid monodisperse aerosol (3 and 5 um
diameter) tagged with uranine generated by a vibrating orifice generator
was supplied into the aerosol chamber. Aerosol sampling through the liquid
lined tube was continued for 5 - 20 min at the same flow rate as the airflow
used for the initial liquid layer formation in the tube. Particles escaping
from the liquid lined tube were collected on a 47 mm diameter membrane filter
(1.0 um pore size) placed immediately after the tube. The amount of particle
mass entering the tube was determined by sampling particles simultaneously
on a filter through a glass tube identical to the entrance section of the
liquid lined tube. Total deposition of the aerosol in the liquid lined tube
was determined by the difference in mass collected between the two filters.
The pressure drop across the liquid lined tube was measured by a pressure
transducer and recorded continuously on a strip chart recorder. Liquid layer
thickness and linear transport speed were also measured.

The liquid layer formed 0.8 - 1.6 mm thickness and moved in average at
0.6 to 2.5 cm/min speed with various types of wave motion, depending upon
aerosol flow rate and the type of liquid used. Aerosol deposition ranged
from 15 to 95% with 3 um particles and from 54 to 97% with 5 um particles.
These were about 30 times higher deposition than in the dry tube. Aerosol
deposition was lower with less viscous and more elastic liquid layer than
with more viscous and less elastic liquid layer and this was related to less

174

pressure drop across the tube with less viscous and more elastic liquid layer.
Althouth deposition of particles of a given size was uniquely related to pressure drop across the tube without regard to the liquid layer conditions, the
deposition curve as a function of pressure drop failed to unify particle size
effect.   Reynolds numbers of aerosol flow ranged from 4000 to 12000 after
a reduction of tube diameter due to liquid layer was taken into account,
indicating that air flows were in the turbulent flow regime.   Despite the wide
variation of Reynolds number, deposition data were well correlated with turbulent deposition parameter, viz dimensionless deposition velocity and dimensionless relaxation time and agreed very well with those of fully developed
turbulent flow in a smooth tube (Friedlander et al, 1957; Liu et al, 1974)
(Figure 1).   This suggests that deposition of aerosol particles in a tube
with irregular liquid lining can be predicted by turbulent deposition theory
with deposition parameters calculated by using a measured pressure drop across
the tube.

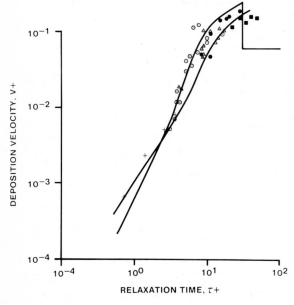

Figure 1: Aerosol depo-
sition data (symbols) ob-
tained from the liquid
lined tube in comparison
with turbulent deposition
theories (solid lines)

References

Friedlander, S.K. and Johnstone, H.F.   (1957) "Deposition of suspended particles from turbulent gas streams", Ind. Eng. Chem. 49: 1151.

Liu, B.Y.H. and Ilovi, T.A.   (1974) "Aerosol deposition in turbulent pipe
flow", Environ. Sci. Technol. 8: 351.

Published 1984 by Elsevier Science Publishing Co., Inc.
Aerosols, Liu, Pui, and Fissan, editors

DEPOSITION IN SAMPLING TUBES

J. B. Stenger and R. A. Bajura
Mechanical and Aerospace Engineering
West Virginia University
Morgantown, WV 26506-6101

EXTENDED ABSTRACT

Introduction

Aspiration probes are often used to obtain samples of flowing aerosol streams. The suspended particles are drawn into the sampling probe inlet and transported to a collection station. An analysis of the particle size and number distribution of the collected sample will lead to erroneous conclusions concerning the characteristics of the aerosol in the free stream unless corrections are applied to the data to account for deposition losses in the transport lines. The purpose of the present research program is to develop effective correlation techniques to estimate the free stream concentration of an aerosol based upon an analysis of the particulate obtained at the collection station, the characteristics of the transport line, and the flow conditions.

Model for Deposition Process

A numerical model was developed to predict deposition in a straight, horizontal sampling line. It is assumed that a uniform velocity profile exists at the probe inlet and the sampling is performed isokinetically. The concentration of solids is considered sufficiently dilute so that the motion of the particle does not affect the flow field. Once a particle is transported to the wall, deposition is assumed to occur without reentrainment. The flow in the sampling tube can be either laminar or turbulent.

The model predicts particle trajectories based upon the combined effects of gravity, Stokes drag, and Magnus and Saffman lift forces. Boundary layer analysis is used to calculate the fluid velocity field from the sampling probe inlet to the location where the velocity profile becomes fully developed. The trajectory of the particle is traced from an arbitrary initial position in the tube cross-section at the probe inlet to the point where deposition occurs on the tube wall.

Experimental Facility

An experimental facility was constructed to generate a dilute monodisperse glass bead aerosol for aspiration into a stainless steel sampling tube. The pipe Reynolds number was varied from 2,000 to 10,000. The probe was lined with an aluminum foil sleeve insert. After testing the foil was removed and sectioned to determine both the weight and size distribution of particles deposited on the wall. The remainder of the particulate was collected on a standard 47 mm absolute filter element. Experiments were performed under isokinetic sampling conditions. The velocity profile in the immediate upstream vicinity of the probe inlet was verified to be uniform by hot wire anemometry studies.

Results

Dimensional analysis and the flow equations show that the deposition process is governed by a particle Stokes number, the pipe Reynolds number, a Froude number based upon the mean velocity and the pipe diameter, the air/particle density ratio and the particle/pipe diameter ratio. A series of parametric studies was completed using the numerical model. Typical results of the study are illustrated in Figures 1 and 2. These results predict the distance particles in a given cross-section of the sample line will travel downstream from the inlet (in terms of pipe diameter) before deposition on the wall. Figure 1 shows the results obtained with the complete flow model compared to the predictions of a less realistic model which assumed fully developed flow at the pipe inlet and considers only Stokes drag acting on the particle. The simple fully developed flow model underpredicts the deposition distance for the particles in the lower half of the tube and overpredicts the deposition distance for particles in the upper half of the tube, compared to the more complete developing flow model. The difference between these two models is primarily attributed to the action of the transverse velocity component in the boundary layer of the developing flow region. Figure 2 illustrates the deposition distances for particles having different Stokes numbers.

The predictions of the numerical model were compared to the experimental results obtained from the test facility. For a pipe Reynolds number of 3300 and a particle Stokes number of 0.23, the experimental results showed that 12% of the entering particulate deposited on the pipe wall before reaching the collection filter. The numberical model predicted that 14% of the entering particulate would deposit on the wall, which shows good agreement with the experimental data.

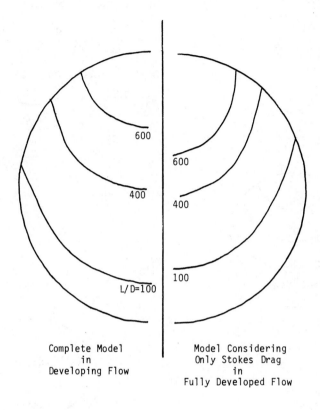

Complete Model
in
Developing Flow

Model Considering
Only Stokes Drag
in
Fully Developed Flow

Re=3300
St=0.3

Figure 1. Comparison of Predicted Deposition Length Contours

178

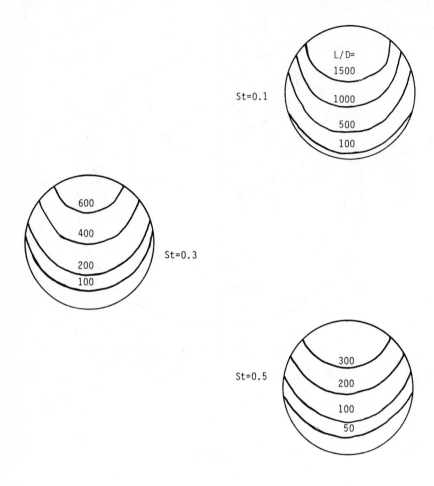

Figure 2. Predicted Deposition Length Contours for Various
Stokes Numbers

Published 1984 by Elsevier Science Publishing Co., Inc.
Aerosols, Liu, Pui, and Fissan, editors

Particulate Sampling Efficiency of Thin and Thick-Walled Probes
in Highly Turbulent Flow Streams

by

S. A. Lutz
U. S. Department of Energy
Morgantown Energy Technology Center
P. O. Box 880
Morgantown, West Virginia 26505

and

R. A. Bajura
Department of Mechanical and Aerospace Engineering
West Virginia University
Morgantown, West Virginia 26506

EXTENDED ABSTRACT

An experimental laboratory study has been performed to determine the effects of
wall thickness and highly turbulent sampling conditions on the collection efficiency of sampling probes for aerosols of various sized particles. The experiments were conducted in a four-inch ID pipe where the average flow velocity was
approximately 16 ft/sec (nominal pipe Reynolds number of 34,000). The aerosols
to be sampled were generated by fluidizing a mixture of glass beads. The polydisperse glass beads had mass mean diameters of 3.1 and 9.0 microns which yielded particle Stokes numbers of 0.03 and 0.24, respectively.

Turbulence in the flow was generated by an orifice upstream from the sampling
station. Streamwise mean velocities and turbulent properties of the carrier gas
were measured by a single channel hot wire anemometer. The integral time scale
was determined at free stream conditions and the resulting computed integral
length scale was on the same order as the nozzle diameter. The amplitude probability density distribution for the free stream sampling conditions showed that
the turbulence became more homogeneous as the streamwise turbulent relative intensity was increased.

Streamwise mean and turbulent velocities were measured upstream from thin and
thick-walled aspiration nozzles at probe sampling velocities corresponding to
conditions below, at, and above the isokinetic value. The measurements were
taken in flows with 19 percent, 12 percent, and 5 percent streamwise turbulent
intensity, respectively. The aspiration probe began to alter the free stream
conditions approximately one probe diameter upstream from the nozzle inlet. The
streamwise turbulence and mean velocities became increasingly distorted as the
flow approached the nozzle leading edge. This effect was more pronounced for
the thick-walled nozzle. All experiments were conducted for speeds in the range
from 30 percent to 300 percent of the isokinetic sampling velocity.

Particle concentration measurements were obtained from filter samples taken
through both thin and thick-walled probes (outer-to-inner diameter ratios of
1.01 and 1.74, respectively. When sampling at the three turbulent conditions

studied, no effect of turbulence on the sampling efficiency of either the thin or thick-walled probes could be observed within the uncertainty range of the measurements. A flow model was proposed which indicates that higher turbulence levels should result in lower sampling efficiency for sampling conditions below isokinetic speeds in view of the increased degree of homogeneity in the free stream conditions which were observed for higher turbulence levels. Trends in the data support this model, but the measurement uncertainty precludes drawing firm conclusions. The measurements, however, strongly support the conclusion that the effect of turbulence on the probe sampling efficiency is minimal, especially when compared to other sources of measurement uncertainty present in typical process stream applications.

Measurements with the thin-walled probe at a Stokes number of 0.24 were in close agreement with the theoretical predictions of Svinos, el al (1), for both above and below isokinetic sampling conditions as shown in Figure 1. Figure 2 illustrates that for a Stokes number of 0.03, the sampling efficiency was markedly below the theoretical predictions for speeds below the isokinetic value. The sampling efficiency observed experimentally was in close agreement with the theoretical predictions for speeds above the isokinetic value.

The probe leading edge thickness produced an observable change in the sampling efficiency. Comparison measurements were made with the thin and thick-walled probes operating simultaneously in identical free stream conditions in the test facility. These results are shown in Figures 3 and 4. When sampling at iso-kinetic speeds, the thick probe sampling efficiency was the same as for the thin probe for the Stokes number 0.03 flows. For the Stokes number 0.24 flows, the sampling efficiency of the thick probe was approximately 12 percent higher than the thin probe. These results indicate that the smaller particles follow the streamlines more faithfully and are not affected by the probe thickness. The larger particles are strongly deviated in their trajectory when the leading edge is thick. A higher sampling efficiency is obtained which is due to the larger effective inlet diameter of the thicker probe and particle rebound effects from the leading edge. Over the range of velocities studied, the overall efficiency of the thick-walled sampling probe is closer to 100 percent than for the thin-walled probe. Therefore, if no corrections are applied to the sampling data, the thick-walled probe yields particle concentration measurements which are closer to the true value for an arbitrary set of sampling conditions than the thin-walled probe.

References

1. Svinos, J. G., Bajura, R. A., and Padhye, A., "Anisokinetic Sampling of Particulate in Duct Flows," Final Report, Department of Mechanical and Aerospace Engineering, West Virginia University, Morgantown, WV, Feb. 1982.

2. Lutz, S. A., "Particulate Sampling with Thin- and Thick-Walled Nozzles in Highly Turbulent Flow Streams," Ph.D. Thesis, West Virginia University, 1984.

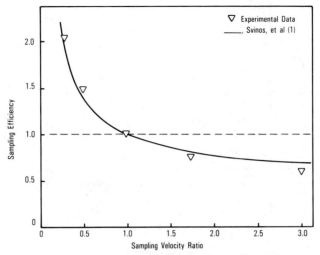

Figure 1. Nozzle Sampling Efficiency Versus Dimensionless
Sampling Velocity for Stokes number, St = 0.24.

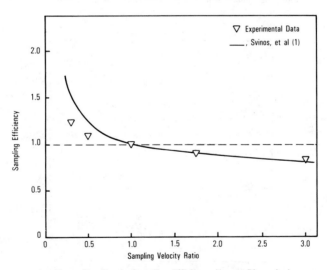

Figure 2. Nozzle Sampling Efficiency Versus Dimensionless
Sampling Velocity for Stokes number, St = 0.03.

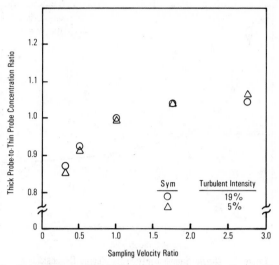

Figure 3. Thick-Walled-to-Thin-Walled Concentration Ratio Versus Dimensionless Sampling Velocity for Stokes number, St = 0.03.

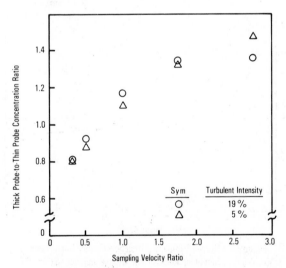

Figure 4. Thick-Walled-to-Thin-Walled Concentration Ratio Versus Dimensionless Sampling Velocity for Stokes number, St = 0.24.

Published 1984 by Elsevier Science Publishing Co., Inc.

Aerosols, Liu, Pui, and Fissan, editors

The Entry Efficiencies of Disc-Like Samplers

K. Y. K. Chung and T. L. Ogden

Health and Safety Executive
Occupational Medicine and Hygiene Laboratories
403 Edgware Road
London NW2 6LN
England

EXTENDED ABSTRACT

Introduction

Extensive research has been conducted on sampling efficiency as a function of particle size, wind speed and sampling flowrate (Badzioch, 1959; Sehmel, 1970; Raynor, 1970; Durham and Lundgren, 1980; Davies and Subari, 1982; and many others). Experimental studies have concentrated on evaluating samplers designs that already existed, rather than trying to elucidate which sampler parameters were most important in general. Consequently, it is not possible to use these results to predict efficiencies of new sampler designs, except in very general terms. This paper describes measurements for a very simple case, a disc with a central circular orifice through which the aerosol was aspirated, facing the wind. Efficiency was measured as a function of disc diameter (10 to 120mm), orifice diameter (5 and 10mm), sampling flowrate (0.5 to 4l/m), wind speed (1.7 to 4.5m/s) and particle aerodynamic diameter (3 to 13μm). The results were compared with recent theoretical studies to obtain empirical formula which might be used to estimate the behaviour of practical samplers of this type of design.

Theoretical Background

Fuchs(1975) summarised the results of Zenker(1971) and Belyaev and Levin(1974) for a thin-walled tube facing the wind, giving sampling efficiency:

$$E = 1 + (W/V - 1)\frac{(2 + 0.62\ V/W)St}{1 + (2 + 0.62\ V/W)St} \qquad \ldots\ldots(1)$$

for ranges $0.18 < W/V < 6$ and $0.18 < St < 2.0$.

Vincent and Mark(1982) generalised this result to a blunt sampler and obtained:

$$E = \left\{1 + \left[\left(\frac{D_D^2\ W}{D_0^2\ V}\right)^{1/3} - 1\right]K_1\right\}\left\{1 + \left[\left(\frac{D_0\ W}{D_D\ V}\right)^{2/3} - 1\right]K_2\right\} \qquad \ldots\ldots(2)$$

where $K_1 = g_1 St\ \left(\frac{D_D^2\ W}{D_0^2\ V}\right)^{1/3}\Bigg/\left[1 + g_1 St\ \left(\frac{D_D^2\ W}{D_0^2\ V}\right)^{1/3}\right]$

$$K_2 = g_2 St \left(\frac{D_D}{D_O} \frac{V}{W}\right)^{1/3} \bigg/ \left[1 + g_2 St \left(\frac{D_D}{D_O} \frac{V}{W}\right)^{1/3} \right]$$

St is here defined from the disc diameter, not the orifice diameter. Vincent(1984) proposed the experimental constants $g_1 = 0.15$ and $g_2 = 1.5$ or 5.0. However, better estimate of these parameters would permit use of equation (2) in predicting the efficiency of disc-like samplers facing the wind.

Experimental

An open-cycle 70x70cm wind tunnel capable of maintaining 0.5 to 4.5m/s low turbulence wind was used in carrying out the experiments. Monodisperse Uvitex-tagged DOP aerosols were generated with a May spinning top generator. The isokinetic sampler was a 10mm diameter ultra fine-edged conical tube attached to a 25mm Gelman filter holder type 1107. The test samplers of various outer diameters and orifice sizes were machined to fit the same filter holder type. Aerosol was captured on a 25mm Millipore type RA membrane filter. The resultant filters were analysed by spectrofluorimetry after dissolution in 15ml acetone. The fluorescence intensity ratio is the measure of sampling efficiency of the samplers.

Results and Discussion

The results are plotted as efficiency against aerodynamic diameter to bring out the effects of various other parameters. The solid lines are derived from Vincent and Mark's(1982) theory. Despite their scatter, a possible +/-20 percentage points, the results aid selection of the best values of g_1 and g_2 in equation (2), giving $g_1 = 0.25$ and $g_2 = 6$. The correlation coefficient between the measured efficiency and the calculated efficiency was 0.7. However the results are fitted almost as well by a much simpler but purely empirical expression:

$$E = 1 + \frac{(2.6 \ V/W)St}{1 + (2.6 \ V/W)St} \left[(W/V)^{1.66} - 1 \right] \quad \ldots\ldots(3)$$

with a correlation coefficient of 0.67.

It follows that in the range of variables studied, efficiency increases as St increases, i.e. as particle aerodynamic diameter or wind speed increases or as disc diameter decreases. Efficiency also increases as orifice diameter increases, or as intake flowrate falls. An 80mm diameter disc with a 5mm orifice aspirated at 2 l/min has an efficiency between 0.9 and 1.2 for particle aerodynamic diameters up to 14μm and windspeeds up to 4 m/s. Its efficiency is not very dependent on disc diameter or flowrate. This form of sampler is therefore to be preferred if isokinetic sampling is not possible.

References

Badzioch,S.(1959).Brit. J. Appl. Phys. 10:26-32
Belyaev,S.P. and Levin,L.M.(1974).J. Aerosol Sci.5:325-338
Davis,C.N. and Subari,M.(1982).J. Aerosol Sci.13:59-71
Durham,M.D. and Lundgren,D.L.(1980).J. Aerosol Sci.11:179-186
Fuchs,N.A.(1975).Atmos. Environ.9:697-707
Sehmel,G.A.(1970).Am. Ind. Hyg. Ass. J.31:758-771
Raynor,G.S.(1970).Am. Ind. Hyg. Ass. J.31:294-304
Vincent,J.H.(1984).Atmos. Environ. (in press)
Vincent,J.H. and Mark,D.(1982).Ann.occup. Hyg.26:3-19
Zenker,P.(1971).Staub 31:30-36

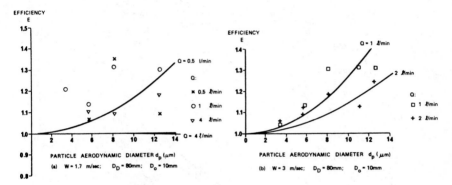

Fig. 1. Effect of sampler flowrate.

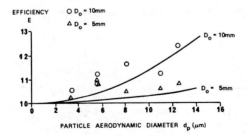

Fig. 2. Effect of varying orifice diameter
(W = 3m/sec; Q = 2l/min; D_D = 80mm).

186

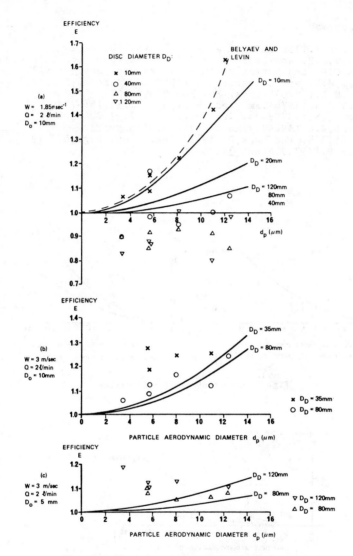

Fig. 3. Effect of varying disc diameter.

CHAPTER 6.

MEASUREMENT AND DATA ANALYSIS

Published 1984 by Elsevier Science Publishing Co., Inc.
Aerosols, Liu, Pui, and Fissan, editors

INVERSION OF AEROSOL PARTICLE SIZE DISTRIBUTION DATA

Kenneth A. Cox
Philip Morris U.S.A.
Research Center
P. O. Box 26583
Richmond, Va 23261
U. S. A.

EXTENDED ABSTRACT

The experimental determination of an aerosol particle size distribution, f, normally requires the solution of an integral equation of the form,

$$\underline{k} = \int_0^1 \underline{K}\, f\, dx.$$

The inversion of this equation to obtain the particle size distribution, f, is complicated by both the presence of experimental error in $\underline{k}$ and by the nonuniqueness of the solution. Frequently f is approximated by a linear combination of basis functions,

$$\hat{f} = \underline{c} \cdot \underline{\Psi},$$

and the coefficients determined by minimization of $s = \int_0^1 |\hat{f}''|^2 dx$ subject to the constraint,

$$||\hat{\underline{k}} - \underline{k}|| / \sqrt{n-1} = \sigma, \qquad \text{where } \hat{\underline{k}} = \int_0^1 \underline{K}\, \hat{f}\, dx.$$

This leads to a set of linear equations for $\underline{c}$. The constant, σ, is chosen so as to achieve a balance between the smoothness of $\hat{f}$ as measured by s and the agreement between $\hat{k}$ and $\underline{k}$. In the simplest case the basis functions are chosen to be step functions and $\hat{f}$ is then a histogram distribution. However, the method is more general. Crump and Seinfeld (1982) have recently discussed the use of a well chosen basis set of continuous functions permitting the determination of more realistic approximations to f.

An alternative to this approach is presented which appears to have several advantages. The integral equation is expressed in the equivalent form,

$$\underline{\phi} = \underline{\Phi}(o) f'(o) - \underline{\Phi}'(o) f(o) + \int_0^1 \underline{\Phi}(x) f''(x) dx,$$

190

where the functions, $\Phi_i(x)$, are mutually orthogonal and simply related to the kernels, K_i. The approximation of f'' by a linear combination of these functions,

$$\hat{f}'' = \underline{c} \cdot \underline{\Phi},$$

leads to explicit expressions for the coefficients $\underline{c}$ in terms of $\underline{k}$ and the boundary values $f(o)$ and $f'(o)$. Determined in this way $\hat{f}$ is the smoothest solution to the integral equations consistent with the boundary conditions $f(o)$ and $f'(o)$. If the boundary conditions are unknown, they may be readily determined by further minimization of s. Furthermore, the smoothing conventionally carried out by proper selection of σ can be achieved here simply by truncating the basis set. Simple rules can be developed for determining the optimum number of functions to be kept.

References

Crump, J. G., and Seinfeld, J. H. (1982) "A New Algorithm for Inversion of Aerosol Size Distribution Data," Aerosol Sci. Technol. 1:15-34.

Crump, J. G., and Seinfeld, J. H. (1982) "Further Results on Inversion of Aerosol Size Distribution Data: Higher-Order Sobolev Spaces and Constraints," Aerosol Sci. Technol. 1:363-369.

Published 1984 by Elsevier Science Publishing Co., Inc.

Aerosols, Liu, Pui, and Fissan, editors

DISTORTION OF SIZE DISTRIBUTIONS BY PARTICLE SAMPLING INSTRUMENTS

P. Biswas, C.L. Jones, and R.C. Flagan
California Institute of Technology 138-78
Pasadena, CA 91125
U. S. A.

EXTENDED ABSTRACT

Introduction

Aerosols which contain volatile species or condensible vapors may be altered by changes in temperature, pressure, and vapor concentration. When such changes occur within aerosol sampling instruments, the measured size distribution may be distorted significantly. An extreme case is the low pressure impactor in which the pressure is reduced intentionally to facilitate collection of small particles. At the lower pressure, a water containing particle may shrink due to evaporation. If, on the other hand, the flow is accelerated to high velocity, aerodynamic cooling may lead to condensation of water vapor and particle growth. Sheath air flows are heated by pumps and electrical dissipation in optical particle counters. This again shifts the vapor equilibrium, possibly leading to a decrease in particle size due to evaporation.

This paper explores both theoretically and experimentally the distortion of particle size distributions in a number of commonly used aerosol instruments including cascade impactors, both conventional and low pressure instruments, and optical particle counters. Ammonium sulfate aerosols under humid atmospheric conditions are taken as a test case for this study.

Theoretical

When the pressure of an aerosol is decreased at constant temperature, the partial pressure of a vapor decreases proportionately. Since the equilibrium vapor pressure at the surface of the particle is determined by the particle composition and temperature, a change in the vapor pressure or in the temperature provides a driving force for vaporization. As vapor is lost, the droplet solution becomes more concentrated. This reduces the equilibrium vapor pressure at the particle surface, allowing the particle to stabilize at a smaller size. The time required for equilibration with changes in water vapor pressures at high relative humidities is relatively short. For situations where the vapor pressure is small, the time required for a new equilibrium to be established may be long compared to the residence time in the instrument. In the general case, the change in the particle size must be evaluated by integrating the particle growth equation for the particle's history as it is transported through the instrument. Such calculations have been made for aqueous ammonium sulfate aerosols in order to examine the distortions of the particle size distribution encountered in various instruments.

Experimental

Experimental studies of instrument induced changes in aqueous ammonium sulfate aerosols have been performed for several instruments. An ammonium sulfate solution was atomized with a constant feed rate atomizer, charge neutralized, dried in a silica gel diffusion column, and then classified with a Thermo Systems Model 3071 electrostatic classifier to produce a dry monodisperse aerosol. This aerosol was then humidified by passing it over a heated water bath. Dry, filtered air was then added to the aerosol stream to reduce the humidity to the desired level. Sample lines were insulated to maintain the aerosol at constant temperature. When a dry aerosol was to be produced, the relative humidity downstream of the diffusion column was not allowed to exceed the deliquescence point of ammonium sulfate, i.e., 81 percent. For impaction studies, the quantity of ammonium sulfate collected on the impactor stages was determined by the flash vaporization/flame photometric detection method of Roberts et al. (1974).

Results

1. Caltech low pressure impactor

The low pressure impactor developed and calibrated by Hering and coworkers (1978, 1979) has the potential to evaporate volatile particles as they enter the low pressure stages of the instrument. The last two stages of that impactor operate with sonic jet velocities, leading to a substantial temperature drop and particle growth. Figure 1 shows the measured size distributions for two different monodisperse, wet ammonium sulfate aerosols. The broken lines represent the distribution as it would be measured if there were no change in particle size. The solid lines present the measured distributions. The 0.2 micron particles of Fig. 1a are observed to shrink, being collected on the 0.075 to 0.11 micron seventh stage, rather than on the fifth and sixth stages as expected from calibration data. By the eighth stage, the pressure is so low that the net effect is a reduction in the relative humidity and a decrease in particle size. At the same relative humidity (95%), 0.024 micron particles which calibration data suggests would be carried through the impactor to the after-filter grow and are collected on the seventh impactor stage. Thus, particles differing by an order of magnitude in diameter are collected on the same impactor stage. This is an extreme case which is, nevertheless, of interest in light of current concern over acid deposition. The particle size distribution is also seriously distorted at relative humidities near the deliquescence point of the aerosol. The effect is less pronounced, but still significant, at lower relative humidities.

2. Conventional cascade impactor

Conventional cascade impactors are designed for operation with jet Mach numbers below 0.3, the range for which the flow may be considered to be incompressible. The temperature decrease corresponding to a Mach 0.3 flow is about $5^{o}C$ for an ambient temperature inlet flow. The pressure drop is about 6 percent. These changes are large enough to distort the particle size distribution significantly, as is shown by the theoretical predictions in Fig. 2. Thus, the conditions which assure that dynamic effects are small from a momentum point of view, do not assure unbiased aerosol sampling.

3. Optical particle counters

Commercial optical particle counters such as the Climet Particle Analyzer and the Royco Laser Particle Counter employ an air recirculation loop to supply the sheath air necessary to direct the flow of the aerosol sample to the focal point of the light collection optics. This air is derived from the sample, and is pumped through high efficiency filters before being introduced into the measurement volume coaxially with the sample flow. The pump, the electronics of the instrument, and the light source heat the sheath air, thereby shifting the equilibrium of the aerosol when the two flows are mixed. Filters supplied with some optical particle counters have been found to absorb water, leading to very slow response of the instrument to changes in the sample relative humidity. These factors again lead to distortions in the particle size distribution upstream of the measurement point.

Discussion

Pressure, temperature, and humidity changes in aerosol analysis instruments have been found to seriously distort the particle size distributions of aerosols which contain volatile species or condensible vapors. The biases are most pronounced for small particles at high relative humidities or near the deliquescence point of the aerosol material. Some of the problems are readily corrected by minor modifications of the instruments or restrictions of the operating conditions, while others such as pressure effects in low pressure impactors are inherent to the instrument design.

References

Roberts, P.S. and Friedlander, S.K. (1976) Atmos. Environ., 10, 403.

Hering, S.V., Flagan, R.C., and Friedlander, S.K. (1978) Environ. Sci. Techn., 12, 667.

Hering, S.V., Friedlander, S.K., Collins, J.J., and Richards, L.W. (1979) Environ. Sci. Techn. 13, 184.

Acknowledgement

Funding for this research was provided by the Office of Research and Development, U.S. Environmental Protection Agency, under Grant No. R-809191010. The conclusions represent the views of the authors and do not necessarily represent the opinions, policies, or recommendations of the U.S. Environmental Protection Agency.

194

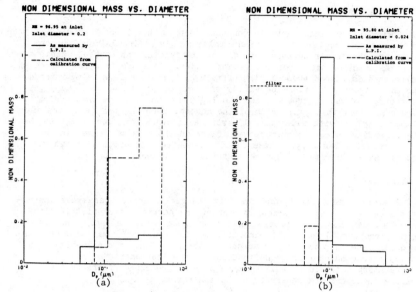

Figure 1.   Size distribution of two different monodisperse
$(NH_4)_2SO_4$ aerosols as measured by the low pressure
impactor.

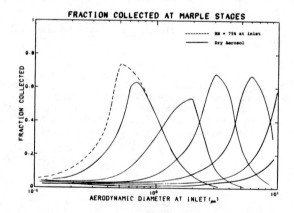

Figure 2.   Fraction of mass collected by six stages of the
Sierra-Marple Impactor.   (——) Dry aerosol.
(———) Aerosol at an inlet relative humidity of 75%.

Published 1984 by Elsevier Science Publishing Co., Inc.

Aerosols, Liu, Pui, and Fissan, editors

THE GENERAL RESULTS OF THE VIENNA WORKSHOPS ON ULTRAFINE AEROSOLS 1979 AND 1980, *) **)

O. Preining, Institute for Experimental Physics, University of Vienna, Strudelhofgasse 4, 1090 Vienna, Austria.

In Minneapoalis the development of the electrical mobility analysers (Whitby and Clark 1966), stimulated further research on ultrafine aerosols and led by the end of the 60ies, to Whitby's size acquisition system combining optical particle counters, an electrical mobility analyzer and a condensation nuclei counter (Whitby et al. 1972). Several systems were assembled and widely used, especially for air pollution studies. One of the very important results were found when averaging over many locally measured size distributions at very different locations and situations, i.e. when forming the "grand average". It simply turned out there are generally three (size) modes present in any (atmsopheric) aerosol: The nucleation, the accumulation and the coarse mode. With this system Whitby had closed for the first time the gap between submicron particles on the one hand and condensation nuclei respectively large molecules on the other(Whitby 1974); it permitted particle size analysis down to about of 10 nm by the mid 70ies.

In Vienna the Size Analysing Nuclei Counter, SANC, which is now a computer controlled and on line operated fast expansion cloud chamber for (among other things) acquiring size distributions in the particle size range between 5 nm and 120 nm was developed (Riediger 1971, Wagner 1973, 1979, Pohl and Wagner 1980). But the Minneapolis as well as the Vienna system relied for interpretation of the results on specific calibrations and assumptions (e.g. the charging probabilities for very small particles, their condensation behaviour and so on), so there were still some doubts about the results. Whitby especially felt an urge to broaden the platform on with the knowledge was based.

During the 1977 GAeF conference at Karlsruhe at the reception at the home of Prof. Schikarski, Christian Junge, Ken Whitby and I, discussed the state of the art of acquiring size distributions of very small particles and there the idea was born to organize a workshop on ultrafine aerosols and compare sizes and concentrations derived with the University of Minnesota system and Vienna SANC from the very same specially prepared and well defined test aerosols.

---

*)This paper is dedicated to the memory of Kenneth Whitby, who contributed so much to aerosol science and who deserves the greatest respect for his humanitarianism.

**) The work was supported by TSI Inc., St. Paul, Minnesota, MN 55164; USA; Max Planck-Gesellschaft, Munich, F.R.G.; University of Minnesota, Minneapolis, MN 55455, U.S.A.; National Bureau of Standards, Washington, D.C. U.S.A.; Fonds zur Förderung der Wissenschaftlichen Forschung, Vienna, Austria, Project No. 3481; EDV-Zentrum der Universität Wien, Abt. Prozeßrechenanlage Physik; and Institut für Experimentalphysik der Universität Wien,Vienna, Austria.

Since the SANC can not be transported the workshop had to be held in Vienna. Unfortunately neither Whitby nor Junge were able to participate; but Ben Liu and R. Jaenicke came instead. Three weeks found the following persons together with a host of instruments at Vienna: B.Y.H. Liu,D.Y.H. Pui, U. of Minnesota, and J.K. Agarwal, TSI, Minneapolis, R.L. McKenzie from NBS, Washington, D.C., R. Jaenicke from the Max Planck Institue,Mainz and F.G. Pohl, G. Reischl, W. Szymanski and P.E. Wagner from the University of Vienna.

The results, published in detail elsewhere(Liu et al. 1982, Liu et al. 1984), were essentially the following.
1) The concentrations measured by the TSI-Aerosol spectrometer and the SANC, which are both absolute instuments, agree well in the whole concentration range from 300 to 300 000 particles/cm^3, but there remains a systematic unexplained calibration difference, a factor of 0,59 much larger than the standard deviations of 6,7 % of the statistical errors. But, very important, the concentration readings of the two measuring systems based on entirely different principles are strictly proportional over the whole mentioned range of concentrations, (see fig. 1) (Liu et al. 1982).

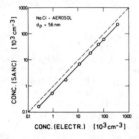

Fig.1: Number concentrations of monodisperse NaCl aerosols (of mean particle diameters 56 nm) measured by the SANC compared with aerosol electrometer measurements. (Liu et al. 1982)

2. Defining a size parameter, which is acquired by the SANC, the Kelvin equivalent diameter, i.e. the diameter of a droplet of pure water which just starts to grow at the supersaturation which is given by the Kelvin equation, opens the possibility of sizing aerosols by an expansion cloud chamber by exposing different samples of the same aerosol to different supersaturations, i.e. operating the cloud chamber consecutively at different expansion ratios. The results are cumulative size distributions of the investigated aerosols, see fig.2 The equivalent diameters of a not soluble and very likely only modestly hygroscopic DOP particle *) agrees well with the diameters determined by the electrical mobility analyzer. The situation is different for NaCl particles, these grow in the humidifier and the Kelvin sizes of particles with electrical mobility sizes of 13 nm become larger than 50 nm (Liu et al. 1983).

---

*) Since the DOP particles were produced by spraying an alcoholic solution (isopropanol) of DOP the surface properties are not really known and the particles can not be considered to be hydrophobic.

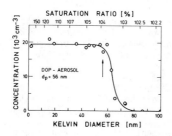

Fig. 2: Cumulative Kelvin-
equivalent size distribution
of a monodisperse DOP-aerosol
with an electrical mobility
particle diameter of 56 nm
(Liu et al. 1983).

This 1979 workshop stimulated the 1980 workshop, again at Vienna. Porstendörfer and H.G. Scheibel from Göttingen and F.G. Pohl, O. Preining, G. Reischl and P.E. Wagner from Vienna used the techniques developed at Göttingen (Scheibel and Porstendörfer, 1983,Porstendörfer et al. 1984) to produce monodispersed aerosols of very small particle sizes of the materials Ag and NaCl in the size range between 6 and 18 nm using N_2 as a carrier gas and the SANC as the analysing system to compare the electrical mobility diameters, d_e, and the Kelvin diameters, d_K, of the respective particles. This comparison is given in the table (Porstendörfer et al. 1984)

T a b l e :

Ag		NaCl		
d_e nm	d_K nm	d_e nm	d_K nm	
6	3,3	6	15	
8	3,9	8	21	
12	4,8	12	28	
18	7,2	18	38	

The elctrical mobility
diameter, d_e, and the Kelvin
diameter, d_K, of monodispersed
silver and sodium chloride
aerosols.
(Porstendörfer et al. 1984)

The electrical mobility diameters agree well with geometric diameters as determined by electron microscopy (Scheibel and Porstendörfer, 1983). These last results for Ag show that the Kelvin diameter for a substance with hydrophobic surface can be considerably smaller than the geometric diameter. (The results correspond to unsoluble particles whose material has an cosine of the contact angle with water of 0.8 according to a theory by Fletcher (1958), however no experimental values of contact angles of silver with water are available in the literature.

These results prove an old observation of Ken Whitby:Condensation nuclei counters do not always "see" all the (hydrophobic) particles.

Literatur:

Fletcher, N.H. (1958)
Size effect in heterogeneous nucleation, J. Chem. Phys. 29, 572

Liu, B.Y.H., D.Y.H. Pui, R.L. McKenzie, J.K. Agarwal, R. Jaenicke, F.G. Pohl, O. Preining, G. Reischl, W. Syzmanski, P.E. Wagner (1982) Intercomparison of Different "Absolute" Instruments for Measurement of Aerosol Number Concentration, (1982) J. Aerosol Sci.,13, 429 - 450

Liu, B.Y.H., D.Y.H. Pui, R.L. McKenzie, J.K. Agarwal, F.G. Pohl, O. Preining, G. Reischl, W. Szymanski, P.E. Wagner (1984) Measurements of Kelvin-Equivalent Size Distributions of Well-Defined Aerosols with Particle Diameters > 13 nm, Aerosol Science and Technology,3, 107-115

Pohl, F.G. and P.E. Wagner (1980), Measurements of Kelvin-Equivalent Size Distsributions of Urban Aerosols, J. Physical Chemistry 84, 1642

Porstendörfer, J., H.G. Scheibel, F.G. Pohl, O. Preining, G. Reischl,P.E. Wagner (1984) Heterogeneous Nucleation of Water Vapor on monodispersed Ag- and NaCl-particles with diameters between 6 and 18 nm, Aerosol Science and Technology, in press

Riediger, G., (1971) Ein automatischer Kondensationskernzähler mit Expansionsprogramm, Staub-Reinhaltung Luft 31, 237-239

Scheibel H.G., J. Porstendörfer (1983) Generation of monodispersed Ag and NaCl aerosols with particle diameters between 2 nm and 300 nm. Journal Aerosol Science 14, 113 - 126

Wagner, P.E. (1973) Optical determination of the size of fast growing water droplets in an expansion cloud chamber, J. Colloid Interface Sci., 44, 181-183

Wagner, P.E. (1979) Droplet growth kinetics and design principles for condensation nuclei counters., from "Aerosol Measurement", ed. D.A. Lundgreen, Univ. Press, Gainsville, Florida, 574-584

Whitby, K.T., W.E. Clark (1966) Electric aerosol particle counting and size distribution measuring system for the 0,0015 to 1 µ size range Tellus XVIII, 573-586

Whitby, K.T., B.Y.H. Liu, R.B. Husar and N.J. Barsic (1972) The Minnesota Aerosol-Analyzing System Used in the Los Angeles Smog Project, Journal of Colloid and Interface Science, 39, 136-164

Whitby, K.T. (1974), Modeling of Multimodal Aerosol Distibutions Aerosole in Naturwissenschaft, Medizin und Technik, V. Böhlau, H. Straubel ed's, Bad Soden/Taunus, Gesellschaft für Aerosolforschung, Kongreßbericht.

Published 1984 by Elsevier Science Publishing Co., Inc.

Aerosols, Liu, Pui, and Fissan, editors

APPLICATIONS OF A PERSONAL DUST MONITOR AND DATA MANAGEMENT WITH A MICROCOMPUTER

Takashi Yamaguchi
Applications Engineer
MDA Scientific, Inc.
1815 Elmdale Avenue
Glenview, IL 60025

EXTENDED ABSTRACT

Introduction

In the industrial hygiene field, personal dust samplers have been widely used to measure worker exposure to dust in the workplace. This information is used to improve environmental conditions and assist in reducing dust exposure levels.

However, because these measurements can only provide average concentration values and often take several days to evaluate, it is difficult for the industrial hygienist to (1) relate high dust levels to a particular process or work procedure and (2) take immediate action on proper engineering controls.

To compensate for this limitation in traditional dust measuring methodology, a continuous direct-reading dust dosimetry system was developed. It consists of a lightweight dust sensor, dosimeter, and portable readout unit and provides both realtime mass concentration readings as well as TWA data.

Personal Dust Monitoring System Description

The system consists of three components - PDS-1 Personal Dust Sensor, MDM-1 Minidosimeter, and Portable Readout Unit. Tables 1 and 2 show the specifications of the sensor and the dosimeter unit.

Table 1
PDS-1 Personal Dust Sensor

Range:	Dual range, switch selectable; 0.01 to 10 mg/m^3 0.1 to 100 mg/m^3
Sensitivity:	0.01 mg/m^3
Particle Size Range:	0.01 to 10 um
Calibration:	Built-in reference board
Power Source:	Rechargeable Ni-Cad Batteries
Operating Time:	9 hours
Dimensions:	100x90x37 mm
Weight:	650 grams

Table 2
MDM-1 Minidosimeter

Input Range:	0-100 mV
Resolution:	1 mV
Precision:	+/-0.5% Full Scale
Data Acquisition:	180 data points per minute; stores 800 1 minute averages
Display:	3-1/2 digit mass concentration from 0 to 10.00 mg/m
Data Output:	Serial (ASCII) TTL Level
Power Source:	Rechargeable Ni-Cad Batteries
Operating Time:	12 hours
Dimensions:	92x45x155 mm
Weight:	450 grams

A portable readout unit based on the battery powered computer system allows the industrial hygienist to easily analyze exposure information gathered with the PDS-1 and MDM-1. Using the menu-driven software program, one minute averages may be printed and minute-to-minute concentration profile and statistical data can be obtained, as shown in Figure 1.

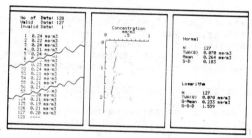

Figure 1

Calibration of the Sensor

The PDS-1 personal dust sensors were dynamically calibrated prior to the evaluation in accordance with the specifications, as shown in Table 3.

Table 3

Calibration Standard:	Arizona Road Dust
Particle Size:	Fine Grade
Concentration:	5.00 mg/m^3 +/-10%
Accuracy:	+/-10%
Ambient Conditions:	22.5 C +/-2.5 R.H. <70%

Lab Evaluation

In order to evaluate the performance of the PDS-1, a series of lab tests was conducted comparing the instrument with two commonly used gravimetric samplers. The three instruments were tested with three different varieties of dust, as shown in Table 4.

Table 4
Test Dusts

Particulate	MMD	CMD
Arizona Road Dust	10.4 micron	1.92 micron
Silica Dust	10.8 micron	2.08 micron
Coal Dust	11.6 micron	2.17 micron

Figure 2 illustrates the relationship between the PDS-1 and a gravimetric sampler using a type 113A elutriator, and Figure 3 illustrates the relationship between the PDS-1 and a personal dust sampler using 10 mm nylon cyclone.

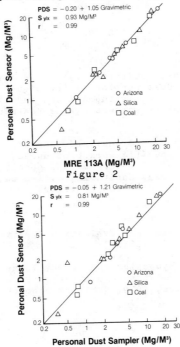

Figure 2

Figure 3

Field Evaluation

The Personal Dust Monitoring System was also tested under field conditions in coal mines. The following X-Y plot was obtained with the combination of dosimeter, portable computer, and X-Y plotter. Figure 4 illustrates the concentration at a drilling area in underground mines.

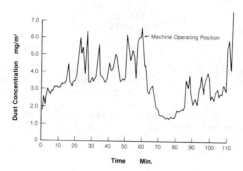

Figure 4

Conclusion

The Personal Dust Monitoring System is a viable alternative to traditional dust monitoring techniques and can, if properly used, significantly improve the industrial hygienist's ability to protect workers against harmful exposures to respirable dusts.

Exposure levels may be analyzed, periods of high concentrations related to specific work procedures, and immediate actions taken on atypical environmental conditions. As a result, the Personal Dust Monitoring System can be a powerful tool for respiratory protection programs.

References

"Air Sampling Instrument For Evaluation Of Atmospheric Contaminants" (5th edition), ACGIH.

"Proceedings of the Symposium on the Development and Usage of Personal Monitors for Exposure and Health Effect Studies", EPA-600/9-79-032, June, 1979, U.S. Environmental Protection Agency.

Published 1984 by Elsevier Science Publishing Co., Inc.
Aerosols, Liu, Pui, and Fissan, editors

Continuous Particulate Mass Emissions Monitor for a Coal-Fired Steam Generator

Gilmore J. Sem, Frederick R. Quant, and Glenn S. Nordell*
TSI Incorporated, P.O. Box 64394, St. Paul, MN 55164 USA
*Northern States Power Co., 414 Nicollet Mall, Minneapolis, MN 55401 USA

EXTENDED ABSTRACT

Introduction

Large coal-fired electric power generating plants must maintain particulate mass emissions below a legally-determined rate. The measurement method in the U.S. is EPA Method 5, a manual gravimetric filter-sampling technique which requires 2 well-trained technicians 4-6 hours to obtain the sample and several days to determine the mass emissions rate. Such time-consuming and expensive methods usually cannot be used to evaluate the effect of various process changes on the emissions rate. While transmissometers measure stack particle concentration in real-time, results often correlate poorly with Method 5.

A new system has been developed for continuous monitoring of particulate mass emissions from a large coal-fired electric power generating plant. The system includes a sample extraction and dilution system which conditions the aerosol for mass concentration measurement by an automated piezoelectric microbalance. A computer controls the system, collects the data, calculates and displays the stack concentration and mass emissions rate, and prints the results. The system has been operating nearly continuously on a large power plant equipped with wet scrubbers since August 1983.

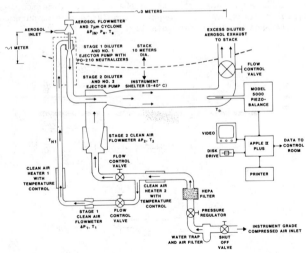

Figure 1. Continuous particulate mass emissions monitor with 2-stage dilution and computerized data acquisition and control.

System Description

The sample extraction and dilution system (Figure 1) isokinetically removes a sample of airborne emissions from the stack (Felix et al, 1981; Tsurubayashi, 1981). A cyclone removes the small fraction (< 4% by mass) of particles larger than 7 μm. A flowmeter measures the undiluted sampling rate (3-4 l/min). The first dilution stage also serves as an air ejector pump to draw the undiluted sample into the system. The first stage uses 45 l/min of dry, particle-free air heated to 65°C, above the dew-point temperature of the emissions. The first stage reduces the particle concentration by about 15X and reduces the dew-point temperature of the sample. A second 10X dilution stage, also an air ejector pump, uses about 450 l/min of dry, 20°C, particle-free air. It reduces the dry-bulb temperature (originally 85°C), the dew-point temperature, and the particle concentration to less than 1 mg/m³. The automatic piezoelectric microbalance (TSI Model 5000) then removes 1 l/min of aerosol isokinetically from the 500 l/min diluted air stream and measures its concentration. The remaining 499 l/min of diluted aerosol passes back into the stack.

An Apple II computer measures air temperatures and air flow rates every 30 seconds. It then calculates the dilution ratio for that set of measurements and also computes the average dilution ratio from the beginning of that run. When the measurement is completed, usually 8 minutes after its start, the computer calculates the average particle concentration and particulate emission rate during the measurement. The repeated calculation of flow rates every 30 seconds means the precise dilution ratio need not remain stable during a single run or during a 24-hour set of runs.

Figure 2 is one of the computer's screen displays showing the most recently measured or calculated parameters next to the corresponding components on the schematic representation. As the computer completes new measurements and calculations, it updates the screen display.

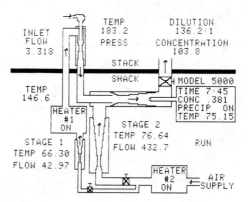

Figure 2. Computer screen display showing current data superimposed on a schematic diagram of the system.

System Performance

The validity of the measurement depends upon
* removal of representative aerosol sample from the stack
* accurate dilution of the aerosol
* minimum and well-characterized particle losses
* accurate sensing of diluted aerosol concentration

The satisfaction of all 4 of these factors would have been formidably more difficult had the particles been larger. Cascade impactor measurements indicated a mass mean diameter of 0.3 μm and less than 4% by mass above 7 μm, thus validating the predicted performance of the scrubber on this power plant.

Emissions enter the vertical stack through a horizontal duct located about 12 diameters upstream of the sampling point. The 10-meter-diameter stack carries highly turbulent effluent with a velocity of about 30 m/sec. The velocity profile at the sampling plane is very flat to within 30 cm of the wall. The aerosol inlet is sized for isokinetic sampling under full-load conditions, assuming straight-line stack flow. As with any "isokinetic" sampler, the high turbulence in the stack results in non-representative sampling of large particles. However, the system should sample the small particle sizes found in this stack representatively.

Accurate dilution requires accurate clean air flow measurement. The system includes a venturi meter with differential pressure and temperature transducers for each of the clean air flows. The computer scans the transducers every 30 seconds and averages these values for the full sampling period. Aerosol inlet flow is sensed by a differential pressure transducer across an expanding section of the isokinetic nozzle. The aerosol inlet flowmeter performs adequately except when high atmospheric winds cause pressure fluctuations in the stack. A better inlet flowmeter is needed. The ejector pumps assure good mixing of the aerosol and clean air by creating turbulent flow upstream of the piezobalance while shielding the aerosol from the walls until it is mixed and diluted.

Large particles can be brought to the wall and lost from the airstream by turbulent deposition, gravitational sedimentation and impaction. Conservative calculations estimate that less than 6% of 3 μm particles will be lost by a combination of these factors. Particles in the 0.3 μm range are not lost as easily as larger and smaller ones. Small particles may be lost by Brownian diffusion, but calculations estimate less than 4% of 0.02 μm particles will be lost by this mechanism. Electrostatic losses are difficult to estimate, but should be negligible because of two Po-210 neutralizers placed in the first ejector pump. Rough penetration experiments using room air verifies that greater than 90% of the particles penetrate the system. Experiments with atomized fluorescein (0.3 μm mass mean diameter) resulted in no concentrated fluorescein deposits in the sampling tubes. A cyclone removes particles larger than 7 μm aerodynamic diameter at the aerosol inlet, preventing them from plugging the system. Continuous operation for 4-8 weeks shows insignificant deposition in the sampling system except in the tube delivering undiluted aerosol from cyclone to first ejector pump. Even those deposits are small compared to the particles which pass through the tube. There are no signs of condensation anywhere within the sampling system.

The diluted aerosol conditions are nearly ideal for accurate measurement of the particle mass concentration by the piezobalance (Sem and Quant, 1983).

206

Figure 3 is a typical set of measurements obtained on 18 October 1983 by
this system, a transmissometer, and a set of 3 Method 5 replicates. Until 1345,
the piezobalance system operated on a cycle of 8-minute measurements followed by
2-minute sensor cleaning periods, resulting in one mass concentration
measurement every 10 minutes. The opacity data is the average measured value
during the corresponding 8-minute period. After 1345, the cycle was changed to
a single 8-minute measurement every half hour. Each Method 5 replicate sample
covered a little more than 1 hour during the periods shown. Opacity has been
arbitrarily scaled so that 25% opacity equals 0.050 lbs/MBTU. Piezobalance
measurements consistently fall a factor of 2 below Method 5 measurements so the
piezobalance data in Figure 3 is plotted twice the actual measured values.
While some of the discrepancy between Method 5 and the piezobalance system may
be caused by the losses in the sampling system, we believe the majority is
caused by more basic differences between the 2 measurement methods. While
piezobalance and opacity measurements rise and fall together for factor-of-two
changes in emission rate, they do not track closely for smaller, short-term
fluctuations. This may be because the piezobalance system measures the mass
concentration of particles reaching it while the transmissometer measures
opacity, a light-scattering technique characteristic of the aerosol.

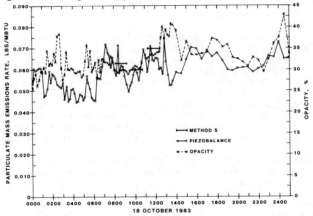

Figure 3. Particulate mass emissions rate (LBS/MBTU = pounds per mega-British
thermal unit) measured by the new system and by Method 5 and opacity
measured by a transmissometer on 18 October 1983.

References

Felix, L.G., R.L. Merritt, J.D. McCain, and J.W. Ragland (1981) "Sampling and
Dilution System Design for Measurement of Submicron Particle Size and
Concentration in Stack Emissions Aerosols", TSI Quarterly VII, Issue 4, p. 3.

Sem, G.J., and F.R. Quant (1983) "Automatic Piezobalance Respirable Aerosol Mass
Monitor for Unattended Real-Time Measurements", Aerosols in the Mining and
Industrial Work Environments, Vol 3: Instrumentation (Edited by V.A. Marple and
B.Y.H. Liu), Ann Arbor Science, Ann Arbor, Michigan.

Tsurubayashi, K. (1981) Personal communication related to ejector pumps as
diluters, Nihon Kagaku Kogyo Co. Ltd., Osaka, Japan.

Published 1984 by Elsevier Science Publishing Co., Inc.

Aerosols, Liu, Pui, and Fissan, editors

PERFORMANCE OF THE AERODYNAMIC PARTICLE SIZER
UNDER DIFFERENT CONDITIONS*

Yung-Sung Cheng, Bean T. Chen and Hsu-Chi Yeh
Inhalation Toxicology Research Institute
Lovelace Biomedical and Environmental Research Institute
P. O. Box 5890
Albuquerque, NM 87185
U. S. A.

EXTENDED ABSTRACT

The aerodynamic particle size analyzer (APS) (TSI, St. Paul, MN) is a real-time sizing instrument based on the principle of accelerating particles in a nozzle and measuring their inertial lag behind the air. The magnitude of the lag which depends on the aerodynamic diameter of the particle, is determined by the transit time of the particles between two split laser beams. A calibration curve relating the transit time and aerodynamic diameter is provided by the manufacturer under a certain set of operating conditions (temperature, pressure, flow rate, and pressure drop across the nozzle) and, furthermore, these conditions are interrelated. For example, at reduced ambient pressure, the flow rate and pressure drop cannot both be maintained at the calibrated values. Thus, a new calibration curve is needed in each operating condition. This paper described performance of the APS under different operating conditions and a way of establishing a new calibration curve without actually performing the calibration. The effect of particle density on the calibration is also studied.

The APS was calibrated by the manufacturer at a total flow rate of 5 L/min (1 L/min of aerosol flow and 4 L/min of sheath flow), a pressure drop of 97 mmHg, ambient pressure of 740 mmHg and room temperature (20°C). Monodisperse aerosols of PSL, DOP, oleic acid, and fused aluminosilicate particles (FAP) were used in the calibration of the APS operated at reduced ambient pressure (620 mmHg) and room temperature (20°C). Three flow rates, which correspond to the same volumatric flow, mass flow, and pressure drop as the manufacturer's calibration (740 mmHg), were performed (Table 1). Calibration curves were different from the manufacturer's with the one operating at the same pressure drop being the closest to the original calibration (Figure 1). Deviation of the calibration curve caused by different densities of the aerosol was not found to be significant in the range of 0.89 (oleic acid) to 2.3 (FAP).

In an effort to combine calibration curves obtained under different operating conditions into a single curve, the Stokes number at nozzle exit and the ratio of the particle velocity to and gas velocity at nozzle exit were calculated. The gas velocity was calculated using Bernoulli's equation of motion for compressible, invicid plug flow of an ideal gas. The particle velocity was estimated from the transit time of an aerosol traversing the

---

*Research performed under U. S. Department of Energy Contract Number DE-ACO4-76EVO1013.

distance between the two-spot laser beam. The transit time was calculated
from the channel number given by the multi-channel accumulator and the
distance between two spots of the laser beam was estimated to be 124 µm.
A single curve unifying all four calibration curves was obtained by plotting
velocity ratio versus Stokes number as shown in Figure 2. With this
dimensionless curve and the manufacturer's calibration, one can generate
accurate calibration curves at different operating conditions without
performing the actual calibrating (Figure 3).

In summary, our results show that calibration of the APS is needed for
each operating condition for accurate determination of the particle
aerodynamic diameter. The calibration curves under different conditions can
be represented by a single curve relating the normalized particle velocity
and Stokes number at nozzle exit. Furthermore, a procedure was described to
generate a calibration curve of a given operating condition without
performing the experimental work.

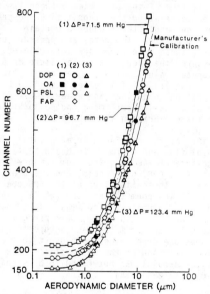

Figure 1.   Calibration curves of the APS for three pressure drops across the
nozzle: (1) ΔP = 71.5 mmHg, (2) ΔP = 96.7 mmHg, (3) ΔP =
123.4 mmHg. Ambient pressure P = 620 mmHg, Temperature θ =
20°C. The particle densities of the materials are 0.986 g/cm^3,
0.894 g/cm^3, 1.00-1.05 g/cm^3, and 2.2 g/cm^3 for DOP, OA,
PSL, and FAP, respectively.

209

Table 1

Three sets of operating conditions were used to calibrate the APS at a
temperature of 20°C. Each set (1, 2, and 3) of conditions corresponds to
the condition (constant mass flow, constant pressure drop, and constant
volumetric flow, respectively) of the manufacturer's calibration.

Set of Conditions	Ambient Pressure (mmHg)	Pressure Drop (mmHg)	Total Flow Rate (L/min)	Dilution Ratio
1	620	123.4	6.3	4:1
2	620	96.7	5.7	4:1
3	620	71.5	5.0	4:1

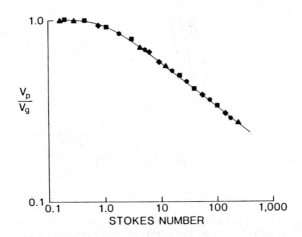

Figure 2.   The generalized curve, which is independent of the flow
conditions, can be used to predict calibration curves for the
APS. ( ) P = 620 mmHg, ΔP = 96.7 mmHg; (0) P = 620 mmHg,
ΔP = 71.5 mmHg; (□) P = 620 mmHg, ΔP = 123.4 mmHg;
(Δ) P = 740 mmHg, ΔP = 96.7 mmHg [conditions used by the
manufacturer].

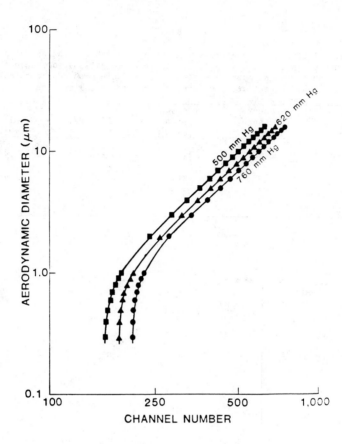

Figure 3. Calibration curves established by the unique curve shown in
Figure 2. Three different ambient pressures (500, 620, 760 mmHg)
were used (ΔP = 96.7 mmHg, θ = 20°C).

Published 1984 by Elsevier Science Publishing Co., Inc.
Aerosols, Liu, Pui, and Fissan, editors

MEASUREMENT OF AEROSOLS IN EXPLOSIVE ENVIRONMENTS*

G. J. Newton[a] and R. P. Sandoval[b]

Lovelace Biomedical and Environmental Research Institute[a]
Inhalation Toxicology Research Institute
P. O. Box 5890
Albuquerque, NM 87185
U. S. A.

Sandia National Laboratory[b]
P. O. Box 5800
Albuquerque, NM 87185
U. S. A.

EXTENDED ABSTRACT

As part of national defense activities, potentially toxic heavy metals and high explosives are assembled into units inside safety enclosures. These safety enclosures are underground reinforced concrete, 34-ft (I.D.), circular rooms with a roof consisting of a cable network that supports a gravel bed 15 to 20 feet thick (Fig. 1). If an untoward event detonates the high explosive, the gravel bed will be expanded and function as a granular bed filter to control the airborne release. A recent safety test at the Nevada Test Site conducted by personnel from Sandia National Laboratory with assistance from the Inhalation Toxicology Research Institute measured the penetration of explosively generated UO_2 through the gravel bed.

One aspect of this research effort required quantification of the source term by obtaining a measurement of UO_2 airborne mass and size distribution inside the underground round room immediately after detonation of several hundred kilograms of high explosives.

We designed a blast-proof aerosol sampling package to measure the source term before collapse of the gravel bed. A total of six gravel aerosol sampling packages (GASP) were located inside the test structure. Four were placed inside the round room and two were placed in storage bays at some distance from the round room.

Each sampling package consisted of a sampling section and a shield designed to withstand blast, fragments and collapse of the gravel bed. The samplers consisted of a 20-L evacuated chamber attached to a measurement section (Fig. 2). Just after detonation, a timer opened a valve that allowed aerosol to enter the evacuated chamber through a critical orifice designed to fill the 20-L chamber within the 3-sec. before collapse of the entire structure. After the 3-sec sample capture period, the valve was closed. Timers then actuated the pump located in the measurement section

---

*Research performed under U. S. Department of Energy Contract Number DE-AC04-76EV01013 and DE-AC04-76DP00789.

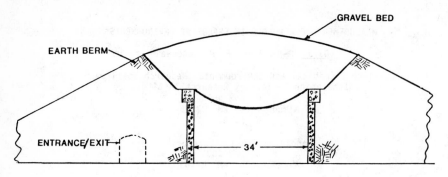

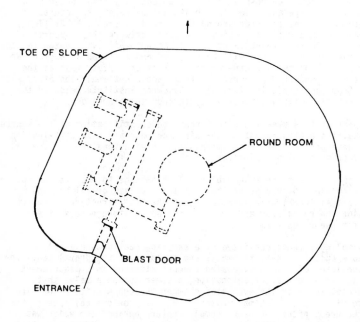

Figure 1. Sketch of the underground containment structure. Upper view is an elevation and lower view is overhead.

and the 20-L aerosol sample was processed through filters and Mercer style cascade impactors (Mercer et al., 1970).

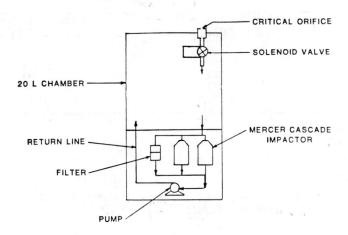

Figure 2.   Schematic diagram of the aerosol sampling package (GASP) Material is 1-in type 304 stainless steel. Blast shield is not shown.

Recovery and analyses of samples occurred several days after detonation when the Gravel Gertie structure was excavated to recover the sampling packages.

Table 1 shows the results of the aerosol measurements obtained with the GASP's. The data were analyzed by two techniques. The critical flow method used the assumptions for mass flow through a critical orifice and the second method used a fraction of the filled volume as determined by the pressure measurements in the individual GASP sampling chambers. Both methods agreed within the errors associated with each type of measurement.

Using this blast-proof sampling package, we determined that approximately 185 g of the uranium dioxide was released as a respirable sized aerosol. This was in agreement with theoretical calculations that predicted approximately 175 g would be aerosolized.

214

Table 1. Results of GASP Source Term Measurements Inside
the Underground Round Room

GASP No.	Critical Flow Method U Mass (g)	Fraction of Filled Volume U Mass (g)
1	217	252
2	230	216
3	180	144
4	147	180
Ave Mass	193 ± 38	198 ± 46

MMAD = 2.5 μm
σ_g = 3.0

Reference

Mercer, T. T., M. I. Tillery and G. J. Newton (1970) "A Multi-Stage, Low Flow Rate Cascade Impactor", $\underline{J. Aerosol Sci.}$ $\underline{1}$:9-15.

Published 1984 by Elsevier Science Publishing Co., Inc.
Aerosols, Liu, Pui, and Fissan, editors

AERODYNAMIC PARTICLE SIZER CALIBRATION AND USE

P. A. Baron
Division of Physical Sciences and Engineering
National Institute for Occupational Safety and Health
Cincinnati, OH   45226
U. S. A.

EXTENDED ABSTRACT

The Aerodynamic Particle Sizer (APS) from TSI, Inc. has been used at NIOSH
for over two years, beginning with a prototype model and more recently with
a commercial model. The APS has been tested and used in a variety of
situations.

Calibration of the APS was originally carried out at the factory using
monodisperse latex particles less than 3 μm diameter and monodisperse DOP
droplets from a Vibrating Orifice Monodisperse Aerosol Generator (VOMAG,
TSI, Inc) at larger sizes. Concern for distortion of the droplets in the
high acceleration flow field of the APS sensor led to measurements of
droplet shape in the sensor region using a laser imaging system (Laser
Holography, Inc). It was observed that oleic acid droplets distorted into
oblate spheroids while water droplets did not noticeably distort at
standard APS conditions. The difference is apparently due to the greater
surface tension of the water droplets. The APS was recalibrated using
monodisperse latex particles at 5, 7, 10, and 15 μm. With the new
calibration, DOP and oleic acid droplets that were calculated to be 15 μm
were indicated by the APS to be approximately 25% smaller.

It had been predicted that particles larger than 10 μm with same
aerodynamic diameter but different density would accelerate differently in
the APS sensor (Wilson and Liu, 1980). Since the APS is calibrated with
essentially unit density particles, measurement of higher density particles
will be somewhat inaccurate. In order to estimate the magnitude of this
effect, monodisperse spherical spherical particles of potassium biphthalate
(KHP) and sodium tungstate were generated. These particles were sampled
and measured with a) a light microscope/image analysis system (the shape
was checked with a scanning electron microscope), b) a settling velocity
measurement system, and c) the APS. The data from these measurements,
combined with the predicted particle size based on the generation
conditions allowed the determination of the aerodynamic size, the physical
size and the density of each type of particle. These particles were
generated from solution so they tended to form a spherical shell with some
solvent retained internally. The percent solvent and air within each type
of particle was also calculated from the above data. The potassium
biphthalate particles were found to have an average density of 1.15 and the
sodium tungstate particles were found to have a density of 2.20. For
particles of the same aerodynamic size, the APS indicated that the sodium
tungstate particles were 0.9 μm larger than 9 μm diameter KHP particles and
1.0 μm larger than 15 μm KHP particles. Thus, a 15 μm particle of density
2.2 measured with the APS will give an apparent aerodynamic size
approximately 7% larger than the true aerodynamic diameter.

216

Techniques have been developed to determine the size dependent penetration of a number of separation devices including the 10 mm nylon cyclone, various impactor samplers, and horizontal and vertical elutriators. A technique was developed for measuring the size dependent response of direct reading instruments. A response curve has been measured for the mini-RAM (GCA/Technology) to volcanic ash aerosol. Other direct reading devices have also been tested.

Field measurements were made using the APS at several locations including a health club whirlpool, a foundry and a bag opening operation.

References

Wilson, J.C. and B.Y.H. Liu (1980) "Aerodynamic Particle Size Measurement by Laser-Doppler Velocimetry", J. Aerosol Sci. 11:139.

Published 1984 by Elsevier Science Publishing Co., Inc.

Aerosols, Liu, Pui, and Fissan, editors

Intercomparison Test of Various Aerosol Measurement Techniques

D. Boulaud*, C. Casselmann**, W. Cherdron, J. B. Deworm***
S. Jordan, J. Mitchell****, V. Prodi*****, G. Tarroni*****

Laboratorium für Aerosolphysik und Filtertechnik I
Kernfoschungszentrum Karlsruhe GmbH
7500 Karlsruhe, W.-Germany

Extended Abstract

Introduction

Although fast reactor accidents are very low probability events, their consequences are to be assessed. During such accidents aerosols can be formed either by mechanical dispersion of molten fuel and fission products or as a result of combustion of sodium. The priority accorded to the study of sodium aerosols, is justified by the fact that they will be the major carrier of radioactive species and hence determine the amount of airborne radioactivity available for release via any leaks in the secondary containment during deliberate venting. Therefore, considerable effort has been devoted to study the different sodium aerosol phenomena.

To ensure that the results obtained by different experimenters can be compared effectively and to assess the compatibilities of different sodium aerosol measurement techniques, an intercomparison of measuring devices was undertaken by interested laboratories from different member states of the European Community. The exercises were conducted in the frame of the containment expert group of the Fast Reactor Coordinating Committee of the European Communities. The aerosol experts participating in the exercises agreed to concentrate on the techniques of measuring aerosol particle size distribution. The tests were performed in the aerosol-loop of FAUNA-test facility at the KfK Laboratory for Aerosol Physics and Filter Techniques.

Aerosol Loop

The aerosol measuring loop is an extension of the original FAUNA facility. The operating conditions for this loop are as follows: gas flow, up to 20.000 m³/h; pipe diameter, 700 mm; maximum gas temperature, 75 °C; relative humidity up to 80 % at 75 °C. The experiments were performed in the open-loop mode. A number of instruments have been installed to monitor the operating conditions. These include the gas temperature in the vessel and in the pipe at the aerosol sampling points (12 in Fig. 1),the relative humidity in the vessel and in the pipe, the flow velocity in the pipe, and the absolute pressure at the aerosol sampling points. A sodium spray fire was chosen as the method of aerosol production in order to keep the mass concentration constant at the sampling points throughout the measurement period. The spray fire offers the following advantages over other methods of sodium fire aerosol generation: a high reaction rate and consequently a low sodium consumption; the ability to change the aerosol yield by appropriate

* CEA-Fontenay-aux-Roses, France; ** CEA-Cadarache, France; *** CEN/SCK Mol, Belgien; **** UKAEA Winfrith, United Kingdom ***** ENEA-Bologna, Italien

218

selection of nozzles and spray pressures; ease of clean-up because only
small amounts of metallic residue are present; simple design.

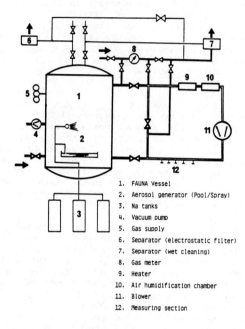

1. FAUNA Vessel
2. Aerosol generator (Pool/Spray)
3. Na tanks
4. Vacuum pump
5. Gas supply
6. Separator (electrostatic filter)
7. Separator (wet cleaning)
8. Gas meter
9. Heater
10. Air humidification chamber
11. Blower
12. Measuring section

Fig. 1: FAUNA aerosol loop modified for the EC workshop

Exp.	Date	Airborne Na Mass Concentration mg/m³	Gas Velocity m/s	Reynolds Number
1	05.10.82	$\sim$ 68	6	280 x 10³
2	05.10.82	$\sim$ 230	2	93 x 10³
3	06.10.82	$\sim$ 220	2	93 x 10³
4	06.10.82	$\sim$ 430	1	47 x 10³
5	07.10.82	$\sim$ 1200	1	47 x 10³

Tab. 1: Test conditions

Experimental conditions and results

The test conditions are placed together in table 1. An important parameter is the aerosol mass concentration which was varied by changing the gas flow rate in the loop and the aerosol generation rate.

The experiments were performed over a period of three days on each of which the aerosol generation rate was kept constant. The aerosol mass concentration was varied within a day by changing the gas flow rate.

The sodium mass concentration was monitored continuously during the tests using the sodium aerosol mass monitor SAMM /1/. Although the primary objective of the workshop was to determine the particle size distributions of the aerosols, measurements of the sodium mass concentration were also made. Comparison of the results of the sizing instruments which had discrete sampling periods varying from 30 s to 60 min with the time averaging SAMM responses show many agreements. However occasionally individual results were much higher than the SAMM responses. Some of these discrepancies may have occured because the time - averaging procedure of SAMM smoths out large concentration increases which lasted for less than 1 min. Also systematic factors such as particle losses, instrument handling may have influenced the measurements.

The six groups of experts used twelve aerosol sizing instruments of which ten were based on the principle of particle inertia (7 impactors, 1 spiral duct centrifuge, 1 inertial spectrometer, 1 sedimentation battery) and two provided data evaluated from aerosol photographs. To ensure consistency all groups used the evaluation program developed by the Italian group /2/. This program enables the experimental data from the different instruments to be compared. The output gives the mass median diameter and the geometric standard deviation.

Table 2 places together a selection of data obtained. The relative standard deviations (S) of the measured AMMD's are given for the different tests.

From the results the following conclusions can be drawn:

- The individual AMMD's were almost always within +20 % and σ_g s were generally within $\pm$ 50 % of the sample mean values. The occasional large differences in individual results were clearly greater than the random variations and can be attributed to systematic deviations. These differences were particularly evident during the first experiment when participants were developing their sampling procedures.

- All of the instruments used in the workshop produced aerosol data that were consistent and in reasonable agreement. Provided that the AMMD's were within the instrument's range of operation, their magnitude was not influenced by either the width or the resolution of the operating range. However, the σ_g s obtained for the impactors were slightly greater than those obtained for the spectrometers.

- The particle sizes obtained by the two methods of image analysis (Zeiss

TGZ 3 particle counter and IBAS image analyser) were in good agreement. Limited information about changes in the particle shape caused by ambient humidity variations were obtained using the automatic image analyser.

Date	Time (h)	Mean AMMD (μm)	$\frac{S}{AMMD}$ (%)	Number of Instruments
05.10.1982	10.20	1.42	18	5
	11.00	1.75	31	6
	12.00	1.41	28	7
05.10.1982	13.40	1.01	23	7
	14.20	0.96	17	8
	15.00	1.02	21	9
06.10.1982	10.48	1.21	16	10
	11.20	1.20	20	5
	11.50	1.14	16	6
	12.17	1.46	16	5
	13.50	1.40	19	10
	14.15	1.54	14	6
	14.30	1.54	14	6
07.10.1982	10.48	1.50	13	9
	11.20	1.89	16	10
	12.00	1.75	18	9

Table 2: Mean AMMD Value

References

/1/   W. Cherdron, Ch. Hofmann, S. Jordan: Experience on Measurement Methods for Sodium Fire Aerosols, 2nd. CSNI Specialist Meeting on Nuclear Aerosols in Reactor Safety, Gatlinburg/Tenn, April 15-17, 1980

/2/   W. Cherdron et al.: Intercomparison Test of Various Aerosol Measurement Techniques, Commission of the European Community, EUR 9203, EN

Published 1984 by Elsevier Science Publishing Co., Inc.

Aerosols, Liu, Pui, and Fissan, editors

221

REAL-TIME MEASUREMENT AND CALIBRATION OF PARTICLE CHARGE

M. K. Mazumder, R. E. Ware, D. Cramer, and G. S. Ballard
University of Arkansas
Graduate Institute of Technology
P. O. Box 3017
Little Rock, Arkansas 72203

EXTENDED ABSTRACT

The properties of aerosol particles can be studied by monitoring their response to acoustic or electric fields applied in the suspending medium (Mazumder, et al. 1983). In the present work, both acoustic and electric fields are applied to determine the aerodynamic diameter and electrical mobility of single aerosol particles. The emphasis of this report is on calibration of the apparatus for electric mobility measurement.

To calibrate the apparatus for measurement of particle electrical mobility, an electric field with sinusoidally varying intensity is applied to test particles of known diameter and mass density. The response to the driving electric field is monitored along the direction of the applied field with a frequency-biased, dual-beam, laser Doppler velocimeter (LDV). Analysis of the particle response provides the values of the needed instrumental parameters.

Using Stokes' law, a one-dimensional equation of motion for a spherical particle in a stationary medium subjected to a sinusoidal electric field can be written as

$$\tau_p \frac{dV_p}{dt} + V_p = Z E_0 \sin \omega t \tag{1}$$

where V_p is the particle velocity, and E_0 and ω are, respectively, the amplitude and angular frequency of the applied electric field. The quantity τ_p is the relaxation time of the particle given by

$$\tau_p = m_p C_c / 3\pi \eta d_p \tag{2}$$

where m_p is the particle mass, C_c the Cunningham slip correction factor referred to the diameter d_p of the particle, and η is the medium viscosity. The quantity Z in Eq. (1) is the particle electrical mobility given by

$$Z = q C_c / 3\pi \eta d_p \tag{3}$$

where q is the net electric charge on the particle.

The steady state solution of Eq. (1) is

$$V_p(t) = Z E_0 (1 + \omega^2 \tau_p^2)^{-1/2} \sin(\omega t - \phi). \tag{4}$$

As a result of this motion, the LDV system produces a frequency modulated signal whose frequency deviation ν_d is proportional to the instantaneous particle velocity and is given by

$$\nu_d = \frac{2V}{\lambda} \sin(\theta/2) \tag{5}$$

where λ is the laser wavelength and θ is the angle of intersection of the two laser beams.

From Equations (3), (4) and (5), the amplitude of the frequency deviation is

$$|\nu_d| = \frac{2qE_0 C_c \sin(\theta/2)}{3\pi\eta d_p \lambda} (1 + \omega^2\tau^2_p)^{-1/2}. \tag{6}$$

Once the amplitude of the frequency deviation is measured for a given particle, Eq. (6) can be used to calculate the electric charge on the test particle provided the remaining parameters are known.

To calibrate the system for electric charge measurement, monodisperse aerosols of polystyrene latex spheres were prepared to have a Boltzmann equilibrium charge distribution. Frequency deviations for several hundred particles in the test aerosol were measured and number of particles counted versus frequency deviation was plotted as shown in Fig. (1). For these measurements, the frequency of the driving electric field was 400 Hz. Measurement time for each particle was about 20 ms.

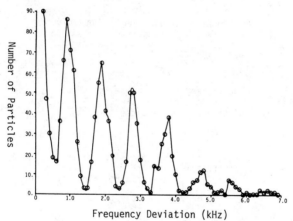

Figure 1.   Measured frequency deviation distribution for a monodisperse aerosol of 0.62 μm diameter polystyrene latex spheres. The aerosol was neutralized to produce a Boltzman equilibrium distribution of particle electric charges.

Periodicity in the plot in Fig. (1) is due to electric charge quantization and the frequency deviation per elementary charge on these test particles is readily determined from the plot. This permits the remaining parameters to be calculated, particularly the electric field amplitude E_0 in terms of the amplitude of the sinusoidal driving voltage applied to the test chamber electrodes.

From Fig. (1), the number of particles in the sample having a given number of elementary charges can be counted. Fig. (2) shows this measured charge distribution compared to the calculated Boltzmann equilibrium distribution. Agreement is sufficient to validate the experimental procedure for measurement of electric charge.

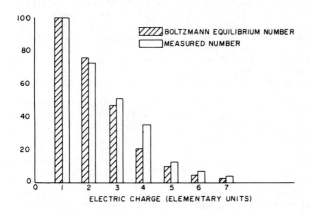

Figure 2.   Measured particle electric charge distribution for a neutralized monodisperse aerosol of 0.62 μm polystyrene latex spheres. Particle count is normalized to 100 at one elementary charge.

Reference

Mazumder, M. K., Ware, R. E., and Hood, W. G. (1983) "Simultaneous Measurements of Aerodynamic Diameter and Electrostatic Charge on a Single Particle Basis," Measurement of Suspended Particles by Quasi-Elastic Light Scattering (Edited by B. Dahneke), Wiley-Interscience, New York.

Published 1984 by Elsevier Science Publishing Co., Inc.

Aerosols, Liu, Pui, and Fissan, editors

224

Simultaneous Measurement of the Diffusion Constant and
Electrical Mobility of Single Aerosol Particles

D. K. Hutchins, D. D. Gregory, and R. A. Sims
University of Arkansas
Graduate Institute of Technology
P.O. Box 3017
Little Rock, Arkansas 72203
U. S. A.

EXTENDED ABSTRACT

Dynamics of fine suspended particles are often dominated by diffusion and, if the particles are electrically charged, by their electrical mobilities. These two transport mechanisms play a signficant role in several aspects of aerosol behavior including coagulation, collection of particles by filters, and deposition of particles in the human respiratory system. This makes it important in many cases to monitor the diffusion constants and electric charges of aerosol particles as well as other features such as aerodynamic diameters and light scattering properties.

The work described employs a variation of the quasi-elastic light scattering technique (Hutchins and Mazumder, 1983) to measure the diffusion constant and electrical mobility of single particles in an aerosol flow stream.

The aerosol to be studied is drawn in a vertical laminar flow through the test chamber parallel to the z-axis of a rectangular coordinate system. A frequency-biased dual-beam laser Doppler velocimeter (LDV), with its probe volume located in the test chamber, is used to monitor the motion of test particles along a horizontal direction, designated as the x-axis of the coordinate system, perpendicular to the aerosol flow stream. An electric field directed along the horizontal x-axis is applied in the LDV probe volume by means of electrodes in the test chamber. In the absence of turbulence in the aerosol flow, particle displacements along the x-axis are due to Brownian motion, and in the case of a charged particle, to the steady drift velocity imparted to the particle by the electric field.

Use of the LDV for monitoring the motion of a test particle can be described in terms of parallel interference fringe planes formed in the LDV probe volume (Durst et al., 1981). The intensity of light scattered by a test particle located in the LDV probe volume is a function of particle position x and time t given by

$$I_s(x,t) = I_0[1 + \cos(\omega t + Kx)] + I_B \qquad (1)$$

where K is the LDV scattering vector magnitude.

Since the intensity of light scattered by a single fine aerosol
particle is low, the method of photon correlation is used to analyze the
scattered light. Assuming a Gaussian distribution of Brownian
displacements, the time autocorrelation function of $I_s(x,t)$ is

$$G(\tau) = (I_0 + I_b)^2 + (1/2)I_0\cos(\omega\tau)\exp(-DK^2\tau) \qquad (2)$$

where D is the particle diffusion constant. Measured autocorrelation
functions are analyzed in accordance with Equation (2) to extract the
values of ω and D. A diffusion diameter is assigned to the particle
through the Stokes-Einstein relation, and the electrical mobility M_E of
the particle is calculated from

$$M_E = (\omega - \omega_0)/KE \qquad (3)$$

where ω_0 is the LDV angular bias frequency and E is the electric field
strength. For a spherical particle, this procedure determines the
diameter and electric charge.

The measured diffusion diameter distribution for a test aerosol is
shown in Figure 1. Measured electric charge distribution for a second
test aerosol is shown in Figure 2.

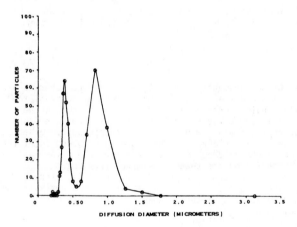

Figure 1. Measured diffusion diameter distribution for an aerosol
sample formed by nebulizing an alcohol suspension containing polystyrene latex spheres of nominal diameters
0.36 and 0.82 μm.

226

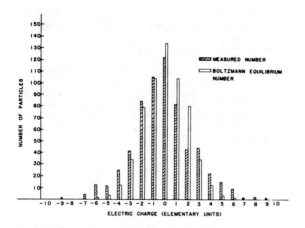

Figure 2.   Measured electric charge distribution for an aerosol
of 0.36 μm nominal diameter polystyrene latex spheres.

In order to make the measurements in real time, a microcomputer is used
to manage data collection and analysis. The microcomputer monitors the
photon count rate via a pulse density detector in order to detect the
presence of a particle in the LDV probe volume. The microcomputer turns
on the digital autocorrelator and the high voltage power supply to set up
the electric field when a particle enters the probe volume, and turns them
off when it leaves the probe volume. Data representing the
autocorrelation function formed are then transferred from the
autocorrelator to the microcomputer for preliminary analysis, with
selected data being output to a desktop computer for further analysis and
storage. This permits measurements to be made at rates up to ten
particles per second.

References

Durst, F., Melling, A., and Whitelaw, J. H. (1981) Principles and
Practice of Laser Doppler Anemometry, Academic Press, New York.

Hutchins, D. K. and Mazumder, M. K. (1983) "Measurement of Diffusion
Diameter Distributions on a Single Particle Basis Using Dual-Beam Laser
Doppler Velocimetry", Measurement of Suspended Particles by Quasi-
Elastic Light Scattering (Edited by B. E. Dahneke), Wiley-Interscience,
New York.

Published 1984 by Elsevier Science Publishing Co., Inc.

Aerosols, Liu, Pui, and Fissan, editors

A TRANSIT-TIME CORRECTED SAMPLING METHOD
FOR THE SPART ANALYZER

J. D. Wilson and M. K. Mazumder
University of Arkansas
Graduate Institute of Technology
P. O. Box 3017
Little Rock, Arkansas 72203

Extended Abstract

Introduction

In the most basic form of the optical particle counter (OPC), the particle size information is derived from the scattered light intensity. The illumination in the sensing volume of an OPC must be carefully controlled to avoid sizing errors. The single particle aerodynamic relaxation time (SPART) analyzer developed at the Graduate Institute of Technology circumvents this requirement by using a technique for particle sizing which is dependent upon the aerodynamic behavior of particles and is therefore independent of the light scattering properties of single particles.

In the SPART analyzer (Mazumder, et. al. 1979) a dual-beam laser Doppler velocimeter (LDV) is used to measure particle velocity in a sinusoidal acoustic field. When the particle velocity is compared with the driving acoustic wave the particle diameter may be deduced, independent of the particle scattering characteristics. Because of the independence of the SPART analyzer method upon particle scattering characteristics, the laser beams used in the LDV are focused to produce a relatively small sensing volume (3.10^{-12} M^3) which gives the instrument the capability of measuring relatively highly concentrated aerosols.

Since the sensing volume formed by the LDV system has a non uniform light intensity, the boundries of the SPART sensing volume are dependent upon the scattering characteristics of individual particles; hence, also upon particle diameter. The size dependence of the sensing volume causes the effective sampling rate to also be size dependent.

A theoretical analysis of the size dependence of the SPART analyzer sampling rate is presented which identifies the important parameters and yields an explicit expression for the effective sample rate. In addition, a method for measuring the size dependence is presented.

Theory

The sensing volume of an LDV is usually described as an ellipsoid defined by the e^{-2} radius of the LDV beams (Farmer 1972). Since the signal strength of scattered light and the signal processing amplifier gain determine the sensing volume boundary within the SPART analyzer, the ellipsoid model was generalized to an arbitrary dimension characterized by the particle detection boundary.

228

Because the geometry of the sensing volume is known, a measurement of one dimension is sufficient to completely describe the sensing volume. The mean thickness of the sensing volume is determined by measuring the average particle transit time through the sensing volume which is multiplied by the carrier gas velocity. It is necessary to measure a mean transit time for each particle size interval of interest and to keep the carrier gas velocity constant. If this is done, the effective sampling rate (Q_s) of the SPART analyzer may be expressed as

$$Q_s = \pi(t_2^2 - t_m^2)U_s^3/2\sin\Theta, \tag{1}$$

where t_m is minimum time required to size a single particle,
$\quad\quad t_2$ is the time required to traverse the center of the sensing volume,
$\quad\quad U_s$ is the carrier gas velocity,
and $\quad\Theta$ is the angle between the beams.

The parameter t_2 is not directly measured but rather is calculated from the mean transit time measurements for each particle size range.

A heterodisperse aerosol of Bis (2-ethylhexyl) sebacate (BES) was sampled by the SPART analyzer and the effective sampling rate, based on transit time measurements, is presented in Figure 1.

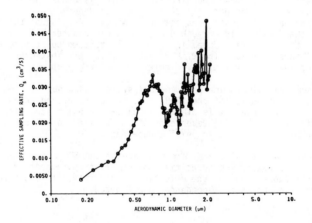

Figure 1. The effective sampling rate of BES particles as computed by means of the mean transit time measurement.

Experimental measurements have demonstrated that the SPART analyzer may be used for concentration measurements of heterodisperse aerosols which have substantially different optical properties when the particle transit time is used to correct the effective sampling rate.

References

Mazumder, M. K., Ware, R. E., Wilson, J. D., Renninger, R. G., Hiller, F. C., McLeod, P. C., Raible, R. W., and Testerman, M. K. (1979) "SPART Analyzer: Its Application to Aerodynamic Size Distribution Measurement," J. Aerosol Sci., Pergamon Press Ltd., Great Britain.

Farmer, W. M. (1972) "Measurement of Particle Size, Number Density and Velocity Using a Laser Interferometer," Applied Optics, 11; 2603-2612.

Published 1984 by Elsevier Science Publishing Co., Inc.

Aerosols, Liu, Pui, and Fissan, editors

THE MEASUREMENT OF THE FIBER-LENGTH DISTRIBUTION FUNCTION BY ELECTRICAL BIREFRINGENCE

M. Yang, M. Hatziprokopiou and D. T. Shaw
Laboratory for Power and Environmental Studies
Department of Electrical and Computer Engineering
State University of New York at Buffalo
4232 Ridge Lea Road
Amherst, New York 14226

EXTENDED ABSTRACT

A new method for the measurement of the size distribution function of aniso-
meric aerosols is developed. Both the rotational diffusion coefficient and
length ratio were determined by the decay of the change in the intensity of
the scattered light in a small forward angle with respect to time after removing
the electrical field which was applied to the aerosol flow in the scattering
volume. This method works well for not only fiber but also disk like aerosols.

Experimental Apparatus and Procedure: The schematic diagram of the
experimental apparatus is shown in Fig. 1. The light beam was formed by
35 mW He-Ne laser, fed through pin hole, lense and diaphragm. The scattering
control volume was the intersection of the light beam and the aerosol flow.
This volume was left in the open air enclosed by a semi-open plastic chamber
with least disturbance from the surrounding and least fluctuation of concentration of aerosol. The pulse electric field was supplied by two blackened
bronze rectangular electrodes, which were parallel to the plane of observation
and were attached to the tubes. Also, the glass tubes were blackened and
separated by a distance of the diameter of the light beam. The scattered
light was collected by lense and was measured only at an optimal angle (5°wrt
forward direction). A photomultiplier was used as a detector and a stabilized
voltage supply was applied to it. A pulse generator and a Dipolar power supply/
amplifier could be connected to the electrodes for control and application of
the electrical field. The pulse generator was for pulse (± 10V. max) with a
length 16ms (on, either + or -), 23ms (off). The continuous output was fed
through a current amplifier with amplification of 10^4 times and then fed into
Boxcar Integrator. The electric pulse and the variation of the photocurrent
from the output of the preamplifier of the Boxcar Integrator were shown on
Double Trace Oscilloscpoe 1. The continuous output of Boxcar Integrator was
sent to and recorded by chart recorder. Unless the DC level of the signal
was stable, the picture of the signal which was taken by camera on the oscilloscope 1 would be more stable than the average output of the Boxcar Integrator.

Fig. 2 shows a typical oscillogram of the relaxation curve of the scattering light of flake particle. The relaxation time and the rotational curve
diffusion coefficient D_r can be estimated from the relaxation curve after
removal of electric field. A computer program called P3R in BMDP is a useful
procedure for finding the solution of the relaxation curve. A computer result
of Fig. 2 would be given in signal 1 of table I. Similarly, this procedure
could also be applied to chain like particle.

Nonlinear regression and some modifications have been applied in the
data analysis. Some of the analysis have been done includes polydispersity,

external disturbance, and fluctuation of the particle concentration from the aerosol generation, provided that the relative time associated with external disturbance, fluctuation of particle concentration were large compared to $T = (1/6D_r)$. If we were carrying experiment for a long time and looking for the average rotational diffusion coefficient $< D_r >$, then the most contributory coefficient would come from slope at $t \to 0$. From the limitation of the set up, necessary truncation of the data should be made.

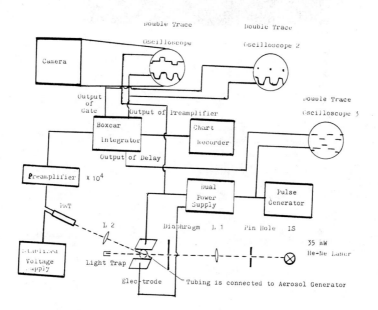

Fig. 1

232

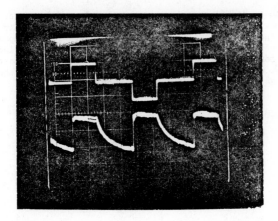

Fig. 2

TABLE 1

Relaxation Times, Rotational Diffusion Constants, Aspect Ratios,
and Particle Diameter of Flake Like Particle

Signal Number	Thickness (um)	Relaxation Time (msec)	D_r (sec^{-1})	Weighting (%)	Aspect Ratio	Diameter (Perrin) (um)
1	.1	2.10	79.2	9.0	8.9	.89
		9.39	17.8	91.0	14.8	1.48
2	.1	.61	272.9	7.7	5.8	.58
		8.57	19.5	92.3	14.4	1.44
3	.1	4.90	34.0	49.7	11.9	1.19
		16.02	10.4	50.3	17.8	1.78
4	.1	3.17	52.5	29.0	10.3	1.03
		12.16	13.7	71.0	16.2	1.62
5	.1	8.79	19.0	100.0	14.5	1.45
6	.1	5.45	30.6	65.1	12.3	1.23
		17.73	9.4	34.9	18.4	1.83

Published 1984 by Elsevier Science Publishing Co., Inc.

Aerosols, Liu, Pui, and Fissan, editors

REAL-TIME PARTICULATE MASS AND SIZE FRACTION MEASUREMENTS IN A PRESSURIZED FLUIDIZED BED COMBUSTOR EXHAUST*

James C. F. Wang
Combustion Research Division
Sandia National Laboratories
Livermore, CA 94550

Introduction

The pressurized fluidized bed combustor (PFBC) is an advanced power generation system for direct combustion of coal in an environmentally acceptable manner. However, its combustion gases contain significant quantities of erosive coal ash and attrited bed material. A key technical objective in the commercialization of this promising concept is to reduce the particulate emissions to achieve reliable gas turbine operation. Hot-gas cleanup components are designed to control mass loading and size distribution of particulate emissions from a PFBC with minimal efficiency losses.

One of the instruments needed to guide the development of suitable hot-gas cleanup eqipment is a real-time particulate mass analyzer, operated at process stream conditions. At Sandia National Laboratories, Livermore, we have previously demonstrated the feasibility of using the tapered element oscillating microbalance for direct particle-mass measurements in real time at room and elevated temperatures up to 500 K.[1,2] Two prototype real-time particle mass analyzers which incorporate an array of these microbalances with a set of cyclones to permit size classification at 1100 K and 6.4 atm were built and tested in the exhaust stream of a pressurized fluidized bed combustor.[3] This paper reports measurements on particle mass loading and size fraction from those prototype analyzers during recent hot-gas cleanup equipment tests at Curtiss-Wright Corporation.

Test Equipment and facility

The test facility at Curtiss-Wright is a 1 MW pilot plant completed in 1977 under the sponsorship of the U.S. Department of Energy. It consists of a coal and dolomite handling facility, a pressurized fluidized bed combustor, an in-bed air heat exchanger, a recycle cyclone, a Dynatherm first-stage separator, a hot gas cleanup system, an Aerodyne cyclone, and a backup exhaust scrubber system.

The hot-gas cleanup equipment tested was the high-temperature electrostatic precipitator developed and built by Research-Cottrell. It consisted of nine cylindrical collection tubes arranged like wheel spokes, with electrodes along the axis of each tube. These vertical collection tubes were housed inside a pressure vessel which was designed to be operated at the combustion exhaust condition. The high-voltage electrodes were surrounded with water-cooled insulators, while the pressure vessel was lined with refractory material.

The Sandia sampling system was installed at the inlet and outlet pipes of the cleanup equipment. The upstream sampling system consisted of one

*Sponsored by Morgantown Energy Technology Center, U.S. Dept. of Energy.

real-time particle mass detector and a total-mass filter, which had been
tested in the laboratory, capable of collecting all particles larger than
0.5 micron in diameter. This system was used to measure the total particle
mass loading of the inlet flow of the precipitator. Because of the high
particle mass loading, the cyclone train arrangement was not used and no
size fraction measurements were performed for the inlet flows.

The downstream sampling system consisted of three cyclones and a
total-mass filter with a particle mass detector coupled to each cyclone and
to the filter. The aerodynamic sizes for particles collected by the
cyclones and filter are confirmed through calibration tests reported in
Ref. 4 as follows: larger than 10 micron in the 10 micron cyclone catch;
between 3 and 10 micron in the 3 micron cyclone catch; between 1 and 3
micron in the 1 micron cyclone catch; less than 1 micron in the filter.
Particle size fraction and total mass loading data were obtained for the
flow exiting from the precipitator. Figure 1 shows a schematic
diagram of the downstream sampling system.

Efficiency of the Electrostatic Precipitator

The on-line precipitator efficiency and corrected particle mass
loading data from the Sandia sampling systems are summarized in Table I.
The mass loading data are presented in g/m^3 (grams per normal cubic meter
measured at room temperature and pressure condition). The power density of
the precipitator is presented in W/m^2 (watts per square meter of the
collector). On the basis of limited data, we found that the precipitator
efficiency changes as the temperature and velocity of the inlet flow and
the power input to the precipitator changes. An average efficiency of 98
percent or better was obtained at 1060 K and dropped to about 83 percent at
1115 K. The latter effect was also partially caused by low precipitator
power, which could not be raised because of a problem of hardware
distortion. Detailed explanations of the precipitator performance are
given in Ref. 5.

Data from the conventional samples, taken simultaneously with the Sandia
measurements, are also included in Table I for comparison. That the Sandia
data appear to yield the higher precipitator efficiencies can be attributed
to the differences between the real-time and the batch-type measurements.
Our real-time measurements usually were taken over a shorter time (10-20
min) than the conventional samples (2 hr). There was also evidence that
the efficiency of the precipitator was not constant, but deteriorated in
time. In addition, the effect of the rapping and the sonic horn used for
electrode cleaning contributed to further increases in particle loading
during a long conventional sampling process. Once these effects were
qualitatively allowed for, a similar dependence of precipitator efficiency
on gas temperature and velocity, and on the power input to the precipitator
was obtained from both measurements.

Transient Observations

Although both the Sandia and conventional sampling systems provide
similar information on the efficiency of the precipitator under various
operating conditions, the unique advantages in having the real-time
information were clearly demonstrated during the tests reported here.

First, we can characterize the performance of the precipitator faster than
the conventional system (as demonstrated in Table I). This facility
allows us to reduce greatly the actual testing time. Second, we can
detect transient changes in precipitator performance in seconds and thus
understand the performance better dynamically.

Figure 2 shows a typical example of the usefulness of a real-time
measurement. It was taken from the total-mass filter of the Sandia
downstream system. The sudden increase shown in the cumulative mass
collected indicates a corresponding increase in the particulate mass
loading when the power of the precipitator was suddenly turned off. The
quick response of our measurements to this abrupt change demonstrates the
capability of the system for observing transient changes in precipitator
performance.

Particle Size-Fraction Measurements

All the measurements on particle size-fraction are for the outlet
flows from the precipitator (see Table II for results). Most of the data
were taken during January at a precipitator inlet temperature of 1115 K and
are the first on-line size measurements ever obtained at the operating
conditions of a pressurized fluidized bed combustor.

At 1115 K, we found the largest portion of the particles to be between
1 and 10 micron in diameter, with a mean size at 3 micron. About 25
percent of the particles were larger than 10 micron, while those smaller
than 1 micron but larger than 0.5 micron were about 30 percent by weight.
At a lower temperature (1060 K), the efficiency of the precipitator
increased from about 86 to 95 percent. The efficiency increase can be
attributed to better collection of the large particles (3 micron and
larger) of the precipitator based on the size distribution measurements.
At 1060 K the small particles (less than 1 micron) constituted about 70
percent of the particles by weight in the precipitator exhaust. This seems
a reasonable observation; however, since we have very limited data on the
low-temperature condition, we need more measurements to substantiate the
findings.

Conclusions

We have demonstrated the capability and utility of the particulate
analyzer developed at Sandia to measure the exhaust of an actual pressurized
fluidized bed combustor. Useful data were obtained during the hot-gas
cleanup tests to assess in real time the functionality of the high
temperature electrostatic precipitator. Constant particle mass loadings at
the inlet of the precipitator were verified by the measurements. The
decrease in particulate loading at the inlet of the precipitator due to the
air-mixing process of hot clean air and cooler particulate laden combustor
exhaust was also detected and agreed with conventional measurements.
During the tests, on-line efficiency of the precipitator was obtained every
two minutes when the two sampling systems were operational simultaneously.
Particulate size-fraction measurements downstream of the precipitator were
obtained for the first time at the process stream conditions and used to
provide additional information about the performance of the precipitator.

References

1. Wang, J. C. F., Patashnick, H. and Rupprecht, G.: JAPCA, 30, p. 1018, September 1980.
2. Wang, J. C. F., Patashnick, H. and Rupprecht, G.: JAPCA, 31, p. 1194, November 1981.
3. Wang, J. C. F. and Libkind, M. A.: "A Real-Time Particulate Mass Analyzer for Pressurized Fluidized Bed Combustion Exhaust Applications," SAND 82-8829, Sandia National Laboratories, Livermore, CA, November 1982.
4. Wang, J. C. F. and Libkind, M. A.: "Particle Collection by Cyclones at High Temperature and Pressure," SAND 82-8611, Sandia National Laboratories, Livermore, CA, August 1982.
5. Kumar, K. S.: "Final Report on Hot Gas Cleanup Tests of the High Temperature Electrostatic Precipitator," DOE/METC Contractors' Meeting, Washington, PA, May 1983.

TABLE I. On-Line Precipitator Efficiency and Corrected Particle Mass Loading from Sandia's Sampling System

DATE	TIME	ESP POWER DENSITY (W/m^2)	INLET TEMP. (K)	PARTICLE LOADING UPSTREAM (g/m^3)	DOWNSTREAM (g/m^3)	ESP EFFICIENCY(%) SANDIA	C-W
11/16/82	10:45-12:00	269.1	1061	4.12	0.024	99.4	99.51
	14:10-15:30	172.2	1061	--	0.0352	99.1	99.57
	20:00-20:35	183.0	1061	--	0.0481	98.8	98.24
11/17	12:20-12:36	134.9	1061	--	0.0792	98.1	98.12
	12:36-12:50	399.6	1061	--	0.0208	99.5	98.12
	17:20-19:30	215.3	1061	--	0.0545	99.3	96.79
11/18	13:00-14:48	317.0	1061	3.57	0.0485	98.6	93.35
11/21	11:10-11:30	179.9	1061	4.0	0.2518	93.5	95.13
	14:00-15:10	179.9	1061	--	0.3616	91.0	97.47
1/13/83	15:00-16:40	290.6	1134	2.2	0.4326	80.0	76.96
1/29	17:40-18:40	20.0	1061	2.33	0.1167	95.0	92.24
1/30	4:58- 5:38	19.4	1117	2.52	0.3410	86.5	83.50

TABLE II. Particle Size-Fraction Measurements at Outlet of Research Cottrell's High Temperature Electrostatic Precipitator

TEMPERATURE (K)	POWER (W/m^2)	SIZE-FRACTION % (10um)	(3-10um)	(1-3um)	(1-0.5um)	EFFICIENCY (%)	REMARKS
1134	732.0	27.0	30.0	13.0	30.0	90.5	
1134	290.6	27.4	30.1	11.6	30.9	80.0	
1117	60.0	27.5	53.7	8.9	9.9	88.1	
1117	101.2	25.1	49.4	4.3	21.2	83.6	
1117	101.2	26.8	52.6	3.9	16.7	78.4	HIGH SPARK
1117	101.2	21.0	41.2	3.2	34.6	83.1	
1117	15.0	21.7	43.4	10.9	24.0	83.2	
1117	12.0	22.5	44.8	9.7	23.0	62.1	
1061	20.0	7.5	15.0	7.0	70.5	95.0	
1061	20.0	12.4	25.6	6.3	54.7	88.4	SONIC HORN
1117	19.4	17.4	34.8	10.5	37.3	90.6	
1117	19.4	22.4	45.0	7.9	24.7	86.5	
1117	19.4	22.7	44.8	8.9	23.6	84.5	HIGH SPARK

237

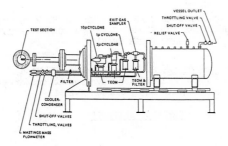

Figure 1. Assembly Sketch of the Sandia Real-Time
Particle Mass Monitor System

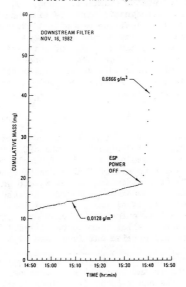

Figure 2. Demonstration of the Real-Time Measuring
Capability of the Sandia Instrument

Published 1984 by Elsevier Science Publishing Co., Inc.

Aerosols, Liu, Pui, and Fissan, editors

AN AERODYNAMIC PARTICLE SIZE ANALYSER TESTED WITH SPHERES,
COMPACT PARTICLES AND FIBRES HAVING A COMMON SETTLING
RATE UNDER GRAVITY

W. D. Griffiths, S. Patrick and A. P. Rood
The Occupational Medicine and Hygiene Laboratories,
403 Edgware Road, London NW2 6LN

EXTENDED ABSTRACT

Introduction

A Timbrell aerosol spectrometer (Timbrell, 1954, 1972) was used in
conjunction with an inertial aerodynamic particle size analyser, APS,
working in the range 0.5 to 15 μm (Wilson and Liu, 1980, Agarwal et al.,
1981) to check the performance of the second instrument in carrying out
measurements of the aerodynamic diameters $(\emptyset_a)$ of a variety of aerosol
particles having the same settling velocities (Vg) under gravity. The
work was carried out first with near unit density spherical particles,
then with non-spherical but still relatively compact particles of higher
density and, finally with glass fibres.

Experimental

The Timbrell spectrometer was modified to extract particles of pre-
determined Vg, and hence $\emptyset_a$, for injection into the APS. Four sampling
points were selected along the base plate of the Timbrell chamber to
coincide with the sedimentation points of four sizes of spheres with
$0.0035 \lesssim Vg \lesssim 0.20$ cm sec^{-1} (corresponding to $1 \lesssim \emptyset_a \lesssim 8$ μm). The modified base plate was designed to accommodate several microscope slides,
and with four holes drilled into it at the appropriate points, to fit
four pipes, which protrude a little inside the chamber. Similar sized
holes were made in the slides to allow them to be fitted to the pipes.
Extraction of the particles was carried out from each pipe separately.

Additional work was carried out with glass fibres which had settled-
out on the microscope slides in the Timbrell spectrometer. Scanning
electron microscope (SEM) size analysis of fibre length and diameter was
carried out on the fibres which had settled-out in areas close to each
of the four holes in the slides. Two separate theories were applied
to these results to calculate $\emptyset_a$ for the fibres under the laminar flow
conditions in the Timbrell spectrometer.

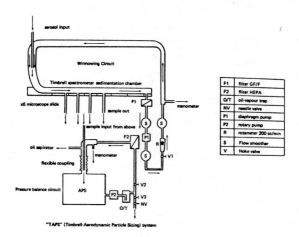

Table accompanying diagram:

F1	filter GF/F
F2	filter HEPA
O/T	oil-vapour trap
NV	needle valve
P1	diaphragm pump
P2	rotary pump
R	rotameter 200 cc/min
S	Flow smoother
V	Hoke valve

"TAPS" (Timbrell-Aerodynamic Particle Sizing) System

Conclusions

The system was a success in carrying out performance checks on the APS with compact non-spherical particles, such as silicon carbide and also with glass fibre. APS measurements of silicon carbide $\emptyset_a$ were fairly accurate up to $\sim$5 μm, but values for glass were perhaps smaller than expected, for smaller than 5 μm.

Calculations made from the SEM values for glass indicated that the fibres were best thought of as prolate spheroids rather than regular cylinders moving perpendicular to their polar axes. This agrees with Davies' interpretation of earlier work carried out by Timbrell, (Davies, 1979 and Timbrell, 1972).

References

Agarwal, J.K., Remiarz, F.J., Quant, F.R. and Sem, G.J. (1982) "Real Time Aerodynamic particle size analysis", J. Aerosol Sci. 13:222.

Davies, C.N. (1979) "Particle-Fluid Interaction", J. Aerosol Sci. 10:477.

Timbrell, V. (1954) "The terminal velocity and size of airborne dust particles", Brit. J. of Appl. Phys. 5, (13): 586.

Timbrell, V. (1974) "An Aerosol Spectrometer and its Applications", Assessment of Airborne Particles (Edited by J.J. Mercer, P.E. Morrow and W. Stober), Charles C. Thomas, Springfield, Illinois, USA.

Wilson, J.C. and Liu, B.Y.H. (1980) "Aerodynamic Particle Size Measurement by Laser-Doppler Velocimetry", J. Aerosol Sci. 7:139.

Published 1984 by Elsevier Science Publishing Co., Inc.

Aerosols, Liu, Pui, and Fissan, editors

REAL-TIME OBSERVATION OF AEROSOL DEPOSITION
BY NEUTRON RADIOGRAPHY

Douglas D. McRae and R. W. Jenkins, Jr.
Philip Morris Research Center
P. O. Box 26583
Richmond, VA 23261
and
J. S. Brenizer, M. F. Sulcoski, and T. G. Williamson
Department of Nuclear Engineering and Engineering Physics
University of Virginia
Charlottesville, VA 22901

EXTENDED ABSTRACT

The University of Virginia Nuclear Reactor Facility and the Philip
Morris Research Center have jointly developed a real-time neutron radiographic system. Neutron radiography is in many ways conceptually similar to
X-ray radiography. When a beam of thermal neutrons (with energies less than
0.3 eV) pass through an object, some are absorbed or scattered out of the
beam. The transmitted neutrons can then be used to create an image of the
object. However, while X rays are strongly absorbed by the higher atomic
weight elements, thermal neutrons are efficiently attenuated by relatively
few elements. For example, isotopes of boron and gadolinium are good atten-
uators as a result of their large neutron capture cross sections while in
hydrogen, neutrons are attenuated primarily as a result of a large scattering cross section. One result of this is that while organic materials and
water are nearly transparent to X rays, they are easily seen in neutron
radiographs as a consequence of their high hydrogen content. Neutron and
X-ray radiography are therefore complementary as neutrons can be used to
image organic materials and strong neutron absorbing isotopes present in a
metallic matrix while X rays can image metallic objects contained in an
organic matrix.

The neutron radiography system has been used to observe the real-time
deposition of a neutron absorbing aerosol in a filter. An aqueous solution
of $GdCl_3$ was nebulized, dried and neutralized to form the aerosol. The
filters used for these experiments were standard cellulose acetate cigarette
filters. From the observations, the depositions were analyzed both as a
function of time and penetration distance into the filter.

The radiographic system consists of a neutron beam, a neutron camera,
and a digital image processor to enhance and analyze the images. The neutron beam is part of the 2 MW research reactor at the University of
Virginia. A beam port has been specially modified to provide a collimated
0.15 m diameter beam of thermal neutrons with a flux of 10^{11} neutrons/m^2sec.
The neutron imaging system is shown schematically in Figure 1. The neutron
camera produced by Precise Optics (Bayshore, N.Y.) converts the neutron
image into a video image. The inside of the front screen of the camera is
coated with a 50 μm layer of gadolinium oxysulfide which absorbs the
neutrons, emitting beta particles (electrons) in the process. These
electrons are amplified in the image intensifier. The output screen of the

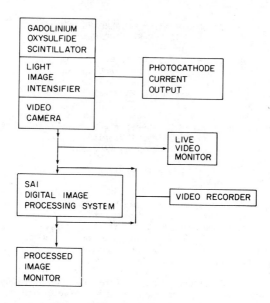

Figure 1. Schematic of the real-time neutron imaging system showing possible video signal and processing paths.

intensifier is viewed by a Newvicon video camera. Because of the high amplification, the system produces a high resolution image in a relatively low neutron flux. The live video signal from the Newvicon camera can be viewed on a monitor and, if desired, recorded on a video tape deck. The recorder allows one to save the real-time images for later analysis. The video image can be analyzed or enhanced using a digital image processor. The SAI Intellect 100 (Science Applications Inc., San Diego, CA) image processing system is an LSI 11/23 based system with a 256K memory plane capable of 512 x 512 pixel resolution. Software functions include recursive noise reduction, picture integration and differencing, a full range of operator selected nonrecursive filtering functions, contrast enhancement and digital image storage. The processed image can also be recorded.

The neutron camera is mounted on a remotely controlled carrier directly facing the neutron beam inside a shielded blockhouse built around the beam port. The video imaging system is located in a nearby room along with the camera controls. In addition to real-time radiography, conventional static neutron radiography can be performed using gadolinium convertor foils and photographic film.

The aerosol used for the deposition experiments was generated from a $GdCl_3$ solution by a TSI Model 3076 nebulizer (Thermal Systems, Inc., St.

Paul, MN) using N_2 as the carrier gas. A gadolinium aerosol was chosen since the gadolinium-155 and -157 isotopes have very large neutron capture cross sections. Amounts less than 0.1 mg in a 2 mm diameter spot were easily detectable using the real-time imaging system. The $GdCl_3$ solution contained 60 kg/m^3 of $GdCl_3 \cdot xH_2O$. The aerosol was dried in a diffusion drier and passed through a TSI Model 3012 aerosol neutralizer to remove the charges left on the particles from the nebulizing process. The output was 53 cm^3/sec with a mass concentration of 1.0 g/m^3. Electron microscopy indicated that the size distribution of the $GdCl_3$ aerosol was log normal with a number mean diameter of 0.63 μm and a geometric standard deviation of 1.75. A second $GdCl_3$ solution containing 600 kg/m^3 was also nebulized to produce an aerosol with a mass concentration of 2.0 g/m^3. This aerosol was used to investigate the rapid clogging of filters.

The filters examined by neutron radiography were standard cellulose acetate cigarette filters. Their physical characteristics are presented in Table 1. All filters were cylindrical with a diameter of 8 mm and a length of 63.5 mm.

TABLE 1

Physical Characteristics of Filters

Designation[1]	Eff. for Cig. Smoke[2] (%)	Fiber Vol. Fraction	Fiber Radius (μm)	Pressure Drop at 17.5 cm^3/sec (kPa)
2.5/32K	49	0.06	8.3	2.2
2.5/48K	58	0.09	8.3	3.0
2.1/60K	79	0.11	7.6	5.4

[1] 2.5/32K ≡ 2.5/32,000. The 2.5 is the denier per filament (the weight in grams of 9000 m). The 32,000 is the total denier (the weight in grams of 9000 m of the fiber bundle used to make the filter).
[2] For a 25 mm filter.

The filters were sealed in a quartz tube and placed directly in front of the neutron camera in the center of the neutron beam. The $GdCl_3$ aerosol was drawn through the filter by a small air pump at a flow rate of 17.5 cm^3/sec. The excess aerosol generated by the nebulizer was exhausted through a high efficiency filter. The aerosol flow through the filters was controlled remotely. A three-way solenoid valve, actuated by a timer, started and ended each run. The runs generally lasted 30 min., unless the filter became clogged.

Figure 2 presents a neutron radiograph of three different density filters. The darkened regions at the left end of the filter are the areas where the $GdCl_3$ aerosol was deposited.

244

Analysis of the image was performed using the line densitometer function of the image processor. Figure 3 shows a comparison of relative luminescence before and after deposition in the 2.5/32K filter. The curve in Figure 3a follows the classic exponential deposition dependence on filter length except for the front end which shows increased deposition due to filter clogging. The curve in Figure 3b exhibits an exponential attenuation as a function of filter diameter both before and after aerosol deposition indicating a uniform deposition across the filter diameter. The same type of video information can be analyzed to show the deposition dependence on time. Non-uniformities in the deposition were also immediately visible. This work is being extended to investigate aerosol deposition in other filters and cascade impactors.

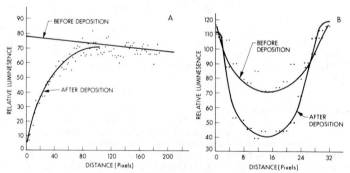

FIGURE 3. Relative liminescence before and after GdCl$_3$ aerosol deposition in a 2.5/32K filter as a function of distance along a) the filter centerline, and b) a selected diameter.

Published 1984 by Elsevier Science Publishing Co., Inc.

Aerosols, Liu, Pui, and Fissan, editors

Development of a New Automatic Aerosol Measuring System for
Determination of Mass Concentration and Mean Particle Diameter

G. Poß

Laboratorium für Aerosolphysik und Filtertechnik I
Kernforschungszentrum Karlsruhe GmbH
7500 Karlsruhe, W.-Germany

Extended Abstract

In monitoring the emission of pollutants it is of increasing interest to
measure not only the mass concentration of the total dust but also the
portion of the finer parts of particulate solids below 10 micron in size.
Most of the usual methods of particle-size analysis are very time consuming
both for calibration and for analysis. The technique presented here gives
the mass concentration and a mean diameter without requiring much expen-
diture in terms of analytical work. The novel method is based on a combined
measurement of mass concentration and extinction caused by the aerosol. The
mass concentration is obtained by simply measuring the increase in weight of
a filter with a microbalance, with the volume flow rate known. The determination of the extinction E is carried out with a photometer. From the two
measured parameters and a constant containing the solid density ρ and the
length l of the extinction path a mean diameter can be calculated.

The whole device consists of 4 components: a measuring cuvette which is the
heart of the system, a microbalance, a minicomputer and a mechanism steering
the various steps of measurements. The analyser itself (Fig. 1) is an
aluminium cuvette with a photometer intergrated. Its design is the result of
minimizing the losses of particles due to depositions on the walls. The
cuvette consists of two parts and can be extended for loading and removing

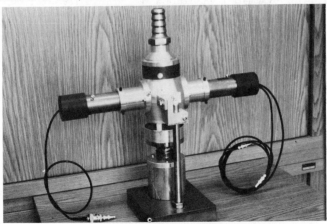

Fig. 1: Measuring cuvette with integrated photometer

246

respectively a membrane or glass fibre filter placed downstream of the photometer measuring volume. The photometer is equipped with a red LED light source with a photodiode placed immediately behind the source as reference and another photodiode as detector opposite the LED for receiving the measuring signal.

The extinction path l is only 40 mm. The windows of the photometer can be flushed with clean air to keep them free of aerosol depositions. The whole gaging process - filter changing, filter weighing, and reinstallation into the cuvette - will be performed by a computer controlled mechanical filter changing system. The microbalance used operates on the principle of electromagnetic force compensation. Its weighing range is down to 1 mg with an accuracy of $\pm$ 1 %.

The idea of estimating a mean diameter with the aid of the two parameters C_m and E starts from a theoretical study comparing the diameter of average mass $d_{\overline{m}}$ (third moment average) with the diameter of average extinction $d_{\overline{ex}}$, both calculated for identical size distributions. These calculations were made for particle size distributions with varying standard deviations and modal values, for different combinations of ligth sources (white light, $\lambda = 0,44$ µm; $\lambda = 0,633$ µm) and light detecting devices taking into account their respective emitting and receiving spectra for particles with different complex refractive indices ranging from 1.05 - 1.96 for the real part and from 0.00-2.21 for the imaginary part and also for bimodal distributions. The results show that $d_{\overline{ex}}$ and $d_{\overline{m}}$ agree well within a well defined often negligible error band. The deviations are up to $\pm$ 15 % for extreme configurations, but in most cases only a few per cent or less for realistic combinations of the parameters considered. Fig. 2 shows an example for the deviations of $d_{\overline{ex}}$ from $d_{\overline{m}}$ calculated for fly ash, a white light source the receiving spectra of a photodetector T 8235 and a standard deviation σ = 1.3 for the given size distributions. The small deviations in comparing $d_{\overline{ex}}$ with $d_{\overline{m}}$ allow to equate them within a well known error band. This is the condition to derive the diameter d_{em} from the equations defining the mass concentration C_M and the extinction E.

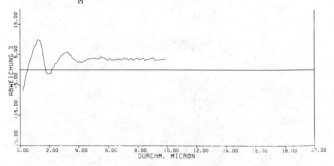

Fig. 2: Deviation (in percent) of $d_{\overline{ex}}$ from $d_{\overline{m}}$ as a function of $d_{\overline{m}}$

One gets

$$C_m = \frac{\pi}{6}\ d_m^3\ \rho\ C_N \qquad (1)$$

for the total mass concentration and

$$E = \frac{\pi}{4}\ C_N\ d_{ex}^2\ Q_e\ 1 \qquad (2)$$

for the extinction, with C_N = number concentration and Q_e = extinction efficiency. Approximating the extinction efficiency Q_e by 2 /1/ which is practicable for particles greater than 2 μm and comparing (1) and (2) under the assumption that $d_{ex} = d_m$

one obtains

$$d_{em} = \frac{3 \cdot 1 \cdot C_N}{E \cdot \rho} \qquad (3)$$

First quantitative results from experimental data comparing the d_{em}-diameter to usual diameters such as the light scattering diameter or the geometrical diameter from TEM and SEM pictures show an agreement within $\pm$ 25 % of d_{em} with the other diameters. These examinations were made for latex and a selection of technical aerosols. Table 1 shows the diameters of various atomized dusts and liquids, measured with different methods.

Material	d_{em}: d	σ/μm	σ/%	C_M / Extinction	Aerodynamic Diameter $\bar{d}_{aero}$	Light Scattering Diameter d_{mass} / d_{number}	σ	$d_{Mod(number)}$	σ	$d_{Mod(mass)}$	Geometric Diameter d_{mass} / d_{number}	σ	$d_{Mod(number)}$	σ	$d_{Mod(mass)}$	Density ρ
Ca(OH)₂	1.51	0.35	22.9	2.9-/1.9-; (2.1-5.771)E-4	2.9	1.6 / 5.8	2.72 / 0.5	6.9 / 8	2.01	1.3	1.34 / 1.80	1.32	1.62	1.3	1.48	2.24
Cement PZ	1.67	0.038	2.8	11.1-7.47; (2.57-56.2)16-3	2.4	1.4 / 2.9	1.91 / 0.7	39	1.77	1.2	1.54 / 1.64	1.34 / 1.2	1.5 / 2	1.28	1.86	3.08
Cement Ra	1.91	0.62	32.4	17.5-; (302-6.71)E-4	2.9	1.5 / 4.4	2.01 / 0.7	4.5 / 4	1.79	1.3	2.42 / 3.25	1.3	2.91	1.27	2.68	2.31
Cocoa Powder	10.45	1.31	12.5													1.1
Diff. Pump Oil	0.69	0.15	21.7			0.60					0.83 / 2.5	1.89	3.3	2.03	0.68	
Fly Ash	2.58	0.44	17	(1.53-2.19)E-3	2.6	1.7 / 5.7	2.25 / 0.6	6.5 / 8	1.95	1.4	2.12 / 3.91	1.27	3.87	1.26	3.27	2.14
Gypsum	2.15	0.54	25	(12.5-1.851)E-3	1.8	1.7 / 4.7	1.96 / 0.7	4.4 / 5	1.73	1.5	1.9 / 2.71	1.34	2.38	1.31	2.14	2.32
Iron	2.08	0.17	8.24	(1.34-2.611)E-3	2.4	1.38	1.74			1.98	0.80 / 0.99	1.28	0.93	1.27	0.84	7.86
Kaolinite	1.7	0.35	20.7	(4.4-5.677)E-4	2.4	1.4 / 4.5	2.01 / 0.65	4.5 / 4.5	1.8	1.2	1.8 / 2.29	1.27	2.1	1.25	2.04	2.52
Latex	0.945	0.16	17.2	(0.76-6.11)E-3	1						1.03 / 1.06	1.07	104	1.07	1.05	1
Solvent Tire Aerosols	0.24	0.09	37.6								0.24 / 0.87	2.06	1.1	2.06	0.19	2
Tin	1.39	0.45	32.5	(1.35-1.54)E-3	1.8	0.98 / 2.5	1.78 / 0.7	2.3 / 2	1.64	0.89	0.63 / 0.78	1.26	0.71	1.27	0.58	6.52

* Whisker - Filter
\# Membran - Filter

Tab. 1:
d_{em} = arithmetic mean from 5-8 measurements; σ/μm/, σ/%/ error in μm and %; C_m = lowest and highest mass concentration in mg/m³ ; Extinction: the respective extinction for the lowest and highest C_M; $\bar{d}_{aero}$ = arithmetic mean aerodynamic diameter; σ = Standard deviation of the measured distributions under the assumption of a log-normal-distribution

Parallel to the measurement of d_{em} additional analyses were made with an
inertial spectrometer "INSPEC" to determine the aerodynamic diameter, a
light scattering device (with a white light source, measuring under 90 °) to
measure the scattering diameter and TEM and SEM to determine the geometrical
diameter. For this latter purpose the droplets were precipitated on a
whisker filter, the solid particles on a membrane filter.
With the exeption of Fe and S powders which have a distinctly higher density
then the other materials $\bar{d}_{em}$ agreed well with the mean arithmetic
geometrical diameter $\bar{d}_{geom}$. $\bar{d}_{em}$ and $\bar{d}_{geom}$ are well correlated (fig. 3)
with a correlation coefficient $r = 0.97$. A correlation between $\bar{d}_{em}$ and $\bar{d}_{aero}$
requires complicated corrections taking into account the dynamic shape factors
of the respective particles, which have not yet been measured.

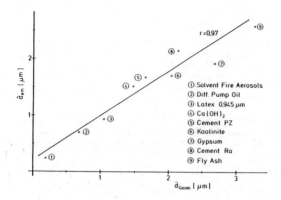

Fig. 3: Correlation between $\bar{d}_{em}$ and $\bar{d}_{geom}$

All measurements were made with a laboratory device still operated manually in
every single step. It is now being automated.

Conclusion

In many applications the determination of an aerosol mass concentration and
the mean particle size is desirable. A method has been developed to provide
information on both parameters. The correlation between the mean geometrical
diameter and the special diameter of the device used shows that the method is
capable to give sufficient information on both mass concentration and particle
size. Therefore a new instrument is available using the mean extinction and
a mass concentration measurement to get additional information about mean
particle size.

References

William C. Hinds (1982): Aerosol Technology, Wiley-Interscience-Publication

PART II.

AEROSOL CHARACTERIZATION

CHAPTER 7.

ATMOSPHERIC AEROSOLS

Published 1984 by Elsevier Science Publishing Co., Inc.
Aerosols, Liu, Pui, and Fissan, editors

AEROSOL CHARACTERIZATION - A SYSTEM ANALYTICAL APPROACH;
AEROSOLS OF THE ATMOSPHERE

O. Preining, Institute for Experimental Physics, University of Vienna,
Strudelhofgasse 4, 1090 Vienna, Austria.

We speak of aerosols and think of solid and/or liquid particulates
suspended in gases, we consider the gas less interesting and describe
the particles only, we consider the suspensions homogeneous and
unlimited and derive from measurements: Particle concentrations, size
distributions, size dependent chemical compositions and similar
descriptors but we hardly ask the question what kind of system do we
investigate, is it sufficiently defined, what are its boundaries, its
total mass, volume, surface, its inhomogeneities and is it stable, at
least stable enough to derive meaningful parameters from our measuring
subsystems so we may predict its future. The programmatic introduction
to Aerosol Microphysics I (Marlow, 1980) devides aerosol physics into
aerosol microphysics and aerosol macrophysics and states explicitely:
"In aerosol macrophysics, bulk, collective, or cooperative properties
of the particle-gas system taken as a whole are treated. An appropriate analogy can be drawn by recognizing that the relationship of
aerosol microphysics to aerosol macrophysics is similar to the relationship between molecular and solid-state physics." In my opinion we
have to go further the realm of aerosol macrophysics has to be
extended to the very large and intrinsically complex systems
comprising even the whole atmosphere or its large subsystems. One
possibility to characterize the "size" of the considered system would
be to specify the total masses involved. A suggestion already made to
describe the extent of air pollution problems (Preining 1974) and
which is also used in climat system analysis (Saltzmann 1983). The
mass range will run from c. 1 kg for a system of a volume of c. 1 m^3
at NTP up to 5.10^{18} kg, the mass of the whole atmosphere.

System analysis developed to a mathematically based interdisciplinary
endeavour between sciences and arts mainly during the past decade. The
concept of system, first extensively used in thermodynamics, is really
at the core of every science (especially physics) but only very simple
systems and the corresponding mathematical models had been its
subjects of study up to very recent times.

Aerosol science is not only in the gap of the "lost dimensions", as
colloid science was christened (Ostwald 1915) but also belongs to the
field of "intrinsic complex systems". In the past few years many
branches of science were forced into modeling of complex systems for
two reasons: The strong pressure to find models applicable to real
world systems and the ubiquitous availability of computers. Suddenly
basic concepts of mathematics, dormant for nearly 50 years and considered unusable and monstrous became tools for modeling and describing quantitatively complexity. Recently the term "Fractal" was coined
(Mandelbrot, 1977, 1982) and in the fall of 83 already the "Third
Conference on Fractals - Fractals in the Physical Sciences", was
organized by the NBS in Maryland. These concepts can be and have to be
applied to aerosls too and they ought to be (and have already occasionally been) used to describe as well the complexities of shapes and

compositions of individual particulates (Kaye 1978) as the inhomogeneities of particulate clouds and such the structure of large aerosol systems. (Preining 1981, Preining and Reischl, 1982) .

We speak of the atmospheric aerosol (Twomey 1977, Pruppacher and Klett 1978) and have to think of a very large, unstable, not stationary, and extremely complex system composed in turn of very large and very complex subsystems like the tropospheric, the stratospheric, the continental, the marine, the arctic aerosols, the anthropogenically produced plumes, the urban, the rural, the background aerosols and so on. All these are ill defined open systems; we should include parameters like the total volume and/or mass of the considered system, e.g. a power plant plume, but we should describe its structures too, the spatial variations of concentrations of the different constituents, of humidity, pressure, temperature and others.But this has not been done so far e.g. the models of the stratospheric aerosols (Turco et al. 1982, Turco 1982) just consider the variations with altitude but assume otherwise a twodimensional uniformly layered structure. Or plumes are still considered having a Gaussian or even a uniform concentration profile (Balko and Peters, 1983). Another example would be shape, extension, composition and structure of the arctic haze layers recently discovered.

Only the use of such structural parameters as fractals (fractal dimensions) will permit to diagnose the applicability of mathematical models to the real world aerosol systems, since all non linear processes, like coagulation, and/or certain chemical reactions will depend strongly on the structure of the system. This is obvious if comparing systems with different structures containing the same masses of reactants. Systems with spatially uniformly distributed masses will differ considerably in their behaviour from systems with the masses of certain reactants occuring in "patches" of high concentrations which only occasionally overlap. The problem is apostrophed in "Turbulent Reacting Flows" (Libby and Williams, 1980);it is a problem between physics, chemistry and engineering and has great relevance to all aspects of aerosol science.

The identity of the structural parameters derived from an analysed system and from its model is a necessary but never a sufficient condition for the applicability of the latter. Such structural parameters can be derived from measurements; comparing samples taken with different resolution (time or space) would be one easy way to derive the parameter "fractal dimension", another method, still hardly explored, would be conditional sampling. Sampling the number of particles arriving at an single optical particle counter in a continous string of time intervalls of equal duration permits to derive a reasonable value of the fractal dimension of the aerosol cloud (Preining 1981). An attempt to derive the fractal dimension from local wind velocities and wind directions led to the basic value of 1,35 (slope of the Richardson plot, Preining and Reischl 1982). These are one dimensional samples of twodimensional sections, so "1" was added, to compare with the two dimensional shapes in Mandelbrot 1982.(pages 266, 267). Analysing the area perimeter relation (i.e. also essentially a scalar parameter) of clouds as seen from a

satellite yields a fractal dimension of 1,35 (Lovejoy 1982). This value of 1,35 for a linear section, or the value of 2,35 for a surface corresponds favorable to recent theoretical predictions. (Hentschel and Procaccia, 1983, 1984).
Another problem which will become more important, especially in this context, is the description of the very subsystems selected by the sampling instrument and its position (sampling location).

The evaluation of the size distribution of aerosols in the range of all possible sizes from molecules up to grains of sands by using several instruments did start here in Minneapolis about 15 years ago, the trimodal distribution of the atmospheric aerosol - the Whitby distribution - was the important result of this endeavour. Now we stand again at a new beginning, not only the sizes and compositions of the aerosol particles have to be acquired, the essential structures of the large systems of macro aearosol physics have to be explored. Let us start again here in Minneapolis and let us hope that in another 15 years, at the and of this century, we know as much about the essential structures of the atmospheric aerosols and their implications as we do now about the size distributions.

Literatur:
Balko J.A. and L.K. Peters (1983)
A modeling study of SO_x-NO_x Hydrocarbon plumes and their transport to the background troposphere, Atmospheric Environment 17 : 1965

Hentschel H.G.F. and I. Procaccia (1983)
Fractal nature of turbulence as manifested in turbulent diffusion, Phys. Rev. A, 27 : 1266

Hentschel H.G.F. and I. Procaccia (1984)
Relative diffusion in turbulent media: The fractal dimension of clouds. Physical review A, 29 : 1461

Kaye B.H. (1978)
Specification of the Ruggedness and/or Texture of a Fine Particle Profile by its Fractal Dimension. Powder Technology, 21 : 1

Libby P.A. and F.A. Williams editors,
Topics in Applied Physics, Vol. 44, Turbulent Reacting Flows, Springer, Berlin

Lovejoy S. (1982)
Area Perimeter Relation for Rain and Cloud Areas, Science 216 : 185

Mandelbrot B.B., (1977)
Fractals, Form, Chance and Dimension. W.H. Freemann and Co. San Francisco

Mandelbrot B.B. (1982)
The Fractal Geometry of Nature. W.H. Freemann and Company, San Francisco

256

Marlow W.H. (1980)
Introductions: "The Domains of Aerosol Physics", Topics in Current
Physics, Aerosol Microphysics I. Particle Interaction edited by W.H.
Marlow, Springer, Berlin

Ostwald W., (1915)
Die Welt der vernachlässigten Dimensionen, Th. Steinkopf

Preining, O. (1974)
Das Aerosol und die Luftverunreinigungen.
Aerosol in Naturwissenschaft, Medizin und Technik, Jahreskongreß 1974
der Gesellschaft für Aerosolforschung e.V. edited by V. Böhlau and H.
Straubel, Bad Soden, BRD

Preining, O. (1981)
"On the use of optical single particle counters to acquire spatial
inhomogeneities of particulate clouds", Atmospheric Sciences Research
Center, State University of New York at Albany, Pub. No. 813.

Preining O. and G. Reischl (1982)
Aerosols, description and descriptors. Journal of the Hung. Met.
Service 86: 104

Pruppacher H.R. and S.D. Klett (1978)
Microphysics of Clouds and Precipitation. D. Reidel, Dordrecht
(Holland)

Saltzmann B. (1983)
"Climatic System Analysis" Advances in Geophysics Vol. 25, Theory of
Climate, edited by B. Saltzmann, Academic Press, New York

Turco R.P. (1982)
Models of Stratospheric Aerosols and Dust. Topics in Current Physics:
The Stratospheric Aerosol Layer, edited by R.C. Whitten, Springer,
Berlin.

Turco R.P., O.B. Toon, R.C. Whitten, R.G. Keesee, P. Hamill (1982)
Importance of Heterogeneous Processes to Tropospheric Chemistry:
Studies With a One-Dimensional Model. Geophysical Monograph 26,
Heterogeneous Atmospheric Chemistry edited by D.R. Schryer. American
Geophysical Union, Washington D.C.

Twomey S. (1977)
Atmospheric Aerosols, Elsevier, Amsterdam.

Published 1984 by Elsevier Science Publishing Co., Inc.

Aerosols, Liu, Pui, and Fissan, editors

A RATIONAL BASIS FOR NATIONAL AMBIENT AIR STANDARDS FOR
PARTICULATE MATTER

James P. Lodge
Consultant in Atmospheric Chemistry
385 Broadway
Boulder, Colorado 80303
U.S.A.

EXTENDED ABSTRACT

Unlike all other ambient standards, particulate matter standards do
not relate to concentrations of chemically defined species. Airborne particulate matter probably contains at least traces of every one of the naturally
occurring elements except the noble gases. In addition, these elements are
not distributed uniformly, but tend to segregate in certain particle size
ranges, and be virtually absent in others. Furthermore, the health and
welfare effects of particulate matter, as they are known, are themselves a
function of particle size, as well as (in some cases) the unevenly distributed
chemical composition. Further to complicate matters, it is not clear that
all effects would be expected to correlate with mass concentration; effects
could instead be functions of particle volume, particle surface areas, or
even particle number.

Additionally, the collected mass, and hence the mass concentration, is
a function of the particular collector employed. All sampling devices discriminate to some degree as a function of particle size. This means that the
reported concentration is in some sense an artifact of the particular
sampling device, and the measured concentration, if that device is optimized
to measure the mass responsible for a particular effect, will not necessarily
be an equally good measure of other effects.

Given this context, let us first examine the available evidence relating
effects and airborne concentrations of particular fractions of the aerosol
mass, then turn our attention to the available sampling devices.

With regard to effects, it is essential that attention be centered on
general, nonspecific particulate matter, and not on individual substances
present in a few locations from particular sources. For example, this
consideration alone would exclude from consideration the studies of Ferris,
et al. (1973, 1976) in Berlin, NH, since the particulate matter in that city
was dominated during the early study by emissions from a pulp mill that are
certainly not present in the vast majority of American urban areas.

There is only one group of studies that the majority of epidemiologists
consider to be the best evidence on which to base ambient standards--the
studies by Lawther and his colleagues in London, beginning in the 1950's
(Lawther, 1958; Lawther, et al., 1970). These workers have unequivocally
stated that their cohort of bronchitic and emphysematous patients did not
respond adversely to concentrations of particulate matter, measured as
British Smoke, until above 250 $\mu g/m^3$, and then only when the sulfur dioxide
concentration was above 500 $\mu g/m^3$. The studies of mortality from all causes
evidenced association with excess mortality deviation when smoke and sulfur

dioxide levels concurrently were markedly higher. There have been numerous
statistical manipulations that have tried to show effects with these data at
lower levels, but all of these have fatal errors. Lawther and his colleagues
are so confident that, now that British smoke concentrations remain below
this level and there are no adverse effects, they have disbanded the MRC Air
Pollution Unit.

The quantity called British Smoke is not a direct gravimetric measure,
but a correlated concentration based on the reflectance of light from the
particle deposit on a white filter. The sampling device collects only
particles less than about 4.5 μm. Reflectance is primarily a function of
the soot content of the particulate matter sampled, and hence the reported
British Smoke concentration is a true measure of total particles less than
4.5 μm when the composition of the particulate matter is constant. This
condition appears to have occurred during the period of the key epidemiological studies; furthermore, the good correlations with associated effects
strongly suggest that the British Smoke concentration was a good surrogate
for the particles that actually caused the effects.

Recently, Bailey and Clayton (1982) have rechecked the calibration, and
find that, by the latter part of the 1970's, the calibration had changed for
all except a single test site in an area that had not yet seriously undertaken smoke control. In comtemporary London, particulate matter smaller than
4.5 μm behaves as though it were only 40 percent British Smoke; the balance
is some sort of colorless fine particulate matter. Thus, the decline in
smoke concentration and health effects did not involve an equal decline in
total fine particulate matter. Bailey and Clayton actually measured days on
which the true fine particle concentrations approached 250 μg/m^3. Statistically speaking, it is thus almost certain that concentrations of total fine
particles exceed 250 μg/m^3 with moderate frequency, yet the most sensitive
populations remain unaffected. The inevitable conclusion is that British
Smoke, not total fine particulate matter, is a measure of this effect on the
well-being of respiratory cripples.

Turning to welfare effects of airborne particulate matter, these would
appear superficially to include the reduction of visibility and the soiling
of surfaces. Soiling is a compound of two widely different effects;
deposition of predominantly black particles on the surfaces independent of
orientation, a function of fine particles, and the deposition of very coarse
dust, larger than roughly 50 μm, on horizontal or near-horizontal surfaces
(Lodge, 1984). For better or for worse, at this time there is no known
quantitative relationship between the concentrations of either of these
particle populations and the resulting soiling. Visibility effects are far
better understood, but the confounding effects of differing chemistry and
humidity result in major regional and temporal differences in the relationship between dry fine particle concentration and visibility degradation. In
addition, it is extremely difficult to specify a nationally acceptable
threshold at which visibility loss becomes adverse. Finally, implementation
of a fine particle standard becomes extremely complex, since the majority of
fine particles are not emitted as such, or locally, but as precursor gases
with sources as much as two or three states distant from the location at which
they cause visibility degradation.

The conclusion is that at this time the sort of quantitative relationship between concentration and effect necessary for the setting of standards does not exist with regard to the welfare effects of airborne particles. However, it seems clear that the precursor gases are likely to be controlled in the near future for other reasons, such as compliance with ambient sulfur oxide and nitrogen oxide standards, and otherwise to decrease precipitation acidity. In addition, energy costs provide an incentive to decrease the discharge of soot that has nothing to do with environmental concerns, but that probably will result in a still further decrease in black particulate matter. It should be observed that the carbon content of American particulate matter is already very similar to that in contemporary London (as contrasted with the high soot content there during the period when the epidemiological studies identified adverse effects).

The employment of the lognormal distribution in both the description of particle size distributions and sampler effectiveness curves makes possible the rapid calculation of theoretically collected particle concentrations. The mathematics are, of course, more precise than the experiments, but the behavior of the resulting equations permits the prediction of sampler behavior, the verification of questionable effectiveness curves and, if enough different samplers are deployed simultaneously, the inference of particle size distributions.

Based on those considerations, the methods proposed by EPA to date (EPA, 1984) for qualifying samplers as Federal Reference Method devices are deficient. Since health effects are, for practical purposes, mediated by the respiration of particles, it is desirable as nearly as possible to approximate the effectiveness curves of the human respiratory tract. These have been best demonstrated by the studies of Swift and Proctor (1982), and would correspond to a sampler with a d_{50} of approximately 6 μm. Such a sampler can be shown to sample approximately the same percentage of coarse particles as the human respiratory tract, and to oversample fine particles. In view of the chemical composition of the respective fractions this appears prudent; the potential toxicity of the fine particles is almost certainly greater than that of the coarse particles, which are predominantly soil. However, political considerations seem to militate in favor of a sampler with a d_{50} of 10 μm. Furthermore, the standards set forth in the discussion draft of the proposed National Air Quality Standards include a test size distribution containing almost no coarse particles, with the result that almost any sampler will collect the same mass within the specified 10 percent. These same samplers, exposed to particles in the arid west, where coarse particles frequently exceed fine particles by a factor of 5 to 10, give resulting concentrations that differ by as much as a factor of 1.5. If differing samplers are to be used as the basis for the crucial decision as to whether a location does or does not attain compliance with the standard, the most extreme distributions must be specified for testing. In addition, better and more uniform methodology must be specified for the determination of effectiveness curves. Finally, field testing under extreme conditions is needed, and it needs to be demonstrated that the samplers actually yield the collected masses predicted by the theoretical calculations.

Finally, assuming that we have samplers that collect in the specified relationship to the mass of particles present, let us put the pieces together and consider what the values of the standard might be. It was noted above that the threshold of effect found in the British studies was about 250 µg/m³ of British Smoke, 24-hour average (in the presence of more than double that concentration of sulfur dioxide). In terms of contemporary fine particulate matter, this corresponds to a fine particle mass 2.5 times as great. Or, to put it another way, if the standard were set on the basis of the 250 µg/m³ of fine particulate matter, it would have a built in safety margin of more than 2.5. Consideration of the relationship between fine particles and total mass less than 10 µm and decoupling the high sulfur dioxide level raises this safety factor to an excess of 3, and other matters deriving from the statistical form of the standard raise this safety factor still higher.

There appears to be no justification for such a large margin of safety. On the other hand, proposing that the standard should be expressed as a value to be exceeded only once annually constitutes an unwarranted incursion into the realm of extreme value statistics. (Almost everyone has heard of areas that have had three or four "100-year floods" in the past decade.) Practically speaking, this results in a major, further increase in the margin of safety. If single years, or even three-year averages are to be utilized, a sensible alternative is rather to decrease this excessive margin of safety by setting the standard in terms of a percentile. If the standard were set at 250 µg/m³ 24-hour mean, at the 95th percentile, there would still be a safety margin in excess of a factor of 2. Since the available evidence suggests that chronic effects are simply the sum of repeated acute exposures, the annual level should merely be set at the corresponding geometric mean.

References

Bailey, D.L.R. and Clayton, P. (1982), "The Measurement of Suspended Particle and Total Carbon Concentrations in the Atmosphere Using Standard Smoke Shade Methods," Atmos. Environ. 16:2683.

Environmental Protection Agency (1984), "40 CFR Parts 50, 53, and 58 Proposed Rules," Federal Register, 49:10408.

Ferris, B.G., Jr., Higgins, I.T.T., Higgins, M.W. and Peters, J.M., (1973) "Chronic Non-Specific Respiratory Disease in Berlin, New Hampshire, 1961-67," A follow-up study. Am. Rev. Respir. Dis. 107:110.

Ferris, B.G., Jr., Chen, H., Puleo, S., and Murphy, R.L.H., Jr., (1976), "Chronic Non-Specific Respiratory Disease in Berlin, New Hampshire, 1967-1973," A further follow-up study. Am. Rev. Respir. Dis. 113:475.

Lawther, P.J., (1958), "Climate, Air Pollution and Chronic Bronchitis," Proc. R. Soc. Med. 51:262.

Lawther, P.J., Waller, R.E. and Henderson, M. (1970), "Air Pollution and Exacerbations of Bronchitis," Thorax 25:525.

Lodge, J.P. (1984), "Laboratory and Field Studies of Soiling," EPA Workshop on Research Strategies to Study the Soiling of Buildings and Materials, April, Environmental Protection Agency, Res. Tri. Pk., NC., p. 19.

Swift, D.L. and Proctor, D.F. (1982), "Human Respiratory Deposition of Particles During Oronasal Breathing," Atmos. Environ. 16:2279.

Published 1984 by Elsevier Science Publishing Co., Inc.

Aerosols, Liu, Pui, and Fissan, editors

MUTAGENIC ACTIVITY AND COMPOSITION OF ORGANIC
PARTICULATES FROM DOMESTIC HEATING AND
AUTOMOBILE TRAFFIC

G.W. Israël, R. Freise and J. Büsing
Technische Universität Berlin
Fachgebiet Luftreinhaltung/Sekr. KF 2
Straße des 17. Juni 135
1000 Berlin 12
        West Germany

EXTENDED ABSTRACT

During the past years we have studied the spatial and temporal variation
of the mutagenic concentration (MC+) of the cyclohexane extractable organic
fraction (CEOM) of the ambient particulates (TSP). The CEOM fraction was
also analysed for several polycyclic aromatic hydrocarbons (PAH). The mutagenicity was determined by the Ames-Salmonella/mammalian microsome test
using strain TA 98 with metabolic activation (1).

In order to gain some understanding as to the sources of the ambient
CEOM fraction and its mutagenic activity we studied the emission of two
types of likely sources, namely those of coal stoves and of motor vehicle
traffic.

The emissions from hard coal (HC) and brown coal (BC) were collected
from the chimney plume near by the chimney exit. The plume collections MIX
were collected from various chimneys of a large appartment building without
knowledge of the solid fuel which was burnt in the stoves. At the point of
collection the emissions were cooled to near ambient temperature. The results
of these plume measurements are shown in table 1. They are reported as total collected material.

The motor vehicle traffic emissions (MV) were determined from samples
taken in a 1 km long one-way traffic tunnel close to its exit. They are
expressed as concentrations in the tunnel air (table 1).

For comparison respective ambient concentrations are shown for average
winter time conditions (WIN) and a period of very high ambient pollution
levels (SMOG). The ambient measurements (RF) were taken on the roof of the
building during the time period when the plume measurements (MIX) were
carried out. The average winter time level of TSP = 100 $\mu g/m^3$; MC+ = 8
Rev/m^3 and BaP = 6.7 ng/m^3 were exceeded by a factor 2 - 6 during the roof-
measurement period (RF) and by a factor of 15 - 25 during the 4-day smog
period.

The ratios of some selected properties are summarized in table 2. The
ambient MC/CEOM-ratio ranges from 0.4 to 0.8 and appears to be a little
lower than the ratio for the investigated sources (range 1.0 - 1.9). There
is no significant difference between the mutagenicity of the CEOM fraction
for motor vehicle traffic and coal combustion. This might explain the fact

that the MC+/CEOM ratio shows no seasonal changes in ambient air in Berlin.

Also the ambient MC+/BaP ratios are lower than those of the source emissions. It could be speculated that this might be due to a greater atmospheric instability of certain mutagenic substances in comparison to CEOM and BaP.

The CEOM fraction of the coal emissions is very high accounting for 36 - 57 % of total particulate emissions. The CEOM fraction of motor vehicle traffic averages 16 %. In Berlin the ambient CEOM fraction of TSP averages 8.4 % with a mean value of 10 % for the winter. This is much lower than for the discussed sources. The reason being that other sources and secondary inorganic aerosol production "dilute" the emitted CEOM fraction. The inhanced CEOM-fraction during the periods SMOG and RF indicate a relative large contribution of coal emissions to the ambient particulates.

At present we are in the process of determining emission factors for these and other sources and hope to be able to conduct source apportionment for the mutagenicity of ambient particulates by chemical element balance in the near future.

References

Israël, G. W. and J. B. Büsing (1983) "Investigation on the Mutagenic Action of Urban Particulates", J. Aerosol Sci. 14:192.

Table 1: Total collected material of plume collection, motor vehicle emissions and ambient levels

	n	TSP	CEOM	MC+	BeP	BkF	PER	BaP	IND	BghiP	ANT	COR
	-	$\frac{mg}{sample}$	$\frac{mg}{sample}$	$\frac{Rev}{sample}$	$\frac{\mu g}{sample}$ →							
a) plume collections												
HC	4	158	66	$15\cdot10^4$*	92	72	11	84	93	92	23	12
BC	5	131	58	$9\cdot10^4$*	11	9	2	12	8	8	3	10
MIX	8	239	132	$15\cdot10^4$	49	22	5	45	34	27	3	24
b) motor vehicle emissions and ambient levels												
	-	$\frac{\mu g}{m^3}$	$\frac{\mu g}{m^3}$	$\frac{Rev}{m^3}$	$\frac{ng}{m^3}$ →							
MV	10	347	56	76	20	4	7	19	14	38	7	24
WI	7	101	10	8	-	4	1	7	-	10	-	-
SMOG		946	265	142	-	-	13	92	-	75	-	32
RF	14	230	58	24	27	17	2	39	22	22	7	13

* one measurement only

263

Table 2: Ratios of selected properties

	$\dfrac{\text{MC+}}{\text{CEOM}}$	$\dfrac{\text{MC+}}{\text{BaP}}$	$\dfrac{\text{CEOM}}{\text{TSP}}$
	$\dfrac{\text{Rev+}}{\mu g}$	$\dfrac{\text{Rev+}}{ng}$	$\dfrac{\mu g}{\mu g}$
HC	1.9*	0.9*	0.40
BC	1.4*	3.8*	0.36
MIX	1.0	2.8	0.57
MV	1.4	4.0	0.16
WI	0.8	1.2	0.10
SMOG	0.5	1.6	0.28
RF	0.4	0.9	0.24

* one measurement only

Published 1984 by Elsevier Science Publishing Co., Inc.
Aerosols, Liu, Pui, and Fissan, editors

HEAVY METAL CONCENTRATIONS IN URBAN AEROSOLS FROM THE CITY OF BERLIN - SIZE DISTRIBUTION, ENRICHMENT, METEOROLOGICAL IMPLICATIONS AND HYGIENICAL RELEVANCE

J. U. Lieback and H. Rüden
Institut für Allgemeine Hygiene der
Freien Universität Berlin
Hindenburgdamm 27
D-1000 Berlin 45
West Germany

EXTENDED ABSTRACT

During the winter of 1981/82 there have been some severe episodes of air pollution in Berlin, with particle mass concentrations reaching up to over 1 mg/m³ within ambient air at some sites (1).

Aerosol (µg/m³)	Elements (ng/m³)	Fe	Pb	Cd	Mn	Ni	V	Cu	Cr	Zn	As
129	x̄	2010	342	4.5	74	32.9	33.8	208	22.0	630	16.2
33	min	450	94	1.1	14	9.2	4.5	74	6.2	90	1.8
412	max	9660	1040	15.1	286	101.4	101.5	932	98.2	4340	67.3

TABLE 1: Medium concentrations and extreme values of the aerosol and elemental mass within particles smaller than 10 µm

Because such pollution episodes occur quite often during the winter time, especially on cold days under the influence of very stable high pressure systems with low mixing layers due to temperature inversion, we decided to start a measurement program in Berlin, gathering the particle mass and determining its compounds continously over a one year period at a rather polluted site of the city (2).
The measurements have been carried out with a six-step-plus-backup sierra high-volume cascade impactor, taking three filter sets per week over 48 hour intervals (2).
16 heavy metals contained within all size fractions seperated have been determined so far by means of the atomic absorption spectroscopy,to get some idea of the hygienical relevance of the ambient particulate material and to recieve some information about possible air pollution sources (2).
Together with meteorologic data and informations on the chemistry of particles from different sources, we hope to support and broaden our recent knowledge on the major air polluters, by contributing the material on heavy metals into an overall balance for air pollutants. This includes an estimate of the transport of gaseous and particulate pollutants inside and outside of the city.
Now there is great convenience among the workers, that especially the lokal coal and oil heating, emitting pollutants low above the ground, is the main source and contributes most to the heavy burden of particulate material and sulfurous smog during the worst pollution episodes.

Elements	Concentration (ng/m³)			Aerosolportion (ppm)			MMD		(µm)
	summer	winter	factor	summer	winter	factor	summer	winter	factor
Iron	1170	3370	0.35	15360	18330	0.84	2.50	3.65	0.70
Lead	305	378	0.18	4050	2650	1.53	0.66	0.85	0.78
Cadmium	3.6	5.7	0.63	51	31	1.65	0.77	0.77	1.00
Manganese	49	116	0.42	692	652	1.06	1.68	1.94	0.87
Nickel	17	51	0.33	244	355	0.69	o.87	0.85	1.02
Vanadium	16	52	0.31	207	340	0.61	0.62	0.85	0.73
Copper	257	206	1.25	3770	1270	2.97	1.27	1.48	0.86
Chromium	15	22	0.71	247	199	1.24	0.54	0.96	0.56
Zinc	311	1170	0.27	4390	6700	0.66	1.10	1.08	1.02
Arsenic	10	21	0.47	135	136	0.99	0.77	1.48	0.52
Aerosol(µg/m³)	81	200	0.40	-	-	-	0.99	0.88	1.13

TABLE 2: Average concentration (ng/m³), portion of the aerosol (ppm) and MMD (µm) for some heavy metals during the summer and during the winter months (Jul., Aug./Dec., Jan., Feb.). Always given is also the summer/winter- factor. For reason of comparison we included the values of concentration (µg/m³) and the MMD for the aerosol mass.

Elements	Stage							Sum of the Stages 2-Backup <10 µm
	Backup <.4 µm	6 <.7 µm	5 <1.3 µm	4 <2.1 µm	3 <4.2 µm	2 <10 µm	1 >10 µm	
Iron	173	125	159	267	417	868	652	2010
Lead	110	64	52	48	34	34	14	342
Cadmium	1.19	0.98	0.84	0.66	0.37	0.50	0.28	4.45
Manganese	7.5	10.2	11.1	12.8	12.0	20.0	14.1	73.6
Nickel	10.2	5.2	3.9	3.7	4.1	5.8	4.3	33.8
Vanadium	10.6	8.1	4.1	2.9	2.7	5.5	4.2	33.8
Copper	31	31	38	40	31	37	19	208
Chromium	6.5	3.2	3.2	3.0	2.7	3.5	2.9	22.0
Zinc	98	118	121	120	99	60	113	625
Arsenic	3.4	3.2	2.6	2.2	2.9	2.9	1.8	16.2
Aerosol (µg/m³)	36	23	19	16	12	24	21	129

TABLE 3: Average immission concentrations of some selected heavy metals (ng/m²) and of the aerosol mass (µg/m²) in each size interval of the impactor. Also given are the sums of the components within particles smaller than 10 µm.

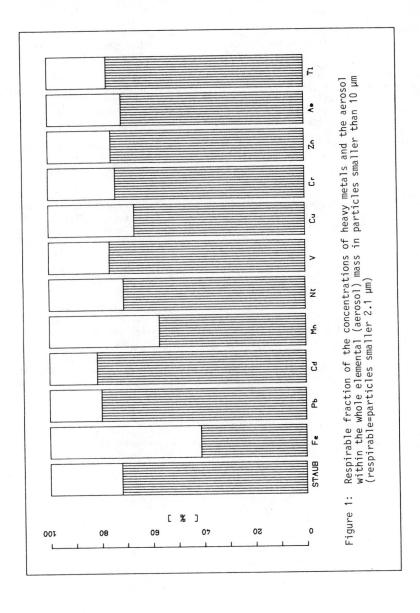

Figure 1: Respirable fraction of the concentrations of heavy metals and the aerosol within the whole elemental (aerosol) mass in particles smaller than 10 µm (respirable=particles smaller 2.1 µm)

268

Table 2 demonstrates that lead (that is mostly contributed from the traffic)
and copper (wich is very low in the coal burned within the city) decrease
on a ppm-base during the winter obviously, whereas zinc, vanadium and nickel
(who are preferably emitted from the burning of coal and oil) increase
markedly.As one can seeout of table 3 and figure 1, there are always 60-70
percent or more of most elements concentrated in the particle size fraction
smaller 2.1 µm, which is known to be certainly respirable.
Those facts in connection with the findings of other authors (1,2,3) demand
a rather quick improvement of the heating habits throughout the western part
of Berlin.
This might though resolve the situation not completely, because of the
rather large influence of air pollutants beeing blown in from the east of
Berlin and having to travel just a few kilometers to pollute the city as a
whole.

1) Israel, G.W., K. Wengenroth und H.-W. Bauer: Schadstoffzusammensetzung
der Luft während Smogsituationen in Berlin (West). Gutachterlicher
Bericht des Fachgebiets Luftreinhaltung der Technischen Universität
Berlin an den Senator für Stadtentwicklung und Umweltschutz Berlin,
Berlin 1983

2) Lieback, J. U. and H. Rüden: Heavy metal concentrations in urban aerosols
of Berlin during episodes of very high particle concentrations in
the winter of 1981/82. J. Aerosol Sci. (in press)

2a) Lieback, J. U., L. Xander, I. Trauer, R. Kneiseler und H. Rüden: Zur
hygienisch-medizinischen Bedeutung von Schwermetallen im Aerosol
des Belastungsgebietes Berlin. Forum Städte-Hygiene 35, 2-10 (1984)

3) Moriske, H.J., M. Wullenweber, G. Ketseridis und H. Rüden: Untersuchungen
zur mutagenen Wirksamkeit von polycyclischen aromatischen Kohlen-
wasserstoffen (PAH) und polaren organischen Verbindungen (POC) im
Stadtaerosol. Zbl. Bakt. Hyg., I. Abt. Orig. B 176, 508-518 (1982)

Published 1984 by Elsevier Science Publishing Co., Inc.
Aerosols, Liu, Pui, and Fissan, editors

PERIODICITIES AND LOCAL FEATURES IN THE
BEHAVIOR OF AEROSOLS IN SOUTHERN FINLAND

R. Hillamo
Air Research Laboratory
Department of Aerology
Finnish Meteorological Institute
P.O Box 50, 00811 Helsinki
Finland

EXTENDED ABSTRACT

Introduction

In most cases the knowledge of the ambient aerosol concentrations based
on integrating sampling methods is insufficient. Therefore it is necessary
to know the time dependence and the local fine structure of the aerosol behavior
if health or non-health effects will be considered. In this work some typical
measurement environments of our laboratory were studied using modern aerosol
research instruments. Size distribution measurements were done using an
electrical aerosol analyzer (TSI Model 3030) and a low pressure impactor
(Hauke low pressure impactor LPI 25/0.015). Besides LPI sample collection
was carried out with multifilter technique (Nuclepore). One of the practical
aims of this work was to study the performance of the conventional sampling
methods our laboratory uses in background areas and in air quality measurements in urban areas.

Experimental

Measurements presented in this paper were carried out in three places:
in Helsinki, in Jokioinen (about 60 km NW from Helsinki) and in Raisio
(about 100 km W from Helsinki). The measurements in Helsinki were done in
an environment which consists of light industry (i.e. printing shops, metal
and plastic industry) forming a diffuse source of aerosols originating from
different processes. The measurement place in Raisio was influenced by
high traffic (in rush time about 3000 cars in hour). Jokioinen represents
an example of rural background areas. The measurement place was in the
sounding station in one of the observatories of the Finnish Meteorological
Institute.

In all cases two different instruments were used: an electrical
aerosol analyzer (EAA) for aerosol concentration and size distribution
monitoring, and a low pressure impactor (LPI) for size distribution
measurement and sample collection. Nuclepore low volume multifilter
technique (polycarbonate membrane filter, $\emptyset$ 25 mm, pore sizes 8.0 μm and
0.2 μm) was used in Raisio and Jokioinen.

In Jokioinen the sample for EAA was sucked through 50 mm diameter
PVC tube with blower (air flow 150 lmin^{-1}). Filter and LPI sampling
were carried out at the height of 1 m from ground level, about 15 m from the
sounding station building. In both methods the collection face was upward
and equipped with circular top and flange to achieve better transport
efficiency for coarse particles as described by Liu and Pui (Liu & Pui, 1981).

In Raisio the measurement was carried out using a mobile laboratory unit. Due to the immediate vicinity of the road (about 15 m) brief fluctuations in the aerosol concentration were typical especially in rush hour. Fluctuations made the size distribution measurement with EAA difficult. They were reduced by using a 20 l stabilizing chamber. After the chamber the EAA sample was sucked through a 2 m long PVC tube, 10 mm in diameter with 50 lmin⁻¹ total flow. The arrangement for the filter and LPI sampling was same as presented before.

In Helsinki the sampling with LPI and EAA was performed in the heart of the 1 km² industrial area. The sampling points were on the roof of a three-floor industrial building on top of a 2 m high mast. For EAA the sample was sucked through 5 m long PVC tube, 10 mm in diameter with 50 lmin⁻¹ total flow. In LPI the same inlet was used as before.

15 μm thick aluminium foils were used as collection surface in the low pressure impactor. In Jokioinen the foils were uncoated. In Raisio the preimpactor (over 16 μm aerodynamic equivalent diameter) was coated with thick (about 50 μm) Apiezon M vacuum grease layer to avoid coarse particle bouncing. In Helsinki all collection surfaces were coated with thin (35 μg/cm²) Apiezon H layer. To make a thin layer the Apiezon vacuum grease was first dissolved in toluene and then spreaded on the foil by wiping.

The impactor foils and the filters were weighed using Mettler M3 electronic microbalance (sensitivity 1 μg). The total error of weighing determined with blank foils and filters was in all cases below 3 μg.

Results

The particle size distributions based on EAA measurements were calculated using DISFITE log normal curve fitting program (Whitby & Whitby, 1982) and constant EAA sensitivities developed by Whitby and Cantrell (Whitby & Cantrell, 1979). The error caused by the use of constant sensitivities was studied by comparing results with one of the data reduction programs presented by Kapadia (EAAKIA; Kapadia, 1980). In wide concentration range 5–15 % differences were observed in the total volume concentration between the constant sensitivity and the iterative data reduction method. In this study EAA was used essentially as monitor for fine particle aerosol concentration. Therefore errors due to the constant sensitivities may be accepted. The accurate size distribution analysis will be performed out later using an iterative data reduction method.

The low pressure impactor results were evaluated by using DISFITE log normal curve fitting program. The impaction spots of fine particle stages (aerodynamic equivalent diameter below 2 μm) were studied with electron probe microanalysis. The mass size distribution parameters presented in table 1 were typical for samples collected in Helsinki. For the sample of table 1 the size distribution parameters of elemental sulfur (assuming log normal distribution) were DG = 0.29 μm with geometric standard deviation σ_g = 2.34.

Table 1.   Mass Size Distribution Parameters For One LPI Sample
Collected In Industrial Area In Helsinki (Sampling Time 24 h)

Mode	DGM (μmAD)	σ_g	C (μgm^{-3})
Nuclei	.033	1.40	1.8
Accumulation	.35	1.85	14.4
Coarse	4.80	2.60	18.0

DGM = geometric mean by mass σ_g = geometric standard deviation
C = calculated concentration of the mode

Conclusions

In this work modern aerosol research instruments have been used to
study the significance of the particle size distribution in different
measurement environments.

In all cases (also in winter when ground was snow-covered) coarse
particle mass was over half of the total mass collected. Therefore total
suspended dust collected with filter sampling without size selective
inlet may often be incorrect in describing air quality.

It is possible to get accurate knowledge of measurement environment
in addition to methodical results by using particle size distribution
measurements. In the case of Jokioinen, local energy production is a
source of high winter concentration of fine particle aerosols. Measurements
with EAA and low pressure impactor showed with wind direction $240^{\circ} \pm 10^{\circ}$
concentrations which were about five times greater than typical background
concentrations measured with other wind directions. Effects of local
sources makes the use of the sounding station in Jokioinen as background
station difficult (Kulmala et al., 1982).

References

Kapadia, A. (1980) "Data reduction techniques for aerosol size
distribution measuring instruments", Particle Technology Laboratory
Laboratory Publication Number 413, University of Minnesota, Minneapolis.

Kulmala, A. (1982) & A. Estlander, L. Leinonen, T. Ruoho-Airola &
T. Säynätkari "Air quality at Finnish background stations in the 1970´s"
(Finnish), Ilmatieteen laitoksen tiedonantoja (Finnish Meteorological
Institute Publication) 36, Helsinki.

272

Liu, B.Y.H. (1981) & D.Y.H. Pui "Aerosol sampling inlets and inhalable particles", Atmos.Environ. 15:589.

Whitby, K.T. (1979) & B.K. Cantrell "Electrical aerosol analyzer constants", Aerosol Measurement (Edited by D.A. Lundgren et al.), University Presses of Florida, Gainesville.

Whitby, K.T. (1982) & E.R. Whitby "DISFITE size distribution and fitting program", Particle Technology Laboratory Publication Number 441, University of Minnesota, Minneapolis.

Published 1984 by Elsevier Science Publishing Co., Inc.
Aerosols, Liu, Pui, and Fissan, editors

HYDRATION OF AEROSOL PARTICLES IN THE LOS ANGELES AIR BASIN

M.J. Rood, D.S. Covert and T.V. Larson
Department of Civil Engineering, FX-10
University of Washington
Seattle, WA 98195

INTRODUCTION

The relationship between fine particle light scattering extinction coefficient (bsp), fine particle chemical composition and relative humidity (RH) were measured during an experiment in Riverside, CA during a five week period in August and September of 1983. These experiments involved the use of an integrating nephelometer as well as a system for controlling the temperature and RH of the air directly upstream of the nephelometer. A modified "humidograph" (Covert et al. 1972) was used to characterize the hygroscopic nature of the aerosol by measuring bsp under instrumentally controlled RH conditions. The "thermidograph" (Waggoner et al. 1983) was used to exploit the aerosol's hygroscopic properties and thermal decomposition characteristics to estimate its NH4+/SO4= molar ratio and sulfate mass concentration. A separate bsp measurement was also performed at ambient RH conditions to determine the state of hydration of the ambient aerosol.

INSTRUMENTATION

The humidograph system is schematically presented in Figure 1a. Initially, the sample airstream was diluted with dry filtered air to reduce its RH. In the humidifier, air was passed through a hollow cylinder whose walls were covered with nylon mesh and wetted with dilute H2SO4 solution. The walls were gradually heated from an external source such that the aerosol experienced a corresponding increase in dew point temperature. The airstream exiting the humidifier was then heated to a peak constant temperature of 60 C. Increased dry bulb temperature decreased the aerosol's RH to < 20 % which was sufficient to effloresce the influent aerosol. Subsequent cooling of the aerosol increased the RH to its final value measured at the nephelometer inlet. Because the cooler was maintained at room temperature, all of the air returned to room temperature but none of the air dropped below that temperature. Therefore, subsequent to leaving the heater section, the highest RH experienced by the aerosol was at the nephelometer inlet. As RH was scanned, bsp and RH values were stored in a microcomputer. The resultant humidogram consisted of a plot of RH versus normalized bsp values (the ratio of measured bsp to that at the beginning of the RH scan).

In the continuous flow thermidograph system, aerosol was preheated, cooled, humidified, heated, cooled and then sensed by an integrating nephelometer (Fig. 1b). The initial thermal treatment consisted of heating the aerosol to a predetermined temperature, usually between 150 and 200 C, to vaporize relatively "volatile" liquid and solid components of the aerosol. As the aerosol exited this "preheater", it was rapidly cooled to room temperature. The vaporized particulate components either recondensed to very small particles that did not efficiently scatter light, adsorbed onto the cooler's walls, or both. The aerosol was then passed through a humidifier that consisted of an annular space between concentric tubes. The inner and outer walls were covered with felt and nylon mesh, respectively, and were wetted with a dilute H2SO4 solution. The outer wall of the humidifier was heated to produce a RH of 65 % at approximately 20 C. However, as the aerosol passed through the humidifier, the particles experienced a RH greater than 80 % because of the suppressed dry bulb temperature caused by evaporative cooling. As the aerosol exited the humidifier, it was heated such that the amount of power applied to the main heater caused the aerosol temperature to scan from room temperature to 380 C in 3.5 minutes. Aerosol exiting this main heater was then rapidly cooled to room temperature in another heat exchanger and then sensed by an integrating nephelometer between 65 and 68 % RH. During a temperature scan, bsp and heater temperature data were stored in a microcomputer until a "thermidogram" was plotted. Such a graph consisted of a plot of heater temperature versus normalized bsp (the ratio of measured bsp to that at approximately 20 C heater temperature).

The state of hydration of the ambient aerosol was studied by comparing bsp values measured in ambient air to bsp measured in air that was dried to 30 % RH. Care was taken to prevent unwanted heating in the nephelometer that was measuring ambient bsp. Data obtained from this test was presented as a ratio of bsp (ambient RH) to bsp (RH = 30 %) as a function of time.

RESULTS

Select humidograph results for ambient (Fig. 2A) and ambient "preheated" to 100 C (Fig. 2B) indicate the presence of hygroscopic aerosols that do not significantly change their hygroscopic properties upon heating to 100 C. Typically, the magnitude of bsp change was between 20 and 70 % for ambient aerosols during RH scans from 55 to 85 %. Such a change in bsp indicates a proportional amount of H2O associated with the aerosol. When the aerosol was heated to 200 C and NH3 was added upstream of the system (Fig. 2C), a sharp deliquescence step appeared at 78 % RH. Analysis of all of the humidogram data indicated that deliquescence was observed in 52 % of the measurements and the deliquescence humidity varied from 73 to 78 % RH with a median value of 76 %. The magnitude of the bsp increase upon deliquescence varied from 10 to 50 %.

Typical thermidograph results (taken at the same approximate time as the above humidograms) for ambient, ambient with NH3 addition and ambient preheated to 200 C with NH3 addition are presented in Figure 3D through 3F, respectively. Note the similiarity in the ambient and ambient with NH3 addition thermidogram curves. Such results indicate that the ambient aerosol was already chemically neutralized with respect to NH3. Upon preheating this neutralized aerosol, the resultant thermidogram curve had the same structure as (NH4)2SO4 test aerosol. Combined results from the humidograph and thermidograph indicate the presence of a neutralized sulfate aerosol mixed with other more volatile compounds, presumably NH4NO3 and organic compounds.

Results from the separate light scattering measurements at ambient and 30 % RH are shown in Figure 3. Such a figure presents the ratio of bsp at ambient RH to that at 30 % RH and is plotted as a function of time. Note the corresponding diurnal response of the bsp ratio with ambient RH. By utilizing results from the humidograms and Figure 3, it was apparent that the particles were hydrated a majority of the time although the ambient RH was below the aerosol's observed deliquescence humidity.

SUMMARY AND CONCLUSIONS

Integration of experimental results from the humidograph, thermidograph
and ambient light scattering measurements indicate that during the time
period of our study, the aerosol was typically hydrated, even at ambient RH
values less than the aerosol's deliquescence humidity. In addition, the
aerosol frequently exhibited deliquescence phenomena that was enhanced by
thermally stripping the more volatile compounds from the particles but was
not enhanced by addition of gaseous NH3. Test results suggest an aerosol
consisting of ammonium sulfate mixed with other more volatile material.

REFERENCES

Covert, D. S., Charlson, R. J. and Ahlquist, N. C. (1972) "A Study of the
Relationship of Chemical Composition and Humidity to Light Scattering by
Aerosols", J. App. Meteor. 11: 968-976.

Waggoner, A. P., Weiss, R. E. and Larson, T. V. (1983) "In-Situ Rapid
Response Measurement of H2SO4-(NH4)2SO4 Aerosols in Urban Houston: A
Comparison with Rural Virginia", Atmos. Environ. 17: 1723-1731.

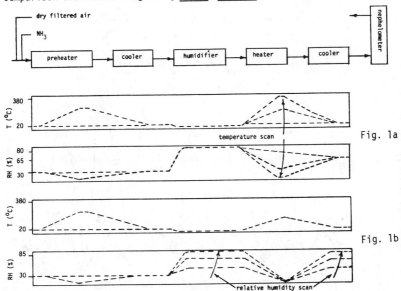

Figure 1. Block diagram of integrated humidograph-thermidograph system with
preheater-cooler assembly. Note the temperature and relative humidity
exposure patterns of the aerosol as it passes through the humidograph (1a)
and thermidograph (1b).

276

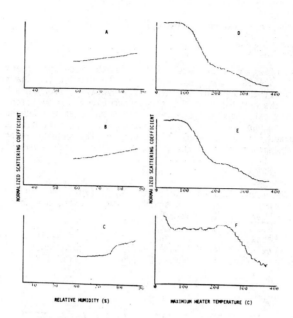

Figure 2 . Humidograms (A to C) and thermidograms (D to F) of ambient Riverside, CA. aerosol. The tests were taken at approximately 7:00 PDT on 09/17/83. Ambient bsp was 2x10-4 m-1.

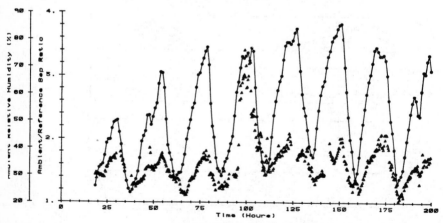

Figure 3 . Relative Humidity and
Bep Data From U.C.R. Experiment
Circle - Relative Humidity
Triangle - Bepa/Bepr Ratio
Time(0.0) - 00.00, 09/11/85
09/11-19/85

.uuu1shed 1984 by Elsevier Science Publishing Co., Inc.

Aerosols, Liu, Pui, and Fissan, editors

THE PACIFIC MARINE AEROSOL AND EVIDENCE FOR A FINE PARTICLE

SULFURIC ACID COMPONENT

Antony D. Clarke, David S. Covert, Norman C. Ahlquist
Environmental Engineering and Science Program
University of Washington
Seattle, Washington 98195

During the past several decades, a large number of investigators have
been involved in measurements of the optical, physical and chemical
properties of atmospheric aerosol. The majority of these efforts have been
directed towards continental aerosol and anthropogenic sources.
Motivations for these efforts have often involved concern over visibility
impairment, chemical conversion processes, burdens, fluxes, deposition,
and transport of pollutant emmisions. Evidence has accumulated that
indicates active transport mechanisms exists for moving primary and
secondary aerosol species over distances exceeding one thousand kilometers
(e.g. Asian dust, Arctic haze). Transport over these and even smaller
spatial scales will frequently involve the influence of natural aerosol
sources on the aerosol composition. In turn, aerosol composition
influences or controls the impact of aerosol on visiblity degradation,
acid rain phenomena, cloud nucleation and scavenging etc. Consequently;
the optical, chemical and physical characteristics of the so called
natural "background" aerosol are important to know in order to better
understand the properties and effects of continental and "polluted"
aerosol.

This motivation was partially responsible for participation in a
joint Soviet-American research cruise (R/V KOROLEV) in the Pacific Ocean
during November-December 1983 and a similar NOAA cruise (R/V RESEARCHER)
in May of 1984. At the time of this writing, the latter cruise is getting

underway and only preliminary data is available for the former. Some of these preliminary observations on the remote Pacific aerosol are reported here.

Aerosol sampling was performed aboard R/V KOROLEV on a continuous basis between 30 N and 30 S latitude along a longitude of ca. 155 W. The sample inlet was located at the top of a forward mast at ca. 14 m above the ocean surface. A vacuum blower brought the aerosol through 25 m of 3 cm diameter vinyl tube to a sampling plenum in less than 6 s. The sample plenum was in the forward laboratory at deck level and was provided with sample ports for each instrument. The majority of the flow exited the plenum to an integrating nephelometer (MRI 1550) before entering the blower. The nephelometer measured the scattering coefficient, b_{sp}, of the aerosol after preheating the ambient air about 5 C in order to reduce sensitivity of the instrument response to fluctuations associated with excursions in relative humidity, RH. Twenty four hour filter samples were also collected on Nuclepore and quartz fibre filters for future light absorption and chemical analysis. A condensation nuclei (CN) counter monitored for high CN values associated with ship contamination and terminated sampling when a programmable threshold value was exceeded. A laser optical particle counter (Particle Measurement Systems ASASP-X) counted particles in the 0.1 to 2.5 um diameter range in sixty channels and was operated under Apple computer control to provide and store three hour averages of the size distribution. Continuous data were collected and stored on floppy disk for b_{sp}, CN, temperature, RH, wind speed, wind direction, ozone, and % solar insolation.

The majority of data were collected in the central Pacific ocean between 30° N and 30° S and 130° W and 155° W. During the KOROLEV cruise, a general symmetry in several aerosol optical and physical parameters was observed for both Northern and Southern hemispheres centered on the Intertropical Convergence Zone (ITCZ), while a minima in aerosol mass concentration was observed for this region (ca. 2° N to 8° N). An exception was observed for the fine particle size distribution as measured by an active scattering aerosol spectrometer probe (ASASP-X). The ASASP-X revealed large increases in the accumulation mode mass (mode diameter centered at ca. 0.2um) immediately south of the ITCZ (Figure 1). A factor of five increase over the typical measured background values persisted between 0° S and 10° S and coincided with regions of CO_2 supersaturation in the relatively colder surface water flowing westward from South America. In general, the accumulation mode mass varied independently from the coarse mode mass and independent of surface wind speed.

While in this region, measurements were made of the changes in the ASASP-X size distribution during preheating the aerosol to 300 C. For certain compounds, the physio-chemical changes due to prehating provide a means to characterize aerosol molecular species. An example of the variation in the size distribution for selected temperatures is included in Figure 2. Changes in the fine mode as a function of temperature are consistent with behavior expected for a mixed aerosol consisting of sulfuric acid and its ammoniated salts. As in most cases, no evidence for a significant contribution of sea salt to this mode is observed.

The size distribution and thermal decomposition data were collected in concert with the additional data mentioned above. The latitudinal variability in the data sets collected under both programs in the mid-Pacific will be discussed and compared.

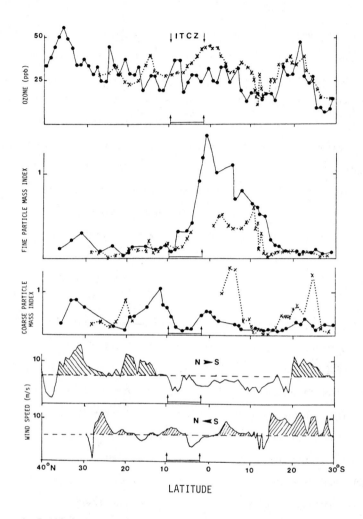

Figure 1. Preliminary unrefined SAGA data for ozone, aerosol mass index, and wind speed collected for the cruise of the AKADEMIC KOROLEV, Nov.-Dec. 1983. Data collected from 40°N to 30°S indicated by solid line (——●——) and data from 30°S to 30°N indicated by dotted line (···×···). Note that high winds speeds are related to coarse mass variability but not to fine mass variability. The large increases in fine mode aerosol mass is clearly seen between 0° and 12°S for both cruise tracks at 155°W and 130°W longitude.

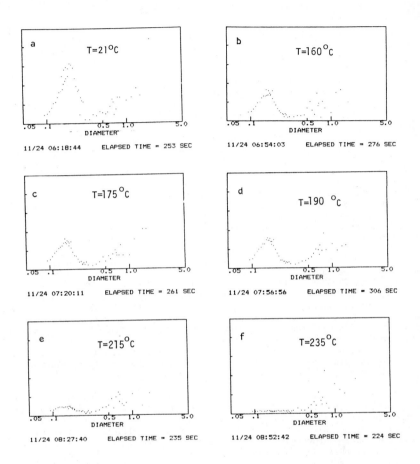

Figure 2. Particle volume distributions dV/d(log d_p) as a function of preheat temperature as measured with the ASASP-X. Data collected during SAGA cruise at 5°S and 130°W. Note rapid decrease in volume of fine mode for temperatures 140-155 °C.(e and f). This behavior is characteristic of ammonium sulfate and ammonium bisulfate in this system while the decrease for 21-160°C is typical of sulfuric acid.

Published 1984 by Elsevier Science Publishing Co., Inc.
Aerosols, Liu, Pui, and Fissan, editors

Analysis of Ambient Aerosol and Gaseous Pollutant Concentrations

John D. Spengler, P. Barry Ryan and Peter W. Severance

Department of Environmental Science and Physiology
Harvard School of Public Health
677 Huntington Avenue
Boston, MA   02115

September 17-21, 1984
Minneapolis, MN

Historically, air pollution measurements have included some measure of total suspended particulate matter (TSP), usually measured by a high volume (hi-vol) sampler and perhaps a measure of sulfur dioxide (SO_2). It was thought that these measures were the most important from both a regulatory and health effect point of view.

The Harvard Air Pollution Health Study was designed as both a cross-sectional (across 6 cities) and as a longitudinal examination of respiratory health in a population of 12,000 children and 10,000 adults. Throughout this study, the traditional measures of TSP and SO_2 have been made to be consistent with standards and the historic record. In addition to these pollutant variables, size fractionated particulates have been collected and analyzed for the water soluble sulfate as well as the elemental component. Ozone and nitrogen dioxide were also measured continuously and reported as hourly values. In the examination of health indices across populations, TSP and SO_2 are used as independent variables. This paper addresses the pollutant covariance structure with each city in an attempt to better define the "pollutant vectors".

In this paper we present analysis of data collected at various sites at which monitors for TSP, size fractionated particulate matter, and various gases were collocated. These data represent a three year period from 1978-1980. Analysis includes bi-variate correlations blocked so that the particulate-particulate and gaseous-gaseous interactions are investigated first. After discussion of these interactions, the "off-diagonal" particulate-gaseous inter-actions are investigated. Comparison with previous particulate-particulate and gaseous-gaseous work is made. The discussion of the interaction between the gaseous and particulate measures with respect to historical TSP and SO_2 measure-ments are made. Finally, suggestions from preliminary multi-variate analysis are presented.

Published 1984 by Elsevier Science Publishing Co., Inc.

Aerosols, Liu, Pui, and Fissan, editors

DEMONSTRATION OF THE MONTE CARLO PARTICLE DISPERSION (MocaPD) MODEL'S

CAPABILITY THROUGH COMPARISON WITH SMOKE WEEK IV DATA

K. H. Huang and Walter Frost
FWG Associates, Inc.
Rt. 2, Box 271-A
Tullahoma, Tennessee 37388

EXTENDED ABSTRACT

Introduction

The objective of this paper is to demonstrate the applicability and validity of the Monte Carlo Particle Dispersion (MoCaPD) Model methodology to war game scenarios and smoke generator designs.

The essential ifnormation required for an optical study of the short-term dispersion of battlefield obscurants is the spatial number density distribution for each particle size at any particular line of sight. These number density distributions should account for ground-level source effects and must be spatially and temporally dependent. The MoCaPD model has unique capabilities to predict these parameters including the effects of temporal mean wind speed variation with height over flat terrain and the qualitative effects of the wind speed variation due to terrain features.

Model Concept

The MoCaPD model computes the individual trajectories of many particles released into a simulated turbulent wind field. The equations governing the motion of the particles are the classical force balance equations with aerodynamic forces included as a function of the Reynolds number plus the Cunningham correction factor to include rarefaction effects. These equations can be written as

$$m\ddot{x} = -\frac{1}{2}\rho_a A C_D (\dot{x} - W_x)|V_a|$$

$$m\ddot{y} = -\frac{1}{2}\rho_a A C_D (\dot{y} - W_y)|V_a| \qquad (1)$$

$$m\ddot{z} = -\frac{1}{2}\rho_a A C_D (\dot{z} - W_z)|V_a| - mg$$

where m is the mass of particle with cross sectional area A, ρ_a is the air density, C_D is the drag coefficient, V_a is the velocity of the particle relative to the air, and g is the accelration of gravity. W_x, W_y, and W_z are the three instantanegus wind velocity components, and are the sum of the mean wind speed $\bar{W}(\vec{r})$ and the turbulence fluctuations $w(\vec{r},t)$, i.e.,

$$\vec{W}(\vec{r},t) = \overline{\vec{W}}(\vec{r}) + w(\vec{r},t) \qquad (2)$$

The horizontal mean wind field is reasonably and accurately described by a log-linear law (Frost et al. 1978, Haugen 1973) in the form:

$$\overline{W}_H = \frac{u_*}{\kappa}\left[\ln\left(\frac{z + z_0}{z_0}\right) + \psi(z/L') \right]$$

$$\overline{W}_z = 0 \qquad (3)$$

where u_* is the friction velocity, κ is von Karman's constant, z_0 is the roughness parameter, and ψ is a stability parameter which is a function of $\zeta = z/L'$, where L' is the Monin-Obukhov stability length. Figure 1 illustrates the method of simulating the turbulent component of the wind (Wang and Frost 1980). A Gaussian white noise is computer generated and passed through a digital filter designed to produce a random signal having the same statistical properties as the turbulence found in the atmosphere. However, the turbulent fluctuations encountered by the adjacent particles in the atmosphere form coherent Lagrangian turbulence. The turbulence generator of MoCaPD model is able to use z-transformation technique to generate a set of non-homogeneous coherent Lagrangian turbulence as the particle moves downstream. Details of this technique are given in Huang and Frost (1983). The form of the filter function is:

$$|H(\hat{n})|^2 = \phi(\hat{n}) \qquad (4)$$

where $\phi(\hat{n})$ is the power spectral density for the atmospheric turbulence being simulated. The Dryden spectrum with Lagrangian time scale is a proper description of Lagrangian turbulence (Huang and Frost 1983, Pasquill 1974).

A computer program which performs the computations described above allows rapid generation of many particle trajectories as functions of time and spatial position for realistic atmospheric conditions and specific source emission characteristics. The number of particles for size i is counted at each line of sight, thus producing the probability distribution of the number density $P_i(\vec{r},t)$ as a function of time and position. The concentration length CL (g/m^2) can be calculated for given source emission rates. According to Farmer (1981), the basic parameter needed to optically characterize an obscurant cloud is the extinction coefficient α which can be written as:

$$\alpha(\vec{r},t) = \frac{3 \sum\limits_{i=1}^{M} Q_{Ei} P_i(\vec{r},t) D_i^2}{2 \rho_0 \sum\limits_{i=1}^{M} P_i D_i^3} \qquad (5)$$

where Q_{Ei} is the extinction efficiency for particles of size D_i and ρ_o is the density of the obscurant. Finally, the transmittance of a line of sight can be calculated from the Beer-Bouguer law as

$$T(\vec{r},t) = e^{-\alpha(\vec{r},t)CL} \qquad (6)$$

Again, the transmittance is a function of position and time.

Model Evaluation with Field Test Data

Three obscurant field trials from Smoke Week IV (Nelson et al. 1982) were used for comparison with the model predictions. Those were: (1) Trial 8, which contained eight L8A1 grenades consisting of 2.88 kg red phosphorus (RP), which were detonated on tops of poles 5.5 m high. The wind speed was 2.25 m/s at the 2-m height, and the relative humidity was 59 percent; (2) Trial 16, which contained Hexachlorethane (HC). Six M116 rounds were simulated by detonating six groups of canisters. The total HC fill weight was 113.8 pounds. The wind speeds were marginally low (1.2 m/s at 2-m height) and the relative humidity was 95 percent; and (3) Trial 2, three Mitey Mite generators dispensing 22.7 kg of Arizona road dust were exploded. The wind speed was 1.7 m/s at a 2-m height and relative humidity was 60 percent.

The MoCaPD model predicted the extinction coefficient α in Trial 8 for RP at 3.4 μm wavelength to be in the range of 0.3 to 0.5 m^2/gm (as shown in Figure 2) when the RH is in the range of 60 to 70 percent. Also, values of α measured from high humidity hydroscopic smoke (H^3S) test (Farmer 1980) remain in the range of 0.2 to 0.5 m^2/gm up to a relative humidity equal to 60 percent. Figure 3 shows the comparison of the transmittance computed by the MoCaPD model and the measured data. The agreement of the magnitudes is good, but the time for the minimum transmittance at LOS has been shifted. Exact correspondence between the measurements and computations cannot be expected since the turbulent wind field cannot be simulated exactly. Figure 4 shows the comparison of the computed α predicted by MoCaPD is bracketed by the reported measured values. The prediction of the transmittance of Trial 16 from the MoCaPD model was of the same order of magnitude as the measured data, as shown in Figure 5. In Trial 2 the obscurant material is Arizona road dust which is non-hygroscopic. The absorption property of this material is so great and the particles are of a sufficient size that the extinction efficiency Q_E can be assumed to be equal to 2. Figure 6 shows the comparison of the computed α at 3.4 μm wavelength for Trial 2 and the averaged experimental data. The predicted α from the MoCaPD model closely corresponds to the PSI (particle size interferometer) measurement. Time history of transmittance at 3.4 μm wavelength for Trial 2 is wavy indicating that there is some chance that holes existed in the smoke. This phemonenon exists for both computed and measured data and might be a general characteristic of nozzle-type generators.

The results of the study indicates that the MoCaPD model has the capability of accounting for the source characteristics and has a unique

286

algorithm to generate coherent Lagrangian turbulence which is necessary
for particle trajectory calculations. The computed and measured data
compare reasonably well, indicating that the MoCaPD model is capable of
predicting the optical properties of the environment as a function of
time and position.

References Cited

Farmer, W. M. (1980). "A Summary and Evaluation of Data Obtained in the
    H^3S Test," Final Report, Gas Diagnostics Division, UTSI, Tullahoma,
    Tenn.

Farmer, W. M. (1981). "The Instrumentation Cluster Concept in Obscurant
    Field Testing," Smoke Symposium V, Harry Diamond Labs., Adelphi,
    Maryland, April 24-26.

Farmer, W. M. (1983). "A Summary and Evaluation of Data from Smoke Week
    IV," Final Report, UTSI, Tullahoma, Tenn.

Frost, W., B. H. Long, and R. E. Turner (1978). "Engineering Handbook
    on the Atmospheric Environmental Guidelines for Use in Wind Turbine
    Generator Development," NASA TP 1359.

Haugen, D. A. (editor) (1973). Workshop on Micrometeorology. Boston:
    American Meteorological Society.

Huang, K. H., and W. Frost (1983). "Monte Carlo Particle Dispersion
    (MoCaPD) Model of Battlefield Obscuration," Final Report for
    U.S. Army, Contract DAAG29-81-D-0100, DO #0563, FWG Associates,
    Tullahoma, Tenn.

Milham, M. E., and D. H. Anderson (1983). "Obscuration Sciences Smoke
    Data Compendium: Standard Smokes," ARCSL-SP-82024, Chemical
    Systems Lab., Aberdeen Proving Ground, Maryland.

Nelson, G. J., W. M. Farmer, V. E. Bowman, and J. E. Steedman (1982).
    "A Summary of Obscurant Characterization in Smoke Week IV:
    Vol. I," DRCPM-SMK-T-005-82.

Pasquill, F. (1974). Atmospheric Diffusion, 2nd edition. New York:
    John Wiley & Sons, Inc.

Wang, S. T., and W. Frost (1980). "Atmospheric Turbulence Simulation
    Techniques with Application to Flight Analysis," NASA CR 3309.

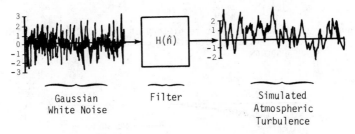

Figure 1   Turbulence simulation technique.

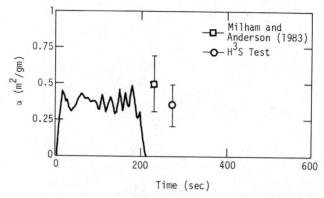

Figure 2   Comparison of the time history of the computed extinction coefficient and the averaged experimental data for red phosphorus.

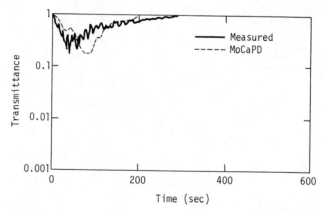

Figure 3   Comparison of the transmittance at 3.4 μm wavelength computed by the MoCaPD model and the measured data from Trial 8, Smoke Week IV, RP.

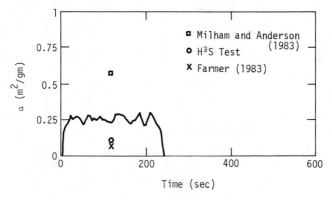

Figure 4   Comparison of the time history of the computed extinction
coefficient and the averaged experimental data of HC.

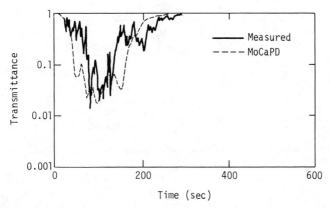

Figure 5   Comparison of the transmittance at 3.4 μm wavelength computed
by MoCaPD model and the measured data from Trial 16, Smoke Week
IV, HC.

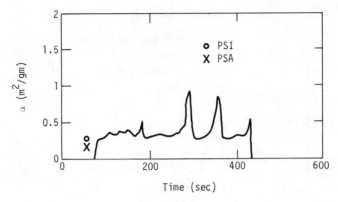

Figure 6   Comparison of the time history of the computed extinction coefficient and the averaged value reduced from measured data of Arizona road dust.

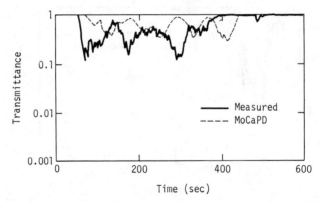

Figure 7   Comparison of the transmittance at 3.4 μm wavelength computed by the MoCaPD model and the measured data from Trial 2, Smoke Week IV, Arizona road dust.

Published 1984 by Elsevier Science Publishing Co., Inc.
Aerosols, Liu, Pui, and Fissan, editors

TRACE-ELEMENTS IN THE NATURAL AND URBAN AEROSOLS OF BRAZIL

C.Q.Orsini, P.A.Netto, M.H.Tabacniks, and V.L.Soares
Group for Studies of Air Pollution (GEPA)
Instituto de Física da USP - Caixa Postal 20516
01498 - São Paulo - Brazil

EXTENDED ABSTRACT

Introduction

Airborne particles play important roles in physical - chemical properties and air pollution problems of any atmosphere. Particularly, the fine and coarse aerosol components (respectively originated from gases, through gas-to-particle conversion, and mechanical pulverization plus dispersion processes) have a very important role in air pollution problems. These two components form the inhalable one, which interacts strongly with the human respiratory system through the respiration process.

An extensive project, which started in August/82, aiming at characterizing atmospheric aerosols, and evaluating air qualities of one natural and seven urban areas in Brazil (see Figure 1) is now being carried out by the GEPA. The investigations are based mainly on the measurement of the fine (< 2.5 μm) and coarse (2.5 - 15 μm) air particle concentrations (denominated parameters FP and CP), and on their elementary constitutions. The results of the measurements performed up to now are herein presented. For comparison purpose, results from the GEPA Pilot-Station (USP campus) and from the Amazon Forest Basin have also been included (Orsini et al.,1983, and Orsini et al., 1982).

FIGURE 1
The geografic location of natural and urban sampled areas (between parenthesis, millions of inhabitants).

292

Experimental

The Nuclepore Stacked Filter Unit, SFU, with filters with
8.0 μm (for CP) and 0.4 μm (for FP) pores are used. Each aerosol
under investigation is sampled weekly, during 2 or 3 days
(depending on its concentration), for one whole year, by means of
one SFU sampler. Changing of filters is provided by local
operators, and exposed filters are exchanged with GEPA by airmail,
in boxes of four SFU units. An intensive short-term (about one
week) sampling is also performed on each site, with two
simultaneous sampling stations, each one with SFU and 10-stage
cascade impactor (Battelle model).

All the SFU filters are gravimetrically analyzed, by means
of a high sensitivity (1μg) analytical balance, to determine the
FP, CP and IP (inhalable particle concentration) values. The
collected samples are analyzed by the PIXE (Particle Induced X-
-Ray Emission) method for trace-elements $(z \geq 11)$ determination
(Orsini et al., 1984). The X-ray spectra are automatically
reduced by means of the HEX-SP computer program. The detection
limit for low z elements is about 30 ng/m^3, and for medium z
elements, about 5 ng/m^3. Reproducibility is within 15% and the
accuracy is better than 25%.

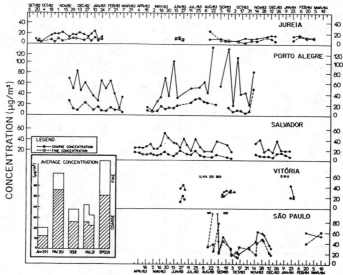

FIGURE 2
Temporal variation of FP and CP (on the left-side, below in
histogram format, the averaged results FP and CP for each sampled
site.

293

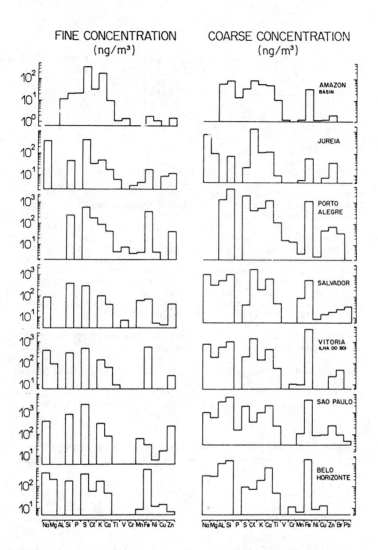

FIGURE 3-The Fine and Coarse Elemental concentrations.

294

Results and Discussions

Figure 2 presents the measured temporal variation of FP and
CP on five sampled sites. The fluctuation of FP values as
shown in Figure 2, is much smaller than for CP (except in São
Paulo). This fact reflects the more stable generation process
for fine particles (gas emissions plus gas-to-particle
conversion).On the same figure, the averaged values of these
parameters are included in histogram format. These averaged
results indicate an unfavorable air pollution picture in São Pau
lo and Porto Alegre.

Results on trace-element concentrations obtained through the
PIXE method are presented as averaged histograms for fine and
coarse particles, for each sampled site, in Figure 3. It is
remarkable that for the coarse particle histograms, the soil and
sea constituent elements (respectively Al, Si, Ca, Fe and Na,Mg,
Cl) are dominant, and with nearly the same ratios as their
respective sources. The significant presence of trace-elements
associated with dominant industrial activies as, for instance,
the iron in Vitória (where the largest Brazilian harbor for
iron-ore is located) and in Belo Horizonte (located close to
iron mines), and the sulfur in São Paulo and Porto Alegre is
also noticeable in these histograms.On the other hand the fine
particle histograms are dominated by sulfur both in natural and
urban areas (except in Vitória and Belo Horizonte).

Another interesting result refers to the average percentage
of the trace-elements to, respectively the FP and CP: 10% and
13% in Juréia; 13% and 17% in Porto Alegre; 7.8% and 25% in Sal-
vador; 14% and 19% in Vitória; 12% and 21% in Belo Horizonte;
9.3% and 18% in São Paulo. Variations is within 8% to 14%; for
FP and 13% to 25% for CP. This means that C and O make up about
90% for the FP and 81% for CP in the investigated aerosols.

References

Orsini, C.Q., Netto, P.A. and Tabacniks, M.H. (1982)
"Preliminary data on atmospheric aerosol of the Amazon basin"
Atmospheric Environment 16: 2177.

Orsini, C.Q.. Artaxo, P. and Tabacniks, M. (1983) "Trace-
Elements in the Urban Aerosol of São Paulo", Proc. of VIth.World
Congress on Air Quality, Poster-Presentation: 69.

Orsini, C.Q., Netto, P.A. and Tabacniks, M.H. (1984) "The
São Paulo - PIXE System and its use in a National Monitoring
Air Quality Program", being published in the Nucl. Instr. and
Meth.

Published 1984 by Elsevier Science Publishing Co., Inc.
Aerosols, Liu, Pui, and Fissan, editors

LARGE PARTICLE SIZE DISTRIBUTION IN FIVE U.S. CITIES
AND THE EFFECT ON A NEW AMBIENT AIR QUALITY STANDARD

Dale A. Lundgren and Brian Hausknecht
Environmental Engineering Sciences
University of Florida
Gainesville, FL 32611

EXTENDED ABSTRACT

Introduction
   A mobile aerosol size classifying sampling system for the collection of
very large (100 μm diameter) particles was designed and constructed by
Lundgren and Rovell-Rixx at the University of Florida.[1] An analysis van was
outfitted to accompany the sampling trailer. A specially designed air
sampling inlet was fitted to a very high flowrate (~40 m^3/min) sampler, which
greatly reduced the large particle sampling errors due to inertial effects, as
described by Lundgren and Paulus.[2] In a 10 km/hr wind, the design criteria
predicted a less than 20% error for sampling 100 μm particles. Test results
indicated that the sampling error was within this design limit.

   Ambient aerosol mass distribution were measured in five cities across the
U.S. and compared with data collected using several conventional ambient
aerosol samplers and size selective inlet (SSI) samplers. The five cities
sampled were Birmingham, Alabama (an industrial area); Research Triangle Park,
North Carolina (a background site); Philadelphia, Pennsylvania (metropolitan
site); Phoenix, Arizona (high fugitive dust area); and Riverside, California
(photochemical aerosol site). These cities provided a variety of sampling
conditions and aerosol compositions. The actual location selected in each
city was at an EPA Inhalable Particulate (IP) Network station where a history
of data for the high-volume air sampler, size selective inlet sampler and the
dichotomous sampler were available.

   Present ambient air quality standards for particulate matter are based on
measurements made by the EPA reference method (High-Volume Method).[3] Weight
gain by the sampler filter media, divided by the volume of air sampled, is
defined as total suspended particulate matter, or TSP. Health effect studies
have correlated this measurement with adverse health effects. However, it is
generally accepted that some of the particulate mass collected by this
reference method sampler is too large to cause health effects. This has
resulted in proposed changes to the primary air quality standard and method of
measurement. If new standards are to be set and the method of measurement
changed it is necessary to determine the relationship between the present
reference method (High-Volume Method) measurements and a size selective
reference method measurement.

Equipment
   Large particle size-distribution data were obtained using the mobile
aerosol-sampling system, called the Wide Range Aerosol Classifier (WRAC). A
schematic diagram of the WRAC is shown in Figure 1. The large (60cm) diameter
aerosol inlet tube leads to a cluster of five individual sampler units, each
of which operates at an actual sampling rate of 1.56 m^3/min (55 acfm). The
center sampler collects what is considered to be a total aerosol mass sample
onto a standard 20.3 by 25.4 cm (8" by 10") glass or quartz fiber filter

media. Four other samplers, placed at 90° intervals around the center sampler, are single stage, rectangular slot impactors. These single stage impactors collect size fractionated samples of the large ambient particles onto grease coated impaction plates. Remaining particles are collected by a standard filter which follows each single stage impactor. Each impactor has a different particle collection efficiency. The particle cutoff diameter (for 50% collection efficiency) for Impactors 1, 2, 3, and 4, are 47, 34, 18.5 & 9.3 μm, respectively. These impactor nozzles were carefully calibrated at the University of Minnesota, Particle Technology Laboratory and the University of Florida to determine their exact cutpoints, as described by Vanderpool.[4]

Particle size-fractionating samplers were operated simultaneously with the WRAC at each site. These samplers included instruments typically found at an Inhalable Particulate (IP) network site such as: high-volume air sampler (HIVOL), a 15 μm type size selective inlet sampler (SSI), and dichotomous sampler (Dicot). These instruments were normally located at the site and were operated by the WRAC sampling team during the special sampling. At each site was at least one high-volume ambient cascade impactor was also used. Most of these samplers were run with a duplicate unit at one or more location to check for repeatability of results.

Results
At each site, samples collected under similar conditions were averaged to determine a representative distribution.

Aerosol Mass Fraction and Particle Size
Collected by the High-Volume Sampler
The total atmospheric aerosol mass fraction and particle size collected by the standard High-Volume Sampler (Hi-Vol) can be inferred by comparison with the aerosol mass and size distribution measurements made using the Wide Range Aerosol Classifier (WRAC).

The grand distribution for all 41 usable WRAC runs produce a total aerosol concentration of 134.0 μg/m³ with 91.0% of the aerosol mass < 34 μm diameter. Most single city average distribution and the 41 day grand distribution average suggests that the standard High-Volume Sampler collected all particles less than ~30 μm diameter (on the average). Calibration data of McFarland, Ortiz and Rodes[5] also suggests a Hi-Vol sampler 50% cut size of about 30 μm. These data were also presented in an article by Watson, Chow, Shah and Pace[6] which discusses the Hi-Vol aerosol collection.

Atmospheric Aerosol Large Particle Mass Distribution
Plots of the total large particle grand average distribution and various city average distributions suggest that the large particle distribution is approximately log-normal. These data also suggests a minimum value between the large and small particle mass modes at about 3 μm (aerodynamic diameter). If one assumes the large particle mass mode is log-normal and that there is a minimum point at 3 μm, several features of the distribution can be determined. A best fitting curve was drawn through the actual mass measurement data plotted as a cumulative distribution curve on log-normal probability paper. Several of these curves were then drawn as histograms in Figure 2.

Discussion

There has been much discussion recently about incorporating an upper size limit for the regulation of particulate matter. A size limit of 10 μm has been suggested. The WRAC measurements reveals that the fraction of particulate matter in ambient air associated with particles less than 10 μm diameter can vary between about 50% and 90% for single run percentages (for Birmingham and Riverside respectively). The average distribution data for the high concentration days in Birmingham and Riverside suggest that a 10 μm size selective sampler would collect 68% and 89% respectively, of what the standard High Volume Air Sampler would collect. An average distribution for all 41 test days from 5 cities suggest the 10 μm sampler would collect 74% of that collected by the Hi-Vol.

Ambient aerosol distributions display a large particle mass mode under a variety of situations. The situations include: relatively clean areas like Research Triangle Park, areas with high small-particle concentrations like Riverside, areas with high large-particle concentrations like Phoenix, and areas with high concentrations of large and small particles like Birmingham. Each of these areas will be affected differently with the implementation of a new health-related particulate matter standard. This will relieve certain areas which have historically been in a non-attainment status because of the presence of a high mass fraction of large particles.

Acknowledgment

This investigation was supported by cooperative agreement CR808606 from the Environmental Monitoring Systems Laboratory, Environmental Protection Agency, Research Triangle Park, North Carolina.

References

1. Lundgren, Dale A. and David C. Rovell-Rixx, 1982. Wide Range Aerosol Classifier, EPA-600/4-82-040, PB82-256264 N.T.I.S.

2. Lundgren, Dale A. and H.J. Paulus, 1975. The Mass Distribution of Large Atmospheric Particles, JAPCA 25 (12):1227.

3. "Reference Method for the Determination of Suspended Particulates in the Atmosphere (High Volume Method)", 40 CFR 50, Appendix B, U.S. Government Printing Office.

4. Vanderpool, Robert, 1983. Particle Collection Characteristics of High Flow-Rate Single Stage Impactors, M.S. Thesis, University of Florida.

5. McFarland, A.P., C.A. Ortiz and C.E. Rodes, 1979. Characteristics of Aerosol Samplers Used in Ambient Air Monitoring, Presented at 86th National Meeting of the American Institute of Chemical Engineers, Houston, TX.

6. Watson, J.G., J.C. Chow, J.J. Shah and T.G. Pace, 1983. The Effect of Sampling Inlets on the PM-10 and PM-15 to TSP Concentration Ratios, JAPCA 33:114.

298

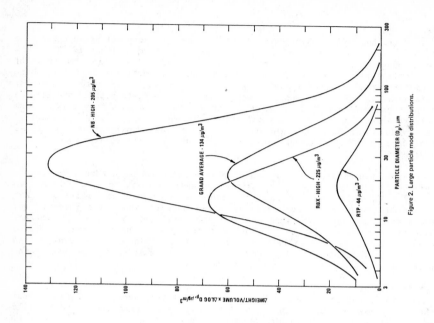

Figure 2. Large particle mode distributions.

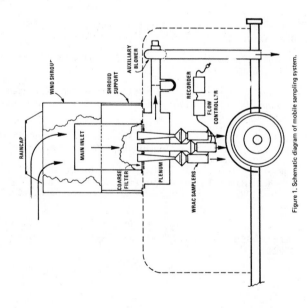

Figure 1. Schematic diagram of mobile sampling system.

Published 1984 by Elsevier Science Publishing Co., Inc.

Aerosols, Liu, Pui, and Fissan, editors

MEASUREMENTS OF AEROSOL DISTRIBUTIONS AND CONSTITUENTS
IN THE BLACK FOREST (SCHAUINSLAND)

Graul[*], R., Georgi[**], B.

[*] Umweltbundesamt Meßstelle Schauinsland,
Postfach 1229, D-7815 Kirchzarten, FRG

[**] Niedersächsisches Institut für Radioökologie,
Herrenhäuser Str. 2, D-3000 Hannover 21, FRG

EXTENDED ABSTRACT

Introduction

In Germany there is increasing public concern about rapidely propagating damages in forests occuring remote from source areas. The air quality in such regions has thus gained high degree of interest (Ulrich, 1983). At the sampling site, Mount Schauinsland (1200 m) in the Black Forest serious damages to conifers have taken place (Schröter, 1983). For several years the total suspended mass, dry and wet deposition has been measured together with meteorological parameters within the routine measuring programme of the Umweltbundesamt. In addition the aerosol size distribution was measured gravimetrically with a low pressure Berner impactor and an Andersen impactor at various time intervals.

Materials and Methods

The sampling site is on a mountain with fairly clean air p.e. TSP yearly average 24 $\pm$3 $\mu g\,m^{-3}$. Therefore appropriate time intervals were chosen to one month. Shorter intervals up to two days are possible for the Berner impactor. The comparison between intervals both impactors required the longer measuring intervals.

The technical data of the low pressure impactor (Berner, *1979*) and the normal one (Andersen, *1966*) are already described.

The samples are measured gravimetrically and analysed by various methods for conductivity, pH-value, Na, K, Mg, Ca.

Results

The meteorological data and the corresponding particle size distributions are shown in Fig. 1 and Fig. 2

300

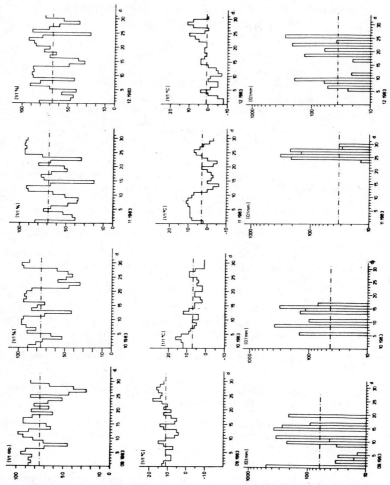

Figure 1. The meteorological date depth of rainfall (mm), temperature
(1/1 °C) and rel. humidity (1/1 %) corresponding with the aerosol
measurements

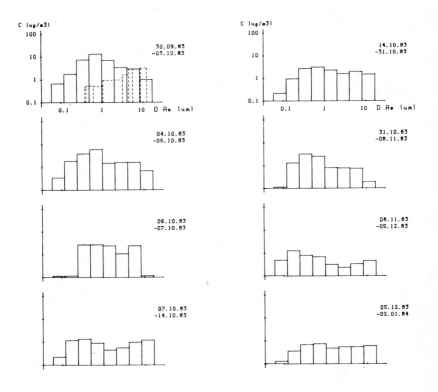

Figure 2. Aerosolconcentration c (µg m^{-3}) in the particle range 0.06 - 16 µm
d a.e at the Schauinsland

302

References

Ulrich, B. and Pankrath, J. (1983) "Effects of Accumulation of Air Pollutants in Forest Ecosystems", D. Reidel Publ. Comp. Dodrecht, Holland.

Schröter, H. (1983) "Krankheitsentwicklung von Tannen und Fichten auf Beobachtungsflächen der FVA in Baden-Württemberg", AFZ, 26/27, 648 - 650.

Berner, A. (1979) In "Aerosol Measurements in the Submicron Size Range", EPA-600/2-79-105; Washington, D.C.

Andersen, A.A. (1966) "A Sample for Respiratory Health Hazard Assessment", Am. Ind. Hyg. Ass. J., 27, 160.

Published 1984 by Elsevier Science Publishing Co., Inc.
Aerosols, Liu, Pui, and Fissan, editors

Composition and Origins of Aerosol at a Forested Mountain in Soviet Georgia

Thomas G. Dzubay and Robert K. Stevens
Environmental Sciences Research Laboratory
U.S. Environmental Protection Agency
Research Triangle Park, North Carolina   27711

Phillip L. Haagenson
National Center for Atmospheric Research
Boulder, Colorado   80307

In an attempt to examine the properties of natural aerosol, investigators from the U.S. and U.S.S.R. conducted the Abastumani Forest Aerosol Experiment (AFAEX-79) during July 1979 in the Georgian Republic of the Soviet Union. The study site was at Abastumani Astrophysical Observatory, which is atop a conifer-covered mountain, 1650 m above mean sea level (MSL). The AFAEX-79 field measurements included concentrations of terpenes and their particulate oxidation products, meteorological conditions, gaseous species, aerosol size distributions, and optical coefficients (Stepanenko and Vdovin, 1982; Vdovin et al., 1982; Shaw et al., 1983).

For our contribution to AFAEX-79, we collected and analyzed ambient aerosol particles in three size ranges: ultrafine (<0.1 μm aerodynamic diameter), fine (<2.5 μm), and coarse (2.5-15 μm). A micro-orifice impactor was used to remove particles larger than 0.1 μm (Kuhlmey et al., 1981) from the sampling air stream so that the ultrafine particles could be collected on a filter. Fine and coarse particles were collected in several dichotomous samplers that were equipped with Teflon, quartz, and Nuclepore filters and were analyzed for the species whose concentrations are shown in Table 1. Figure 1 shows time-series plots of sulfur and manganese concentrations in the fine-fraction. Figure 1 also shows mean resultant wind vectors for the layer -- 1500 to 2100 m MSL. We computed these from rawinsonde ascents made three times daily (Vdovin et al., 1982).

Sulfate and ammonium accounted for 50% of the fine-fraction mass. X-ray fluorescence (XRF) and ion chromatography (IC) analyses together indicated within experimental error that all of the sulfur occurred as sulfate. No hydrogen ion was detected, and sufficient ammonium was measured to indicate that most of the sulfate occurred as ammonium sulfate -- a conclusion that was confirmed by x-ray diffraction (XRD) analysis (Thompson et al., 1982) of fine-particles. The amount of particulate carbon that Shaw et al. (1983) could relate to terpene emissions is less than 4% of the fine-fraction mass. In the coarse-fraction montmorillonite, quartz, gypsum, and calcite were determined by XRD. Nitrate, a labile species, contributed 3.5% to the average coarse-fraction mass and may have been retained in the sample by reaction with the calcite. In general, the air we sampled was purer than that found in rural areas of eastern U.S.A. (Stevens et al., 1984). The exception was manganese, which appeared in unusually large concentrations.

To determine the nature of the particles containing manganese, the aerosols collected during the night of July 10 on 0.3-μm pore size Nuclepore

304

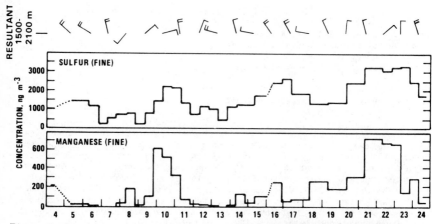

Figure 1. Species concentrations and resultant wind vectors during July 1979.

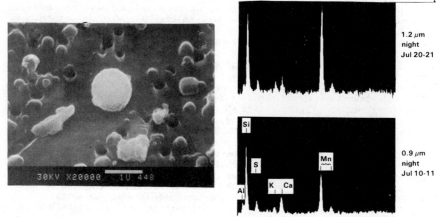

1.2 μm
night
Jul 20-21

0.9 μm
night
Jul 10-11

Figure 2. Micrograph and typical X-ray spectra of manganese-rich spheres.

Figure 3. Trajectories in 12-h segments for air parcels arriving at Abastumani.

filters were analyzed by scanning electron microscopy. Although most of the particles in the fine-fraction contained only sulfate, spherical particles containing manganese and silicon and ranging in diameter from 0.5 to 1.5 μm were very prominant (Figure 2). The high manganese content and spherical shape of these particles indicate that they were emitted from a furnace or smelter. The mean resultant wind vectors shown in Figure 1 indicate that air flowed to our sampling site from the north, where manganese mines at Chiatura and a manganese smelter at Zestafoni are located ~50 km from our sampling site. Manganese production at Chiatura and Zestafoni is second largest in the U.S.S.R. Isoentropic air parcel trajectories showed that the Abastumani air was not affected by emissions from the vicinity of Nikopol, 940 km to the northwest, where the U.S.S.R.'s largest manganese production occurs.

Particulate sulfur (Sp) concentrations measured in South America (Lawson and Winchester, 1979) are typically ~0.1 μg m^{-3} and represent an upper limit for the natural component from continental and oceanic sources. If the natural component of Sp in Europe is comparable to that in South America, then anthropogenic sources accounted for more than 90% of the 1.6 μg m^{-3} particulate sulfur at Abastumani. The major anthropogenic source of particulate sulfur are sulfur dioxide emissions that are transformed to sulfate in the atmosphere. According to an evaluation of emissions in Europe, SO_2 emissions in Georgia, where Abastumani is located, are among the lowest in Europe (Dovland and Saltbones, 1979). Vdovin et al. (1982) measured SO_2 at the study site. Their average value expressed as sulfur is Sg = 0.05 μg m^{-3} and is among the lowest in Europe during July 1979 (Skjelmoen et al., 1981). The Sp/Sg ratio for Abastumani is 32 ± 6 -- much higher than for any of the other sites in Europe (Skjelmoen et al., 1981). Such a high Sp/Sg ratio indicates that the aerosol at Abastumani is well aged.

We can estimate the aging time for an ideal situation in which the SO_2 reaching our site is assumed to have originated from a single source and to have been transported via a single path. Use of an equation and rate constants by Trägardh for this situation indicates that the aerosol reaching Abastumani had aged for 147 ± 74 h. Three episodes of high sulfate concentrations at Abastumani were evaluated by isoentropic trajectory analysis. Two are shown in Fig. 3. One episode could be explained in terms of rapid transport of an air parcel from Central Europe. For the other two episodes, the trajectories indicate four days of stagnation, which obscures the identity of any distant anthropogenic source of the sulfur reaching Abastumani.

Although the research described in this article has been conducted at the U.S. Environmental Protection Agency, it has not been subjected to Agency review and therefore does not necessarily reflect the views of the Agency, and no official endorsement should be inferred.

References

Dovland, H.; Saltbones, J. (1979) "Emissions of Sulfur Dioxide in Europe in 1978," EMEP/CCC-Report 2/79, Norwegian Institute for Air Research, N-2001 Lillestrøm, Norway.

Kuhlmey, G.A.; Liu, B.Y.H.; Marple, V.A. (1981) Am. Ind. Hyg. Assn. J., 42:790.

Lawson, D.R.; Winchester, (1979) J.W. Science, 205:1267.

Shaw, R.W.; Crittenden, A.L.; Stevens, R.K.; Cronn, D.R.; Titov, V.S. (1983) Environ. Sci. Technol. 17:389.

Skjelmoen, J.E.; Dovland, H.; Schaug, J. (1981) "Data Report, October 1978 - September 1979," EMEP/CCC-Report 4/81, Norwegian Institute for Air Research, N-2001 Lillestrøm, Norway.

Stepanenko, V.D..; Vdovin, B.I. (1982) Atmosfernaya Diffuziya and Zagryaz-neniya Vozdukha 450:52.

Stevens, R.K.; Dzubay, T.G.; Lewis, C.W.; Shaw, R.W. (1984) Atmos. Environ. 18:261.

Thompson, A.C.; Jaklevic, J.M; O'Connor, B.H.; Morris, C.M. (1982) Nucl. Instr. and Meth. 198:539.

Trägardh, C. (1982) Atmos. Environ. 16:1451.

Vdovin, B.I.; Tulchinskaya, Z.G.; B.V. Pastukhov. (1982) in Monitoring Fonovovo Zagryazneniya Prerodnoi Credie, State Committee of U.S.S.R. for Hydrometeorology, Leningrad, p. 95.

Table I. Average Size Fractionated Aerosol Composition (ng m^{-3}) at Abastumani Astrophysical Observatory between July 5 and July 24, 1979.

species	ultrafine	fine	coarse	Method	Filter
mass	250 ± 1400	12050 ± 3600	5350 ± 3100	β-gauge	Teflon
C$_{or}$		1710 ± 730	750 ± 660	pyrolysis	quartz
C$_{el}$		980 ± 150	108 ± 120	pyrolysis	quartz
H$^+$ion		<5	<5	titration	Teflon
ammonium		1390 ± 180	19 ± 23	electrochem.	Teflon
nitrate		<70	188 ± 22	IC	Teflon
sulfate		4600 ± 340	73 ± 85	IC	Teflon
Al	19 ± 31	98 ± 79	690 ± 270	XRF	Teflon
Si	11 ± 12	275 ± 36	1290 ± 430	XRF	Teflon
S	51 ± 6	1570 ± 115	81 ± 84	XRF	Teflon
Cl	0 ± 3	5 ± 7	28 ± 12	XRF	Teflon
K	9 ± 2	129 ± 10	100 ± 23	XRF	Teflon
Ca	2 ± 2	78 ± 7	327 ± 59	XRF	Teflon
Mn	3 ± 1	187 ± 14	38 ± 5	XRF	Teflon
Fe	2 ± 4	72 ± 10	196 ± 20	XRF	Teflon
Zn	2 ± 3	15 ± 7	6 ± 6	XRF	Teflon
Br	0.3 ± 0.1	3 ± 1	1 ± 1	XRF	Teflon
Pb	1.8 ± 0.5	18 ± 2	1 ± 1	XRF	Teflon

Published 1984 by Elsevier Science Publishing Co., Inc.
Aerosols, Liu, Pui, and Fissan, editors

NATURAL ATMOSPHERIC RADIOACTIVE AEROSOLS IN THE ATMOSPHERE OVER
THE PACIFIC OCEAN OF THE NORTHERN HEMISPHERE

S.Mochizuki, T.Tanji, M.Okino, K.Orikasa and N.Matsumura
            Muroran Institute of Technology
            27-1 Mizumoto, Muroran, Hokkaido, JAPAN

EXTENDED ABSTRACT

Introduction

    In the recent years, investigation of progressive transition character of
aerosols including anthropogenic heavy metals and their impacts to the global
scale oceanic environment, they are moving from land to over ocean, has increased
its importance in relation with grasping the progressing property of back ground
pollution.
    Development of this investigation has increasingly emphasized that the in-
vestigation of radon and its daughters dispersed to wide region in the atmosphere
over ocean is urgent topics to be made, because of its usefulness as a tracer to
see the dilution rate of dispersing aerosols and to evaluate their impacts to
the ocean.(M.Misaki,et al. 1975; K.K.Turekian,et al. 1977)
    We have repeatedly made the natural radioactive aerosol observation by the
aid of ship in the Pacific Ocean near Japan Islands from 1975.(S.Mochizuki,1978;
1982)
    To know the distribution trend of radon, its decay products and aerosol par-
ticles dispersing from land to more wide region over the Pacific Ocean, we have
made the observation by the aid of research vessel Hakuho-maru of Tokyo University
when she made the round voyages in the Pacific Ocean of Northern Hemisphere from
1982 to 1983 and in 1983.
    We will present some observational evidences found in the wide region over
the Pacific Ocean from near Japan Islands to near North and South American continent, and discuss the distribution trend of natural atmospheric radioactive
aerosols over ocean and their behaviors in the atmosphere near seasurface, and
also particularly discuss the aerosol residence time estimated from the measurement results of short lifed decay products of radon.

Measuring method of radioactive aerosols and other items measured simultaneously

    We adopted silicon semiconductor detector (HORIBA Ltd. Silicon Surface Ba-
rrie Type, 300 SB 60 L) for alpha ray detection of radon daughters collected on
the filter. The detector output was fed to multichannel analyzer that displayed
relative integral alpha events as a function of energy. Alpha ray spectrometory
was adopted for the measurement of RaA(Po-218) and RaC'(Po-214).
    The measurement was made at an interval of about 8 hours or 6 hours through
the full period of cruisings. Radon daughters collection time and analysis time
were both set at 4000 sec. Radon daughters were collected on the filter with
suction pump with flow rate of 60 liter/min.
    Number concentration of Mie particle and its size distribution, number con-
centration of Aitken particle and atmospheric electric conductivity were measured
simultaneously.
    Furthermore, using an eight staged Andersen air sampler, we made aerosol
sample collections for their composition analysis.

308

Results of radon daughter measurement

The cruise routes of the expedition voyage máde during the period from 22
Nov. 1982 to 24 Feb. 1983 are shown in Fig.1.
On the ship position and meteorological data, we reffered the data obtained
by the officers on the deck.
Wind direction, wind force(by Beaufort wind scale) and weather along the
cruise routes are shown in the upper part of each figures.
The radon daughters measured over the middle latitude of the North Pacific
are shown in Fig.2.
In this figure the abscissa shows the longitude of the ship position along
the cruise route, and weather is shown by Beaufort's weather notations. In the
middle part of this figure, the ordinate shows an equivalent radon concentration
computed from the results of Po-218 and Po-214 measurement.
As clearly seen in the figure, radioactive aerosols showed considerable
variation even in mid-ocean. Generally speaking, in the case of northerly or
southerly wind, excepting near the North American Continent, radioactive aerosols
showed low levels. But, on the observation when the vessel was stopped or float-
ing on the ocean around the spot 40°N, 130°W, a remarkable variation was seen
that radioactivity concentration levels were found to be distributed from 0.052
$\times 10^{-17}$ Ci/cc to 0.61 x 10^{-17} Ci/cc.
The radon daughters measured over the eastern North Pacific and the eastern
tropical Pacific are shown in Fig.3 and 4. In these two figures, concerning to
wind direction and route chart, the rightward denotes north. Since the cruise
routes were consisted of two or three segments, to present the measuring spot
clearly, the base lines were corresponded upon the expanded broken segments of
the cruise routes.
As clearly seen in Fig.3 and 4, when the wind was being blown from the continent, the radon daughters were found to increase with considerable high levels.
A lowest concentration of radon daughters was found at around a spot 2.3°S,117°W
shown in Fig.4 and its value was found to be 0.021 x 10^{-17}Ci/cc.
We would like to discuss more detail on the variation trend and its reason
at the conference presentation, and also report an interesting phenomenon obtained
from similar observation which has made in the considerably wide western zone of
the North Pacific Ocean during the period from 11 Aug. to 13 Sept. in 1983.
Further, using the method reported by Misaki(Misaki, et al. 1975), as the
time constant for radon concentration decrease due to dilution, the time constant
was found to be distributed from 34 hours to 63 hours based on several observation results which were obtained in the atmosphere over the Pacific Ocean near
Japan Islands.

Reference

Misaki,M. et al.(1975) "Deformation of the size distribution of aerosol particles dispersing from land to ocean". J. Meteor. Soc. Japan. 53:111.
Turekian,K.K. et al.(1977) "Geochemistry of atmospheric radon and radon decay
products". Ann. Rev. Earth Planet Sci. 5:227.
Mochizuki,S.(1978) "Comparison of natural atmospheric radioactivity appeared
over Hachijo Island and Miyake Island in the early spring". J. Meteor. Soc.Japan
56:52.
Mochizuki,S.(1982) "Radon and its daughters in the maritime atmosphere near Japan
Islands". J. Meteor. Soc. Japan. 60:787.

309

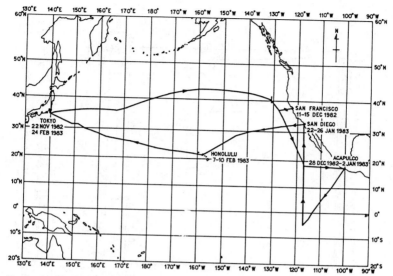

Fig. 1. Cruise routes of the expedition voyage KH-82-5.

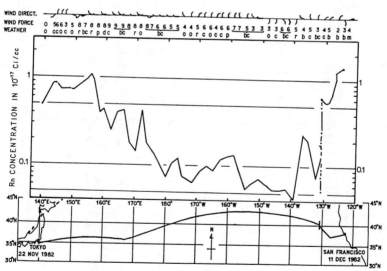

Fig. 2. Radioactive aerosol distribution trend found over the North Pacific Ocean
from Tokyo to San Francisco.

310

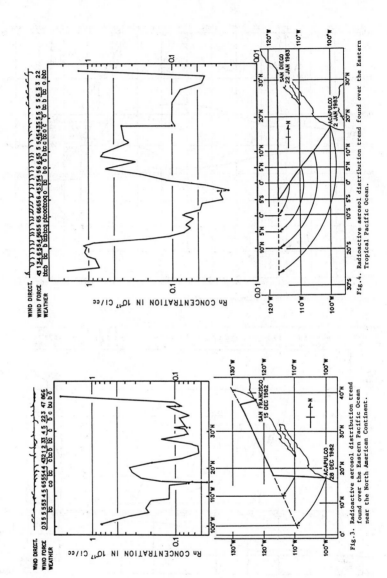

Fig.3. Radioactive aerosol distribution trend
found over the Eastern Pacific Ocean
near the North American Continent.

Fig.4. Radioactive aerosol distribution trend found over the Eastern
Tropical Pacific Ocean.

Published 1984 by Elsevier Science Publishing Co., Inc.

Aerosols, Liu, Pui, and Fissan, editors

MULTIVARIATE STATISTICAL ANALYSIS OF ATMOSPHERIC
AEROSOL COMPOSITION DATA FOR SANTIAGO, CHILE

A. Trier and C. Silva Z.

Facultad de Ciencia, Universidad de Santiago de Chile,
Casilla 5659, Santiago 2, Chile.

Introduction

Air pollution is an acute problem in the metropolitan area of Santiago,
Chile. The area has a population of four million and an urban spread estimated at 1200 Km^2. Mean altitude is 600 m. Topography and meteorology favor
conditions of altitude atmospheric thermal inversion. Average annual rainfall
stays below 400 mm. Concentrations of TSP exceeding WHO guidelines for the
protection of public health are frequently observed by a local high-volume
sampling network (Silo et al., 1982). Other pollutants monitored are NO_x, CO,
SO_2, hydrocarbons, dry deposited matter (Silo et al., 1982). Observations by
academic research groups confirm the overall picture of heavy air pollution
(Préndez et al., 1984; Trier, 1983). Government intervention is planned for
which more precise and reliable information on sources contributing to pollution and local pollutant reaction and dispersion mechanisms is required. Exist
ing information on atmospheric aerosol size distribution and its elemental and
chemical analysis is still inadequate. The present work es intended as a contribution towards a better understanding of air pollution problems in Santiago.

Experimental

Samples of atmospheric particulate matter were obtained at a fixed station
on the Universidad de Santiago de Chile campus between July, 1981 and March,
1983 using commercial equipment. Total (TSP) samples for periods of one week
at a time were taken from July, 1981 through June, 1982 (mid-winter to mid-win
ter) at an average flow rate of 0.66 (15) $m^3 h^{-1}$. Upper sampling cutoff size
was not determined. Samples were obtained both on polycarbonate membranes (PC
filters) and on ashless paper (W filters). Thus two groups of TSP data are
available. Samples were analyzed by X-ray fluorescence (XRF) with an energy-
dispersive spectrometer at an excitation of 45 KV. Mass concentrations for
the PC filter data set were derived. Net XRF K α or Lα line intensities were
extracted from the spectra. Dichotomous samples of atmospheric particulates
were obtained from October, 1981 (mid-spring) through March, 1983 (mid-autumn)
at a flow rate of 1 $m^3 h^{-1}$. Sampling time ranged from 6 to 10 hours; some data were obtained during night hours. The sampling apparatus has a cut-point
at 2.5μm aerodynamic diameter and an inlet cutoff at 15μm. Membrane filters
were used for most of this work but some samples were collected on ashless paper. Mass concentrations were derived and XRF analysis was carried out. Net
XRF line intensities were extracted from the spectra. Because of variations
in sampling and analytical procedures the data have been grouped into three
daytime sample groups and two nighttime groups. Each group comprises a fine
particle and a coarse particle subgroup. The data have been subjected to statistical analysis which is being continued.

Results

1. TSP data

The distributions of characteristic line intensities and of net filter mass were tested for approach to normality. Transformations of the data (SQRT and LOG) tend to improve approach to normality. Pair correlations were examined. Pair correlations in the group (Al, Si, S, K, Ca, Ti, Fe, Pb, mass) were found to be highly significant (i.e. at the 99% level of confidence, or better). The Br/Fe, Pb, mass and the Zn/S, Ca, Fe, Pb, Br, mass pair correlations were also highly significant. Multiple regression analysis was performed for some cases. No-intercept models were found to be more satisfactory in all cases tested. On the W filters there is a clear association of Br with Pb and of S with Fe. On the PC filters Br is associated with Pb and S, and S is associated with Ca and Zn. Principal component analysis was carried out on SQRT and LOG transformations of the TSP data sets. Examination of the structure of the two first eigenvectors shows that differences are mainly related to pollutants S, Zn, Pb. No general conclusions were derived from this analysis.

2. Dichotomous data

The distributions of characteristic elemental line intensities and of net filter mass were tested for approach to normality. It was found that SQRT and (especially) LOG transformations of the data improve approach to normality. Pair correlations have been examined for both fine and coarse particles in each of the five data groups as well as fine/coarse pair correlations for each element and for net filter mass. For two groups of daytime samples (141 datum points from October, 1981 through December, 1982, on membrane filters) highly significant pair correlations both in the fine and in the coarse particle fraction were: Si/Fe, Zn, Pb, mass; S/Pb, Br; Ca/Fe, Zn, Pb, mass; Fe/Si, Ca, Zn, Pb, mass; Zn/Pb, Br; Pb/Br. A number of other pair correlations were highly significant in one of the fractions only. The fine/coarse particle correlations show that more than 90% of the Si and between 80 and 90% of the Ca and the Fe are found in the coarse particle fraction, on the average. On the other hand, two to three times as much S is found in the fine fraction as in the coarse fraction, two to four times as much Zn and about three times as much Pb, on the average. One third to one half of the collected mass is found on the fine particle filter. While multiple regression anlysis is still in progress principal component analysis has been carried out. The number of principal components retained in each case in shown below, where in the group designations F stands for fine, G for coarse, D for daytime, N for nighttime:

1FD	2	1GD	2
2FD	4	2GD	2
2FN	4	2GN	3
3FD	3	3GD	2
3FN	2	3GN	2

Analysis of the G data generally shows a clear split between groups (Al, Si, K, Ca, Ti, Fe, mass) and (S, Pb, Br) while elements Cu and Zn occupy an intermediate position. Elements in the first group, which accounts for about half of the variance, are recognized as of "natural" origin. Analysis of the F data does not yield simple general conclusions.

Discussion

Although the information content of the available data sets has not yet been fully explored the results so far indicate:
i) coarse fraction mass is mainly determined by the contribution of elements of "natural" origin (with the caveat that elements C, N, O escape XRF detection). The XRF intensities of elements Si, Ca, Fe could be used as predictors of coarse particle fraction mass;

ii) as expected, pollutants S, Zn, Pb are mainly found in the fine particle
fraction. The Br/Pb correlation is highly significant in both size frac
tions indicating that at our station the main contribution to Pb comes
from automotive exhaust products;
iii) the behavior of elements Cu and Zn is intermediate between that of the
"natural" group (Al, Si, K, Ca, Ti, Fe) and that of the "anthropic"
group (S, Pb, Br), as judged from principal component analysis.

Acknowledgements

The authors want to thank C. Rojas, J. Valdés and M. Eugenia Cantillano for
assistance with the XRF analysis and the sampling. The assistance of M. A.
Cumsille with the statistical analysis is also appreciated. One of us (A.T.)
wants to thank A. Seguel M. for aid and encouragement. This work is being
supported by DICYT/USACH.

References

Prendez, M.M. et al. (1984) J.A.P.C.A., January 1984.
Silo, C. et al. (1982) "Suspended matter in the urban area of Santiago, Chile",
Proceedings of the 5 th International Clean Air Conference, p. 321, Buenos
Aires, Argentina.
Trier, A. (1983) "Observations on inhalable atmospheric particulates in
Santiago, Chile", 11 th Annual Conference, Association for Aerosol Research,
Munich, FRG.

Published 1984 by Elsevier Science Publishing Co., Inc.

Aerosols, Liu, Pui, and Fissan, editors

314

A 2-D MODEL OF GLOBAL AEROSOL TRANSPORT INCLUDING SULPHUR SPECIES

J. Rehkopf [1], M. Newiger [1], and H. Grassl [2]

1) Max-Planck-Institut für Meteorologie, Hamburg, FRG
2) Institut für Meereskunde, Kiel, FRG

EXTENDED ABSTRACT

Introduction

Man's activities have already led to changes in local and regional climate. Aerosols are known to influence the earth radiation budget both directly by interaction with incoming solar radiation and indirectly inside clouds, where aerosol particles serve as condensation nuclei thus changing optical cloud properties. However, even the sign of the energy budget change due to man-made aerosols is still questioned. For an assessment of the possible effects of anthropogenic aerosols on global climate information is needed about particle number density and particle size distribution as a function of height and geographical location for natural as well as man-made particles.

The main aim of the present study is to investigate transport and size dependent removal of natural and man-made aerosol particles and to derive seasonal steady state distributions of particle number density as well as particle size distribution for climate studies.

Model

Tropospheric aerosol transport is calculated with a 2-D grid model using seasonal means of climatological data. The vertical resolution of 100 m in the planetary boundary layer is gradually increased to a 2 km step for upper layers. A horizontal grid distance of 10° has been adopted, however, resolution is improved down to 2° for the domain between 30-60°N where 90% of the man-made emissions are released. The aerosol size spectrum has been divided into three size classes (r - particle radius) nucleus mode: $r < 0.1$ µm, accumulation mode: $0.1 \lesssim r \lesssim 1.0$ µm, coarse mode $r > 1.0$ µm. Each of these classes is given a mean radius according to Whitby (1978). The following processes are considered removing aerosol particles from the atmosphere: rainout, washout, dry deposition, gravitational settling, and coagulation. Thereby rainout is understood as removal of aerosol particles by becoming cloud condensation nuclei CCN and falling out. The CCN are assumed to grow to a mean droplet radius according to cloud typ and particle number density. The amount of precipitating cloud water and thus particle removal, is determined by the zonal mean precipitation distribution. In consequence of this rainout concept, washout is understood to be the collision between precipitation elements and aerosol particles both within clouds and below cloud base. In addition to directly emitted particles the products from gasphase reactions of SO_2 and H_2S have been considered as a secondary particle source independent of the ground. Thus the continuity equation has to be solved for five compounds at each time step. For details about the numerical solution as well as production and removal processes of particles see Rehkopf et al. (1984).

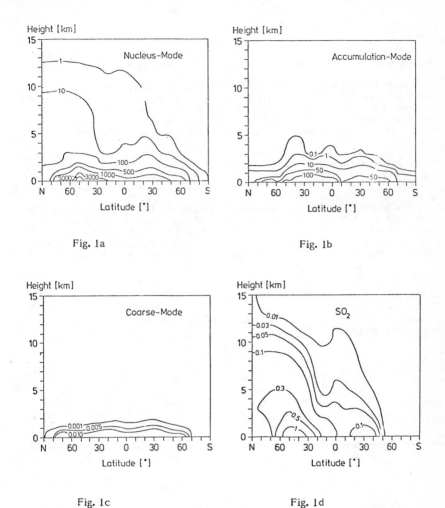

Fig. 1a

Fig. 1b

Fig. 1c

Fig. 1d

Fig. 1a-d: Steady state distribution for natural plus man-made emissions after 10 weeks model time: particle number densities (cm^{-3}) for nucleus-, accumulation-, and coarse-mode (a-c) and mixing ration (ppbv) for SO_2 (d). Results refer to January.

316

Results

In Fig. 1 (a-d) the steady state distribution of particle concentration is given resulting from natural plus man-made emissions of particles (a-c) and sulphur dioxide (d).

Nucleus mode particles show a maximum concentration of 9000 cm^{-3} at the ground between at ~ 50° N. This number density is decreased by two orders of magnitude within the first 2 km of the atmosphere. Accumulation mode particles have a maximum concentration of 200 cm^{-3} at ~ 20°N at the ground. There is hardly any transport of accumulation mode particles into higher levels, since rainout is an effective sink for them.

Wet deposition is the prevailing sink for all particles with rainout being more effective than washout. However, near the earth's surface particle concentration in the nucleus mode is reduced stronger by coagulation than by washout. With increasing height the effectiveness of coagulation decreases, since it depends on the square of the particle concentration.

It is found that at latitudes of maximum man-made influence (~ 50°N) particle concentration is increased by nearly one order of magnitudes due to anthropogenic aerosol particles. Since this increase is mainly caused by small particles, the exponent of an assumed truncated power law size distribution is changed from $\langle$ = 3.2 for natural particles to $\langle$ = 3.7 for natural plus man-made particles near the ground.

References

Whitby, K.T. (1978) "The physical characteristics of surlfur aerosols". Atm. Environm. 12:135.
Rehkopf, J., M. Newiger, and H. Grassl (1984) "A 2-D model of global aerosol transport" submitted to Atm. Environm.

Published 1984 by Elsevier Science Publishing Co., Inc.

Aerosols, Liu, Pui, and Fissan, editors

COUPLING OF RADIOACTIVITY AND ELEMENTAL COMPOSITION DATA IN
AEROSOL TRANSPORT ANALYSIS

J. Jokiniemi, T. Raunemaa, R. Mattsson*, A. Reissel** and
A. Reponen

Department of Physics, University of Helsinki
Siltavuorenpenger 20 D, SF-00170 Helsinki 17, Finland

*Finnish Meteorological Institute
PL 503, SF-00100 Helsinki 10, Finland

**Department of Radiochemistry, University of Helsinki
Unioninkatu 35, SF-00170 Helsinki 17, Finland

EXTENDED ABSTRACT

1. Introduction

Measurements collected by the Finnish Meteorological Institute have
revealed interesting facts about the concentrations of sulphur
dioxide and short lived ^{222}Rn progeny in the atmosphere. The
concentration of sulphur dioxide shows good correlation with ^{222}Rn
progeny activity, which is an approximate measure of ^{222}Rn
concentration. Trajectory analysis of the air mass generally shows
that the air has passed through some heavily industrialized areas
on its way to the observer. Because sulphur dioxide can in no
way be connected with ^{222}Rn emitted from the earth's crust, the
good correlation is most probably due to enhanced mixing in the
troposphere.

The influence of local emissions on the ^{222}Rn and SO_2
concentrations of an air mass should not be overestimated. In
the afternoon, when the troposphere often is well mixed, the
concentration of ^{222}Rn is practically the same over a large
region, e.g. throughout southern Finland. Let us suppose that
radon has accumulated during 1-2 weeks in an air mass in some
continental region to the south-east of Finland. If this air
mass moves over some region of Finland within one hour, any
local concentration is too small to interfere. This is more
obvious if, instead of gaseous SO_2 and radon, we look at their
particle-bound daughter products -SO_4^{2-} and ^{210}Pb. Most aerosol
sulphur is formed from gaseous SO_2 at a rate of 1-2 % per hour
similarly to the rate at which ^{210}Pb is formed from ^{222}Rn
($T_{\frac{1}{2}}$ = 3.8 d). When considered together, the sulphur and ^{210}Pb
concentrations in the atmosphere can therefore give valuable
information on long range transport.

2. Experimental

Aerosol samples for this study were collected at the radio-

318

Figure 1. The sampling localities.

activity registration stations of the Finnish Meteorological
Institute. Sampling time for each sample was one week, at the
sampling localities shown in Figure 1. The filters used in
collection were Whatman 42 cellulose fibre filters.[1] Samples chosen
for analysis were collected weekly in February and in August in
even-numbered years from 1966 to 1982.

Pieces (25 x 10 mm^2) from the filters were used for elemental
analysis by particle induced X-ray emission, PIXE. The PIXE
facility at the Accelerator Laboratory of the University of Helsinki[2] was employed for the analysis of 15 trace elements from
Al to Pb. The analysis was carried out using a 1.5 mm diameter
internal beam of 2.0 MeV protons. Low 2-8 nA currents were used.
Elemental concentration was corrected by assuming an exponential
aerosol depth distribution in the Whatman filters, and the
quantitative analysis was performed by computer.

The ^{210}Pb content of the filters was analyzed radiochemically
by plating on silver foils and counting the alpha emissions of
^{210}Po, the 138 d alpha-active daughter of ^{210}Pb.

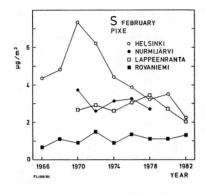
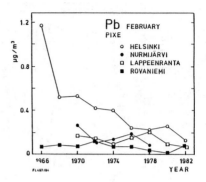

Figure 2. Average monthly
sulphur concentrations in
the February samples.

Figure 3. Average monthly lead
concentrations in the February
samples.

3. Results and discussion

Stable climatic conditions generally occur in mid-winter in
northern Europe, at a time when the heating energy consumption
is at its highest, usually in February. Whereas energy consumption
in the 1970's increased, the mean sulphur content in the
February samples from Helsinki decreased exponentially by 9 %
per year (Figure 2). The sulphur level in Helsinki, the
capital of Finland, is presently approaching that of Nurmijärvi
and Lappeenranta. In the north, in Rovaniemi, the sulphur
concentration is presently 50 % of that in southern Finland.
This is in accordance with the observations for SO_2 at several
background stations.[3] The sulphur levels in August samples were
about 50 % lower than in February samples. The decrease of the
weekly mean of sulpnur and also of several other combustion-
related elements is obviously due to higher stacks on power
plants and rapidly decreasing amounts of low-level surface
sources in the cities.[4]

The decrease in airborne lead concentration in central Helsinki
in the 1970's can be seen in Figure 3 and has been established
also elsewhere.[4] There are many reasons for this decrease. At
the beginning of the 1970's lead pollution from the lead
smelters north of Helsinki decreased, the number of on-site
incinerators was reduced, overall lead combustion decreased and
the lead content of gasoline was lowered.

Large amounts of particle sulphur and ^{210}Pb relative to the
gases, sulphur dioxide and radon, are often indicators of long
range transport. Figure 4 shows the variations of the sulphur

320

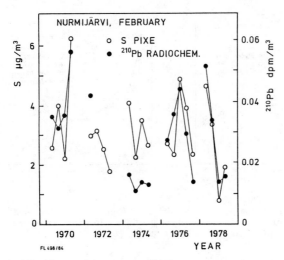

Figure 4. Weekly sulphur and ^{210}Pb concentrations in Nurmijärvi in February. The radiochemically determined ^{210}Pb is expressed as disintegrations per minute (dpm)

and ^{210}Pb contents in the weekly collected samples in Nurmijärvi. Both variations display similar patterns. The correlation coefficient between sulphur and ^{210}Pb values is 0.65. The ^{210}Pb level in 1974 is considerably lower than in the other years, due no doubt to the exceptionally warm winter in 1974 with prevailing maritime air masses from the south-west.

References

1. Observations of Radioactivity 1960 to 1982, Nos. 1-21, The Finnish Meteorological Institute, Helsinki 1960-1982

2. Raunemaa, T., Vaittinen, M., Räisänen, J., Tuomi, T., Gerlander, M., Nucl. Instr. and Meth. 181 (1981) 43-48

3. Kulmala, A., Estlander, A., Leinonen, L., Ruoho-Airola, R., Säynätkari, T., Air Quality at Finnish Background Stations in the 70's, Finn. Met. Inst. Reports. No. 36, Helsinki 1982

4. Mattsson, R., Jaakkola, T., An Analysis of Helsinki Air 1962 to 1977 Based on Trace Metals and Radionuclides, Geophysica 16:1, pp. 1-42, Helsinki 1979

Published 1984 by Elsevier Science Publishing Co., Inc.
Aerosols, Liu, Pui, and Fissan, editors

A PARTICLE SAMPLER FOR TETHERED BALLOON SYSTEMS

Robert G. Flocchini
Land, Air and Water Resources
University of California
Davis, CA   95616
U. S. A.

EXTENDED ABSTRACT

Introduction

Many researchers have noted the importance of obtaining vertical information on the concentration and composition of atmospheric pollutants. A review by Armstrong et al. (1981) describes studies in which balloons have been used to investigate meteorological parameters in the lower atmosphere as well as particulate and gas concentrations. A number of particulate samplers were also described in that paper. Described in this paper is an independent development of a light-weight particle sampler that enables collection of samplers in the vertical at 5 levels at heights up to 500 meters. The system utilizes a Astro Flight ASTRO CHALLENGER COBALT 05 MOTOR, a heavily modified GAST pump, a Nuclepore sequential filtration system and Lithium cells. The samplers are attached to a commercially available tethered balloon.

Sampling Device

The electric motor utilized in the system is a ASTRO CHALLENGER COBALT 05. This motor is approximately 7 cm long and 3 cm in diameter. The motor is most efficiently used from 5 to 10 volts input and from 5 to 25 amperes load. In the present configuration with weight as the prime criteria this motor is driven with one 2.8 Volt Lithium cell.

Several GAST rotary vane vacuum pumps were evaluated. It was determined that the GAST Model 0333 would provide adequate performance if overdriven at 150% of design rpm. Due to the nature of the application it was felt that the reduction in operating lifetime would be secondary. The Model 0333 GAST vacuum pump is constructed of castiron and would normally be unacceptable for tethered balloon sampling because of its excessive weight.

A redesign of the stator and end bells was undertaken. The rotor and vanes were retained.

New end bells were machined from 6061-T6 aluminum. Aluminum bronze plates were pressed into the inboard ends to minimize friction with the GAST rotor. One of the aluminum bronze plates was machined to allow proper port passages for the inlet and outlet tubes. The driven end of the rotor retained the standard sealed ball bearing while the other end was fitted with a miniature precision stainless steel instrument bearing.

Inlet and outlet tubes were constructed of .050 mm diameter aluminum tubing and pressed into the rear bell.

The stator was constructed of 2.54 cm diameter and 0.15 cm wall D.O.M. 4130 aircraft tubing. The ends were machined parallel. The whole assembly was fastened with two small studs and sealed with Loctite #601. The finished product was run at 1000 rpm and a mixture of Bon Ami and Dexron ATF was used to lap all internal surfaces. The figure below shows the combination motor and pump. Total weight of this system is 394 grams.

The configuration chosen for the filtration system which is described in detail elsewhere (Cahill, 1977, 1979) is shown below. Basically two 25 mm stacked Nuclepore filters are used to separate the coarse and fine mass. The first filter which is a 8 μm pore size Nuclepore membrane collects the coarse particulates while the 0.3 μm membrane collects fine particles. A flow rate of approximately 5 ℓ/min results in a cut point of approximately 2.0 μm. The single lithium cell can maintain this flow for approximately two hours.

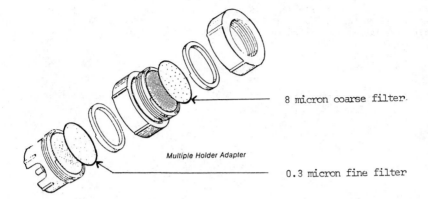

8 micron coarse filter.

Multiple Holder Adapter

0.3 micron fine filter

Tethered Balloon

A commercially available balloon system is utilized with five samples to provide a system capable of gathering particle information at six levels. The bright orange blimp shaped ballon is 6.6 meters long, has a volume of 7.50 m^3. The balloon when filled with helium has a free left of 5.7 kg. The balloon system is best operated when wind speeds are below 10 m/s.

A second tethered balloon is utilized to provide meteorological information during the particle sampling. The basic meteorological data set includes pressure, temperature, wind speed and wind direction. Vertical information is obtained by raising and lowering this system.

Summary

The system described in this abstract provides an effective method to obtain information on the vertical distribution of particulates.

References

Armstrong, J. A., P. A. Russell, L. E. Sparks and D. C. Drehmel (1981) "Tethered Balloon Sampling Systems for Monitoring Air Pollution", JAPCA 31:735.

Cahill, T.A., L. L. Ashbaugh, J. B. Barone, R. A. Eldred, P. J. Feeney, R. G. Flocchini, C. Goodart, D. J. Shadoan, and G. W. Wolfe (1977) "Analysis of Respirable Fractions in Atmospheric Particulates via Sequential Filtration", JAPCA 27:675.

324

Cahill, T. A., R. A. Eldred, J. B. Barone, and L. L. Ashbaugh "Ambient
Aerosol Sampling with Stacked Filter Units", Report FHWA-RD-78-178
Federal Highway Administration Office of Research and Development,
Washington, D.C. 20590

Published 1984 by Elsevier Science Publishing Co., Inc.

Aerosols, Liu, Pui, and Fissan, editors

Size Distribution of Particles That May
Contribute to Soiling of Material Surfaces

Fred H. Haynie, USEPA, Research Triangle Park, NC, 27711

In November 1983, EPA held a workshop entitled Research Strategies to Study the Soiling of Buildings and Materials. One of the main subjects of discussion was what size of particles contributes most to soiling and related to that, what ambient air measurement technique provides the best proxy for soiling potential. Because of this interest and its implications on future research and possible regulations, the following "quick and dirty" experiment was performed.

Two thin pieces of glass, each having a one side area of approximately 10 mm^2 were taped to glass microscope slides. In February, one of the slides was mounted vertically on a post under a mailbox. The mailbox was approximately two meters from an uncurbed asphalt suburban roadway with a traffic rate of approximately 13,000 vehicles per day. The other was placed on a window sill at an angle of approximately 70° to the horizontal at approximately 20 m from the same road. One month later, the pieces of glass were removed and examined by both light and scanning electron microscopy.

Light micrographs were taken at magnifications of 93X, 236X, 370X, and 590X, and scanning electron micrographs were taken at 100X, 600X, 2000X, 5000X, 10,000X and 20,000X. Particle size counts were made on each micrograph. For sizing purposes, nonspherical shapes were assumed to have a diameter equal to the square root of the product of two primary dimensions.

A different technique was used to determine whether the amount deposited on a small piece of glass is representative of the amount accumulated on windows over a long period of time. Clear paint was sprayed on a garage window that was approximately 40 m from the road. The window had not been washed in over 10 years and had been exposed to blowing rain. The dried film was stripped from the glass and examined with a light microscope (particles were embedded in the film and could not be seen with the scanning electron microscope).

Figure 1 is a composite of scanning electron and light micrographs at comparable magnifications for the glass exposed near the road. The central spherical particle in the low magnification light micrograph is pollen (deformed when bombarded with electrons). Pollen particles were not counted. The dark irregular shapes in the light micrographs are particles of tire rubber. Most of the other nonspherical particles are earth materials.

Figure 2 shows the distribution of submicrometer particles deposited on the glass near the road. Distribution was not uniform on the samples.

Figure 3 is a composite of low magnification micrographs of particles deposited on the glass at the window sill. The glass was not protected from rain, and the very large particles tended to wash to the bottom. Because the glass was at an angle of approximately 70°, gravitational settling contributed to the deposition of the larger particles. Although most of the irregular shapes are crystalline earth materials, some are particles of rubber from the road.

Figure 4 shows that the distribution of particles accumulated on a window is similar to that deposited on a small piece of glass. It indicates that particles larger than 10 µm adhere to vertical surfaces and contribute significantly to soiling.

Particle counts in size intervals of one-third the logarithm of decade diameter were made on each micrograph. Counts were adjusted between magnifications by multiplying by factors to give the number of particles per square millimeter of glass surface. These counts were then correlated to the middiameters of the intervals by using a fourth degree polynomial regression.

$$N = Exp(A_o + A_1 \ln(D) + A_2(\ln(D))^2 + A_3(\ln(D))^3 + A_4(\ln(D))^4) \qquad (1)$$

where N = number of particles per mm^2 within the interval

D = mid diameter of the interval in micrometers

A_i = regression coefficients

The regression equation for the two samples were then used with the middiameters of each interval to calculate the number of particles, the limits of one standard deviation range, particle cross sectional area per square millimeter of glass surface area, the percent of total coverage, and the percent of total particle volume. The results are given in Table II. Figures 5 and 6 show the bimodal character of the soiling effect.

These calculations reveal several facts. Most significant is that a few very large particles deposited by gravitational settling can dominate both particle volume and cross-sectional area measurements. By considering surface coverage as soiling, 99% of the volume and 63% of the soiling on the glass at the window sill was caused by only 39 particles, with a diameter greater than 10 µm. Even on the vertical glass near the road, where gravity did not contribute to deposition, large particles were dominant. Particles with diameters greater than 10 µm account for 90% of the volume and 32% of the soiling. The total soiling was significantly greater on the glass near the road than on the glass on the window sill, 0.138 mm^2/mm^2 compared to 0.070 mm^2/mm^2. This was consistent with the expectation that the ambient concentration of particles is greater near the primary source.

In conclusion; (1) Coarse particles dominate soiling of smooth surfaces such as glass, (2) A sampler designed to collect particles with diameters less than 10 µm would be a very poor indicator of soiling on these surfaces, (3) A Hi-Vol sampler may provide a fair indicator of soiling on vertical surfaces but would be poor for horizontal surfaces, where settling dust is the dominant fraction, and (4) More definitive research is needed, especially for rough surfaces for which deposition velocities of submicrometer particles are expected to be two to three orders of magnitude higher than for smooth surfaces.

I greatly appreciate John L. Miller's assistance with the scanning electron microscopy.

Table I. Fourth degree polynomial regression results of ln(N) as a function of ln(D).

	Near road		Window sill	
	coefficient	standard error	coefficient	standard error
A_0	9.7968	0.2193	8.6002	0.2060
A_1	-2.0986	0.1345	-2.1632	0.1326
A_2	0.0413	0.0853	-0.0304	0.0667
A_3	0.0390	0.1193	0.0476	0.0113
A_4	-0.0116	0.0523	-0.0043	0.0038
Standard error of estimate	1.3191		0.9602	

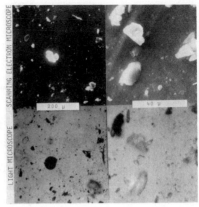

Figure 1. Particles deposited on glass near a busy roadway.

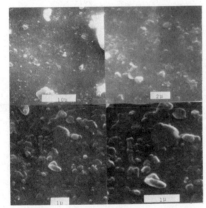

Figure 2. Scanning electron micrographs of small particles on glass near a busy roadway.

Table II. Particle size distribution on glass.

Location	Diameter Range (µm)	Number per mm^2 median	standard deviation Lower limits	Upper limits	Area (mm^2/mm^2)	Area (percent)	Volume (percent)
Near road	0.010-0.021	3.54×10^5	9.47×10^4	1.32×10^6	0.0001	0.0	0.0
	0.022-0.045	1.59×10^6	4.26×10^5	5.96×10^6	0.0013	0.9	0.0
	0.046-0.099	1.75×10^6	4.67×10^5	6.53×10^6	0.0064	4.6	0.0
	0.100-0.214	7.61×10^5	2.03×10^5	2.85×10^6	0.0129	9.4	0.1
	0.215-0.463	1.96×10^5	5.25×10^4	7.34×10^5	0.0154	11.2	0.3
	0.464-0.999	4.04×10^4	1.08×10^4	1.51×10^5	0.0147	10.7	0.5
	1.00-2.14	8.10×10^3	2.17×10^3	3.03×10^4	0.0137	9.9	1.1
	2.15-4.63	1.76×10^3	4.71×10^2	6.59×10^3	0.0138	10.0	2.3
	4.64-9.99	4.20×10^2	1.12×10^2	1.57×10^3	0.0153	11.1	5.5
	10.0-21.4	1.00×10^2	2.68×10	3.75×10^2	0.0170	12.3	13.1
	21.5-46.3	2.00×10	5.35	7.48×10	0.0157	11.4	26.2
	46.4-99.9	2.52	6.73×10^{-1}	9.41	0.0092	6.7	33.0
	100-215	1.37×10^{-1}	3.65×10^{-2}	5.11×10^{-1}	0.0023	1.7	17.9
Window sill	0.010-0.021	2.05×10^5	7.85×10^4	5.35×10^5	0.0000	0.0	0.0
	0.022-0.045	5.04×10^5	1.93×10^5	1.32×10^6	0.0004	0.6	0.0
	0.046-0.099	4.62×10^5	1.77×10^5	1.21×10^6	0.0017	2.4	0.0
	0.100-0.214	2.00×10^5	7.95×10^4	5.43×10^5	0.0035	5.0	0.0
	0.215-0.463	5.81×10^4	2.22×10^4	1.52×10^5	0.0046	6.5	0.0
	0.464-0.999	1.24×10^4	4.74×10^3	3.23×10^4	0.0045	6.4	0.0
	1.00-2.14	2.36×10^3	9.05×10^2	6.18×10^3	0.0040	5.7	0.1
	2.15-4.63	4.61×10^2	1.77×10^2	1.21×10^3	0.0036	5.2	0.3
	4.64-9.99	1.01×10^2	3.86×10	2.64×10^2	0.0037	5.2	0.6
	10.0-21.4	2.62×10	1.00×10	6.84×10	0.0041	6.3	1.5
	21.5-46.3	8.22	3.15	2.15×10	0.0065	9.2	4.7
	46.4-99.9	3.07	1.18	8.03	0.0112	16.0	17.7
	100-215	1.30	.50	3.39	0.0220	31.4	74.9

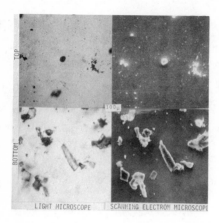

Figure 3. Particles deposited on glass at 70° to the horizontal on a window sill.

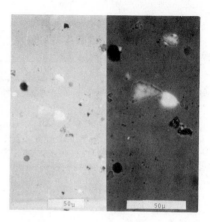

Figure 4. Particles in a clear paint film applied to and stripped from a dirty garage window (partially polarized light).

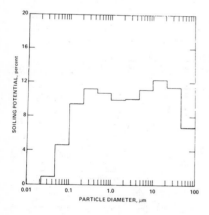

Figure 5. Distribution of soiling potential as a function of particle diameter on glass near roadway.

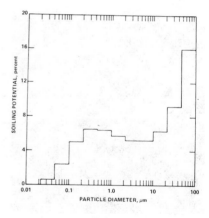

Figure 6. Distribution of soiling potential as a function of particle diameter on glass at window sill.

Published 1984 by Elsevier Science Publishing Co., Inc.

Aerosols, Liu, Pui, and Fissan, editors

Therapy of damaged forests through pyrotechnically
generated, neutralizing and fertilizing aerosols

Georgi, B.[+], Kühn, W.[+] and Krone, U.[++]

Forests are damaged by airborne pollution. In the same way
that the negative effects took place, calcium, magnesia,
potassium and their compounds together with trace elements
for neutralisation and fertilization are applied. The
aerosol is generated pyrotechnically and the suspended
matter consists of particles 50 % $\leq$ 1 µm d a.e.. Particles
of this size have a neglectable sedimentation velocity and
are able to enter the leaves through the stomata. The intake
has been proved by tracer experiments and microprobe analysis. The change of the pH value according to deposition
was continously measured by a specially constructed set up.
Starting with 3.5 pH after application of the aerosol > pH 8
was reached and lasted for several hours. Transpiration and
plant toxicity experiments showed no differences to the
control plants. Experiments combined with the application of
suspended acids are under investigation as well as field
trials. So far the application in selected areas has not
shown negative results, but further vegetation time is
needed to prove the positive effects.

[+] Niedersächsisches Institut für Radioökologie, Herrenhäuser
Str. 2, D-3000 Hannover 21, FRG

[++]Fa. Nico Pyrotechnik, Bei der Feuerwerkerei 4,

CHAPTER 8.

RECEPTOR MODEL

Published 1984 by Elsevier Science Publishing Co., Inc.
Aerosols, Liu, Pui, and Fissan, editors

X-RAY DIFFRACTION AND SOURCE APPORTIONMENT OF DENVER AEROSOL

Briant L. Davis
Cloud Physics Group
Institute of Atmospheric Sciences
South Dakota School of Mines and Technology
Rapid City, South Dakota 57701-3995
U. S. A.

EXTENDED ABSTRACT

Introduction

Low level urban air pollution in Denver has received much recent attention, particularly with regard to fine particulate characterization (Dzubay et al., 1981; Wolff et al., 1981), although very little information on the coarse particulate composition is available in the Denver "Brown Cloud."

X-ray diffraction is one of the analytical techniques able to provide quantitative information on both fine and coarse particulate concentrations, although mass requirements on the filter requires extended sampling periods to achieve the additional sensitivity needed for analysis of minor components. We here summarize a quantitative XRD study of dichotomous and low volume samples obtained over a continuous 3-1/2 day sampling interval at the Mile High Kennel Club in Commerce City (northeast Denver).

Field and Laboratory Methods

Sampling was accomplished using a low volume sampler having a 20 μ diameter upper cut point and a manual Sierra dichotomous sampler with a 15 μ diameter upper cut point and 2.5 μm diameter separation of fine and coarse particles. Sampling began on 2010 on January 25 and was concluded at approximately 0600 on 29 January 1982. Relatively light southwest to northwest winds prevailed during the period, and only minor precipitation occurred early on the morning of 29 January.

Samples were analyzed periodically for the following eight months using x-ray diffraction, x-ray transmission, and polarizing optical microscopy. Samples were collected on glass fiber filters, supposedly of high quality and artifact resistant; later tests, however, revealed artifact nitrate formation with these filters.

While field sampling progressed, several sets of street dust, soil dust, and industrial samples were obtained for quantitative XRD analysis. Using these analyses and emission inventory data for transportation and combustion compositions, an effective variance source apportionment (Davis, 1982) was completed.

Results and Discussion

Table 1 presents the quantitative analyses for the bulk soil and street residue materials. In these analyses, we observed a remarkable complexity of the crystalline components that may be airborne in the Denver metropolitan

TABLE 1: Quantitative component analyses for soil, street residue, and sand stock.

	Street/Soil Source (A) (GS 1-4)		Street/Soil Source (B) (GS 7-11)		Street Sand Stock (C) (GS-12)	
	Wt. %	σ_i	Wt. %	σ_i	Wt. %	σ_i
Biotite	0.80	1.6	0	--	0	--
Muscovite	0	--	16.16	3.13	8.0	1.1
Chlorite	0.25	0.29	0	--	0	--
Kaolinite	4.10	1.95	2.64	1.68	0.5	0.2
Montmorillonite	14.68	1.74	8.46	4.40	5.8	0.7
Quartz	32.58	4.56	27.52	4.60	19.2	2.4
Microcline	21.45	2.86	20.20	7.39	23.4	3.4
Oligoclase	15.35	3.59	11.54	3.21	10.4	1.3
Calcite	1.45	2.90	1.52	1.69	0	--
Hematite	0.23	0.45	0.20	0.45	0	--
Magnetite	0.70	0.62	1.36	0.63	0.8	0.4
Hornblende	0	--	0.62	0.67	0.5	0.5
Epidote	0.13	0.25	0.24	0.43	0	--
Serpentine	0	--	1.94	4.34	1.5	0.9
Limonite	5.23	1.83	2.16	0.76	2.4	2.9
Opal	0.70	0.25	2.74	4.08	0	--
Carbonac. matter	1.90	2.37	1.38	1.90	0	--
Fly Ash	0.58	1.15	0	--	0	--
Halite	0	--	1.30	1.78	25.5	4.3
Zircon	0	--	0.10	0.22	0	--
Gypsum	0	--	0	--	2.0	0.4
	100.13		100.08		100.0	

area. Table 2 presents the actual analysis of the ambient filter samples showing many of the source materials present observed in Table 1, as well as a number of anthropogenic compounds either formed as atmospheric reaction products or as filter artifacts. Subsequent laboratory testing of the filters used for the dichotomous sampling demonstrated that the sodium of the $NaNO_3$ component was a positive artifact provided by the filter substrate material, whereas the ammonium sulfate and ammonium lead sulfate components appear to be topochemical artifacts.

Table 3 presents a summary of selected elemental and ionic components measured by the EPA laboratories using x-ray fluorescence (XRF) and ion chromatography (IC) compared with values reduced from the component compositions of the XRD analysis. The additional nitrate observed in the fine mode by our measurements (IAS) suggests the trapping of additional nitrate on the alkaline filter sites and subsequent reaction to $NaNO_3$.

TABLE 2: Analyses of ambient filter samples.

Component	Lo-Vol Sampler Wt. %	Lo-Vol Sampler s(Wᵢ)	Dichotomous Sampler Coarse Wt. %	Dichotomous Sampler Coarse s(Wᵢ)	Dichotomous Sampler Fine Wt. %	Dichotomous Sampler Fine s(Wᵢ)
Biotite	0.5	0.2	0.6	0.3	--	--
Muscovite	12.9	1.9	12.1	1.8	--	--
Kaolinite	3.1	1.1	3.3	1.1	10.6	2.7
Quartz	13.6	2.0	19.6	2.8	6.8	0.8
Microcline	9.2	1.7	19.0	3.3	--	--
Oligoclase	6.5	1.0	14.5	2.1	--	--
Calcite	1.6	1.0	2.3	1.3	--	--
Magnetite	--	--	<0.01	--	--	--
$(NH_4)_2SO_4$	--	--	--	--	4.7	1.2
$(NH_4)_2Pb(SO_4)_2$	--	--	--	--	3.2	0.4
$CaSO_4 \cdot 2H_2O$	3.8	0.9	1.4	0.1	--	--
$NaNO_3$	--	--	--	--	21.7	1.8
Limonite	7.3	3.1	15.4	4.4	--	--
Carbonac.	37.7	6.9	4.6	5.6	52.9	20.0
Fly Ash	1.0	0.3	1.5	0.4	--	--
Halite	1.5	0.4	6.0	1.4	--	--
Anthophyllite	1.7	0.5	--	--	--	--

TABLE 3: Selected elemental and ionic components analyzed by EPA (XRF and IC) and compared to values computed by IAS from XRD analysis.

DICHOTOMOUS SAMPLE

$\mu g\ m^{-3}$

Component	Coarse Mode EPA	Coarse Mode IAS	Fine Mode EPA	Fine Mode IAS
Si	11.4	9.1	0.4	1.3
Al	4.3	2.7	0.5	0.5
K	1.1	1.3	0.09	0.01
Pb	0.1	ND	0.3	0.4
$SO_4^=$	0.5	0.3	1.2	1.2
NO_3^-	0.4	ND	1.9	3.9
NH_4^+	<0.01	ND	0.2	0.4
C	NM	1.4	11.1	10.0

NM = Not Measured   ND = Not Detected

336

Figure 1 presents two diagrams representing the source apportionment of
the dichotomous samples based on the soil and street sand analyses of 1 April,
and vehicle emission compositions of 0.7% $SO_4^=$, 14.8% lead (Pb), and 56.3%
carbon for lightweight vehicles; and 21.5% $SO_4^=$ and 78.5% carbon representing
diesel emissions. The apportionment of the low-volume analysis was very
similar to that of the coarse dichot analysis.

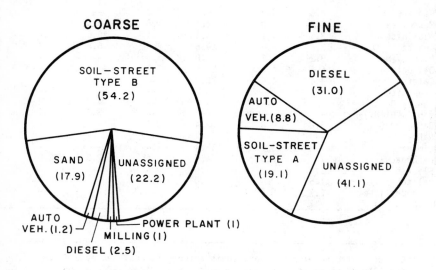

COARSE

SOIL—STREET
TYPE B
(54.2)

SAND
(17.9)

UNASSIGNED
(22.2)

AUTO
VEH. (1.2)

POWER PLANT (1)

MILLING (1)

DIESEL (2.5)

FINE

DIESEL
(31.0)

AUTO
VEH.(8.8)

SOIL—STREET
TYPE A
(19.1)

UNASSIGNED
(41.1)

Figure 1: Source apportionment of dichotomous sample, Jan 25-29, 1982,
Commerce City, CO.

References

Davis, B. L. (1982) "Source characterization and apportionment of the Houston
aerosol by means of x-ray diffraction techniques," Proc. Specialty Conf. on
Receptor Models Applied to Contemporary Pollution Problems, Air Poll. Control
Assoc., 84-95.

Dzubay, T. G., R. K. Stevens, W. J. Courtney, and E. A. Drane (1981) "Chemical
element balance analysis of Denver aerosol," In Electron Microscopy and X-ray
Applications to Environmental and Occupational Health Analysis (Edited by
P. A. Russell), 2:23-42.

Wolff, G. T., R. J. Countess, P. F. Groblicki, M. A. Ferman, S. H. Cadle,
and J. L. Muhlbaier (1981) "Visibility-reducing species in the Denver "brown
cloud" - II. Sources and temporal patterns," Atmos. Environ. 15:2485-2502.

Published 1984 by Elsevier Science Publishing Co., Inc.
Aerosols, Liu, Pui, and Fissan, editors

PARTICLE COLLECTION BY A FIXED COLLECTOR CONSIDERING THE
COMBINED EFFECTS OF INERTIA AND CHARGE

H. C. Wang , J. J. Stukel, K. H. Leong, and P. K. Hopke
Department of Civil Engineering
University of Illinois at Urbana-Champaign
Urbana, IL 61801
U. S. A.

EXTENDED ABSTRACT

Introduction

To increase the removal efficiency of fine particles in abatement
devices, it is a common practice today to impose electrostatic forces in
conventional devices to enhance the overall collection efficiencies. In
most cases charge enhancement is utilized to improve systems in which
inertial impaction is the primary collection mechanism. Therefore, the
understanding of the competition between these two mechanisms on
particle collection is necessary. Recently, the combined effects of
inertial impaction and electrostatic attraction were theoretically
examined by George and Poehlein (1974), Nielsen and Hill (1976), and
Beizaie and Tien (1980). All the above authors assumed that the drag on
the particles could be approximated using Stokes law. However in actual
practice, this may not be true especially for large particles and/or
high velocity. In addition, no experimental measurements of particle
collection considering the combined effects of particle inertia and
charge are available in the literature, probably because of the
difficulty in experimentally matching these two mechanisms by carefully
controlling the particle size, charge, and velocity. This study extends
the theoretical model proposed by previous investigators to include
non-Stokesian correction. To insure the validity of the modified model,
well-defined experiments were performed with selected experimental
conditions parallel to those frequently encountered in industrial
applications. The experimental results are compared with the
theoretical prediction and the significance is discussed.

Theoretical Model

The dominant collision mechanisms of interest in this study are
inertial impaction and Coulombic attraction. By equating the inertial
force, Coulombic force (F), and the drag force, the equation of motion
for a spherical particle can be non-dimensionalized and expressed as

$$St \frac{dV}{dt} = C_D (U - V) + Kc F \tag{1}$$

where V , t, and U are dimensionless particle velocity vector, time, and
flow velocity vector. St and Kc are Stokes number and Coulombic
parameter as defined by Nielsen and Hill (1976), C_D and Re_p are the drag

coefficient and the particle Reynolds number, respectively. The value of C_D for $0 < Re_p < 3x10^5$ is given by Brauer and Sucker (1978) as

$$C_D = 0.49 + 24/Re_p + 3.73/Re_p^{0.5} - 4.83 \times 10^{-3} Re_p^{0.5} /(1+3x10^{-6} Re_p^{1.5}) \qquad (2)$$

Substituting Eq.(2) into Eq.(1), and assuming potential flow field around the collector, particle trajectories can be obtained by integrating Eq.(1) twice. The limiting trajectory is determined by trial and error in conjunction with an interval-halving search technique. The collision efficiency can then be calculated from

$$E = [X_c/(R_c + R_p)]^2 \qquad (3)$$

where R_c and R_p are the radii of the collector and particle and X_c is the initial displacement of the limiting trajectory from the centerline in the undisturbed flow region.

Experimental Method

A 4-inch diameter stainless-steel pipe, approximately 3.4 m in length, was used as a wind tunnel with the flow induced by a blower on the exit end and exhausted through a ventilation hood. The vibration caused by the blower was damped out using a flexible connection between the wind tunnel and the blower. The flow velocity, temperature and pressure were measured by means of a flow measuring system located 61 cm upstream from the blower. An absolute filter was installed in the inlet section to remove the particles in the intake air. A vibrating orifice aerosol generator (Berglund and Liu, 1973), modified for induction aerosol charging (Reischl et al., 1977), was used to obtain an aerosol of uniform charge and size. Monodisperse charged aerosol thus generated was introduced into the wind tunnel immediately after the filter. The charge flux was continuously monitored by an aerosol electrometer equipped with an isokinetic sampling probe located 137 cm downstream of the inlet. The aerosol then passed through the aerosol test section in which a stainless-steel ball of known size was located. This ball was electrically connected to a high voltage supply and served as the charged collector. In this study, 5 different balls were used with diameters of 2.311, 3.937, 4.737, 5.537, and 6.345 mm, respectively. The amount of particle mass collected on the ball was determined by fluorescence spectroscopy and was used to calculate the collection efficiency.

Results and Discussions

Since the competition between inertial and electrical effects and the non-Stokesian behavior of the particles were the major concerns of this investigation, two experimental schemes were planned. In the first experiment, the dependence of collision efficiency on Kc was examined at a constant value of St. It was carried out by fixing R_p, R_c, and the flow velocity to give constant values of St, while the value of Kc was

varied by adjusting the voltage supply to the collector. The particle
Reynolds number selected was 5.084 corresponding to the common
industrial applications to allow an examination of the non-Stokesian
behavior of the particles. The results were shown in Figure 1. The
theoretical results with and without the non-Stokesian correction are
also included. The experimental efficiencies increased smoothly with
increasing values of -Kc as predicted by both theories. Obviously, the
theory that includes the non-Stokesian correction gives a better
prediction. Estimates using Stokes law overestimate the efficiency by
approximately 15% under this experimental condition. Therefore, the
non-Stokesian correction whenever applicable is necessary. With this
correction, the theoretical model can provide good estimates of the
collection efficiency.

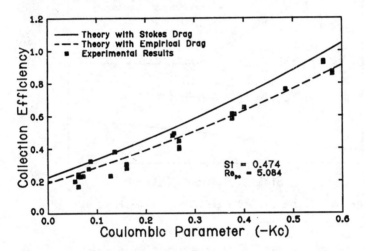

Figure 1. Collection efficiencies as a function of Kc.

In the second experiment, the variation of collision efficiency with
St was examined at a constant value of Kc. This experiment was more
complicated in terms of experimental control because the variation of
each adjustable physical variable involved in St also caused Kc to vary
simultaneously. Taking advantage of the ease of controlling the applied
voltage in this setup, the variation of St was obtained by varying the
ball size while Kc was kept constant by changing the applied voltage
accordingly. In order to cover the minimum efficiency point as
indicated by Nielsen and Hill (1976), a wide range of St values must be
used. It was not feasible to cover the whole range of interest with a
single experimental configuration. Instead three configurations were
necessary because of the limitation of the ball sizes that were used to
vary St. Therefore, three experiments were performed separately using

balls ranging in diameter from 2.311-6.345 mm. These three sets of data together with the corresponding theoretical curves that include the non-Stokesian correction are shown in Figure 2. Good agreement was obtained between experiment and theory. Thus, the competition between inertial and electrical forces in the collection of particles by a spherical collector can be predicted very well by means of a theoretical model which includes the non-Stokesian correction and assumes potential flow conditions.

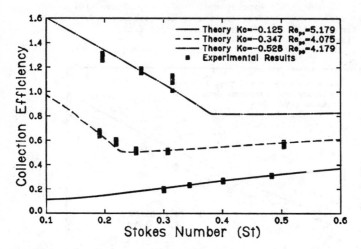

Figure 2. Collection efficiencies as a function of St

References

Beizaie, M. and Tien, C. (1980) "Particle Deposition on a Single Spherical Collector." Canadian J. Chem. Eng. 58:12.

Berglund, R. N. and Liu, B. Y. H. (1973) "Generation of Monodisperse Aerosol Standards." Environ. Sci. Technol. 7(12):147.

Brauer, H. and Sucker, D. (1978) "Flow about Plates, Cylinders, and Spheres." Int. Chem. Eng. 18(3):367.

George, H. F. and Poehlein, G. W. (1974) "Capture of Particles by Spherical Collectors." Environ. Sci. Technol. 8(1):46.

Nielsen, K. A. and Hill, J. C. (1976) "Capture of Particles on Spheres by Inertial and Electrical Forces." Ind. Eng. Fundam. 15(3):157

Reischl, G., John, W., and Devor, W. (1977) "Uniform Electrical Charging of Monodisperse Aerosol." J. Aerosol Sci. 8:55.

Published 1984 by Elsevier Science Publishing Co., Inc.

Aerosols, Liu, Pui, and Fissan, editors

RECEPTOR MODEL APPLICATION TO DENVER AEROSOL

Charles W. Lewis and R. K. Stevens
Environmental Sciences Research Laboratory
U.S. Environmental Protection Agency
Research Triangle Park, NC   27711
U. S. A.

EXTENDED ABSTRACT

Introduction

In Denver, CO the total suspended particulate (TSP) often exceeds 100 $\mu g/m^3$ in the winter time and these events are invariably associated with what has been termed the "Denver Brown Cloud." Even if the National Ambient Air Quality Standard for TSP is revised to reflect a modified upper size limit, it is anticipated that the Denver brown haze is likely to persist, since it is the fine particles which are primarily responsible for visibility degradation.

The most comprehensive study aimed at determining the origins of the Denver winter haze was performed in November and December, 1978 under the auspices of the Motor Vehicle Manufacturers Association (Heisler et al., 1980) and General Motors Research Laboratory (Wolff et al., 1981). During January 1982 a follow-up study was performed by EPA. The latter study attempted to exploit some of the advances in sampling and analytical methods that have occurred since the earlier experiments, and to develop auxiliary source signature data for sources previously shown to be of importance.

Experimental

The measurement site for the EPA January 1982 study was located at the Mile High Kennel Club (MHKC) which was approximately 10 km northeast of downtown Denver and was the same site used by General Motors during their 1978 study. The air quality measurements and extinction coefficient measurements were essentially identical to those measurements made in an earlier visibility study conducted in Houston (Dzubay et al., 1982). A principal difference between the EPA Houston study and the EPA 1982 Denver winter haze study was the manner in which the absorption extinction coefficient bap of fine particle aerosol was deduced in Denver (from telephotometric, NO_2 and light scattering measurements). Further differences were the employment of an artifact free method for the separate measurement of fine particle nitrate and gaseous nitric acid (Shaw et al., 1982); and the determination of volatile and elemental (black, graphite-like) aerosol carbon through a two step thermal analysis (650° and 850° in 100% helium and 98% helium/2% oxygen atmospheres, respectively) of the aerosol collected on quartz filters. The 1982 study also included a comprehensive experimental investigation on the relationship between human judgment of visual air quality and the chemical composition of the air. The latter results have already been reported by Middleton et al. (1984).

Conclusions

(1) The central feature of the Denver fine particle aerosol is the large percentage of the mass represented by carbonaceous material. While the EPA and GM results for total fine particle carbon are in qualitative

342

agreement (Table 1), the EPA result (54 ± 11%) is even larger than the GM result (36 ± 3%). A more important difference is the much greater amount of volatile (non-black) carbon, as a percentage of total fine particle mass, found by EPA (42 ± 8%), compared with GM (20 ± 4%). The latter difference is made more understandable by the recent recognition (Cadle et al., 1983) that the method GM used for their Denver study tends to overestimate the elemental component of the carbonaceous aerosol.

Table 1. Denver Fine Mass ($\leq$ 2.5 µm) Comparison

	Day ng/m^3	EPA, January, 1982 Night ng/m^3	24-h Ave. ng/m^3	Mi/Mass %	GM Nov-Dec, 1978 24-h Ave. ng/m^3	Mi/Mass %
Mass	18220	21950	20080	100.	33700	100.
SO4$^-$	2280	1840	2060	10.3	3000	8.9
NH4$^+$	654	505	579	2.8	2500	7.4
NO3$^-$	2570	1860	2210	11.0	4100	12.2
Cvol	6460	10270	8360	41.6	6600	20.0
Cel	2160	2890	2520	12.6	5500	16.3
Al	389	399	394	2.0	700	2.0
Si	264	290	277	1.4	800	2.3
P	38	47	42	0.2	--	--
S	755	662	708	3.5	2240	6.6
Cl	46	57	52	0.3	--	--
K	59	98	78	0.4	100	0.3
Ca	43	51	47	0.2	80	0.2
Ti	--	--	--	--	10	0.0
Mn	8	13	11	0.1	5	0.0
Fe	77	81	78	0.4	110	0.3
Cu	9	10	9	0.0	--	--
Zn	38	54	46	0.2	60	0.1
Br	78	128	103	0.5	490	1.5
Pb	268	383	325	1.7	1680	5.0

(2) The fine particle sulfur measured in the EPA study is entirely accounted for by the observed sulfate concentration. This is in sharp contrast to the 44% "excess sulfur" seen in the GM results (Table 1). On the basis of the new results there seems to be little difference between the sulfur species present in Denver compared with what has been measured in many other U.S. cities.

(3) Measured amounts of nitric acid were small in comparison to fine particle nitrate (303 vs. 2210 ng/m^3 for the 24-h average).

(4) The telephotometer-based measurements of bap resulted in a value of 0.13 for the average of the ratio bap/bext. The corresponding result during the GM study was 0.29. The latter was determined using the Integrating Plate method, which has been criticized in the past as having a systematically high bias. The EPA result suggests a reduced role for aerosol absorption in the Denver atmosphere.

(5) Table 2 shows a <u>preliminary</u> source apportionment of the EPA-measured fine particle aerosol. The method employed was that of Kleinman et al. (1980), using the single element tracers Pb, S, Si, and (K - 0.68 x Fe) for mobile sources, coal combustion, crustal material and wood smoke, respectively. The results from the GM study are also given for comparison. Additional steps are planned to improve the EPA source apportionment. These include (a) analysis of Denver soil samples to increase the accuracies of the crustal contribution and the wood smoke correction term, and (b) using profiles of vehicle emissions measured in Denver to confirm the vehicle contribution shown in Table 2.

Table 2. Comparison of Source Apportionments for Denver Winter Haze Studies Conducted in 1978 and 1982.

Source	U.S. EPA January 1982 % of Fine Mass	GM Nov.-Dec. 1978 % of Fine Mass
Motor vehicles	46 ± 7	26
Coal combustion	29 ± 5	20
Crustal/Flyash	9 ± 13	14
Woodburning	6 - 12	12
Natural gas combustion	U	9
Fuel oil combustion	U	7

U = unquantified at present

References

Cadle, S.H., Groblicki, P.J. and Mulawa, P.A. (1983) "Problems in the Sampling and Analysis of Carbon Particulate," Atmospheric Environment, 17:593-600.

Dzubay, T.G., Stevens, C.W., Hern, D.H., Courtney, W.J., Tesch, J.W., and Mason, M.A. (1982) "Visibility and Aerosol Composition in Houston, TX," Environmental Science and Technology, 16:514-525.

Heisler, S.L., Henry, R.C., Watson, J.G., and Hidy, G.M. (1980) "The 1978 Denver Winter Haze Study, Volume II, Final Report", Environmental Research and Technology, Inc. Report No. P-5417-1.

Kleinman, M.T., Pasternack, S., Eisenbud, M. and Kneip, T.J. (1980) "Identifying and Estimating the Relative Importance of Sources of Airborne Particulates," Environmental Science and Technology, 14:62-65.

Middleton, P., Stewart, T.R., Ely, D. and Lewis, C.W. (1984) "Physical and Chemical Indicators of Urban Visual Air Quality Judgments," Atmospheric Environment (in press).

Shaw, R.W., Stevens, R.K., Bowermaster, J., Tesch, J.W., and Tew, E. (1982) "Measurements of Atmospheric Nitrate and Nitric Acid: The Denuder Difference Experiment," Atmospheric Environment, 16:845-853.

344

Wolff, G.T., Countess, R.J., Groblicki, P.J., Ferman, M.A., Cadle, S.H., and Muhlbaier, J.L. (1981) "Visibility-reducing Species in the Denver "Brown Cloud" II Sources and Temporal Patterns", <u>Atmospheric Environment</u>, 15:2485-2502.

Published 1984 by Elsevier Science Publishing Co., Inc.
Aerosols, Liu, Pui, and Fissan, editors

ELECTRON MICROSCOPY AND RECEPTOR MODELING
APPLICATION TO URBAN AND RURAL AEROSOLS

Yaacov Mamane* and Thomas Dzubay
Environmental Sciences Research Laboratory
U.S. Environmental Protection Agency
Research Triangle Park, NC 27711, U.S.A.

EXTENDED ABSTRACT

Introduction

Aerosol receptor modeling assesses the impact of major pollution and natural sources using "fingerprints" of particles as measured at their sources, measured elemental concentrations at a receptor site, and a mathematical model (Dzubay, 1980). The use of electron microscopy techniques to study morphology and chemical content of individual particles provides a direct way to deduce the contribution of various sources to the aersol make-up.

Experimental

In this study two sets of samples were analyzed. One from a rural site in West Maryland at the Deep Creek Lake (summer 1983), and one from an urban region -- Philadelphia (summer 1982). In the Deep Creek Lake transmission electron microscopy and micro spot techniques (Mamane and Puschel, 1980) were used to study individual sulfate (Mamane and Pena, 1978) and nitrate particles (> 0.1 μm), while a scanning electron microscope (SEM) equipped with an x-ray analyzer is used to study both morphology and elemental content of particles (> 0.2 μm). In the Philadelphia study particles were only collected on Nuclepore filters using dichotomous sampler and analyzed by SEM.

Results

In the DCL sulfates, mostly as acidic sulfates, dominate the fine fraction (< 2.5 μm) while crustal material and biological sources, such as pollen, dominate the coarse fraction. Fly ash particles from coal power plants were found mostly in the 1 to 10 μm size fraction. Other few particles of industrial sources, transported over long distances, were also found. Nitrates were not found in the submicrometer size range, say 0.4 to 1 μm, but only in the upper size range of the fine fraction, 1 to 2.5 μm. Even then, the percent of particles containing nitrate was between 2 to 6 percent. Particles of soil and biological origin seem to be enriched in sulfur, probably as a sulfate layer on the particle surface.

In the Philadelphia study chemical plants, power plants, a refinery and an incinerator were sampled at the source. Scanning electron microscopy characterization of these sources provide the "fingerprints" to be used later in determining their contribution at the receptor site. The main results of these analyses, as related to receptor modeling, are as follows: (a) Fly ash particles are highly variable in their iron content, less so far the Al, Si elements. Figure 1 shows x-ray spectra of two fly ash particles, one (top) has an "average" composition and one highly enriched in iron. (b) Incinerator and the catalytic cracker (refineries) particles vary in composition and morphology as a function of size. (c) Antimony from a chemical plant was a simple tracer easily identified with the scanning microscopy in ambient sam-

---

* NRC Research Associate. On leave from the Department of Environmental Engineering, Technion, Haifa, 32000, ISRAEL.

ples in Philadelphia when the flow was to the sampling site. An example is given in Figure 2, which is a photomicrograph of particles collected in the fine fraction. The sample is loaded with sulfate particles -- the grey round particles. Antimony (probably as oxide) particles are the bright square ones.

Analysis of the ambient samples revealed the dominant presence of sulfates in the fine fraction, and small contributions from large number of industrial sources. In the coarse fraction natural sources were dominating however contributions from chemical and power plants, and motor vehicles in some cases were significant.

Disclaimer:

This paper has not been subjected to the Environmental Protection Agency review and therefore does not necessarily reflect the views of the Agency, and no official endorsement should be inferred.

References:

Dzubay, T. G. (1980) "Chemical element balance method applied to dichotomous sampler data," Ann. N.Y. Acad. Soc., 328, p. 126.

Mamane, Y. and R. F. Pueschel (1980) "A method for the detection of individual nitrate particle," Atmos. Environ., 14, 629.

Mamane, Y. and R. de Pena (1978) "A quantitative method for the detection of individual submicrometer size particles," Atmos. Environ., 12, 69-82.

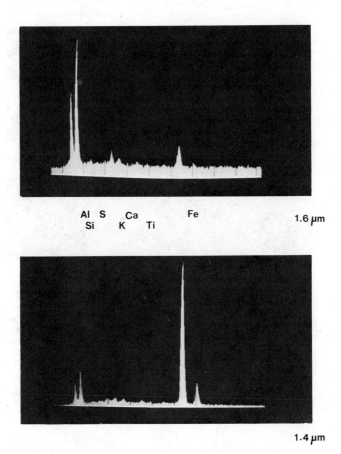

Figure 1. X-ray spectra of two fly ash particles collected at the source. While the "typical" particle (top) contain mostly Al, Si and some Fe, K, Ca, Ti, and S, the atypical is almost made entirely of Fe (bottom).

348

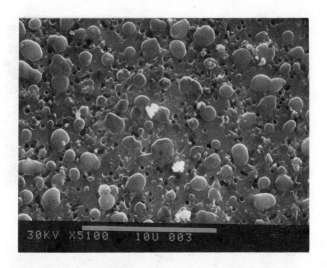

Figure 2.   A typical photomicrograph of particles of the fine fraction
            collected in Philadelphia. Most particles contain sulfur, and
            probably are made of sulfate -- the round grey particles. The
            very bright rectangular particles are made of antimony oxide,
            and were found only in episodic cases where the contribution from
            the plant was significant.

Published 1984 by Elsevier Science Publishing Co., Inc.

Aerosols, Liu, Pui, and Fissan, editors

CAN PAH SOURCE EMISSION PROFILES BE USED TO IDENTIFY SOURCES
OF AIRBORNE PARTICULATE POLLUTANTS?

J. M. Daisey, J. L. Cheney[*], and P. J. Lioy
Institute of Environmental Medicine
New York University Medical Center
550 First Avenue
New York, New York 10016
*Environmental Sciences Research Laboratory
U.S. Environmental Protection Agency
Research Triangle Park, N.C. 27711

EXTENDED ABSTRACT

Introduction

There is a need to identify different types of source emissions tracers
due to the impending loss of Pb as a tracer for motor vehicle emissions and
our inability to distinguish certain types of sources through trace elemental
profiles. Published data on polycyclic aromatic hydrocarbons (PAH) and unique
organic tracers in source emissions have been reviewed to determine their
potential usefulness for this purpose. Emphasis was placed on emissions from
source types suspected to contribute most significantly to particulate organic
matter in urban air - motor vehicles, space heating, power production and
certain industrial processes. For the purpose of comparison among sources,
ratios of measured PAH compounds to benzo(e)pyrene (BeP) were computed; BeP
was selected because it is almost wholly in the particulate phase, appears to
be relatively stable in the atmosphere and it is frequently measured.

Factors to be Considered in Comparing Source Emissions Data

Several major experimental factors can cause variations in the measured
PAH profiles which make comparisons among published source emissions data
difficult. 1) Published emissions data are usually for only a few individual
sources, measured under limited conditions for short sampling intervals and
do not represent an average profile for a given source type; 2) source
emissions sample collection methods generally differ from those used for
ambient sampling and such differences can affect sample combustion; 3) differences in extraction and analytical methods can affect PAH profiles. Within
these limitations, two questions were considered: 1) How well do the reported
profiles for a given source type agree within and among laboratories? 2) Do
there appear to be differences in the reported PAH profiles from different
source types which might be useful for distinguishing sources?

Results and Discussion

Adequate PAH data do not exist for certain sources of interest, e.g.,
residential oil burning, resuspended soil. In addition, sets of PAH compounds
measured by different investigators often differ and other composition
variables of interest (trace elements, carbon, particle mass) have rarely been
measured simultaneously in a source test. However, the data reported by
Grimmer et al. (1977; 1980) for motor vehicles and single-source dominated
urban aerosols, by Bjorseth (1979) for coke and aluminum plants, and by

Burlingame et al. (1981) for industrial boilers, indicate that the profile of
a given source is reproducible within a factor of 2-3, when the same sampling and analytical method is used.

A summary of the best available PAH profiles is presented (Daisey et al.,
1984) in Table 1. Some of the data gaps mentioned above are apparent in
Table 1. For tunnel samples, the ratios of many compounds agree within a
factor of 3 to 5 despite differences in tunnels (and traffic mixes) and
sampling and analytical methods. Similar variations exist in the ratios of
some trace elements to Pb among three U.S. tunnels (Pierson and Brachaczek,
1983). The variation in PAH ratios was similar (3-5) for anthracite and
bituminous coal burning for 5 of the 7 industrial boilers and for the coke
ovens; ratios varied more widely for the wood burning, but each reported
profile was for a different combustion condition. For samples which included
collection of PAH vapors, the ratios of PAH within a given profile generally
varied over two orders of magnitude. Such a range of variation is useful
for source resolution and the data in Table 1 suggest some variations among
sources.

The data in Table 1 suggest that it may be possible to distinguish motor
vehicle emissions from other sources using cyclopenta(cd)pyrene and benzo(k)-
fluoranthene. The compounds benzo(ghi)perylene and coronene, used commonly
as indicators of motor vehicle emissions, are probably useful tracers in the
absence of oil, coal and wood combustion. Under such conditions, it may be
possible to further distinguish between diesel and spark-ignition engines
using BNT as an indicator of diesel emissions. The higher proportion of
alkylated PAH reported for coal combustion (Bartle et al., 1974) may be useful for distinguishing this source from resuspended soil and from oil and gas
combustion emissions but there are few data available. Retene (Ramdahl, 1983)
has been suggested as a unique marker for combustion of resinous soft wood
and levoglucosan (Hornig et al., 1983) as a unique marker for wood combustion
in general.

Although the data are severely limited, they provide some indications
that PAH profiles used in combination with trace elements may be helpful in
resolving combustion source emissions using receptor models. The recent work
of Thrane and Wikstrom (1983), who have reported distinguishing several PAH
sources through PAH profiles via cluster analysis, supports this. However,
confirmation of this will require more comparable and extensive source emissions testing for major sources of interest.

References

Bartle, K. D., M. L. Lee and M. Novotny (1974) "High-Resolution GLC Profiles
of Urban Air Pollution Polynuclear Aromatic Hydrocarbons," Int. J. Environ.
Anal. Chem. 3:349-356.

Bjorseth, A. (1979) "Determination of Polynuclear Aromatic Hydrocarbons in
the Working Environment," In: Polynuclear Aromatic Hydrocarbons (Edited by
P. W. Jones and P. Leber), Ann Arbor Science Publishers, Inc., Ann Arbor, MI.,
pp. 371-381.

Burlingame, J. O., J. E. Gabrielson, P. L. Langsjoen and W. H. Cooke (1981) Field Tests for Industrial Coal Stoker Fired Boilers for Inorganic Trace Element and Polynuclear Aromatic Hydrocarbon Emissions, EPA-600/7-81-167.

Daisey, J. M., P. J. Lioy and T. J. Kneip (1984). A complete set of references to the sources of the data in Table 1 may be found in the report Receptor Models for Airborne Organic Species. Submitted to the U.S. Environmental Protection Agency, Contract No. CR810300-01-0, January, 1984, Project Officer, J. L. Cheney.

Grimmer, G., H. Bohnke and A. Glaser (1977) "Investigation on the Carcinogenic Burden by Air Pollution in Man. XV. Polycyclic Aromatic Hydrocarbons in Automobile Exhaust Gas - An Inventory," Zbl. Bakt. Hyg., I. Abt. Orig. B 164: 218-234.

Grimmer, G., K.-W. Naujack and D. Schneider (1980) "Changes in PAH Profiles in Different Areas of a City During the Year," In: Polynuclear Aromatic Hydrocarbons: Chemistry and Biological Effects (Edited by A. Bjorseth and A. J. Dennis), Battelle Press, Columbus, OH. pp. 107-125.

Hornig, J. F., R. H. Soderberg, A. C. Barefoot and J. F. Galasyn (1983) "Woodsmoke Analysis: Vaporization Losses of PAH from Filters and Levoglucosan as a Distinctive Marker for Woodsmoke. Paper presented at the Eighth International Symposium on Polynuclear Aromatic Hydorcarbons, Battelle Columbus Laboratory, Columbus, OH.

Pierson, W. R., and W. W. Brachaczek (1983) "Particulate Matter Associated with Vehicles on the Road. II," Aero. Sci. Technol. 2:1-14.

Ramdahl, T. (1983) "A Molecular Marker of Wood Combustion in Ambient Air," Nature 306:580-582.

Thrane, K. E. and L. Wikstrom (1983) "Monitoring of Polycyclic Aromatic Hydrocarbons in Ambient Air," Paper presented at the Eighth International Symposium on Polynuclear Aromatic Hydrocarbons, Battelle, Columbus Laboratories, Columbus, OH.

Acknowledgments

This study was supported by the U.S. Environmental Protection Agency, CR810300-01-0 and is part of a Center Grant Program supported by the National Institute of Environmental Health Sciences, Grant No. ES00260 and the National Cancer Institute, Grant No. CA13343.

This work has not been subject to the Agency's peer and administrative review and therefore does not necessarily reflect the views of the Agency and no official endorsement should be inferred.

Table 1. Summary of Best Available PAH Source Emissions Profiles

Compound	Tunnel Measurements	Ranges of Ratios to Benzo(e)pyrene Residential Heating Anthracite Coal	Bituminous Coal	Wood[c]	Industrial Coal Fired Boilers	Coke Ovens
Fluoranthene	1.0-2.0	16-54	2.0-3.2	1.9-124	16-161	3.0-11.4
Pyrene	0.6-1.7	12-39	1.8-5.5	1.3-69	1.6-25	2.8-8.2
Cyclopenta(cd)pyrene	5.3[a].	0.02[b].	–	–	–	–
Benzo(b)fluoranthene	0.6-1.0	1.4[b].	–	{1.9-6.9	{1.0-2.4	{1.0-1.2
Benzo(k)fluoranthene	0.3-0.4	3-4.3	1.9-3.3	n.r.	n.r.	
Benzo(j)fluoranthene	0.2-1.4	–	–			
Benzo(a)pyrene	0.2-2.0	0.6-2.0	0.6-1.4	0.5-3.2	1.0	1.5-2.3
Benzo(ghi)perylene	0.4-3.0	0.3-1.9	0.5-0.8	0.1-5.9	0.5-5.0	0.5-1.3
Indeno(1,2,3-cd)pyrene	0.3-1.5	0.6-2.3	0.1-1.8	0.6-5.3	0.5-4.4	0.3-1.0
Coronene	0.3-1.1	0.2-1.3	0.2-0.6	0.2-2.7	1.0-1.8	–
BNT[d].	0.5[a].	0.6[b].	–	–	–	–
% BaP in particles	0.008-0.037	0.001-0.002	0.02-0.06	–	–	0.1-0.58
Sampling Method	Filter	Modified EPA Method 5	Modified EPA Method 5	Modified EPA Method 5	SASS train	Filter with backup cold trap
Maximum # of Samples	17 (5 tunnels)	4	4	6 (3 stoves)	9 (7 boilers)	32 (2 seasons)

n.r. = not resolved chromatographically.
a. Measurements in one tunnel, compound not present in automobile exhaust suggesting diesel emissions source (Grimmer et al., 1980).
b. Ambient measurements in a residential coal burning area of Federal Republic of Germany (Grimmer et al., 1980).
c. Combustion conditions varied widely; oak and pine.
d. BNT = benzo(b)naphtho(2,1-d)thiophene.

Published 1984 by Elsevier Science Publishing Co., Inc.
Aerosols, Liu, Pui, and Fissan, editors

INVESTIGATION OF ELEMENTAL AND ORGANIC TRACERS TO
IDENTIFY THE SOURCES OF IPM IN NEWARK, N.J.

M. T. Morandi, P. J. Lioy and J. M. Daisey
New York University Medical Center
Institute of Environmental Medicine
550 First Avenue
New York, New York 10016

EXTENDED ABSTRACT

Introduction

The multivariate techniques of factor analysis and regression analysis
have been used to identify sources and their tracers and estimate source contributions of different fractions of the New York City aerosol (Kneip et al.,
1983; and references therein). More recently, these methods have been used
to develop receptor models for inhalable particulate matter (IPM; particles
$\leq$ 15 µm diameter) in Newark, N.J. (Morandi et al., 1983). This study suggested that due to the complex mix of sources in Newark, several traditionally
used tracers had multiple sources. A new approach which permits the use of
these multiple source tracers to estimate source contributions and assist in
source identification in Newark is presented here.

Sampling and Analytical Methods

The sampling site description, sampling and analytical procedures and
statistical methods have been described in detail elsewhere (Morandi et al.,
1983). The only variation from the previous study is the addition of more
samples. The total data base consisted of 154 daily samples obtained during
the periods of 7/6-8/14/81, 1/18-2/25/82, 7/6-8/14/82 and 1/17-2/25/83.

Data Analysis

Data screening was done with a CDC Cyber Computer using BMDP-81.
SPSS-6000-8.3 was used for the factor analysis and multiple regression procedures. The main source types and their tracers were identified using R
type common factor analysis with iterations, varimax rotation and listwise
deletion of missing data. Outliers greater than 4 S.D. from the mean were
excluded. All factors with a minimum eigenvalue of 0.2 and at least one
variable with a 0.3 loading or higher were retained. Selected tracers, were
entered stepwise in a multiple regression equation. Source contributions
were then estimated using the regression coefficients for each tracer.

Results and Discussion

The results of the factor analytical procedure as well as their interpretation and selected tracers are presented in Table 1. Source identification was based on previous work done in this laboratory. Source tracers for
the IPM model were chosen so as to minimize multicollinearity and analytical uncertainties (Morandi et al., 1983). Neither CO nor Pb was satisfactory as
a motor vehicle tracer in the regression model developed for IPM due to correlation with tracers for other sources of Pb. Benzo(ghi)perylene and benzo-

354

(a)pyrene were tested as possible motor vehicle tracers but had non-significant regression coefficients.

Table 1. Factor Analysis of the Newark, N.J. Data Set

Factors	F1	F2	F3	F4	F5	F6	F7
Variables	CDD(0.85)	Mn(0.82)	V(0.82)	Pb(0.73)	Cu(0.81)	Zn(0.65)	$SO_4^=$(0.48)
	HDD(-0.83)	Fe(0.75)	Ni(0.77)	COM(0.70)	Cd(0.44)	Pb(0.38)	
	Fe(0.40)	Pb(0.34)	Pb(0.34)	Cd(0.37)	Fe(0.37)		
	$SO_4^=$(0.36)	Cd(0.34)	HDD(0.31)				
Source	Seasonal	Soil	Oil	Motor Vehicle	Industrial	Zinc Source	Sulfate/ Secondary
% of Variance	42.5	30.7	7.3	6.3	5.3	3.6	2.2
Tracer		Mn	V	COM	Cu	Zn	SO_4

COM = maximum 1 hr. carbon monoxide concentration.
CDD = cooling degree days.
HDD = heating degree days.
Number of cases = 137.
Only variables with loadings $\leq$ 0.3 are presented.

An estimate of motor vehicle lead contributions was obtained using a new approach. A multiple regression equation for Pb as a function of its various source tracers was developed (Table 2., eqn. (1)). The concentration of motor vehicle related Pb (Pba) was then estimated as per eqn. (2), and used as the motor vehicle related tracer in eqn (3). The actual source contribution estimates to IPM and total Pb based on these equations are presented in Table 3.

Table 2. Apportionment Models for IPM, Pb and Fe

$Pb = (0.04\pm0.01)COM + (1.2\pm0.4)Cu + (1.9\pm0.4)V + (6\pm1)Mn + (0.04\pm0.01)Zn$ (1)
$+ (0.01\pm0.03)$

$Pba = Pb - (1.2Cu + 1.9V + 5.8Mn + 0.04Zn)$ (2)

$IPM_1 = (2.3\pm0.2)SO_4^= + (113\pm30)Cu + (474\pm100)Mn + (22\pm6)Pba$ (3)
$+ (73\pm27)V + (6\pm3)$

$Fe = (23\pm3)Mn + (4.4\pm0.8)Cu + (0.3\pm0.05)$ (4)

$FeR = Fe - (23.7Mn + 4.4Cu)$ (5)

$IPM_2 = (2.1\pm0.1)SO_4^= + (86\pm27)Cu + (24\pm3)FeR + (445\pm86)Mn$ (6)
$+ (110\pm24)V + (18\pm5)Pba + (1.3^*\pm0.8)Zn + (1\pm2)$

Number of cases = 137; coefficients are shown ± S.E.
F for IPM eqns. $\leq$ 90.
R^2 for IPM eqns. $\leq$ 0.77.
*p=0.076.

The residuals of eqn. (3) were found to be positively correlated with Fe and dichloromethane soluble organics ($p\leq0.001$) indicating an additional source of Fe related aerosol. To apportion Fe among its known sources, Fe was re-gressed against other related variables (eqn. (4)). The residuals of eqn. (3) were strongly correlated with the residuals of eqn. (4) but not with the fraction of Fe accounted for by Mn and Cu. The residual Fe (FeR, eqn. (5)) was then used as another tracer to apportion IPM (eqn. (6)). This Fe related source was the third most important contributor to IPM. Although the identity of this source is as yet unconfirmed, there is some evidence that it is related to emissions from a number of auto body repair and painting shops which sur-round the site.

Table 3. Estimates of Source Contributions to IPM

Dependent Variables	Sulfate Secondary	Soil	Motor Vehicle	Industrial	Oil-burning/ Space heating	FeR Related	Zn Smelter	Res.
Pb $\mu g/m^3$	–	0.1	0.2	0.04	0.07	–	0.02	–
%	–	26.3	46.7	10.3	16.8	–	4.4	–
IPM_1 $\mu g/m^3$	27.9	8.8	4.1	4.0	2.8	–	–	5.8
%	52.6	16.6	7.6	7.5	5.2	–	–	10.8
Fe $\mu g/m^3$	–	0.4	–	0.2	–	0.3	–	–
%	–	51.0	–	17.8	–	31.0	–	–
IPM_2 $\mu g/m^3$	26.1	8.3	3.2	3.0	4.2	6.6	0.65	–
%	48.9	15.5	6.1	5.7	7.8	12.4	1.2	–

% - As % of dependent variable.

This work indicates that regression of multiple source tracers against other related variables can assist in source identification and apportionment of sources which could not otherwise be resolved with factor analysis - multi-ple regression methodology.

References

Kneip, T. J., R. P. Mallon and M. T. Kleinman (1983) "The Impact of Changing Air Quality on Multiple Regression Models for Coarse and Fine Particle Fraction," Atmos. Environ. 17:299.

Morandi, M. T., J. M. Daisey and P. J. Lioy (1983) "A Receptor Source Apportionment Model for Inhalable Particulate Matter in Newark, N.J.," Paper No. 83-14.2. Proceedings of the 76th Annual Meeting of the Air Pollution Con-trol Association, Atlanta, GA.

Acknowledgments

This work was supported by the Office of Science and Research of the New Jersey Department of Environmental Protection and is part of a Center Program supported by the National Institute of Environmental Health Sciences, Grant No. ES00260 and the National Cancer Institute, Grant No. CA13343.

Published 1984 by Elsevier Science Publishing Co., Inc.

Aerosols, Liu, Pui, and Fissan, editors

356

CARBON ISOTOPES AS TOOLS TO DETERMINE COMBUSTION SOURCES OF
ELEMENTAL AND ORGANIC CARBONACEOUS AEROSOLS

J.S. Gaffney, R.L. Tanner, J. Forrest
Dept. of Applied Science

and

R.W. Stoenner
Dept. of Chemistry
Brookhaven National Laboratory
Upton, NY U.S.A. 11973

EXTENDED ABSTRACT

Introduction

Carbonaceous aerosols can be derived from wide variety of different sources, and are complex in nature. Methods which can simplify the characterization of these materials are needed if we are to evaluate their potential and current impacts on the environment. As pointed out previously, carbon isotopes (^{14}C and ^{13}C) can be useful tools in evaluating source terms for carbonaceous aerosols produced primarily from combustion (so-called elemental carbon).[1-7] As well, these same isotopes can aid in determining sources of organic carbon associated with the aerosol which can be primary or secondary in nature. This work describes preliminary efforts to use ^{14}C and $^{13}C/^{12}C$ determinations on organic and elemental fractions of carbonaceous aerosols to deduce source terms.

Experimental

Thermal evolution is used to separate organic from elemental fractions[8]. A scaled-up version of a system previously described is used to process 10-20 mg of total carbon collected on quartz fiber filters[8]. The organic and elemental carbon fractions as CO_2, are purified by passing the sample through a series of cold traps and chemical filters to remove NO_x, sulfur dioxide and any other impurities. The carbon-13 isotope ratio is then measured using standard isotope ratio mass spectrometry[10] and miniature gas proportional counters are used to determine the amount of ^{14}C present in the samples[9].

Applications to Specific Problems

When coupled to other available measurements, carbon isotope techniques enable one to more readily evaluate the importance of specific sources. An overview of isotopic mapping will be presented and two specific examples will be discussed.

1) Differentiation of wood burning and agricultural slash burn elemental carbon from fossil fuel combustion elemental carbon.

2) Evaluation of the contribution of secondary organic aerosol to the "haze" over forested areas such as the Great Smokey Mountains (data collected near Oak Ridge, Tennessee).

Carbon-13 measurements can be used to differentiate between sources (C3 or C4)[10]. Their use when coupled to C-14 measurements allows the differentiation of C-4 (corn, sugar cane) slash burning from C-3 biomass burning (wood) and from fossil fuel combustion processes (No C-14).

In the case of evaluating secondary organic aerosol formation from natural reactive hydrocarbons (isoprene, α-pinene, etc.), carbon isotopic data and thermal evolution data along with gas phase measurements of olefins and aldehydes, indicate that the natural contribution to the aerosol is much less than suggested by Went[13]. Measurements of aerosol components support the conclusion that most of this haze is due to acidic sulfate although some local carbonaceous soot (elemental carbon) can be a contributing factor, consistent with other previous work[11,12].

Use of carbon isotopes in future studies will require further refinement of measurement techniques such as use of laser infra-red spectroscopic techniques or Tandem van der Graaf mass spectrometry in order to reduce sample size requirements. As well these measurements are most useful when coupled to other aerosol component measurements and analyses to gaseous pre-cursors in question.

Acknowledgment

The authors wish to thank Mary Phillips and Dan Spandau for their assistance in collecting the Oak Ridge samples. We are grateful to Dr. Bruce Hicks of Atmospheric Turbulence and Diffusion Laboratory and Dr. Steven Lindberg of Oak Ridge National Laboratory for use of their facilities.

References

1. T. Novakov, ed. Conference on Carbonaceous Particles in the Atmosphere, Mar. 20-22, 1978, Lawrence Berkeley Laboratory, LBL-9037 (1979), and references therein.

2. G.T. Wolff, J. Air Pollut. Control Assoc. 31 (1981) 935-938, and references therein.

3. J.S. Gaffney, Organic air pollutants: setting priorities for long-term research needs. Report BNL 51605 UC-11 (Environ. Control Technol. and Earth Sci., TIC-4500) April 1982, 76p.

4. S.H. Cadle, P.J. Groblicki and P.A. Mulawa, Atmos. Environ. 17 (1983) 593-600.

5. P.J. Groblicki, S.H. Cadle, C.C. Ang, and P.A. Mulawa, Atmos. Environ. 17 (1983) in press; GMR-4054, ENV 152, and references therein.

6. J.A. Cooper, L.A. Currie and G.A. Kleidal, Environ. Sci. Technol. 15 (1981) 1045-1050, and references therein.

7. J.S. Gaffney, R.L. Tanner and M. Phillips, Separating carbonaceous aerosol source terms using thermal evolution, carbon isotopic measurements, and C/N/S determinations. The Science of the Total Environment (1984) in press.

358

8.   R.L. Tanner, J.S. Gaffney and M.F. Phillips, Anal. Chem. 54 (1982) 1627-1630.

9.   G. Harbottle, F.V. Sayre, and R.W. Stoenner, Science 206 (1979) 683-685.

10. J.S. Gaffney, A.P. Irsa, L. Friedman and D.N. Slatkin, Biomed. Mass. Spectrom. 5 (1978) 494-497, and references therein.

11. R.W. Shaw, Jr., A.L. Crittenden, R.K. Stevens, D.R. Cronn and V.S. Titov, Environ. Sci. Technol. 17 (1983) 389-395.

12. R.K. Stevens, T.G. Dzubay, R.W. Shaw, Jr., W.A. McClenny, C.W. Lewis and W.E. Wilson, Environ. Sci. Technol. 14, (1980) 1491-1498.

13. F.W. Went, Nature, 187, (1960) 641.

This work was supported by the Office of Health and Environmental Research, U.S. Department of Energy and performed under the auspices of U.S. DOE under Contract No. DE-AC02-76CH00016.

Published 1984 by Elsevier Science Publishing Co., Inc.
Aerosols, Liu, Pui, and Fissan, editors

FUNDAMENTAL LIMITATIONS OF
FACTOR ANALYSIS RECEPTOR MODELS

Ronald C. Henry
Environmental Enginnering Program
Civil Engineering Department
University of Southern California
Los Angeles, CA 90089-0231
U.S.A.

EXTENDED ABSTRACT

Introduction

Receptor models based on a factor analysis of sequential elemental particulate data have been applied to estimate the source composition matrix and the source contributions to the aerosol by several groups. Space limitations make referencing them all impossible, two examples are Henry (1977) and Alpert and Hopke (1981). In order for these models to be consistent with basic physical principles certain constraints must be placed on the source composition matrix and source contributions inferred by the model from the data.

These constraints are:

1) The data must be accurately reproduced by the model; the model must explain the observations.

2) the model predicted source compositions must be nonnegative; a source cannot have a negative percentage of an element.

3) the sum of the elemental composition for each source must be less than or equal to 1; the whole cannot be greater than the sum of its parts.

4) the source strengths predicted by the model must all be greater than 0; a source cannot produce negative mass.

5) the sum of all the predicted source strengths on any day must be less than or equal to the observed total mass on that day, given some allowance for error.

This paper shows by example that even in the simplest nontrivial case that these constraints are not sufficient to define a unique physically consistent factor model. Additional physical constraints are necessary.

360

An Example

Table 1 gives the assumed source composition matrix and
source contributions (strengths) used to create the set of
artificial data also given in the table. There are only two
sources each with only three elements.

Based on the paper of Lawton and Sylvestre (1971) on a
totally different application, one can calculate the range of
physically feasible factor analysis solutions using the
artificial data in table 1. This is best shown on a plot in
the space spanned by the two nonzero eigenvectors of the
data, as shown in the Figure.

Those sources which satisfy all the constraints given
above must lie on line 5 in the Figure. One source must lie
between lines 1 and 3 and the other must lie between lines 2
and 4. This defines a fundamental range of uncertaity on the
sources. Transforming the results from eigenvector space, it
is easily shown that the fundamental uncertainties in the
determination of the source compositions are:

	Range for Source 1	Range for Source 2
Element 1	0.0001 - 0.1997	0.5003 - 0.9332
Element 2	0.4401 - 0.5600	0.0000 - 0.2598
Element 3	0.3601 - 0.4400	0.0666 - 0.2398

These uncertainties are for perfect error free data. No
factor analysis model can estimate the true source
compositions with less inherent uncertainty without making
addional physical assumptions. This implies that factor
analysis models should only be used in conjunction with to
other receptor models.

References

Alpert, D. J. and P. K. Hopke (1981) "A Determination of the
Sources of Airborne Particulate Matter Collected During the
Regional Air Pollution Study", Atmos. Emviron., 15:675

Henry, R. C. (1977) "Application of Factor Analysis to Urban
Aerosol Source Identification", in Proceedings of the Fifth
Conference On Probability and Statistics In Atmospheric
Sciences, Am. Meteor. Soc., Boston, MA

Lawton, C. L. and E. A. Sylvestre (1971) "Self Modeling Curve
Resolution", Technometrics, 13:617.

TABLE 1

TRUE SOURCE MATRIX AND SOURCE STRENGTHS
AND COMPUTED ARTIFICIAL DATA

TRUE SOURCE MATRIX

	SOURCE 1	SOURCE 2
ELEMENT 1	0.1000	0.6000
ELEMENT 2	0.5000	0.2000
ELEMENT 3	0.4000	0.2000
SUM	1.0000	1.0000

TRUE SOURCE STRENGTHS

	DAY1	DAY2	DAY3	DAY4	DAY5
SOURCE 1	1.00	2.00	3.00	1.00	0.50
SOURCE 2	1.00	0.50	1.00	3.00	2.00
TOTAL	2.00	2.50	4.00	4.00	2.50

COMPUTED ARTIFICIAL DATA

	DAY1	DAY2	DAY3	DAY4	DAY5
ELEMENT 1	0.70	0.50	0.90	1.90	1.25
ELEMENT 2	0.70	1.10	1.70	1.10	0.65
ELEMENT 3	0.60	0.90	1.40	1.00	0.60

362

FACTOR SOLUTION FEASIBLE REGIONS

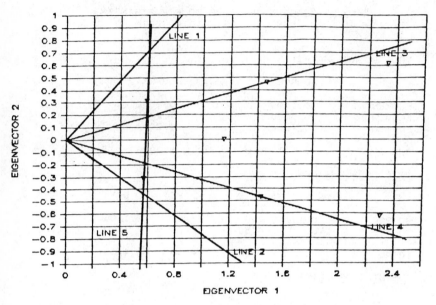

FIGURE   **Feasible sources lie on Line 5 between Lines 1 and 3 and between Lines 2 and 4. They satisfy all physical constraints.**

Published 1984 by Elsevier Science Publishing Co., Inc.
Aerosols, Liu, Pui, and Fissan, editors

363

Study of the Use of Linear Programming Methods
in Receptor Modeling

Meng-Dawn Cheng and Philip K. Hopke

Department of Civil Engineering and
Institute for Environmental Studies
1005 W. Western Avenue
Urbana, Il 61801

Several authors have suggested that Linear Programming (LP) might
be a useful receptor modeling approach for source identification and
aerosol mass apportionment. LP might be an alternative linear model to
replace the chemical mass balance (CMB) least-squares model. Assumming
adherence to the principle of mass conservation, the CMB model resolves
the total sample mass collected on a filter at a receptor site into
several components. Each identified component represents the
contribution of aerosol mass from one identified source. Ordinary or
weighted least squared methods and ridge regression have been used to
estimate the source strengths (1). In addition to these statistical
methods, LP had been proposed to resolve aerosol mass by Henry (2) and
Hougland (3). However, only limited development and no detailed
evaluation studies have been performed.

The CMB model can be formulated as follows:

$$Ax = b + e \tag{1}$$

where the matrix A has emission source profile in each column and is of
dimension m measured elements by n identified sources; the vector x is
the unknown strength of each source and has dimension n by 1; the
ambient concentration vector, b, has the dimension of m by 1; the error
vector of the linear system, e, is m by 1. LP is used to estimate the
unknown emission strengths, x.

The problem can be written in typical LP terminology as:

$$Z = \max \{C'x \text{ subject to } Ax < b, \text{ b and } x > 0\} \tag{2}$$

where C' is the transpose of vector C (n by 1). C is the coefficient
vector of the objective function, Z,

$$Z = C'x \tag{3}$$

The conditions

$$Ax < b \text{ and } x > 0 \qquad (4)$$

are called the constraints on the problem.

Two distinct LP approaches have been used to estimate the unknown vector, x, the vector of source strengths. The first one was initially proposed by Henry (2). This approach maximizes the functional value of the objective function to estimate x subject to the contraints of equation (1). To maximize the linear sum of source strength is the basic idea and also the objective function of this method. This approach will be designated as the MASSS method. Alternatively, Fisher (4) noted a LP method to fit data using the criterion of least sum of deviations. Following this idea, the sum of absolute residuals can be minimized (MISAR) as the second criterion of aerosol mass apportionment. Hougland (3) first applied the method to apportion the aerosol mass of the Washington, DC data set and obtained better results than those obtained by least squares.

Two different options of the LP methods have been tested using National Bureau of Standards (NBS) simulated data sets (5). These data sets consist of data for 40 samples with 19 or 20 elemental concentrations provided for each sample. Two different approaches will be used to obtain the average properties of the data set. By analyzing the average composition of the 40 samples in the data sets, these values will be called the Average Concentration (AC) results. The Average Prediction (AP) values are computed using the concentrations for each individual sample and then the results are averaged over the whole data set to obtain the average prediction.

Only the results for NBS data set one are presented here. These data were derived from nine sources operating in a multiple source dispersion model (5). Table 1 shows the results of AP option of two LP approaches, and the Table 2 shows the results of AC option. MISAR's results are very stable for both AC and AP approaches in terms of average L/S ratio. Further, the number of identified sources is significantly improved by the AP approach in an overall judgement. However, MASSS is extremely poor compared to MISAR for the AP approach. Both methods gave similar results for the AC approach.

Table 1. Comparison of Average Prediction MISAR and MASSS Mass Values
(ng/m3) for NBS data set 1.

Source	True	MISAR	L/S	MASSS	L/S
Steel-A	51	19	2.68	0	-
Incinerator	1312	952	1.38	898	1.46
Oil-B	1997	1482	1.35	582	3.46
Sandblast	4254	3472	1.23	2203	1.93
Basalt	4732	1723	2.75	897	5.28
Soil	8034	7238	1.11	4352	1.85
Road	6954	5324	1.30	1438	4.84
Wood Combu.	3205	2449	1.31	18	178.06
Coal-B	2349	1199	1.96	3	783.00
Average L/S			1.67		122.48

Table 2. Comparison of Average Concentration MISAR and MASSS Mass
Values (ng/m3) for NBS data set 1.

Source	True	MISAR	L/S	MASSS	L/S
Steel-A	51	0	-	0	-
Incinerator	1312	1151	1.14	1031	1.27
Oil-B	1997	1925	1.04	1760	1.14
Sandblast	4254	2746	1.55	2193	1.94
Basalt	4732	0	-	0	-
Soil	8034	12115	1.51	12442	1.55
Road	6954	5946	1.17	6073	1.15
Wood Combu.	3205	2054	1.56	2074	1.55
Coal-B	2349	597	3.93	314	7.48
Average L/S			1.70		2.30

Reference cited

1). Henry, R. C., Lewis, C. W., Hopke, P. K. and Williamson, H. J. "Review of Receptor Model Fundamentals", Atmospheric Environment, in press (1984).

2). Henry, R. C. "A Factor Model of Urban Aerosol Pollution: A New Method of Source Identification". Ph. D. dissertation, Oregon Graduate Center, Environmental Sciences, 1978.

3). Hougland, E. S. "Chemical Element Balance by Linear Programming", Air Pollution Association Paper No. 83-14.7, 1983.

4). Fisher, W. D. "A Note on Curve Fitting with Minimum Deviations by Linear Programming", J. American Statistical Association, 56(294):359, (1961).

5). Currie, L. A., Gerlach, R. W., Lewis, C. W., Balfour, W. D., Cooper, J. A., Dattner, S. L., De Cesar, R. T., Gordon, G. E., Heisler, S. L., Hopke, P. K., Shah, J. J., Thurston, G. D., Williamson, H. J., "Interlaboratory Comparison of Source Apportionment Procedures: Results for Simulated Data Sets", Atmospheric Environment, in press (1984).

Published 1984 by Elsevier Science Publishing Co., Inc.
Aerosols, Liu, Pui, and Fissan, editors

ESTIMATES OF THE CONTRIBUTIONS OF VARIOUS SOURCES TO
INHALABLE PARTICULATE MATTER IN DETROIT

G. T. Wolff and P. E. Korsog
Environmental Science Department
General Motors Research Laboratories
Warren, MI 48090
U.S.A.

EXTENDED ABSTRACT

Introduction

As part of the 1981 Detroit Summer Air Pollution Study, General Motors
Research Laboratories (GMR) conducted an aerosol characterization study in
southeastern Michigan in an area that is in violation of both the secondary
annual and 24 h total suspended particulate (TSP) standards. The study was
conducted during June, July and August because the highest TSP values are
generally observed during these months (Wolff et al. 1983). The purpose of
this paper is to identify the major sources of inhalable particulate matter.

Methodology

The sampling site was situated in an open field which is approximately
3 km north of Detroit's central business district and 10 km north of the large
downriver industrial areas. Two Beckman automatic dichotomous samplers were
used to collect 4 h filter samples in two size fractions: diameter < 2.5 μm
(FPM) and 2.5 μm-15 μm (CPM). The sum of the CPM and FPM is the inhalable
particulate mass (IPM). The filters were analyzed for: mass, organic and
elemental carbon, sulfate, nitrate, ammonium, Al, Si, S, K, Ca, Ti, Mn, Fe,
Zn, Br, Pb, and Ni.

The source apportionment method employed was absolute principal component
analysis (APCA) (Thurston, 1983). In this procedure the principal components
(PC) are first determined and then rotated using the varimax procedure. Only
components with eigenvalues $\geq$ 1 after rotation were retained (Hopke, 1982).
The absolute zero component scores are then calculated by separately scoring
an extra observation wherein all the elemental concentrations are zero. The
absolute principal component score (APC) is then equal to the principal component score minus the absolute zero component score. The particulate mass
can then be regressed on the APC scores to determine the multiple regression
coefficients which convert the APC scores into estimates of mass concentrations in μg/m^3.

Sampling every four hours for seven days resulted in 42 sampling periods.
However, due to less than 100% data capture for any given variable, the correlations, which were used to generate the factor patterns, were generally based
on 41 data pairs for the FPM and 35 for the CPM. The number of complete
observations needed to do the subsequent regressions was less: 27 for the fine
mass and 30 for the coarse mass.

Results

The rotated factor pattern for the FPM species contained 8 components
which explained 95% of the total variance. The first component is the motor

vehicle component since it has very high loadings for Pb and Br and a moderate loading for elemental carbon (EC). The second factor has very high loadings for SO_4 and Al and moderate loadings for K and Fe. This component represents coal burning. The third and fourth factors are highly loaded in crustal materials. The fifth component, which has a high loading for V and moderate loadings for Ca and Ni, most likely represents fuel oil combustion. A high loading for Zn and a moderate loading for Fe in the sixth component suggests that this is an incineration term. The seventh component is highly loaded in Mn, which could represent emissions from iron and steel-making processes. The last component is highly loaded in only OC. Since OC from combustion sources would be expected to be associated with EC also, this component may represent the regional, perhaps secondary, source of OC previously identified by Wolff et al. (1984).

The FPM was then regressed on the eight components using multiple regression analysis. Only the coefficients for components 2, 5, and 6 were significant at the 95% confidence level. However, since the coefficient for component 1 was significant at the 80% confidence level, and since we know motor vehicles contribute to FPM, this component was retained. The individual source estimates in $\mu g/m^3$ are the products of the means of the absolute principal component scores and the regression coefficients. These are given in Table 1. The R^2 for the regression equation was 0.70.

Table 1. Source Estimates of FPM from the APCA Method

Component	Source	$\mu g/m^3$ (± Uncertainty*)	% Contribution to FPM
1	Motor Vehicles	2.2 (± 1.6)	5.1
2	Coal Burning	23.5 (± 3.0)	55.0
5	Oil Burning	1.3 (± 0.5)	3.0
6	Incinerator	1.6 (± 0.7)	3.8
	ꓱidentified	14.1 (± 8.3)	33.1

* Uncertainty is calculated from ⸻ ⸻ ⸻s of the regression coefficients.

The rotated factor pattern for the CPM species contained 6 components which explained 95% of the total variance. The first factor, the crustal component, is highly loaded in Si, Al, K, and Ti and moderately loaded in Ca. Fe and Mn are highly loaded in factor 2, while EC has a moderate loading. This factor is believed to be iron and steel-manufacturing emissions. The third component represents motor vehicle emissions as it is highly loaded in Pb and Br. Zn, the incinerator tracer, is the only heavily-loaded species in factor four. The fifth factor is highly loaded in coarse S and moderately loaded in Ca, and we will label it an unknown S-rich source. The sixth factor is heavily loaded in OC.

Regressing the CPM on the APC scores of the first six factors resulted in a negative intercept of -3.9. Since a negative intercept is physically impossible, the regression procedure was repeated with the equation forced through zero. Except for components 4 and 5, all regression coefficients were significant at the 90% confidence level and R^2 was 0.79. The resulting apportionment is shown in Table 2.

Table 2. Source Estimates of CPM from APCA Method

Component	Source	$\mu g/m^3$ (± Uncertainty)*	% Contribution to CPM
1	Crustal	16.9 (± 1.6)	63.9
2	Iron & Steel	3.1 (± 0.8)	11.6
3	Motor Vehicles	4.2 (± 1.4)	16.0
6	Organic Carbon	1.5 (± 0.8)	5.6
	Unidentified	0.7 (± 3.4)	2.7

* Uncertainty is calculated from the standard errors of the regression coefficients.

Discussion

The APCA procedure clearly identifies coal combustion and crustal material as major sources of FPM and CPM, respectively. By examining the distribution of aerosol species from the various sources, we can compare the chemical composition of the estimated source profiles with ones reported in the literature. This is accomplished by regressing each element or aerosol species on the absolute principal component scores. The results of this procedure are discussed below.

The coal component predicts an $SO_4^=$ concentration of 16.1 $\mu g/m^3$ which compares well to the observed value of 15.4 $\mu g/m^3$. In addition, the sum of the above coal-related constituents is then 22.5 μ/m^3, which is in good agreement with the 23.5 $\mu g/m^3$ estimated by the APCA. The estimated vehicular contribution to fine Pb of 0.16 $\mu g/m^3$ corresponds to 7.3% of the fine vehicular particles estimated from the APCA. This is in good agreement with literature values. The small contributions of incineration and fuel oil combustion to the FPM are consistent with the small number of sources listed in the Wayne County Emissions Inventory.

The APCA method failed to determine the crustal contributions. Previous studies have indicated that Si represents 20-28% of the crustal aerosol component (Countess, et al. 1980). Assuming that all of the observed fine Si (0.66 $\mu g/m^3$) is of crustal origin, the total crustal component would be 2.46-3.45 $\mu g/m^3$ or 5.8-8.1% of the FPM. In the absence of better information, however, we feel that the 5.8-8.1% range is reasonable. The reason for the failure of the APCA to quantify this source is not obvious, but we believe it is due to the fact that the crustal species were not all heavily loaded in a single component. This did not occur because there appears to be many other important sources besides soil dust of some of the crustal species.

The regression of the FPM on the APCA's did not identify a statistically significant contribution of the OC APC, component 8, to the FPM. The regression of OC on the APC's, however, indicated that component 8 accounts for 5.17 ± 0.33 $\mu g/m^3$ or 73.5% of the total fine OC. This would also account for 12.1% of the FPM. Because EC, an ubiquitous particulate product from combustion, is absent from the component, secondary production seems to dominate this component (Wolff et al., 1984).

The uncertainty associated with the coarse crustal factor is relatively small, and the calculated ratios of the various crustal species are in good agreement with crustal reference material. Assuming that Si accounts for 20-28% of the coarse crustal material, the crustal fraction is 13.5-19.0 $\mu g/m^3$. This represents 52.2-73.5% of the CPM and is in good agreement with the APCA results. The marker elements for the iron and steel source are Fe and Mn which occur in a ratio of 49:1. Dzubay (1980) found that at a site in St. Louis, where the coarse material was dominated by steel processing emissions, the Fe:Mn ratio was 49.4:1. In addition, the APCA predicts that Fe represents 23.5% of the coarse particles from this source. This compares to Dzubay's number of 23.8%. Consequently, we feel that the 11.6 ± 3.0% estimate is reasonable for the percent of the coarse particulate mass attributable to the iron and steel source. In the coarse motor vehicle fraction, the calculated coarse Pb represents only 1.2% of the mass. This low percentage is due to high loadings for the crustal components in this factor. Consequently, it appears that this term represents coarse particles both emitted and suspended by motor vehicles.

References

Countess R. J., Wolff G. T. and Cadle S. H. (1980) "The Denver Winter Aerosol: A Comprehensive Chemical Characterization," J. Air Pollut. Control Assoc. 30:1194-1200.

Dzubay T. G. (1980) "Chemical Element Balance Method Applied to Dichotomous Sampler Data," Annal. NY Anal. Sci. 338:126-144.

Hopke P. K. (1982) "Comments on 'Trace Element Concentration in Summer Aerosols at Rural Sites in New York State and Their Possible Sources'," Atmos. Environ. 16:1279-1280.

Thurston G. D. (1983) "A Source Apportionment of Particulate Air Pollution in Metropolitan Boston," Ph.D. Dissertation, Harvard School of Public Health, Boston, MA.

Wolff G. T., Stroup C. M. and Stroup D. P. (1983) "The Coefficient of Haze as a Measure of Particulate Elemental Carbon," J. Air Pollut. Control Assoc. 33:746-750.

Wolff G. T., Korsog P. E., Stroup D. P., Ruthkosky M. S. and Morrissey M. L. (1984) "The Influences of Local and Regional Sources on the Concentrations of Inhalable Particulate Matter in Southeastern Michigan," Atmos. Environ. (In Press).

Published 1984 by Elsevier Science Publishing Co., Inc.
Aerosols, Liu, Pui, and Fissan, editors

ELEMENTAL COMPOSITION OF SIZE-SEGREGATED SUBMICROMETER AEROSOL
PARTICLES: IMPLICATIONS FOR SOURCE APPORTIONMENT

J.M. Ondov
Martin Marietta Environmental Systems
9200 Rumsey Rd.
Columbia, MD 21045
U. S. A.

EXTENDED ABSTRACT

Introduction

High-temperature combustion sources, such as coal and oil-fired power
plants, municipal incinerators, motor vehicles, smelters, cement kilns,
etc., account for virtually all of the anthropogenic emission of primary
atmospheric fine particles. Particles from these sources differ chemically
and are emitted in distinct and narrow peaks in the submicrometer aerosol
accumulation region. Computer simulations and experimental evidence
suggest that these peaks remain intact during plume transport and should
be observable in the ambient air.

Experimental

To determine if submicrometer aerosol properties could be exploited
for source identification and apportionment, over 50 size-fractionated
ambient aerosol particulate samples were collected in August of 1983 at
Deep Creek Lake (DCL) in northwestern Maryland using a uniform-deposit
five-stage micro-orifice impactor (MOI) fitted with a precyclone separator
and an outboard afterfilter. The rural DCL site was chosen for study
because of its proximity to many large coal-fired power plants. Fifty-
percent cut diameters for the cyclone and five impactor stages were 2.5,
1.1, 0.7, 0.5, 0.2, and 0.1 μm. Additional samples were collected in the
plumes of two coal-fired power plants with two five-stage MOIs equipped
with four-stage pre-impactors of conventional design. The D50s of the preimpactor stages were 5, 2.5, 1.0, and 1.0 μm. Two 1- μm stages were used
to reduce the potential contamination of the submicrometer particulate
fractions by supermicrometer particles not collected in the previous stages.
The air sampling rate for each of the impactors was 30 lpm. Teflon filters
(37-mm) were used as impaction substrates because of their low blank for
most chemical elements. Substrates used in the first stage of the ground-
based MOI and those used in the conventional pre-impactors were coated
with Apiezon L vacuum grease to reduce particle bounce and reentrainment.

Two of the ground based MOI samples (run for 6 and 60 hrs) were analyzed for up to 26 elements by instrumental neutron activation analysis
(INAA). Irradiations were performed at the National Bureau of Standard's
10 Mw research reactor at a flux of approximately 6×10^{13} $n/cm^2 \cdot s$. Short-
lived neutron capture products were counted on large Ge(Li) detectors for
5- and 20-minute periods following a 5 to 7 minute decay. The same samples

were later reirradiated for 4 hours and counted for 4 and 24 hours after respective decay times of 3-days and 1-month.

Results

   In the 60-hr sample, the concentrations of most of the elements, especially Al, Ca, and Na which are predominantly associated with large supermicrometer particles, increased on the backup filter. This behavior is typically an artifact of severe particle bounce, and probably indicates that the substrates became overloaded with particles during the long 60-hr sampling time. Results of the analysis of the 6-hr sample are listed in Table 1. Despite the rather lengthy counting times, relatively few elements could be determined in the 6-hr sample, and analytical uncertainty often exceeded 25%. Despite their low concentrations in the teflon substrates, Zr, Cs, and Sc could not be detected above blank. Uncertainties associated with the concentrations of As, Sb, Se, V, Mn, Na, and Al typically range from 7 to 20%.

   Elemental concentration vs particle size distributions for elements detected in the 6-hr sample were fit to single, bimodal, or trimodal log-normal distribution functions. The distributions of As and Se were best fit to a unimodal lognormal function with mass median aerodynamic diameter (mmad) of 0.2 μm. Zinc and Al were bimodal, with mmads occurring at 0.4 (for both elements) and at about 0.1 μm for Al, and less than 0.05 μm for Zn. The geometric standard deviations (σ_g) ranged from about 1.4 to 1.5; values consistent with the distribution parameters of submicrometer aerosols emitted from pulverized coal-fired power plants. Resolution of source contributions by receptor modeling techniques is currently achieved through statistical treatment of whole-filter, or dichotomous sampler data. The application of resolved submicrometer aerosol particle analysis to receptor-based source apportionment should improve the identification and resolution of the impacts of major high-temperature combustion sources.

Table 1.   Concentrations of Elements in Fine Particles Collected in a
           Micro-Orifice Impactor run for 6 hours at Deep Creek Lake,
           August 1984, ng/m^3

Element	
As	0.39 ± 0.02
Sb	1.20 ± 0.1
Sc	1.61 ± 0.07
Zn	28 ± 2
Fe	9.4 ± 2.4
Co	0.023 ± 0.005
W	0.049 ± 0.004
Br	1.71 ± 0.06
Cl	7.0 ± 0.8
V	0.22 ± 0.02
Mn	0.75 ± 0.04
Cu	2.0 ± 0.4
Na	15.9 ± 0.4
K	30.8 ± 3.6
I	0.53 ± 0.05
Ca	15 ± 6
Al	12.3 ± 0.6
Nd	< 2.0
Sm	0.0011 ± 0.0001
La	0.14 ± 0.0015
Hf	< 0.02
Eu	< 0.002
Cr	< 0.6
Zr	NDB*
Cs	NDB*
Sc	NDB*

* Not detected above the substrate blank.

Published 1984 by Elsevier Science Publishing Co., Inc.
Aerosols, Liu, Pui, and Fissan, editors

374

Composition of Haze at a Site on the East Coast of the U.S.

by

George T. Wolff, Martin A. Ferman, Nelson A. Kelly,
David P. Stroup, and Martin S. Ruthkosky

Environmental Science Department
General Motors Research Laboratories
Warren, Michigan 48090

Abstract

During August, 1982 and February, 1983, General Motors Research Laboratories
conducted an extensive aerosol characterization study on the Atlantic Coast
near Lewis, Delaware. Dichotomous samplers were used to collect aerosol
samples in two size fractions. The measured species included: mass, $SO_4^=$,
NO_3^-, NH_4^+, organic carbon, elemental carbon, and trace elements (by XRF).
Optical measurements were obtained using two integrating nephelometers --
one operated at ambient conditions to obtain the light scattering due to
ambient aerosols and one operated at 10C° above ambient to obtain the
light scattering of the dried aerosols. The relationship between light
scattering and the chemical composition of the fine particles was determined
using the multiple regression techniques which we successfully employed at
previously studied urban and rural continental sites. A total light extinction budget is calculated using the nephelometer, NO_2, and elemental carbon
measurements. The results of the extinction budgets are presented and the
seasonal variations are discussed.

Published 1984 by Elsevier Science Publishing Co., Inc.
Aerosols, Liu, Pui, and Fissan, editors

^{14}C AS A TRACER FOR CARBONACEOUS AEROSOLS:
MEASUREMENT TECHNIQUES, STANDARDS AND APPLICATIONS

Lloyd A. Currie
Gas and Particulate Science Division
National Bureau of Standards
Washington, DC 20234
U. S. A.

EXTENDED ABSTRACT

For the past half dozen years radiocarbon has been used for distinguishing fossil from biogenic carbon components in atmospheric particles. Assessment of the extent of woodburning contributions during winter periods and fossil fuel contributions to urban areas, for example, was made possible through the development of small gas proportional counters capable of measuring ^{14}C in 5-10 mg-carbon. Important recent advances in instrumentation, standards, and chemical approaches have greatly increased the information which can be gained through ^{14}C. Two such advances will be discussed in detail: source resolution enhancement by combining isotopic (^{13}C, ^{14}C) and chemical data, and "micro-radiocarbon dating" via Accelerator Mass Spectrometry. The power of these approaches is illustrated with data from urban, rural (forested), and remote regions including samples containing as little as 100 µg-carbon.

Ambient Concentrations; ^{14}C Measurement

The first indication of fossil carbon depletion of atmospheric ^{14}C was the so-called "Suess Effect", after the ~2% decrease of ^{14}C/^{12}C in CO_2 reported by H. E. Suess (1955). At a concentration in excess of 150 mg-C/m^3 (as CO_2) there was little difficulty in obtaining sufficient material for conventional low-level ^{14}C counting. Carbonaceous aerosols, however, occur at concentrations ranging from ~1-100 µg-C/m^3, depending upon the level of pollution. Early attempts to ^{14}C "date" such particles in polluted U.S. cities therefore required somewhat heroic sampling efforts, but suspected major fossil fuel combustion contributions were confirmed (Clayton et al., 1955; Lodge et al., 1960).

Important advances in ^{14}C measurement have come about in the last few years, based on miniature gas counters (Currie and Murphy, 1977) and high energy tandem accelerators (Gove, 1978). The former technique has been applied to atmospheric samples containing as little as 5 mg-carbon; whereas the latter, in which ^{14}C atoms are counted directly, has been successful with samples containing only 100 µg-carbon (Klouda et al., 1984). Both measurement techniques require critical attention to recovery, contamination, and stability of the blank. Means for monitoring these items will be described, together with a discussion of the limits they impose on the measurement process. (At present, for example, chemical processing yields an accelerator mass spectrometry blank equivalent to about 15 µg of contemporary carbon.)

376

Summary of Results

Table 1 gives a condensed summary of our results on contemporary ("living") carbon in source, reference, and ambient samples.

Table 1. Median Results for Atmospheric ^{14}C

Sample Class	Contemporary Carbon (%)
Source particles	
wood smoke	92 ± 4
vehicle tunnel	4 ± 2
Urban particles	
reference material	61 ± 4
U.S. cities	32
Arctic Haze (Barrow)	27
Wood-burning communities (C-particles; CO)	73
Forest, desert aerosol	88
Atmospheric methane	82

For the first two categories specific samples were involved; therefore measurement uncertainties (Poisson standard deviations) are stated. Other results are presented as medians for a number of ambient samples, where (except for the last two entries) significant environmental variations were observed. These variations in isotopic as well as the inorganic and organic composition gave important clues as to the origins of the particles in question. This topic will be discussed in some detail.

Highlights from Table 1 include the following:

•Source particle measurements are consistent with the dichotomous ("living", "dead") assumption on which the atmospheric ^{14}C method is based.
•Ambient urban particles are, on the average, primarily fossil. (The urban standard reference material will be discussed below.) The remarkable urban particle characteristics of the soot-like Arctic Haze (Rosen et al., 1982) are supported by the large fossil carbon contribution we observed in a series of early spring samples.
•The burning of wood as a supplemental fuel is causing alarming levels of particulate pollution in the U.S. and Europe (Cooper, 1981). In a number of detailed chemical and meteorological studies, ^{14}C has served as the primary direct tracer for carbonaceous particles and carbon monoxide from woodburning (Ramdahl et al., 1984).

•In preliminary studies of atmospheric methane and rural areas (desert, forest) anticipated to be largely free from anthropogenic sources, we observed the carbon to be mostly biogenic (80-90%) with little variation in ^{14}C content among samples.

The [13]C-^{14}C plane; Standards

Figure 1 is presented to indicate the type of two-dimensional isotopic information which can be derived if both stable (^{13}C) and radioactive (^{14}C) isotopes are measured.

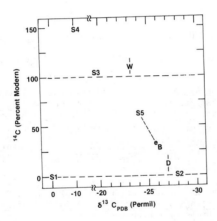

Figure 1. [14]C vs [13]C: Ambient Samples, and Isotopic and Urban Particulate Standards. (See text for explanation of symbols.)

The fossil/biogenic discrimination power of ^{14}C is thus complemented by the discrimination among certain classes (e.g. C_3 vs C_4 vegetation) afforded by ^{13}C. The primary reason for presenting Fig. 1, however, is to indicate some standard materials for isotopic carbon measurements. S1 (Pee Dee Belemnite, a fossil carbonate) and S2 (graphite, RM 21) are specific for ^{13}C, whereas S3 (oxalic acid, SRM 4990B) and S4 (Australian National University Sucrose) are used for ^{14}C. S5 is one of two Urban Particle Reference Materials (SRM 1648); its elemental carbon fraction ("e") demonstrates the significance of isotopic variation with chemical species. W, B and D represent particles from woodburning, Barrow (Arctic Haze) and diesel exhaust, respectively.

Conclusion

Measurement capabilities for carbon isotopes in carbonaceous aerosols, combined with modern techniques for inorganic and organic characterization, make possible important advances in our understanding of sources of fossil and biogenic carbon in the atmosphere.

References

Clayton, G. D., Arnold, J. R. and Patty, F. A. (1955) Science 122, 751.

Cooper, J. A., Currie, L. A. and Klouda, G. A. (1981) Environ Sci. Tech. 15, 1045.

Currie, L. A. and Murphy, R. B. (1977) "Origin and residence times of atmospheric pollutants: application of ^{14}C" in Methods and Standards for Environmental Measurement, (W. H. Kirchoff, ed.), NBS Spec. Pub. 464, National Bureau of Standards, Washington, D.C., p. 439.

Gove, H. E., ed. (1978) Conference on radiocarbon dating with accelerators, 1st Proc: Univ. Rochester. (See also Currie, L. A., " Environmental radiocarbon measurements", p. 372).

Klouda, G. A., Currie, L. A., Donahue, D. J., Jull, A. J. T. and Zabel, T. H. (1984) in Proceedings of the 3rd International Conference on Accelerator Mass Spectrometry (in press).

Lodge, J. P., Bien, G. S. and Suess, H. E. (1960) Int. J. Air Pollution 2, 309.

Ramdahl, T. Schjoldager, J., Currie, L. A., Hanssen, J. E., Moller, M., Klouda, G. A., and Alfheim I. (1984) Sci. Total Environ. 36:81.

Rosen, H., Hansen, A. D. A., Dod, R. L., Gundel, L. A., Novakov, T., (1982) in Particulate Carbon, Atmospheric Life Cycle, Wolff, G. T. and Klimisch, R. L. Eds., (Plenum Press, New York, p. 273).

Suess, H. E., (1955) Science 122, 415.

Published 1984 by Elsevier Science Publishing Co., Inc.

Aerosols, Liu, Pui, and Fissan, editors

RECEPTOR MODELING FOR CARBONACEOUS AEROSOLS
PART I: SOURCE CHARACTERIZATION

S.V. Hering, A.H. Miguel[1], R.L. Dod[2], and J.M. Daisey[3]
Department of Chemical Engineering, University of California
Los Angeles, CA 90024

EXTENDED ABSTRACT

Two types of sources, freeway vehicles and a commercial airport, were sampled with the objective of determining source profiles suitable for receptor modeling of carbonaceous aerosols. A threefold approach is used: 1) measurement of polycyclic aromatic hydrocarbon (PAH) concentrations, 2) bulk characterization of the carbon component by thermal analysis and 3) determination of elemental composition by X-ray fluorescence. Additionally the relative stability of the polycyclic aromatic hydrocarbons upon exposure to filtered urban air was evaluated. Data were collected at a receptor site, which by virtue of its location is expected to be impacted by these two source types.

Sampling Sites and Methods

Vehicular emissions were collected from the airduct of a 1 km freeway tunnel on California State Route 24, east of Berkeley, CA. The individual bores of the tunnel each carried two lanes of unidirectional traffic with a 2^0 grade. Samples were collected for both uphill and downhill traffic for which the traffic mix consisted of less that ½% heavy duty diesels, and for uphill traffic with 2-4% heavy duty diesel trucks. The non-catalyst fraction of the vehicle fleet was estimated by manual counts to be 15-25%. Samples were collected at Los Angeles International Airport 15 meters from the taxiway adjacent to the main runways. The sampling took place in the mornings and evenings under near stagnant conditions with winds less than 5 mph. Samples collected at the airport represent a mix of jet aircraft, prop planes (18% of total aircraft), airport ground vehicles (mostly diesel), and some urban background. Measured soot/Pb ratios at the airport were 12 ± 1 as opposed to 3 ± 1 for the tunnel samples, indicating the dominance of a non-auto combustion source for these airport samples.

Filter samples were collected with three cyclones operated in parallel at a flow rate of 52 lpm to give a precut of 1.3 μm. One of the cyclones used a teflon downstream filter for gravimetric and X-ray fluorescence analyses. The other filters were quartz fiber; one was prefired for thermal analyses, the other unfired for PAH. A HiVolume sampler, employing 20x25 cm quartz filters at $1.4m^3$/min, was also used for the polycyclic analyses. PAH were analyzed by reverse phase high pressure liquid chromatography with fluorescence or absorbance detection. The carbon thermal analyses were performed by heating the sample in an oxidizing atmosphere, and measuring the CO_2 evolved. Simultaneous measurement of the filter opacity during combustion allows the high temperature ($\sim500^0$ C) peak to be identified as "black" (presumably soot) carbon. (Ellis and Novakov, 1983)

[1] Pontificia Universidade Catolica do Rio de Jۮneiro, Brazil
[2] Lawrence Berkeley Laboratory, Berkeley, CA 94720
[3] New York University Medical Center, New York, NY 10016

380

Carbon Thermal Analyses (Thermograms)

The data collected here can be catogorized into 5 classes: (1) uphill freeway traffic with 2-4% heavy duty diesel (HDD), (2) uphill freeway traffic of 0-½% HDD, (3) downhill traffic with 0-½% HDD, (4) airport sampling and (5) receptor site sampling. Thermograms from each of these five classes are shown in Figure 1. Although not shown here, these thermal profiles are quite consistent within a specified source type. There is no significant difference between the uphill freeway samples with 2-4% truck diesels as opposed to those with 0-½% truck diesels. For the downhill traffic, the percentage of carbon associated with the 500°C "black carbon" peak is smaller. The airport thermogram has a more rounded black carbon peak as does that of the receptor site.

Although the thermograms offer encouragingly different signals for the tunnel and airport sources, there are several questions which must yet be addressed. First, the stability of the thermograms with respect to atmospheric reactions should be established. For source resolution one must know how similar the thermograms measured at the source are to the aged source aerosol seen at the receptor site. Secondly, one must show whether the principle of linear superposition, tacitly assumed in all receptor modeling, is valid for the thermograms.

Polycyclic Aromatic Hydrocarbon Profiles

One method of presenting PAH profiles would be to divide the concentration of each compound by the total fine mass loadings. This would be analogous to the elemental profiles developed for elemental mass balance analysis. The difficulty is these compounds one known to react in a polluted atmosphere (eg. Grosjean et al, 1983). Thus, the mass fraction of a primary polycyclic hydrocarbon observed in the aged aerosol of the receptor site will not be the same as at the source. Lifetime studies conducted by A. Miguel (1984) on our Caldecott freeway tunnel samples showed 35-100% PAH losses upon 4 day exposure to filtered Los Angeles air. However, five of the compounds chrysene, benzo(b)fluoranthene, benzo(k)fluoranthene, indenopyrene and benzo(ghi)perylene where found to decay at a similar rate, so that the percentage of a specified PAH of the total PAH considered is relatively constant (Figure 2).

PAH profiles expressed as a fraction of the sum of PAH loadings $(PAH_i/\Sigma_i\ PAH_i)$ are shown in Figure 3 for our five data types. Unfortunately we do not see any marked differences here. The profiles of the different source types, and of the receptor site are quite similar. We note that these four PAH compounds have previously been identified in the exhausts of aircraft turbines (Spitzer and Dannecker, 1983) and automobiles (National Research Council, 1983).

Acknowledgements: The authors thank Jay Turner, Sotiris Pratsinis, Debra Leucken, Lara Gundel, A.D.A. Hansen, James Butler and James Cheney for their assistance in the sampling and analysis, Sheldon Friedlander and Tihomir Novakov for their support and comments, and James Kelley of TWA for his cooporation with the airport sampling. This work was supported by the U.S. Environmental Protection Agency Grant #CR 810455-01.

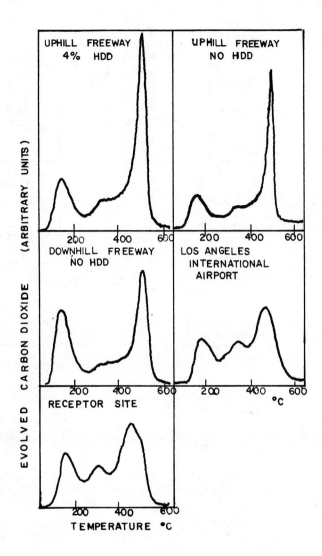

Figure 1.   Relative CO_2 evalued as a function of temperature for five samples corresponding to (1) uphill freeway traffic with 4% HDD, (2) uphill freeway no HDD, (3) downhill freeway no HDD, (4) LA International Airport, (5) receptor site near freeway and airport.

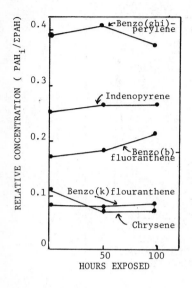

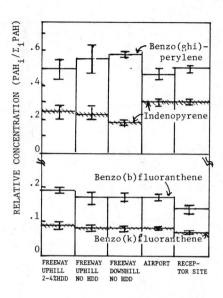

Figure 2. Relative concentrations of PAHs collected in a freeway tunnel and exposed to filtered Los Angeles air (data from Miguel, 1984).

Figure 3. Average relative concentration of some PAHs for four types of source samples and one receptor site. Error bars represent standard deviations among successive measurements.

References

Ellis, E.C., Novakov T. (1982) "Applications of Thermal Analysis to the Characterization of Organic Aerosol Particles", Science of the Total Environment 23 227-238.

Grosjean, D., Fung, K. Harrison, J. (1983) "Interactions of Polycyclic Aromatic Hydrocarbons with Atmospheric Pollutants", Environ Sci Technol 17, 673-679.

Miguel, A.H. (1984) "Atmospheric Reactivity of Particulate Polycyclic Aromatic Hydrocarbons Collected in a Urban Tunnel", Science of the Total Environment, in press.

National Research Council (1983) Polycyclic Aromatic Hydrocarbons: Evaluation of Sources and Effects (National Academy Press, Washington, D.C.)

Spitzer T., Dannecker W. (1983) "Glass Capillary Gas Chromatography of Polynuclear Aromtic Hydrocarbons in Aircraft Turbine Particulate Emissions" Journal of Chromatography 267, 167-174.

CHAPTER 9.

VISIBILITY

Published 1984 by Elsevier Science Publishing Co., Inc.

Aerosols, Liu, Pui, and Fissan, editors

A SIZE CLASSIFYING ISOKINETIC AEROSOL SAMPLER
DESIGNED FOR APPLICATION AT REMOTE SITES

C.F. Rogers and J.G. Watson, Desert Research Institute,
Atmospheric Sciences Center, P.O. Box 60220, Reno, NV 89506;
and C.V. Mathai, AeroVironment, Inc., Pasadena, CA 91107

EXTENDED ABSTRACT

Research quality aerosol monitoring projects require an aerosol filter
sampling device with the following characteristics:

● Measurement of inhalable (0 to 10 or 15 µm) and fine (0 to 2.5 µm)
  size-classified particulate matter, with acceptable sampling effectiveness.

● Simultaneous sampling on two different substrates, one amenable to
  elemental and the other amenable to carbon analysis.

● Sequential sampling, without operator intervention, at greater than
  75 l/min flow rates to obtain continous samples over 4- to 24-hour
  sample durations with sufficient deposits for chemical analysis.

● Simple and reliable field and laboratory operation at an affordable
  price.

A Size Classifying Isokinetic Sequential Aerosol Sampler (SCISAS) combines the best features of the SURE/ERAQS (Mueller and Hidy et al., 1983)
sequential filter sampler and the WRAQS (Allard et al., 1982) and Henry
(1977) isokinetic sampling manifolds to meet these requirements.

Ambient air is continuously drawn at a rate of 1100 l/min into a ten
inch diameter PVC stack through a 10 or 15 µm McFarland size-selective
inlet (SSI). Particles are then drawn from this stack, at a velocity close
to that flowing through the main stack to sample 0 to 10 or 15 µm particles
on two different filter media. Each of the two 0 to 2.5 µm aerosol samples
is withdrawn from the main stack at a flow rate of 113 l/min through a
single two-inch internal diameter tube which leads to a cyclone for exclu-
sion of particles larger than 2.5 µm diameter. The outlet of the cyclone
leads to a simple rectangular supply manifold into which six 47 mm
Nuclepore filter holders are mounted. A general view of the SCISAS is
shown in Figure 1.

The basic configuration of the SCISAS includes fourteen two-inch internal diameter supply tubes clustered inside the main ten-inch stack. The
flow rate of approximately 80 l/min within each inhalable particle two-
inch supply tube was chosen empirically to provide very nearly isokinetic
matching between the average velocities in these tubes and that in the main
stack with 1100 l/min flow. An alternative design draws the 0 to 10 or

386

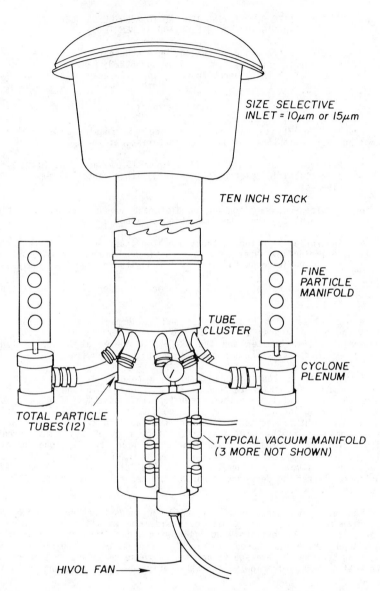

SIZE SELECTIVE
INLET = 10μm or 15μm

TEN INCH STACK

FINE
PARTICLE
MANIFOLD

TUBE
CLUSTER

CYCLONE
PLENUM

TOTAL PARTICLE
TUBES (12)

TYPICAL VACUUM MANIFOLD
(3 MORE NOT SHOWN)

HIVOL FAN →

FIGURE 1: General view of SCISAS, not showing Nuclepore filter holders, three of four vacuum manifolds, four suction pumps, connecting tubing and stand.

15 µm sample at isokinetic velocities through a four inch diameter tube into a 46" long sampling plenum. Filters along the side of the plenum draw the sample from it.

The following particle loss mechanisms were theoretically evaluated and predicted to be negligible (less than 1%):

● Electrostatic capture in the main PVC stack.

● Turbulent diffusion in the main stack and two inch sampling tubes.

● Brownian diffusion losses.

● Inertial impaction losses in the two inch tubes.

Sedimentation losses in the inhalable particle sampling tubes are calculated to be a maximum of 7% for 15 µm particles. The maximum bias in the 0 to 10 or 15 µm mass estimation is much less than 5%.

In ten tests of a 15 µm cut-point SCISAS prototype, aerosol mass concentrations measured in Reno, NV, ranged from 6 µg/m^3 to 32 µg/m^3; the maximum difference between any two of the four SCISAS 0 to 15 µm particle sampling tubes operated simultaneously in each of these tests was 4%. More typically, any two sampling tubes agreed to better than 3%. At the same sampling site, three comparisons of the SCISAS prototype to a collocated hivol outfitted with an identical 15 µm size selective inlet were conducted. Ratios of the mass concentrations measured by the SCISAS, to those measured by the hivol/SSI, were 0.98, 0.95, and 0.96 for these three preliminary tests.

Further tests of the SCISAS are now scheduled and will include extensive comparisons with other sampling devices. Other tests include 1) measurement of passive deposition inside the SCISAS sampling tubes, 2) measurements of re-entrainment of large particles inside the SCISAS, 3) evaluation of virtual impaction into non-operating sampling tubes in the SCISAS tube cluster, 4) quantitative evaluation of the effects of non-isokinetic mismatches at the entrance to the tube cluster, 5) and measurement of the effects of flow rate variations in the main stack and SSI. Velocities inside the main stack at the approach to the tube cluster will be mapped with a hot-wire anemometer. A theoretical evaluation of the possible effects of aerosol particle deliquesence or shrinking, due to heat transfer in the SCISAS, will also be performed.

388

REFERENCES

Allard, D.W., Tombach, I.H., Mayrsohn, H., and Mathai, C.V., 1982. "Aerosol Measurements: Western Regional Air Quality Studies" Air Pollution Control Association Annual Meeting, New Orleans, LA.

Henry, R., 1977. "A Factor Model of Urban Aerosol Pollution," Ph.D. Dissertation Oregon Graduate Center, Beaverton, Oregon.

Mueller, P.K., Hidy, G.M., Baskett, R.L., Fung, K.K., Henry, R.C., Lavery, T.F., Warren, K.K. and Watson, J.G., 1983. "The Sulfate Regional Experiment: Report of Finding Volume 1 Report EA-1901, Electric Power Research Institute, Palo Alto, CA.

Published 1984 by Elsevier Science Publishing Co., Inc.
Aerosols, Liu, Pui, and Fissan, editors

VISIBILITY REDUCTION AS RELATED TO ATMOSPHERIC AEROSOL CONSTITUENTS

BY

B. R. Appel, Y. Tokiwa, J. Hsu, E. L. Kothny and E. Hahn

Air & Industrial Hygiene Laboratory
California Department of Health Services
2151 Berkeley Way, Berkeley, California 94704
U.S.A.

EXTENDED ABSTRACT

Introduction

The contribution to visibility reduction of both light-scattering and absorption by air pollutant particles and gases was assessed. Emphasis was placed on minimizing sampling artifacts for nitrate and sulfate since previous visibility studies were generally subject to substantial errors from these sources. The aerosol species measured were fine and coarse particulate mass, sulfate, nitrate, and elemental carbon, plus organic carbon and ammonium ion. The gases measured were nitric acid, NH_3, SO_2, NO_2, and O_3. Sampling was done at San Jose, Riverside and downtown Los Angeles.

The atmospheric extinction coefficient, b_{ext}, may be expressed as the sum of the scattering coefficients for dry aerosol, $(b_{sp})_{dry}$, liquid water, b_{sw}, and gases, b_{sg}, plus the absorption coefficients for particles, b_{ap}, and gases, b_{ag}. Omitting Rayleigh scattering, the extinction coefficient will be termed b'_{ext}. To minimize sampling errors, fine particle sulfate was sampled using Teflon filters. Fine particle nitrate was sampled with an acid gas diffusion denuder followed by tandem Teflon and NaCl-impregnated cellulose filters (Appel et al, 1981).

Results

Table 1 lists mean concentrations of the aerosol and gaseous pollutants and mean visibility parameters measured at each site. At all sites, scattering by dry particles, b_{sp}, contributed over half of the b'_{ext}. b_{ap} showed the largest percentage contribution to b'_{ext} at San Jose, apparently because of the relatively low concentration of sulfates and nitrates at this site. The extinction coefficients correspond to mean visual ranges of 43, 15 and 13 Km at San Jose, Riverside and Los Angeles, respectively.

The proportion of the aerosol species in the fine particle fraction were 92% for sulfate, 62% for nitrate, 80% for elemental carbon, and 62% for total carbon. A scatter diagram of elemental carbon concentrations against the corresponding b_{ap} values showed high correlation ($r = 0.92$), negligible intercept and a slope of 12.0 m^2/g, in excellent agreement from results with other investigators.

The coefficient b_{sw} was subjected to multiple regression against the concentration of the hygroscopic aerosol constituents sulfate and nitrate, with alternative forms for humidity dependence:

$$b_{sw} = a + \frac{b\ (SO_4)}{(1-\mu)^v} + \frac{c\ (NO_3)}{(1-\mu)^w} \tag{1}$$

$$b_{sw} = a + b\ (SO_4)\ \mu^x + c\ (NO_3)\ \mu^y \tag{2}$$

Where μ = % RH/100 and v, w, x, and y were varied from 0 to 2.

The choice of equation as well as specific values for the coefficients in the range 0.5 to 2 was relatively insignificant; Pearson's correlation coefficient, r, remained in the range 0.78 to 0.81. For simplicity, equation (2) with x = y = 1, yielding r = 0.81, was employed.

To assess the contribution of aerosol-phase pollutants to light scattering, multiple regression analyses were performed yielding:

$$b_{ap} = -0.14 + 0.051\ (FSO_4) + 0.064\ (FNO_3) + 0.049\ (FC_e) + 0.11\ (CSO_4)$$

$$r = 0.95$$

Where F = fine, C = Coarse, C_e = elemental carbon and R = residual fine mass.

Notable features of this equation include: (1) Higher scattering efficiency for fine nitrate compared to fine sulfate, in contrast to most previous studies; (2) A high and physically unreasonable coefficient for coarse sulfate (we suspect that this material may be functioning as a surrogate for light scattering species whose concentrations vary linearly with that for coarse sulfate, e.g. sea salt aerosol); (3) Fine elemental carbon is similar in scattering efficiency to fine sulfate.

The coefficient b'_{ext} can now be expressed by the sum of the corresponding regression equations for b_{sp}, b_{sw}, b_{ap}, and b_{ag}:

$$b'_{ext} = -0.14 + 0.051\ (FSO_4) + 0.064\ (FNO_3) + 0.049\ (FC_e) + 0.11\ (CSO_4) +$$
$$0.081\ (FSO_4)\ \mu + 0.062\ (FNO_3)\ \mu + 0.116\ (FC_e) + 3.3 \times 10^{-3}\ NO_2$$

Where NO_2 is in ppb and b'_{ext} in $10^{-4}\ m^{-1}$.

Using the mean 4-h average concentrations for the chemical species judged to be contributors to b_{sp} and b'_{ext}, the mean proportions due to each species are shown in Table 3. The contributions due to nitrate- and sulfate-related water are included as part of the contributions of these species. Nitrate was the largest single contributor to both optical parameters, providing, on average, over half of the light scattering and over one third of the extinction. Sulfate was second in important. Fine elemental carbon provided a total of nearly 20% of the extinction. Coarse sulfate (or species proportional to it) appeared to contribute nearly 14% of the light scatterng and 7% of the extinction. Absorption due to NO_2 represented 8% of the mean b'_{ext}.

Table 1. Mean Values of Pollutant Concentrations, Meteorological and Optical Parameters ($\mu g/m^3$)

Species[a]	San Jose	Riverside	Los Angeles
FSO_4	1.66	6.97	13.0
FNO_3	3.38	13.5	7.88
FC_e	1.54	2.50	3.34
CSO_4	0.61	1.46	1.80
CNO_3	2.11	8.28	7.93
NH_4	0.30	6.09	5.18
CC_e	0.25	0.60	0.86
C_o	9.44	17.58	18.3
FMass	22.3	47.5	61.5
TSP	51.5	115.7	107.2
b_{ap}[b]	0.17	0.21	0.36
b_{sp}	0.47	1.44	1.61
b_{sw}	0.050	0.58	0.82
b_{ag}	0.090	0.20	0.21
b'_{ext}	0.78	2.43	3.00

---

[a] F = Fine, C = Coarse
[b] Units $10^{-4} m^{-1}$ for all optical parameters.

Table 2. Average Percent Contributions of Chemical Species to the Particle Scattering Coefficient, b_{sp}, and to the Extinction Coefficient, b'_{ext} [a]

Species	Mean % of b_{sp}	Mean % of b'_{ext}
FSO_4	36.0	30.0
FNO_3	52.1	36.2
CSO_4	13.9	6.7
CNO_3	ca.0	ca.0
FC_e	11.9	5.7 (scattering)
		13.5 (scattering)
CC_e	ca.0	ca.0
C_o	ca.0	ca.0
R	ca.0	ca.0
NO_2	--	8.0

a. b'_{ext} refers to the total extinction coefficient minus the Rayleigh scattering coefficient.

Reference

Appel, B.R., Tokiwa, Y., and Haik, M. (1981) Sampling of Nitrates in Ambient Air, Atmos. Environ. 15:283-289 (1981).

Published 1984 by Elsevier Science Publishing Co., Inc.
Aerosols, Liu, Pui, and Fissan, editors

THE ALBUQUERQUE WINTER VISIBILITY STUDY

B. D. Zak, W. Einfeld, H. W. Church,
G. T. Gay, and A. L. Jensen
Applied Atmospheric Research Division 6324
Sandia National Laboratories
Albuquerque, NM 87185
U. S. A.

EXTENDED ABSTRACT

In the fall of 1982, the City of Albuquerque sought the assistance of
Sandia National Laboratories in planning and carrying out a study of winter
visibility impairment in Albuquerque. The purpose of this study was to
determine the composition of the haze over the city, and from that
information, ascertain the relative contribution of the various types of air
pollution sources to visibility impairment in Albuquerque in winter.

Three existing City of Albuquerque air quality monitoring stations were
augmented with additional instrumentation. The base site was Zuni Park, near
the population center of the city. A second site known as Volandia was
located nearer the northern edge of the city. The third site was atop a City
water tank in the Rio Grande Valley. Meteorological, heated nephelometer, and
continuous concentration data for all regulated pollutants were acquired at
the Zuni and Volandia sites. Dichotomous samplers were operated at all three
sites on a 12-hour schedule, beginning at 7 am and 7 pm. Aerosol sampling
began on December 28, 1982, and continued until February 23, 1983. The
aerosol samplers were not operated during periods of forecast rain, snow, or
high winds.

On most weekdays, the continuous monitors showed a morning peak in
concentration of pollutants known to be associated with vehicles: CO and NO.
Light scattering also peaked during the same period. Vehicle-related
pollutant concentrations typically fell to low levels during the middle of the
day, before beginning to rise again after 3 pm. These pollutant concentrations usually peaked again between 7 and 11 pm, before falling to low levels
by 4 or 5 am. For CO and NO, morning and evening peak concentrations were
approximately equal. For light scattering, evening peaks were on average
three times as intense as morning peaks.

Meteorological measurements were made aloft with an AIR tethersonde system
which gave wind speed, wind direction, temperature, "wet bulb" temperature,
and pressure as a function of altitude. Another tethered balloon system was
used to collect size-segregated aerosol samples and to measure carbon monoxide
concentrations at selected heights above ground. The tethersonde system
revealed that under the clear sky, low wind conditions which favor visibility
impairment, a temperature inversion typically formed in late afternoon, grew
more intense during the evening and night, and only slowly broke up in the
morning. Results from the aerosol and carbon monoxide sampling balloons
confirmed that under inversion conditions, pollutants were being trapped in a
very shallow layer close to the ground.

394

The elemental composition of the dichotomous filter samples was determined by X-ray fluorescence. Light absorption measurements were made using the integrating plate methods. Samples taken at Zuni Park were also analyzed for volatile and for elemental carbon, as well as for sulfate and nitrate. A detailed breakdown of the fine particle aerosol content is given in Figure S-1.

Apportionments of fine particle mass and visibility impairment to local sources were calculated by using tracer element and multiple linear regression techniques. Results of the visibility impairment apportionment from the Zuni Park day, night and overall data sets are given in Figure S-2.

It was found that on average, approximately 78% of the daytime visibility degradation was attributable to mobile sources -- essentially vehicles. About 19% was attributable to wood burning, with the small remaining amount attributable to dust and natural gas combustion. During the night, on average, 59% of the visibility degradation was attributable to wood combustion. On nights with severe visibility impairment, the contribution from wood burning was higher. On one night examined in detail, it was found to be 70%. The uncertainties on the apportionments are estimated to be about ±20% of the magnitude of each term.

Overall visibility impairment apportionments for the Zuni, Volandia, and Candelaria sites were compared to get an indication of how site-specific such apportionments are. The Volandia apportionment was similar to that for the Zuni site, but with a slightly heavier contribution from wood combustion. The Candelaria site was quite different. Vehicles dominated both day and night. Thus the Valley may form an air pollution domain significantly different in character from the Heights.

During the study, remote sampling stations throughout the Southwest recorded fine particle mass concentrations averaging about 10% of the concentrations measured in Albuquerque. Thus, approximately 90% of Albuquerque's winter visibility impairment is locally generated. On average, sulfur compounds accounted for less than 5% of the Albuquerque fine particle mass. Based on the remote station elemental composition data, more than half of that is attributable to local sources, rather than to distant coal-fired power plants and smelters.

Heavy-duty diesel trucks and buses contribute disproportionately to visibility impairment. Per vehicle mile traveled, such vehicles exhaust on average more than 100 times as much fine particle mass as catalyst-equipped passenger vehicles. Using national averages for the vehicle fleet composition, we calculate that about 60% of the visibility impairment apportioned to mobile sources is attributable to diesels. There is evidence, however, that at least at the Zuni Park site, this calculation moderately overestimates the impact of diesels.

A more complete description of this study is available (Zak et al., 1984).

References

B. D. Zak, W. Einfeld, H. W. Church, G. T. Gay, A. L. Jensen, M. D. Ivey, P. H. Homann, and J. Trijonis (1984) The Albuquerque Winter Visibility Study - Volume I: Overview and Data Analysis, Sandia National Laboratories Report SAND84-0773.

OVERALL

AVERAGE FINE PARTICLE MASS = 37.7 µg/M^3

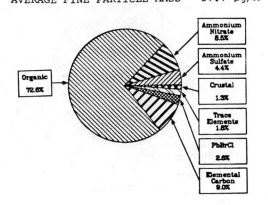

DAY

Average FPM - 18.6 µg/m^3

NIGHT

Average FPM = 54.1 µg/m^3

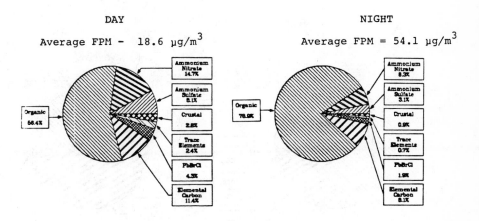

FIGURE S-1.   Fine Particle Mass Balance for the Zuni Park Site.

396

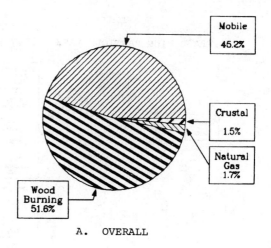

A. OVERALL

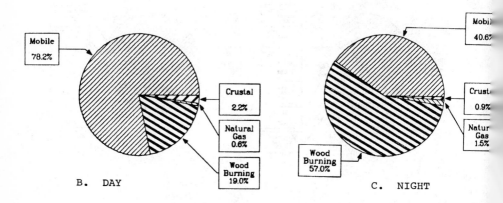

B. DAY

C. NIGHT

FIGURE S-2. Total Visibility Impairment Apportioned by Source Type
for the Zuni Park Site.

Published 1984 by Elsevier Science Publishing Co., Inc.
Aerosols, Liu, Pui, and Fissan, editors

AEROSOL OPTICAL ABSORPTION: ACCURACY OF FILTER
MEASUREMENT BY COMPARISON WITH IN-SITU EXTINCTION

R. E. Weiss and A. P. Waggoner
Environmental Engineering and Science
Department of Civil Engineering
University of Washington
Seattle, WA   98195
U. S. A.

EXTENDED ABSTRACT

Introduction

Several methods have been used to measure the optical absorption
properties of suspended particles. Most techniques require removing the
particles from the suspension by collection on filters or plates to amplify
the absorption to measurable levels. The effect of concentrating the
particles and the influence of the collection surfaces on the accuracy of the
techniques is complex and most methods are calibrated by argument rather than
using aerosols of known absorption. We are describing a procedure to
calibrate ambient methods for measuring absorption using in-situ measurements
on laboratory generated aerosols in a 24 meter, continuous flow extinction
cell.

Aerosols with controlled optical and size properties are injected into
the cell and absorption is measured from the difference between total and
scattering extinction. With a time response of about 1 second, total
extinction is determined from the direct attenuation of light through the
known cell path length as the cell is alternately filled with aerosol and
particle free air; scattering extinction is measured with a three wavelength
nephelometer as the aerosol flows from one end of the cell. By varying the
composition and size of the particles, the effect of single scattering albedo
and particle size on the accuracy of the various ambient absorption techniques
can be tested by comparison with the cell in a controlled manner.

We are reporting preliminary results testing the Integrating Plate method
(IPM) for absorption (Lin, 1973, Weiss, 1980) measured on nuclepore filters.

Instruments

A schematic of the extinction cell is shown in figure 1. The cell
consists of a 24 meter enclosed path, about 7.5 cm in diameter. There
are internal baffles at 2 meter intervals to eliminate reflected light from
the walls, reducing the potential viewing diameter to about 4 cm. At one end
there is a diffuse light source (two H-3 quartz-halogen 55 watt lamps behind a
7 cm diameter milk glass window) and a multi-wavelength radiometer at the
other. The lamp current is regulated to 1 part in 10,000. The radiometer
uses a 600 mm lens and fiber optic light path to a photomultiplier, to reduce
the actual field of view to 0.0004 radians. The combination of noise and
drift in the radiance measurement is equivalent to about 10^{-5} m^{-1} over
several minutes when the cell is filled with filtered air. Since the
calibration aerosols are generated at extinction coefficients greater than

398

10^{-3} m^{-1}, the uncertainty in the extinction measurement is less than one percent. The output from a typical aerosol injection is shown in figure 2. In addition to the cell transmittance, the nephelometer signal is also shown.

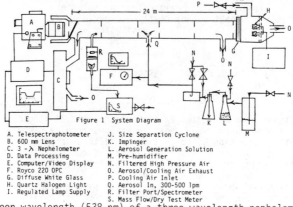

Figure 1 System Diagram

A. Telespectraphotometer	J. Size Separation Cyclone
B. 600 mm Lens	K. Impinger
C. 3 - λ Nephelometer	L. Aerosol Generation Solution
D. Data Processing	M. Pre-humidifier
E. Computer/Video Display	N. Filtered High Pressure Air
F. Royco 220 OPC	O. Aerosol/Cooling Air Exhaust
G. Diffuse White Glass	P. Cooling Air Inlet
H. Quartz Halogen Light	Q. Aerosol In, 300-500 lpm
I. Regulated Lamp Supply	R. Filter Port/Spectrometer
	S. Mass Flow/Dry Test Meter

The green wavelength (538 nm) of a three wavelength nephelometer was used to measure scattering extinction. The nephelometer was calibrated with particle free air and CCl_2F_2 prior to an after each sampling session using values (Ruby and Waggoner, 1981) adjusted to system temperature and pressure and a wavelength of 538 nm. Nephelometer truncation error was corrected using curves generated by varying the size distribution of non-absorbing aerosols and measuring the ratio of scattering to extinction. In figure 3 the results of four distributions are shown as well as a measure of system linearity. In the small particle limit, the ratio of uncorrected scattering to extinction approached 1 for non-absorbing particles.

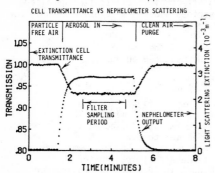

CELL TRANSMITTANCE VS NEPHELOMETER SCATTERING

Figure 2 Nephelometer and extinction cell response to aerosol injection and clean air purge. The aerosol exchange period is about 1 minute in the cell. The signal from the extinction cell is in radiance and is normalized to 1 with filtered air; the nephelometer signal is in extinction coefficient for particles only. Filter samples are taken during the period of constant scattering and transmittance.

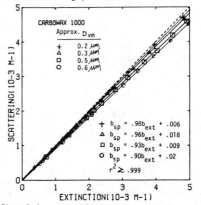

Figure 3 Comparison of light scattering extinction measured with a nephelometer and total extinction from the extinction cell for nonabsorbing aerosol. Aerosol concentration was varied by addition of dilution air. The separate lines are for different size distributions. The difference in slope is due to truncation error in the nephelometer which increases with increasing particle size.

For use with the Integrating Plate Method, an IPM spectrometer was constructed such that diffuse transmission through the filter could be measured continuously as the particles were deposited on the filter. Samples were taken on 37 mm, 0.4 um pore Nuclepore filters. Volume flow was measured using a mass flow meter as well as a dry test meter.

The signals from the cell spectrometer, nephelometer, IPM spectrometer and flow devices were monitored and stored in an Apple II based data aquisition system as approximately 1 second averages.

Results

Absorbing particles were produced from solutions of polyethylene glycol, water, ethyl alcohol and India ink. By varying the ratio of polyethylene glycol and india ink and the particle size, the single scattering albedo could be varied from about 0.5 to 1. The particles were generated using a sonic jet of filtered air submerged below the surface of the particle forming solution. The aerosol produced above the solution was diluted with 1 part dry filtered air, passed through an impactor and tandem cyclones to control the size distribution. Approximately 10 parts dry air was mixed with the aerosol in a turbulent flow system prior to injection at the center of the cell. Stable concentrations of submicron aerosol at 1 mg/m^3 and flow rates of 500 lpm can be maintained by this method. The size distribution was monitored using the wavelength dependence of scattering as a sensitive measure of mean size parameters. Sufficient filter mass loadings for testing the IPM method could be obtained with sampling periods of 1 to 5 minutes.

Examples of recorded cell and nephelometer extinction data for a non-absorbing and absorbing aerosol respectively are shown in Figures 4 and 5. For the non-absorbing case, extinction appears to be slightly greater than scattering. The difference is because scattering is not corrected for truncation error in the direct signal output. The difference increases for larger particles and approaches zero for small particles.

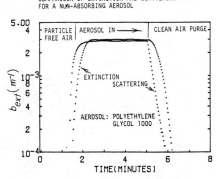

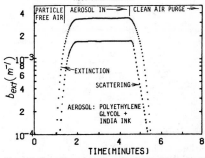

Figure 4. Same as figure 3 except that cell transmittance is converted to extinction coefficient. The non-absorbing aerosol is generated from a mixture of polyethylene glycol 1000, H$_2$O and ethyl alcohol. The slight difference between scattering and extinction is caused by some loss of forward scattered light in the nephelometer due to truncation of the near forward angles.

Figure 5 Comparison of light scattering to total extinction for an absorbing aerosol generated from an internal mixture of poly-ethylene glycol and india ink. For this aerosol, scattering and absorption are about equal. The ratio of absorption to scattering can be changed by varying the fraction of ink in the mixture.

The effect of particle size on the single scattering albedo is shown in figure 6. The wavelength dependence parameter, α, is used as an indication of mean size distribution. An α range of 1.8 to 3 represents a range in volume mean diameters for log normal size distributions of about 0.15 to 0.4 um. The size distribution was changed by varying the cyclone and impactor flow rate and the solution concentration. The symbols represent two mixtures of polyethylene glycol and india ink; the solid lines are Mie calculations for a series of complex refractive indices. The scattering values were corrected for truncation error, about one to five percent for the range of values measured.

IPM tests were done on 37 mm, 0.4 um Nuclepore polycarbonate filters. Aerosols with volume mean diameters of about 0.25 um and scattering to extinction ratios of 0.63, 0.76 and 0.84 were tested. The results of the three tests are shown in figure 7. For each aerosol, the IPM overestimated absorption extinction compared with that measured in the cell. The more highly absorbing particles presented the greatest overestimation; on the order of 30 percent compared with 15 to 20 percent for the less absorbing particles. The overestimation appears to be caused by the particles affecting the internal transmittance properties of the substrates (Clark, 1982). Further tests are being done to better understand and correct this effect.

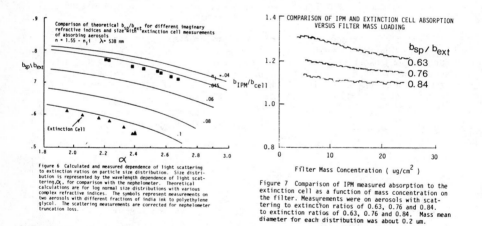

Figure 6 Calculated and measured dependence of light scattering to extinction ratios on particle size distribution. Size distribution is represented by the wavelength dependence of light scattering, α, for comparison with the nephelometer. Theoretical calculations are for log normal size distributions with various complex refractive indices. The symbols represent measurements on two aerosols with different fractions of india ink to polyethylene glycol. The scattering measurements are corrected for nephelometer truncation loss.

Figure 7 Comparison of IPM measured absorption to the extinction cell as a function of mass concentration on the filter. Measurements were on aerosols with scattering to extinction ratios of 0.63, 0.76 and 0.84. to extinction ratios of 0.63, 0.76 and 0.84. Mass mean diameter for each distribution was about 0.2 um.

References

Clarke, A. D. (1982), App. Opt., 21, 3021.
Lin, C.I, Baker, M. B., Charlson, R. J. (1973), App. Opt., 12, 1356.
Ruby, M. G. and Waggoner, A. P. (1981), Env. Sci. & Tech, 14, 109.
Weiss, R. E. (1980), Ph. D. dissertation, University of Washington, Seattle.

Published 1984 by Elsevier Science Publishing Co., Inc.

Aerosols, Liu, Pui, and Fissan, editors

A COMPARISON OF OPTICAL METHODS FOR THE MEASUREMENT OF ELEMENTAL
CARBON IN COMBUSTION AEROSOLS

Steven M. Japar
Research Staff
Ford Motor Co.
P.O. Box 2053
Dearborn, Mi. 48121

Extended Abstract

The elemental (or black) carbon formed during combustion has been the subject of increasing study recently for two reasons. First, the material is
emitted from most combustion sources in the accumulation mode (mmd $\simeq$ 0.2μm),
thus permitting elemental carbon a very long lifetime in the atmosphere. In
fact, the material is a ubiquitous component of atmospheric aerosol, being
found in such remote regions as the Arctic (Rahn, 1980; Rosen et al, 1981).
Second, elemental carbon is a very efficient absorber of visible light, thus
making it possible for elemental carbon in the atmosphere to degrade visibility seriously (Charlson et al, 1978), and to perturb the radiation balance in
the atmosphere.

One of the major areas of interest has been the development of methodology
to monitor and quantitatively measure elemental carbon both in the atmosphere
and at the combustion source. Here too the optical properties of elemental
carbon play an important role, and have been the basis of a number of techniques, some capable of "real-time" applications.

Optical methods which have been used with some degree of success are listed
in Table 1, where they are separated into bulk (or filter-collection) and in-
situ methods. A bibliography of articles important in the development and
intercomparison of the techniques appears in Table 2.

Bulk methods have been used for the last ten years. These methods generally involve collecting an aerosol sample on a suitable filter and subjecting
the filter to analysis. The approach which has received the most study is
the integrating plate (IP) method developed by Charlson and coworkers (Lin et
al, 1973). This technique involves measuring the transmission through a filter and a diffuse reflector, such as opal glass, before and after sample collection, and attributing the increase in light attenuation to light absorption
by the impacted aerosol. This approach, and a number of related ones, such as
laser transmission through a filter-collected sample (Rosen et al, 1980), have
been extensively used to measure carbon in atmospheric samples.

More recently there have been a number of efforts to develop in-situ techniques capable of real-time measurement of elemental carbon in the laboratory
and in the atmosphere. The most extensive studies have involved the use of
photoacoustic spectroscopy (Japar and Szkarlat, 1981; Roessler and Faxvog,
1979) to measure aerosol light absorption, without interference from light
scattering. However, the use of total light extinction (light scattering plus
light absorption) (Japar and Szkarlat, 1981) and light scattering (Flower,
1982) have also been described. A new method, photothermal deflection spectroscopy, which is capable of spatial resolution, has also been reported
(Campillo et al, 1980).

This paper is intended to review the progress which has been made regarding optical methods for the measurement of elemental carbon. The relative merits of various techniques will be described, based on literature data where possible (see Table 1); and the applicability of these techniques, in terms of sensitivities and potential interferences, to the measurement of elemental carbon in urban and rural atmospheres will be discussed.

References

Campillo, A.J., Lin, H.-B., Dodge, G.J. and Davis, C.C.(1980). Stark-effect-modulated phase-fluctuation optical heterodyne interferometer for trace gas analysis. Opt. Lett. 5, 424-426.
Charlson, R.J., Waggoner, A.P. and Thielke, J.F.(1978). Visibility protection for Class I areas: the technical basis. PB-288 842, U.S. Department of Commerce (August, 1978).
Flower, W.L.(1982). Optical measurements of soot formation in pre-mixed flames. Sandia Report Sand82-8812, Sandia National Laboratory, Albuquerque, N.M.
Japar, S.M. and Szkarlat, A.C.(1981). Real-time measurement of Diesel vehicle exhaust particulate using photoacoustic spectroscopy and total light extinction. Society of Automotive Engineers Transactions 90, 3624-3631.
Lin, C.-I., Baker, M.B. and Charlson, R.J.(1973). Absorption coefficient of atmospheric aerosol: a method for measurement. Appl. Opt. 12, 1356-1363.
Rahn, K.A. and McCaffrey, R.J.(1980). On the origin and transport of winter Artic aerosol. Ann. N.Y. Acad. Sci. 338, 486-503.
Roessler, D.M. and Faxvog, F.R.(1979). Optoacoustic measurement of optical absorption in acetylene smoke. J. Optical Soc. Amer. 69, 1699-1703.
Rosen, H., Hansen, A.D.A., Dod, R. and Novakov, T.(1980). Soot in urban atmospheres: determination by an optical absorption technique. Science 208, 741-744.
Rosen, H., Novakov, T. and Bodhaine, B.A.(1981). Soot in the Artic. Atmos. environ. 15, 1371-1374.

Table 1 - Optical Methods for the Measurement of Elemental Carbon in
Combustion Aerosols

Bulk (Filter-Collection Methods	Intercomparisons
Light Transmission - Integrating Plate (IP)	LT, PAS-B, PAS-A
- Laser Transmission (LT)	IP, PAS-B, RSS
Reflectance spectroscopy (RS)	PAS-B, IP
Photoacoustic spectroscopy (PAS-B)	IP, RS, LT
Raman scattering spectroscopy (RSS)	LT
In-Situ Methods	
Photoacoustic spectroscopy (PAS-A)	TLE, IP
Photothermal deflection spectroscopy (PDS)	
Total light extinction (TLE)	PAS-A
Light scattering (LS)	

Table 2 - Bibliography of Optical Methods and Intercomparisons

BULK METHODS

Integrating Plate
Lin, C.-I., Baker, M.B. and Charlson, R.J.(1973). Absorption coefficient of atmospheric aerosols: a method for measurement. Appl. Opt. 12, 1356-1363.
Clarke, A.D. and Waggoner, A.P.(1982). Measurement of particle optical absorption, imaginary refractive index, mass concentration, and size at First International LAAP Workshop. Appl. Opt. 21, 398-402.
Szkarlat, A.C. and Japar, S.M.(1981). Light absorption by airborne aerosols: comparison of integrating plate and spectrophone techniques. Appl. Opt. 20, 1151-1155.
Patterson, E.M. and Marshall, B.T.(1982). Diffuse reflectance and diffuse transmission measurements of aerosol absorption at the First International Workshop on Light Absorption by Aerosol Particles. Appl. Opt. 21, 387-393.

Laser Transmission
Hansen, A.D.A., Rosen, H. and Novakov, T.(1982). Real-time measurement of the absorption coefficient of aerosol particles. Appl. Opt. 21, 3060-3062.
Rosen, H., Hansen, A.D.A., Dod, R.L. and Novakov, T.(1980). Soot in urban atmospheres: determination by an optical absorption technique. Science 208, 741-744.
Rosen, H. and Novakov, T.(1983). Optical transmission through aerosol deposits on diffusely reflective filters: a method for measuring the absorbing component of aerosol particles. Appl. Opt. 22, 1265-1267.

Photoacoustic Spectroscopy
Bennett, C.A., Jr. and Patty, R.R.(1982). Monitoring particulate carbon collected on Teflon filters: an evaluation of photoacoustic and transmission techniques. Appl. Opt. 21, 371-374.
Rohl, R., McClenny, W.A. and Palmer, R.A.(1982). Photoacoustic determination of optical properties of aerosol particles collected on filters: develop-

ment of a method taking into account substrate reflectivity. Appl. Opt. 21, 375-380.

Raman Scattering Spectroscopy

Blaha, J.J., Rosasco, G.J. and Etz, E.S.(1978). Raman microprobe character- ization of residual carbonaceous material associated with urban airborne particulates. Appl. Spect. 32, 292-297.

Rosen, H. and Novakov, T.(1978). Identification of primary particulate carbon and sulfate species by Raman spectroscopy. Atmos. Environ. 12, 923-927.

Rosen, H., Hansen, A.D.A., Gundel, L. and Novakov, T.(1978). Identification of the optically absorbing component in urban aerosols. Appl. Opt. 17, 3859-3861.

IN-SITU METHODS

Photoacoustic Spectroscopy

Faxvog, F.R. and Roessler, D.M.(1979). Optoacoustic measurements of diesel particulate emissions. J. Appl. Phys. 50, 7880-7882.

Faxvog, F.R. and Roessler, D.M.(1982). Mass concentration of diesel particle emissions from photoacoustic and opacity measurements. Aerosol Sci. Tech. 1, 225-234.

Japar, S.M. and Killinger, D.K.(1979). Photoacoustic and absorption spectrum of airborne carbon particulate using a tunable dye laser. Chem. Phys. Lett. 66, 207-210.

Japar, S.M. and Szkarlat, A.C.(1981). Measurement of diesel exhaust particu- late using photoacoustic spectroscopy. Combust. Sci. Tech. 24, 215-219.

Japar, S.M. and Szkarlat, A.C.(1981). Real-time measurement of diesel vehicle exhaust particulate using photoacoustic spectroscopy and total light ex- tinction. Society of Automotive Engineers (SAE) Paper No. 811184; SAE Transactions 90, 3624-3631.

Killinger, D.K., Moore, J.A. and Japar, S.M.(1980). The use of photoacoustic spectroscopy to characterize and monitor soot in combustion processes. Amer. Chem. Soc. Symp. Ser. 134, 457-462.

Roessler, D.M. and Faxvog, F.R.(1980). Photoacoustic determination of optical absorption to extinction ratio in aerosols. Appl. Opt. 19, 578-581.

Roessler, D.M. and Faxvog, F.R.(1980). Optical properties of agglomerated acetylene smoke particles at 0.5145um and 10.6um wavelengths. J. Opt. Soc. Amer. 70, 230-235.

Roessler, D.M.(1982). Opacity and photoacoustic measurements of diesel parti- cle mass emissions. SAE Paper No. 820460.

Roessler, D.M.(1982). Diesel particle mass concentration by optical techni- ques. Appl. Opt. 21, 4077-4086.

Terhune, R.W. and Anderson, J.E.(1977). Spectrophone measurements of absorp- tion of visible light by aerosols in the atmosphere. Opt. Lett. 1, 70-72.

Truex, T.J. and Anderson, J.E.(1979). Mass monitoring of carbonaceous aerosol with a spectrophone. Atmos. Environ. 13, 507-509.

Photothermal Deflection Spectroscopy

Campillo, A.J., Lin, H.-B., Dodge, G.J. and Davis, C.C.(1980). Stark-effect- modulated phase fluctuation optical heterodyne interferometer for trace- gas analysis. Opt. Lett. 5, 424-426.

Jackson, W.B., Amer, N.M., Boccara, A.C. and Fournier, D.(1981). Photothermal deflection spectroscopy and detection. Appl. Opt. 20, 1333-1344.

Rose, A., Pyrum, J.D., Muzny, C., Salamo, G.J. and Gupta, R.(1982). Applica- tion of the photothermal deflection technique to combustion diagnostics. Appl. Opt. 21, 2663-2665.

Total Light Extinction
Faxvog, F.R. and Roessler, D.M.(1982). See listing under Photoacoustic Spec-
    troscopy.
Japar, S.M. and Szkarlat, A.C.(1981). See listing under Photoacoustic Spec-
    troscopy.
Roessler, D.M.(1982). See listing under Photoacoustic Spectroscopy.
Roessler, D.M. and Faxvog, F.R.(1981). Changes in diesel particulates with
    engine air/fuel ratio. Combust. Sci. Tech. 26, 225-231.

Light Scattering
Dalzell, W.H., Williams, G.C. and Hottel, H.C.(1970). A light scattering me-
    thod for soot concentration measurements. Combust. Flame 14, 161-170.
Flower, W.L.(1982). Optical measurements of soot formation in premixed flames
    Sandia Report Sand82-8812, Sandia National Laboratory, Albuquerque, N.M.

METHOD INTERCOMPARISONS

Integrating Plate vs Laser Transmission
Sadler,   M., Charlson, R.J., Rosen, H. and Novakov, T.(1981). An intercom-
    parison of the integrating plate and laser transmission methods for deter-
    imination of aerosol absorption coefficients. Atmos. Environ. 15, 1265-68.

Integrating Plate vs Photoacoustic Spectroscopy (Bulk)
Bennett, C.A., Jr., Patty, R.R. and McClenny, W.A.(1981). Photoacoustic de-
    tection of carbonaceous particles. Appl. Opt. 20, 3475-3477.
Bennett, C.A., Jr. and Patty, R.R.(1982). Monitoring particulate carbon col-
    lected on Teflon filters: an evaluation of photoacoustic and transmission
    techniques. Appl. Opt. 21, 371-374.

Integrating Plate vs Photoacoustic Spectroscopy (Ambient)
Szkarlat, A.C. and Japar, S.M.(1981). Light absorption by airborne aerosols:
    comparison of integrating plate and spectrophone techniques. Appl. Opt.
    20, 1151-1155.

Reflectance Spectroscopy vs Photoacoustic Spectroscopy (Bulk)
Delumyea, R.D. and Mitchell, D.(1983). Comparison of reflectance and photo-
    acoustic photometry for determination of elemental carbon in aerosols.
    Anal. Chem. 55, 1996-1999.

Reflectance Spectroscopy vs Integrating Plate
Patterson, E.M. and Marshall, B.T.(1982). Diffuse reflectance and diffuse
    transmission measurements of aerosol absorption at the First International
    Workshop on light absorption by aerosol particles. Appl. Opt. 21, 387-393.

Laser Transmission vs Photoacoustic Spectroscopy (Bulk)
Yasa, Z., Amer, N.M., Rosen, H., Hansen, A.D.A. and Novakov, T.(1979). Photo-
    acoustic investigation of urban aerosol particles. Appl. Opt. 18, 2528-30.

Laser Transmission vs Raman Scattering Spectroscopy
Rosen, H., Hansen, A.D.A., Gundel, L. and Novakov, T.(1978). Identification
    of the optically absorbing component in urban aerosols. Appl. Opt. 17,
    3859-3861.

Photoacoustic Spectroscopy (Ambient) vs Total Light Extinction
Japar, S.M. and Szkarlat, A.C.(1981). Real-time measurement of diesel vehicle
    particulate using photoacoustic spectroscopy and total light extinction.
    SAE Transactions 90, 3624-3631.

Published 1984 by Elsevier Science Publishing Co., Inc.

Aerosols, Liu, Pui, and Fissan, editors

406

EFFECT OF DIESEL VEHICLES
ON VISIBILITY IN CALIFORNIA

by

John Trijonis

Santa Fe Research Corporation
Route 7, Box 124K
Santa Fe, New Mexico 87501

ABSTRACT

The effect of diesel vehicle elemental carbon on visibility in California is estimated for 1980 and projected to the early 1990's. An emissions budget model indicates that heavy-duty diesel trucks contributed nearly one-half of statewide elemental carbon emissions in 1980 and about 5-15% of statewide light extinction (visibility reduction). A lead tracer model indicates somewhat larger extinction contributions from heavy-duty diesel trucks, about 5 to 25% statewide in 1980. Even greater visibility impacts (too large to be reasonable) are suggested by a CO tracer model. Because of increased diesel usage -- due to both overall traffic growth and partial conversion of the vehicle fleet to diesels (10% of light-duty, 20% of medium-duty, and 60% of currently gasoline heavy-duty) -- visibility in California is projected to decrease significantly (about 9 to 35%) from 1980 to the early 1990's under a "no control" scenario. Diesel vehicle elemental carbon would then contribute 13 to 40% of statewide visibility reduction. Even in the early 1990's, heavy-duty trucks would account for about three-fourths of the emissions from the entire diesel fleet. The most uncertain aspect of the conclusions is the overall magnitude of the visibility effect predicted by the emission budget and Pb tracer models. The partition of visibility impacts among light-, medium-, and heavy-duty vehicles is more definite. The conclusions are rather insensitive to reasonable changes in the assumptions regarding emission factors, traffic growth, and dieselization percentages. The findings indicate that, in order to be effective in protecting visibility, the current California standards for light- and medium-duty vehicles would have to be extended to include regulations for heavy-duty trucks.

Published 1984 by Elsevier Science Publishing Co., Inc.
Aerosols, Liu, Pui, and Fissan, editors

SIZE DISTRIBUTIONS OF ELEMENTAL CARBON IN ATMOSPHERIC AEROSOLS

[1] Antonio H. Miguel_ and Sheldon K. Friedlander [2]
[1] Department of Chemistry, PUC/RJ, R.Marques de S.Vicente,225
Rio de Janeiro 22453, Brazil
[2] Department of Chemical Engineering, UCLA, CA, 90024, USA

EXTENDED ABSTRACT

Introduction

Environmental problems caused by atmospheric aerosols are well documented in the specialized literature. Studies reporting on the role of dense clouds of soil particles in past mass extintions of life on Earth and, more recently (Turco et al.,1983), on calculations of potential global effects of dust and aerosol clouds (generated in a nuclear war) have received the attention of the public through oral and written reports by the news media. Because of the importance played by particulate elemental carbon (PEC) on climatic impact, a study was conducted on the size distribution of PEC in samples collected: i) in an urban area, ii) inside the air duct of an urban tunnel, and iii) next to the take-off runway of a busy airport. Besides providing needed data to improve the precision of calculations of potential local and global environmental effects, the PEC size distributions reported in this study help understand the mechanism of association of PEC and polycyclic aromatic hydrocarbons (PAHs), a class of organic compounds that include several mutagens.

EXPERIMENTAL

Sampling

The urban area sample was collected on the roof (7th floor) of Boelter Hall on the UCLA Campus. Sampling was carried out for a 24-hour period in the summer. Averaged-fleet aerosol emissions were sampled inside the air duct of the Caldecott Tunnel, on a freeway commuter route east of Berkeley. The tunnel is 1.1 Km long, carrying two lanes of uphill traffic with a 2 degree grade. Sampling was conducted in the summer for a period of 226 minutes. Aircraft-dominated aerosols were collected about 50 meters from the take-off area of large aircrafts on Runway 26 of Los Angeles International Airport (LAX). Sampling time was 336 minutes. All samples were collected using a calibrated eigth-stage low-pressure impactor (LPI) with 50% efficiency cut-offs of 4.0, 2.0, 1.0, 0.5, 0.26, 0.12, 0.075, and 0.050 μm aerodynamic diameter, dp (Hering et al., 1978,1979). During sampling at both the airport and in the tunnel, a cyclone (Johns and Well, 1981) was connected to the inlet of the LPI to remove particles resulting from resuspension which could overload the impaction discs of the LPI. Particle bounce in the LPI was reduced by a layer of Vaseline about 2 μm thick by about 3 mm diameter (Miguel, 1982) deposited on unbaked pre-cleaned stainless steel (ss) strips mounted onto 2.54 cm diameter glass discs.

Instrumentation and Analytical Procedure

All PEC determinations were made using a Dohrmann Carbon Analyser (DCA) model DC-50A/52A operated in the total carbon mode and manual boat injection. The DCA's original sample port was modified to allow passage of the ss strips. Manganse dioxide powder was used as oxidant in the PEC determination of the ambient air sample. For the other two samples, only zero air was used to oxidize PEC. The first step in the analysis procedure was to vaporize the Vaseline and all non-elemental carbon compounds (low boilers) contained in the sample. This was done under anaerobic (Helium) conditions by sliding the Pt boat containing the ss strips onto the vaporization zone , operated at 120 ± 10 degrees C for a period of 148 seconds and then onto the pyrolysis zone operated at 530 ± 15 degrees C for a period of 298 seconds. Vaseline blanks were carried out in the same manner. The temperature in the pyrolysis zone was then raised to 825 ± 20 degree C and allowed to stabilize for about 2 hours. At this point zero air was introduced into the sample port at a rate of 18 cc/min. Potassium acid phathalate (KHP) standards were run to set the appropriate instrument span. Activated carbon (Aldrich 24,227-6, Darco 6-60, 100-250 mesh) suspensions prepararced in water (using an ultrasonic device) deposited on ss strips were used to validate the method. Further details of the analytical procedure employed can be found elsewhere (Miguel, 1984).

At 825 degrees C, under aerobic conditions, PEC is oxidized to carbon dioxide which is then reduced to methane by a bed of nickel catalyst. The methane formed is quantitated by the flame ionization detector of the DCA. The digital readings of the detector (ppm) are converted to mass and the PEC mass distribution functions calculated in the usual manner. The measured PEC corresponds to the portion of the sample that is oxidized by manganese dioxide or air.

RESULTS AND DISCUSSION

The major fraction of "sooty" aerosols is comprised of a mixture of elemental carbon (designated here as PEC), tars and oils (Turco et al., 1983). Among the minor constituents, perhaps the most important from a toxicological point of view are the PAHs and their -nitro and -methyl derivatives. In the ambient sample (Fig.1c), observations of the deposit on the first stage (dp>4µm) using an optical microscope revealed the presence of large quantities (perhaps larger than that of particles) of transparent, oily droplets. Upon pyrolysis, this deposit became black, indicating that the oily material was partly converted to elemental carbon, which resulted in an artificial reading of PEC concentration. For this reason, data obtained from the first stage were discarded. Since the airport and tunnel samples were collected with a cyclone connected to the LPI inlet, first stage data for this samples were also discarded.

Data on the concentration of PEC as a function of particle

size are presented in Figures 1a, 1b, and 1c, respectively for the
tunnel, the airport, and the ambient air samples. In all samples,
most of the PEC was found in aerosols with aerodynamic diameter
below 1.0 μm. For the tunnel and airport samples, the mass of PEC
associated with aerosols of aerodynamic diameter less than 0.5μm
was 87 and 77%, respectively. For the ambient air sample it was
65%. These values agree qualitatively with the expected relative
"age" of the aerosols in the three collection sites, that is, the
tunnel aerosols should be less aged than the others. The ambient
air aerosols should be aged more that the airport aerosols, since
the latter are closest to an important source of PEC (aircraft
exhaust) and thus had less time to coagulate and agglomerate.

While the distribution of PEC with respect to particle size
was unimodal (centered between 0.05 and 1.0 μm) in the tunnel
sample (Fig. 1a) it was bimodal in both the airport and the
ambient air samples (Figs. 1b and 1c). The additional PEC mode
in the size range between 0.5 and 1.0 μm, may have resulted from
coagulation of fine PEC onto nitrates, sulfates and chlorides
which, in mixed polluted/maritime air masses, have a major mode in
this size range (Harrison and Pio, 1983). PEC in this size range
are responsible for significant attenuation of the solar flux with
attended lowering of ambient temperatures and decreased
visibility. In addition, due to their large surface to volume (or
mass) ratio, submicron PEC may catalyse specific chemical
reactions with entrained gases near the combustion process.

The mass distribution of PEC as a function of particle size
for the ambient air sample (Fig. 1c) was similar in form to
typical distributions of benzo(a)pyrene (BaP) with respect to
particle size (Fig.1d) in Rio de Janeiro aerosols (Miguel, 1982).
This observation is consistent with an adsorption mechanism for
the PAH-elemental carbon association.

Acknowledgements

The authors thank Susanne V. Hering for collecting the
tunnel sample and for her invaluable help in the collection of the
others. The ready cooperation of James M. Kelly, supervisor of
facilities, maintenance and technical services of TWA at LAX is
greatly appreciated.

This research has been funded in part by the National
Research Council of Brazil (CNPq), FINEP(Brazil), and the U. S.
Environmental Protection Agency under Grant CR-810455-01, and does
not necessarily reflect the view of the agency and no official
endorsement should be inferred.

References

Harrison, R.M, and Pio, C.A. (1983) "Size Differenciated
Composition of Inorganic Atmospheric Aerosols of Both Marine and
Polluted Continental Origin", Atmos.Environ. 17,1733.

Hering, S. V., Flagan, R. C., and Friedlander, S. K. (1978)

410

"Design and Evaluation of a New Low-Pressure Impactor I". _Environ. Sci. Technol._ _13_, 667.

Hering, S. V., Friedlander, S. K., Richards, L. W., and Collins, J. J. (1979) "Design and Evaluation of a New Low-Pressure Impactor II." _Environ. Sci. Technol._ 13,184.

Johns, W., and Wall, S. (1981) "Size Selective Sampler for Particulate Monitoring in California". Air Industrial Hygene Lab., California State Dept. Health Services, _Technical Report No. CA/DOH/AIHL/SP-24_, October.

Miguel, A. H. (1982) "On the Determination of Benzo(a)Pyrene in Atmospheric Aerosols Fractionated by a Low Pressure Impactor, by TLC Separation and in situ Spectrophotofluorometry at the Picogram Level". _Intern. J. Environ. Anal. Chem._ ,12,17.

Miguel, A. H. (1984) in preparation.

Turco, R. P., Toon, O. B., Ackerman, T.P., Pollack, J. B., and Sagan, C. (1983) "Nuclear Winter: Global Consequences of Multiple Nuclear Explosions". _Science_ ,222,1283.

Wolff, G. T., Groblicki, P. J., Cadle, S. H., and Countess, R. J. (1982) "Particulate Carbon at Various Locations in the United States". _Particulate Carbon: Atmospheric Life Cicle_ ,(Edited by G. T. Wolff and R. L. Klimisch), Plenum Press, New York.

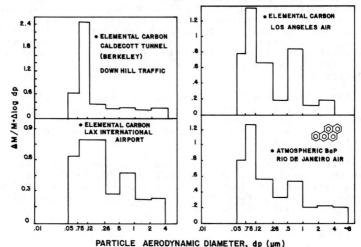

Figure 1. Mass distributions of PEC and benzo(a)pyrene as a function of particle size. PEC loadings,ΔM,($\mu g/m^3$): 34.3, 14.4, and 2.13, respectively for 1a, 1b, and 1c. BaP loading,ΔM=1.62 ng/m^3.

Published 1984 by Elsevier Science Publishing Co., Inc.
Aerosols, Liu, Pui, and Fissan, editors

THE SCENES PROGRAM

Prem Bhardwaja, Ray Kelso,
Bill Malm, Peter Mueller and Marc Pitchford

Rob Farber
Southern California Edison
Rosemead, CA 91770

EXTENDED ABSTRACT

     The SCENES program is a multi-year visibility research
monitoring program which is addressing the layered and regional
haze problem in the southwestern U.S. The program is unique in
many aspects including the cooperative nature of the study.
SCENES is an acronym for Subregional Cooperative Electric
Utility, National Park Service, and Environmental Protection
Agency Study. Electric utilities participating in this study
are Salt River Project, Southern California Edison and Electric
Power Research Institute. The other participant is the
Department of the Defense. The goal of SCENES is to establish
the relative contributions of various source categories to
atmospheric aerosols in the desert southwest, with emphasis on
the national parks and recreation areas in Arizona and southern
Utah and restricted airspace in the western Mojave desert.

     To achieve this goal each participant is operating
monitoring sites in their area of interest. Over a dozen sites
comprise the network which stretches from the western Mojave
desert in California southward to central Arizona and north and
eastward into southern Utah. Generally, these sites are very
isolated, designed to monitor the regional atmospheric constituents in terms of both optical and aerosol measurements.
Additionally, they have been carefully selected with the idea of
monitoring layered hazes and corridors of various long-range
transport plumes.

     In general, because of the research nature of visibility
monitoring, new instruments have been developed for SCENES to
reflect both the latest technology, and because of the length of
the monitoring program, instruments are automated whenever
possible. For example SCE and NPS have developed automated
teleradiometers. Both SCE and NPS are field testing two types
of transmissometers. The latest in aerosol sampling technology
is also being deployed in the network. Isokinetic sampling
devices are being used which have precisely measured cuts at
2.5 µm Dp and 10 or 15 µm Dp. These devices are automated,
operating for several days unattended. The participants are

also evaluating three different types of small submicron
particle multi-size cut aerosol samplers. Also included at the
various monitoring sites are meteorological instruments,
integrating nephelometers specially designed with an LQL of one-
fiftieth of Rayleigh light scattering coefficient, and camera
equipment. The camera is recording the sky conditions and in
many cases will serve as a record for future perception studies.

Because many participants are using different optical and
particulate collection devices, an extensive intercomparison and
equivalency program is currently underway. Criteria are being
defined and all those instruments deployed in the field will
meet these criteria. In general, there are intercomparison
studies for the various teleradiometers and aerosol samplers.
Because of the research nature of visibility and aerosol
monitoring devices, newly developed instruments which meet these
criteria will be deployed, and in some cases replace the
existing instrumentation. Various participants are also
conducting several special studies in the context of SCENES.
These include determining the most appropriate method for
inherent contrast (C$_O$) of a target, the importance of water
associated with hygroscopic aerosol, examination of appropriate
measurement techniques for volatile compounds, including
nitrates and organics, developing methods for determining
changes in scene color/texture from changes in air quality,
techniques for measuring particles >15μm Dp.

Field measurements which began in 1984, will continue for a
minimum of five years. This amount of time is necessary in
order to establish a statistically significant data base on
aerosol and visibility climatology in the region. The first
year is devoted to exploratory sampling. During this period the
various sampling devices are being tested and the data is being
analyzed to determine the necessary parameters to be sampled
during the remaining years. During the first year daily samples
will be obtained, but chemical analyses will be performed every
third day. Every third day was selected to randomize the data
meteorologically. A fixed time interval was selected in order
to statistically examine the lead lag relationships of the many
variables affecting visibility. Analyses will include nitrates,
sulfates, total and elemental carbon, elements, size speciated
sulfates, ammonium and visibility parameters such as the
radiance at three wavelengths and the three luminance response
values, etc.

The following years will include this routine sampling plus
intensive measurements. Intensives will last for up to one
month and will occur four times a year, once each season. The
purpose of these intensive periods is to conduct source
apportionment analyses on a case study basis. In addition to
the routine measurements, there will also be extensive size
distribution measurements and an array of pibals for airmass
trajectory analyses. Tracers unique to smelters (e.g.,

arsenic and molydenum) power plants (e.g., cenospheres, mullite and total fluorides) vegetive emissions (e.g., terpenes), soils (e.g., silicon) and urban areas (e.g., fluorocarbons) will be sampled and subsequently analyzed. During intensive periods, daily analyses will be conducted for light extinction budgets, trajectory analysis, mass budgets, and source apportionment.

At the conclusion of this study, the participants hope to have a firm understanding of the causes, source categories and effects of both regional and layered hazes in the southwestern U.S. The frequency, spatial extent and duration of those hazes will also be examined. In addition, how regional and layered hazes relates to human perception will also be quantified during the study.

Published 1984 by Elsevier Science Publishing Co., Inc.

Aerosols, Liu, Pui, and Fissan, editors

414

ABSTRACT

Western Regional Air Quality Studies:
Accuracy and Precision of Atmospheric Aerosol and Visibility Measurements

C.V. Mathai, D.H. Bush, R.C. Nininger, D.W. Allard
and I.H. Tombach

AeroVironment Inc.
145 Vista Avenue
Pasadena, California 91107

An eleven-station network called "Western Regional Air Quality Studies (WRAQS): Visibility and Air Quality Measurements" was operated during 1981 and 1982 in rural areas of nine Western States under the sponsorship of the Electric Power Research Institute. Visibility data were obtained from teleradiometers and integrating nephelometers. Fine and coarse aerosol data included mass concentrations and their chemical components. This paper is concerned with determination of accuracy and precision of WRAQS visibility and aerosol data.

Uncertainties in the aerosol data depended on sample volume measurements, maintaining the integrity of the sample between sampling and analyses, and the laboratory analytical techniques. Accuracy of these data was calculated using information from repeated calibrations, performance audits and analyses of "controls." The precision and lower quantifiable limits for various analytical techniques were determined from replicate analyses of sample and blank filters. A number of special experiments were conducted to quantify factors relating to sample integrity.

Accuracy and precision of visibility data were derived from repeated calibration and performance audit data. Specifically, accuracy of teleradiometer radiance and contrast measurements were obtained using a 'contrast standard' and a reference teleradiometer. Evaluations of the teleradiometer measurements of several distant targets along a viewing path show that the precision of these data varied from site to site depending on the skill of the operating technician. Teleradiometer flare was found to introduce errors of 10% or more in visibility data for less than 1% of the time. Details of these evaluations will be presented in the paper.

Published 1984 by Elsevier Science Publishing Co., Inc.

Aerosols, Liu, Pui, and Fissan, editors

THE OPTICAL EFFECT OF FINE-PARTICLE CARBON ON URBAN ATMOSPHERES

L. Willard Richards
Sonoma Technology, Inc., Santa Rosa, CA 95401 USA

Robert W. Bergstrom, Jr.*
Systems Applications, Inc., San Rafael, CA 94903 USA

Thomas P. Ackerman
NASA Ames Research Center, Moffett Field, CA 94035 USA

EXTENDED ABSTRACT

The effects of fine-particle carbon, such as diesel soot, on the
optical properties of urban haze in the visible wavelength range were
explored to determine the dominant effects and to see if simple parameters
(such as visual range in green) provide an adequate measure of these
effects. It is known that fine-particle carbon absorbs more strongly in
the blue than in the red, so that when it is mixed with a white pigment,
the resulting gray can appear somewhat brown. Detailed radiative transfer
calculations were done to investigate the possibility of a similar effect
in urban hazes.

Figure 1 is a chromaticity diagram showing the sky colors calculated
for a viewer on the ground looking upward through a 1 km thick haze layer
with each of the specified compositions. The view elevation angles are
indicated, and the angle of 1.5^o in the plane-parallel geometry of the
calculation corresponds to a horizontal view on the round earth. The solar
zenith angle in these calculations was 13°. The atmospheric composition
above the haze layer is represented by four layers containing air and
non-absorbing fine aerosol. In Figure 1, blue is to the lower left and
yellow to the upper right. The solid line is derived from experimentally
measured sky colors, which typically fall quite close to the line.

The data in Figure 1 show that the sky colors produced by mixtures of
non-absorbing and fine-particle carbon aerosol when the sun is overhead can
also be produced by higher concentrations of non-absorbing aerosol alone.
Thus, yellowing of the midday sky by fine-particle carbon appears to be no
more of a problem than the yellowing of the sky by non-absorbing aerosol,
except that a given mass concentration of carbon causes a larger effect.

Figure 2 shows data for the horizon sky brightness determined by the
radiative transfer calculations. These results show that absorbing
aerosols make the horizon sky darker, and cause a qualitative change in the
radiance profile near the horizon. Non-absorbing aerosol causes the sky to
become lighter as the horizon is approached. These effects should be
included in computer-generated images showing the visual effect of adding
absorbing aerosol to the atmosphere.

*Current address: McCutchen, Doyle, Brown, and Enerson, San Francisco, CA
94111 USA.

416

A simple formula for the radiance of the horizon sky $I_s(\Omega)$ is derived under the assumption that the upwelling F_+ and downwelling F_- radiant fluxes are diffuse, but not necessarily equal:

$$I_s(\Omega) = \frac{\omega_0}{4\pi} \quad [2 F_+ + 2 F_- + F_0 P(\Omega,\Omega_0)]$$

Here F_0 is the solar flux, ω_0 is the single scattering albedo, $P(\Omega,\Omega_0)$ is the light scattering phase function, Ω the direction opposite the direction of view, and Ω_0 the direction of the solar flux. At wavelengths or aerosol

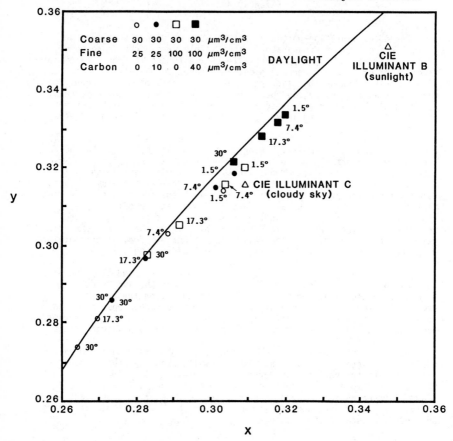

Figure 1. CIE tristimulus values at four view elevation angles for four haze layer compositions.

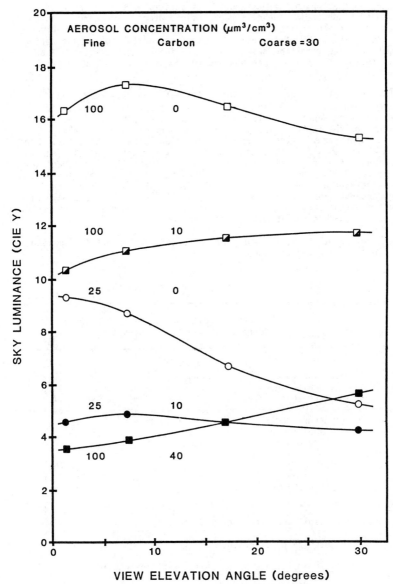

Figure 2. Sky luminance as a function of view elevation angle for five haze layer compositions.

concentrations at which the horizon sky is optically thick, this equation can be used to determine the effect of various atmospheric components on the radiance of the horizon sky. For the case represented by open circles in Figure 1, about 30% of the horizon brightness is due to the scattering of light reflected from the ground into the line of sight, about 30% due to the scattering of skylight, and 40% due to the scattering of direct sunlight.

Fine-particle carbon is a highly efficient light absorber, and thus has large absorption and extinction cross sections per unit mass. The extinction cross section per mass of fine-particle carbon is three or four times that of the typical fine-particle component of the atmospheric aerosol. Therefore, carbon is very effective at limiting the visual range. Table 1 shows results derived from the radiative transfer calculations for the contrast of distant white and black objects. The limnal contrast was taken to be ± 0.02. The results show that white objects can be seen at a somewhat greater distance than black objects for the range of ambient carbon concentrations considered. The convention of using black objects to define the visual range may therefore slightly overemphasize the effect of fine-particle carbon on visibility in urban hazes.

Acknowledgement --- This work was supported by the CAPA-13 Committee of the Coordinating Research Council.

Table 1. Visual range for green light.[a]

Fine aerosol concentrations ($\mu m^3/cm^3$)	Carbon aerosol concentrations ($\mu m^3/cm^3$) and corresponding visual ranges (km)				
25	Carbon conc.[b]	0.	2.5	5.	10.
	Visual range	24.0	19.4	16.3	12.3
	White object dist.[c]	28.7	24.6	21.4	17.1
50	Carbon conc.	0.	5	10.	20.
	Visual range	13.6	10.6	8.8	6.5
	White object dist.	15.1	12.9	11.2	8.9
100	Carbon conc.	0.	10.	20.	40.
	Visual range	7.2	5.6	4.6	3.4
	White object dist.	6.7	6.1	5.4	4.4

a. The coarse aerosol concentration is 30 $\mu m^3/cm^3$.

b. For black object (zero reflectance).

c. Maximum distance at which a white object (90% reflectance) can be perceived.

Published 1984 by Elsevier Science Publishing Co., Inc.

Aerosols, Liu, Pui, and Fissan, editors

THE DETERMINATION OF INHERENT CONTRASTS TO BE USED
IN THE REDUCTION OF TELERADIOMETER MONITORING DATA

L. Willard Richards and Nancy L. Alexander
Sonoma Technology, Inc.
3402 Mendocino Ave.
Santa Rosa, California 95401
U. S. A.

William C. Malm
National Park Service
Air and Water Quality Division
301 South Howes
Fort Collins, Colorado 80521
U. S. A.

EXTENDED ABSTRACT

Contrast teleradiometers have been widely used for monitoring visibility
in the western United States, and several additional studies which will use
these instruments are being planned. The physical quantity measured is the
contrast of a distant target such as a mountain or a ridge against the horizon
sky, where contrast = (target radiance/sky radiance) - 1. Measurements are
typically made at several visible wavelengths, but it is common to reduce only
data for the green wavelengths. The calculation of the visual range or the
atmospheric extinction coefficient from the measured contrast requires
selecting a value of the inherent contrast, which is the contrast of the
target against the horizon sky at the target. The visual ranges and
extinction coefficients calculated from the measured contrasts are sensitive
to the inherent contrasts selected.

Data from one brief and two extended monitoring programs have been
examined to explore methods for extracting values for the inherent contrasts
of the targets from the monitoring data themselves. The general approach was
to use one of several methods for selecting very clean days, when the dominant
cause of extinction is the scattering of light by air molecules. The
magnitude of this component of extinction is well known, so the total
extinction coefficient of the atmosphere can be estimated reasonably reliably.
Then the data reduction calculation can be reversed to determine the inherent
contrasts from the estimated extinction coefficients, the measured contrasts,
and the sight path geometry.

When integrating nephelometer data were available, readings of light
scattering by particles were used to select the clean days. When only
teleradiometer data were available, computer searches were used to find days
in which the daily average visibility (calculated from all teleradiometer
measurements on one day at one monitoring site) was very good at the
monitoring site and at several neighboring monitoring sites. It is likely
that on these days a large clean air mass surrounded the monitoring site.
Cloudy days were omitted from consideration.

On the selected clean days, the second most important source of extinction in the sight path is light scattering by particles, which can be 1/3 or less of the scattering by air. The particle scattering over the sight path was estimated from integrating nephelometer data when available. Otherwise, it was estimated from an average of the extinction coefficients determined from all teleradiometer targets at that site. When available, data from neighboring monitoring sites were considered when estimating the extinction coefficients.

The procedure of using monitoring data derived from assumed inherent contrasts to calculate extinction coefficients and then to use these extinction coefficients to calculate inherent contrasts is somewhat circular. However, the reliable selection of clean days minimizes the magnitude of the particle scattering and hence the importance of the value assumed. Averaging the input extinction data yields a set of consistent inherent contrasts for all targets at a monitoring site.

Numerical data will be presented showing how well the inherent contrasts determined in this study agreed with those used in the monitoring programs, and the effects of differences in the inherent contrasts on the monitoring program results.

A general procedure for determining inherent contrasts for use in future manual teleradiometer monitoring programs is recommended. A key element is having the observer use the apparent texture of distant targets to select very clean days and to estimate the atmospheric extinction coefficient on those days. If there are no clouds in the sun-target-observer plane and no other non-standard conditions, the measured contrasts on these days can be used to determine the inherent contrasts of the targets.

Published 1984 by Elsevier Science Publishing Co., Inc.
Aerosols, Liu, Pui, and Fissan, editors

FORMATION OF LIGHT SCATTERING AEROSOL IN SMELTER AND POWER PLANT PLUMES

L. W. Richards, J. A. Anderson, D. L. Blumenthal, J. A. McDonald
Sonoma Technology, Inc., 3402 Mendocino Ave., Santa Rosa, CA 95401, USA

S. V. Hering
Chemical Engineering, University of California, Los Angeles, CA 90024, USA

William E. Wilson, Jr.
US Environmental Protection Agency, Research Triangle Pk., NC 27711, USA

EXTENDED ABSTRACT

Concern about the effects on visibility in the National Parks of
increased electric power generation near coal fields in the desert Southwest
led to the development of mathematical models to estimate the near-source
optical effects of plumes. The EPA sponsored VISTTA study (Visibility
Impairment due to Sulfur Transport and Transformation in the Atmosphere)
obtained field data to evaluate plume optics models. This paper summarizes
and comments on the results of airborne measurements during the VISTTA field
programs of the formation of light scattering aerosol in smelter and power
plant plumes. The plume sampling included: in Arizona, a coal-fired power
plant in summer and winter and two smelters in the summer, and in the Midwest,
two coal-fired power plants in the winter and a different one in the summer.

The light scattering by particles b_{sp} was measured with a
three-wavelength nephelometer, aerosol size distributions were measured with
optical particle counters and an electrical aerosol analyzer, and cascade
impactors were used to measure the particle size distributions of several
elements, especially sulfur. Richards et al. (1981) give additional
information on the experimental procedures.

Significant differences in the aerosol formation rates and size
distributions were observed. Wilson (1981) has reviewed SO_2 to sulfate
conversion rates in plumes and has shown that most data scatter around a rate
of 3.7% per kW h m^{-2} of solar radiation dose, which corresponds to about 3.7%
h^{-1} at midday in the summer. Conversion rates in the Arizona smelter plumes
were close to this value. Winter conversion rates per kW h m^{-2} in the Midwest
plumes were about three times smaller, and summer and winter rates in the
Arizona power plant plume were ten times smaller.

It is believed that the dominant factor controlling the near-source rates
is the ozone concentration in the plume, which in turn is determined by the
rate of plume dilution. Data from this and other studies indicate that the
relative humidity also plays a significant role, with faster conversion rates
at higher humidities. Smelters emit little NO, so ozone concentrations are
not depressed near the source and sulfur oxidation begins immediately. These
general conclusions have been confirmed by model calculations (personal
communication, C. Seigneur and D. A. Latimer).

The regression slopes in Table 1 give the plume excess b_{sp} normalized to
plume excess NO_x or SO_2 concentrations of 1 ppm at the specified downwind
distance. The values observed in the Navajo Generating Station plume in
Arizona are consistent and low. The negative alpha (which specifies the

Table 1. Normalized plume-excess aerosol light scattering coefficients.

Plant & date	Tape/Pass	Distance km	Regression slopes vs. NOx bsp per ppm NOx (Mm⁻¹ ppm⁻¹) [a]			Regression slopes vs. SO2 bsp per ppm SO2 (Mm⁻¹ ppm⁻¹)			alpha=Δlog bsp/Δlog λ [b]	
			Blue	Green	Red	Blue	Green	Red	Blue-Green	Green-Red
Navajo										
27 Jun 79	66/3	29	27	29	32	18	20	22	-0.51	-0.44
9 Jul 79	78/2	32	82	89	97	52	56	62	-0.45	-0.37
13 Jul 79	82/2	59	46	50	56	21	23	25	-0.35	-0.49
5 Dec 79	93/1	32	41	44	46	32	35	37	-0.40	-0.18
9 Dec 79	96/2	29	39	44	48	37	42	46	-0.64	-0.36
12 Dec 79	99/9	45	64	68	74	49	52	56	-0.33	-0.34
13 Dec 79	101/2	25	40	43	46	25	28	29	-0.46	-0.24
Kincaid										
25 Feb 81	225/2	60	1200	900	570	420	340	2		
La Cygne										
1 Mar 81	223/12	32	2100	1400	760	1900	1300	690	2.2	2.7
	223/18	58	500	260	74	530	270	58	3.4	5.5
	226/26c	100	480	140	-46	610	210	-3	6.4	---
Labadie										
23 Aug 81	278/1	5	450	260	130	130	78	39	2.8	3.0
30 Aug 81	282/1c	40	150	34	-15	40	6	-7	7.9	---
2 Sept 81	283/1	30	1800	1300	750	500	290	110	1.6	2.4
3 Sept 81	284/1	25	240	140	46	50	30	10	2.9	4.8
San Manuel										
6 Sept 81	285/1	10	---	---	---	1395	1037	619	1.56	2.23
	285/3	13	---	---	---	1871	1360	806	1.68	2.23
	285/5	14	---	---	---	2590	1851	1067	1.77	2.38

a. 1 Mm = 10^6 meters.
b. Calculated from slopes vs. NOx before rounding.
c. The bsp values determined from the regression slopes for this pass do not fall in a reasonable range.

wavelength dependence of b_{sp}) indicates the plume excess b_{sp} is mostly due to fly ash particles larger than 1 μm. This and other data indicate that very little aerosol was formed in the Navajo plume in the particle size range most effective for scattering light. The data from the Midwestern power plants are more variable. The San Manuel smelter data show that b_{sp} in this plume was large and increased rapidly with downwind distance.

The data in Figures 1 and 2 show that the particles formed in the smelter plume are larger and hence more effective per unit mass at scattering light than the particles formed in the power plant plume. Figure 1 shows the total (plume excess plus background) aerosol size distribution measured 89 km and 4.5 h downwind in the Navajo plume. In this case only 1.6% of the plume excess sulfur was sulfate. Plume excess aerosol was detectable only at particle sizes smaller than about 0.1 μm, and the geometric mean diameter D_{gv} of the background accumulation mode aerosol was < 0.2 μm. Figure 2 shows the size distribution of the plume excess aerosol in the San Manuel plume for which D_{gv} > 0.3 μm. The fraction of the plume excess sulfur as sulfate was 3.5% at 10 km and 1.5 h downwind and 7.7% at 14 km and 3.1 h. Sulfate was formed more rapidly in the smelter plume and the resulting particles were larger and much more effective at scattering light. Comparison with theoretical models indicated that the particles in the smelter plume grew by condensation rather than liquid phase reactions. Even if all particles were liquid, the liquid volume was about 1000 times too small for the observed sulfate formation to be accounted for by liquid phase reactions. Iron and manganese concentrations were measured and catalysis by these species included in the calculations.

The values of D_{gv} for the plume excess accumulation mode aerosol in the Midwestern plumes fell in the range between 0.1 μm and 0.3 μm in agreement with prior results.

It is difficult to see the particles formed in plumes in a hazy environment because the plumes disappear in the haze close to the source. In the Midwestern summer experiments, the background b_{sp} ranged from 50 to 300 Mm^{-1} and the Labadie plume could not be seen during the hazier periods. When a wildfire increased the background b_{sp} to 110 Mm^{-1} in Arizona, the Navajo plume could not be seen from the ground observation locations.

Although the research described in this abstract has been funded in part by the United States Environmental Protection Agency through contract 68-02-3225, it has not been subjected to Agency review and therefore does not necessarily reflect the views of the Agency and no official endorsement should be inferred.

REFERENCES

Richards L. W., Anderson J. A., Blumenthal D. L., Brandt A. A., McDonald J. A., Waters N., Macias E. S. and Bhardwaja P. S. (1981) The chemistry, aerosol physics, and optical properties of a western coal-fired power plant plume. Atmospheric Environment 15, 2111-2134.

Wilson W. E., Jr. (1981) Sulfate formation in point source plumes. Atmospheric Environment 15, 2573-2581.

424

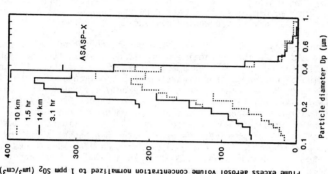

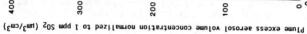

Figure 2. Plume excess aerosol size distribution measured 6 September 1981 in the San Manuel smelter plume.

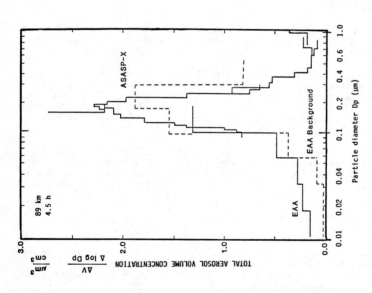

Figure 1. Total (plume excess plus background) aerosol size distribution measured 13 July 1979 in the Navajo Generating Station plume.

Published 1984 by Elsevier Science Publishing Co., Inc.
Aerosols, Liu, Pui, and Fissan, editors

INCORPORATION OF A HUMAN VISUAL SYSTEM MODEL
IN THE PLUME VISIBILITY MODEL PLUVUE

Ronald C. Henry
Environmental Engineering Program
Civil Engineering Department
University of Southern California
Los Angeles, CA   90089-0231
U.S.A.

John F. Collins
Environmental Research & Technology
975 Business Center Circle
Newbury Park, CA   91320

EXTENDED. ABSTRACT

Introduction

The goal of protecting visibility in Federal Class I areas conflicts
to some extent with the national need to develop energy resources in the
American West. Quantitative measures of visibility are absolutely
necessary in an analysis of the benefits of improved visibility. Review
of current visibility indices (visual range, ΛE, plume to sky contrast,
blue-red ratio) indicated that they are all inappropriate and inadequate
for predicting or assessing plume perceptibility under typical conditions
in the West (Henry and Collins 1982). The primary flaw in all these
indices is their inability to incorporate in their calculation any
consideration of size, shape or other spatial characteristics of the plume
that are known to greatly influence perception. Application of current
visibility indices is also unwarranted because none have been shown to
apply to perceptibility of plumes in the natural environment. All are
based on extensions of laboratory studies and their predictions have not
been verified against on-site measurements of humanly perceived brightness
and color.

This study concludes that modern visual system models can be used to
improve predictions of plume perceptibility by including spatial effects
ignored by existing indices. An example calculation of plume perceptibility
for a 2250 MW coal-fired power plant showed that the effect of spatial
characteristics such as size and diffuse edges is to lower predicted plume
perceptibility by a factor of three from a level that was very perceptible
to the perceptual threshold. Several versions of a model that incorporates
the spatial and color response of the eye have been published. The model
of Faugeras (1979) has been evaluated in terms of human perception and has
been designed to be compatible with the existing C.I.E. methodology. For
these reasons, the Faugeras model was chosen for application in a plume
visibility model. In this model, C.I.E. colorimetric data and spatial
information are processed through several stages to simulate the response
of the human visual system. These stages are:

426

- conversion of brightness and chromaticity data to human eye receptor responses

- log transformation of the receptor response

- Fourier transformation of the spatial image into the frequency domain

- weighting the frequency domain image by the spatial frequency response function of the human visual system

- inverse Fourier transformation of the weighted frequency domain image back to the spatial domain

- calculation of perceptability measure with the resulting spatially filtered image data.

Application

The model briefly described above was implemented in the plume visibility model PLUVUE. The PLUVUE model calculates the perceptibility of plumes from smokestacks based on standard Gaussian modeling of plume dispersion and chemistry, simplified radiometric calculations, and standard C.I.E. colorimetric methods (Johnson et al 1980). The plume perceptibility values calculated by PLUVUE are based on a view through the plume centerline. The spatial extent and distribution of plume color and brightness effects are ignored by the standard PLUVUE. We have extended the PLUVUE model to calculate plume effects for views ranging over a ten degree vertical scan of the plume. The spatial information in this ten degree scan is the starting point for processing by the Faugeras model. The vertical scan of plume effects is plotted both before and after spatial filtering has been applied. This allows direct evaluation of how the spatial extent and distribution of a plume affects the plume's perceptibility.

The modified model was applied to a sample case from the PLUVUE users manual. The test case is characteristic of a 1600 MW power plant operating in the American Southwest. The effect of the diffuse distribution of the plume is to reduce its perceptibility. Reduction in perceptibility ranged from 13 to 50 percent depending the angular extent of the plume and choice of spatial filtering model. Work remains to be done to quantitatively relate human visual system model predictions to field observations of pollutant plumes.

References

Faugeras, O.D. (1976). Digital Color Image Processing and Psychophysics within the Framework of a Human Visual Model. Ph.D. dissertation. Dept. of Computer Science, University of Utah, Salt Lake City, UT.

Johnson, C.D., D.A. Latimer, R.W. Bergstrom and H. Hogo. 1980 "User's Manual for the Plume Visibility Model (PLUVUE)."

Henry, R.C. and J.F. Collins 1982. "Visibility Indices: A Critical Review" prepared for Western Energy Supply and Transmission Associates, ERT Doc. No. P-A771.

CHAPTER 10.

CHEMICAL CHARACTERIZATION

Published 1984 by Elsevier Science Publishing Co., Inc.

Aerosols, Liu, Pui, and Fissan, editors

IN-SITU MEASUREMENT OF PARTICLE $SO_4^=$, NH_4^+, AND CONTRIBUTION TO SCATTERING BY WATER USING A MODIFIED INTEGRATING NEPHELOMETER

A. P. Waggoner, R. E. Weiss, T. V. Larson
Civil Engineering Department
University of Washington
Seattle, WA 98195

INTRODUCTION

Ambient particles in the rural eastern US are typically 50% sulfate and the sulfate molecular form varies from sulfuric acid to ammonium sulfate. The molecular form affects the exchange of water molecules between the particle and its surroundings. The effects of particles on visual range or human health will also depend on molecular form either through composition related effects or because the relation of hydrated size to relative humidity (RH) is a function of composition. The molecular form also may contain information about the chemical formation mechanism and/or the transport history of the air mass. However, variation in particle concentration and composition may take place on a shorter time scale than can be resolved by traditional particle sampling on filters for later chemical or gravimetric analysis.

We have operated a set of instruments in rural Virginia and urban Houston that use an integrating nephelometer to measure particle scattering to determine change in particle diameter in response to a change in either sample air RH or to upstream heating and cooling. These techniques allow us to infer the fine particle mass, the sulfate ion concentration and the molar ratio of sulfate to ammonium ions by an in-situ measurement with response time of a few minutes. We found the particles to be largely sulfate and to be more acid in the afternoon and less acid at night. In Houston we found the aerosol to retain liquid water at an RH below that which would dry the particles at equilibrium conditions.

THEORY

Hydrated particle diameter is a function of sample air relative humidity and can be altered by evaporation/decomposition of particle mass at elevated sample air temperature. A particle of ammonium sulfate in equilibrium with the partial pressure of water molecules changes in diameter in response to change in relative humidity as shown in Figure 1a. For increasing RH, an initially dry particle of ammonium sulfate changes to a saturated solution droplet at its deliquescent RH, about 79%. As RH increases above 79%, the droplet increases in size and becomes more dilute. When the RH of air surrounding the droplet is reduced, the droplet loses water and decreases in size. The particle can retain liquid water at an RH<79% forming a supersaturated solution droplet with respect to dissolved salt concentration but with a partial pressure of water in equilibrium with its surroundings. The behavior of diameter versus RH is shown in Figure 1b-d to depend on molar ratios of ammonium to sulfate ions.

Two systems of instruments were used in our measurements. The "Humidograph" system first exposes particles to RH<30% to evaporate water associated with ammonium sulfate particles, then exposes the particles to a test RH that is scanned over several minutes from 30 - 90%. A nephelometer senses change in diameter by measurement of particle scattering extinction as a function of test RH. The Humidograph system allows measurement of the role of liquid water associated with the particles in terms of impact on extinction over a range of test RH. The system can detect ammonium sulfate by change in slope of scattering versus RH at RH=79%. Acid sulfate aerosol is detected by adding 100 ppb ammonia gas to the sample air stream to convert acid sulfate to $(NH_4)_2SO_4$.

The "Thermidograph" exposes particles to RH>80%, then heats particles and sample air to a peak temperature scanned in time from ambient to 380 C, and finally cools sample air to approximately ambient temperature and RH 65-70% for measurement of particle scattering. Data is a plot of scattering versus peak temperature, shown in Figure 2A & 2B. The trace is interpreted as follows (Larson et al., 1982). The particles enter the heater as droplets at RH>80% and are exposed to a peak temperature that is increased from ambient to 380 C over 3-4 minutes. The scattering at (a) is that of particles plus water of hydration at RH=65% in the nephelometer. The drop, (a)-(b), is the change in scattering due to loss of liquid water when peak temperature is sufficient to dry the entering droplet. The particle remains in the dehydrated condition because RH in the measurement nephelometer is not high enough to reform a droplet. The scattering (b) to (c) is flat because the dry particle is not altered by temperatures up to about 175 C. The scattering increases, (c) to (d), as ammonium sulfate cracks to ammonium acid sulfate with loss of ammonia. Ammonium acid sulfate is hydrated at RH=65% in the measurement nephelometer. The decrease in scattering, (d) to (e) occurs because the ammonium acid sulfate decomposes at temperatures above 200 C. The decomposition products either condense on the walls of the cooler or form sub-optical size particles. In either case, the decomposition removes particle mass from the size region of efficient optical scattering. The Thermidograph trace can be interpreted using the marked points on Figure 2B as follows: (a) minus (b) equals scattering by particulate water; (b) equals scattering by dry particles; (a) minus (d) equals scattering by non-water volatiles; (d) equals scattering by ammonium sulfate plus water (alternate runs have ammonia gas added to reach concentration of 100 ppb) and (e) equals scattering by refractory materials such as silica, inorganic carbon or sea salt.

MEASUREMENTS

The humidograph and thermidograph systems were operated in a program sponsored by EPA at a rural site in Shenendoah National Park, VA and in urban Houston, TX in August and September of 1980. Fine particle mass and PIXE atomic analysis on 8 hour samples, particle scattering at ambient RH and at RH<30% were measured in addition to operating the humidograph and thermidograph systems. We compared the thermidograph determinations of sulfate and sulfate/ammonium ion ratio to that determined from analysis of filter samples. We also compared three measures of the contribution of scattering by particulate water: 1) the increase in scattering extinction as a function of RH measured in the humidograph; 2) the thermidograph measurement of scattering by particulate water and 3) the ratio of scattering extinction as measured by the nephelometer when alternately operated at ambient RH and at RH<30%.

RESULTS

The instrument and results are described in detail by Waggoner et al. (1983). At both the Virginia and Texas sites, comparison of fine particle mass concentration to dry particle scattering extinction found these two parameters highly correlated R = 0.96 and in the ratio of scattering to mass concentration was 4.2 and 4.4 meters squared per gram at Virginia and Texas, respectively.

Using the measured scattering extinction efficiency and the analysis of thermidograms for sulfate concentration as described earlier, we found the thermidograph sulfate and PIXE sulfur to be correlated R = 0.98 in Virginia and 0.95 in Texas. The slopes were 2.85 and 3.03 in Virginia and in Texas, respectively, nearly equal to the expected value of 3.00.

Our Thermidograph data on ammonium/sulfate was compared to measurement of the two ions from filter samples taken by T. Dzubay of EPA. Agreement was found except during periods of variable composition over the filter sampling interval.

We found RH, sulfate concentration, ammonium/sulfate ratio and particle scattering extinction to all be highly variable in time as shown in Figure 3. Presented are: A) the hourly thermidograph and filter measured sulfate concentrations; B) the sulfate fraction of fine particle mass from thermidograph and filter measurements; C) the ammonium/sulfate molar ratio; D) the acid/salt character from the humidograph; E) the ambient and low RH nephelometer scattering extinction. Excellent agreement was found with the Thermidograph system detecting variation not resolved in the filter measurements.

The average diurnal behavior is shown in Figure 4 with the aerosol nearly chemically neutral in the evening and becoming more acid during the afternoon at both sites. These observations are consistent with either of two models: 1) the acid sulfate is formed photochemically or 2) the acid aerosol is neutralized by ammonia gas released by biological processes at the surface and composition is determined by solar driven mixing. The particles were found to be more acid in Virginia than in Texas.

Particle chemical composition will control the exchange of water with variation in RH. The deliquescence RH for particles with ammonium/sulfate molar ratios of 2.0, 1.5 and 1.0 are 79%, 69% and 40%, respectively (Tang, 1980). At ambient RH<79%, hydrated ammonium sulfate exists as supersaturated droplets; the other compounds can exist as unsaturated solutions to lower RH. We have compared the three measures of the retention of water in the particles at ambient conditions as mentioned earlier to explore interactions of RH, particle composition and particulate water in the ambient atmosphere. We were especially interested in retention of water in supersaturated droplets of ammonium sulfate. In Virginia, the particle ammonium/sulfate molar ratio averages 1.4 and acid periods and the lowest RH occur during in the afternoon. These conditions produce an aerosol in equilibrium with water at ambient RH. In Houston TX, the aerosol has higher ammonium/sulfate molar ratio and the RH is lower during the periods with high molar ratio. The composition of particles in Houston and ambient RH combine to produce conditions under which the particles could retain water in excess of that expected from equilibrium conditions. In comparison of the three measures of particulate water, we found that in Virginia, the particles are in not supersaturated in terms of water content. In Houston, the particles are supersaturated about 1/3 of the time. This result is consistent with differences between sites in terms of aerosol composition and ambient RH. We find that sulfate and ammonium ions plus water caused 50 to 70% of visibility reduction during our measurments.

REFERENCES

Larson, T. V.; Ahlquist, N. C.; Weiss, R. E.; Covert, D. S.; Waggoner, A. P.; "Chemical Speciation of H_2SO_4 - $(NH_4)SO_4$ Particles Using Temperature Controlled Nephelometry"; Atmos. Env.; 16, 1587-1590 (1982)

Tang, I. N. in "Generation of Aerosols and Facilities for Exposure Experiments",; ed. by Willeke, K.; Ann Arbor Science, Ann Arbor, (1980)

Waggoner, A. P.; Weiss, R. E.; Larson, T. V.; "In-Situ Rapid Response Measurment of $H_2SO_4/(NH4)_2SO_4$ Aerosol in Urban Houston: A Comparison to Rural Virginia", Atmos. Env., 17: 1723-1731 (1983)

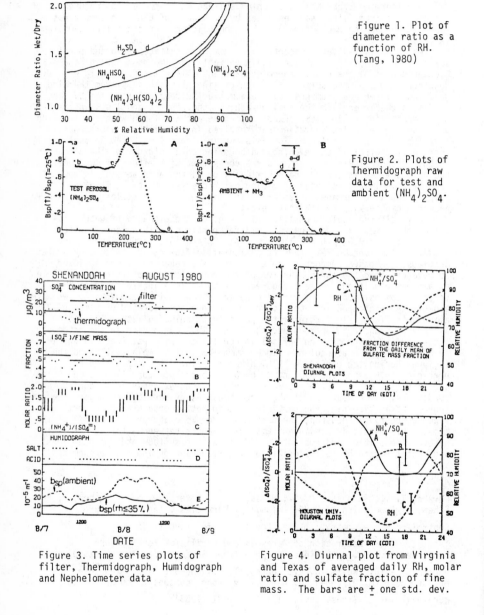

Figure 1. Plot of diameter ratio as a function of RH. (Tang, 1980)

Figure 2. Plots of Thermidograph raw data for test and ambient $(NH_4)_2SO_4$.

Figure 3. Time series plots of filter, Thermidograph, Humidograph and Nephelometer data

Figure 4. Diurnal plot from Virginia and Texas of averaged daily RH, molar ratio and sulfate fraction of fine mass. The bars are ± one std. dev.

Published 1984 by Elsevier Science Publishing Co., Inc.
Aerosols, Liu, Pui, and Fissan, editors

PHYSICOCHEMICAL MEASUREMENTS WITH SINGLE SUSPENDED PARTICLES

I.N. Tang, H.R. Munkelwitz, and Ning Wang
Brookhaven National Laboratory
Upton, New York 11973

EXTENDED ABSTRACT

Introduction

Atmospheric aerosols often contain a mixture of inorganic salts, which absorb or give off water as the relative humidity in the atmosphere changes. As a result, the nature and size of an aerosol particle are governed by the chemical and thermodynamic properties of the aqueous electrolyte solution composing the particle. A knowledge of the behavior of concentrated multicomponent electrolyte solutions is also necessary for the rational design and optimum operation of many industrial processes involving crystallization, evaporation, or phase separation. The extensive chemical and thermodynamic data required for both atmospheric and industrial applications are often not available at the present time because of the laborious efforts needed to obtain them.

Experimental

An experimental technique has recently been developed which involves the suspension of an electrostatically charged solution droplet in an environment of controlled humidity and temperature. The great stability with which the suspended particle is held in position by an AC field superimposed on the DC field makes it possible to alter the gaseous environment without disturbing the particle. Consequently, it is possible to study droplet growth, evaporation, phase transformation, or gas-particle interaction with single aerosol particles.

The technique has been successfully employed in our laboratory to investigate nucleation and crystallization in electrolyte solution droplets relevant to atmospheric processes. Extensive results obtained for several inorganic salt-water systems including NaCl, KCl and $(NH_4)_2SO_4$ demonstrate that the technique is ideally suited for measuring thermodynamic properties such as water activity, activity coefficient, partial molar enthalpy, solubility, and critical supersaturation for electrolyte solutions of all compositions. It is particularly valuable for concentrated and supersaturated solutions, for which data are extremely scarce due to conventional measurement difficulties.

The experimental results are used to fit the semi-empirical relations currently available in the literature for predicting the water activities as a function of concentration. The coefficients derived from the present work are valid for the whole concentration range up to the crystallization point, whereas in the past they were only applicable to the saturation point.

Published 1984 by Elsevier Science Publishing Co., Inc.

Aerosols, Liu, Pui, and Fissan, editors

436

IN SITU SURFACE ANALYSIS OF COATED PARTICLES

H. Burtscher°, R. Niessner* and A. Schmidt - Ott°

° Laboratorium für Festkörperphysik ETH - Hönggerberg, CH - 8093 Zürich, Switzerland
* Universität Dortmund, Abteilung Chemie, Postfach 500 500, D - 4600 Dortmund 50, F.R.G.

Extended Abstract

Introduction

Surface coating of photoelectrically inactive particles with photo-electrically active material (PAM) is a phenomenon frequently found in ambient aerosols, especially in those originating from combustion of organic material (Burtscher and Schmidt - Ott, 1984). The PAM presumably consists of polyaromatic hydrocarbons (PAH), at least in part. The technique described in the following yields coating parameters and is applied to combustion aerosols as well as artificially PAH - coated particles.

Experimental Setup

A set - up applied is shown in Fig. 1. After all charged particles have been removed in the electrostatic filter EF 1, the aerosol is irradiated with the full spectrum of a high pressure Xe flash lamp. Sub-sequently, the particles charged by photoemission are removed in EF 2 and the size spectrum of the remaining aerosol is determined in an electrostatic aerosol analyzer (EAA). The difference between the spectrum with and without light exposure leads to the fraction N^+/N of photo-electrically charged particles for each size class D_p. The total energy W of light impinging on the particles is varied either by grey filters or by the number of pulses applied.

N^+/N well followed the following W - dependence for various aerosols :

$$I \quad N^+/N = A (1 - e^{-W\varepsilon})$$

Here, A is the fraction of particles chargeable and ε is the photoelectric activity of the particles with respect to the lamp spectrum. Equation I follows

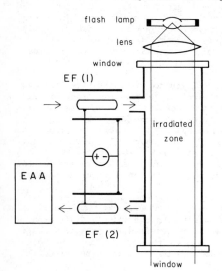

Figure 1. Experimental Setup

from $dN^+/dt \propto N_0$ (Burtscher and Schmidt – Ott,1984), N_0 is the number of uncharged particles. For PAM – coated particles, A represents the fraction of particles having provided at least one condensation nucleus for the PAM, and ε is a measure for the surface area of PAM.

By fit of the measured N^+/N (D_p) to formula I, ε and A can be determined. For all aerosols from combustion of organic material the measurements led to A = 1 and $\varepsilon \approx$ const $\cdot D_p^x$, with $1 \lesseqgtr x \lesseqgtr 2$. For an extremely O_2-deficient butane flame one obtained x = 2, i.e. the emission per unit surface area of all particles is totally covered with the same mixture of PAM. In general, however, the distribution of PAM on the particle surface shows to be size-dependent, i.e. x < 2, indicating partial coverage of the surface. For auto-exhaust aerosols, for instance one obtains x ≈ 1.

Besides the described technique, another set-up was applied to combustion aerosols : The aerosol first passes through a differential mobility analyzer (DMA) and subsequently the aerosol photoconductivity (Schmidt – Ott and Federer, 1981) is measured. The results obtained were the same as above. Additionally, the photo – threshold of the particles can be measured by this method.

The technique of Fig. 1 was further applied to NaCl particles produced in an air jet solution nebulizer/dryer configuration and coated with PAH in a LaMer arrangement (Niessner,1984). The coating thickness was varied by the temperature of the LaMer generator up to 15 % of increase of the mean diameter $\bar{D}_p$.

In Fig. 2, ε is plotted versus D_p for Perylene on NaCl (5 % growth); ε is proportional to D_p rather than D_p^2, indicating partial coverage. A ≈ const = 1 shows that all particles have nucleated. For Anthracene on NaCl, ε (D_p) is similar, while A is not always unity. Fig. 3 shows A (D_p) for 5 % and 15 % growth of $\bar{D}_p$ by coating : For " thin " coatings and/or small particles, not every particle provides a nucleus. As expected, the probability of nucleation rises with the size.

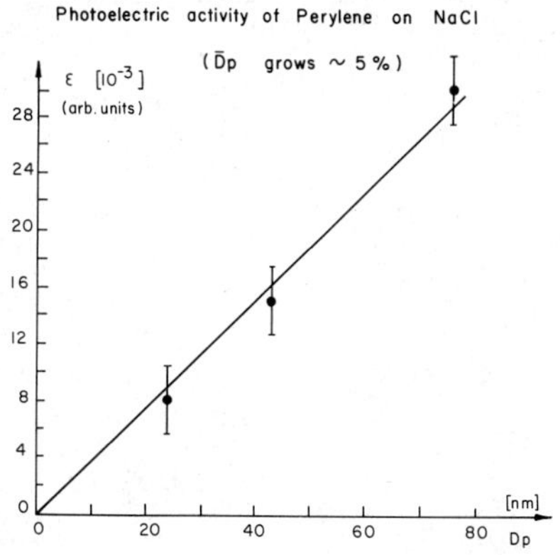

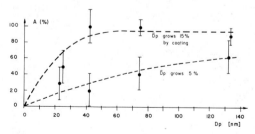

Figure 2

Figure 3

Conclusions

In conclusion, A and ε provide information on the surface of particles in situ. For particles from combustion of organic material the measurement yields that the particle surfaces are only partly covered with PAM in general. Partial coverage means that the mass of the material exposed on the surface, and determining the biological, chemical or physical properties of the particle, may be extremely small compared to the bulk mass. In the case of artificially coated particles for laboratory studies A and E enable monitoring of the coating process.

References

Burtscher, H. and A. Schmidt – Ott (1984) " Surface Enrichment of Soot Particles in Photoelectrically Active Trace Species ", <u>Sci. Total Environ.</u>, in press.

Niessner, R. (1984) " Coated Particles : Preliminary Results of Laboratory Studies on Interaction of Ammonia with Coated Sulfuric Acid Droplets or Hydrogensulfate Particles ", <u>Sci Total Environ.</u>, in press.

Schmidt – Ott, A. and B. Federer (1981) " Photoelectron Emission from Small Particles Suspended in a Gas ", <u>Surface Science</u> 106 : 538 – 543.

Published 1984 by Elsevier Science Publishing Co., Inc.
Aerosols, Liu, Pui, and Fissan, editors

APPLICATION OF RAMAN SCATTERING TO AEROSOL ANALYSIS

R. Rambau and G. Schweiger
Fachgebiet Thermodynamik
Fachbereich Maschinenbau
Universität Duisburg
4100 Duisburg 1
F.R.G.

EXTENDED ABSTRACT

Introduction

Very refined and highly sensitive methods have already been developed to analyse even trace concentrations. The main drawback of about all of these methods is that the particle has to be removed from its original neighborhood before it can be analysed. This can result in a whole bunch of complications (particle stability, particle deformation, particle reaction).

An established technique used since decade to circumvent the problems generated by the removal of aerosolparticles from the bearing gas before analysis ist the measurement of elastically scattered light (Mie scattering, Rayleigh scattering). This technique provides usually no information on the particle composition. Such an information is available in the inelastically scattered light, e. g. in the Raman spectrum of the particles. It is not surprising, therefore, that Raman scattering on microparticles is already used successfully for the analysis of microparticles [1,2] since the powerful monochromatic light sources needed for this technique are available. Most of the work, however, was done on particles deposited on a suitable carrier substrate to avoid particle heating.

The work presently under way in Duisburg concentrates on Raman scattering by liquid aerosol particles in their natural environment to avoid any noteworthy disturbance caused by the measurement technique and to get real time information about the particles. In addition to that, this concept offers the possibility to apply the theory of Raman scattering developed by Kerker and co-workers [3] which provides quantitative information for simple particle geometries as outlined briefly in the next section. The experimental problems and the set-up for Raman scattering under construction in Duisburg will be described in the last section. A comparable experiment, but without any attempt for comparison with theoretical predictions, was reported by [4].

Theory

There are several fairly good models for calculating the inelastically scattered light from microparticles. We give preference to that introduced by Kerker et. al., because there the simple spherical symmetry of the particle is taken into account in a quite sufficient way. This is exactly the geometry we are dealing with in our experiment. On the other hand one can expect a still tolerable amount of computing time by treating this relatively simple case.

In this model an incident light wave, denoted by its electromagnetic field vectors $(\vec{E}_2, \vec{B}_2)$, causes an e. m. field $\vec{E}_1(\vec{r})$, $\vec{B}_1(\vec{r})$ within the microparticle, which has to be calculated in its local dependence. The local field strength forces the molecules for induced dipole oscillations. The contributions of all molecules are superimposed. The whole problem is a typical mathematical boundary-value problem, which is written in spherical coordinates. In this case a most powerful method to achieve a solution has proved to be the expansion in spherical orthonormal functions. As a result of this one gets the general solution of the appropriate set of Maxwell's vector equations as an expansion in vector spherical harmonics, e. g. for the E-field

$$\vec{E}(\vec{r}) = \sum_{l,m} \frac{iC}{n^2 w} \, a_E(l,m) \, \vec{\nabla} \times [h_l^{(1)}(kr) \, \vec{Y}_{llm}(\hat{r})] + a_M(l,m) \, h_l^{(1)}(kr) \, \vec{Y}_{llm}(\hat{r}) \qquad (1)$$

with the expansion coefficients a_E, a_M. $\vec{Y}_{llm}$ are the vector spherical harmonics and $h_l^{(1)}$ denote the spherical Hankel-function of the first kind.

By applying the appropriate boundary conditions for the e. m. fields, the expansion coefficients a_E, a_M can be calculated. All the e. m. fields involved $(\vec{E}_2, \vec{B}_2, \vec{E}_1, \vec{B}_1, \vec{E}_{dip}, \vec{B}_{dip}, \vec{E}_{scat}, \vec{B}_{scat})$ are described as multipole expansions of the given form (1). However, it should be mentioned, that some numerical problems are unavoidable in practical computation, especially with regard to a wide range of parameter variation.

Experiment

The number of photons n_s scattered per laserpower density I_o can be factorized into a proportionality factor F, a factor P dependent only on the particle material, a quantity E, which is determined by the experimental set-up and the size d of the particle (diameter for spherical particles)

$$n_s/I_o = F \cdot P \cdot E \cdot d^3; \quad F = N_A \cdot K \cdot \pi/6, \quad P = (\partial\sigma/\partial\Omega) \cdot \rho/M_p, \quad E = \Omega \cdot T \cdot \eta \qquad (2)$$

N = Avogrados number, K = photons per unit laserpower, $(\partial\sigma/\partial\Omega)$ = average differential scattering cross section, ρ = particle density, M = molecular weight of the particle material, I_o = laserpower density, Ω = scattering aperture angle, T = total transmission of the optical system, η = quantum efficiency of the light detector.

More important than the number of scattered photons per particle in unit time is the number of photons scattered per particle transit through the scattering volume. The transit time has a lower limit determined by the acceptable particle heating. If we assume, for simplicity, a rectangular scattering volume of the size $\vec{a} \times \vec{b} \times \vec{c}$ (a points into the flow direction of the particle and c into the propagation direction of the laser beam), the temperature rise of the particle during the transit time $\tau = a/v_p$ (v_p = particle velocity) is

$$\Delta T = I_o \cdot V \, \alpha \, \tau/\rho \, c_p \, V = P_o \, \alpha/(b \cdot \rho \cdot c_p \cdot v_p) \qquad (3)$$

V = particle volume, α = absorption coefficient, c_p = specific heat, P_o = laserpower

Using equation (3), the number of photons scattered inelastically into the solid angle Ω per particle passage through the scattering volume can be calculated from equation (2)

$$n_{ST} = n_s \cdot \tau = F \cdot P \cdot E \, \frac{\rho \cdot c_p}{\alpha} \frac{\Delta T}{} \, d^3 \tag{4}$$

This number has an upper limit determined by the particle properties and the tolerable temperature rise and can be calculated by (4). Fig. 1 shows n_{ST} for some materials calculated by using the differential scattering cross section of the bulk material.

The total number of scattered photons per unit time caused by a flow of particles with concentration ρ_p (particles per volume) and a velocity v_p is given by

$$N = n_{ST} \, b \, c \, v_p \, \rho_p = F \, P \, E \, P_o \, c \, \rho_p \, d^3 \tag{5}$$

This is exactly the same relation as known from scattering on bulk material. The number of scattered photons is for a given particle material proportional to the laserpower, length of the scattering volume (in laser beam direction) and to the mass concentration of the particle phase in the aerosol.

Fig. 2 shows the experimental set-up which is under development for Raman scattering on aerosols containing liquid particles. As light source serves an Ar-Ion laser and the collection optics allows a continuous variation of the observation angle to facilitate a comparison between theoretical prediction and experimental findings. The present experiment is designed for high aperture single particle scattering. It is, therefore, expected, that this set-up will be relatively unsensitive to particle morphology of form but responds mainly to the mol mass of the particle under investigation.

References

(1) Etz, E.S. (1979) "Raman Microprobe Analysis: Principles and Applications", Scanning Electron Microscopy

(2) Etz, E.S., Rosasco, G.J., Blaha, J.J. (1978) "Observation of the Raman Effect from Small, Single Particles: Its Use in the Chemical Indentification of Airborne Particulates", Environmental Pollutants, Toribara, ed., Plenum Press, New York

(3) Kerker, M., Chew, H., McNulty, P.J. (1976) "Model for Raman and Flourescent Scattering by Molecules Embedded in Small Particles", Physical Review A, Vol. 13, 1, p 396

(4) Benner, R.E., Dornhans, R., Long, M.B., Chang, R.K. (1979) "Inelastic Light Scattering from a Distribution of Microparticles", Dale E. Newburg, ed., Microbeam Analysis

442

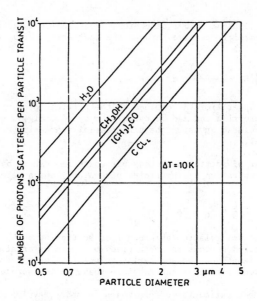

Fig. 1: Scattered photons per particle transit for
various materials as a function of particle
diameter

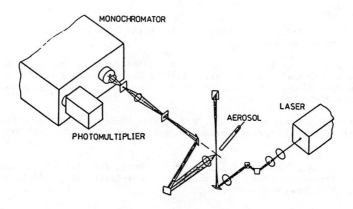

Fig. 2: Sketch of the optical set-up of the Raman
scattering experiment

Published 1984 by Elsevier Science Publishing Co., Inc.

Aerosols, Liu, Pui, and Fissan, editors

IN SITU DETECTION BY PHOTOELECTRON EMISSION OF PAH ENRICHED ON PARTICLES SURFACES

H. Burtscher°, R. Niessner* and A. Schmidt - Ott°

° Laboratorium für Festkörperphysik, ETH - Hönggerberg, CH - 8093 Zürich, Switzerland
* Universität Dortmund, Abteilung Chemie, Postfach 500 500, D - 4600 Dortmund 50, F.R.G.

EXTENDED ABSTRACT

Introduction

The application of photoelectron emission as a tool for surface analysis of suspended particles has been described already by Joffé (1913). For the first time Burtscher et al. (1982) made use of this principle for monitoring continuously particles in the atmosphere. For detection of chargeable particles they used a low-pressure Hg-discharge lamp in combination with an aerosol electrometer. It was suspected that polycyclic aromatic hydrocarbons (PAH) enriched at the surface of the particles are responsible for the distinct capability of photoelectron emission even at a low photon energy of 4.9 eV. This study was undertaken to receive qualitative informations on the photoelectron emission of different organic substances deposited on photoelectrically inactive condensation nuclei. The application of gas-phase reactions for selective conversion of the photoemitting surface was investigated too.

Experimental Set-up

The whole system used is given in Fig. 1. Generation of coated particles was accomplished by means of a condensation apparatus already described by Niessner (1984). A Collison atomizer or a graphite aerosol generator

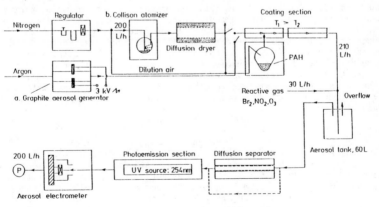

Figure 1. Schematic diagram of the system used

(Burtscher and Schmidt - Ott, 1982) was employed for production of condensation nuclei. The compounds under study were as listed in Table 1.

Table 1. Gases and substances used

Condensation nuclei	Coating material	Gases
NaCl ($d_{50} \sim 0.1$ µm; $\delta_g \sim 1.8$; N $\sim 10^5$ cm^{-3}) Graphite ($d_{50} \sim 20$ nm; $\delta_g \sim 1.9$; N $\sim 10^6$ cm^{-3})	n - Decanol n - Hexadecanol Toluene Anthracene Fluoranthene Pyrene Perylene Chrysene	Nitrogen dioxide Bromine Ozone

The relative humidity of the aerosol was kept close to 0 %, because a water
layer prevents the photoemission process at 4.9 eV. Ozone as a reactive gas
was generated by use of a Welsbach O_2-discharge unit (520 - 1550 ppb). Bro-
mine (5.4ppm) and NO_2 (0.38 ppm) were delivered from a permeation system
(Teckentrup and Klockow, 1978) with known permeation rates. The ozone concentration was determined by iodometric analysis. The PAH content of the
aerosol was quantified by fluorimetric analysis after filter sampling and
extraction of the filters with n - Heptane.
After having passed the condensation section the coated particles were mixed
with the reactive gas. The mixture was conducted through a plastic chamber
(60 L) allowing a residence time of about 10 min. From this vessel a stream
of 200 L/h was pulled through the photoemission section. In some experiments
a diffusion separator (loaded with MnO_2 for O_3 and charcoal for Br_2, resp.)
was inserted between vessel and photoemission section. The denuder removed
the reactive gas from the aerosol stream without a significant loss of particles, thus avoiding gas-phase reactions under the influence of UV-light.
After having left the photoemission section the charged particles were detected by an aerosol electrometer.
The test procedure was as follows : 1. The background photoemission activity,
expressed as the electrometer current, caused by the inactive condensation
nuclei was observed. 2. The temperature of the reservoir containing the organic material to be adsorbed onto the particle surfaces was increased thus
increasing the vapour pressure of the organic material. The temperature was
raised until a significant photoelectrical activity was indicated by a positive current. 3. The reactive gas was now added to the aerosol stream. If a
reaction between gas and surface results in formation of products with a
higher ionization threshold a derease of the photoelectric current should be
observed.

Results

The results obtained are listed in Table 2. A typical plot is given in
Fig. 2. Only PAH's with an extended π-electron system are good photoemitting
substances. The electron-rich graphite does not emit electrons under the
applied photon energy of 4.9 eV. No reaction between the PAH's and Br_2 or NO_2,
with and without UV-irradiation applied could be observed. Ozone and PAH's
did react significantly under irradiation of UV-light. Without UV-light only
Anthracene showed a change of the photoemitting surface. This was confirmed

Table 2.　Photoelectrical activity of different organic substances.
Influence of reactive gas with and without UV-light

Aerosol (Core/Surface)	Photoemission at 4.9 eV	Photoemission at 4.9 eV after reaction with Br_2	NO_2	O_3	O_3+UV
Graphite	−				
Graphite/Toluene	−				
Graphite/n − Decanole	−				
Graphite/n − Hexadecanole	−				
NaCl	−				
NaCl/Anthracene	+		+	+	−	−
NaCl/Fluoranthene	+		+	+	+	−
NaCl/Pyrene	+		+	+	+	−
NaCl/Perylene	+		+	+	+	−
NaCl/Chrysene	+		+	+	+	−

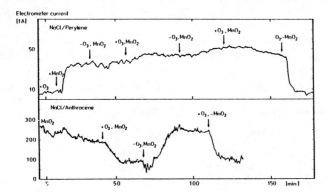

Figure 2.　Upper part : Readout of electrometer current caused by photoelectron emitting NaCl/Perylene particles (272 ng Perylene/m^3).
In a mixture of Ozone and NaCl/Perylene aerosol (1.55 ppm O_3)
photoelectric activity is depressed. Insertion of a MnO_2 diffusion separator for removal of O_3 prevents the UV-assisted O_3-
Perylene reaction.
Lower part : Readout of electrometer current caused by irradiated
NaCl/Anthracene particles (66 μg Anthracene/m^3). The conversion
of Anthracene by O_3 is achieved already in the mixing tank. The
O_3-removing MnO_2 diffusions separator has no influence on the decrease in photoelectric activity.

by measurements of the Ozone and Anthracene concentration. Laser Micro Mass
Analysis (LAMMA) of selected particles indicated the conversion of Anthracene
into Anthrachinone (Niessner et al., 1984). In contrast to Anthracene the
PAH's Perylene, Chrysene, Fluoranthene and Pyrene were not chemically converted. This could be confirmed by LAMMA and chemical analysis. The behaviour
of other PAH's and reactive gases are under study.

Conclusions

Polycyclic aromatic compounds adsorbed onto particles are responsible for a distinct photoelectric effect. Br_2 and NO_2 gave no reaction even under the influence of UV-radiation at 254 nm. Without assistance of UV-light only Anthracene reacted with Ozone to yield a product with a higher ionization threshold. All PAH's were converted by Ozone under the influence of mono-chromatic UV-light.

References

Burtscher, H., L. Scherrer, H. C. Siegmann, A. Schmidt - Ott and B. Federer (1982) " Probing Aerosols by Photoelectric Charging ", J. Appl. Phys. 53 : 3787.

Burtscher, H. and A. Schmidt - Ott (1982) " Enormous Enhancement of van der Waals Forces between Small Silver Particles ", Physical Review Letters 48 : 1734.

Joffé, A. (1913) " Beobachtungen über den photoelektrischen Elementareffekt ", Sitzungsberichte der königlichen und bayerischen Akademie der Wissenschaften : 19.

Niessner, R. (1984) " Coated Particles : Preliminary Results of Laboratory Studies on Interaction of Ammonia with Coated Sulfuric Acid Droplets or Hydrogensulfate Particles ", Sci. Total Environ., in press

Niessner, R. D. Klockow, F. Bruynseels and R. van Grieken (1984), unpublished results

Teckentrup, A. and D. Klockow (1978) " Preparation of Refillable Permeation Tubes ", Analytical Chemistry 50 : 1728.

Published 1984 by Elsevier Science Publishing Co., Inc.

Aerosols, Liu, Pui, and Fissan, editors

CALIBRATION AND SOME NEW APPLICATIONS
OF A SCINTILLATION PARTICLE COUNTER FOR
MEASUREMENT OF AIRBORNE ALKALI-NITRATES

Th. Waescher[1], R. Heger, H. Barth, and T. Timke

Kernforschungszentrum Karlsruhe GmbH
Laboratorium für Aerosolphysik und Filtertechnik II
Postfach 36 40, 7500 Karlsruhe 1

[1] Kraftanlagen AG
Im Breitspiel 7
6900 Heidelberg 1

F. R. G.

EXTENDED ABSTRACT

Introduction

Methods of flame scintillation spectral analysis for counting individual airborne particles according to size and composition were developed on different levels of technical design (Binek et.al., 1967; Crider, 1968; Pueschel, 1969). This method is very well suited for the measurement of test aerosols which are injected into the atmosphere concerned to simulate the behaviour of highly toxic aerosols. In most cases these test aerosols contain a selected chemical composition for an easy excitation in the flame.

As excitation of individual particles in a flame is a very complex process, which is dependent not only on the two main parameters, particle mass (size) and metal content of the aerosol substance, but also on the flame conditions (local composition and temperature) and kind of anion of the particle substance (e.g., Cl^-, Br^-, SO_4^- or NO_3^-), it is necessary for quantitative results to use a homogeneous substance that has no cross sensitivity to other particles present in the atmosphere (Pueschel, 1969). Accurate and reproducible calibrations require continuous control of flame background-radiation and stability of the negative pressure inside the aerosol inlet-system and the connected flame chamber.

Several investigations in the field of aerosol entrainment phenomena during nuclear fuel reprocessing steps were carried out at the Laboratorium für Aerosolphysik und Filtertechnik II (Heger, Waescher, 1982). To provide an instrument for on-line measurement of the relevant behaviour of the simulated high toxic aerosol, we developed a modified version (MSST) of the Sartorius Szintillations Teilchenzähler, (SST). The basic principles of this instrument (double-flame system, sensitive pressure control of the aerosol inlet) create the necessary conditions for accurate and reproducible calibration.

Description of the MSST

Because of the very high detection sensitivity, most instruments are calibrated for use of a NaCl-aerosol. To provide a similar sensitivity to other substances like $LiNO_3$, $NaNO_3$, KNO_3, and $CsNO_3$, we made some system modifications mainly with respect to pulse amplification and pulse height analysis.

The scintillation pulses emitted from the excited atoms and molecules are detected by the photomultiplier via an optical system which furthermore consists of a narrow bandpass interference filter with a peak transmission wavelength tuned to the line spectrum of the element to be measured. Current-pulse output of the PM-tube has a dynamic range of 5 to 6 orders of magnitude. For contraction of range the pulses were fed directly into a fast-response logarithmic amplifier. Additional pulse conditioning and amplification provided unipolar pulses in the range between 0.03 and 4.00 volts, corresponding to channel 2 through 128 of the multichannel analyser (MCA). Dependent on particle size, pulse widths were found to be in the range between 1 and 2 milliseconds at baseline. Background noise which results from unavoidable microturbulences in the flame produced pulse heights between 10 an 20 millivolts. These pulses could be sufficiently discriminated by the 30 millivolt threshhold of the second channel of the MCA. The input stage of the nuclear MCA employed was modified to process the comparatively long pulses without errors in counting. Long term particle counting was realized by extending the time base of the MCS-mode of the MCA.

Calibration

Calibration of the MSST was performed by the use of a monodisperse droplet generator (Binek and Dohnalova, 1967). Spherical-shape $MeNO_3$ particles were reliably generated by evaporation of the 10 um-diameter droplets produced by means of the vibrating quartz-fiber in a definite geometric arrangement and with a constant frequency. Figure 1 shows the calibration curves of the MSST for different alkali nitrate particles of various sizes.

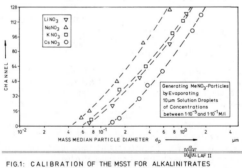

FIG.1: CALIBRATION OF THE MSST FOR ALKALINITRATES

To avoid significant aerosol-loss the droplet generation chamber was radiated by a Po 210 -source of 4,5 uCi. This radiation produced bipolar ions in a sufficient number to bring the droplets to a state of charge equilibrium in the air flow of 200 cm^3/min.

For calculation of the size (diameter) of the completely
dried MeNO$_3$-salt particles we used the expression

$$d_p = d_d \left(\frac{c_0 \, \rho_0}{100 \, \rho_p} \right)^{1/3}$$

which is evaluated from the mass bilance of the salt content of
the droplet and the particle. Hereby, d_p is the salt-particle
diameter, d_d is the original droplet diameter, c_0 and ρ_0 respectively are the content of the solute (wt%) and the solution
density, and ρ_p is the density of the salt particle. SEM-photos
of the carefully evaporated droplets showed the uniform spherical appearance and the nearly pore-free surface structure of
the residual salt-particles. Therefore with neglegible error we
set equal the value of density of the salt particle with that
of the homogeneous substance.

Figure 2 shows the response of the MSST to the monodisper-
sely generated droplets of CsNO$_3$-solutions with various salt
content. Compared to optical particle-counters the MSST revealed a remarkable good size-resolution capability with regard
to the homogeneous salt particles. Included into the values
stated (7.1 to 2.7 % relative size resolution) is the limited
degree of the monodispersity of the droplets, which is characterized by values of σ approx. between 0.05 and 0.15.

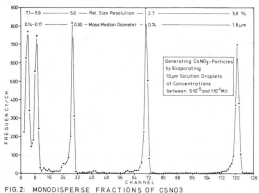

FIG.2: MONODISPERSE FRACTIONS OF CSNO3

Droplet entrainment by bursting of air bubbles

One of the main sources of droplet entrainment into natural and technical atmospheres is the bursting of air-bubble
films at the surface of liquids. To detect size and frequency
of the small fraction of droplets produced by the film-cap rupture, many investigations have been made by means of diffusion
or expansion cloud-chamber experiments (e.g. Day, 1964). In a
first application we used the MSST to determine the size and
frequency of these droplets with the help of the experimental
arrangement shown in figure 3. To provide a constant surface
tension, which is an important parameter of droplet formation,
the liquid surface at the top of the glass tube is renewed continuously by the constant flow of the solution feed.

450

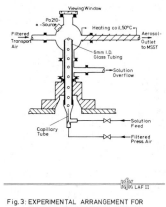

Fig. 3: EXPERIMENTAL ARRANGEMENT FOR
SINGLE BUBBLE BURSTING AEROSOL
(from Day, 1964, modified)

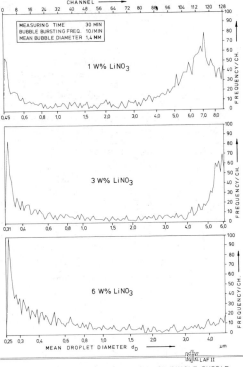

FIG. 4: DROPLET SIZE DISTRIBUTIONS OF SINGLE BUBBLE
BURSTING IN 1,3 AND 6 W% LiNO₃-SOLUTIONS

Figure 4 shows the droplet size distribution at various contents of LiNO₃. At higher concentrations the peak of large droplets is drifting to the right whereas the small fraction appear on the left. This small fraction represents the salt nuclei of sizes less than 0.1 um which are generated by the film-cap rupture.

References

Binek,B and B.Dohnalova (1967) "Die Anwendung des Szintillationsspektralanalysators für Aerosole in Forschung und Technik", Staub-Reinh. d. Luft 9:379

Crider, W. L. (1968) "Spectrothermal Emission Aerosol Particle Analyser", Rev. Sci. Instrum. 39(2):212

Pueschel, R. F. (1969) "Principles and Applications of the Flame Scintill. Spectral Analysis", Ph. D. diss., Univ. of Wash.

Heger, R. and Th. Waescher (1982) "Bestimmung der Quelltherme von Aerosolen bei verfahrenstechnischen Operationen im Wiederaufarbeitungsprozeß", KfK-Nachr. 14(3):143

Binek, B. and B. Dohnalova (1967) "Ein Generator zur Erzeugung monodisperser Aerosole aus der flüssigen Phase", Staub-Reinh. d. Luft 27:492

Day, J. A. (1964) "Production of Droplets and Salt Nuclei by the Bursting of Air-Bubble Films", Quart. J. Roy. Met. Soc. 90:1964

Published 1984 by Elsevier Science Publishing Co., Inc.
Aerosols, Liu, Pui, and Fissan, editors

MEASUREMENT OF MASS DISTRIBUTION OF CHEMICAL SPECIES IN AEROSOL PARTICLES

M. P. Sinha, Jet Propulsion laboratory, California Institute
of Technology, Pasadena, CA 91109 and S. K. Friedlander,
University of California, Los Angeles, CA 90024

Introduction

Aerosols are often generated by nebulizing solutions and evaporating the
solvent to leave the dry solute particles. The particle size distribution is
then measured with an optical counter which is usually calibrated with polystyrene latex particles. Hence, the diameters of the particles measured in
this way are the optical equivalent of the polystyrene particles. This leaves
much uncertainty associated with the particle shape and the refractive index, and
therefore may not be used for the mass distribution of particles. Other
methods of measurements based on aerodynamic size and ion mobility also leave the
uncertainty associated with the particle shape and density. None of the above
methods possesses any chemical specificity, i.e., they provide no information on
the chemical nature of the particles.

A method has been developed in our laboratory for the direct determination
of masses of chemical species in individual aerosol particles on a continuous,
real-time basis. This is accomplished by applying the method of Particle
Analysis by Mass Spectrometry (PAMS). Some of the preliminary results on the
mass measurements of particles using PAMS are described in this paper.

Experimental

PAMS involves a combination of the techniques of particle beam generation
and mass spectrometry. Its details have been described elsewhere (Sinha et
al., 1982, 1984). In this method, the aerosol particles are introduced into
the ion source of a quadrupole mass spectrometer (MS) in the form of a beam
that is produced by the expansion of the aerosol sample into a vacuum through a
capillary nozzle and a skimmer system. The particles impinge on a hot rhenium
filament in the ion source of the MS. The resulting plume of vapor is then
ionized by the electron bombardment. For particles composed of materials with
low ionization potential, ionization may be performed on the surface of the
rhenium filament (Stoffels, 1981). A pulse of ions is produced from each particle. The intensities of different masses in the ion pulses are measured by
the MS.

Aerosols containing sodium chloride particles were selected for mass
determination in the present work. These aerosols are important because of the
fact that aerosols of various active compounds used in studies employing animal
or human subjects are routinely generated from their saline solutions. Sodium
chloride is also an important constituent of marine aerosols.

In order to determine the salt content in the particles, a calibration
curve for the MS signal versus mass of sodium chloride needs to be generated.
Such a calibration curve has been determined by using monodisperse aerosols
generated from aqueous solutions of sodium chloride by a vibrating-orifice

452

aerosol generator. Nitrogen gas used for the dilution and dispersion of
the aerosol was heated for complete drying of the particles. The particles
were introduced into the ion source in the form of a beam for their volatilization and ionization on the surface of the rhenium filament. Surface-ionization
is particularly efficient for alkali metal compounds.

In our earlier studies on particle analysis, the rhenium filament was
made in the form of a V-groove. This facilitated the volatilization of particles. It was found that for particles (.8 to 2.0 μm diameter) composed of
different organic compounds, the ion signal (electron impact ionization) increased linearly with the particle mass showing a complete volatilization of
particles within the filament. Complete volatilization is important for
quantitative measurements. However, when sodium chloride particles were analyzed, the linear relation did not hold. This may be due to particle-bounce
from the the filament surface before complete volatilization (Dahneke, 1973,
1975). Sodium chloride has a high boiling point. To overcome this problem,
the V-filament was replaced by a Knudsen cell oven (5 mm diameter) and was
situated in the ion-optics region of the MS. The oven was fabricated of zone-
refined rhenium metal sheet (0.018 mm thick) and was resistively heated to
~1600 K during operation. An opening of ~2 mm diameter was made in the oven
for the introduction of the particle beam. A particle inside the oven is trapped and consequently it is likely to be completely volatilized. The vapor molecules are thermally dissociated and a fraction of the resulting sodium atoms
are ionized on the surface before effusing out of the oven. The intensity of
the Na^+ ion signal was measured by integrating the area under the ion pulse
from the electron-multipier of the MS. The ion pulse width varied from 4 to
12 ms for the different size particles. This may be contrasted with a typical pulse width of 200 μs when the volatilization was made with a V-type
rhenium filament. The increase in the pulse width may be attributed to the
slow effusion rate of ions from the oven.

Results and Discussion

Table I lists the relative Na^+ ion intensities from particles of different
masses. Also listed in the table are the equivalent spherical diameters of the
particles.

Table 1
Mass spectrometer signal for
different particles

Particles Mass $(x10^{-2}g)$	Equivalent Spherical Particle Diameter (μm)	Na^+ Ion Intensity
1.2	1.0	2.2
3.1	1.4	5.2
6.2	1.8	8.9
12.5	2.2	16.9
54.1	3.6	55.0

Figure 1 is the log–log plot of the particle mass with the ion intensity, I.

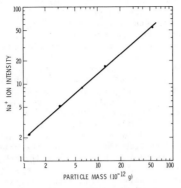

Fig. 1. Variation of Na$^+$ ion intensity with particle mass

A straight line with a scope of 0.86 is obtained, showing that I α (mass)$^{.85}$. We believe that the particle is completely volatilized in the oven. The departure of the exponent from 1 may be attributed, in part, to the effusion profile of the ions from the oven and to the beam collimator located just above the oven.

The MS intensity distribution from particles of 1.0 μm size was also measured. The ion pulse is fed into a charge integrator that intergrates the signal pulse from a preset time interval depending on the pulse width. Following this, the background is integrated for the same length of time and substracted from the stored signal. The distribution of the peak voltages of the integrated signal pulses is then measured with a pulse height analyzer. Figure 2 shows the signal intensities (proportional to channel number) on the probability-graph paper.

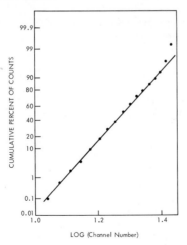

Fig. 2. Pulse height distribution of the mass spectrometer signal at 23 amu resulting from 1.0 μm diameter sodium chloride particles.

454

A linear plot shows a log-normal mass distribution of the MS signal. The departure from the straight line in the highest channels may be due to the doublet particles present in the input aerosol from the vibrating orifice aerosol generator. The geometric standard deviation, σ_g of the distribution is found to be 1.19, indicating that the mass spectrometric measurement does not introduce much broadening of the aerosol size/mass distribution.

The above results show that a miniature oven in the ion source of the MS provides for the complete vaporization of aerosol particles over a wide size range. This is important for the quantitiative mass measurements of different chemical species present in airborne particulate material. These particles may be heterogeneous in composition. The design seems to overcome the particle bounce problem from the filament even with relatively high boiling point material. The calibration curve is found to correspond to a single power law of 0.86. These observations, together with the fact that the mass spectrometeric measurements do not introduce broadening in the distribution, make the PAMS method suitable for the mass determination of chemical species in individual particles. Work on measurements of polydisperse particles are underway in our laboratory. For particles composed of chemical species which cannot be efficiently surface-ionized, electron ionization is being used.

The work was performed at the Jet Propulsion Laboratory and was supported by the National Science Foundation under Grant CHE-8008676.

References

B. Dahneke (1973), J. Colloid Interface Sci., 45, 584; (1975), 51, 58.

M. P. Sinha, C. E. Giffin, D. D. Norris, T. Estes, V. Vilker and S. K. Friedlander (1982) J. Colloid Interface Sci. 87, 140.

M. P. Sinha, R. M. Platz, V. L. Vilker and S.K. Friedlander (1984) Int. J. Mass Spectrom. Ion Processes 57, 125.

J. J. Stoffels (1981), Int. J. Mass Spectrom. Ion Phys. 40, 243 and reference therein.

CHAPTER 11.

NUCLEAR AEROSOLS

Published 1984 by Elsevier Science Publishing Co., Inc.
Aerosols, Liu, Pui, and Fissan, editors

Recent Developments in Nuclear Aerosol Research

W. O. Schikarski

Laboratorium für Aerosolphysik und Filtertechnik I
Kernforschungszentrum Karlsruhe GmbH
7500 Karlsruhe, FRG

Extended Abstract

Introduction

There is a common understanding on the definition of nuclear aerosols, namely

"nuclear aerosols are airborne particles of solid or liquid nature, which may be formed by various processes occurring in nuclear installations, which are usually composed of a variety of chemical compounds including radionuclides and which pose a certain health risk when released into the environment".

This implies that nuclear aerosols are not simply identical with radioactive aerosols which are formed by natural processes in the environment too. Since abnormal conditions in nuclear installations are the most important source of nuclear aerosols, there are typical differences compared to other aerosol sources (e.g. environmental aerosols) nuclear aerosols are usually formed under extreme conditions like high temperature, high pressure or high humidity. They can be formed at high initial concentrations and their chemical composition corresponds to the nuclear or other material of nuclear installations involved. Finally nuclear aerosols are formed in confined spaces representing a three-phase aerodisperse system (Fuchs /1/) which is generally dynamic in space and time depending on aerosol processes and carrier gas convection involved.

Nuclear aerosols and their behaviour are strongly related to the nuclear installation considered, namely nuclear power plants and nuclear fuel reprocessing plants. This paper reviews recent developments in nuclear aerosol research related to LMFBR's, LWR's and RP's.

2. Sources of nuclear Aerosols

2.1 LMFBR

Generally, there are two major sources of nuclear aerosols in LMFBR's which are called the primary and the secondary source term. Analysing the primary source term, particularly, hypothetical core disruptive accidents (HCDA) are investigated. In the case of an HCDA melting and vaporization of major parts of the core inventory is assumed. A bubble may be formed which expands into the sodium pool and transfers kinetic energy to the coolant and to the structures. A sodium slug may impact the reactor vessel head, causing finally material leakage into the inner containment space. The primary source term is defined as the amount of radioactive material released from the reactor vessel. The majority of this mass material consists of aerosols. The secondary source term is related to the usually delayed emissions of nuclear aerosols from hot sodium carrying fission and activation products into the inner containment space.

Considerable progress has been made in the investigation of both, the primary and the secondary aerosol source term. For the primary source term main emphasis has been placed on the aerosol retention during bubble rise in the core and upper plenum region in the coarse of an HCDA. Most interesting experiments have been carried out in the U.S. (ORNL-FAST programme) and the FRG (KfK-FAUST programme) showing a high aerosol removal capability of the moving bubble. An aerosol in a fast moving gas bubble is removed from the gas mainly due to impaction and coagulation. Consequently, all FAUST tests (so far water tests) show high aerosol retention factors in the order of $RF > 10^4$ /2/. A typical result of a model calculation of the particle motion relative to the motion of the bubble surface is given in Fig. 1, confirming the experimental results of a removal already in the first oscillation of a rising bubble. Fuchs's theory /1/ of a relatively small aerosol absorption by bubbling seems to be valid only for slow bubble rise due to buoyancy.

For the case of the secondary source term new data are available from the KfK-NALA II-tests /3/. Liquid sodium has a high retention capability for UO_2- and SrO-particles. Retention factors (RF) are in the order of 10^3 to 10^4 for UO_2 and about $5 \cdot 10^2$ for SrO. The RF for NaI was found to be betweeen 1 an 10 indicating a different release mechanism; RF values were measured for sodium pool temperatures of 550 °C. A dependence of the RF's on particle size was measured. Most experiments were carried out with UO_2-particles with sizes between 10-20 μm. UO_2-particles with 200 μm showed a much higher RF than the smaller particles.

2.2 LWR

THe LWR aerosols source term is of interest only for the case of hypothetical core melt accident scenarios. In the SASCHA-programme /4/ releases of fission products as aerosols have been investigated by simulation of core melt releases and determination of the relevant aerosol parameters. The initial size distribution was mesured for a core melt aerosol to be lognormal with a median geometrical diameter of 0.1 μm and a standard deviation of $\sigma = 2.1$.

3. Aersosol transport and deposition in confined systems

Early experiments on nuclear aerosol behaviour in confined volumes have been started before 1965, mainly in the US and in the FRG. A review of the various experimental programmes, their main results and the related code development are given in /5/. Tab. 1 shows the experimental programmes and corresponding code development of our KfK-programme from 1968 up to now. There was essentially a parallel development of aerosol behaviour models for LMFBR's and for LWR's. Whereas the PARDISEKO code /6/ was aimed for LMFBR accident scenarios (mixed aerosols consisting of mainly sodium with fuel and fission products impurities), the NAUA code /7/ describes the aerosol decay in LWR containment systems after severe accidents (mixed aerosols consisting of mainly Ag, Fe and U plus some other materials including condensed steam). Both codes comprise all important aerosol processes, namely coagulation

(Brownian and gravitational) sedimentation, diffusion. Additionally, in PARDISEKO IV thermophoresis and turbulent deposition, and in NAUA Mod5 condensation and diffusiophoresis are taken into account. It has been shown that aerosol codes should have the possibility to treat arbitrary particle size distributions in order to avoid errors for long term calculations of aerosol mass concentration and other parameters. This implies relatively large computing times. Due to some recent numerical improvements both codes have been improved considerably in that respect. Nuclear aerosol codes use the homogeneous mixing assumption, which has been confined in the range of experimental errors for small and medium large vessels by several authors. However, there may occur an inhomogenity of the aerosol mass concentration in the confined volumes as long as a strong local aerosol source exists. Although principally the compartmentalisation of a large volume and the corresponding transformation of the present single-compartment aerosol codes into multi-compartment-codes is possible, this would increase the computing time extraordinarily. Therefore, this problem is of major interest and closely related to the large scale aerosol behaviour tests currently carried out in Europe and USA. I may refer to the DEMONA program /5/ and to another paper of this conference /8/, indicating that after source shut-off the homogeneous mixing assumption can be considered as a reasonable approximation.

Another problem has been investigated more thoroughly, namely the shape factor concept. The application of nuclear aerosol codes has shown already several years ago that the dynamic shape factor and the coagulation shape factor are of great significance in predicting aerosol behaviour in confined systems /9/. Using the wrong shape factors could lead to large deviations of the aerosol concentration in the long term. Whereas the dynamic shape factor κ can be measured easily by normal aerosol measuring techniques (i.e. centrifuge etc.), the determination of coagulation shape factors f needs a more sophisticated laboratory set-up consisting of a coagulation tube and aerosol generation and measuring devices. There is also a special evaluation method necessary. Measured shape factors for PtO ($d_m = 0.058$ μm) and UO_2 particles ($d_m = 0.27$ μm) are given in /10/. An interesting result (Table 2) of the first direct measurement of a coagulation shape factor is that for both aerosols κ and f have comparable values.

Three major nuclear aerosol behaviour test programs are currently under way, namely DEMONA (FRG), MARVIKEN V (Sweden) and LACE (USA). Together with the results of the small and medium scale volume test programs already performed in the recent years, they will contribute to the better understanding of nuclear aerosol behaviour and the improvement of the aerosol codes, already mentioned, and other codes internationally under development.

4. Air cleaning and nuclear aerosols

Although industrial air cleaning related to various types of aerosols is in an advanced stage, some special problems for nuclear aerosols exist and have been discussed already in the beginning of nuclear aerosol research. Two aspects are to be mentioned, namely the necessity for high efficiency of off-gas filter systems and the (to some extent) chemically aggressive nature of nuclear aerosols. This required new air cleaning devices, in particular for LMFBR's and nuclear fuel reprocessing plants (RP).

460

Sodium aerosols together with high mumidity can destroy the ordinary
glass fibre HEPA filters. Therefore, a special multilayer sand bed filter
suitable for the off-gas system of an LMFBR's and with HEPA quality had been
developed /11/. This device has recently been changed and optimized for
Kerosene fire aerosols which might occur in RP accidents. The loading capacity and the filter efficiency were improved by using crushed lava which has
a high porosity together with large surface area of the sand. Removal efficiencies of 99.9% (integral) and 98.5% (initial) were achieved. The known
excelllent loading capacity is about 1500 g per m² at a pressure drop lower
than 30 mbar. Wet scrubbers of diffferent types and with different numbers
of stages were also tested for sodium aerosol removal. A disc sprayer type
scrubber showed an efficiency of 70% per stage and 98% with three stages. A
combination of a Venturi scrubber, a wet deep bed filter and an electrostatic precipitator were also tested for sodium aerosols. Efficiencies of about
99.9% were observed.

Conclusion

Progress has been made in the investigation of nuclear aerosol formation,
nuclear aerosol behaviour and air cleaning related to nuclear aerosols. Primary and secondary aerosol source terms in LMFBR's and LWR's have been studied. Except for the LMFBR primary source term a good quantification of the
aerosol emissions and their characteristics is possible. Aerosol behaviour
in confined systems can be predicted reasonably well by codes like PARDISEKO
and NAUA. Necessary input parameters (shape factors) have been measured. New
air cleaning devices for nuclear aerosols have been tested successfully.

References

/1/ N.A. Fuchs: "The Mechanics of Aerosols", Pergamon Press (1964)

/2/ W. Haenscheid, W. Schütz: "Particle formation, transportation and
removal in a rapidly expanding and rising bubble"
Proc. GAeF Munich, J. Ae. Sci. (to be published)

/3/ H. Sauter, W. Schütz: "Aerosol- und Aktivitätsfreisetzung aus
kontaminierten Natriumlachen in Inertgasatmosphäre", KfK-3504 (1983)

/4/ H. Albrecht, H. Wild: "Freisetzungsquellterme bei hypothetischen
LWR-Störfällen", KTG-Fachtagung 'Freisetzung und Transport von
Spaltprodukten bei hypothetischen Störfällen von LWR, Karlsruhe (1982)

/5/ W. Schikarski: "DEMONA-Forschungsprogramm zur Demonstration nuklearen
Aerosolverhaltens - Grundlagen, Ziele, Auslegung
KfK-Report 3636 / EIR-Report 502 (1983)

/6/ H. Bunz: "PARDISEKO IV - Ein Coumputerprogramm zur Berechnung des
Aerosolverhaltens in geschlossenen Behältern", KfK-Report 3545 (1983)

/7/ H. Bunz, M.Koyro, W. Schöck: "NAUA-Mod4 - A code for calculating
Aerosol behaviour in LWR Core Melt Accidents",
Code description and Users' manual, KfK-Report 3554 (1983)

/8/ H. Bunz, W. Schöck, R. Taubenberger: "Spatial Distribution of an Aerosol in a large structured Building", this conference

/9/ H. Jordan, W. Schikarski, H. Wild: "Nukleare Aerosole im geschlossenen System, KfK-Report 1989 (1974)

/10/ W. Zeller: "Direct Measurement of aerosol shape factors", KfK-Report 3560 (1983), to be published in J. Ae. Sci. & T.

/11/ L. Böhm, S. Jordan: "On the filtration of Sodium Oxide Aerosols by multilayer sans bed filters", J. Ae. Sci. 7, 311

Code	Improvement	Development time	Experiment Volume	Experimental Program · Time
PARDISEKO	coagulation sedimentation diffusion	1968-69	-	-
PARDISEKO II	+ thermophoresis	1971-73	TUNA 0.2...2.2 m³	1968-73
NAUA Mod 1	condensation	1973	-	-
NAUA mod 2	+ coagulation + sedimentation + diffusion	1974-75	-	-
PARDISEKO III	+ time dependent source	1973-75	TUNA 0.2...2.2 m³ NABRAUS 4.7 m³	1968-73 1973-75
PARDISEKO IIIb	+ improved numerics	1976-79	FAUNA 220 m³	1975-
NAUA Mod 3	+ improved numerics	1976-80	NAUA 3.7 m³	1976-
NAUA Mod 4	+ application to PWR-containments	1981-82	-	-
PARDISEKO IV	+ turbulent deposition	1982	FAUNA 220 m³	1983
NAUA Mod 5	+ diffusiophoresis	1982-83	NSPP 38 m³ DEMONA 580 m³	1981-83 1983-85

Table 1: Development of aerosol codes and verification at KfK

462

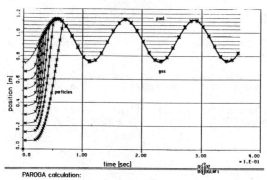

PAROGA calculation:

1.1 MPa discharge of 0.02 mm iron particles at FAUST geometry

PtO-particles			
d_M (µm)	$\kappa(t_o, d_m)$	$\bar{\kappa}(t)$	$\bar{f}(\bar{t})$
0.038	1.?	$\bar{\kappa}(t_o) = 1.6$	–
0.041	2.0	$\bar{\kappa}(t_1) = 2.1$	–
0.055	2.9	$\bar{\kappa}(\bar{t}) = 1.85$	1.8
0.071	3.6		
UO_2-particles			
d_M (µm)	$\kappa(t_o, d_m)$	$\bar{\kappa}(t)$	$\bar{f}(\bar{t})$
0.19	2.6	$\bar{\kappa}(t_o) = 2.0$	–
0.24	2.8	$\bar{\kappa}(t_1) = 2.5$	–
0.33	3.3	$\bar{\kappa}(\bar{t}) = 2.25$	3.2
0.55	1.2	–	–

Table 2: Dynamic (κ) and coagulation (f) shape factors for PtO and UO_2 particles as function of mass diameter and time

Published 1984 by Elsevier Science Publishing Co., Inc.
Aerosols, Liu, Pui, and Fissan, editors

STUDY OF HEAVY-METAL TRANSPORT BY SMOKES

K. Buijs, B. Chavane de Dalmassy, H.E. Schmidt
Commission of the European Communities
Joint Research Centre, Karlsruhe Establishment
European Institute for Transuranium Elements
Postfach 2266
D-7500 Karlsruhe
Federal Republic of Germany

Radioactive materials, in particular alpha-emitting elements such as plutonium, are handled in a closed system, e.g. a glove-box, when the quantities exceed trace levels.

During the fabrication of plutonium-containing nuclear fuels, the interiors of such glove-boxes become heavily contaminated with plutonium dioxide, a refractory oxide. Fires in such installations are expected to give rise to plutonium-contaminated smoke.

The degree and nature of Pu-contamination of smoke from burning glove-boxes was studies in a series of test-fires involving polymethymetacrylate (PMMA), a material frequently used for the construction of glove-boxes. As a simulant for plutonium dioxide (which in practice usually contains up to one percent of americium), a co-precipitated oxide of 84% of cerium and 16% of europium was used.

Smoke particles were sampled on nuclepore filters as well as in an inertial spectrometer. The samples collected were analysed by neutron-activation analysis, scanning electron microscopy with energy dispersive analysis and laser microprobe mass analysis. Up to 6 kg of PMMA contaminated with 0.1 mg/cm^2 of Ce-Eu-oxide were burnt. The maximum heat release rate was of the order of 0.5 MW. More than half of the heavy metals became airborne and the heavy-metal concentration in the smoke reached a few hundred $\mu g/m^3$. Possible mechanisms of entrainment and transport are discussed.

Published 1984 by Elsevier Science Publishing Co., Inc.

Aerosols, Liu, Pui, and Fissan, editors

THE DESIGN AND OPERATION OF
A SEMI-AUTOMATED SAMPLING SYSTEM
FOR SIZE CLASSIFICATION OF SIMULATED
REACTOR CORE AEROSOLS

J. Mäkynen, W. Piispanen, and L. Ström
The Marviken Project
S-610 24 Vikbolandet
SWEDEN

Introduction

In tests designed to demonstrate the behavior of aerosols released in a degraded nuclear reactor core accident, a semi-automated sampling system has been developed to provide representative size fractionated samples of simulated fission product aerosols. The aerosols consist of vaporized or condensed species of cesium, iodine, and tellurium. The aerosols are transported through a reactor system in an atmosphere of steam, and nitrogen, with a small quantity of hydrogen to maintain reducing conditions. Temperatures of the gas at various measurement stations range from 170°C to 500°C with resulting gas viscosities from approximately 16×10^{-6} to 30×10^{-6} kg/m·s.

Sampler Particle Sizing Design

The sampling train has been designed to provide size fractionated samples of the aerosol at various locations. The sampling train is constructed entirely of stainless steel and is designed for sampling up to 500°C. A single piece nozzle/probe assembly is attached to a 4 stage cyclone train. The cyclone cut points (D_{50}) for the last three cyclones were determined by the designer as Cyclone 280-1 D_{50} = 8.4 μm, Cyclone 280-3 D_{50} = 2.4 μm cut Cyclone 280-5 D_{50} = 0.87 μm at a flow rate of 237 cm^3/s air (25°C) (Smith et al., 1979).

Determination of Sampling Conditions

In actual sampling operations, the test conditions will specify the approximate gas temperatures, gas flow rate and gas composition. Since most all of the aerosols are less than 15 μm aerodynamic diameter it is desirable to set the flow rate of the first cyclone to provide approximately a 15 μm cut point.

By knowing the sample gas temperature, the individual gas component viscosities can be calculated. Knowing the molecular weight and mole fraction of each gas component, the mean gas density, mean free path and mean gas viscosity can then be calculated. The sampling flow rate required for the approximately 15 μm D_{50} cut point is then determined based on calibrations by Smith et al. (1982). The calculated mean viscosity and sampler flow rate are used to determine the approximate performance of the sampling train based on individual cyclone fitting equations.

It should be noted that the test conditions provide a significantly different environment than the "as calibrated" conditions due to the gas mixture, high temperatures, and particle form and density. The viscosity relation may not be valid at these conditions. Calculated cut points are reported as a spherical partile of unit density but the actual particles may be liquid or solid with an approximate density of 4 g/cm^3. Additional calibration work and research is underway to determine the effect of these conditions on cyclone performance.

Sampling System Design

Each aerosol sampling station consists of one fractionating sampler and 2 total particulate (filter plus probe) sampling units. Each sampling unit is installed on a moving cradle inside a temperature controlled oven. The entire unit is connected to a drive mechanism which rotates at 2 rpm to provide continuous traversing along a single 300 mm duct diameter.

The sample gas exiting the filter passes first through a heat exchanger. The heat exchanger is designed to heat or cool as necessary to maintain gas temperature at $\simeq$ 140 °C. In order to prevent transient temperature feed back hunting, the cooler has a large thermal mass and is enclosed in a temperature controlled insulated housing. Thermal stabilization is assured before sampling and also an LED readout of the sample gas temperature is monitored continuously. A heated sample line transports the filtered sample gas to the flow control system.

Sampling Control

A single flow monitoring station is connected to the three incoming heated lines and parallel solenoid switching valves (air cooled) allow sequential operation of samplers. A ganged valve allows simultaneous back purging of all 3 systems with argon when not in use.

Sampler flow control is monitored using a turbine flow meter along with a pressure transducer and a gas thermocouple in a temperature controlled cabinet. The temperature is maintained above 100° C in order to measure the steam component. A computer program determines duct gas velocity and sampler probe velocity and then relays an "isokinetic voltage set". A flow control valve is then manually adjusted to meet the set point. Data is monitored as real time on a visual display monitor. The turbine flow meter output voltage and the sampler pressure drop are monitored on a strip chart recorder for a system performance record.

System Performance

The sampler system has been used in four actual tests to date. The number of sampling stations operating simultaneously under a single remote computer control has been three. Only minor problems have been experienced such as 1) one case of corrosion induced leakage 2) a misalignment of a traversing machine push rod causing damage to the machined surface, 3) leakage in threaded fittings and soldered joints and 4) initial problems in maintaining non-condensing steam temperatures throughout the system. These problems and other potential problems have been addressed in a detailed "Sampling Methods Manual" which includes specifications for calibrations, leak checks, Quality Assurance procedures, and an operations check list. Vulnerable system components have been replaced with improved designs.

The cyclone and filter samplers have performed very well for the collection of both liquid and solid aerosols but there is a question regarding the applicability of the manufacturer calibration curves to the as sampled conditions. Recent simulated condition calibrations indicate that both gas density and gas viscosity may be major parameters in predicting the size classification of aerosols rather than gas viscosity alone as traditionally assumed. Preliminary correlations of D_{50} versus Reynolds number (Re) indicate a possible inverse relationship. Calibrations are being made for the cyclones and the probe for final calculation of aerosol data.

References

Smith, W. B., Cushing, K. M., Wilson, R. R., and Harris, D. B. (1982) "Cyclone Samplers for Measuring the Concentration of Inhalable Particles in Process Streams", J. Aerosol Sci. 13:3.

Smith, W. B., Wilson Jr., R. R., and Harris, D. B. (1979)" A Five Stage Cyclone System for in situ Sampling", ES&T. 13:11.

Published 1984 by Elsevier Science Publishing Co., Inc.

Aerosols, Liu, Pui, and Fissan, editors

RELEASE OF OVERHEATED LIQUIDS FROM
PLUTONIUM-NITRATE TRANSFER CONTAINERS

H.D. Seehars, D. Hochrainer, M. Spiekermann
Fraunhofer-Institut
für Toxikologie und Aerosolforschung
Grafschaft, Germany

EXTENDED ABSTRACT

Introduction

In the Federal Republic of Germany Plutonium-nitrate transfer containers consist of a Titanium vessel, a cladding material of asbestos and a special resinaceous foam material. The Titanium vessel itself contains the nitric (5m) solution of Pu-nitrate with a concentration of 250 g Pu/l.

Potential traffic accidents of those containers during road transportation may induce an exposure of the Titanium vessel itself to a fire due to the ignition of the leaking fuel. The heat energy transfer leads to an increasing temperature and overpressure of the solution within the vessel up to a critical value, causing the burst of the vessel and the more or less complete release of the contents in form of liquid and/or solid aerosol particles. A high percentage of the released aerosol may belong to the respirable particle fraction, leading to a potential hazard of the surroundings.

Theoretical

Here it is reported on experiments with a Cerium(IV)-nitrate-solution used as a substitute with similar physico-chemical specification as the radioactive solution. There are several arguments for the selection of this substitute:

1. Lanthanides and Actinides indicate a certain chemical similarity.

2. The valences of Ce and Pu are:
   Ce: 3, 4
   Pu: 3, 4, 5, 6

3. As well Pu(IV)-nitrate- as Ce(IV)-nitrate-compounds show a very high degree of solubility in salpetric acid.

4. As well Pu(IV)- as Ce(IV)-nitrate are decomposed at temperatures between 200 and 220 °C into the respective 4-valent oxides.

5. Conversion of a droplet into a solid salt particle after the release does not result in serious particle mass differences of the both phases if the Pu- and Ce-nitrate solutions have the same metallic concentrations.

468

The spontaneous release of liquid material with high initial velocities leads to the formation and breakup processes of droplets (Wolfe et al, 1964). The breakup processes as well as the generated drop size spectrum are described by the Weber-number, which is defined as the ratio of disruptive hydrodynamic force and the stabilizing surface tension force. The mass median diameter of droplets produced by the breakup of the original drop, is highly dependent on the initial drop velocity.

Experiments

Because of the high costs only three original Titanium vessels were used during the experiments, but most of the experiments were performed with an equivalent composed of stainless steel. To have defined bursting conditions the equivalent was closed by a bursting stainless steel diaphragm. The bursting properties of the diaphragm due to the experiments with the equivalent were derived from the maximum bursting pressure of the Titanium vessel. Heat energy was produced by a propane burner. Initial droplet velocities were measured by a high speed camera with some 1000 frames per second (Fig. 1).

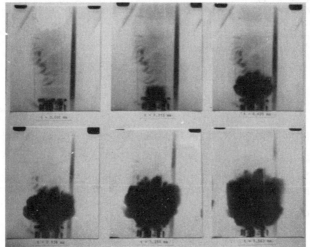

Fig. 1: Example of a droplet cloud after the spontaneous release of the solution at a pressure of 7 MPa and a temperature of 240 °C. Time between the single frames: 0.31 ms.
Lower part of the frame: top of the equivalent vessel.

An array of horizontal elutriators and membrane filters was used to measure as well the respirable particle fraction as the concentration of the pollutant, dependent on the degree of acidity of the solution, the Ce-concentration per liter acid, the orientation (horizontal or vertical) and degree of filling of the vessel (40 or 80 %).

All experiments took place within a tent of plastic foil of 450 m^3 volume to prevent disturbances by meteorological influences.

Results

Initial droplet velocity

Measured initial droplet velocities covered a range between 300 and 510 m/s and decreased with increasing degree of filling. An influence of the orientation of the vessel as well as the acid concentration was not discernible. However, an increasing concentration of Cerium in the solution resulted in an increase of the droplet velocity.

Released respirable and total particle fraction

Maximum release of the respirable and the total mass fraction of particles were determined in connection with the vertical orientation of the vessel and a degree of filling of 40 %. The release rate, defined as the ratio of the mass of elemental Cerium and the total mass of Cerium in the solution amounted to 1.8 %.

The totally released fraction of mass of the respirable aerosol increased as well with the acid concentration as with the concentration of Cerium in the solution. However, the relese rate decreased with the Cerium concentration. The concentration of the totally released mass fraction of aerosol covered the range between 40 and 60 mg/m^3. The investigation of the collected particles on the filters with methods of analytical chemistry demonstrated, that there was no conversion of the aerosol particles from the soluble Ceriumnitrate to the insoluble Ceriumoxide. This result is important due to the different properties of deposition and transfer in the human body.

Reference

Wolfe, H.E., H.W. Andersen (1964)
"Kinetics, Mechanisms and Resultant Droplet Sizes of the Aerodynamic Breakup of Liquid Drops", Aerosol-General Corporation, Research and Engineering Division, California.
Defense Documentation Center for Scientific and Technical Information, Cameron Station, Alexandria, Virginia
Rep.No. 0395-04(18)SP

Published 1984 by Elsevier Science Publishing Co., Inc.
Aerosols, Liu, Pui, and Fissan, editors

470

THE GRAVEL BED CONFINEMENT VERIFICATION PROGRAM*

R. P. Sandoval, R. E. Luna, J. M. Taylor, N. Grandjean

Transportation Technology Center
Sandia National Laboratories
P. O. Box 5800
Albuquerque, NM 87185
U.S.A.

and

G. J. Newton
Lovelace Biomedical and Environmental Research Institute
Inhalation Toxicology Research Institute
Albuquerque, NM 87115
U.S.A.

EXTENDED ABSTRACT

Introduction

The Gravel Bed Confinement Verification Program (GBCVP) was initiated in early 1982 to provide information on the ability of gravel bed structures to contain respirable-sized toxic aerosols produced in detonation of a high explosive device containing heavy metals. Earlier tests of the containment structure design during the 1950s were primarily concerned with the degree to which the structure reduced ground contamination by radioactive materials. However, those tests did not address the exposure to respirable material of persons nearby and at site boundaries which is the principal concern, under current radiation protection standards, of the hazard to man.

The full GBCVP contained three major components:

1. Scaled Tests--To explore in a quick and inexpensive manner the possible impact of structural design features which could not be duplicated in the full-scale test and which were not included in earlier tests,

2. Confinement Vessel Tests--To verify the high explosive uranium aerosol generator (HEAG) design in terms of the amounts of respirable aerosol produced, and

3. Full-Scale Test--To verify in full scale the capture efficiency of a gravel bed structure for aerosolized uranium produced.

*Research performed under U. S. Department of Energy Contract No. DE-ACO4-76DP00789 and DE-ACO4-76EV01013.

Subscale Test Program

The subscale test series confirmed the predicted scaling law behavior for overall gravel bed movement. The dynamic behavior of the gravel bed basically followed a $W^{1/3}$ rule, where W is the mass of the gravel bed. In addition, the variability introduced by either an off-center detonation or the inclusion of a dropped ceiling were found to be insignificant.

Confinement Test Program

A series of subscale explosive tests were performed in a spherical steel confinement chamber to verify the hydrocode predictions of aerosol quantity produced by the HEAG as well as to determine a particle size distribution for the aerosols produced. The high explosive aerosol generator used for these experiments was scaled from those designed for the full-scale test.

Full-Scale Test Program

The site chosen for the full-scale gravel bed test was the same as that used for the tests performed in the mid 1950s. The structure was fully excavated and, after inspection by structural consultants, was found to be suitable for additional testing. The staging area used in the 50's test series did not simulate actual geometry and volume of current designs. As a result, a new staging area was designed and constructed using appropriate lengths of 10-ft-diameter steel culvert.

Since the principal goal of the full-scale test was to determine the capture efficiency of the gravel roof of the structure, aerosol measurement systems were designed to characterize the aerosol inside the structure before gravel bed collapse as well as any aerosol escaping from the structure. Depleted uranium metal was selected to be the primary tracer and was generated by four high explosive aerosol generators.

Within the structures were a set of six blast-hardened and fragment-shielded samplers. Four were in the round room and two in the staging area to sample respirable aerosols prior to gravel bed collapse. Outside the structure were a number of air sampler systems designed to obtain data to characterize the aerosol emissions by quantity, size distribution and location. These included filter, impactors, electrostatic precipitators and fallout trays.

Pressure measurements were made at four locations in the round room, and various points in the staging area. In addition, a series of pressure measurements were made inside and outside the structure by Southwest Research Institute.

The test was documented by still photographs, video tapes, and motion pictures spanning construction, test, and recovery phases of the event.

Extrapolation of the aerosol concentration in the mass of gas captured by inside samplers to the total mass of gas in the structure provided an estimate of the total uranium aerosol mass produced inside the gravel structure. The total quantity of respirable uranium aerosol inside the structure was measured to be 185g ± 40 g.

Estimating the mass of respirable aerosol penetrating the structure was done in two ways. One involved fitting air sampler data with a diffusion model. The second involved using air sampler results together with photographic measurements of cloud volumes sampled to form a concentration-volume product which yielded aerosol mass. Both processes indicated a released aerosol mass of 0.5 to 0.9 grams.

The fallout tray data indicated five locations where significant deposits were received within 125 meters of the test structure; the remaining values were generally indistinguishable from background.

The test result of principal interest is the respirable aerosol release fraction. This quantity was obtained by dividing the estimates of uranium aerosols external to the structure by the measured quantity inside before roof collapse. This yields a total aerosol release fraction of $4 \times 10^{-3} \pm 1.5 \times 10^{-3}$ with a respirable component of $3 \times 10^{-3} \pm 1.1 \times 10^{-3}$.

The results obtained in the test were not fully consistent with the results of the 50's test. In that test no fallout tray received deposits above that received by the same trays uncovered for a comparable time before the test. In addition, qualitative review of test films also indicate that gravel bed behavior was somewhat different in the two tests. In searching for possible reasons for these discrepancies, several differences between the tests were noted. The three most important relate to the quantity of aerosol produced, the construction of the gravel bed, and the volume and configuration of the staging areas. Since the Department of Energy's specifications for these types of structures includes about 20% more gravel and greater gravel bed depth in the areas above the round room periphery than used for this test, the use of 4×10^{-3} as a release fraction for assessing gravel bed confinement capability is clearly conservative.

The Gravel Bed Structure Confinement Verification Program has provided a significant increase in the understanding of the behavior of granular filtration beds under high explosive detonation conditions.

Published 1984 by Elsevier Science Publishing Co., Inc.
Aerosols, Liu, Pui, and Fissan, editors

TEMPORAL AND SPATIAL AEROSOL CONCENTRATION MAPPING IN A GLOVE BOX

by

V. A. Marple and D. J. Rader
Particle Technology Laboratory
University of Minnesota
Mechanical Engineering Department
111 Church Street Southeast
Minneapolis, MN 55455-0111

and

Dr. Hans Schmidt and Dr. Steve Pickering
European Institute for Transuranium Elements - Karlsruhe Establishment
Postfach 2260
D-7500 Karlsruhe
Federal Republic of Germany

EXTENDED ABSTRACT

In the machining of radioactive materials in glove boxes, it is desired to
know how long the air flow must be maintained after a machining process has
terminated to clear the aerosol from the box. This time will depend upon the
air flow rate, glove box ventilation system, machining process, and geometric
shapes of the objects in the box. The mapping of the aerosol concentration in
the glove box was accomplished by first defining the air flow pattern with the
solution of the Navier-Stokes equations using numerical techniques. This
technique involves the defining of a grid over the entire volume of the glove
box and solving the finite difference form of the Navier-Stokes equations
using an iterative relaxation procedure. The result is the velocity vector
components and pressure at each of the node points (grid line intersections)
of the grid. The only boundary condition necessary is the flow velocity at
the inlet or outlet. The obstacles in the glove box are defined by setting
the viscosity at node points inside or at the surface of the obstacle to a
very large value. The Navier-Stokes equations are solved at all the node
points in the glove box but since obstacles are defined by node points with
very high (essentially infinite) viscosity the air flows around instead of
through these obstacles.

After the velocity vectors have been determined, the particle concentration equations are solved considering the gravitational, diffusional and
convective motion of the particles. The only boundary conditions for this
routine are vanishing particle concentration at the walls and the strength of
the source of aerosol generation. The particle concentration equation is
expressed in transient form so that both temporal and spatial concentration
are calculated. By stopping the routine at specific times, concentration maps
can be printed.

474

The problem is being analyzed in both two- and three-dimensions by using either the two- or three-dimensional form of the Navier-Stokes equations. An example of the flow field is shown in Figure 1 for a two-dimensional solution of the glove box with an obstacle present. The fluid velocities are presented at each node point with the direction and length of the arrow being the direction and relative magnitude of the flow, respectively.

The particle concentration for the flow filed shown in Figure 1 is presented in Figure 2. The particles are generated at the source at the rate of 10^{-8} g/min and exit at the outlet or settle on the walls. Isopleths are shown for aerosol concentrations of 2, 5 and 7 particles/cc.

An experimental program has also been undertaken to verify the results of the theoretical study. The boundary conditions for the experimental and theoretical studies are matched as closely as possible, with the specifications of the glove box size (160 cm wide by 100 cm high by 110 cm deep), flow rate (3.8 l/sec), air velocity at the inlet (190 cm/sec) and geometric obstructions.

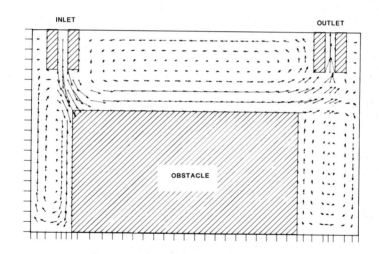

Figure 1.   Two-dimensional flow field in a glove box.

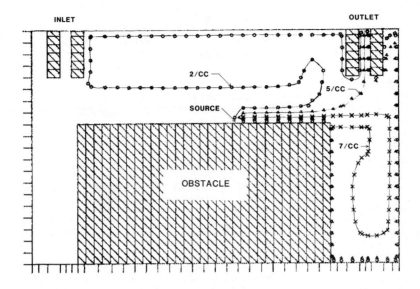

Figure 2. Steady-state two-dimensional particle concentration for
6 μm aerodynamic-diameter particles.

Published 1984 by Elsevier Science Publishing Co., Inc.

Aerosols, Liu, Pui, and Fissan, editors

476

CONDENSATION OF C_sI ON FERRIC-OXIDE PARTICLES

M. T. Cheng, W. Chan, D. T. Shaw
Laboratory for Power and Environmental Studies
State University of New York at Buffalo
Buffalo, New York

and

Mati Merilo
Electric Power Research Institute
Palo Alto, California

EXTENDED ABSTRACT

    Cesium iodide is a fission product generated in large amount in a nuclear reactor degraded-core accident. The effect of C_sI condensation on the shape factor of ferric-oxide particles is the subject of the present project. Experiments are carried out in a two-oven evaporating-condensing system in which C_sI vapor generated in a ceramic boat is condensed onto Fe_2O_3 particles fed into the oven. It was found that C_sI vapor can be condensed in three modes: (1) At relatively low vapor pressure, uniform condensation on the Fe_2O_3 chain-aggregate to form amorphous structure; (2) Localized crystallization occurs at higher vapor pressure; and (3) At even higher pressure, rhombic crystals of C_sI are formed around small particles. Figure 1 shows a plot of the ratio of the aerodynamic diameter to the volume equivalent diameter versus the ratio of the condensed vapor mass to the particle mass. The aerodynamic diameters of for (1) and (2) are calculated based the uniform-coating and string-of-pearls models. The effects C_sI condensation on particle dynamics in the reactor primary system are presented.

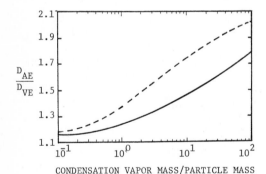

CONDENSATION VAPOR MASS/PARTICLE MASS

Figure 1.

Ratio of the aerodynamic diameter to the volume equivalent diametric versus the ratio of the condensed vapor mass to the particle mass for uniform coating and localized condensation.

Published 1984 by Elsevier Science Publishing Co., Inc.
Aerosols, Liu, Pui, and Fissan, editors

Spatial Distribution of an Aerosol in a Large Structured Building

H. Bunz, W. Schöck
Laboratorium für Aerosolphysik und Filtertechnik
Projekt Nukleare Sicherheit
Kernforschungszentrum Karlsruhe GmbH
7500 Karlsruhe, Germany

R. Taubenberger
Eidgenössisches Institut für Reaktorforschung
Würenlingen, Switzerland

Introduction

Computer codes for calculating natural aerosol deposition in closed systems are frequently used to calculate the time dependent behavior of radio-active particles in nuclear power plant buildings to determine the risk associated with hypothetical accidents. For economical reasons these codes use certain simplifying assumptions which are correct only for the special application and cannot be generalized. One of the more important assumptions is that the aerosol be well mixed and homogeneously distributed over a control volume at all times. It is quite obvious that in a system with sources and sinks this assumption cannot be true in a mathematical sense, which of course has never been postulated. What is really meant is, that the relaxation time of any inhomogeneity is short compared to the times needed to change the state of the system by other processes. Or, in other words, there are stirring mechanisms vigorous enough to distribute the aerosol all over the control volume before a noticeable effect of space dependent interaction or depletion occurs. The mixing of aerosols in a large cylindrical vessel during a fire was recently investigated /1/. In this paper we present results of measurements in a structured building of more complex shape.

Experimental methods

The measurements were performed in the frame of the large scale experiment DEMONA /2/. The experiments are conducted in the model containment facility at Battelle, Frankfurt, which is a 1/4-scale model of a 1300 MWe KWU type pressurized water reactor containment. The volume of the facility is 640 m³, schematic sections are shown in Fig. 1. The aerosol is composed of metal oxide particles with concentrations of several grams per cubic meter. The aerosol generation technique has been described earlier /2,3/, the aerosol is fed into the center of the building and is distributed by natural convection.

The time dependent behavior of the aerosol is measured by various methods which are described elsewhere /2,4/. Additionally in order to look for the spatial distribution of the aerosol, ten identical photometers are distributed at representative locations in different compartments of the facility as is shown in Fig. 1. These photometers have been developed especially for this application, to measure in the concentration range of several g/m³, and give an on line information of the state of the aerosol system. It is self explaining that spatial inhomogeneities, should they occur, are expected and will be important in the early phase of the experiment when the concentration is still very high. Four of the photometers are placed close to filter sampling devices and can be absolutely calibrated after the experiment.

478

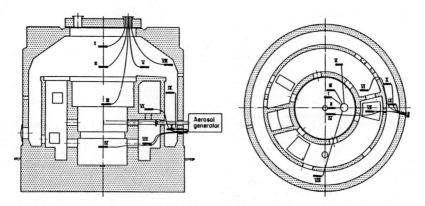

<u>Fig. 1:</u> Sections of the facility showing the photometer positions

<u>Results</u>

The following results have been obtained during shake down tests at
ambient temperatures. During aerosol generation the pressure increased to
3 Bar and decreased slowly afterwards due to leakages in the facility. The
temperature of the atmosphere was about 40 °C at the end of the aerosol
production period and decreased to ambient temperature within one hour.

In Fig. 2 the measured mass concentration of the aerosol is drawn as a
function of time. The data are evaluations of filter samples taken at two positions, one close to photometer X and one between photometers II and V. The initial mass concentration at the end of the aerosol production was about 1 g/m³.
It is noted that the mass concentrations at the upper and lower position follow
the same time dependence and that no local differences can be noticed.

The initial rate of decrease of the mass concentration was significantly
higher than anticipated. This was due to an unexpected fraction of coarse particles in the aerosol which consisted of molten powder grains that had not been
completely vaporized. The aim of the aerosol generation is to condense the
primary particles from the vapor phase which consequently leads to very small
primaries in the size range below 0.2 μm. Due to their high concentrations the
primaries coagulate rapidly to larger aggregates, these aggregates, however,
should have average diameters well below three microns in all experiments. The
inital depletion rate in Fig. 2 corresponds to settling of particles with an
aerodynamic diameter of about 6 μm. This result was confirmed by measurements
with a cascade impactor and by microscopic investigation of the samples showing
a significant fraction of coarse particles.

Since gravitational settling is proportional to the square of the particle
diameter the larger particles are removed from the aerosol quickly and the
slope of the mass concentration curve decreases more and more. After 20 hrs the
slope corresponds to an aerodynamic diameter of 2 μm which agrees well with the
expectations.

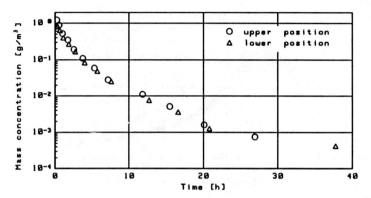

Fig. 2: Aerosol mass concentration as function of time

To look at the spatial distribution of the aerosol the signals from the ten photometers have been evaluated. In Fig. 3 a simplified diagram of the photometer recordings vs time is shown for the first four hours of the experiments. It was found that within the range of fluctuations of the individual signals seven of the ten photometers read identical values at all times. These are represented in Fig. 3 by the shaded band.

The only clearly different signals were obtained from photometers IV, VII and X and are represented as separate curves in Fig. 3. These photometers are in the lowest positions in the building, almost in the dead zones below the openings connecting the different compartments. The coupling of these locations to the convectional currents is certainly weaker than elsewhere in the building. The total volume of these dead zones is about 40 m³ or 6 % of the total volume.

Additionally, these locations are at the bottom of the building and a coarse particle fraction existed. Therefore, the deviation of the signals from photometers IV, VII and X can be explained by an enrichment of the coarser particle fraction due to gravitational settling and a lack of convectional stirring. On the other hand it was clearly seen that the signals from photometers III, VI and IX at medium elevation are the same as those from photometers I, II, and VIII in the top of the building. All these are within the large convectional loops and permanently undergoing mixing.

Another detail in Fig. 3 confirms that picture. At around one hour the signals from the lower photometers return into the range of the other signals. At that time the aerosol production rate was decreased but the heat input from the generators continued. The gases from the generator enter the building with a temperature of 250 °C and enhance the large convection loop considerably, so that obviously the locations of the lower photometers are coupled to the stirring, too. When the generators are shut down at 1.5 hrs the forced mixing stops and the lower photometer locations decouple from the main volume again.

480

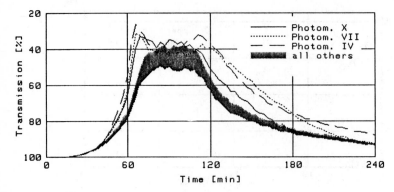

<u>Fig. 3</u> Signals from the ten photometers during the first four hours

An additional proof for the good initial mixing by the hot generator plume is obtained from sedimentation samples. They were taken from all over the building and exhibit identical amounts of deposited material irrespective of their location. This shows that the aerosol is distributed evenly over the main part of the volume during its generation as well as at later times.

Conclusions

The aerosol behavior in a large compartmentalized building was investigated with special emphasis on the spatial ditribution. It was seen that due to forced stirring by energy input during the aerosol generation and, later on, by normal convection the aerosol system is well mixed. Only in dead volumes which amount to 6 % of the total volume a temporary enrichment of very large particles was observed. The large particles were an unexpected by-product of the aerosol generation technique and are not typical for aerosols that are generated by evaporation and re-condensation of material. After the large particles had been removed by settling the system was well mixed at all times.

References

/1/ W. Cherdron, Berechnung der Konvektionsströmung bei Natriumflächenbränden in geschlossenen Behältern, Jahrestagung der Kerntechnischen Gesellschaft, Frankfurt, 22. - 24. May, 1984

/2/ W. Schikarski et al., DEMONA-Forschungsprogramm zur Untersuchung nuklearen Aerosolverhaltens, KfK 3636/EIR 502 (1983)

/3/ W. Schöck et al., Generation and properties of highly concentrated metal oxide aerosol in a steam atmosphere, Annual Meeting 1983 of the Association for Aerosol Research, München 14 - 16. September 1983

/4/ G. Friedrich et al., Aerosol measurement system for the DEMONA Experiment, CSNI Specialist Meeting on Nuclear Aerosols in Reactor Safety, Kalsruhe, 4. - 6. September 1984

Published 1984 by Elsevier Science Publishing Co., Inc.
Aerosols, Liu, Pui, and Fissan, editors

481

THEORETICAL EVALUATION OF SOURCE CONCENTRATION
IN A NUCLEAR REACTOR DEGRADED-CORE ACCIDENTS

Sushil Patel, D. T. Shaw and Mati Merilo*
Laboratory for Power and Environmental Studies
State University of New York at Buffalo
Buffalo, New York

*Electric Power and Research Institute
Palo Alto, California

EXTENDED ABSTRACT

The source term evaluation for degraded-core accidents is the subject
of many reports on nuclear safety analysis. Controversies currently exist
on the maximum aerosol concentration in such accidents. The dynamic behavior
of a highly dense coagulating aerosol flowing in a straight tube is investigated with a computer program which determines the time dependent particle
size distribution function. In addition to Brownian, gravitational and turbulent coagulation, such removal mechanisms as the gravitational settling,
laminar/turbulent deposition and laminar/turbulent thermophoretic deposition
are included. The computer program covers the changes of the particle size
over three orders of magnitude (typically 0.01 to 30 μm) without assuming
any fixed form of the distribution function (such as log-normal).

The dynamic behavior of an initially high density aerosol flowing through
a straight pipe is numerically computed. Figures 1, 2 and 3 show the time
change in the mass, number concentration and particle mean diameter respectively for various initial mass loadings and pipe diameters. Figure 1 shows
that the mass concentration decreases more rapidly for smaller diameter pipe.
This is expected because of the larger surface to volume ratio per unit
length of pipe. Also shown is that a higher initial mass loading has a higher
rate in decrease in mass concentration. This is due to rapid coagulation and
simultaneous removal. At low mass loadings, removal is dominant. Since
removal is proportional to concentration, and in the absence of coagulation,
the three curves for different initial mass loadings would be similar and
would not intersect. But the fact that they will intersect at some time indicate the degree of coagulation and their consequent removal by sedimentation,
turbulence and thermophoresis.

Similar is the reasoning for Figure 2 which shows the time change in
number concentration. It is seen that the coagulation only curves follow the
curves with removal up to a certain time. This indicates the time until which
coagulation is dominant. This time is longer for higher mass loadings again
showing the relative effects of coagulation and removal.

Figure 3 shows that the higher initial mass loading has a faster increase
in mean diameter. Coagulation is primarily responsible for this increase in
particle diameter. Particle diameter would normally increase faster in the
case of coagulation only relative to coagulation and removal. But this is not
the case in Figure 3. The smaller particles are removed more rapidly by
thermophoresis in comparison to the removal of larger particles by sedimentation and turbulence, resulting in an increase in the mean particle diameter.

482

Results such as shown in Figure 1, 2 and 3 indicate that very high mass loadings (1 kg/m³) can exist only for a period of 0.01 to 1 second. For such high concentration aerosol, the coagulating size distribution function depend sensitively on the initial size distribution function and the so called 'quasi-equilibrium' condition does not exist in the reactor primary system.

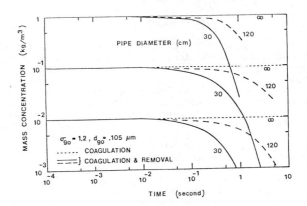

Figure 1. Mass concentration vs time with pipe diameter and initial mass concentration as parameters. Gas velocity = 10 m/s, gas temperature = 500°K, wall temperature = 300°K.

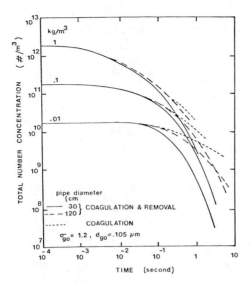

Figure 2.   Total number concentration vs time with pipe diameter
and initial mass concentration as parameters: gas
velocity = 10 m/s, gas temperature = 500°K,   wall
temperature = 300°K.

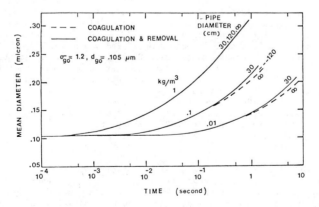

Figure 3.   Geometric mean diameter vs time with pipe diameter and
initial mass concentration as parameter: gas velocity = 10 m/s,
gas temperature = 500°K, wall temperature = 300°K

Published 1984 by Elsevier Science Publishing Co., Inc.

Aerosols, Liu, Pui, and Fissan, editors

484

WIND RESUSPENSION OF TRACE AMOUNTS OF PLUTONIUM PARTICLES
FROM SOIL IN A SEMI-ARID CLIMATE

G. Langer
Rockwell International
Rocky Flats Plant
Post Office Box 464
Golden, Colorado   80401
U.S.A.

EXTENDED ABSTRACT

Introduction

This study of resuspension of soil containing minute amounts of plutonium (Pu-239) has been in progress at the Rocky Flats (RF) Plant since
1978. It is one of several studies initated after wind relocated small
amounts of soil-borne Pu-239 during cleanup of an outdoor storage area
(Sehmel, 1980). The Pu-239-settled field is now sparsely covered with
prairie grass typical of the area. Past studies were limited to comparisons of bulk soil activity with total activity in the airborne dust. This
work covers the physics of the particle resuspension process.

This report covers the following: (1) Pu-239 resuspension rate versus
wind speed, (2) mechanisms of soil particle resuspension, (3) vertical concentration profile of Pu-239 particles, (4) Pu-239 and host particle size
distribution and activity concentration.

Experimental Methods

A small wind tunnel provided data on dust resuspension under controlled conditions. The sampling train for the resuspended dust consisted of a
cyclone, cascade impactor and a back-up filter. The dust size fractions
were analyzed for Pu-239 by alpha spectroscopy. Individual Pu-239 particle
size was determined by autoradiography using CR-39 plastic foil. Airborne
dust samples were collected with high-volume samplers using size selective
inlets (SSI) with a D-50 of 15 μm. The SSI was followed by a cascade impactor and back-up filter. The size cuts were analyzed as above. Hivols
were installed at 1 and 10 m above ground on a tower to measure the vertical Pu-239 and dust concentration gradient about 100 m from the source
region. These data serve to predict Pu-239 transport.

Results and Discussion

The wind tunnel is necessary to develop a predictive model for the
Pu-239 resuspension at RF. Data cover the range of RF wind speeds (0 to
140 mph). This empirical approach is used because we are not dealing with
wind erosion from a plowed field, where the classical saltation theory is
applicable. The data are highly variable with respect to windspeed. Also,
there was no clear relation between resuspension rates and soil surface
conditions, such as soil type and grass cover. The effect of soil moisture

was not discernible until the moisture content exceeded 15% in the top 1 cm
and resuspension ceased. A statistical analysis provided the following
relationships between resuspension of dust and Pu-239 with windspeed based
on wind tunnel data: Dust (μg/m^2-min) = $-18 + 0.45$ X Windspeed and Pu-239
(nCi/m^2-min) = $-13 + 0.59$ X Windspeed. The equations were derived from a
linear regression analysis. A power relationship between the two variables
had been expected from past studies of soil erosion by wind. But, the
variance for each data point was too large to give such a fit. The physical conditions in the Pu-239 source area are very complex. An extensive
wind tunnel study would be necessary to accurately model resuspension by
that approach.

The above equations cannot be used directly to predict resuspension
over prolonged periods such as several hours. The data are based on 1 min
exposures of a given soil area to the wind tunnel. An exploratory test has
shown that dust release drops off two orders of magnitude in a 15 min exposure compared to the release in the first minute. After the first 15 min,
the resuspension rate levels off.

An important result of the wind tunnel work was that dust is resuspended even at low windspeeds. The classical theory of wind erosion is
predicated on the saltation process. That is, particles a few tenths mm or
larger start to move in the wind. These sand-like particles either hop or
roll along the surface and their impact suspends smaller (<100 μm) particles. These particles become airborne when projected above the boundary
layer. However, windspeeds over 30 mph are necessary to initiate the
saltation process. This was verified at RF by observations with a ribbon-
like laser beam aimed along the surface of bare soil areas. Also, a
Bagnold Catcher was operated for several months in the Pu source area
during dry weather periods. No measurable amounts of creeping or saltating
particles were collected. Windspeeds during this period did not exceed
40 mph. Research was undertaken to define the physics of soil particle
resuspension during wind <30 mph, especially from grass covered areas.

The source area has less than 25% of patchy, bare soil spots. Therefore, the release of Pu-239 from prairie grass was considered. Archer
(1982) reports that at RF root uptake of Pu-239 is minimal. In that case,
any Pu-239 particles that are resuspended come directly off the surface of
the grass blades. Grass samples were taken and rinsed to remove any dust
particles. The activity of the dust was about the same as that of nearby
soil. Then, 37 grass samples were collected over a larger area and
analyzed directly for Pu-239 without the tedious rinsing process. The
ashed grass residue averaged 47 pCi/g Pu-239. The activity again was close
to that of the soil, when compensating for the grass residue contribution
to the sample mass. The litter on the ground was also examined for Pu,
because it is representative of decaying grass blades that have not yet
disintegrated. Again, the activity was near that of the soil. Overall,
the plant-to-soil activity ratio was 0.015. Dreicer (1984) reports a ratio
of 0.05 for tomato plants raised in an arid climate. Future studies are
necessary to determine how Pu-239 activity is released from the grass. At
low windspeeds, wind tunnel resuspended dust was about 50% organics,
indicating that the flexing of decaying grass blades releases particles as

the plant breaks up. The release mechanism from a green plant surface, if there is one, is not clear. Use of a scanning electron microscope would provide basic information on the attachment of the dust particles.

Pu-239/dust particles are transferred to the grass by rain splash, an effective mechanism for transferring soil contaminants to plant surfaces. Dreicer (1984), Gregory (1973). During high winds (often over 100 mph at RF) the grass blades would also intercept soil particles released by the saltation process.

If the saltation and resuspension from grass, which should predominate at low winds, were the principal resuspension mechanisms, a better correlation with windspeed would be expected. Data from earlier sampling were examined for Pu-239 resuspension versus windspeed during actual field conditions. An ultra high-volume (7.65 m^3/min) sampler allowed weekly sample collections for Pu-239 analysis. The sampler was operated from June 1980 through May 1981 in the Pu-239 source field. The data failed to show a significant relation between resuspension and windspeed for either Pu-239 or dust. Closer examination showed that resuspension was unimpeded during weekly periods when the ground was always wet. No Pu-239 was released during only two weeks in March 1981, when the ground was always covered with snow. At that time, the Pu-239 concentration dropped 30-fold to fallout levels. It is clear that we are dealing with another resuspension mechanism that is not windspeed dependent: resuspension by rain splash. More work is necessary to quantify this process for a predictive model.

For modeling the Pu-239 dispersal, the following information is required: particle size and activity distribution of the Pu-239-carrying soil particles versus height as the plume leaves the source region. The data from the vertical dust flux tower, 100 m east of the source area are in Table 1. The original data were reported in 6 size fractions; Table 1 reports the respirable, inhalable and coarse size ranges. The respirable size presents the real risk from Pu-239. The inhalable particles enter the gut and are quickly eliminated. The coarse particles are no threat, except during windy conditions when coarse particles could be forced into the nasal passages and mouth (Vincent, 1982). The Pu-239 concentration of 4.5±3.3 and 9.3±11 aCi/m^3 Pu-239 in the respirable dust at 1 and 10 m respectively is somewhat above the fallout level of 2.2±1.8 aCi/m^3 Pu-239 for this 16-month period. For the present, the higher averge at 10 m is not statistically different from the value at 1 m because of its high variance. Fifty respirable dust samples taken in the source area at 1 m during June 1980 to July 1982 were twice fallout levels, like the 1 m samples at the tower.

The isotopic ratio of Pu-240/Pu-239 in the respirable dust was 0.07 versus 0.05 for the inhalable and coarse particles. This ratio should be higher for the respirable dust, since it contains fallout plutonium, which has a ratio of 0.16. The RF weapons grade plutonium has a ratio of 0.05. The fact that the particles >3 µm had only local plutonium shows that good particle separation was achieved by the cascade and SSI. Fallout is attached to the small (<3 µm) particles.

Table 1. Vertical Concentration Profile of Dust and Pu-239 Particles

| | Dust Concentration, $\mu g/m^3$ | | | | | | | |
| | 1 m | | | | 10 m | | | |
Date	Respi-rable	Inhal-able	Coarse	Total	Respi-rable	Inhal-able	Coarse	Total
Nov-Dec 82	8.1	7.3	20	35	6.8	3.8	11	22
Jan-Feb 83	8.9	7.1	36	52	7.9	5.4	19	32
Mar-Apr	11	9.6	22	43	10	6.1	14	30
May-Jun	6.4	8.8	30	45	6.4	8.0	18	32
Jul-Aug	8.0	10	27	45	9.3	9.0	28	46
Sep-Oct	7.4	14	24	45	8.4	12	23	43
Nov-Dec	9.3	7.0	21	37	7.3	4.5	10	22
Jan-Feb 84	8.7	11	47	67	8.0	7.7	26	42
	8.5±1.4	9.4±2.4	28±9.2	46±9.9	8.3±1.3	7.9±2.8	19±6.7	34±9.3

| | Pu-239 Concentration, aCi/m^3 | | | | | | | |
| | 1 m | | | | 10 m | | | |
Date	Respi-rable	Inhal-able	Coarse	Total	Respi-rable	Inhal-able	Coarse	Total
Nov-Dec 82	4.5	4.6	55	64	26	26	44	96
Jan-Feb 83	11	8.0	190	210	7.9	3.6	40	51
Mar-Apr	5.5	8.6	45	59	28	9.1	10	47
May-Jun	5.6	29	51	86	8.9	4.5	28	41
Jul-Aug	4.9	19	52	76	2.4	5.8	19	27
Sep-Oct	1.4	19	27	47	0.9	7.0	49	57
Nov-Dec	0.0	0.2	31	31	0.0	0.0	4.9	5
Jan-Feb	3.1	20	320	350	0.0	0.0	35	35
	4.5±3.3	14±9.7	96±104	110±110	9.3±11	7.0±8.3	29±16	45±26

The respirable dust loading is quite constant and statistically the same at both levels. This is expected because the respirable particles are mostly combustion pollutants from the metropolitan area and vehicle exhausts. This size fraction is deep black in color, while the other fractions are the color of the local soil (light brown). The dust loadings at 10 m for the inhalable, coarse particles and total dust are statistically significantly less than at 1 m. The difference is based on a 95% confidence level. Also, the inhalable dust fractions are highly correlated ($r^2 = 0.83$). The dust concentration should drop with height due to larger particles settling.

The above trend is not yet significant for the corresponding Pu-239 data, which are much more variable. About 10 more sample periods are required before the apparently lower Pu-239 concentration measurements at 10 m are statistically significant at the 95% level for the coarse and total dust fractions. The inhalable Pu-239 fraction requires 20 more

488

samples to establish this difference, which is numerically smaller for the inhalable particles.

The Pu-239 carried by the dust particles ranged from 0.1-3 μm as determined by autoradiography. The soil contains Pu-239 aggregates up to 10 μm in size, but these break up at the surface due to weathering. The Pu-239 was originally dispersed in oil as it leaked into the ground.

Activity concentration versus particle size in the resuspended dust from the wind tunnel was compared against the distribution measured in the airborne dust collected at the tower. This serves to verify wind tunnel information on the resuspension processes. In the wind tunnel resuspended dust, 60-80% of the activity was in the >75 μm fraction, while the airborne dust at the tower had only 4-12% of its activity in particles >75 μm. The airborne dust had most of its activity in the 15-45 μm fraction. These two peaks are also the corresponding peaks in dust loading. This means that activity concentration is proportional to mass concentration, an important relation for modeling.

References

Archer, W. J. (1982), Importance of plutonium contamination on vegetation surfaces at Rocky Flats, Colorado. Environmental and Experimental Botany 22,33-38.

Dreicer, M. (1984), Rain splash as a mechanism for soil contamination of plant surfaces. Health Physics 46, 177-187.

Gregory, P. H. (1973), The Microbiology of the Atmosphere, 2nd ed., Chapter V, John Wiley & Sons, N.Y.

Sehmel, G. A. (1980), Transuranic and tracer simulant resuspension. Transuranics in the Environment (Ed. Hanson, W. C. DOE/TIC-2280), p. 262.

Vincent, J. H. (1982) Inhaled Particles (Ed. Walton, W. H.), Pergamon Press, Oxford, p. 21.

Published 1984 by Elsevier Science Publishing Co., Inc.
Aerosols, Liu, Pui, and Fissan, editors

AEROSOL GENERATION BY PRESSURIZED MELT EJECTION
SAND84-0246A

J. E. Brockmann*
Sandia National Laboratories
Albuquerque, New Mexico 87185
U.S.A.

W. W. Tarbell
Ktech Corporation
Albuquerque, New Mexico
U.S.A.

EXTENDED ABSTRACT

Introduction

Pressurized ejection of molten core material from the reactor pressure vessel has recently been suggested (Commonwealth Edison Co, 1981) as an important element of severe nuclear reactor accidents. On the order of three-quarters of the accident sequences that lead to core melting are predicted to result in vessel failure while the vessel is pressurized. Phenomena associated with the high pressure (1.3 to 17 MPa) ejection of melts (Tarbell, Brockmann and Pilch, 1983) are under investigation at Sandia National Laboratories as part of an ongoing research program in reactor safety.

The following observations have been made.

1) Aerosol particles may be formed by condensation of melt vapor and also by physical breakup of the melt.

2) The extent of melt disruption affects the aerosol generation in the melt jet.

3) Effervescence of gases dissolved in the melt may play a part in melt jet disruption.

4) Ducted melt jets expel fragmented melt debris from the duct exit.

Aerosols, which we have defined as particles with an aerodynamic diameter less than 30μm, are of interest in reactor safety because they have a radionuclide content and constitute a radiological source term to the reactor containment building atmosphere. The melt debris, which we have defined as particles larger than 30μm aerodynamic diameter, do not remain suspended in the containment atmosphere. The debris is important because as it is expelled from the reactor cavity instrumentation tunnel, it may travel several meters through the atmosphere producing a heat load by cooling and chemical reaction of unoxidized metal with the oxygen in the containment atmosphere. If a large portion of the core debris is relocated outside the cavity by pressurized ejection, it may cool more readily reducing the level of concrete attack.

*This work was supported by the U.S. Nuclear Regulatory Commission under Fin. No. A1247 and performed at Sandia National Laboratories which is operated for the U.S. Department of Energy under contract number DE-AC04-76DP00789.

Experimental Set-up

For all tests, 2.5 or 10 kilogram melts were ejected from a pressure vessel at 1.3 to 17 MPa. The melts were generated by an iron oxide-aluminum metallothermic reaction which yielded iron (55 w/o) - aluminum oxide (45 w/o) melts. Two types of tests have been conducted. The first type ejected the melt as a free jet directed downwards for a distance of 1 m into a gravel bed which caught and quenched the melt. These were free jet tests and conducted at pressures from 1.3 to 17 MPa with CO_2 or N_2 as the pressurizing gas (Brockmann and Tarbell, 1983).

The second type of test, the cavity tests, ejected 10 kilogram melt charges into a 1/20th linearly scaled model of a reactor cavity and instrumentation tube access way. The tests were pressurized with N_2 to about 10.7 MPa. Two of these tests (18 and 19) were conducted inside a 45 m^3 steel chamber. In these tests, the melt was ejected downward about 23 cm to the bottom of a 25 cm diameter circular cavity. The cavity opened on the side to a horizontal duct about 40 cm long, 16 cm high, and 12 cm wide which turned upward 63° from horizontal for about 41 cm. Catch pans were used to collect the melt debris thrown out of the cavity into the test chamber. The debris was sized by sieving.

Flash x-ray photographs were taken of the melt jets in the free jet tests and of the melt debris exiting the model cavity, to provide information on the structure of the jet and expelled melt. Anderson MkIII cascade impactors were used to measure the size distribution of the aerosol.

Results and Discussion

Copius aerosol generation was observed in all tests. Measurements from the cavity tests show that 0.5% to 1% of the 10 kg melt mass was aerosolized as particles under 10µm aerodynamic diameter. These results are consistent with the estimated aerosol production from the free jet tests.

The free jet tests allowed study of jet behavior. Flash x-ray photographs of a CO_2 driven jet showed it to be disrupted by hydrodynamic forces; a solid core with ligaments of melt being stripped away. Flash x-ray photographs of a N_2 driven jet showed a much more uniform disruption similar in appearance to a jet of flashing liquid or gas effervescence in a liquid jet. Solubility calculations (Powers, 1983) indicate that N_2 is much more soluble in the melt used than is CO_2. This supports the possible explanation of effervescence as the disruptive mechanism of the N_2 driven jet.

The character of the aerosol size distribution depended on the pressurizing gas. Figure 1 shows the normalized mass distributions of the aerosol for the two tests mentioned above. The CO_2 driven test, Test 12, yields a distribution shifted toward the larger sizes while the N_2 driven test, Test 8, displays a noticable submicron aerosol mode. Electron micrographs of the collected particles suggest that the submicron particles are a condensation aerosol formed by nucleation, condensation, and agglomeration. The submicron particles are comprised of small (<0.1µm) primary particles which are too small to be collected individually on the impaction stage. Their appearance is similar to electron micrographs of steel fumes formed by vapor condensation (Ellis and Glover, 1971). The electron micrographs indicate that the micron sized and larger particles are formed by fragmentation of the melt into discrete droplets which cool into spheres.

The greater disruption of the N_2 driven jet provided greater surface area for vaporization and any effervescing gas would act to convect this vapor out of the jet. This jet could produce an enhanced vapor condensation aerosol which would be consistent with observation. The higher level of disruption is also more likely to produce small micron sized droplets from film fragmentation while the coarser fragmentation in the CO_2 driven jet would more likely produce droplets of a larger size from the action of gas flow on melt fragments (Pilch, 1981).

The mass distribution of melt debris collected in the cavity tests was fit by a lognormal distribution function. The geometric mass mean diameters (based on sieve size) were 0.75 mm and 0.43 mm for Test 18 and 19 respectively. On both tests, the geometric standard deviation was about 3.5. In Test 18 about 60% and in Test 19 more than 90% of the melt was expelled from the cavity and duct. The differences in mean size and expelled melt can be attributed to lower gas velocities in Test 18. Transport through the duct may have an influence on the distribution of droplets produced by melt fragmentation and the aerosol produced by vapor condensation. The aerosol distribution from one of these tests is shown in Figure 2. It is not significantly different from the observed distributions for N_2 driven free jets.

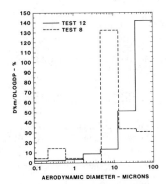

Figure 1.   Aerosol mass distribution   Figure 2.   Aerosol mass distribution
            from free jet Tests 8 and 12.   from cavity Test 18.

Figure 3 contains three micrographs of particles collected from these tests. Figure 3a is typical of the submicron vapor condensation aerosol. Figure 3b is typical of the melt fragmentation aerosol collected in the impactors. Figure 3c is typical of the sieved debris collected in the cavity tests.

Figure 3. Electron micrographs of a) vapor condensation aerosol, b) melt fragmentation aerosol and, c) melt debris collected in melt ejection tests.

Conclusions

Pressurized melt ejection produces aerosols from two sources: vapor condensation and melt fragmentation.

Effervescence may play a role in jet breakup. It could drive the film fragmentation mechanism and enhance melt vaporization.

Melt fragmentation occurs during jet disruption, blowdown of the melt generator, and transport thorugh the cavity and duct.

More research is required to determine if ducting the melt jet produces aerosols of significantly different character than those produced by free jets.

References

Brockmann, J. E. and W. W. Tarbell (1983), "Aerosol Source Term in High Pressure Melt Ejection", First proceedings of Nuclear Thermal Hydraulics, American Nuclear Society 1983 winter meeting.

Commonwealth Edison Company (1981), "Zion Probabilistic Safety Study", Chicago, IL.

Ellis, A. F. and J. Glover (1971) "Mechanism of Fume Formation in Oxygen Steelmaking" Journal of the Iron and Steel Industries. p. 593.

Pilch, M. (1981), Acceleration Induced Fragmentation of Liquid Drops, PhD. Dissertation, University of Virginia.

Powers, D. A. (1983) "Solubility of Gases in Molten Materials Produced During Reactor Core Meltdown", SAND83-1796, in press.

Tarbell, W. W., J. E. Brockmann, and M. Pilch (1983), "High Pressure Melt Streaming (HIPS) Program Plan", SAND82-2477, NUREG/CR-3025, on press.

Published 1984 by Elsevier Science Publishing Co., Inc.

Aerosols, Liu, Pui, and Fissan, editors

MEASUREMENT OF SPECIFIC SURFACE AREAS OF SMALL POWDER SAMPLES OF URANIUM, PLUTONIUM, AND MIXED URANIUM AND PLUTONIUM OXIDES USING [85]Kr AS ADSORBATE*

S. J. Rothenberg,[1] J. A. Mewhinney,[1] A. F. Eidson,[1] D. K. Flynn,[1]
G. J. Newton,[1] and R. Scripsick[2]
[1]Inhalation Toxicology Research Institute
Lovelace Biomedical and Environmental Research Institute
P. O. Box 5890
Albuquerque, NM 87185
U. S. A.

[2]Los Alamos National Laboratory
P. O. Box 1663
Los Alamos, New Mexico 87545
U. S. A.

EXTENDED ABSTRACT

Introduction

Important factors determining the rate of dissolution of a particle deposited in the lung are the chemical composition of the particle, the crystalline structure and the amount of surface interacting with body fluids. This paper describes work designed to allow measurements of specific surface area for extremely small samples of radioactive particles. Several commercial methods are available for determination of specific surface area of particles, but these are limited to sample sizes exceeding 10 mg (1). Particulate samples from filters or cascade impactors are normally available in quantities between 0.1 mg and 10 mg, and thus development of a method suitable for such small samples was necessary.

This paper describes the adaptation of a method for measurement of the specific surface area described by Chenebault and Schurenkamper (2) for powder samples. Comparison of the method with other methods of measurement of specific surface area is described. Application of this new method to measurement of specific surface area of particulates taken from simulated tank penetrator tests and of mixed U and Pu oxides of industrial origin are described.

Materials and Methods

Samples of Al_2O_3 and TiO_2 were obtained from Duke Standards, Palo Alto, California. Powder samples produced during normal operation of nuclear fuel pellet fabrication facilities were collected (3,4), aerosolized for animal exposures, and characterized by density determination, TEM and [85]Kr adsorption. Samples were collected at the Hanford Engineering and Development Laboratory (HEDL) during centerless grinding of $(U,Pu)O_2$ fuel pellets. These pellets had been heat-treated at 1750°C for several hours prior to the finish grinding of the pellet to exact dimensions. A second sample collected at HEDL represented mechanically mixed powders of PuO_2 and UO_2 treated at 750°C. A sample of "pure" PuO_2 was collected at the Babcok and Wilcox (B&W) mixed oxide fuel fabrication facility prior to mixing the PuO_2 and UO_2 powders. A second sample was collected at B&W representing particulates produced by centerless grinding of $(U,Pu)O_2$ treated at 1750°C. The respirable fraction from combustion of depleted uranium metal, a mixture of oxides, was collected on filters in vitro dissolution measurements made using lung serum simulant. At the beginning and end of the study, samples were removed for ^{85}Kr adsorption study.

A vacuum microbalance (Cahn RG2000) was used for gravimetric determination of adsorption of research grade nitrogen (Mattheson). A standard Quantasorb instrument was employed, using 1%, 5% and 10% research grade nitrogen in helium mixtures (Scientific Gas Co.) and research grade nitrogen (Mattheson Gas Co.). The apparatus used for ^{85}Kr adsorption consisted of a vacuum manifold to which adsorption cells and a reservoir of ^{85}Kr were attached and a pumping assembly capable of producing a vacuum of 10^{-6} Torr.

Results

Results are shown in Tables 1-3.

Table 1. Specific Surface Area Values for Standard Powders Determined by Three Different Methods

| Sample | Specific Surface Area, $m^2 g^{-1}$ | | | |
	Micro-Balance	Quantasorb	^{85}Kr [a] Adsorption	Manufacturer's Specifications
Al_2O_3, A	2.6	2.53	3.23	3.06
Al_2O_3, B	81.9	76.0	90	81.4
Al_2O_3, C	13.0	12.6	12.6	14.0
TiO_2	8.85	8.95		10.3

[a]TiO_2 used as a standard with value 8.9 m^2g^{-1} at 250°C.

Table 2. Comparison of Specific Surface Areas of Two Samples of Uranium Oxide at the Commencement and end of a 30 day Dissolution Study in Lung Serum Simulant

| Sample | Surface Area ($m^2 g^{-1}$) | | Decrease (%) |
	Predissolution	Post-dissolution	
A	1.84	0.64	65
B	3.85	2.36	39

Table 3. Specific Surface Area of Particulate Samples of Materials Used
in Fabrication of U and Pu Mixed Oxide Nuclear Fuels

HEDL (U,Pu)O$_2$ 1750°C		HEDL UO$_2$ + PuO$_2$ 750°C		B&W (U,Pu)O$_2$ 1750°C		B&W "Pure PuO$_2$ 850°C	
Cumulative Outgassing Time (hr)	Specific Surface Area (m^2/g)	Cumulative Outgassing Time (hr)	Specific Surface Area (m^2/g)	Cumulative Outgassing Time (hr)	Specific Surface Area (m^2/g)	Cumulative Outgassing Time (hr)	Specific Surface Area (m^2/g)
16	2.78	16	2.82	16	3.76		
32	2.90	32	3.35	32	8.44	16	19.1
48	2.66	48	4.79	48	15.9	32	19.0
		64	5.82	64	20.9	48	20.3
		80	12.7	80	24.1		
				96	28.3		
16	3.06			112	47.1		
32	3.08						
48	2.97						

Discussion

The values in Table 1 demonstrate good agreement between the three methods. The apparent precision of the [85]Kr adsorption method is somewhat less satisfactory (± 10%) than that of the other two methods (± 5%). There is no evidence of systematic error in the [85]Kr adsorption method. Since the [85]Kr adsorption of samples as small as two mgs (S $\geq$ 1 m^{2}g^{-1}) can readily be determined, this makes it the method of choice for small samples.

The data for the uranium oxide samples (Table 2) demonstrate that, in accord with theoretical expectations, surface roughness decreases with time. This implies that the shape factor of Mercers theory changes slowly with time.

Replicate samples of HEDL (U,Pu)O$_2$ 1750°C gave concordant results. The high values for the B&W PuO$_2$ sample suggest the sample is porous, which confirms previous inferences from electron microscopy and density determinations (4). The pattern of increasing surface area with successive 250°C outgassing for two samples suggests slow annealing of these samples, but the difference in behavior of the B&W α HEDL 1750°C samples is inexplicable. Additional samples are under study.

Summary

An apparatus suitable for the determination of the specific surface area characterization of small masses of particulates requiring containment has been constructed. The method gives self-consistent data for samples between 2 and 50 mg. The method has been tested against other methods and applied to (Pu/U) oxide samples.

References

1. Rothenberg, S. J., P. B. DeNee, Y. S. Cheng, R. L. Hanson, H. C. Yeh, and A. F. Eidson (1982) "Methods for the Measurement of Surface Areas of Aerosols by Adsorption," Adv. Colloid Interface Sci. 15:233-249.

2. Chenebault, P. and A. Schürenkämper (1965) "The Measurement of Small Surface Areas by the B.E.T. Adsorption Method," J. Phys. Chem. 69:2330.

3. Raabe, O. G., G. J. Newton, C. J. Wilkinson, and S. V. Teague (1978) "Plutonium Aerosol Characterization Inside Safety Enclosures at a Demonstration Mixed-Oxide Fuel Fabrication Facility," Health Phys. V35:649-661.

4. Newton, G. J. (1978) "Mixed Oxide Fuel Fabrication," in Radiation Exposure and Risk Estimates for Inhaled Airborne Radioactive Pollutants Including Hot Particles - Annual Report for Period July 1, 1976 - June 30, 1977, NUREG/CR-0010.

CHAPTER 12.

WORK ENVIRONMENT AND INDOOR AEROSOLS

Published 1984 by Elsevier Science Publishing Co., Inc.
Aerosols, Liu, Pui, and Fissan, editors

ENVIRONMENTAL TRACERS OF AIRBORNE FIBROUS AEROSOLS

K.R. Spurny

Fraunhofer-Society
Institute for Toxicology and Aerosol Research
D-5948 Schmallenberg 11 - Grafschaft
West Germany

Introduction

The application of modern physical methods in analytical
chemistry makes possible to identify the origin organic and
inorganic of particulate air pollutants. Analytical data of
inorganic and organic components from aerosol emission samples
often contain complex information about the sources contributing to the particulate air pollution (Puxbaum, 1984). A
seven-element tracer system showed that regional pollution
aerosols in North America and Europe have characteristic signatures that can be followed into remote areas up to several
thousand kilometers. Furthermore molecular marker analysis
provided promising information about sources of carbonaceous
particulates in tropospheric aerosols (Simoneit, 1983). The
analysis of many individual particles from a coal fly ash
source and the categorization of these particles based upon the
relative abundance of energydispersive spectra types provided a
method for assigning a signature to each fly ash source
(Minnis, 1982).

In our investigation tracer elements of atmospheric fibrous
aerosols were identified by means of single particle analysis.

Measurements

Electron microscopy (EM), as well as the associated analytical procedures EDXA and SAED, micro Raman spectroscopy, mass
spectroscopy (LAMMA procedure), SIMS and some other modern analytical methos are suitable for analyzing and characterizing
single solid aerosol particles. Different trace elements and
molecules associated with particles can be used as natural
tracers, that are characteristic for the composition of the
pollutants original matrix.

In our investigation tracer elements of atmospheric fibrous
aerosols were identified by means of EM, EDXA and LAMMA. The
preliminary results have shown, that it could be possible to
find characteristic trace elements originated from matrices of
asbestos-cement products, brake linings, etc. Figure 1 shows an
example of analyzed fibers released from an asbestos-cement
roofing tile. X-ray fluorescence analysis of fibers indicates

500

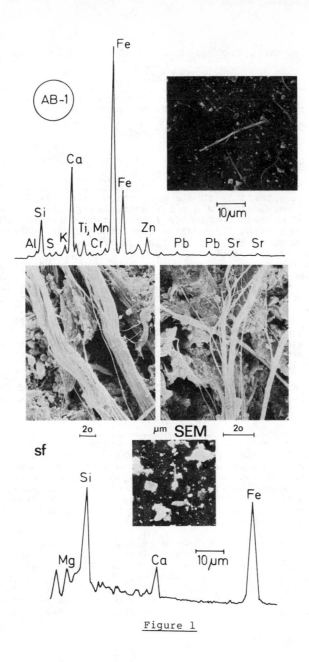

Figure 1

Figure 1. Analysis of released chrysotile fibers from the
surface of an asbestos-cement roofing tile:
<u>AB-1</u> is the element spectrum (x-ray fluorescence
spectroscopy) of the released material, the <u>SEM</u>
means and electron micrograph of the weathered tile
surface and <u>sf</u> is the element spectrum (Scanning
electron microscopy and EDXA) of a single chrysotile
fiber from ambient, air contaminated with calcium.

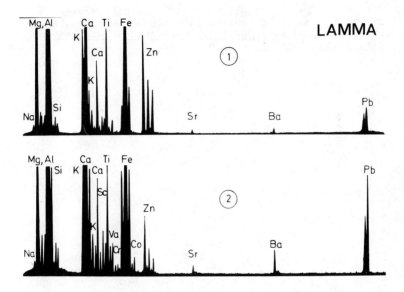

Figure 2. Element spectra (LAMMA-analysis) of two single
asbestos fibers released from the surface of a
weathered asbestos-cement tile. Contamination by Zn,
Ba, Ti, Pb etc.

many trace elements originating from the cement matrix. Using
SEM and EDXA analysis of single fibers could be done also. Ca-
peak seemed to be characteristic for chrysotil fibers released
from cement matrices.

Other measurements indicated, that asbestos fibers released
from brake linings of automobiles are labeled with metal elements, e.g. with Cu and Cr.

References

Minnis, M.M. (1982) "A Method for Assigning to Coal Fly Ash Samples by an Individual Particle Analysis Technique", Dissertation, State University N.Y., Syracuse, N.Y., USA.

Puxbaum, H. and B. Wopenka (1984) "The application of Receptor Models for Aerosol Source Analysis", Fresenius Z. Anal. Chem. 317:278.

Simoneit, B.R. (1983), "Application of Molecular Marker Analysis to Reconcile Sources of Cabonaceous Particulates in Tropospheric Aerosols.", 2nd Int. Conference on Carbonaceous Particles in the Atmosphere, Linz, Austria.

Published 1984 by Elsevier Science Publishing Co., Inc.
Aerosols, Liu, Pui, and Fissan, editors

THE STATIC ELECTRIFICATION OF AIRBORNE DUSTS IN INDUSTRIAL WORKPLACES

J.H. Vincent, A.M. Johnston and A.D. Jones
Physics Branch, Institute of Occupational Medicine,
8 Roxburgh Place, Edinburgh, U.K.

Introduction

A growing body of work now exists to demonstrate that the electrical forces which an airborne dust particle can experience as a consequence of its carrying an electric charge can be of considerable practical importance; for example, in relation to sampling and filtration (Liu, Pui, Rubow and Szymanski 1984) and, perhaps most interestingly, in relation to lung deposition after inhalation (Yu 1984, Prodi 1984). In order that such work can be placed in a practical context, it is desirable that measurements of the states of charge of industrial workplace aerosols be carried out. With this in mind, experiments were conducted at a total of 48 workplace locations in 13 factories, 2 quarries and 1 coal mine, encompassing textile, extraction and manufacturing industries.

Instrumentation

A measurement system was required which could provide the full charge distribution for dust particles of defined size and distinguish between positive and negative. It also needed to be portable and robust enough to be transported within and between industrial locations. The apparatus which was developed and used in most of the studies reported is the 'split-flow electrostatic elutriator' system described by Johnston (1983). Only for the coal mine parts of the investigation was it necessary, for reasons of intrinsic safety, to develop a separate system. This was based on the one which we used in our earlier laboratory studies with asbestos dust (Vincent, Johnston, Jones and Johnston 1981). The operation of both systems is based on analysis (by the method of tangents) of the curve relating penetration through an electrostatic elutriator to the potential difference applied between its plates.

Results and Discussion

For all the dusts examined (with the exception of asbestos), open filter samples showed that the airborne particles were basically isometric, even for those derived from bulk material which itself was fibrous. For such particles, it was found that the median magnitude of charge per particle ($\bar{q}$) may be expressed to a fair approximation as a function of particle diameter (d) by

$$\bar{q} = Ad^n \tag{1}$$

where, if d is in microns, A is therefore the median magnitude of charge
carried by a 1 μm particle.   For fibrous particles of asbestos, it was
found to be more appropriate to describe the charge carried as being proportional to length.

Examination of the charge distribution of aerosols at the workplace
locations visited showed that, without exception, the distribution was
symmetric;   that is, to within the limits of experimental error, the
distributions of positive and negatively charged particles were the same.

The results for the median magnitude of charge, obtained in the form
consistent with Equation (1), showed for isometric particles that values of
A ranged from 2 to 50, with the largest values apparently being associated
with industrial processes involving the greatest transfer of energy or with
the 'freshest' dust.   Values of n ranged from 0.8 to 2.0.   For airborne
dust in an asbestos textile factory, charge per unit length of airborne
fibre ranged from 6 to 13 electrons/micron, including one particularly
interesting result for dust near the spinning process where it was found that
the level of charge was significantly lower for material which had been made
damp prior to working.   In general, the results were comparable with those
obtained for airborne dust clouds generated in our laboratories for animal
inhalation studies, etc.

If these workplace results are considered together with the information
which is available from studies with animals and humans (summarised by
Prodi 1984) and from the theory of lung deposition (summarised by Yu 1984),
then it becomes apparent that, for most such dusts, the contribution of
electrostatic forces to the lung dose received during exposure should not
be large enough to warrant special consideration by industrial hygienists.
For asbestos fibres, however, the matter might not be dismissed so readily.
Here particles with small aerodynamic diameter (governed predominantly by
fibre diameter) and so capable of penetrating to the deep lung, may carry
a relatively large amount of charge (governed by fibre length) and so
experience correspondingly strong forces towards the lung wall.   This
interpretation is consistent with the high electrostatic enhancement of
lung burden which was observed in our earlier experiments with rats in-
haling both amosite and chrysotile asbestos (Vincent, Johnston, Jones and
Johnston 1981, Jones, Johnston and Vincent 1982).

References

Johnston, A.M. (1983).   "A semi-automatic method for the assessment of
electric charge carried by airborne dust", J. Aerosol Sci. 14: 643-655.

Jones, A.D., Johnston, A.M. and Vincent, J.H. (1982).   "On the static
electrification of airborne asbestos dust", Aerosols in the Mining and
Industrial Work Environment (Edited by V.A. Marple and B.Y.H. Liu),
Ann Arbor Science, Ann Arbor (Mich.), 613-632.

Liu, B.Y.H., Pui, D.Y.H., Rubow, K.L. and Szymanski, W.W. (1984). "Electrostatic effects in aerosol sampling and filtration", Ann.occup.Hyg. (to be published).

Prodi, V. (1984).   "Electrostatic lung deposition experiments with humans and animals", Ann.occup.Hyg. (to be published).

Vincent, J.H., Johnston, W.B., Jones, A.D. and Johnston, A.M. (1981). "Static electrification of airborne asbestos: a study of its causes, assessment and effects on deposition in the lungs of rats", Am.Ind.Hyg.Ass.J. 42: 711-721.

Yu, C.P. (1984).   "Theories of electrostatic lung deposition of inhaled aerosols", Ann.occup.Hyg. (to be published).

Published 1984 by Elsevier Science Publishing Co., Inc.
Aerosols, Liu, Pui, and Fissan, editors
506

The Growth of a Welding Fume Aerosol as a Function of Relative Humidity

Mats S Ahlberg[1], Hans-Christen Hansson[2] and Ingrid Fängmark[1]

[1] NBC department, National Defence Research Institute,
S-901 82 Umeå, Sweden

[2] Department of Nuclear Physics, Lund Institute of Technology,
Sölvegatan 14, S-223 62 Lund, Sweden

The lung deposition of hygroscopic aerosols is known to be affected
by the growth of the particles due to the water uptake that occurs
when they are exposed to the high relative humidity in the respiratory
tract.

In this study the growth of a manual metal arc (MMA) welding fume
as a function of relative humidity is investigated using two electrostatic
classifiers in series. The first classifier extracts a well defined
mobility fraction of the welding fume aerosol. This mobility fraction
is then exposed to humidified sheath air in the second classifier and
the growth is determined. The growth of the water soluble fraction of
the welding fume (about 25 % in this case) is also studied in the same
way by generating an aerosol of the water soluble material with a
Collison atomizer.

The welding fume particles show different degrees of growth. From no
growth at all to that for the pure water soluble material. This behaviour
suggests that the welding fume aerosol is an external mixture consisting
of particles containing different concentrations of hygroscopic material.
From particles containing no hygroscopic material at all to particles
consisting of only such material.

The implications of these findings on the lung deposition of the
welding fume is discussed.

Published 1984 by Elsevier Science Publishing Co., Inc.

Aerosols, Liu, Pui, and Fissan, editors

TRANSPORTATION, DEPOSITION AND ANALYSIS OF FINE AND COARSE PARTICLES IN AIR-CONDITIONED SPACES

T. Raunemaa, K. Karhula, M. Kulmala, A. Reponen

University of Helsinki, Department of Physics
Siltavuorenpenger 20 D, SF-00170 Helsinki 17, Finland

EXTENDED ABSTRACT

1. Introduction

Transportation of aerosols to indoor air in buildings has been investigated. Particle concentrations are determined by fractionated sampling with the cut-off size around 1 μm. PIXE is used for elemental identification, which gives the concentrations of several source derived elements / 1 /. Fine particle distributions below 1 μm were obtained by EAA (TSI 3030). The two kinds of analyses, when considered together, reveal detailed particle transport characteristics to indoors and for various ventilation systems. These characteristics can be applied also in health risk analyses.

2. Particle concentrations

Particle concentrations indoors are strongly influenced by human activity, e.g. by smoking. The present study concerns only the indoor air quality in buildings where there are no cigarette or pipe smokers. Particle studies were carried out in these environs both in the day and at night.

The study houses are situated in urban and suburban Helsinki within much the same regional climate. During the experiments the outside temperature varied between 2-4 °C, barometric pressure between 89-101 kPa, relative humidity was > 60 % and surface wind ≤ 5 m/s. During the experiments only once in no. 2 house a rain occurred during the night. The sampling took place in October-November, 1983.

The average indoor concentrations of fine particles ($D_p \leq 1$ μm) were 17-30 μg/m^3, with corresponding I/O-ratios (indoor/outdoor) varying between 0.50 and 0.29. Average indoor concentrations for coarse particles ($D_p > 1$ μm) varied between 27-16 μg/m^3, while the corresponding I/O-ratios were 0.48-1.7.

The coarse particle concentrations were obtained in the first instance, to be affected by human indoor activities. High I/O coarse particle ratios are accordingly coupled with low fine particle I/O-ratios. The I/O-ratios were seen to be lowest when particle concentrations were highest.

3. Elemental analyses

Transport of aerosols indoors was documented by the presence of cumbustion derived aerosols in indoor air. E.g., lead and sulphur can be detected. Exceptionally high I/O values were obtained for K and Pb. Outdoor concentrations have no strong influence on the elemental I/O-ratios (see Figure 1). It is the ventilation rate which was observed to affect elemental fine particle concentrations with an apparent minimum at 1-2 h^{-1}. The elemental I/O-ratios can be presented as relative transmission values / 2 /, which reveal that lead behaves differently from the other elements.

In a recent study in automated telecommunication exchanges / 2 /, where mechanical filtration (F 85) was shown not to be enough for removing combustion originated aerosols, similar results resulted.

4. Fine particle distributions

The fine particle distribution is shown to comprise at least two modes, the nuclei and accumulation mode particles / 3 /. The nuclei mode particle group around 0.02 μm was studied outdoors overnight during the climatically fairly stable period. The number of nuclei mode particles Nn decreased smoothly during the night, until early morning rush hours (Fig. 2). The indoor particle number followed the outdoor change with a slight time delay. Early in the morning Nn indoors was seen even to exceed Nn outdoors.

Similar results as for nuclei mode particles were obtained also for accumulation sizes at around 0.1 μm. Mode parameters indicate that the changes apparently are delayed relative to the nuclei mode, a situation which may be attributed to aerosol growth / 3 /.

5. I/O relationships and size transformations

When I/O relationships for Nn are considered for different outdoor concentrations, a semiempirical behaviour I/O = $1.4 \cdot 10^3 (Nn)^{-0.8}$ is obtained (Fig. 3). The indoor concentration nearly equals outdoor concentration when Nn ≅ 10^4 cm^{-3} (and Va ≅ 1 μm^3 cm^{-3}). Both mean sizes, DGNn and DGVa, the have their largest values, i.e. these aerosols are aged.

Average aerosol concentrations and mean geometric sizes (Table 1) reveal several aerosol characteristics; a.o. the following ones:

The semiempirical relationship for Nn versus DGNn (2) holds for outdoor as well as for ventilation inlet aerosols.

The mean size DGNn is larger by 60 % when indoor aerosols are compared with outdoor aerosols.

Indoor concentration Nn reaches a uniform level of on average 2600 cm^{-3}. Thus the I/O-ratios in some instances may reflect more the effect of indoor sources rather than aerosol transport to indoors.

The accumulation mode mean sizes DGVa, as well as aerosol concentrations Va, are smaller in urban than in suburban areas; this is attributable to there being more fresh aerosols in the former case.

References

1. T. Raunemaa, A. Hautojärvi, Aerosols in the Mining and Industrial Work Env., Vol. 3, Eds. V.A. Marple, B.Y.H. Liu, Ann Arbor Sci., Ann Arbor 1983, 971

2. T. Raunemaa, A. Hautojärvi, V. Lindfors, A. Reponen, Characterization of Fine Particle Aerosols in Indoor and Outdoor Air in Telecommunication Exchange Buildings, to be published in Nucl. Instr. & Meth., 1984

3. Whitby, K.T., Atmospheric Environment <u>12</u> (1978) 135-159

ID	Nn_{OUT} $DGNn_{OUT}$	Nn_{VENT} $DGNn_{VENT}$	Nn_{IN} $DGNn_{IN}$	Va_{OUT} $DGVa_{OUT}$	Va_{VENT} $DGVa_{VENT}$	Va_{IN} $DGVa_{IN}$
1 Urban	65000 0.016	3580 0.015	2000 0.028	9.3 0.23	1.03 0.23	1.8 0.24
2 Suburban	30000 0.020	2050 0.020	2200 0.028	11.1 0.28	1.16 0.22	2.4 0.31
3 Suburban	28500 0.026	3010 0.018	3500 0.028	14.8 0.026	0.78 0.25	1.4 0.26

Nn in units cm^{-3}, Va in $\mu m^3 \, cm^{-3}$, DGNn and DGVa in μm

Table 1. Average nuclei mode numbers, Nn and DGNn, as well as accumulation mode numbers Va and DGVa, in the three study regions.

510

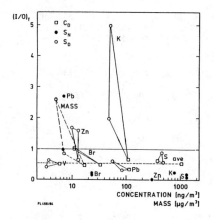

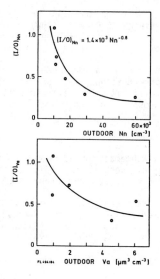

Fig. 1. Elemental indoor-to-outdoor ratios for fine particles $D_p \leq 1$ μm in suburban areas (o day o night) and in urban Helsinki (□ day). Average value 0.55 is depicted by a dotted line.

Fig. 3. I/O-ratios for nuclei size (in Nn) and accumulation size (in Va) particles as a function of outdoor concentration.

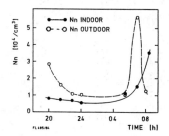

Fig. 2. Nuclei mode size particle numbers in outdoor and indoor air at study environs in urban Helsinki.

Published 1984 by Elsevier Science Publishing Co., Inc.
Aerosols, Liu, Pui, and Fissan, editors

Experiment and Data Interpretation for Fiber Mobility
Distribution from Penetration Measurements

P.Y. Yu, S.Y. Hwang and J.W. Gentry
Department of Chemical and Nuclear Engineering
University of Maryland, College Park, MD 20742
U.S.A.

EXTENDED ABSTRACT

Introduction

Electrostatic properties of both isometric and non-spherical aerosols have
been a subject of intensive investigation for the last ten years. Electrostatic mobility measurements have been applied to determine indirectly the aerosol
size distribution for ultrafine aerosols (Liu, et al, 1975). However, there is
no analogous instrument for fibrous particles or non-spherical particles. This
is due to the fact that the electrostatic properties of fibers are not fully
understood.

An objective of this study was to measure the fiber mobility distribution
in a variable voltage mobility analyzer (VVMA). Before entering the mobility
analyzer, the fibers were charged in a unipolar diffusion charger in which the
ion concentration was varied. By measuring the particle concentration before
and after the mobility analyzer, the penetration was determined as a function
of the voltage applied in the analyzer. The mobility distributions were re-
trieved from the penetration measurements using an inversion algorithm. Those
experimental results were compared with theoretical results of Laframboise et
al. (1977).

Equipment

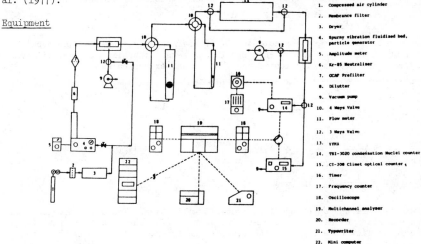

1. Compressed air cylinder
2. Membrane filter
3. Dryer
4. Spurny vibration fluidized bed, particle generator
5. Amplitude meter
6. Kr-85 Neutralizer
7. GCAF Prefilter
8. Dilutter
9. Vacuum pump
10. 4 Ways Valve
11. Flow meter
12. 3 Ways Valve
13. VVMA
14. TSI-3020 condensation Nuclei counter
15. CI-208 Climet optical counter
16. Timer
17. Frequency counter
18. Oscilloscope
19. Multichannel analyzer
20. Recorder
21. Typewriter
22. Mini computer

The schematic diagram of the mobility measurement is shown in Fig. 1. The
fibrous particles were generated from a Spurny vibrating fluidized bed. Then,
by passing through a Kr-85 neutralizer and a glass capillary array filter (GCAF),

agglomerates could be removed. The particles were, also, relatively uncharged. The aerosol flow rate was finely controlled by a diluter (Yu, et al., 1983).

As shown in Fig. 2, the VVMA is a modified TSI model 3030 EAA (electrical aerosol analyzer). The charging parameters can be controlled individually and the particle numbers before and after the mobility analyzer can be measured. It was necessary to bypass the Faraday cage of EAA, since the concentration of particle is too low to be counted accurately. An additional Kr-85 neutralizer was installed at the exit of VVMA to remove the charges on fibers. The particle concentrations were measured by two particle counters - CLIMET model 208 and ROYCO model 203.

Experiment

Amosite and actinolite served as the source for the fibrous particles. The sample was placed in the Spurny vibrating fluidized bed with a frequency of 55 Hz, an amplitude of 7,000 μm and a superficial flow rate of 10 cm3/sec. After 10-20 minutes the concentrations of the aerosol became stable. By changing the aerosol and the charger sheath air flow rates, the residence time in the charger can be controlled. The ion density is dependent on the screen voltage which is in the range of 300 to 1200 volts. The penetration was calculated as the ratio of the concentration after to average concentration before the mobility analyzer. The experiments were carried out for several different operation conditions - residence time (t) and ion density (N_o).

Data Interpretation

The penetration measurement, $\overline{Pt}_i$, can be expressed by the equation,

$$\overline{Pt}_i = \int_0^\infty Pt(h_i, Z_p) \, F(Z_p) \, dZ_p \tag{1}$$

where, h_i is the operation condition, Z_p is the mobility, $F(Z_p)$ is the mobility distribution function, and $Pt(h_i, Z_p)$ is the kernel function of the mobility analyzer. The kernel function of the VVMA can be expressed as

$$Pt = \begin{cases} 0 & , \text{ if } Z_p \geq Z_{pu}, \\ 1 - \dfrac{Z_p - Z_{p\ell}}{Z_{pu} - Z_{p\ell}} & , \text{ if } Z_{p\ell} \leq Z_p \leq Z_{pu}, \\ 1 & , \text{ if } Z_p \leq Z_{p\ell} \end{cases} \tag{2}$$

where

$$Z_{p\ell} = \frac{1}{2\Pi VL} (Q_{tm} - Q_{am}) \cdot \ln\left(\frac{R_2}{R_1}\right) , \tag{3}$$

and

$$Z_{pu} = \frac{1}{2\Pi VL} Q_{tm} \ln\left(\frac{R_2}{R_1}\right) \tag{4}$$

Z_{pu} and $Z_{p\ell}$ are the upper and lower cut-off mobilities. Those are dependent on three operation conditions: the applied voltage (V), the total flow rate (Q_{tm}) and aerosol flow rate (Q_{am}) in the mobility analyzer. L is the length of the analyzer.

The apparent mobility (or size) method (AMM) (Park, et al, 1980) was applied to retrieve the mobility distribution. It is described in detail elsewhere.

Results and Discussions

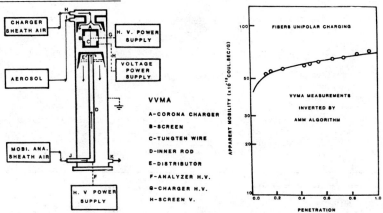

Fig. 3 shows the typical AMM inversion from penetration measurements. The operation conditions were: $N_o t = 4.2 \times 10^6$ ions.sec/cm^3, $Q_{am} = 22.2$ cm^3/sec and $Q_{tm} = 183.3$ cm^3/sec. The mean and deviation of the mobility distribution, (for a log-normal distribution) were 6.44×10^{-11} coul.sec/g and 0.70. Using the same inversion method, other penetration measurements were interpreted for different charging conditions.

The theoretical charging theory by Laframboise et al, (1977) was applied to compare the experimental results. In Fig. 4 the mean mobility of the fibers is plotted as a function of the residence time in the unipolar diffusion charger. When the fiber aspect ratio (B) or the charging time (t) increases, the mean mobility increases. Consequently the electrostatic classification by fiber length is possible.

The experimental and theoretical results are compared in Fig. 5. Theory gave mobilities that were 15% higher than those obtained from experiment. This could be due to the assumption of neglecting image force in the theory or to particle loss in the charger.

Conclusions

The principal conclusions of this study are summarized below:
1. A variable voltage mobility analyzer was developed to measure the fiber mobility distribution.
2. The "Apparent mobility method" can be applied for retrieving the fiber mobility distribution, from penetration measurements.
3. The fiber mobility distributions were measured for unipolar charging at different ion fluxes N_{ot}. These experimental values were compared with the Laframboise theory.
4. Electrostatic classification is a feasible method for measuring the fiber size distribution including both the mean length and diameter.

514

References

1. Liu, B.Y.H., and D.Y.H. Pui, J. Aerosol Sci., 6, 249-264 (1975).
2. Yu, P.Y., C.A. Bohse, and J.W. Gentry, J. Aerosol Sci., 14, 334-339 (1983).
3. Park, Y.O., W.E. King, and J.W. Gentry, I&EC Prod. Res. Dev., 19, 151-157 (1980).
4. Laframboise, J.G., and J.S. Chang, J. Aerosol Sci., 8, 331-338 (1977).

Acknowledgement

We wish to acknowledge the support of the State of Maryland Department of Natural Resources and the National Science Foundation for their support under grants P 67 80 04 and CPE 80 11269.

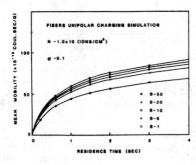
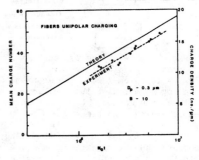

Published 1984 by Elsevier Science Publishing Co., Inc.
Aerosols, Liu, Pui, and Fissan, editors

RESUSPENSION OF DUST FROM WORK CLOTHING
AS A SOURCE OF INHALATION EXPOSURE

B.S. Cohen and R. Positano
Institute of Environmental Medicine
New York University Medical Center
New York, New York 10016
U.S.A.

EXTENDED ABSTRACT

Introduction

Previous studies of the accuracy of conventional aerosol sampling techniques for the estimation of occupational exposures showed that significant amounts of dust could be resuspended from work clothing into lapel mounted monitors, and that this was the single most important factor for differences among exposure estimates made by different sampling methods (Cohen, et al., 1984). Bohne (1982) conducted experiments with cotton and Nomex[R] aramid fabrics which are used for work clothing at the Be plant. He used lengths of new fabric obtained from the manufacturer of work uniforms for the Be production company (Brush Wellman, Inc.). The fabrics were prelaundered and then exposed to the Be containing atmosphere. Subsequently, the fabric was tested to measure the Be dust load and determine whether the dust would resuspend into filter samplers placed either directly in front of, or above the mechanically agitated fabric samples. Bohne's (1982) results showed that a significant amount of Be containing dust could be resuspended from the fabrics into monitors mounted right near (0.5 cm) the fabric surface. Since these experiments were carried out at the Be plant, it was not clear whether the resuspended dust also contributed to the air concentration measured by samplers placed at some distance above the fabrics to simulate the nose or mouth. In order to assess some of the factors that may influence the amount of resuspended dust that is actually available for inhalation by a worker, we initiated a pilot study using Nomex[R] workshirts that were worn at the Be refinery and we measured the amount of dust resuspended in a clean laboratory glove box.

Experimental Methods

Six shirts, three new (two unwashed, one washed) and three well worn (two unwashed, one washed) were obtained from Brush Wellman, Inc. These shirts were worn for one full workshift by employees at the Be plant, and either laundered or not as indicated. The shirts were selected by plant personnel who classified them as "old" or "new" by observation. The classes refer to worn "old" looking shirts compared with shirts that appeared to be fairly new. A complete history for each shirt (i.e., age, number of times worn, number of times laundered) was not attainable. The shirts were received at our laboratory enclosed in individual plastic bags. Four 8.5" diameter circular sections were cut from each shirt. The sections were taken from the upper front and back as those areas were most likely to have been exposed to Be dust and were

516

near the breathing zone. Each of the 8.5" diameter sections was placed
in a 7" diameter embroidery hoop. The excess material about the perimeter of each hoop was removed, weighed, and saved for subsequent
analysis. Three additional sections were cut from each shirt and measured and weighed to establish conversion factors for mass per unit area
of the fabric. The hoops were suspended in a sealed 0.87 m^3 glove box.

Four filter cassette monitors with 0.8 μm pore size membrane
filters were placed 3" in front of the suspended hoops. Two pairs of
monitors, one of each pair having a cyclone precollector, were placed
near the fabric array, one set on each side of the glove box. The flow
rates through the filter samplers was maintained at 2 Lpm with critical
orifi. Flow rates were checked before and after each sampling run. The
fabrics were agitated for 1.25 hours by a motor driven apparatus which
repeatedly struck the back of the fabric to simulate the motion of a
worker. The air sampling continued for an additional 15 minutes. The
volume of air sampled was 180 liters. The glove box was decontaminated
between each experiment.

Filters were wet ashed with nitric and sulfuric acids and analyzed
for Be by Inductively Coupled Plasma Spectrometry (ICP) at a wavelenth
of 313.04 nanometers. Fabric samples were dry ashed at 450°C for 48
hours then wet ashed and handled in the same manner as the filter samples. Blanks and spikes were included for each experimental trial. The
lower limits of detection (LLD) for an individual sample were calculated
according to Gabriels (1970). The LLD for the filter samples was .015
μg per sample. This converts to an LLD for air concentration of .08
$μg/m^3$. For the fabric samples the LLD was 15.4 $μg/m^2$.

Results

The average initial concentration for the nearly new washed shirt
was 0.2 ± 0.3 mg/m^2 compared to 30 ± 3 mg/m^2 for the old washed shirt.
The mean concentrations of Be remaining on the washed new and old shirts
after agitation were 0.4 ± 0.2 and 27 ± 6 mg/m^2. All error terms
reported here are ± 1 standard deviation. The mean initial concentration of the two unwashed nearly new shirts was 16 ± 14 mg/m^2, and 29 ±
16 mg/m^2 for the two unwashed old shirts. These values are not significantly different at the p ≤ 0.05 levels*. The concentration of Be
remaining after agitation for both new and old unwashed shirts (8 ± 10,
and 26 ± 13 mg/m^2) were not significantly different either.

When the initial concentrations in all samples of the three nearly
new shirts are combined and compared with all samples from the three old
shirts, the older shirts demonstrate significantly higher initial concentrations (p ≤ .001). Concentrations left on the shirts after agitation are also significantly different (p ≤ .001).

The average air concentrations of Be measured in front of the three

*Means were compared by testing the null hypothesis by Student's
t-test, to determine whether there were significant differences.
The t-test was used for both fabric and air samples.

new shirts was very low (0.03 ± 0.04 µg/m^3) and 15 of the 18 filter samples were below the LLD. For the old shirts concentrations were higher (0.28 ± 0.33 µg/m^3). These measured air concentrations for the three old and three new shirts are significantly different ($p \leq .001$). The mean measured air concentrations for the two unwashed new and old shirts were 0.04 ± 0.05 and 0.39 ± 0.36 µg/m^3. These too are significantly different ($p \leq .001$). The mean air concentrations for the washed new (0.01 ± 0.01 µg/m^3) and washed old (0.08 ± 0.03 µg/m^3) shirts were also significantly different ($p \leq .001$).

Based on a paired t-test, there was no significant difference between air concentrations measured with and without cyclone pre-collectors for the three old shirts. For the three new shirts, air concentrations measured were too low to permit meaningful comparison.

Discussion and Conclusions

While this pilot study was limited by the small number of samples and lack of detailed history of the work shirts, some interesting conclusions are apparent. Very high concentrations of Be were found in the work shirt fabric. This was particularly notable considering that the air concentration of Be measured at the refinery was found, on repeated occasions, to be only a fraction of the threshold limit value (TLV) of .002 mg/m^3. Except for the new washed shirt, all values measured were much greater than expected from the values measured by Bohne (1982). He reported that fabric samples in motion on an oscillating frame collected $.075 \pm .009$ mg/m^2/hour at the Be plant. A few stationary samples which he exposed for up to three days in a die stripping hood collected about 1 mg/m^2 of Be. The mean Be values for the shirts tested here, omitting the washed new shirt, ranged from 12 to 37 mg/m^2.

The washed new shirt had less Be initially than did the unwashed new shirts which would be expected. Also, it appears that washing a nearly new shirt significantly reduces the amount of Be released on agitation. Washing of the old shirt however, did not make a difference in the initial Be concentration when compared to the unwashed old shirts and had little or no effect on the amount of Be released from the fabric. This is surprising, because washing is supposed to cleanse clothing and remove particles. It is possible that repeated washing may contribute to the Be load seen in the old shirts. The amount of Be released into the glove box by the shirt samples when they were agitated was very variable. No difference was detected between the new and old unwashed shirts, but the old shirt which had been washed released significantly more than the washed new one. There was no correlation between the amount released and the measured air concentrations. This suggests that some of the released material did not remain suspended.

For some sample sections in five of the six shirts, the calculated mass of Be per unit fabric area were higher from agitated shirt sections (circular hoops) than from adjacent sections which were not agitated (perimeter). The magnitude of these discrepancies was greater than that which could be caused by analytical error based on the analysis of the blank and Be spiked fabric samples. These discrepancies indicate that

518

the Be concentration on the shirts was not uniformly distributed. However, it was necessary to assume uniform distribution, at least within each original 8.5" diameter circle, to obtain an estimate of original Be concentration on the circular samples and the amount of Be released. This must be considered when interpreting the results.

When the measured air concentrations are examined separately for either the new or the old shirts, no significant differences resulted from agitation of unwashed shirts as compared with those from washed shirts. This may be because measured air concentrations were very low, particularly for the experiments with new shirts, thus it would be difficult to observe significant differences with this limited data set. When all old and all nearly new shirts are grouped, agitation of old shirts resulted in more Be in the air as compared to the new shirts (p $\leq$ .001). In addition, some fraction of this dust is clearly respirable because no difference was observed between air concentrations measured with and without cyclone precollectors for experiments with the old shirts.

From the results of this pilot study, it may be concluded that these old shirts suspended a significantly greater amount of Be into the air than did the new shirts and that the suspended particles can contribute to inhalation exposure of the worker.

Acknowledgements

The authors thank Dr. Morton Lippmann who suggested a study of the used workshirts, Mr. Martin Powers of the Brush Wellman Co. who provided the shirts, and Mr. Robert Morse who did the spectrometric analysis. This study is part of Center Program supported by the National Institute of Environmental Health Sciences Grant No. ES 00260. Mr. Positano was supported by Training Grant IT 15 OH 07125 from the National Institute for Occupational Safety and Health.

References

Bohne, J.R. (1982) "The Effects of Particle Resuspension on Beryllium Exposure Estimates as Measured by Garment Mounted Personal Monitors". M.S. Thesis, Department of Environmental Health Sciences, New York University.

Cohen, B.S., Harley, N.H., and Lippmann, M. (1984). "Bias in Air Sampling Techniques Used to Measure Inhalation Exposure." Am. Ind. Hyg. Assoc. J., 45: 187-192.

Gabriels, R. (1970) "A General Method for Calculating the Detection Limit in Chemical Analysis." Anal. Chem. 12: 1439-1440.

Published 1984 by Elsevier Science Publishing Co., Inc.
Aerosols, Liu, Pui, and Fissan, editors

ROTATIONAL ELECTRODYNAMICS OF AIRBORNE FIBERS

Pedro Lilienfeld
GCA/Technology Division
Bedford, Massachusetts   01730
U.S.A.

EXTENDED ABSTRACT

Introduction

One of the most effective methods of selective detection and sizing of
airborne fibers is based on the alignment and rapid rotation of these elongated
particles by means of applied electric fields. Electrodynamic alignment in
combination with light scattering or transmission has been used to study
acicular-shaped macromolecules in liquid suspensions (Jennings, 1968). Birefringence effects have been applied to study rotational diffusion properties
and to determine the particle size distribution of such colloids (Oakley, 1982).
Within the development effort of a fibrous aerosol monitor instrument, relevant
electrical and rotational dynamic properties of airborne acicular particles
were investigated and several electrical alignment/rotation configurations were
explored (Lilienfeld, 1979). The theory underlying the field-induced alignment
of acicular particles and the critical parameters influencing the electrodynamic behavior of such particles are discussed herein.

Theoretical

Acicular particle rotational electrodynamics involves two steps: a) induced electrical polarization, and b) particle axial alignment along the
electric field lines. These two steps are essentially concurrent. Forces
acting against the alignment of such particles are associated with gaseous
viscosity (Stokes regime), Brownian molecular bombardment, induced turbulence
(non-Stokesian regime), and rotational inertia. For non-conducting fibers,
exposure to an electric field results in the partial alignment of polar
molecules without appreciable transport of charges. Conducting fibers in a
field exhibit rapid free charge migration resulting in electric polarization.
In practice, most aerosol particles can be considered conducting (Fuchs, 1964)
and this was confirmed experimentally during the development of the above
mentioned instrument (Lilienfeld, 1979). Dielectric particles, such as glass,
asbestos and polymer fibers behave as conductive ones at ambient relative
humidities exceeding approximately 30%, evidently as a result of adsorbed water-
enhanced surface conductivity. Furthermore, the small dimensions of most airborne fibers of interest (i.e. asbestos) result in relatively short charge
transport time constants (e.g. 10^{-5} seconds for a fiber 1 μm in diameter and
10 μm in length with a resistivity of 10^{13} ohms/square).

Polarized fibers then undergo axial alignment parallel to the external
field as a result of the torque exerted by this field. The torque on the fiber
depends on the particle volume, the length-to-diameter ratio, the field intensity and the angle between the fiber axis and the direction of the field. In
the Stokes regime, and for a continuously rotating field, the phase lag θ between those two angles (for fiber length-to-diameter ratios exceeding 10,
approximately) can be approximated by:

$$\theta = \tfrac{1}{2} \sin^{-1} (4\text{x}10^5 \omega /E^2)$$

where E is the field intensity (volts/meter) and ω is the rotational frequency. Although the maximum theoretical value that θ can attain for rotational synchronism is 45°, experimental evidence shows that much smaller angles must be maintained (i.e. $\theta < 2°$) for stable rotation of typical asbestos and glass fibers at $\omega = 800\pi \text{ sec}^{-1}$.

As fiber length and rotational frequency are increased, the Reynolds number at the fiber ends can increase well into the non-Stokesian regime resulting in an increase in the phase lag. This behavior can be modeled by integrating the velocity over the entire fiber length in order to obtain an average Reynolds number. The phase angle θ can thus be used to determine fiber dimensions, especially in combination with the light scattering signal associated with particle rotation.

Several different electric field configurations can be applied to align and/or rotate airborne fibers: a) pulsed d.c., b) a.c., c) rotational, and d) oscillating (by superposition of d.c. and rotational fields). Applied field intensities are limited by inter-electrode breakdown in air which, for practical geometries limits the average intensity to a maximum of about 10,000 V/cm.

References

Fuchs, N.A. (1964) "Mechanics of Aerosols", Pergamon Press, New York.

Jennings, B.R. and Plummer, H. (1968) "A Study of the Light Scattered by Hectorite Solutions when subjected to Electric Fields", J. Coll. & Interface Science, 27:377.

Lilienfeld, P., Elterman, P.B., and Baron, P. (1979) "Development of a Prototype Fibrous Aerosol Monitor", Am. Ind. Hyg. Assoc. J., 40:270.

Oakley, D.M. et al (1982) "An Electro Optic Birefringence Fine Particle Sizer", J. Phys. E. Sci. Instrum. 15:1077.

Published 1984 by Elsevier Science Publishing Co., Inc.
Aerosols, Liu, Pui, and Fissan, editors

521

DUST MEASUREMENT ON WORK SITES : THE CIP 10 DUST SAMPLER

P. Courbon
Centre d'Etudes et Recherches de Charbonnages de France (CERCHAR)
B.P. 2 F60550 Verneuil en Halatte, FRANCE

J.F. Fabriès, R. Wrobel
Institut National de Recherche et de Sécurité (INRS)
B.P. 27 F54501 Vandoeuvre, FRANCE

EXTENDED ABSTRACT

Introduction

Knowing aerosol parameters that may be effectively inhaled by workers at their work place requires measurements in the vicinity of their upper respiratory airways. The most adapted method consists in sampling airborne particles using a small sized sampling head. In order to meet such a measurement, an instrument of a new type has been developed, utilizing the capabilities offered by polyurethane foams : the personal air sampler CIP10.

Description

The instrument, as indicated in Figure 1, is made of a cylindrical sampling head, and a parallelepipedic box including components that are necessary for instrument operation.

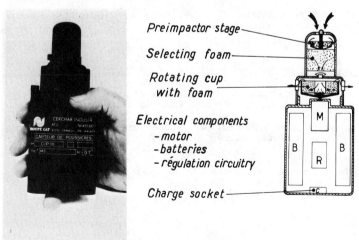

Figure 1. General view and scheme of the CIP10 instrument

[Apparatus manufactured under CdF CERCHAR license by MSA FRANCE - B.P. 617 - 95004 CERGY PONTOISE CEDEX]

The suction of dust-laden air is realized by means of the high speed rotation of the lower cup, about 7000 rpm. The ambient air is sucked in through six circular orifices of 3.5 mm diameter, distributed along a circle on the upper part of the head. Coarsest particles are retained in the preimpactor, made of a small cup containing a large pore foam. It aims at limiting the amount of dust reaching the selecting foam, thus allowing it to keep its separation characteristics during sampling. As previously shown (Brown, 1980 ; Gibson and Vincent, 1981), such porous plastic foams can be utilized for selecting particles according to their size. After having passed through the first static foam, fine particles are then collected by the rotating foam contained in the lower cup. Working conditions of the instrument do not depend on its orientation, and the motions it could be subject to.

Manufacturing characteristics

The main manufacturing characteristics of the CIP10 instrument are :
flow rate : 10 l/min. This high value allow collection of appreciable amounts of dust during a work shift
time of independent operation : more than 10 hours
height : 165 mm
maximum head diameter : 45 mm
box width : 70 mm
box thickness : 25 mm
weight : only 280 g
intrinsic safety for explosible atmospheres, according to european standards.

Sampled dusts can be extracted from the foams by washing for further analysis. Foams can be thus reutilized.

Sampling characteristics

The particle size selection of the instrument was investigated in laboratory for various types of selecting foams in a wind tunnel (calm air conditions).
During measurements the instrument was placed in a reference aerosol having the following characteristics :

nature : aluminium oxide (Aloxite 50, Ets Mercier, France)
time monitored mass concentration close to 20 mg/m3
mass median aerodynamic diameter : 8.95 μm
geometric standard deviation : 1.4

The overall selection efficiency, which is the probability for an airborne particle of given diameter in ambient air to be collected in the rotating foam, was calculated for several size intervals. This calculation was made from the amount of dust collected by each stage and from the particle size analysis of each fraction : preimpactor, selecting foam, rotating foam. These analyses were performed with a 16-channel Coulter Counter TA II (Coulter Electronics, Inc). The main results are reported in Figure 2 and Table I.

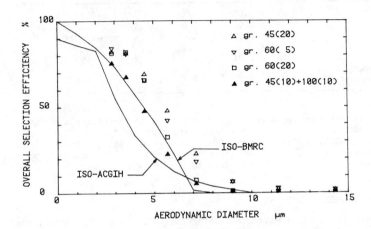

Figure 2. Overall selection efficiency for various selecting foams :
grade, pores per in. (thickness, mm).

It can be pointed out that the instrument selectivity depends on the type of foam utilized. The results obtained, using the same experimental conditions, with the 10 - mm nylon cyclone are also reported in Table I for comparison purposes.

Table I. CIP10 Sampling characteristics with various selecting foams
MMAD : mass median aerodynamic diameter ; σg : geometric st. dev.

Selection foam		Concentration ratio fine fraction/ ambient aerosol	Fine fraction		Flow rate (l/min.)
Grade	thickness mm		MMAD µm	σ_g	
45	20	0.21 ± 0.02	5.2	1.5	10.1 ± 0.2
60	5	0.19 ± 0.02	5.0	1.6	10.3 ± 0.2
60	20	0.14 ± 0.02	4.4	1.4	9.8 ± 0.2
45 + 100	10 10	0.12 ± 0.01	4.2	1.4	9.9 ± 0.2
10 - mm nylon cyclone		0.110 ± 0.004	3.8	1.5	1.72 ± 0.01

ISO conventional curves, representing the product inspirability x alveolar deposition probability, have been also reported in Figure 2 for both conventions, ACGIH and BMRC (ISO, 1981). It appears that the experimental data points obtained for a selecting foam grade 60, 20 mm thick or a combination 45(10) + 100 (10) are close to the ISO - BMRC curve.

Example of utilization

The instrument, equipped with a selecting foam grade 45, 20 mm thick, was utilized in coal mines, in peculiarly hard sampling conditions. When worn on a shoulder-belt, it did not give any opportunity of observing some possible discomfort ; its utilization for systematic research about exposition of miners has been retained in French coal mines. It is contemplated to utilize it for industrial hygiene purposes, although the choice of the selecting foam is being presently discussed.

References

Brown, R.C. (1980) "Porous Foam Size Selectors for Respirable Dust Samplers", J. Aerosol Sci. 11 : 151.

Gibson, H., J.H. Vincent (1981) "The Penetration of Dust through Porous Foam Filter Media", Ann. occup. Hyg. 24 : 205.

ISO International Standards Organization (1981) "Size Definitions for Particle Sampling", Am. Ind. Hyg. Assoc. J. 42 : A64.

Published 1984 by Elsevier Science Publishing Co., Inc.

Aerosols, Liu, Pui and Fissan, editors

525

Effects of Aerosol Concentrations on Radon Progeny

submitted by
Richard G. Sextro*, Francis J. Offermann, William W. Nazaroff,
and Anthony V. Nero

Building Ventilation and Indoor Air Quality Program
Lawrence Berkeley Laboratory
Berkeley, CA 94720

Radon decay products, the radioactive isotopes ^{218}Po, ^{214}Pb and ^{214}Bi, are chemically active, unlike their inert, gaseous parent ^{222}Rn. These species can attach to surfaces, such as aerosols, walls or other indoor surfaces and can also deposit on lung tissue, either attached to aerosols or as unattached molecular species. The associated health hazards depend, in part, on the concentrations of these attached or unattached isotopes. Characteristics of indoor aerosols are thus important determinants of radon progeny behavior.

Experiments have been conducted at the LBL Indoor Air Quality Research House to determine the effects of indoor aerosol concentrations on radon decay product concentrations. Simultaneous real-time measurements of radon and radon progeny concentrations have been made for aerosol concentrations ranging over nearly four orders of magnitude. Aerosol concentrations and size distributions were measured for particle diameters from 0.01 to 3 μm.

The principle modes of radon progeny behavior, e.g. attachment to airborne particles, deposition of unattached and attached progeny on macro surfaces, ventilation, and filtration can be estimated from these measurements. Radon progeny concentrations, unattached fractions and the Potential Alpha Energy Concentrations (PAEC) can be determined as a function of particle concentration. These data indicate that deposition of unattached progeny is the most significant removal mode, even at moderate particle concentrations. The effect of aerosol concentration on these removal mechanisms and on the potential health effects of attached and unattached progeny concentrations is discussed.

* author correspondent: Building 90, room 3058, LBL

Published 1984 by Elsevier Science Publishing Co., Inc.

Aerosols, Liu, Pui, and Fissan, editors

526

Behavior and Control of Indoor Tobacco Smoke Particles

Francis J. Offermann[*], John R. Girman, Richard G. Sextro

Building Ventilation and Indoor Air Quality Program
Applied Science Division
Lawrence Berkeley Laboratory
University of California
Berkeley, California   94720

ABSTRACT

     Concern has been increasing regarding the potential health effects
resulting from chronic indoor exposure to tobacco smoke contaminants.
Since an estimated one third of the general population are smokers, it is
apparent that indoor air pollution from tobacco smoke is extensive. It is
also known that many of the chemical constituents present in sidestream
tobacco smoke are toxic and may pose significant health risks to non-
smoking occupants. In this paper we examine the effects of smoking
density, ventilation, and air cleaning on the indoor contaminant concentrations generated by sidestream tobacco smoke. In the past various steady-
state, mass-balance models have been utilized to predict indoor contaminant
concentrations. Following an examination of these models, we present a
general mass-balance model which has been extended to include the concept
of ventilation efficiency. Ventilation efficiency has been receiving increasing attention in the fields of ventilation and indoor air quality and
is an important consideration when comparing various ventilation control
strategies. Also introduced in this model are experimentally determined
surface deposition rates for tobacco smoke particulate phase contaminants.
Particulate deposition rates, as determined from decay measurements made in
an indoor environment, ranged from 0.05 to 0.8 hr^{-1} as a function of
particle size. In contrast to these measurements the deposition rates of
indoor tobacco smoke particulates have been observed by others to be in the
range of 1.0 to 6.0 hr^{-1}. A discussion of this discrepancy is included.

     Following the development and discussion of the steady-state, mass-
balance model we present several sets of field data acquired in various
indoor settings where there is tobacco smoking and compare these results
with the model predictions. In the final section of the paper, we compare
the effects of various control strategies on the indoor concentrations of
tobacco smoke generated contaminants. Based on these calculations we
estimate the exposure of non-smoking occupants for various smoking
densities and recommended ventilation rates.

This work was funded by the Director, Office of Energy Research, Office of
Health and Environmental Research, Human Health and Assessments Division
and Pollutant Characterization and Safety Research Division of the U.S.
Department of Energy under Contract No. DE-AC03-766F00098, and the
Bonneville Power Administration, Portland, Oregon.

*author correspondent. Building 90 Room 3058

Published 1984 by Elsevier Science Publishing Co., Inc.

Aerosols, Liu, Pui, and Fissan, editors

INDOOR AEROSOL IMPACTOR

W.A. Turner*, V.A. Marple**, J.D. Spengler*

*Department of Environmental Science & Physiology
Harvard School of Public Health
Boston, MA 02115 USA

**University of Minnesota
Particle Technology Laboratory
Mechanical Engineering Dept.
Minneapolis, MN 55455 USA

Extended Abstract

Introduction

An aerosol impactor suitable for indoor sampling has been developed as part of the ongoing Harvard air pollution epidemiology study (Spelzer, et al.). This impactor was developed to meet several design criteria. The design criteria and results of initial testing are presented. Because of its unique features of convenience and versatility, the impactor will be suitable for a wide variety of sampling needs.

Design Criteria

The impactor was designed to meet several criteria. The primary intended use of the impactor is for the sampling of fine fraction (<2.5 um) or inhalable fraction (<10 um) particle sampling in the indoor environment. The flow rate chosen was 4 lpm. The rational for this flow rate was that it was technically achievable with both battery powered flow controlled pumping systems and line power. This flow rate allows the researcher to maximize the amount of material which can be collected during a 24-168 hour sampling period. This will facilitate improved accuracy in gravimetric measurements, Beta-gauging operations, and when using XRF automated elemental analysis techniques. Previous studies have indicated that 24 hr indoor home concentrations can range between 5 and 300 ug/m^3 (Spengler et al., 1981; Ju and Spengler, 1981). A flow rate of 4 lpm will sample approximately 5m^3 of air per day. Thus, a mass loading of 25 ug/m^3 will be collected. This the minimum detectable limit of either the beta-gauge or the gravimetric method.

The impactor has been designed to use dichotomous sampler type 2" x 2" PTFE membrane filters. Both the rigid mount type (supplied by Membrana, Inc., Ghia Div.) and removable mount type filter slides supplied by Beckman, (42 mm filter disks supplied by Ghia can be used in the present configuration). We are also developing the option of using a standard 37mm filter in a removable type 2" x 2" slide holder.

A disposable impactor plate has been developed for use with the impactor and will be reported elsewhere in the literature (Marple, 1984).

Uniform Deposition

In order to utilize automated weighing and analysis procedures which have previously been developed by the US EPA for dichotomous filter samples, a uniform deposition criteria was established (Dzubay and Stevens, 1975). This criteria was defined as the deposit density per CM^2 of the filter shall not exceed 10% of the averaged deposit of filter.

Results

To date the impactor size cut and uniformity of deposition has been characterized at both Harvard School of Public Health and the University of Minnesota. We are currently making arrangements for a third independent calibration with researchers at Florida State University. The results of the 2.5 um and the 10.0 um Minnesota size cut characteristics tests are presented in Figures 1 and 2. The Minnesota tests were performed using fluorescent dyed techniques (Stober and Flachsbart, 1973). The results of the Harvard test are similar for the 2.5 um impactor and show a discrepancy for the 10.0 um impactor. The Harvard testing was performed using a Particle Measuring System (PMS) laser unit and polydispersed mineral oil aerosol. Difficulties in generating and counting larger sized particles may account for the discrepancy.

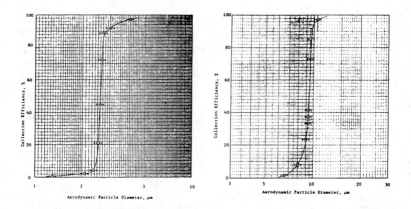

Figures 1 and 2   Size cut characteristics – Minnesota

Further testing is being conducted by all three universities to
confirm the initial results. Testing with solid particles and investigation of the influence of flow rate on the size cuts is also planned. The
results of the uniform deposit testing by both universities have confirmed
the need for a spacing distance between the impactor plate and filter
media; especially for the 10 um impactor. From the initial results it can
be concluded that a plate to filter distance of approximately 7 cm will
assure uniform deposition for both size cuts.

Conclusions

The cut points of the impactors has been determined to be at or close
to the original design criteria of 2.5 um, and 10.0 um. Further testing
will be done to confirm these results and allow us to move out of the
prototype stage and into the production of 100 units by September, 1984.
These impactors combined with a programmable, flow controlled, low power
consumption sampling unit will allow us to begin testing in 300 homes in
Watertown, Massachusetts during the fall of 1984.

Acknowledgements

The authors wish to thank Ken Rubow of the University of Minnesota
and Jack Price and Janet Macher of Harvard School of Public Health for
aerosol testing and Caroline Granfield and Allison Maskell for manuscript
preparation. This work was performed under National Institute of Environmental Health Science Grant No. ES011K80.

References

Dzubay, T.G. and Stevens, R.K. Ambient air analysis with dichotomous
sampler and x-ray fluorescence spectrometer. Environmental Science &
Technology, 9:7, July 1975.

Ju, C. and Spengler, J.D. Room-to-room variations in concentration of
respirable particles in residences. Environmental Science and Technology
15(5): 595-598, May 1981.

Marpel, V. Verbal communication to William A. Turner, April 1984.

Speizer, F.E., Bishop Y. and Ferris, B.G., Jr. An epidemiological approach
to the study of the health effects of air pollution. Proceedings of the
4th Symposium on Statistics and the Environment, March 3-5, 1976,
Washington, D.C. pp. 56-68.

Spengler, J.D., Dockery, D.W., Turner, W.A., Wolfson, J.W. and Ferris,
B.G., Jr. Long-term measurements of respirable sulfates and particles
inside and outside homes. Atmospheric Environment 15:23-30, 1981.

Stober, W. and Flaschsbart, H. Evaluation of nebulized ammonium fluorescein as a laboratory aerosol. Atmospheric Environment 7, 1973.

Published 1984 by Elsevier Science Publishing Co., Inc.
Aerosols, Liu, Pui, and Fissan, editors
530

Electrostatically Charged Water Droplets
for Smoke Clearing in Shipboard Fires

John S. Kinsey
Midwest Research Institute
Kansas City, MO

Joseph Leonard
Naval Research Laboratory
Washington, DC

ABSTRACT

Accidental fires occuring below decks of naval surface vessels create a
dense aerosol of black smoke particles which severely hampers visibility.
There is good theoretical and experimental evidence to indicate that the
introduction of electrostatically charged water droplets into such an aerosol can remove a significant portion of the particles from suspension thus
diminishing the attenuation of visible light. In this paper the results
of a recent experimental program will be presented which evaluates the use
of charged water droplets (fog) for smoke clearing in shipboard fires.

During the program, a test aerosol similar to that created in shipboard
fires was established in a 3.45 m^3 test chamber into which both charged and
uncharged fog was introduced. Real-time measurements of concentration,
particle size, and other important parameters were made throughout the
clearing process. These data were then used to compare the overall time
rate of change of concentration, particle size distribution, and visibility
resulting from the introduction of charged fog over that which occurs by
natural coagulation, deposition, etc. The charged fog generator used in
these experiments is of a novel design developed specifically for this application by Midwest Research Institute. A model of the overall clearing
process is also presented in the paper based on experimental data.

Published 1984 by Elsevier Science Publishing Co., Inc.
Aerosols, Liu, Pui, and Fissan, editors

Measurement of the Deposition of Aerosols in an Iron Ore Mine

W. Koch, W. Behnke, H.P. König, W. Holländer
Fraunhofer-Institue for Toxicology and Aerosol Research
Hannover

F. Lange, R. Martens
Society of Reactor Safety
Cologne

Introduction

Published measurements (Koepe, 1952; Madisetti, 1958) of dust concentrations in galleries of coal mines have shown that, under turbulent flow conditions, there is an exponential decrease of the dust concentration with the distance from the source. However, particle size dependent deposition velocities V_D [cm/sec] and the wall loss coefficient β [sec^{-1}], respectively, were not extracted from these data. But then, fitting available theories to simple wind tunnel experiments may not adequately apply to aerosol deposition in a complex system such as an underground mine.
Thus, in order to assess the order of magnitude of the wall loss coefficient and to reveal the mechanisms responsible for the wall deposition of inhalable aerosol particles, we made a propagation experiment in an iron ore mine.

Experimental

A typical gallery of the mine was selected and a test aerosol of Ytterbium chloride particles of sizes between 0,5 to 20 microns in diameter was fed into this gallery. At four sampling stations down-stream of the aerosol generator, the aerosol was sampled by 9-stage Berner impactors. The mass fractions of Ytterbium were measured by atomic absorption spectrometry. Figure 1 shows a schematic drawing of the test gallery and the position of the aerosol generator and the sampling stations. The arrows indicate the wind direction. S means the distance of the sampling stations from the source and $\dot{V}$ [m³/min.] the flow rate of the ventilation. The average height, cross section and circumference of the gallery is 3.9 m, 24.3 m² and 19.5 m respectively. The relative humidity was 35 % and the temperature was 28°C.

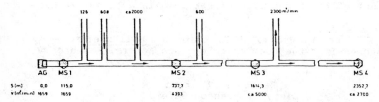

Fig. 1: Schematic drawing of the gallery

The test aerosol was produced by a pneumatic nebulizer (Schlick nozzle, Type 940). Figure 2 shows the mass distribution of the ytterbium chloride aerosol.

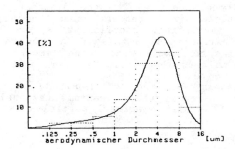

Fig. 2: Size distribution of the test aerosol

Results

Within the experimental error, the results show an exponential decrease of the Yb masses with time t of propagation for each size range.

$$\frac{m_i}{m_1} = \exp(-\beta t) \, . \qquad (1)$$

From these data wall loss coefficients were calculated by linear regression methods for each range (see for example Fig. 3).

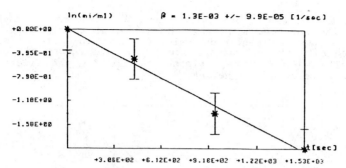

Fig. 3: Correlation between Yb masses and time of propagation for the size range between 2 and 4 microns (i = number of the sampling station)

Note that we have normalized the data to the data of the first sampling station rather than to the generator. The reason is that we took the aerosol measured at the first station as our actual source aerosol to be sure to have a homogeneous spatial aerosol distribution all over the cross section of the gallery.
In addition the wind profile was measured to derive estimates of typical values for the friction velocity u^* and the roughness height k_S in the gallery. According to the theory of Wood, the particle losses are supposed to be influenced by both of these variables. Applying the estimated values of these two parameters to the theory of Wood (Wood, 1981) values for the wall loss coefficient are obtained which are of the same order of magnitude as those measured directly.
In figure 4 we have plotted the measured wall loss coefficient β against the particle diameter. The dashed line is a theoretical curve obtained by the formula

$$\beta = \frac{1}{V} \int_{surface} ds \; \frac{-\overline{V}_{sed} \cdot \overline{K}}{1 - \exp[\overline{V}_{sed} \cdot \overline{K} \cdot Int]} \quad , \qquad (2)$$

534

where V is the volume of the gallery and $\int$ds denotes the integral over its surface. $\vec{K}$ is the vector normal to the surface and $\vec{V}_{sed}$ is the sedimentation velocity. The so-called resistance integral Int is evaluated following the theory of Wood (Wood, 1981) and using the estimated values for the friction velocity and the roughness height.

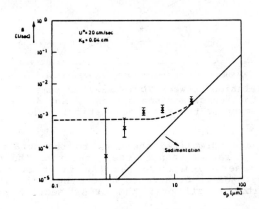

Fig. 4: Wall loss coefficients

References:

Koepe, F. (1952) "Gesetzmäßigkeiten des Staubfluges und der Staubablagerung in Bergwerksstrecken", Staub 28: 38

Madisetti, R.A. (1956) "Ablagerung von feinem Kohlenstaub in einem Staubversuchsraum unter Tage", Staub 47: 653

Wood, N.B. (1981) "A simple method for the calculation of turbulent deposition to smooth and rough surfaces", J.Aerosol Sci. 12: 275

Published 1984 by Elsevier Science Publishing Co., Inc.
Aerosols, Liu, Pui, and Fissan, editors

SUB-MICRON PARTICLE FORMATION FROM
PULVERIZED COAL COMBUSTION

Hisao Makino
Combustion Engineering Section
Environment Department
Energy & Environment Laboratory
Central Research Institute of Electric Power Industry
Komae-shi, Tokyo   201
JAPAN

EXTENDED ABSTRACT

1. Objective

Sub-micron particles (less than 1μm in diameter) exhausted from thermal
power stations do not easily become deposited on the ground and penetrate
our respiratory organs together with other contaminated gases, resulting
health problems.

From these circumstances, the need to investigate the actual conditions
and technology to reduce sub-micron particles exhausted from coal-fired power
plants has increased recently. Based on our investigation, collection efficiency has a very low value for particles whose diameter is from 0.05μm to
5μm, and has a minimum value between 0.1μm and 1μm in particle diameter   in
particular. Thus, in order to reduce sub-micron particle emission from coal-
fired power stations, it is necessary to not only develop collection technology
but also technology to reduce particles generated by coal combustion.

In this study, by using the Pulverized Coal Combustion Research Facility
in CRIEPI, we conducted combustion tests on nine kinds of coal having various
fuel ratios and components. We measured particle concentration and particle
size distribution at the furnace exit of the Pulverized Coal Combustion Research Facility. We investigated the influence of coal characteristics and
combustion conditions for the generation of sub-micron particles.

2. Experimental Facility

The main components of the experimental facilities are shown in Fig. 1.
The experimental furnace was a horizontal cylinder with an inner diameter of
0.85m and a length of 8m.   The inside of the furnace is lined with a 75mm-
thick refractory, and its outside was water-cooled. The burner capacity at
standard load is 100 kg/hr of coal or its equipment and the heat release rate
is 1.48×10^5 kcal/m^{3}hr. Almost all of the tested coal are bituminous coal,
but one was sub-bituminous coal. Tested coal properties are shown in Table 1,
calorific value is 6200 - 7490 kcal/kg, fuel ratio is 0.92 - 2.44, and ash
content 7.65 - 22.03 wt%. Particle size distribution is measured by the Andersen Stack Sampler (so called A.S.S.) of particle diameter exceeding 0.75μm and
the Electrical Aerosol Size Analyzer (so called E.A.A.) for particle diameter
less than 1μm.

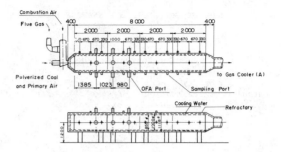

Fig. I Schematic of Combustion Test Furnace

Table. 1 Tested Coals Properties

No.		1	2	3	4	5	6	7	8	9
Item	Coal	Lemington	Witbank	Millmerran	Ermelo	Blair Athol	Fuhsin	Datong	Huaipei	Beluga
Calorific Value (kcal/kg)		6920	6390	6320	6650	6880	6200	7490	6360	5720
Proximate Analysis	Moisture (%)	3.2	3.0	3.2	3.3	6.2	8.3	4.1	2.0	14.5
	Volatile Matter (%)	31.8	24.1	30.3	29.0	28.6	32.8	28.1	25.3	45.4
	Fixed Carbon (%)	56.4	59.1	47.7	56.6	60.3	52.3	64.3	52.8	41.6
	Ash (%)	11.8	16.8	22.0	14.4	11.1	14.9	7.6	21.9	13.0
F.C./V.M.		1.77	2.45	1.58	1.95	2.11	1.59	2.29	2.09	0.92
Ultimate Analysis	Carbon (%)	73.30	69.00	64.93	70.70	72.30	65.33	77.92	67.22	59.65
	Hydrogen (%)	4.88	3.95	4.23	4.34	4.14	4.12	4.30	4.30	3.64
	Nitrogen (%)	1.43	1.46	1.16	1.54	1.44	0.73	0.67	0.97	0.76
	Oxygen (%)	8.23	8.25	6.99	8.46	10.65	14.35	8.85	5.34	22.90
	Sulfur (%) (Combustible)	0.38	0.54	0.66	0.56	0.36	0.49	0.61	0.21	0.05

3. Experimental Results and Considerations

3.1 Influence of Excess O_2 Concentration on Sub-micron Particle Formation

Fig. 2 shows the differential mass particle size distribution for Witbank
coal combustion gas. The sub-micron particle increased when reducing excess
O_2 concentration in flue gas from 6% to 3%. In particular, particles whose
diameter ranged from 0.02μm to 0.3μm increased. Thus, these particles are
assumed to be gaseous phase carbon formed in the volatile matter combustion
area where the excess air ratio is very low. The total concentration of submicron particles whose diameter ranged from 0.02μm to 0.3μm is shown Fig. 3
in comparison to the excess O_2 concentration. For all coal, the concentration of these particles increases 5 times when reducing the excess O_2 concentration from 6% to 3%.

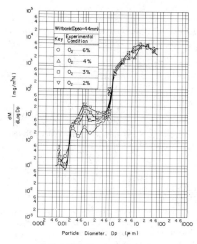

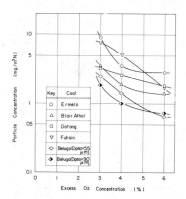

Fig. 2 Differential Mass Particle Size Distribution

Fig-3 Excess O₂ Versus Particle Cncentration
(Particle concentration means the integrated weight of particles of which diameter were from 0.02μm to 0.3μm)

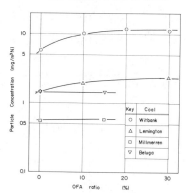

Fig-4 Influence of Two Stage Combustion Method on Particle Concentration
(Particle concentration means the integrated weight of particles of which diameter were from 0.02μm to 0.3μm)

3-2. Influence of Two Stage Combustion for Sub-micron Particle Concentration

Fig. 4 shows the influence of two stage combustion for total concentration of particles whose diameter are 0.02 - 0.3μm. For some coal, particle concentration increases 2 - 3 times by two stage combustion, but for other coal, particle concentration is not influenced by two stage combustion. From these results, the influence of two stage combustion for sub-micron particle concentration is less than that of the excess O_2 concentration. Also, we measured the O_2 concentration distribution in the furnace for Leminton coal combustion gas and Beluga coal combustion gas. For Leminton coal combustion gas, the combustion area in the furnace where the excess air ratio is low expands by two stage combustion, but for the Beluga coal combustion gas, expansion is not seen. Thus, it appears that the formation of gaseous phase carbon is influenced the local excess air ratio in the furnace.

3-3. Influence of Coal Characteristics

The influence of the particle diameter of pulverized coal for differential

538

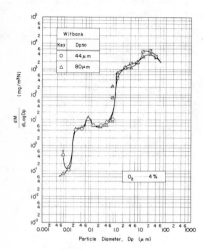

Fig. 5 Differential Mass Particle Size Distribution

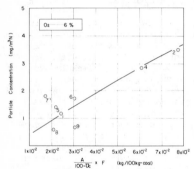

Fig. 6 Influence of Coal and Ash Characteristics for the
Particle Concentration
(Particle concentration means the integrated weight
of particles of which diameter were from
0.02 μm to 0.3 μm)
A   Ash (%)
U_c Unburned carbon in fly ash (%)
F   Weight of the smaller particles than
    Dco in Coal (kg/100kg-coal)
$Dco = (\frac{A}{100-Uc})^{-1/3} \times 0.3$ (μm)

mass particle size distribution is shown
in Fig. 5. In our experiment, the particle
diameter of pulverized coal virtually does
not affect sub-micron particle formation.
And, the influence of coal properties,
for example ash content, volatile matter
and fuel ratio, is not seen.

Now, on the assumption that one
pulverized coal particle generates only
one coal ash particle, and gaseous phase
carbon is almost nonexistance under the
condition where excess O_2 is 6%, sub-micron
particle weight at 6% of excess O_2 is calculated from the pulverized coal characteristics.

The calculation method is explained as
follows. The diameter D_{co} of pulverized
coal which makes coal ash of 0.3μm diameter is calculated by the equation below.

$$D_{co} = (\frac{A}{100-U_c})^{-1/3} \times 0.3 \ (\mu m) ....(1)$$

A: Ash (%), U_c: Unburned carbon in coal
ash (%)

The weight W of coal ash smaller than
0.3μm is expressed by the next equation.

$$W = \frac{A}{100-U_c} \times F \ ......(2)$$

F: Weight of smaller particles than D_{co} in
coal (kg/100kg-coal)

As shown in Fig. 6, the sub-micron
particle concentration measured by E.A.A.
at 6% of excess O_2 correlates with W.

4. Conclusion

Particles formed by pulverized coal
combustion are classified as particles generated by ash, and gaseous phase carbon
whose diameter range from 0.02μm to 0.3μm.
Particle concentration generated by ash is
estimated by pulverized coal characteris-
tics, but the concentration of gaseous phase carbon is affected by the local
excess O_2 concentration and does not correlate with the coal characteristics.
In order to reduce gaseous phase carbon emission, it is necessary to mix pulverized coal and combustion air sufficiently in the volatile matter combustion
area.

PART III.

GAS CLEANING; CLEAN ROOM; PARTICLE
PRODUCTION; COMBUSTION AEROSOLS

CHAPTER 13.

FILTRATION

Published 1984 by Elsevier Science Publishing Co., Inc.
Aerosols, Liu, Pui, and Fissan, editors

ENHANCED DEPOSITION OF SUSPENDED PARTICLES TO FIBROUS SURFACES
FROM TURBULENT GAS STREAMS

Chaim Gutfinger* and S. K. Friedlander
Department of Chemical Engineering and National Center
for Intermedia Transport Research
University of California
Los Angeles, CA 90024, USA

EXTENDED ABSTRACT

Introduction

Prediction of deposition rates of aerosols from turbulent flows is of importance in such diverse areas as atmospheric dry deposition, particle deposition in ducts of gas cleaning equipment, losses in aerosol sampling lines, and the deposition of radioactive particles on walls in coolant circuits of gas cooled nuclear reactors.

Most of the studies that have appeared in the literature deal with deposition from fully developed turbulent flows inside tubes or annuli with smooth walls. Friedlander and Johnstone (1957) assumed that particles were carried by turbulent eddies to within on stopping distance from the wall, and derived an expression for the dimensionless deposition velocity, as a function of the Moody friction factor and dimensionless stopping distance. Several modificaions and extensions of the Friedlander – Johnstone deposition theory have appeared in the literature (Davies, 1966; Owen, 1961; Liu and Ilori, 1973).

There have been fewer studies, under controlled conditions, of turbulent deposition to rough surfaces. In a previous work (Fernandez de la Mora and Friedlander, 1982) a mechanism of deposition to fully rough surfaces based on filtration theory was advanced which successfully correlated the experimental data.

Wells and Chamberlain (1967) reported results for particle deposition to the walls of an annulus lined with filter paper. The wall roughness created by the paper fibers was small and had only a slight effect on the frictional loss, but particle deposition rates were several orders of magnitude greater than those to smooth walls. The goal of the present paper was to develop a theory for particulate deposition from turbulent streams to surfaces covered by fibrous elements of small roughness. Such a theory would bridge the gap between the smooth surface theory of Friedlander and Johnstone (1957), and the theory for deposition to fully rough surfaces of Fernandez de la Mora and Friedlander.

---

*On leave of absence from the Faculty of Mechanical Engineering, Technion, Israel Institute of Technology, Haifa 32000, Israel.

Deposition to Fibrous Surfaces.

Aerosol transport within the turbulent core is extremely efficient due to the turbulent eddies. Closer to the wall, however, the effect of the eddies is very small and the resistance to aerosol transport is drastically increased. The fibrous roughness elements protruding from the surface enhance deposition by providing deposition sites away from the wall. If the fibers are contained within the viscous sublayer they have almost no effect on the resistance to flow. Fibers protruding into the transition region do result in some increase in pressure drop. Wells and Chamberlain (1967) report a 10% increase in pressure drop for flow conditions where the fibers protrude up to y+ = 10.

The beneficial effect of the fibers on deposition is partially offset by the fact that not all the particles approaching the fibrous surface are caught. The fiber canopy may be cosidered as a filter, which for larger particles approaches 100% removal efficiency. For particles in the size range 0.2 - 0.7μm, for which the filtration efficiency is lowest, the fiber canopy may act as an additional resistance for deposition. Thus, deposition to fibrous surfaces can be considered to consist of two resistances in series, namely, that of eddy transport outside the roughness layer and that of filtration by the fibrous elements. The deposition velocity is given for this case by

$$k_{dep} = \frac{1}{1/k_{eddy} + 1/k_{filt}}$$

where k_{eddy} is the eddy deposition velocity on to the roughness layer and k_{filt} is the deposition velocity on the fibrous filter elements. The eddy deposition velocity was found by adapting the Friedlander - Johnstone theory to rough surfaces. The deposition inside the filter canopy was analyzed applying the principles of filtration on cylindrical collectors.

Enhancement of Deposition

Particle deposition from turbulent flows to smooth walls is due mainly to eddy transport. In the viscous sublayer eddy transport is very slow. Fibrous surfaces provide deposition sites away from the wall, thus markedly enhancing the deposition process. Wells and Chamberlain (1967), who performed deposition experiments in the same system first with smooth walls then with filter-paper lined walls, have shown that the fibrous layer of the filter paper does indeed enhance deposition. To characterize this efect quantitatively we define an enhancement factor, F:

$$F = \frac{\text{deposition velocity to rough wall}}{\text{deposition velocity to smooth wall}}$$

In Fig. 1 experimental data from the paper by Wells and Chamberlain (1967) are plotted in terms of enhancement factors vs Reynolds numbers, for two aerosol diameters. The curves were calculated from the present theory. The figure shows good agreement between theory and experiments. As can be seen, the enhancement of deposition by a fibrous surface is impressive indeed, especially at Reynolds numbers of 20,000 – 40,000, where deposition by the fiber elements is enhanced by a factor of 200 – 1000.

We hope this paper will focus attention on this interesting and potentially useful phenomenon.

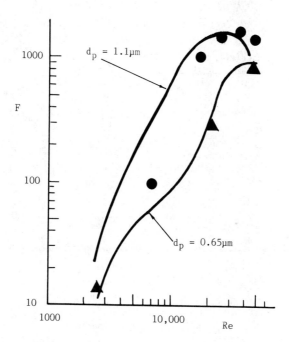

Fig. 1. Enhancement factor as a function of Reynolds number; theoretical curves with experimental points from the data of Wells and Chamberlain (1967).

References

Davies, C.N. (1966). Deposition of aerosols from turbulent flow through pipes. Proc. Royal Soc. A., 289, 235–246.

Fernandez de la Mora, J. and Friedlander, S.K. (1982). Aerosol and gas deposition to fully rough surfaces: Filtration model for blade shaped elements. Int. J. Heat Mass Transfer 25, 1725–1735.

Friedlander, S.K. and Johnstone, H.F. (1957). Deposition of suspended particles from turbulent gas streams. Ind. Eng. Chem., 49, 1151–1156.

Liu, B.Y.H. and Ilori, T.A. (1974). Aerosol deposition in turbulent pipe flow. Environ. Sci. & Technol. 8, 351–356.

Owen, P.R., (1961). Dust deposition from a turbulent airstream. Int. J. Pollut. 3, 1.

Wells, A.C. and Chamberlain, A.C. (1967). Transport of small particles to vertical surface. Brit. J. Appl. Phys., 18, 1793–1799.

Published 1984 by Elsevier Science Publishing Co., Inc.
Aerosols, Liu, Pui, and Fissan, editors

EVALUATION OF AEROSOL SPECTROMETERS FOR QUALITY ASSURANCE
FILTER PENETRATION MEASUREMENTS*

R. C. Scripsick, S. C. Soderholm, and M. I. Tillery
Aerosol Science Section
Industrial Hygiene Group
Los Alamos National Laboratory
Los Alamos, NM 87545
USA

Extended Abstract

Introduction

Every high efficiency aerosol filter used in the United States
Department of Energy (USDOE) facilities is quality assurance (QA) tested
at one of the USDOE filter test facilities prior to installation. This
testing presently includes measurement of filter penetration at rated
airflow using a "hot DOP" aerosol generator, an "Owl" aerosol size
analyzer, and a scattered-light photometer concentration monitor.
Alternative penetration measurement methods for testing high efficiency
aerosol filters with rated airflow capacities of ~28 m^3/min (1000
cubic feet/min) are being studied at Los Alamos National Laboratory.
These methods are intended to take advantage of currently available
aerosol instrumentation.

As part of this program, certain aerosol spectrometers are being
evaluated for their ability to accurately and reliably provide information
on penetration as a function of particle size. The aerosol spectrometers
considered in this evaluation include 1) a laser aerosol spectrometer
(LAS, PMS Inc. Model LAS-X) and 2) a differential electric mobility
analyzer (DEMA). The DEMA incorporates an electrostatic aerosol
classifier (EC, TSI Inc. Model 3071) to segregate aerosol by size and
either a LAS or a condensation nuclei counter (CNC, TSI Inc. Model 3030)
to monitor the EC output.

Early operating experience with the DEMA system (using either the LAS
or the CNC as the EC output monitor) indicated that penetration
measurements could not be made by this system within the time requirements
of the USDOE filter test facilities. For this reason, the DEMA system was
eliminated from consideration as an alternative penetration measurement
device. However, because the size and concentration sensitivities of the
EC and the CNC include the ranges of interest for high efficiency aerosol
filter penetration testing, these devices were used in evaluation of the
LAS performance.

---

*Work performed at Los Alamos National Laboratory under the auspices of the
U. S. Department of Energy, Airborne Waste Management Program Office,
Contract No. W-7405-ENG-36.

548

Size Comparison

A comparison of aerosol sizing by the LAS and the EC was performed over the range from ~0.1 μm to ~0.4 μm by generating nearly monodisperse aerosols with the EC. Di(2-ethylhexyl) phthalate (DEHP) aerosols with concentrations in the vicinity of 1000 particles/cm^3 were used in this comparison. The LAS response was corrected for refractive index effects. Atmospheric pressure and slip factor effects were accounted for in calculating the particle size produced by the EC.

Results of this comparison are presented in Figure 1. A linear least squares fit to these data resulted in the equation:

$$Y = -0.019 + 1.016X,$$

where, Y = LAS measured size in μm, and X = EC measured size also in μm. The coefficient of determination, r^2, for this fit was 0.995. The per cent difference in aerosol size indicated by the two instruments was <10% for aerosols with diameters from ~0.15 μm to ~0.4 μm. The per cent difference increased for aerosol diameters below ~0.15 μm until at an EC size of ~0.12 μm this difference was ~20%. Resolution of the size response discrepancy between the instruments may be possible by modifying the operation of the LAS. This comparison is planned to be repeated using monodisperse polystyrene latex aerosol independently sized by electron microscopy.

Concentration Comparison

Comparisons of number concentration measurements made by the LAS and the CNC were performed using polydisperse and monodisperse DEHP aerosols. Sample airflows for each instrument were checked during the concentration measurements. CNC concentration measurements made in the count mode were corrected for coincidence losses.

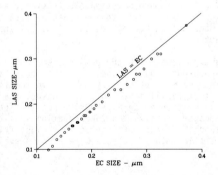

Figure 1. Data points (O) indicate results of a comparison of aerosol sizing by the LAS and the EC. The line denotes theoretical equal response by the instruments.

A polydisperse aerosol with a count median diameter of ~0.25 μm and a geometric standard deviation of ~1.5 was used in the first concentration comparison. Results of this comparison for aerosol concentrations from ~50 particles/cm³ to ~5000 particles/cm³ are displayed in Figure 2. On the average the LAS concentration measurements were ~10% lower than those of the CNC. Some of this bias is thought to be the result of the LAS size range not being broad enough to include the entire aerosol size distribution. This is a possible explanation for the fractional difference between measurements being about the same magnitude over the concentration range studied. The degree of bias that was observed is not believed to be significant in terms of the concentration accuracy requirements for penetration measurements.

Monodisperse aerosols for the second concentration measurement comparison were produced by the EC. Comparisons were conducted at seven aerosol sizes with aerosol diameters ranging from ~0.2 μm to ~0.5 μm. Concentration was varied at each size from 10^2 - 10^3 particles/cm³ up to <10^4 particles.

A typical example of the results from this set of comparisons is shown in Figure 3. In general, good agreement was observed at each aerosol size. Over all sizes studied, the LAS concentration measurement tended to be slightly greater than the CNC measurement with the CNC operating in the count mode. In this operation mode, all LAS concentration measurements were within 10% of the CNC measurements. With the CNC operating in the photometric mode the LAS measurement tended to be the lower of the two measurements. The fractional differences observed in the measurements made with the CNC operating in the photometric mode tended to be greater than the differences observed with the CNC operating in the count mode. This was especially true for CNC measured concentrations greater than ~3000 particles/cm³ where LAS

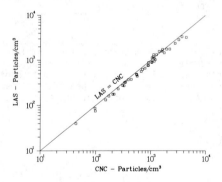

Figure 2.   Data points (□) indicate results of a comparison of number concentration measurements made by the LAS and the CNC using a polydisperse aerosol. The line denotes theoretical equal response by the instruments.

550

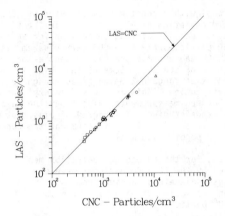

Figure 3.   Example of typical results from comparison of monodisperse
            aerosol concentration measurements made by the LAS and the
            CNC. The triangles (Δ) indicate measurements made on
            0.21 µm diameter aerosol. The squares (□) indicate
            measurements made on 0.31 µm diameter aerosol. The circles
            (o) indicate measurements made on 0.42 µm diameter aerosol.
            The line denotes theoretical equal response by the instruments.

concentration measurements were in some cases less than 70% of the
corresponding CNC concentration measurements. The concentration
measurement differences observed above CNC measured concentrations of
~3000 particles/cm³ may be the result of LAS coincidence losses. No
aerosol size dependent effects were observed.

Conclusion

    The general conclusion drawn from the evaluation completed thus far is
that both aerosol spectrometers have potential for making accurate
penetration measurements. The LAS is currently the instrument of choice
because it has promise of making the required penetration measurements
within the time limitations necessary for efficient operation of USDOE
filter test facilities.

Published 1984 by Elsevier Science Publishing Co., Inc.
Aerosols, Liu, Pui, and Fissan, editors

A FILTER PRESSURE LOSS MODEL FOR UNIFORM FIBERS

Donald R. Monson
Principal Scientist
Donaldson Co., Inc.
Minneapolis, MN. 55440

EXTENDED ABSTRACT

Introduction

In a previous paper Schaefer and Liu (1984) established the need for a
practical pressure loss model which can predict pressure loss in randomly
arrayed filter structures of known fiber size over a wide range of solidity
for the viscous and no-slip flow regime. Further, using test data from media
having essentially uniform fibers but manufactured in various ways and covering a solidity range from 0.0080 to 0.808, it was shown that there were no
models in the previous literature which could accurately predict pressure loss
over such wide solidity ranges. In the present paper it will be shown how
such a model using a phenomenological semi-empirical approach with the Schaefer
and Liu data as a guide can be derived in closed form.

Theoretical

Our objective is to find a regular geometric structure which most closely
represents the pressure losses occurring in a real fibrous structure.

The total pressure loss is related to the dimensionless fiber drag F^* by
means of the Kirsch-Fuchs relation, Fuchs and Kirsch (1967). The dimensionless fiber drag is defined as

$$F^* = \frac{\text{total fiber drag/total fiber length}}{\mu\, U_\infty} = \frac{n\, D_f}{\mu\, U_\infty\, L_f} \qquad (1)$$

where n is the number of fibers, D_f is the drag on an individual fiber, and
L_f is the length of fibers in the matrix, μ is the fluid viscosity and U_∞ is
the mean approach velocity.

It has long been known that in the viscous non-slip regime F^* is independent of Reynolds number and depends only on solidity, see e.g. Davies
(1973). This suggests that the effective length L of the fibers is finite
and that on the average, the length varies with solidity. This is consistent
with the observed fact that random fibers in a matrix touch or cross at various points and that the number of crossings increases as the fibers are compressed to higher matrix solidities. Unlike an infinitely long fiber whose
shear stress and pressure distribution is the same everywhere along its
length but varies with Reynolds number, the matrix fibers have secondary flows
and modifications to their shear stress and pressure distributions at the
crossing points as well as effects from neighboring fibers (boundaries). It
is assumed this produces drag characteristics similar to a finite length fiber
in a bounded viscous flow. This drag varies with fiber length to diameter
ratio $L/d_f = R$ (or aspect ratio) and boundary proximity.

We concentrate on defining a matrix geometry which will allow us to find the unknown relation for n, D_f and L_f as a function of c in Eq. (1). Chen (1955) assuming the matrix to be equivalent to layers of screen having a square mesh of length L and fiber diameter d_f with each successive layer spaced on a center to center distance L from the previous layer. He did not detail the geometry of the intersection which is necessary for accuracy at high solidity. Two modifications of this geometry were considered which seemed more representative of real matrices:

1. Layers of screen formed by a crossed rod matrix. Each layer is 2 d_f thick and spaced 2 d_f apart on centers in the flow direction. Successive layers are randomly oriented with a maximum solidity c = $\pi/4$.

2. Layers of screen with all cylindrical elements mitered at the juncture. Layers are spaced d_f apart on centers in the flow direction. Successive layers are randomly oriented with a maximum solidity c = 0.904.

For the same solidity each geometry will have a different value of R and thus a different fiber drag. If one imagines the real matrix to be divided into layers one d_f thick along its thickness t and the resulting material in each layer is reconstituted into circular fibers arranged into a square screen of mesh length L, it was found that the latter mitered cylinder geometry most closely reproduces this average mesh length. This geometry is shown in Figures 1a & b.

The following relations can be derived from this geometry: The mean fiber aspect ratio R is related to c by Eq. (2), (see equation summary, Table I). The equivalent spherical diameter d_s of the mitered cylinder, Eq. (3). A boundary proximity parameter λ defined as the ratio of d_s to the hydraulic diameter D_h of the boundary formed by the six surrounding fibers (Figure 1a), Eq. (4); other definitions of this parameter yield quite different results. The ratio of total length of constant diameter fiber L_f to total length of mitered cylinders n L in the matrix, Eq. (5).

The viscous drag on any finite length particle (mitered cylinder) in a bounded fluid can be given by Eq. (6), Monson (1983), in which K_∞ is the shape factor of the mitered cylinder and K_{b2} is a boundary correction factor. Substituting eq's. (3),(5) and (6) into Eq. (1) gives F* in terms of parameters relative to the present mitered cylinder model, see Figure (3). Dupuit's relation $U_\infty/(1-c)$ for the mean velocity in the matrix has also been used.

The shape factor of the mitered cylinder could not be represented by that for a cylinder of the same aspect ratio as the mitered cylinder because the fibers in the real matrix are curved between intersections. The ideal matrix was thus modified as shown schematically in Figure 2. Curved fibers have less drag than straight fibers of the same aspect ratio. In this model it was assumed that the shape factor was similar to that for segments of a torus of the same aspect ratio as the mitered cylinder would have at a given solidity. The shape factor is given by a linear transition as a function of R between that for a full torus at c = 0 to that for a cylinder when R < π (c > 0.43), Eq. (7). Semi-empirical shape factor relations for a torus, $K_{\infty t}(R)$ and $K_{\infty c}(R)$ for a cylinder were given by Monson (1983).

When the surrounding effective boundary is porous (other fibers) rather than solid, K_{b2} is assumed to also be a function of porosity (1-c) in addition to the parameters K_∞, λ and k, the boundary shape parameter in Brenner's (1962) general first order boundary theory. Furthermore, K_{b2} was assumed separable into two parts. The porosity effect was modeled after the form given by Maude and Whitmore's (1957) general sedimentation theory, i.e., $(1-c)^m$. The exponent m was empirically found to be 2.0 ± 0.25 for best fit to the data. The second part was assumed to take the form of Brenner's first order boundary theory $K_{b1}(K_\infty, \lambda, k)$ except the boundary shape constant k was replaced by an empirical relation $k_e(\lambda)$ to account for higher order terms and the probability that the effective boundary shape varies with $\lambda(c)$. The experimental values F_{exp}^* and the capillary tube model (where applicable) were used in obtaining the curve fit for k_e. Thus $K_{b2} = (1-c)^2 K_{b1}$ with K_{b1} given by Eq.'s (8) and (9).

The final relation is compared in Fig. 3 with the data of Schaefer and Liu (1984). Also shown are the highest values of F* which can be achieved with a regular infinite fiber array (Sangani and Acrivos (1982)) and the predicted effect on F* if we have a transverse log-normal distribution in solidity throughout the media depth given by values of geometric standard deviation in solidity $\sigma_g > 1.0$. If we had assumed straight fibers, the dashed curve would have resulted.

Table 1. Equation Summary

$$R = \left[\frac{3}{8}\pi - \sqrt{\frac{9\pi^2}{64} - \frac{3}{2}c}\right]^{-1} \quad (2)$$

$$d_s = d_f \left[\frac{3}{2}R - \frac{2}{\pi}\right]^{1/3} \quad (3)$$

$$\lambda \equiv \frac{d_s}{D_h} = \frac{\frac{1}{2}\left[\frac{3}{2}R - \frac{2}{\pi}\right]^{1/3}}{(R-1)(2R-1)+1/2}(3R - 3 + \sqrt{2}) \quad (4)$$

$$\frac{L_f}{nL} = 1 - \frac{4}{3\pi R} \quad (5)$$

$$D_f = \frac{3\pi\mu U_\infty d_s}{K_\infty K_{b2}} \quad (6)$$

$$K_\infty = K_{\infty t}(1 - \frac{\pi}{R}) + \frac{\pi}{R}K_{\infty c}, \quad c \leq 0.4325 \quad (7)$$
$$K_\infty = K_{\infty c} \qquad\qquad , \quad c > 0.4325$$

$$K_{b1} = 1 - \frac{k_e}{K_\infty}\lambda_i \quad (8)$$

$$k_e^{-1} = 0.19 + 1.11\,(\lambda_i + 0.1)^{1.17} \quad (9)$$

where $\lambda_i = \lambda$, c < 0.8737
$\lambda_i = 2.7758 - \lambda$, c > 0.8737

554

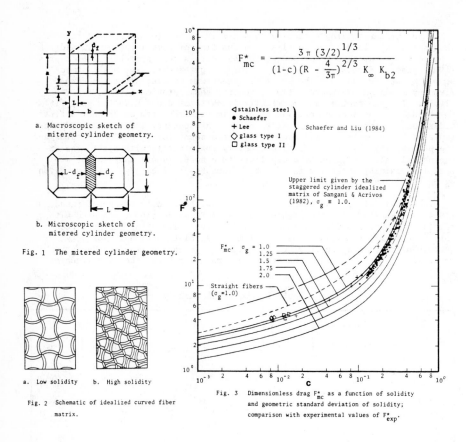

a. Macroscopic sketch of
   mitered cylinder geometry.

b. Microscopic sketch of
   mitered cylinder geometry.

Fig. 1   The mitered cylinder geometry.

a. Low solidity   b. High solidity

Fig. 2   Schematic of idealized curved fiber
         matrix.

$$F^*_{mc} = \frac{3\pi \, (3/2)^{1/3}}{(1-c)(R - \frac{4}{3\pi})^{2/3} K_\infty \, K_{b2}}$$

◁ stainless steel
● Schaefer
+ Lee   Schaefer and Liu (1984)
◇ glass type I
□ glass type II

Upper limit given by the
staggered cylinder idealized
matrix of Sangani & Acrivos
(1982), $\sigma_g \equiv 1.0$.

F^*_{mc}, $\sigma_g = 1.0$
              1.25
              1.5
              1.75
              2.0

Straight fibers
($\sigma_g = 1.0$)

Fig. 3   Dimensionless drag F^*_{mc} as a function of solidity
         and geometric standard deviation of solidity;
         comparison with experimental values of F^*_{exp}.

References

Brenner, H. (1962). "Effect of Finite Boundaries on the Stokes Resistance of an Arbitrary Particle," J. Fluid Mech. 12:35-48.

Chen, C.Y. (1955). "Filtration of Aerosols by Fibrous Media," Chemical Review 55:595-623.

Davies, C.N. (1973). Air Filtration, Academic Press, New York.

Fuchs, N.A. and Kirsch, A.A. (1967). "Studies on Fibrous Aerosol Filters - II. Pressure Drops in Systems of Parallel Cylinders," Ann. Occup. Hyg. 10:23-30.

Maude, A.D. and Whitmore, R.L. (1957). "A Generalized Theory of Sedimentation," British Journal of Applied Physics 9:477-482.

Monson, D.R. (1983). "The Effect of Transverse Curvature on the Drag and Vortex Shedding of Elongated Bluff Bodies at Low Reynolds Number," J. Fluids Eng. 105:308-322.

Sangani, A.S. and Acrivos, A. (1982). "Slow Flow Past Periodic Arrays of Cylinders with Application to Heat Transfer," Int. J. Multiphase Flow 8:3:193-206.

Schaefer, J.W. and Liu, B.Y.H. (1984). "An Investigation of Pressure Loss Across Fibrous Filter Meida," Paper No. 63, First International Aerosol Conference, AAAR and GFA, Mpls., MN., Sept. 17-21.

Published 1984 by Elsevier Science Publishing Co., Inc.
Aerosols, Liu, Pui, and Fissan, editors

AN INVESTIGATION OF PRESSURE LOSS ACROSS FIBROUS FILTER MEDIA

James W. Schaefer   and   Benjamin Y. H. Liu
Corporate Research   Particle Technology Laboratory
Donaldson Co., Inc.   Mechanical Engineering Department
Minneapolis, MN. 55440   University of Minnesota
                                  Minneapolis, Minnesota 55455

EXTENDED ABSTRACT

Introduction

All hydrodynamic filtration models are based upon a geometrical description of the filter matrix. The effectiveness of the matrix description is generally measured in terms of the correlation between the theoretical prediction of pressure loss and the actual experimental data. Many theories are derived and tested for a specific set of boundary conditions covering flow, particle and media geometrical regimes. From the practical perspective of filter media design and filtration performance analysis, a practical pressure loss model will provide an essential tool for general design, analysis, normalization and characterization. Therefore, it is important to know how the available filter pressure loss theories compare to experimental data covering a wide range of practical media geometries composed of known fibers approaching ideal properties for viscous and no-slip flow.

Theoretical

Theoretical models of fibrous filter media can generally be classified as describing either flow between or flow around fibers. The volume fraction of solids (solidity) determines which classification is applicable. For low solidity media fibers are far apart and the flow disturbance is confined to the vicinity of the fiber with minimal fiber - fiber interference. For high solidity media, with fibers close together, the flow disturbance is everywhere in the field and fiber - fiber interference dominates. The media solidity is then a primary media parameter for determining the appropriate conceptual visualization.

Schaefer (1976) analyzed several of the most widely applied pressure loss models including capillary tube and fiber type models. The capillary tube model of Kozeny-Carman is based on the flow between fibers being modeled as a series of circular capillaries, with the number of capillaries being a function of the solidity. An empirical tortuosity factor is employed to account for the length of a flow path being greater (tortuous) than the thickness of the media. The capillary tube model is generally applied for high solidities greater than 0.2-0.3. Fiber type models include cell and drag types. The cell models are based on analyzing the flow in a fluid envelope surrounding a fiber, the fluid envelope diameter being a function of solidity. In 1942, Langmuir examined flow parallel to the fiber axis as flow between two coaxial cylinders, while separate studies in 1959 by Happel and Kuwabara examined flow perpendicular to the fiber axis, each employing a different boundary condition at the fluid envelope surface. The 1955 drag model of Chen employs the analysis of the effective drag on a fiber surrounded by neighboring fibers in a screen matrix using the correlation of the drag of an isolated fiber falling in a bounded fluid for different fiber to wall geometries. Kirsch and Fuchs (1968) developed an empirical fan model where fibers are arranged parallel in each layer, with subsequent layers rotated through an arbitrary angle. Spielman and Goren (1968) developed a model for four matrix orientations that cover all fibers normal to flow but random (case 1), all fibers parallel to flow (case 2), fibers random in two dimensions parallel to flow (case 3), and fibers random in three dimensions (case 4). Their model employs the traditional forces on a fiber interior to the matrix with the addition of a damping force that results from the interference effects of neighboring fibers. The fiber type models are generally applied for solidities less than 0.2-0.3. The theories are all based on the assumption of a single fiber size and a uniform solidity. It is of interest to examine the aforementioned models for viscous and no-slip flow over a broad range of media solidities with each mat containing a single fiber size, to determine which theory is best to generally apply for practical filtration applications.

Experimental

Experiments were conducted for viscous and no-slip flow using filter media representing practical random formations composed of fibers with ideal properties. Schaefer (1976) utilized 13.4 μm Dacron® polyester fibers with a fibrid binder manufactured by a wet handsheet process. Through hot press calendering media solidities from 0.0978 to 0.4446 were obtained. Commercially available Owens-Corning glass media were used for solidities from 0.0080 to 0.0136. Calendered and diffusion bonded fibrous 316L stainless steel media made by Michigan Dynamics was used for solidities from 0.642 to 0.808.

Lee (1977) utilized Dacron® polyester fibers of 11.0 μm for loose packed mats and 13.4 μm with fibrid binder for calendered media and covered solidities from 0.0086 to 0.420. Lee's data is included to fill out portions of the solidity range from 0.0080 to 0.808 covered in this study.

Table I gives a summary of fiber properties listing fiber diameter (D_f) and geometric standard deviation (σ_g).

Table I. Fiber Properties

Fiber	D_f (μm)	σ_g	Fiber	D_f (μm)	σ_g
Dacron [R]			Glass		
• Schaefer (1976)	13.4	1.05	• Type I	45.7	1.45
• Lee (1977)	11.0	1.06	• T.I.W. Type II	10.3	1.47
			Stainless Steel	13.9	1.13

The resulting media composed of the fibers listed in Table I generally satisfied the theoretical assumption of a single fiber size, except for surface fiber deformation (flattening) in the stainless steel media, but did not satisfy the assumption of a uniform solidity. The media formation covered the spectrum of possible nonuniformities encountered in actual media. Solidity variations through the media depth arise from the fiber deposition during manufacturing, as well as, structural deformations during calendering. Lateral solidity variations can be attributed to non-uniform fiber dispersion during manufacturing. Attempts were made to minimize these variations as much as possible. The experimental results yielded 168 data points for media generally having uniform fiber diameters and sufficient thickness to minimize lateral solidity variation effects.

Comparison of Theory to Experimental

The data is compared to theory in Figure 1 in the form of a dimensionless drag parameter (F*) versus solidity, as defined by Fuchs and Kirsch (1967), with a dashed line illustrating the theory extended past the assumed solidity bounds. As illustrated, the Langmuir model covers the broadest solidity range ($0.0080 \leq c \leq 0.808$) the best, with the Spielman and Goren case 3 covering low solidities ($c < 0.1-0.2$) and the Kozeny-Carman model covering high solidities ($c \geq 0.2$) the best. The experimental data falling below the Happel, Kuwabara, and Kirsch and Fuchs models confirms the results of Fuchs, Kirsch and Stechkina (1973) for ideal versus actual media geometries. The Chen model falls below the data for $c < 0.1$ and above for $c > 0.3$. The Spielman and Goren model has case 1 falling above the data, case 2 falling below the data, and case 4 falling above the data for $c < 0.15$ and below for $c > 0.4$. We conclude that no model adequately covers the entire solidity range for actual media geometries encountered in practical filtration applications.

References

Fuchs, N.A. and Kirsch, A.A. (1967). "Studies on Fibrous Aerosol Filters - II. Pressure Drops in Systems of Parallel Cylinders," Ann. Occup. Hyg. 10:23-30.

Fuchs, N.A., Kirsch, A.A. and Stechkina, I.B. (1973). "A Contribution to the Theory of Fibrous Aerosol Filters," Faraday Division of Chemical Society, Symp. No. 7, "Fogs and Smokes."

Kirsch, A.A. and Fuchs, N.A. (1968). "Studies on Fibrous Aerosol Filters-III. Diffusional Deposition of Aerosols in Fibrous Filters," Ann. Occup. Hyg. 11:299-304.

Schaefer, J.W. (1976). "An Investigation of Pressure Drop Across Fibrous Filter Media", M.S. Thesis, Mechanical Engineering Department, University of Minnesota, Minneapolis, MN.

Spielman, L. and Goren, S.L. (1968). "Model for Predicting Pressure Drop and Filtration Efficiency in Fibrous Media", Environmental Science and Technology, V. 2, No. 4:279-287.

558

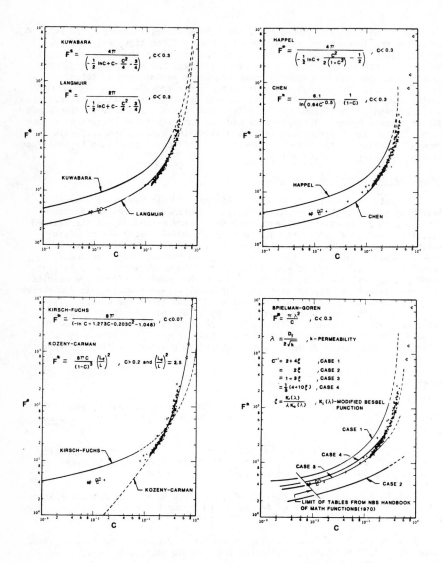

Figure 1. Dimensionless drag as a function of solidity for:
● Schaefer (1976), + Lee (1977), ◇ glass type I,
□ glass type II, and ◁ stainless steel.

Published 1984 by Elsevier Science Publishing Co., Inc.

Aerosols, Liu, Pui, and Fissan, editors

ION LEAKAGE TEST OF AN ELECTROSTATICALLY AUGUMENTED
MEMBRANE FILTER COMPRIZING CORONA PRECHARGER

S. Masuda* and N. Sugita**
* Department of Electrical Engineering
Faculty of Engineering, University of Tokyo
7-3-1, Hongo, Bunkyo-ku, Tokyo, Japan 113

** Midori Anzen Kogyo Co., Ltd.
5-4-3, Hiroo, Shibuya-ku, Tokyo, Japan 150

EXTENDED ABSTRACT

Introduction

An electrostatically augumented membrane filter developed by the authors
(MESA-Filter) comprizes a corona precharger in front of an electrostatically
augumented HEPA membrane, as illustrated in Figure 1, and it indicates a very
high air cleaning performance (Masuda and Sugita, 1981 and 1982). The
conductive separaters, b, inserted in the upstream and downstream pathways of
the folded HEPA-membrane, a, serve as the postive and negative electrodes, with
the upstream electrodes grounded and the downstream ones connected to a negative
dc power supply. The airborn particles, corona-precharged positively, are
mostly collected by Coulombic force on the upstream electrodes, while the
escaped particles, very small in size, are effectively collected in the inside
of the electrified membrane. However, a question arises in this device whether
corona induced positive ions penetrate the electrified membrane to appear in the
clean room and produce electrostatic troubles in the IC or LSI elements being
processed. Tests are made to make this problem clear, using a novel high sensitivity negative ion detector developoed by Gosho (Y. Gosho and A. Hara, 1982).

High-Sensitivity Negative Ion Detector

The detector consists of a needle-to-grid electrode system and an ion
filter attached in front of it (Figure 2). The grid is grounded through a
resistor, while the needle applied with a positive voltage carefully adjusted to
a Geigerr-Counter region just below corona-onset. Then, a burst-like positive
streamer occurs from the needle tip every time when a negative ion penetrating
the ion filter and grid arrives at an active region around the tip. This
streamer produces a current pulse which is detected as a voltage pulse signal
across the grounding resistor. This phenomenon is caused by the field induced
electron detachment from the negative ion, occuring in the high field active
region close to the tip. The detached electron produces an electron-averanche
to cause a large charge multiplication. The ion filter is to distinguish the
effect of large ions and charged aerosols from those of molecular ions, based on
the difference in mobility. Although the present MESA-Filter uses primarily
postive corona for precharging, the present tests are made using a negative
corona precharger, as this ion detector works only for negative ions. No difference is likely to exist, however, when considered the mobility of negative and
positive ions being almost the same.

The present detector does not provide any quantitative information of negative ion concentration, since its counting efficiency is not clearly defined.

560

Hence, the measured pulse counting rate must be compared with its baseline value due to background negative air ions with an equilibrium concentration in the order of 10^3 ions/cm^3. These ions are produced by cosmic ray and natural radioactive elements. Figure 3 shows such background pulse signals measured in an experimental setup shown in Figure 4 with the ion filter detached. The signal occurs very randomly with its average counting rate being 0.6 pps.

Experimental Setup

Three different tests are made, using the experimental setups as shown in Figures 5 (a), (b) and (c). The first and second tests are the reference tests made with the detector being exposed to a much higher ion concentration produced by alpha-ray and negative corona. The last test is the main test made at the outlet of the present MESA-Filter with its corona precharger operated with a reverse (negative) polarity. The polarity of the downstream electrodes is also reversed accordingly.

Results of Tests and Discussions

The results obtained are tabulated in Tables 1, 2, 3 and 4. A great increase in the pulse counting rate as well as its ratio to the background level are observed when the negative ion concentration is enhanced by alpha-ray and negative corona (Tables 1 and 2). However, the penetration of corona-induced negative ions is greatly diminished by insertion of the HEPA-membrane, even if no electrification is made (Table 3). When the membrane is electrified by applying 1 kV positive dc voltage to the downstream electrodes, the pulse counting rate drops down to the background level (Table 4), indicating the ion leakage being completely stopped.

This last result is expected from a simple calculation assuming that ion mobility = 2 cm^2/Vs, contact-to-contact electrode distance across the membrane = 4 mm (average field intensity = 2.5 kV/cm, and electrical ion migration velocity = 50 m/s), membrane thickness = 0.5 mm and membrane velocity of air = 2.5 cm/s (residence time = 20 ms). This gives rise to 1 m movement of ions within 50 ms residence time, far beyond the pore dimension in the HEPA-membrane, so that all of the ions should be completely eliminated.

Conclusion

The penetration of corona induced negative ions can be substantially reduced even by a non-electrified HEPA-membrane, and its level of counting rate becomes almost the same as that of background air ions when the membrane is electrified. Hence, it is concluded that no ion leakage is likely to occur in the present MESA-Filter. A more detailed test is being undertaken for investigating a possibility of detecting charged submicron particles based on the principle of the present ion detector.

References

Masuda, S. and Sugita, N. (1981) "Electrostatically Augumented Air Filter for Producing Ultra-Clean Air", Proc. 74th Ann. Meet. of APCA, Paper No. 81-45.3 (Philadelphia, June 21-26, 1981).

Masuda, S. and Sugita, N. (1982) "Electrostatically Augumented Air Filter for

561

Producing Ultra-Clean Air and Its Performance", Proc. 6th Int. Symp. Contaminat. Control, pp. 243-246 (Tokyo, Sept. 13-16, 1982).

Gosho, Y. and Hara, A. (1982) "A New Technique for Detecting Initial Electrons by Using Geiger Counter Region of Positive Corona", Proc. 7th Int. Conf. Gas Discharge (Sponsored by IEE), pp. 193-195 (Aug. 31-Sept. 3, 1982).

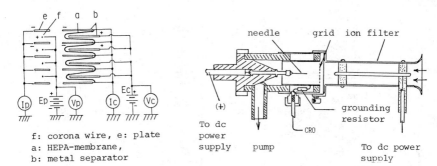

f: corona wire, e: plate
a: HEPA-membrane,
b: metal separator

Fig. 1 Construction of MESA-Filter with Corona Precharger

Fig. 2 High-Sensitivity Negative Ion Detector

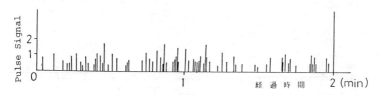

Fig. 3 Pulse Signal by Background Negative Ions in Air (see Fig. 4)

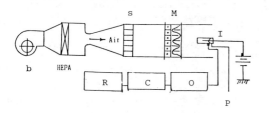

Fig. 4 Experimental Setup - I

s: straightner, b: blower
M: MESA-Filter, P: pump
I: ion detector, O: CRT
C: pulse counter,
R: pen recorder

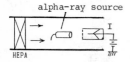
(a) Exposure to alpha-ray induced ions

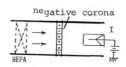
(b) Exposure to negative corona induced ions

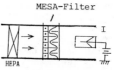
(c) Downstream of MESA-Filter (see Fig. 4)

Fig. 5 Experimental Setup - II for Negative Ion Detection Tests

Table 1 Alpha-Ray Test

Cond.	Alpha Ray	Pulse Rate (pps)	Ratio to Backg.
I	no	0.86	0.3
I	yes	22.8	7.4
I	no	5.29	1.7
II	no	2.13	1.6
II	yes	23.92	18.0
II	no	0.53	0.4

I: In room air with blower and detector pump stop.
II: In clean air with blower and detector pump on. (duct air vel. 0.76 m/s; 32 l/min)

Table 2 Negative Corona Test

Cond.	Corona Curr. (mA)	Pulse Rate (pps)	Ratio to Backg.
I	1.0	6.77	20
I	1.0	3.00	0.9
II	0.0	3.32	1.0
II	1.0	38.9	11.0
I	1.0	79.3	23.0
I	0.0	3.94	1.1

I: In room air with blower off and detector pump on (32 l/min)
II: In room air with blower and detector pump on (duct air vel. 0.76 m/s; 32 l/min)

Table 3 HEPA-Membrane Block Test with Precharger OFF

Cond.	Pulse Rate (pps)	Ration to Background Pulse Rate
see below	12.7	1.2
see below	10.1	0.9
see below	9.18	0.9

In clean air with blower and detector pump on (duct air vel. 0.65 m/s; 32 l/min)

Table 4 MESA-Filter Leakage Test

Cond.	Corona Curr. (mA)	Corona Volt. (kV)	Pulse Rate (pps)	Ratio to Backg.
see below	1.0	1.0	20.7	1.1
see below	0.0	0.0	19.2	1.0
see below	1.0	1.0	12.4	1.1
see below	0.0	0.0	11.0	1.0

In clean air with blower and detector pump on (duct air vel. 0.65 m/s; 32 l/min)

Published 1984 by Elsevier Science Publishing Co., Inc.
Aerosols, Liu, Pui, and Fissan, editors

EFFECT OF INTER-FIBER DISTANCE ON COLLECTION EFFICIENCY
OF A SINGLE FIBER IN A MODEL FILTER COMPOSED
OF PARALLEL FIBERS IN A ROW

Chikao Kanaoka, Hitoshi Emi and Atsushi Deguchi
Department of Chemical Engineering
Kanazawa University
Kanazawa 920, Japan

EXTENDED ABSTRACT

Introduction

Collection efficiency and pressure drop of a filter are affected by
inter-fiber distance as well as filtration conditions such as filtration
velocity, fiber and particle diameters. The effect of inter-fiber distance
on a filter performance is usually taken into account in the calculation of
flow field around a fiber as the change in the domain size which is related
to the average fiber distance, i.e., packing fraction of a filter. However
this treatment is not always adequate, because the flow field around a fiber
changes by the fiber arrangement in a filter.

In this study, to elucidate the effect of filter structure on its
performance, the collection efficiency and pressure drop of model filters
with various combinations of longitudinal and transverse directional
inter-fiber distances were measured in interceptional and/or inertial
collection region using DOP particles. Based on the experimental results,
approximate expressions for the estimation of a single fiber collection
efficiency and pressure drop were proposed.

Experimental

Figure 1 shows the experimental apparatus used in this study. A set of
model filter layers was used as a test filter. DOP particles generated by
vaporization-condensation method (D_p=1.6 m,σ_g=1.2) were used as test
aerosols. They were introduced to the test section. The collection
efficiency of the filter was determined by measuring the concentrations at
the inlet and outlet of the filter continuously by light scattering
photometers. Pressure drop across the filter was also measured continuously
by a pressure difference transducer. The experiment was carried out by
varing approaching velocity and inter-fiber distances in the longitudinal and
transverse direction to the air stream. Table 1 shows the experimental
conditions.

Results and discussion

Single fiber collection efficiency

Single fiber collection efficiency η was calculated by Equation (1) and
the typical result was plotted in Figure 2.

$$\eta = h_T/D_f\{1 - (1 - E)\}^N \qquad (1)$$

where, h_T is the transverse distance.

The experimental single fiber collection efficiency decreases as
longitudinal distance increases, and gradually approaches to a constant

value which depends on the transverse distance. In the previous work, the interference effect of neighboring fiber was treated by taking account of the average inter-fiber distance of a filter, i.e.,

1) to add the correction term, which is a function of packing fraction of the filter, to the collection efficiency of an isolated fiber (Chen 1955), and

2) to change the domain size of the calculation of flow around a fiber (Kuwabara, 1959).

The experimental efficiency suggests that the interference effect depends upon the relative location of neighboring fiber. Hence, to see the effect of the relative location of neighboring fiber, the exerimental single fiber collection efficiency at a given longitudinal or transverse distance was plotted against packing fraction of filter, defined by Equation (2), in Figure 3.

$$\alpha = \frac{\pi D_f^2}{4 h_T h_L} \tag{2}$$

The collection efficiency for a constant transverse distance is higher than that for a constant longitudinal distance at any filtration velocity. The difference of efficiency by the change in filter structure becomes siginificant at lower filtration velocity. This means that the transverse distance has stronger interference effect on the single fiber collection efficiency than the longitudial.

Pressure drop

Figure 4 shows a typical experimental pressure drop as a function of longitudinal distance h_L. For a given filtration condition, the pressure drop changes much by the change of the transverse distance but is insensitive to the longitudinal. Hence, it was converted to drag coefficient by using Equation (3) to see the effect of the transverse distance and plotted in Figure 5. In the same figure, the theoretical drag coefficient for Miyagi flow around parallel cylinders in a row and that of an isolated cylinder are also shown.

$$C_D = \Delta P \frac{h_T}{D_f} \frac{2}{\rho v^2} \tag{3}$$

The drag coefficient decreases with Re in a similar manner for the case of isolated fiber but it increases as h decreases. In small Re region, the experimental drag coefficient approcaches to the corresponding Miyagi's coefficient.

Conclusion

The experiment revealed that although both single fiber collection efficiency and drag coefficient increased as the fiber distance decreased, the transverse distance to the air flow affected them stronger than the longitudinal distance.

References

Chen, C.Y. (1955) "Filtration of Aerosols by Fibrous Media", Chem. Revs, 55, 595

Kuwabara, S. (1959) "The Forces Experienced by Randomly Distributed Parallel Circular Cylinders and Spheres in a Viscous Flow at a Small Reynolds Numbers", J. Phys. Soc. Japan, 14, 527

Table 1   Conditions for the experiments with fan model filter

---

Particle diameter	D_p	[μm]	1.6 (σ_g=1.16)
Concentration of mist	C_i	[mg/m³]	5 ∿ 20
Filtration velocity	v	[m/sec]	1,2,4,7
Transverse distance between the axis of fibers in a row	h_T	[μm]	135,266,525,1020
Longitudial distance between the axis of fibers	h_L	[μm]	260,410,560,1260,2260
Fiber diameter	D_f	[μm]	60
Number of model filters	n_M [layers]		8 ∿ 20

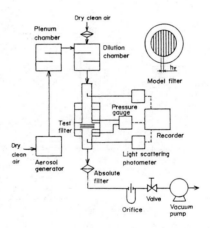

Figure 1 Experimental setup

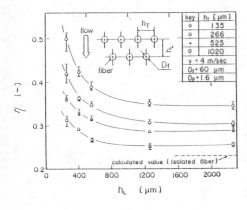

Figure 2 Relationship between single fiber collection efficiency η and h_L

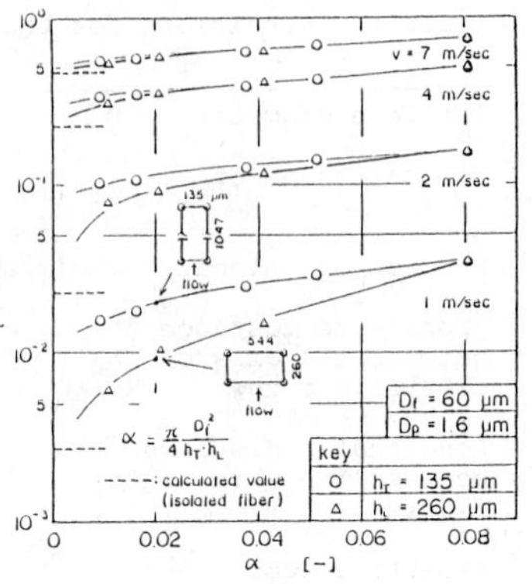

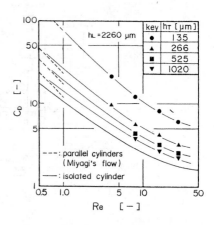

Figure 3 Effect of fiber arrangement on a single fiber collection efficiency η

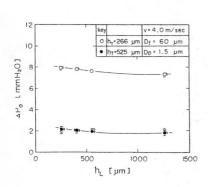

Figure 4 Relationship between pressure drop ΔP and h_L

Figure 5 Relationship between drag coefficient and Re

Published 1984 by Elsevier Science Publishing Co., Inc.
Aerosols, Liu, Pui, and Fissan, editors

MOST PENETRATING PARTICLE SIZE IN ELECTRET FIBER FILTRATION

Hitoshi Emi, Chikao Kanaoka, Yoshio Otani and Takashi Iiyama
Department of Chemical Engineering
Kanazawa University
Kanazawa 920, Japan

EXTENDED ABSTRACT

Introduction

Electret filter is composed of permanently charged electret fibers and is capable of collecting fine particles effectively. In the previous work (Emi et al.), the collection performance of an electret filter was investigated theoretically and experimentally. As a result, the collection efficiency of a single electret fiber was well correlated by a function of Coulombic force and induced force parameters, when electrostatic effects are prevailing.

In actual filtration, Brownian diffusion is also effective to the collection of fine particles. Among three effective mechanisms for the collection of small particles, Brownian diffusion and Coulombic force become stronger as particle size decreases, while induced[force becomes more effective as size increases. This means that maximum penetration appears in the transient region of these effects, i.e., between Brownian diffusion and induced force effect for uncharged particles, and between induced and Coulombic force effects for charged particles.

In this study, collection performance of an electret filter was measured over a wide range of filtration condition using very fine particles (0.03 to 0.2 μm) in different charging states (uncharged, singly, doubly and triply charged, and charged in charge equilibrium). Based on the experimental result, an expression for the collection efficiency of a single electret fiber, which is applicable to the collection of particles in any charging state, was obtained by taking account of Brownian diffusion, and Coulombic and induced force effects simultaneously. The most penetrating particme size obtained from this expression was compared with the experimental result.

Experimental

Figure 1 shows the experimental setup. NaCl particles generated by a Collison atomizer and classified by a differential mobility analyzer (DMA) were used as test aerosol particles. Since particles from DMA are singly charged, charging state of test aerosol was controlled by the following methods, i.e.,
1) Singly charged; particles from DMA (DMA-particle)
2) Doubly or triply charged; DMA-particles classified by the 2nd DMA after passing them over a radioactive isotope Am-241.
3) Charged in equilibrium; DMA-particles passed over Am-241 again ("equilibrium" particles)
4) Uncharged particles; "equilibrium" particles penetrating through an by an electrostatic precipitator.
Experiment conditions are shown in Table 1.

568

Result and discussion

Penetration of an electret filter

Figures 2, 3 and 4 show the experimental penetration of particles in different charging states against particle size with filtration velocity as parameter. The maximum penetration appears below 0.1 μm for uncharged and over 0.1 μm for singly charged particles and the maximum size shifts to smaller size as filtration velocity increases. The penetration curve of "equilibrium" particles is different from those of charged and uncharged particles, i.e., the minimum appears at high filtration velocity in addition to the maximum penetration. Although Brownian diffusion, Coulombic force and induced force work simultaneously for the collection of fine particles, the magnitude of their effects depends upon particle size and charging state of particles. The maximum penetration takes place between Brownian diffusion and induced force effect in Figure 1 and between Coulombic force effect and induced force effect in Figure 2. For the case of "equilibrium" particles, the situation is somewhat different, because they are a mixture of charged and uncharged particles and their composition depends upon particle size. The penetration of "equilibrium" particles will be discussed later.

Single fiber collection efficiency

For uncharged particle, a single fiber collection efficiency is expressed by a function of dimensionless induced force parameter K_{In} (ratio of induced force and air drag), neglecting mechanical collection mechanisms. The experimental single fiber collection efficiency was plotted against K_{In} in Figure 5. In the same figure, the previously obtained semi-empirical line, which estimates the single fiber collection efficiency due to pure induced force effect is also shown. The experimental efficiency coincides with the previous result, except in small K_{In} region for each particle size. Since filtration velocity is small in small K_{In} region for a given particle size, the collection efficiency in this area is significantly affected by the Brownian diffusion of particles. Taking account of Brownian diffusion and induced force effect simultaneously, the following approximate expression for uncharged particle was obtained.

$$\eta_{InD} = 3.2Pe^{-2/3} + 0.18K_{In}^{2/5} \tag{1}$$

For charged particle, Brownian diffusion effect is negligibly small in comparison with Coulombic force even in small size range. Therefore, the two electrostatic forces are predominant collection mechanisms. To see the Coulombic effect, the induce term in Equation (1) was subtracted from an experimental single fiber collection efficiency for charged particles and was plotted against Coulombic force parameter K_C in Figure 6. A semi-empirical expression for the collection of charged particles due to pure Coulombic force effect was also shown in the same figure. $\eta_{exp} - \eta_{In}$ is in good agreement with the semi-empirical expression, which was obtained for singly charged particles. Since number of charges on a particle is varied from 1 to 3 in this experiment, the proposed expression is applicable also to carrying any number of charges. However, the experimental efficiency deviates from the semi-empirical expression in small K_C regime because of interference effect between Coulombic and induced force effects, which was also observed for large particles in the previous work. Finally, the following semi-empirical expression for the estimation of a single fiber collection efficiency was obtained taking account of Brownian diffusion and two electrostatic effects simultaneously.

$$\eta_{InC} = 0.18K_{In}^{2/5} + 0.2K_C^{3/4} + 3.2Pe^{-2/3} - 0.05(K_{In}K_C)^{1/2} \qquad (2)$$

Penetration of particles in charges equilibrium.

Aerosol particles prepared by passing DMA-particles over Am-241 carry various number of positive or negative charges on them. Their charge distribution does not change with time and usually satisfies the following Boltzmann distribution

$$f(i) = \exp\{-(ie)^2/d_p\kappa T\}/ \sum_{i=-\infty}^{\infty} \exp\{-(ie)^2/d_p\kappa T\} \qquad (3)$$

The pentration of the "equilibrium" particles are determined by the following equation.

$$P = f(0)p(0) + 2 \sum_{i=1}^{\infty} f(i)p(i) \qquad (4)$$

$$p(i) = \exp - \frac{4}{\pi} \frac{\alpha}{1-\alpha} \frac{L}{d_f} \eta(i) \qquad (5)$$

where, $f(i)$ and $\eta(i)$ denote a frequency and a single fiber collection efficiency for particles with i charges.

The penetration obtained from Equations (2) and (4) is shown in Figure 7 together with the experimental results. The predicted curves show maximum and minimum penetration at high filtration velocity and follow the experimental results satisfactorily.

Particle size at maximum penetration

Since the value of first derivative of Equation (2) with respect to particle size is zero at maximum penetration, particle size at the maximum was obtained by substituting the physical and electrical properties of fiber and paraticles listed in Tables 1 and 2 as,

$$d_{p,min} = 2.7\times10^{-8}v^{-2/13} \qquad (d_{p,min} < 6.52\times10^{-8}) \qquad (6)$$

$$d_{p,min} = 0.668v^{-1/5} \qquad (7)$$

In above equations, dimension of d_p and v was given by MKS unit.

The experimental most penetrating particle size obtained from Figures 2 and 3 were compared with the predicted values in Figure 8. As can be seen, for each charging state of particle, the experimental and predicted particle sizes are in good agreement with each other and decrease as filtration velocity increases. It is remarkable that the most penetrating particle size of uncharged particle appears around 0.05 m, which is extremely small comparing the value of 0.3 - 1 μm for mechanical filtration.

Conclusion

Collection efficiency of an electret filter was experimentally measured by using various charging states. The major conclusions obtained in this study are;
1) Semi-empirical expressions for a single fiber collection efficiency and particle sizes at the maximum penetration were proposed.

2) Penetration of particles in charge equilibrium was found to be expressed
   by the summation of the single fiber collection efficiency taking account
   of charge distribution.

<u>Reference</u>

Emi, H., Kanaoka, C., Otani, Y. and Ishiguro, T. (1984) "Collection Mechanisms
of Electret Filter", Submitted to <u>Particulate Science and Technology</u>

Table 1 Experimental conditions

Aerosol	dielectric constant ε_p [-]	particle diameter d_p [μm]	number of charges n [-]	number concentration C [1/cm^3]	filtration velocity v [cm/s]
NaCl	5.9	0.03-0.2	0 - 3	1 - 1000	5 - 200

Table 2 Properties of electret filter

filter material	density of fiber ρ_p [g/cm^3]	width of fiber d_{fw} [μm]	hickness of fiber d_{fT} [μm]	packing density α [-]	filter thickness L [mm]	charge density Q [nC/m]
poly propyrlene	0.91	38	10	0.075	5.0	34.2

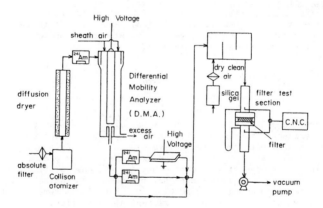

Figure 1 Experimental setup

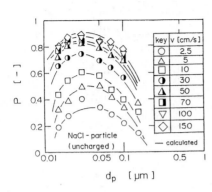

Figure 2 Experimentl penetrarion of
uncharged particles from
an electret filter

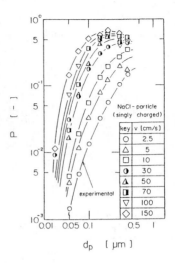

Figure 3 Experimental penetration
of singly charged particles
from an electret filter

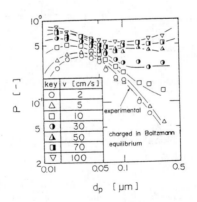

Figure 4 Experimental penetration of
particles in charge equilibrium
from an electret filter

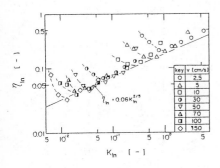

Figure 5 Single fiber collection
efficiency of uncharged
particles

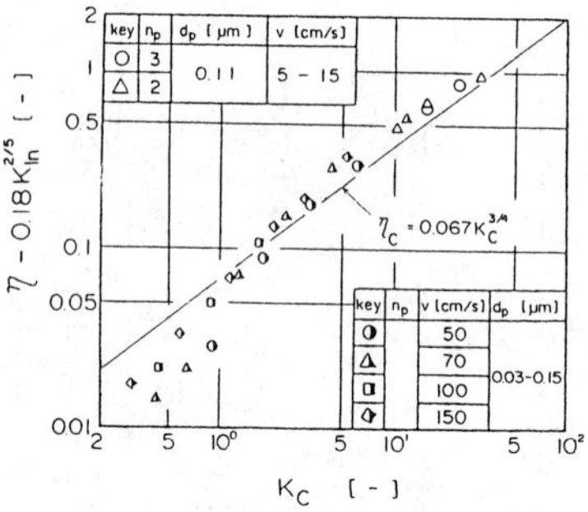

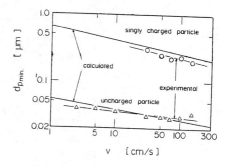

Figure 6 Single fiber collection
efficiency of charged
particles

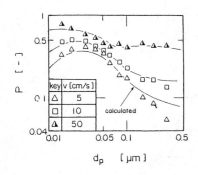

Figure 7 Comparion of calculated
penetration of particles
in charge equilibrium with
experimetal ones

Figure 8 Comparison of calculated
maximum penetrating particle
size with experimental ones

Published 1984 by Elsevier Science Publishing Co., Inc.
Aerosols, Liu, Pui, and Fissan, editors

A SEMI-AUTOMATED SYSTEM FOR MEASURING THE PENETRATION OF RESPIRABLE SIZED AEROSOLS THROUGH FILTERS

D. B. Blackford.*, R. C. Brown and D. Wake
Health and Safety Executive
Red Hill
Sheffield S3 7HQ
England

EXTENDED ABSTRACT

Introduction

Filters are used extensively in respirators giving personal protection against respirable dust; and recently interest has developed in the use of filters as experimental size-selectors in respirable dust samplers (Brown, 1980; Gibson and Vincent, 1981). The degree of protection given by the former, and the accuracy of size-selection of the latter depend on the capture efficiency of the filters with respect to aerosols of all sizes in, and close to, the respirable size range.

The efficiency of filters can vary with the dust load that they hold, either increasing because of clogging or, in the case of electrostatic filters, decreasing because of reduction of effective electric charge. In order to measure filter performance, and its variation with dust load, a system that allows quick and accurate measurement of aerosol penetration is needed.

Any system used in aerosol penetration measurements consists of three fundamental parts: a test aerosol generator; a device for holding the filter while it is exposed to the aerosol; and an aerosol detector. In our system a microcomputer is used to govern the flow of aerosol through the filter and an empty line, to read the counts from the detector, and to compute the aerosol penetration and its likely error.

Test Aerosol Generator

The aerosol generator is a modified Berglund-Lui generator. The syringe pump, which feeds a suitably dilute solution with a volatile solvent to the aerosol generator, has been replaced by a 200 cc capacity pressure vessel. Liquid is forced through a 47 mm, 0.2 µm pore size filter, fitted at the base of the pressure vessel, by the regulated output from a bottle of compressed air. Between the air supply and the pressure vessel is a 20 litre capacity reservoir, which damps any pressure fluctuations. The low-pressure side of the pressure vessel is connected to the aerosol generator through a flow measuring device, consisting of a

---

* Current address: TSI Incorporated, 500 Cardigan Road, PO Box 43394, St Paul, MN 55164, USA.

calibrated horizontal capillary tube of uniform diameter, approximately
30 cm long, and a small hypodermic syringe connected to the main flow line
via a T-piece. The flow of liquid through the system is measured by
injecting a bubble of air into the line and timing its passage along the
tube. The flow can be adjusted using the regulator on the compressed air
cylinder, and it may be checked at any time during an experiment.

This modified liquid feed system overcomes the problem of irregular
liquid flow caused by the intermittent sticking of the plunger in the
barrel of the syringe. In addition filter clogging is much less of a
problem with the pressure feed system than with the syringe pump, because
the area of the filter is about thirteen times as large.

Filter Test Boxes

In order to measure the penetration of aerosols through a filter it is
necessary to compare counts between filtered and unfiltered aerosols. Our
system uses two identical boxes in parallel: the test box which contains
the filter, and the control box which is empty. The path taken by the
challenge aerosol is governed by two solenoid valves. The resistance of
the box containing the filter under test will be higher than that of the
empty box, and so an abrupt change in flow rate will occur when the aerosol
is switched from one box to the other. This is corrected by using a split
air line on the inlet side of the Berglund-Liu generator, fitted with
solenoid valves synchronised with the main solenoid valves. One of the
lines contains a variable constriction, and this can be set to ensure that
the flow rate does not vary when line-switching occurs. The flow rate is
measured by an orifice plate, connected to a micromanometer and chart
recorder.

It is important that the aerosol should flow through the system at a
constant rate. For this reason the transporting air is supplied by a
"Gast" rotating vane air compressor. The air is passed through a 20 litre
capacity reservoir before it reaches the aerosol generator, to provide
additional smoothing.

The test boxes have long inlet cones with a half angle of
approximately seven degrees; this prevents flow separation and its
associated turbulence, which would tend to occur if the angle were larger.
A correct aerodynamic design of the test boxes is critical, especially when
relatively coarse, low-resistance filters are tested. If separation
occurs, the velocity of the aerosol through the front face of the filter
will not be uniform and since filtration efficiency is velocity dependent,
a spurious surface filtration effect will be observed.

Aerosol Detector and Microcomputer

The aerosol is detected by a Royco 225 particle counter with a 518
interchangeable module. A Rockwell AIM 65 microcomputer, interfaced with
the Royco counter, controls the sample time, which can vary from one fifth
of a second to one hour. The microcomputer also reads and stores data from
the five channels of the Royco which have previously been adjusted to

accommodate the challenge aerosol into one channel. When the lines are switched the microcomputer delays counting for the time taken for stagnant aerosol to be flushed out of the system. These dead times have been measured for a variety of airflow rates corresponding to face velocities of 1 cm/sec to 60 cm/sec at the filter.

The microcomputer controls the number of samples of filtered and unfiltered aerosol that are taken, and the relative time spent on each. The lines should be switched frequently if error due to any drift of mean particle count is to be reduced; but frequent switching will increase the error caused by increased dead time. The microcomputer chooses the number of switches that will avoid the high error associated with either extreme. Statistical considerations show that error is less when rather more time is devoted to the filtered aerosol than when the filtered and unfiltered aerosols are counted for equal times, and that this is particularly important for high-efficiency filters. The microcomputer adjusts the sampling routine accordingly.

References

Brown, R. C. (1980) "Porous foam size selectors for respirable dust", J. Aerosol Sci. 11:151.

Gibson, H. and Vincent, J. H. (1981) "The penetration of dust through porous foam filter media", Ann. Occ. Hyg. 24:205.

Published 1984 by Elsevier Science Publishing Co., Inc.

Aerosols, Liu, Pui, and Fissan, editors

576

ELECTRICAL ENHANCEMENT OF FILTRATION BY NUCLEPORE FILTERS

Gentry, J., Spurny, K., Boose, C., Schormann, J.

Electrostatic attraction has frequently been suggested as a principal mechanism for particle separation. Presented in this paper are experimental measurements which quantify the enhancement in filtration efficiencies for nuclepore filters as a function of pore charge and aerosol charge.

Nuclepore filters were coated with gold in order to minimize charge charge build up on the filter. Comparison of experimental measurements with theory are presented.

Published 1984 by Elsevier Science Publishing Co., Inc.

Aerosols, Liu, Pui, and Fissan, editors

REVIEW OF ELECTROSTATICALLY AUGMENTED FILTRATION MODELS

D. W. VanOsdell and A. S. Viner
Center for Aerosol Technology
Research Triangle Institute
Research Triangle Park, North Carolina   27709
U. S. A.

EXTENDED ABSTRACT

Introduction

Electrostatically augmented (EA) filtration has been the subject of considerable research interest over the past several years because of its potential for improving the performance of commercial filters. Experimental data for a number of dust sources and with various electrical configurations have consistently shown that an EA filter operates with improved collection efficiency and reduced pressure drop.

A number of mechanisms have been proposed to account for the effects observed with EA filtration. These mechanisms include increased upstream particle collection, nonuniform filter dust deposits, and a reduction of dust cake porosity. All have successfully been used to explain some of the effects of EA filtration. Understanding the mechanisms which control EA filtration is necessary to allow extrapolation to new conditions and sources.

Each mechanism is examined in turn below to evaluate its applicability to industrial filtration, with the emphasis placed on cake filtration such as is present in industrial fabric filters.

Upstream Particle Collection Mechanism

The upstream surface of a filter medium is, normally, less densely packed with fibers than is the interior of the filter. The "upstream particle collection" mechanism postulates that, due to the presence of electrical forces, particles in an EA filter are more likely to be collected in the low density surface region than would be true of a conventional filter relying solely on physical collection forces. Because more particles are collected in the low density region, the overall permeability of the filter is greater even if the particles pack on the individual fibers with the same porosity as is found in conventional filtration. This mechanism was first suggested by Lamb and Costanza (1980) for the results obtained with an electric field applied parallel to the fabric surface. The principal forces involved would be the Coulombic force between the fiber and a charged particle and gradient forces between particles and fibers. Short range forces dominate.

This mechanism has been modeled and investigated experimentally by Lamb (1982). His model filter consisted of three layers, each having a different porosity. The upstream or surface layer had a porosity of 0.02, and the other two layers were assigned the more typical (for fabric) value of 0.2. The first two layers accounted for 40 percent of the filter thickness, and the final layer made up the balance. Using this model, Lamb (1982) showed that increasing the fraction of the total mass collected in the upstream layer from

20 to 60 percent led to a 50-percent reduction in calculated pressure drop, a pressure drop reduction which has been observed experimentally with surface field EA filters.

Lamb (1982) tested his model experimentally with a filter having an upstream layer porosity of 0.015 and the bulk of the filter having a porosity of 0.23. Application of the electric field to this filter caused an increase in the dust fraction collected in the upstream fabric layer from 40 to 80 percent and reduced the pressure drop in the layer to essentially zero.

This mechanism explains the effect of EA filtration directly for the early stages of fabric filtration while the dust is collecting on the fabric fibers. The mechanism also appears to be valid for fibrous filtration. This mechanism is also consistent with the laboratory observation that fabrics with low surface density respond better to EA than those that have a high surface density (Lamb and Costanza, 1980).

Application of this mechanism to cake filtration is less direct. The reduced pressure drop associated with EA filtration has been shown (VanOsdell et al., 1982) to persist under conditions of cake filtration in a pilot-scale pulse-jet baghouse, and in cake filtration the upstream surface has a relatively high density.

Nonuniform Dust Distribution across Filter Surface

A nonuniform distribution of the dust mass across the surface of a filter allows that filter to operate with a reduced overall pressure drop when compared to a uniformly distributed dust load, even if the flow resistance per unit of dust mass is the same for both cases.

This model was advocated by Chiang et al. (1982). Conceptually, it is analogous to electrical resistances in parallel. The total resistance to flow is less than the sum of the individual resistances. For the model to hold, electrical forces which act over distances of a few centimeters must dominate the trajectory of the particles as they approach the filter. Coulombic forces, which are present when a charged particle enters an electric field, operate over these distances, and, thus, this mechanism requires that the particles be charged.

Chiang et al. (1982) show that with the proper choice of the adjustable parameters of the model, the observed pressure drop behavior of an EA filter can be modeled.

Very nonuniform dust deposits, suggestive of this model, have been reported for a number of laboratory investigations. Such deposits have been observed in the RTI laboratory for naturally charged dust collected in a parallel field EA filter. Penney (1978) and Lamb (1982) present similar results obtained with a precharger immediately before the fabric filter. In these experiments, the dust was observed to deposit along the electrodes, forming deposits which were easily visible.

Nonuniform deposits were not visible at the field pilot unit when no precharger was in use (VanOsdell et al., 1982a). The principle difference between the pilot unit and the laboratory work was that the laboratory experiments did not utilize as high a dust loading and did not clean the bags as regularly. Surface nonuniformity was not observed by Lamb (1982) unless a precharger was used.

Dust Deposit Porosity Reduction

This mechanism is based on a reduction in the porosity of the dust deposit of an EA filter when compared to a conventional filter. In its pure form, this mechanism does not depend on the character of the filter surface. The dust deposit on the filter is uniform, and the EA filter deposit has an increased porosity which leads to a pressure drop reduction. The porosity increase occurs because the electrical forces cause a change in particle deposition patterns. For this mechanism, the electrical forces act over short distances (on the scale of the fiber diameter), and both gradient and Coulombic forces are probably significant. Evidence for this mechanism comes from three principal sources--(1) dendritic particle collections, (2) observation of the surface of EA filters, and (3) a reduction in the density of hopper fly ash of an EA fabric filter.

The presence of an electric field in the vicinity of a single fiber has been shown both theoretically and experimentally to cause increased collection of particles in dendrites. Because of their shape, dendrites would not be expected to pack as well as spheres and, thus, should cause a higher porosity deposit.

In addition, examination of some EA filter surfaces has suggested that an EA filter deposit is less dense than a conventional deposit. Evidence of this type has been presented by a number of authors. This change in appearance was not visible at a pulse-cleaned EA fabric filter pilot unit (VanOsdell et al., 1982a), but the nature of the pulse cleaning may have obscured this effect.

VanOsdell et al. (1982b) have shown that dust cakes collected and maintained under the influence of an electric field have more resistance to being collapsed by increasing pressure drop than conventional dust cakes.

In addition, the dust which was cleaned off the EA pilot unit bags and collected in the hopper had a bulk density 80 percent of that found in the hopper of the conventional baghouse. This difference persisted throughout the entire test program.

Conclusions

Experimental evidence is available for each of the cited mechanisms, and each has been shown to be capable of producing the observed effects of EA filtration. However, in some cases, each mechanism has been shown to be unlikely as a primary cause. In practice, all of these mechanisms might be expected to act at the same time. In particular laboratory situations, single mechanisms appear to dominate, but the field experience suggests that more than one mechanism is operating in that case.

The interacting nature of the forces involved makes the separation of effects difficult. Careful experiments will be necessary to isolate the mechanism in any particular case.

580

References

Chiang, T. K., E. A. Samuel, and K. E. Wolpert. (1982) "Theoretical Aspects of Pressure Drop Reduction in a Fabric Filter with Charged Particles," in: Third Symposium on the Transfer and Utilization of Particulate Control Technology, Volume III, EPA-600/9-82-005c (NTIS PB 83-149609). Pp. 250-260.

Lamb, G. E. R., and P. A. Costanza. (1980) "A Low-Energy Electrified Filter System," Filtration and Separation, 17:319-322.

Lamb, G. E. R. (1982) "Influence of Particulate Precharging on the Performance of an Electrically Stimulated Fabric Filter (ESFF)," in: Proceedings of the World Filtration Congress III, Volume I, Downington, Pennsylvania, September 13-17, 1982. Pp. 165-168.

Penney, G. W. (1978) "Electrostatic Effects in Fabric Filtration: Fields, Fabrics, and Particles (Annotated Data), Volume I," EPA-600/7-78-142a (NTIS PB 288576).

VanOsdell, D. W., et al. (1982a) "Electrostatic Augmentation of Fabric Filtration: Pulse-Jet Pilot Unit Experience," EPA-600/7-82-062 (NTIS PB 83-168625), November 1982.

VanOsdell, D. W., et al. (1982b) "Permeability of Dust Cakes Collected under the Influence of an Electric Field," in: Proceedings of the Fourth Symposium on the Transfer and Utilization of Particulate Control Technology, Houston, Texas, October 11-15, 1982.

Published 1984 by Elsevier Science Publishing Co., Inc.
Aerosols, Liu, Pui, and Fissan, editors

ELECTROSTATIC ENHANCED FILTRATION

H.J. Fissan and S. Neumann
Prozess- und Aerosolmesstechnik
Universität Duisburg, FRG

G. Schürmann
TSI-Deutschland, Aachen, FRG

EXTENDED ABSTRACT

Introduction

The requirements for clean air are becoming increasingly stringent. Small, submicron particles at low concentrations often must be eliminated, because even small particles can cause damage to products in clean rooms or to humans in nuclear power plants.

Fiber filters are often used for the collection of submicron particles. The collection of particles is caused by diffusion, interception and inertia. As a result of these overlapping effects one finds a maximum of particle penetration in the submicron size range.

One way of decreasing the particle penetration through a filter is to reduce the fiber diameter, but this has the disadvantage of causing an increase in pressure drop. Another possibility is to make use of electrostatic effects which can reduce the particle penetration without causing an increase in pressure drop. This is the case with electret filters.

In this paper we report on the investigations of electrostatic effects in commercially available fiber filters (1) carrying charges from the production process.

Experiments

Generation and Characterization of Aerosols

The fractional penetration was determined using DOP-particles and NaCl-particles in the size range, $0.035 \ \mu m \leqslant Dp \leqslant 0.7 \ \mu m$. The aerosol generator consists of an atomizer followed by different types of conditioners (2). Nearly monodisperse DOP-aerosols ($\sigma_g = 1.3$) were produced. The geometric standard deviation of NaCl-particles was $\sigma_g = 1.8$. The aerosols differed with respect to particle size and charge. The particle number concentration of DOP-particles upstream and downstream of the filter was determined with an electrical aerosol detector. The NaCl-particle number concentration was measured with a continuous flow condensation nuclei counter.

Description of filters

Three different fibrous filters (Freudenberg/Germany) were investigated (table 1).

Filter	Fiber Diameter D_f [µm]	Solidity α [-]	Filter Thickness L [cm]	Velocity v_a [cm/s]	Pressure Drop Δp [pa]	Reynold-Number Re [-]	Knudsen-Number Kn [-]
FA-Filter Polycarbonate Fiber	1,8	0,03	0,06	2	16	0,024	0,087
				5	35	0,006	
FB-Filter Polycarbonate Fiber	1,5	0,067	0,07	2	130	0,002	0,073
				5	350	0,005	
VS-Filter Glass Fiber	1,3	0,17	0,08	2	175	0,00173	0,1
				5	430	0,00433	

Table   1.   Dimensions,   Pressure Drop   and   Flow   Regime   of
Investigated Filters

The FA-filter is made of polycarbonate fibers with diameter D_f =
1.8 µm. This filter layer of which further data are given   in
table 1 is supported by another filter layer of fibers with
diameter D_f = 20 µm, which does not contribute to the   collection of submicron particles. To eliminate large particles a third
layer with fiber diameter D_f = 10.5 µm is placed in front of
the FA-filter which again does not contribute to the   collection
of submicron particles. The FB-filter has the same three layer
structure. The main layer differs with respect to fiber diameter
and   solidity.   Both filters carry charges as a result of the
production   process   The charge of the fibers is in the range of
8 · 10^{-12} C/m to 8 · 10^{-11} C/m (3). The VS-filter consists of
uncharged glass fibers with diameter D_f = 1.3 µm. It is a
single layer filter. The penetration was determined for two face
velocities and the corresponding pressure drops are given also in
table 1. Also included are the Reynold- and Knudsen numbers.

Discussion of the Results

In fig. 1 the fractional penetration P of the FA-filter is
plotted for different charge conditions of filter and   particles.
The results shown refer to a face velocity of 5 cm/sec.
The upper curve refers to the case where no electrostatic
effects are involved. The shape of this curve is determined by
the mechanical effects of diffusion and interception. The penetration decreases slightly for neutralized aerosols because   of
electrophoretic effects. A strong decrease in penetration is
observed, if the fibers are charged. For uncharged particles the
decrease is caused by dielectrophoretic forces. The rapidly decreasing penetration with increasing particle diameter occurs, as
a consequence of the dielectrophoretic force, which increases
with the square of particle diameter. In the case of charged
fibers and neutralized aerosol the penetration is further decreased because of Coulomb effects. The Coulomb forces for a
particle with a certain charge decreases with increasing particle
diameter, but on the other hand, the bigger particles carry more
charge.

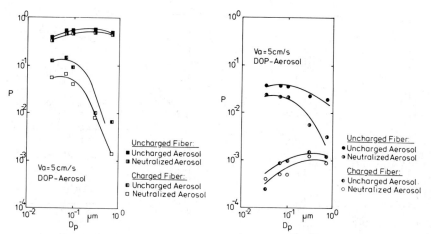

Penetration as a Function of Particle Diameter
Figure 1. FA-Filter   Figure 2. FB-Filter

Fig. 2 shows the corresponding fractional penetrations of the FB-filter. The reduction of fiber diameter and the increase of solidity results in a considerable decrease of fractional penetration by a factor of ten. The reduction in penetration caused by electrostatic effects are comparable to those of the FA-filter (Fig. 1). The reduction of the face velocity reduces the fractional penetration of all the investigated filters under all conditions.

In fig. 3 the fractional penetrations of the three filters for a face velocity of 5 cm/sec and neutralized DOP-particles are compared. For small particles the penetration of the FB-filter is smaller than that for the VS-filter, in spite of the fact that the pressure drop is higher in the case of the VS-filter. In the case of the VS-filter the collection of particles is only caused by mechanical forces, whereas in the case of the FB-filter, electrostatic effects are also involved.

For practical applications the dynamic behavior of a filter is most important. Thus far the particle loading of the filters was negligible. In order to determine the changes in filter behavior with loading, the FA-filter was loaded with liquid DOP-particles and solid NaCl-particles. The particle mass in the filter has been calculated from the measured number of collected particles and their size. The results are shown in fig. 4. In the case of DOP-particles the penetration increases with increasing particle mass loading. The penetration of neutralized aerosol is shifted to lower values because of the strong Coulomb forces acting on the particles. We believe that the increase in penetration is caused by a change in the charge loading of the fibers by particle deposition. Also changes in the structure may also occur.

584

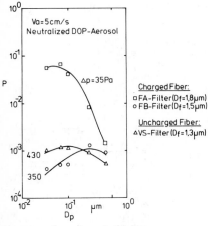

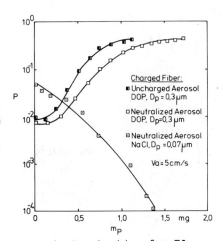

Figure 3. Penetration against Particle Diameter for Different Filters

Figure 4. Penetration for FA-Filter against Mass Loading of Liquid and Solid Particles

The NaCl-particles show a totally different behavior. The penetration drops drastically with particle mass loading. We believe that this effect is mainly caused by structural changes of the filter by the deposition of solid particles.

Summary

The experimental results obtained on the fractional particle penetration for the different filters agree qualitatively with the results of theoretical investigations. They show the importance of electrical effects in charged fiber filters in the submicron size range and that the penetration can be decreased considerably using electrostatic effects without an increase in pressure drop. With respect to the dynamic behavior of the filters theoretical considerations are thus far not very conclusive, because the knowledge of the changes in the filter with particle loading is insufficient at present.

References

(1) Schmidt, K. (1980)
    Melliand Textilberichte 61: 495-497

(2) Schürmann, G. and Fissan, H.J. (1984)
    Aerosol Science, Vol. 15, No. 3, 1984

(3) Schürmann, G. (1983)
    Interner Bericht Nr. 106, Prozess- und Aerosolmesstechnik
    Universität Duisburg

Published 1984 by Elsevier Science Publishing Co., Inc.

Aerosols, Liu, Pui, and Fissan, editors

ELECTRIC STIMULATED HEPA-FILTER FOR PARTICLES 0.1 μm

Dumitru Cucu and H.-J. Lippold

DELBAG-LUFTFILTER GmbH, Holzhauser Str. 159, 1000 Berlin 27, W.-Germany

EXTENDED ABSTRACT

Introduction

This paper presents the results and conclusions following the researches on
an electric stimulated HEPA-Filter. Depending on the utilized filtering media and on the electric configuration, it has been obtained an efficiency
between 99.99 % up to 99.99995 % for 0.1 μm particles, with a pressure loss
dramatically smaller than that resulted by using mechanical HEPA-Filters
without electric stimulation. Achieving a Super-HEPA-Filter for 0.1 μm level
particles or smaller ones and, at the same time, reducing Δp are, in fact,
two contradictory requirements for a mechanical filter. More than that, in
the near future it will be surely necessary to filter particles under 0.1 μm
very efficiently, for example about 0.04 - 0.05 μm and this performance is
impossible for a mechanical filter with a reasonable Δp.

The answer to these special requirements can be given only by the electrically stimulated Super-HEPA-Filters.

Working Principle

By passing through the first stage, the ionizer, the particles are electrically charged. Together with the air that flows through the filter, the
dust particles will pass from the ionizer to the second stage, called
collector. At here, the in the first stage already electrically charged
particles lie in an electrical field, that favours the arrest of the particles (due to the Coulomb Forces) on the separators and in the filtering
medium on fibres. The filtering medium lies, in fact, in the electrical
field of the separators which are connected to a power supply. Under these
circumstances the dielectric material fibres will polarize very strongly.
Between the electrically charged particles and the polarized fibres very
strong attraction forces will apear, so that the particles will be
attracted to and fixed to the fibres with an efficiency which depends on
the applied electrical value and on the filter geometry.

The Collector seen as a Condenser

The collector of the Electro-HEPA-Filter is, actually, a multiple condenser
containing several dielectric sheets. In fact, to simplify the problem we
can consider the filtering medium to be a first dielectric sheet and the
insulating material of the separators to be a second dielectric sheet.
Under these circumstances the dielectric permittivity will be different in
the space between the electrodes. The system considered above changes in
the course of time with the increase of the aerosol quantity retained in
the medium and on the separators. Accordingly, the permittivities and the
intensities of the electric field in the two sheets will change also.
Theoretically, the problem of this "dynamic" condenser is much more complicated and this paper does not deal with it.

The Influences of the relative Humidity and of the deposited Aerosol

The increase of the filtered air relative humidity up to 90 % has been tested. It has been found out that the efficiency increases as the relative humidity increases. We have found an explanation to this phenomenon in the fact that, as the relative humidity increases, the surface resistivity of the dust particles decreases; some of these dust particles are difficult to be ionizated at a low relative humidity.

This kind of particles is easy to be electrically charged when the relative humidity is increased, i. e. they can receive more elementary charges. In these conditions the forces to arrest and retain them in the filter are stronger and, consequently, as a final result, the filter efficiency increases as the relative humidity increases. But at the same time, when very high relative humidities are achieved, for example up to 98 - 99 %, in case the system is not insulated well enough, it is possible that "current leakages" occur which bring about a high voltage decrease in the collector and, consequently, a decrease of the efficiency. In order to keep the filter reliability there have been built separating insulated electrodes which allow the filter running at very high relative humidities.

Analysis of the Pressure Loss of the two Systems - The Mechanical and the Electrical one

The fact that the initial pressure loss (Δp_i) of an Electro-HEPA-Filter is much lower than that of a mechanical filter of similar efficiency results, in part, out of the above mentioned facts and it can be summed up as follows: in order to obtain a certain efficiency it can be used an inferior filtering medium with a much lower pressure loss, due to the electrostatic filter of the same efficiency.

As regards to the slow Δp increase of the Electro-HEPA-Filter it is to be explained with the dust particles depositing structure within the electric field. The study under the microscope of the aerosol depositing process has given us the answer to this problem. The fact is known that, in case of the mechanical filters, the particles deposit themselves on the fibres making them thicker. When the fibres lie in an electric field, i. e. when they are polarized, and the dust particles are electrically charged or polarized, the depositing structure is very much different: The thickening of the fibres happens only partially and the so-called dendrites appear. These dendrites reduce also the possibilities of the so-called "cake" to develop, that obturates the filter very quickly. Because of the fact that these dendrites develop, the filter keeps a better "transparency" for the air flow for a longer time.

Another reason is the way particles deposit on the filter's building structure. In case of the mechanical filter, the aerosol is laying down almost exclusively in the filtering medium. The separators remain almost clean after a long running time. In case of the Electro-HEPA-Filters on contrary a big aerosol quantity is laying down on the separators due to the electric field on the one hand, and, on the other hand, between the separators and the filtering medium, i. e. in the free space through which the air flows, filiform structures develop which also contribute to the prolonged "transparency" and thus to a longer lifetime of the Electro-HEPA-Filters.

a   b

Fig. 1 Dust deposition in a Mechanical HEPA-Filter (a)
in an Electro-HEPA-Filter   (b)

Conclusion

The advantages of the Electro-HEPA-Filter as compared to a mechanical
Super-HEPA-Filter:

- a dramatic increase of the efficiency for very small particles of 0.1 μm
  level (or smaller) by using a filtering medium much cheaper than any
  mechanical Super-HEPA-Filter of same efficiency.

- a reduction of the pressure loss (Δ p) up to 40 - 50 % in comparison
  with any mechanical Super-HEPA-Filter of same efficiency.

- due to the facts mentioned above a lifetime of 2 - 2.5 times longer than
  that of a mechanical Super-HEPA-Filter of proper efficiency.

- due to the much lower pressure loss (Δ p) the vibrations and the noise
  of the ventilators made by the air flowing through the filter are much
  diminished as compared to the proper mechanical Super-HEPA-Filter.

- due to some ultraviolet radiations to an intense ionic bombardment and
  to a small ozone quantity existing in the ionizer stage as well as to
  the electrical field existing in the collector stage, an Electro-HEPA-
  Filter has not only a filtering effect but also a sterilizing one, that
  no mechanical filter has. This sterilizing effect is especially wished
  by such fields as medicine, biology etc.

- through a suitable choosing of the electrical values and of the polarities applied to the Electro-HEPA-Filters it is possible to reduce or
  totally cancel the air electrical charging due to the friction between
  the air passing through the filter and the filter fibres (this troubling
  phenomenon with very negative effects upon the integrated circuits is to
  be met by all the mechanical HEPA-Filters).

References

1. Masuda, S. and Sugita, N., 1982, Electrostatically Augmented Air Filter
   for Producing Ultra Clean Air and its Collection Performance, ICCCS,
   Tokyo.

2. Bergman, W. a o., Electrostatic Filters Generated by Electric Fields,
   Second World Filtrations Congress, London, 1979.

3. H. H. Stiehl and D. D. Cucu, Remarks and Conclusions on the Practical
   Application of Dielectric Medium Filters under imposed Electric Fields,
   World Filtration Congress III, Sept. 1982, Pa, USA.

588

4. Cucu, D. a. o., Technic and Economic Aspects of Ultra High Efficiencies of Electrostatic HEPA-Filters, Filtech Conference, London Sept. 1983.

5. Cucu, D. and Lippold, H.-J., Electrostatic HEPA-Filter, 30th IES Annual Technical Meeting "Environmental Integration Technology Today for a Quality Tomorrow", Orlando, Florida, USA, May, 1984.

6. Cucu, D. and Lippold, H.-J., Electrostatic Ultra Filter, Electrical and Magnetic Separation and Filtration Technology, Antwerp, Belgium, May, 1984.

Published 1984 by Elsevier Science Publishing Co., Inc.

Aerosols, Liu, Pui, and Fissan, editors

589

EFFECT OF DOP HETERODISPERSION ON HEPA FILTER PENETRATION MEASUREMENTS*

W. Bergman and A. Biermann
Lawrence Livermore National Laboratory
P. O. Box 5505, Livermore, CA 94550
U.S.A.

Extended Abstract

Introduction

The accuracy of the standard U.S. test method for certifying High Efficiency Particulate Air (HEPA) filters has been in question since the finding by Hinds, et al, 1978, that the dioctyl phthalate (DOP) aerosol used in the test is not monodisperse as had been assumed. Hinds, et al, 1978, showed that the owl, a particle size analyzer, could not distinguish among the infinite particle size distributions having the same owl reading. Figure 1 shows three different particle size distributions that yield the same owl reading. The concern was that these different particle size distributions would yield different filter penetrations as measured by the light scattering photometer. To address this concern, we have conducted a theoretical and experimental study of filter efficiency for different DOP size distributions having the same owl reading.

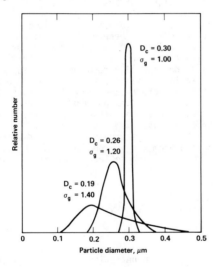

Fig. 1 Three different particle-size distributions that produced identical measurements of average size by the "owl"

*This work was performed under the auspices of the U.S. Department of Energy by Lawrence Livermore National Laboratory under contract No. W-7405-ENG-48.

590

Results

The filter penetration measurement, P_M, using a light scattering photometer is determined from the ratio of the photocurrent produced by aerosols measured before I_B, and after, I_A, the filter.

$$P_M = \frac{I_A}{I_B} = \frac{\displaystyle\int_0^\infty N(r)\,P_F(r)i(r)dr}{\displaystyle\int_0^\infty N(r)i(r)dr} \qquad (1)$$

where $N(r)$ is the number of aerosols having radius, r
$P_F(r)$ is the filter penetration for aerosols of radius, r
$i(r)$ is the photocurrent produced by a single aerosol of radius, r

The filter penetration measurement, P_M, equals the filter penetration, $P_F(r)$, for aerosols of radius, r, only for monodisperse aerosols. If the aerosol size distribution is not monodisperse then the measured filter penetration may deviate from the actual filter penetration. Moreover, the repeatability of the filter test measurement is also questionable since the owl analyzer that is used for verifying the aerosol size distribution gives the same reading for an infinite number of distributions.

We have computed the effect of increasingly heterodisperse aerosols, all having the same owl reading, on the measured filter penetration using Eq. 1. The aerosol size distributions, $N(r)$, were selected from Table 6 of Hinds, et al, 1978. The photocurrent, $i(r)$, was taken from Tillery, et al, 1982. $P_F(r)$ was experimentally determined as a function of aerosol diameter for a standard HEPA filter using a TSI differential mobility analyzer coupled to a condensation nucleii counter and a PMS LAS-X aerosol spectrometer. Figure 2 shows the experimental results. The calculated filter penetration for increasingly heterodisperse aerosols that have a constant $29°$ owl reading is shown in Fig. 3. Note that the penetration measurement is relatively insensitive to the heterodispersion of the aerosols provided the owl reading is constant. However, if the count median diameter is constant, then Fig. 3 shows that changes in aerosol heterodispersion produce large changes in filter penetration measurements.

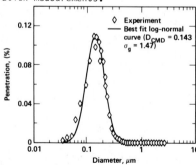

Fig. 2 Experimental penetration of a HEPA filter as a function of DOS particle diameter.

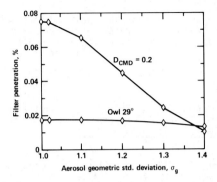

Fig. 3 Calculated filter penetration using Eq. 1 as a function of increasing heterodispersion for various particle size distributions, $N(r)$.

The surprisingly small effect of the DOP heterodispersion on HEPA filter penetration is due to the response function of the owl which is similar to the response of the photometer. This is shown in Fig. 4.

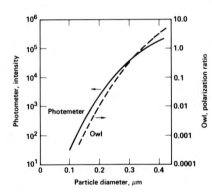

Fig. 4 Response of light scattering photometer and owl to different particle sizes.

Conclusion

We have found that the degree of heterodispersion of DOP aerosols in the U.S. HEPA filter certification tests has little effect on the measured filter penetration. This occurs because the owl (which is used for measuring particle size) and the photometer (which is used for measuring filter penetration) are both based on light scattering from a cloud of particles.

Changes in the aerosol size distribution produce similar changes in the owl and in the photometer measurements. However, if a particle size analyzer is based on a different principle than the light scattering photometer, large variations in the filter penetration will occur with variations in the aerosol size distribution.

References

Hinds, W., First, M., Gibson, D. and Leith, D., (1978), "Size Distributions of 'HOT DOP' Aerosols Produced by ATI Q-127 Aerosol Generator," CONF-780819, pp. 1130-1144, Proceedings of the 15th Nuclear Air Cleaning Conference, Boston, MA.

Tillery, M., Salzman, G. and Ettinger, H., (1982), The Effect of Particle Size Variation on Filtration Efficiency Measured by the HEPA Filter Quality Assurance Test," CONF-820833, pp. 895-908, Proceedings of the 17th DOE Nuclear

Published 1984 by Elsevier Science Publishing Co., Inc.

Aerosols, Liu, Pui, and Fissan, editors

Electric Air Filtration Movie*

W. Bergman and R. Jaeger
Lawrence Livermore National Laboratory
P. O. Box 5505, Livermore, CA 94550
U.S.A.

Extended Abstract

Introduction

The use of electrostatics to improve the performance of conventional air filters has gained considerable attention in recent years. This interest is due to the higher efficiency and reduced pressure drop of electrically enhanced filters compared to conventional fibrous filters. This 30-minute movie presents a state of the art review of electric air filters in the United States with major illustrations provided by the research and development program at the Lawrence Livermore National Laboratory sponsored by the Department of Energy. The electric air filters described in this movie are mechanical air filters to which electrical forces have been added.

Fundamentals

The fundamental principles of electric air filtration are reviewed by means of an art animation sequence and small-scale laboratory experiments. Conventional air filters can be converted into electric air filters by either precharging the aerosols or polarizing the filter media or for maximum efficiency, a combination of both. Precharging the aerosols is generally accomplished by passing aerosols across a corona discharge. The charged aerosols then deposit on the filter media thereby building up a charge on the media itself which consequently attracts more particles by electrical attraction. In the polarization approach, an electric field is applied across the filter media, thereby polarizing the media. The polarized media immediately attracts charged aerosols and larger polarized aerosols.

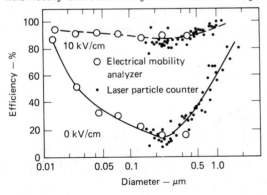

Fig. 1 Filter efficiency as a function of particle size with and without an electric field.

*This work was performed under the auspices of the U.S. Department of Energy by Lawrence Livermore National Laboratory under contract No. W-7405-ENG-48.

594

Once the filter medium is electrified, the filter efficiency increases dramatically. Figure 1 shows the increase in filter efficiency that occurs when an electric field of 10 kV/cm is applied across a typical fibrous filter. Note that the increased efficiency is most pronounced over the particle size range from 0.05 to 0.5 μm diameter where mechanical filters have their lowest efficiency.

In addition to the higher efficiency, an electrified filter also has a lower pressure drop than a conventional filter during aerosol loading. Figure 2 shows the increase in pressure drop across two similar filters as a function of aerosol loading. Note that the filter with an applied electric field has a significantly lower pressure drop than the nonelectrified filter. There are two reasons for the decreased pressure drop of the electric filter. First, on a microscopic scale, the electrical forces cause the aerosols to form a relatively uniform and compact deposit around the entire filtering element thereby reducing the resistance to air flow (Fig. 3). In the absence of electrical forces, the particles form a loose dendritic structure that projects from the front side of the fiber and restricts the air flow past the fiber (Fig. 3).

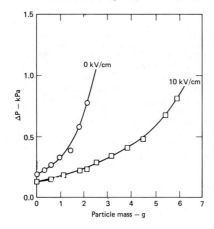

Fig. 2 Pressure drops across filters operating with and without an electric field during aerosol loading.

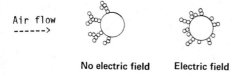

Air flow
------>

No electric field   Electric field

Fig. 3 Schematic of particle deposits on a fiber with no electric field and with an electric field

Second, on a macroscopic scale, the decreased pressure drop in electric fibrous filters is illustrated in Fig. 4 and is due to the natural tendency of most fibrous filters to have a gradation of fiber packing density, the outside portion of the filter mat being loosely packed and the interior portion tightly packed. Without electrical enhancement, the heaviest deposits form on the closely spaced interior fibers, thereby accelerating the blockage of air flow. With an applied electric field the fibers become more efficient and the heaviest deposit on the filter mat forms on the widely spaced outside fibers thereby minimizing the restriction to the air flow. Greater collection efficiency and reduced air flow restrictions are the two major advantages of electric air filtration.

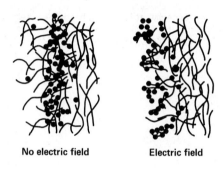

No electric field   **Electric field**

Fig. 4 Schematic of the filter mat cross section showing the particle deposits with no electric field and with an electric field.

Application

The principles of electric air filtration have been used to design and develop prototype models for evaluation in industrial environments. The movie shows a number of experimental units developed by LLNL and by the Environmental Protection Agency, EPA, as well as various commercially available units illustrating the diversity of designs and applications of electric air filters. A rolling electric filter is shown that automatically passes a continuous supply of fresh filter media across a ventilation duct whenever the media becomes plugged. The operation of this filter is illustrated in a ventilation duct containing smoke aerosols. Other LLNL designs include electric prefilters for removing toxic or radioactive aerosols in the exhaust of glove boxes, a pleated 2' x 2' x 1' electric filter for removing radioactive aerosols from burning uranium scrap, and a recirculating electric air filter installed in a plutonium powder-handling cell. Two electric bag filters for removing fly ash from coal-burning power plants are also illustrated. One unit developed by the Apitron Company uses a corona precharger to electrify the bag. A second unit, developed by the EPA, uses a series of high voltage electrodes to polarize the filter media. The final example of an electric air filter is a moving granular bed filter developed by the Combustion Power Company and is shown cleaning 250,000 cfm from a large wood burning boiler. This illustration differs considerably from the previous ones because the filter media consists of pea gravel.

Conclusion

This 30-minute film provides an overview of the electric filter technology in the United States. The recent interest in electric air filters is due to their higher efficiency and decreased pressure drop compared to conventional filters. We have reviewed the primary methods for generating electric filters by precharging the aerosols, applying an electric field across the filter medium or a combination of both. A variety of experimental as well as commercial electric air filters were presented to show the diversity of both their designs and applications.

Published 1984 by Elsevier Science Publishing Co., Inc.

Aerosols, Liu, Pui, and Fissan, editors

REAL INFLUENCES ON THE COLLECTION OF SOLID PARTICLES IN
TECHNICAL FILTERMATS

F.Löffler and H.Jodeit

Institut für Mechanische Verfahrenstechnik und Mechanik,
Universität Karlsruhe, D-7500 Karlsruhe, Germany

Extended Abstract

Fiberfilters are used in a wide variety of fiber volume fraction,
fiber diameter, fiber material and filtration velocity for the
collection of liquid and solid particles in a particle size range
from about 0.01 µm up to larger than 50 µm. Scientific investigations for elucidation of the physical phenomena and for mathematical description of the filter performance started about
50 years ago.

At the time being many observations are well enough understood
but there remain still some important questions which have to be
solved. Recently Lee and Liu |1| reported about studies in the
range where diffusion mechanisms and interception are dominant
for particle collection, i.e. for particle sizes below approximately 1 µm and for filtration velocities below about 25 cm/s.
These authors found good agreement between their theoretical and
experimental results by adopting a constant numerical factor.

On the other hand there are numerous remarks in the literature
about discrepancies in the range where inertia forces controll
particle collection, i.e. roughly for particles of 1 µm and
larger and filtration velocities above 10 cm/s. The gap between
filtration theory and experimental results increases when not
single fibers or model filters but technical filter mals are
dealt with.

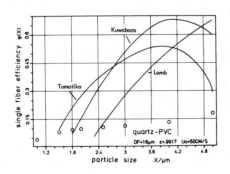

Figure 1. Theoretical (curves)
and experimental
(dots) single fiber
efficiencies for a
fiber mat with
fiber diam. 16 µm,
fiber vol.fraction
0.99,filtration
velocity 50 cm/s

598

Fig. 1 e.g. shows theoretical single fiber collection efficiencies
based on particle trajectory calculation using three different
flow fields. Inertia forces are assumed to be the only transport
mechanism. Single fiber efficiency φ is assumed to be the product
of transport efficiency η and adhesion probability h, i.e.
φ = η · h. Adhesion probability h takes particle bouncing into
account and was calculated according Hiller and Löffler |2|.
Apparently the different flow fields yield different results
mainly in the range where particle bouncing gains more influence.
But the discrepancy to the experimental results (dots) is nearly
more important than the difference between the flow fields. Theories overestimate remarkably the reality even if we introduce
adhesion probability. This gap cannot be explained by the in-
fluence of electrostatic forces |3|. Also it cannot be bridged
by taking into account porosity. This is shown in Fig. 2 where
porosity ε is variied using Kuwabara's flow field.

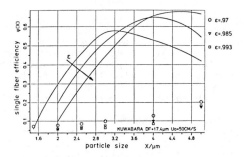

Fig. 2. Influence of filter porosity ε on single
fiber efficiency φ

The theoretical models assume either an isolated fiber or an uniform array of fibers. Of course this presupposition is not
fulfilled with technical filter mats with a more or less inhomo-
genious distribution of the fibers. The question is in how far
the real filter structure influences the collection process.

Structural effects on filter efficiency have already been looked
upon by Benarie |4| about 15 years ago. As a first approach we
followed Benarie's concept for calculation of the influence of
structural inhomogenity on collection efficiency.

Benarie assumed that the distribution of the characteristic
parameter for description of filter geometry, e.g. interfiber
distance, local fiber volume fraction, porosity etc. can be
approximated by a log.-normal function. This distribution
results in a reduction of collection efficiency which does not

depend on the absolute size of the pores but which is only a
function of the geometric standard deviation σ_g, as is shown by
equation (1):

$$\log \frac{\eta_{sf}}{\eta_c} = 4,6 \log^2 \sigma_g$$

(η_{sf} = single fiber efficiency
η_c = corrected efficiency)

Fig. 3 shows an evaluation of eq. (1). For a given filter with
a given σ_g all T(x)-values should fall on a perpendicular line.

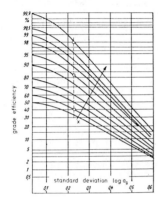

grade efficiency

standard deviation log σ_g

Fig. 3. Relation between grad efficiency T(x) and standard
deviation σ_g according Benarie
(x = particle diameter)

We calculated the reduction of collection efficiency according
eq. (1) for two flow fields as shown in fig. 4. In this graph
we plotted experimental values for one filter type and one
filtration velocity. These values fall fairly well on a perpendicular line as predicted by Benarie. For both flow fields
the range of log σ_g is about 0.3 - 0.4. Therefore we took as an
mean value log $\sigma_g \approx 0.35$. (Direct optical determination of σ_g
is in progress.)

Using the empirical value log σ_g = 0.35 we corrected the theoretical single fiber efficiencies as calculated with different
flow fields for different filter types and different filtration
relocities. Fig. 5 shows as an example the comparison between
experimental values (dots) and corrected and non-corrected
theoretical curves. Obviously the agreement is much better for
the corrected curves. Furthermore the differences between the

600

different flow field loos importance. This observation was confirmed for all our different filtration experiments with different
filters.

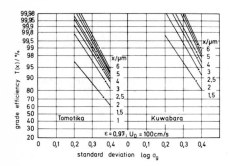 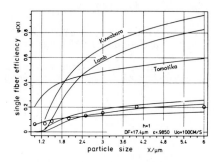

Fig.4. Determination of σ_g by com-   Fig.5. Comparison of experiments
       paring theoretical values   (dots) with noncorrected
       (eq.1) with experiments for   and corrected theoretical
       a filter with D_F = 17 µm,   single fiber effieiencies
       ε = 0.97..., v = 100 cm/s

Conclusion: Obviously structure of the fiber arrangement is a
very important property of technical filter mats which are used
for the collection of particlesin the range with predominant
inertia forces. This influence could be allowed for by using
Benarie's concept.

Experimental and theoretical investigations of this influence
have to be continued. Furthermore there remain some additional
questions open, e.g. consideration of the influence of porosity
or of the adhesion probability in real filter mats.

References

|1| Lee, K.W. (1982), Liu, B.Y.M. "Theoretical Study of Aerosol
    Filtration by Fibrous Filters", Aerosol Sci. and Tech.1:147

|2| Hiller, R., Löffler, F. (1980) "Der Einfluß von Auftreffgrad
    und Haftanteil auf die Partikelabscheidung in Faserfiltern",
    Staub-Reinhaltung der Luft 40:405

|3| Jodeit, H., Löffler, F. (1984) "The influence of electro-
    static forces upon particle collection in fibrous filters",
    J.Aerosol Sci.

|4| Benarie, M. (1969) "Einfluß der Porenstruktur auf den Ab-
    scheidegrad in Faserfiltern",
    Staub-Reinhaltung der Luft 29:76

Published 1984 by Elsevier Science Publishing Co., Inc.

Aerosols, Liu, Pui, and Fissan, editors

THE EFFECT OF IRREGULARITIES OF PORES ON THE LOADING
OF NUCLEPORE FILTERS

A H Leuschner and J Kruger

Nuclear Development Corporation of South Africa (Pty) Ltd
Private Bag X256
PRETORIA
0001
SOUTH AFRICA

EXTENDED ABSTRACT

Introduction

Nuclepore filters with large pores can be used to segregate non-respirable aerosols from respirable aerosols. The purpose of this study is to determine whether local irregularities and imperfections cause deviations in microscopic deposition patterns of particles from the average or theoretical patterns.

The pores of a Nuclepore filter are all approximately of the same size. Random positions of the pores result in some pores being fairly isolated and others being closely packed, and in the worst case, some pores overlapping each other. Pore axes are not always perpendicular to the filter surface.

Theoretical

The irregularity which was expected to have the greatest influence on the microscopic deposition pattern was overlapping pores (also called clusters). Random spread of pore positions allows Poisson statistics to be used to calculate the fraction of the pores forming clusters. About 10 % of the pores on a 12 μm Nuclepore filter form clusters. This was verified experimentally.

In order to predict deposition positions of particles at an idealised pore, a theoretical model using the Navier-Stokes equation has been used. This is a two-dimensional model.

Experimental

The porosity of 12 μm Nuclepore filters has been determined microscopically and found to satisfy Poisson statistics.

Cuts through filters were analysed under a scanning electron microscope to determine the angles that the pore axis makes with the normal to the filter surface.

Pores were selected at regular intervals along a diametrical line on a Nuclepore filter for analysis. For each pore analysed, the following were noted: Is it a cluster? If so, what is the number of pores in the cluster and what is the separation of the pore centres? What is the local porosity P about the pore (i.e. the number of pores within a distance equal to the average inter-pore spacing from the pore)? What is the distance R_1 to the nearest other pore? What is the deviation of the pore axis from right angles to the filter surface expressed as the displacement (D) of the bottom of the pore relative to the top? Does the pore size differ by more than about 2 μm from 12 μm?

The filter was then exposed to an aerosol of monodispersed particle sizes (σ_g=1,10). These spherical particles of polystyrene were generated with a spinning top generator. They were labelled with ^{51}Cr to determine the overall filter efficiency. A control filter was exposed simultaneously to the test aerosol in order to ensure that the analytical procedures, like coating the nuclepore filter with gold, do not influence the filter functioning. The same pores were then analysed under a scanning electron microscope and the number of deposited particles counted in each region as designated in Figure 1.

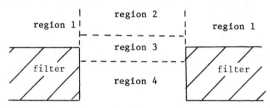

Figure 1. Sectional view of filter showing regions of deposition

The results were grouped per pore characteristic. This procedure of exposure of the filter and analysis as described above was then repeated at least twice, using the same particle size. Such tests were performed for particle sizes of 1 μm, 2 μm and 4.5 μm.

Results

For 1 μm and 2 μm particles, good correlation was obtained between the theoretical efficiency and the overall experimental filter efficiency. For 4.5 μm particles, the experimental efficiency was lower than the theoretical efficiency, owing to particle bounce. Good correlation was achieved between the analysed filter and the control filter.

For the interpretation of the results, the following parameters were defined:

$r_{n,i}$ is the ratio of the number of particles per pore after the ith loading for pores of a certain type to the average number of particles per pore. This parameter gives an indication of the efficiency of the particular type of pore relative to the overall filter efficiency;

$f_{i,a}$ is the fraction of the particles deposited at a particular region for pores of a particular type after a particular loading. This parameter gives an indication of the loading pattern of the particular type of pore.

The variation of $r_{n,i}$ is shown as a function of the distance R_1 to the nearest other pore in Figure 2 for 1 μm particles for good pores.

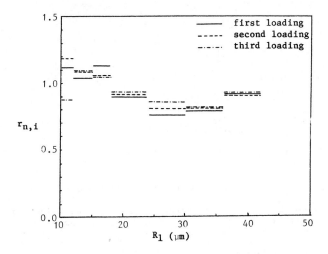

Figure 2. $r_{n,i}$ as a function of distance R_1 to nearest other pore

The results for other particle sizes follow very much the same trend unless particle bounce occur.

No noticeable difference could be detected between analysis with the distance R_1 to the nearest other pore and with the localised porosity P.

The ratio $r_{n,i}$ of the number of particles per cluster of doublet pores to the number of particles per average pore was found to be nearly the same as the ratio of the area of the cluster to the area of a single pore. This indicates that the efficiency of a cluster of double pores may be close to the efficiency of a single pore.

For 1 μm particles, the variation of $f_{i,a}$ with the loading of the filter for all pores and for clusters of pores consisting of doublet pores is shown in Figure 3. For other particle sizes the correlation is also good.

604

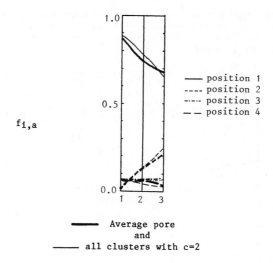

$f_{i,a}$

— position 1
---- position 2
—·— position 3
— — position 4

▬▬▬ Average pore
and
——— all clusters with c=2

Figure 3. $f_{i,a}$ shown as a function of the loading number

The variation of $f_{i,a}$ with loading for different values of the displacement D of the bottom of the pore follows the same trend and is comparable to that of the average pore.

Discussion and Correlation

Variations in interpore spacings, and clusters of pores as well as pore defects like pores axes not perpendicular to the filter surface have only limited effect on the investigated filter characteristics, except for the most extreme cases, which have a very low frequency of occurrence.

Published 1984 by Elsevier Science Publishing Co., Inc.

Aerosols, Liu, Pui, and Fissan, editors

ON THE NAVIER-STOKES EQUATIONS USED IN THE MODELLING OF FILTRATION MECHANISMS OF AEROSOLS THROUGH LARGE-PORE NUCLEPORE FILTERS

J W Rossouw and A H Leuschner
Radiation Physics Division
Nuclear Physics Department
NUCOR
Private Bag X256
PRETORIA
SOUTH AFRICA
0001

EXTENDED ABSTRACT

Introduction

Stream-function values for air incident on 12 and 8 μm filters with different face velocities were calculated from the Navier-Stokes equations, using an existing model (Parker, 1975). From these stream-function values the air streamlines, i.e. lines for constant stream-function values, were calculated.

By letting Stokes' drag forces act on a particle in this stream-function field, the actual trajectories of particles were calculated.

The degree to which the particle trajectories deviate from the streamlines, and the effect of this on the filter efficiency, were determined for particles of different sizes.

In an attempt to describe the loading characteristics of a Nuclepore filter, an investigation has been made of the impaction efficiency of the filter after a single particle has deposited on the filter surface near the pore.

Efficiencies

If the interception efficiency is defined as that efficiency of the filter for particles that follow the air streamlines, i.e. when all external forces are neglected, then this efficiency can be calculated directly from the stream-function values and particle sizes.

The impaction efficiency is defined as the efficiency of the filter where inertial forces cause the particles to deviate from the streamlines, and is calculated from the particle trajectories and particle sizes.

From the definitions of the impaction and interception efficiencies given above, the quantitative difference between these efficiencies can be formulated. Interception is the basic deposition mechanism, that is, a particle that touches the filter due to its geometrical size and movement along a streamline adheres to the filter. In the calculation of the impaction efficiency, an additional parameter is taken into

606

consideration. Therefore the impaction efficiency of the filter includes the interception efficiency as well as the effect of the inertia of the particles.

These two types of efficiencies were calculated for an 8 μm filter with a face velocity of 6 cm s^{-1} for particles ranging from 0.5 to 7.5 μm. The results are given in Figure 1 below.

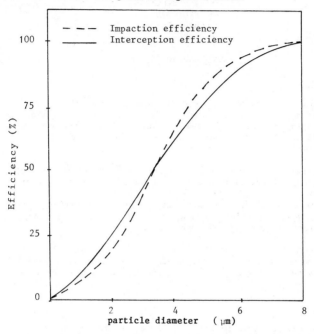

Figure 1.   Comparison of impaction and interception
efficiencies

From Figure 1 it is clear that there is a difference, although less than 10 %, between the two types of efficiencies. The contribution of inertia to the impaction efficiency is less than 10 %, and interception can thus be taken as the dominant filtration mechanism. An interesting point to notice is that for smaller particles, the impaction efficiency is lower than the interception efficiency. Thus the inertia of the particle causes it to pass through the filter rather than to deposit on the filter surface. Comparing the air streamline and particle trajectory of a 2 μm particle in Figure 2 confirms this phenomenon. In this case the difference between the efficiencies is 7.5 %.

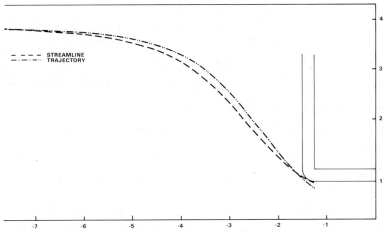

Figure 2. Comparison between air streamline and particle
trajectory of a 2 µm particle

Impaction efficiency after particle deposition

Instead of taking up the task to change the governing partial differen=
tial equations, it is assumed that the same equations and thus the same
stream-function values and particle trajectories can be used. Only the
criterion for particle deposition on the filter was changed. This
criterion is dependent on the size of the deposited particle, the
incident particles and the radial deposition coordinate (Rdep). Figure
3 illustrates the configuration near the pore after particle deposition.
The sizes of the incident particles and of the deposited particles were
assumed to be equal.

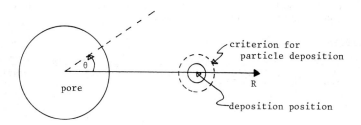

Figure 3. Configuration after particle deposition

From the above figure it is clear that the local impaction efficiency will change with θ. Thus the region $0 < \theta < 2\Pi$ was divided into sub-intervals and the local impaction efficiency calculated. The total impaction efficiency Eff(dep) was then obtained by numerical integration over θ. Table 1 represents some of the calculations done with this procedure. The deposition coordinates were experimentally obtained by investigating the deposition pattern after one loading with particles. Theoretically determined deposition positions with the highest probability of occurring correlated well with the experimental results unless particle bounce occurred.

Table 1. Change in efficiency after particle deposition

Dp (μm)	Eff (clean) (fraction)	Rdep (μm)	Eff(dep) (fraction)	Rdep–Dp (μm)	Eff(dep)/ Eff (clean)
1	0.053	6	0.056	5	1.0589
2	0.174	6.35	0.190	4.35	1.0907
4.68	0.7663	9.75	0.7665	5.07	1.0003

Table 1 shows that the presence of a particle on the filter surface does have an effect on the efficiency of the filter. For larger particles, the presence of a particle does not influence the efficiency as much as for smaller particles at the same deposition position. In practice however, larger particles will deposit further from the pore than smaller particles. For 4.68 μm particles, the efficiency of the filter after particle deposition is within 5 % of that of the clean filter with Rdep as small as 6.5 μm.

Conclusions

Since there is less than 10 % difference between the impaction and interception efficiencies of an 8 μm Nuclepore filter, it is concluded that interception makes the dominant contribution to the impaction efficiency.

For larger particles, the presence of a single deposited particle does not have much influence (less than 5 %) on the efficiency of the filter. For smaller particles, the influence of the particles is greater and highly dependent on the deposition position.

References

Parker, R.D. (1975) "A fundamental study of particle deposition onto large pore nuclepore filters." Ann Arbor, Mich. University Microfilms 1980 -Thesis (Ph.D) - Duke University.

Published 1984 by Elsevier Science Publishing Co., Inc.
Aerosols, Liu, Pui, and Fissan, editors

THE INFLUENCE OF AN ELECTROFILTER
ON RADON DECAY PRODUCTS

Rajala,M.[1], Janka,K.[1], Lehtimäki,M.[2] , Graeffe,G.[1] and Kulmala,V.[1]

(1) Department of Physics, Tampere University of Technology,
P.O. Box 527, 33101 Tampere 10, Finland

(2) Occupational Safety Engineering Laboratory, Technical
Research Centre of Finland, P.O. Box 656, 33101 Tampere 10,
Finland

EXTENDED ABSTRACT

Introduction

Previous studies made by our laboratory in a single-family house have shown that air cleaning devices have a significant influence on the behaviour of radon decay products in indoor air. An electrofilter was found to produce ultrafine particles in some circumstances. The concentration of these condensation nuclei became high enough to eliminate the wall deposition of free radon daughters thus high concentrations of airborne radon decay products were measured (Lehtimäki et.al. 1983).

This paper deals with laboratory measurements concerning the formation of small particles by an electrofilter and their effect on radon daughters.

Measurement system

The measurement system (fig.1.) consists of a test chamber (28 m^3) equipped with a recirculation system having a high-efficiency mechanical filter for air cleaning. To prevent the penetration of unwanted particles to the chamber a small amount of air is led into it from a compressed air line so that in the room there is a small excess of pressure. The air flow also determines the air-change rate of the chamber. A part of the flow may be led through a humidifier, and generated particles are also fed to the air stream. Particles used are dioctyle phthalate (DOP) particles generated by a condensation generator.

As a radon source we use a 1 mCi ^{226}RaCl$_2$ salt dilution. Radonconcentration during the experiments has been about 6000 Bq/m^3. Radon in the chamber is measured by an ionization chamber (Janka and Lehtimäki 1982). Radon daughters are measured by two WL-monitors, the other one for unattached decay products and the other for the total activity of airborne radon daughters. For the calculation of the WL-value we use the modified Kusnetz method (Nazaroff 1981).

Aerosol particles are measured by an electrical aerosol monitor (Lehtimäki 1983). This monitor is also useful for measuring the attachment rate of the decay products to aerosol particles (Lehtimäki et.al. 1983).

The temperature and relative humidity of chambers air are also measured. All results are acquired by a microcomputer.

610

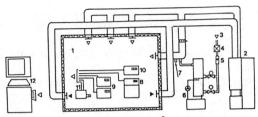

Fig.1. The measurement system. (1) test chamber (28 m^3), (2) recirculation system, (3) compressed air line, (4) filters, (5) flow meter, (6) humidifier, (7) particle, SO$_2$, and ^{222}Rn feeding, (8) r don monitor, (9) two WL-monitors (one for unattached daughters and one for total activity), (10) electrical aerosol monitor, (11) temperature and relative humidity meter, (12) microcomputer

Results

During the experiments an electrofilter with an air throughput rate of approximately 4 hr^{-1} was situated in the chamber. The filter was able to be controlled outside the testroom.

When the chambers air was fed with particles the particle concentration in the room reached equilibrium after a few hours from the starting of the aerosol feeding. Connecting the electrofilter caused a rapid decrease in the particle concentration as well as in the total concentration of radon decay products. Excluding the particles the chambers air was very pure (no gaseous impurities) and dry. In this experiment particles generated by the corona discharge of the electrofilter were not measurable.

When the chambers air was moisten those particles could be observed during the use of the electrofilter (fig.2.). In this experiment no particle feeding was used.

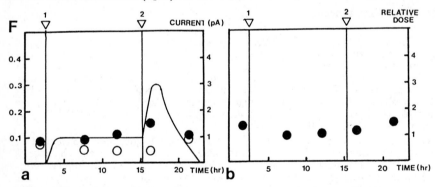

unattached radon daughters ○
total airborne activity ●
the current of the electrical aerosol monitor ——

Fig.2.a. The formation of condensation nuclei and their effect to equilibrium factor F when the chambers air has been moisten (R.H. about 50 %). 1: electrofilter on, 2: elctrofilter off.
2.b. Relative dose calculated using the Jacobi-Eisfeld model and the data seen in figure 2.a.

When the filter was switched off the current of the aerosol monitor increased rapidly. This is probably due to two things. Firstly, the condensation nuclei produced by the electrofilter are presumably HNO_3 products from the reaction between NO_x and O_3 generated by the electrostatic precipitator (Peyrous and Lapeyre 1982). These particles are then grown in the room air due to condensation of water or impurities. Condensated particles are effectively removed from the room air by the filter, and stopping it causes that HNO_3 particles produced just before stopping will be condensated, but not anymore eliminated from the air. As a result of this the particle concentration in the room will increase. Secondly, after disconnecting the filter condensation nuclei tend to grow and the aerosol monitor used is more sensitive for larger particles. The size of these condensation nuclei was measured with an electrical aerosol analysator (EAA), and found to be under 0.02 μm when the electrofilter was on, and increase to over 0.03 μm after stopping the filter.

The formation of aerosolparticles can be observed even more clearly when sulphur dioxide (SO_2) is fed to the chamber (fig.3.)

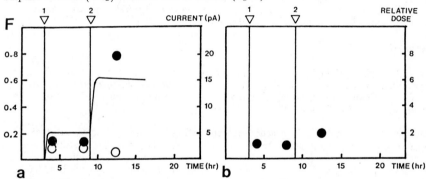

unattached radon daughters   O
total airborne activity   ●
the current of the electrical aerosol monitor ———

Fig.3.a. The effect of SO_2-feeding to the formation of particles by the electrofilter and further to radon daughters. The filter has been on for several hours before SO_2 input. 1: SO_2-feeding (0.1 ppm), 2: the electrofilter off.

3.b. The calculated relative dose using the Jacobi-Eisfeld model and the data from figure 3.a.

Before SO_2-feeding (about 0.1 ppm) the electrostatic precipitator was on for several hours. Sulphur dioxide tends to increase the particle size and consequently the attachment rate of radon daughters to aerosol particles. Due to this the total activity of decay products rised rapidly as can be seen from the figure.

When a charcoal filter was inserted in the electrofilter the formation of particles became unnoticeable. The charcoal filter effectively removes ozone generated by the electrofilter. Without it the ozone production was measured to be 1.7 cm^3/h. When the charcoal filter was used, it not only adsorbed O_3 produced by the filter but also decreased the original ozone concentration in the room.

The influence of discussed particles to the health risk caused by radon daughters is quite complicated. In figures 2.b. and 3.b. the sum of the relative doses to the bronchial basal cell layer and the pulmonary epithelium has been calculated using the Jacobi-Eisfeld model (Jacobi and Eisfeld 1981) and the data presented in figures 2.a. and 3.a. In this model the dose/exposure to the bronchial (B) and pulmonary (P) areas are calculated from the equations

$$H_B/E \approx 0.075(1 + 9.4f_P) \; Sv \; per \; WLM$$

$$H_P/E \approx 0.022(1 - f_P) \; Sv \; per \; WLM$$

where f_P is the unattached fraction of the total inhaled potential α-energy. The dose from unattached decay products may be somewhat overestimated (the diffusion coefficient used for unattached daughters has been 0.05 cm^2/s. In indoor air and especially in respiratory airways it is probably smaller due to clustering). However the effect of condensation nuclei born in the filter to the health risk is presumably smaller than to WL-value because the activity of unattached daughters decreases when the attachment rate increases. In some situations the resulting dose may even decrease inspite of the increasing total activity.

References

Jacobi,W. and Eisfeld,K.: Internal dosimetry of radon-222, radon-220 and their short-lived daughters. Proceedings of the *Second Special Symposium on Natural Radiation Environment*held at Bhabba Atomic Research Centre Bombay, India during January 19–23, 1981

Janka,K. and Lehtimäki,M.: Method of eliminating the effect of decay products in continuous measurement of ^{222}Rn. *Rev.Sci.Instrum.* **53**(4) 523–527 (1982).

Lehtimäki,M.: Modified electrical aerosol detector. Marple,V.A. and Liu,B.Y.H. (editors) *Aerosol in the Mining and Industrial Work Environments*, Vol.3 Instrumentation. Ann Arbor 1983, 1135–1143.

Lehtimäki,M., Graeffe,G., Janka,K., Kulmala,V. and Rajala,M.: On the behaviour of radon daughters in air. International Seminar on Indoor Exposure to Natural Radiation and Related Risk Assessment, October, 3–5, 1983, Anacapri, Italy.

Nazaroff,W.W.: An Improved technique for measuring working levels of radon daughters in residences. *Health Phys.* **39**(4) 683–685 (1980).

Peyrous,R. and Lapeyre,R-M.: Gaseous products created by electrical discharges in the athmosphere and condensation nuclei resulting from gaseous phase reactions. *Atm.Environ.* **16**(5) 959–968 (1982).

Published 1984 by Elsevier Science Publishing Co., Inc.
Aerosols, Liu, Pui, and Fissan, editors

TIME DEPENDENCY OF COLLECTION PERFORMANCE OF ELECTRET FILTER

Chikao Kanaoka, Hitoshi Emi and Tomoaki Ishiguro
Department of Chemical Engineering
Kanazawa University
Kanazawa 920, Japan

EXTENDED ABSTRACT

Introduction

Electret filter is composed of "permanently" charged electret fibers and is capable of collecting fine particles at a high efficiency in the begining of the filtration because of strong electrostatic effects, but its collection performance is reported to decrease with time. However, the time dependency of the collection performace and the reason of the decrease in collection efficiency are not well understood yet.

In this study, the stability of charges in the fiber was first tested by measuring the collection performances of chemically treated electret filters. Then the evolution of collection efficiency and pressure drop of an electret filter was measured using solid and liquid particles. Finally, a countrmeasure for the decrease in the collection efficiency was discussed.

Property of electert filter

An electret filter used in this experiment is composed of polypropylene fibers with rectangular cross section. They carry positive and negative charges inside each fiber as shown in Figure 1. Its charging density is reported to be 34.2 nC/m-fiber. However, it is unknown that all charges stay stably inside the fiber. Furthermore, the value of charge density itself is still uncertain. Other relevanat properties are listed in Table 1.

Collection performance of treated electret filter

To prepare the filters with different charging states, electret filters were treated by dipping for 24 hours into distillated water or 1N-NaCl solution or C_2H_5OH, and then their collection performances were measured using 0.5 μm PSL particles at various filtration velocities. Penetration of the filter was determined from the concentrations at filter inlet and outlet. They were measured by light scattering photometers. Figure 2 shows the experimental penetration at the initial stage as a function of filtration velocity. In the same figure, an estimated penetration of a fibrous filter composed of uncharged cylindrical fibers and with the same structure as the test filter is also shown by a solid line. As seen from the figure, the penetrations increase, i.e., the collection efficiency decreases, as filtration velocity increases. However, the penetration of the electret filter is lower than that of estimated at any filtration velocities. Among the electret filters, the penetration decreases in the order of the treated filter with C_2H_5OH, the treated filters with H_2O and NaCl, and the untreated filter. The penetration does not change by the chemical treatment of the filter, if charges were firmly attached in the fibers. The experimental result shows that some charges on electret fibers are removed by chemical treatment, especially it seems that most of charges are removed from the filter by the C_2H_5OH treatment, since the penetration becomes comparably large as uncharged filter. This is probably caused by the difference in the affinity of of chemicals to the filter material.

Time dependency of filter performance

With proceeding of the filtration, the captured particles accumulate on fibers in the filter and the electrostatic effect becomes weak so that the collection efficiency of the filter decreases with time. However, its decreasing trend is quite different depending on whether solid or liquid particles are filtrated. Figures 3 and 4 show the time dependencies of collection performances of the elctret filters by the collections of solid stearic acid and liquid DOP particls. For the case of solid particles, the collection efficiency of the untreated filter gradually decreases with time in the early stage of the filtration because captured particles form dendritic agglomerates and cover the fiber surface so as to weaken the electrostatic effect of the fiber. After a certain period of time, the collection efficiency of the filter increased again, because the effect of the agglomerates on the collection of particles prevails the degradation of electrostatic effect. On the other hand, the collection efficiency of the treated filter increases with time and finally reaches to 100%, because the previously captured particles always positiviely contributes to the collection of upcoming paraticles.

For the case of liquid particles, the collection efficiency of the untreated filter decreases but that of C_2H_5OH treated fitler increases somewhat with time in the early stage of the filtration. On the contrary to the solid particles, the colleciton efficiencies of both filters finally reach to the almost same asymptotic efficiency, because captured particles never form dendritic agglomerates unlike solid particles but cover the fiber surface with droplets so that electrostatic effect decreases with time but mechanical collection dose not increase so much with time. If electrostatic effect is proportional to the uncovered area with liquid particles and collection efficiency in covered area is given by the mechanical collection efficiency, a collection efficiency of a mist laden fiber is approximated by the following equation.

$$\eta = \eta_e - (\eta_e - \eta_k)m/m_s \quad (0 \leqq m \leqq m_s) \tag{1}$$

where, η_e and η_k denote the single fiber collection efficiencies due to electrostatic effect and mechanical effect respectively. m_s expresses the value of the mass of captured mist per unit filter volume, m, when mist covers the whole fiber surface. Broken line in the Figure 4 shows the solution of a unsteady filtration equation using the above relation. It agrees very well with the experimental collection efficiecny.

Decrease of the collection efficiency is unfavorable for tne filter performance. Hence it is desirable for its practical use to find some countermeasure, which can prevent the decrease or decreasing rate of the filter efficiency with time. Appling the external electric field is considered to be one of the methods. Figure 5 shows the time dependency of the filter performances for untreated and treated filters. As seen from the figure, both efficiencies dose not change so much with time in compared with the efficiencies shown in Figure 4.

Table 1 Properties of electret filter

filter material	density of fiber $\rho\,[\mathrm{g/cm^3}]$	width of fiber $d_{fw}\,[\mu m]$	thickness of fiber $d_{fT}\,[\mu m]$	packing density $\alpha\,[-]$	filter thickness $L\ [mm]$	charge density $Q\,[nC/m]$
poly propyrlene	0.91	38	10	0.075	5.0	34.2

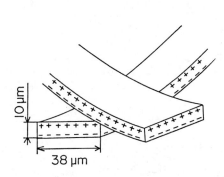

Figure 1 Schematic of an electret fiber

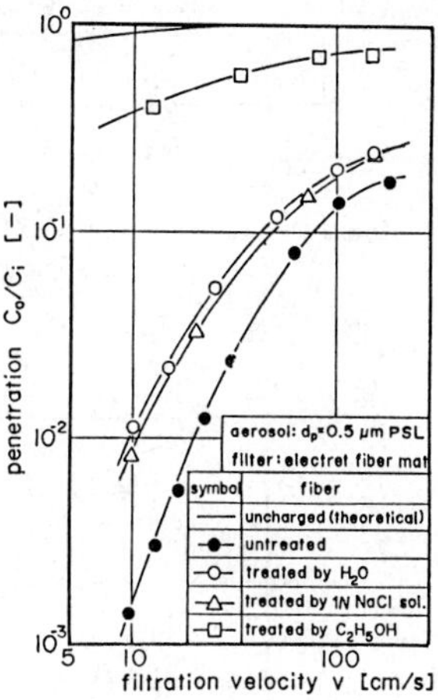

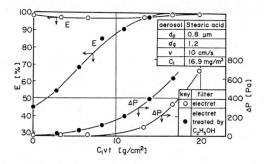

Figure 2 Penetration of variously treated electret filters

Figure 3 Time dependency of collection efficiency and pressure drop of an electret filter, when solid stearic acid particles are filtered

616

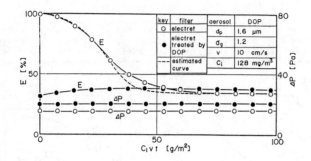

Figure 4 Time dependency of collection efficiency and pressure
drop of an electret filter, when liquid DOP particles
filtered

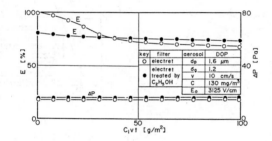

Figure 5 Time dependency of collection efficiency and pressure
drop of an electret filter with external electrical
field, when liquid DOP particles are filtered

Published 1984 by Elsevier Science Publishing Co., Inc.
Aerosols, Liu, Pui, and Fissan, editors

INFLUENCE OF AN ACOUSTIC FIELD ON THE AEROSOL COLLECTION
EFFICIENCY OF A GRANULAR BED

H. Tavossi*, A. Renoux*, D. Boulaud**, C. Frambourt** and G. Madelaine**

* Laboratoire de Physique des Aérosols et de Transfert des Contaminations,
  Université Paris XII, avenue du Général de Gaulle, 94010 Créteil Cedex,
  France

** Laboratoire de Physique et Métrologie des Aérosols, IPSN/DPT/SPIN,
   Commissariat à l'Energie Atomique, B.P. n° 6, 92260 Fontenay aux Roses,
   France

EXTENDED ABSTRACT

Introduction

The study of cleaning of the air containing particulate impurities
(solid or liquid aerosol particles) by a granular bed has numerous
applications. A granular bed has a number of advantages over traditional
filters of high efficiency. First of all, it can be use high temperatures
(500°C if it is metallic, and higher if it is made of a ceramic material)
and it resists to corrosive agents. Furthermore, the mass loading capacity
can be very important with the possibility of automatic recycling of the
granular bed medium.

We know that all types of filters have a minimum of collection
efficiency for aerosols with diameters between 0.1 μm and 1 μm and for many
applications it is necessary to remove these so-called, "maximum penetration
aerosols".

In this paper, the effects of an acoustic wave on the collection of
aerosols by a bed of spheres is studied. The results of the present
investigation indicate that the aerosol collection efficiency of this device
can be increased, especially in the particle size range where the maximum
filter penetration occurs.

Another advantage of the application of acoustic waves is to allow to
reduce the depth of the granular bed for the same collection efficiency,
that is to have the more compact filters which could facilitate the
recycling process.

Since 1931 (Patterson and Cawood) it is known that acoustic waves can
accelerate the coagulation of fine aerosols of high concentrations (Mednikov
1965) by increasing the collision efficiency of the aerosol particles.
However, at low concentration the coagulation is ineffective.

In this study, the increase of aerosol collection efficiency of a
granular bed by means of acoustic waves has been investigated also for low
concentrations.

618

1. BACKGROUND

It is well known that a minimum of collection efficiency exists for aerosols of diameters between 0.1 μm and 1 μm, for all types of filters. Several separation mechanisms could act on aerosol of this size range, but these mechanisms are not sufficiently effective to collect the aerosols.

In therories elaborated for the determination of the total efficiency of a spherical collector, the flow model, the separation mechanisms involved and their interactions are taken into account in a different way. If one mechanism is much more important than the others, the total efficiency is controlled by it. When several separation mechanisms of comparable importance are present, two cases are to be considered :

(1) If these mechanisms are non interacting, that is to say, if they are independent of one another, the collection efficiency is the sum of the individual efficiency of each one of these acting mechanisms.

(2) In the opposite case, when these separations mechanims interact, the total collection efficiency (ηt) is given by :

$$\eta_t = \eta_{M_1} + \eta_{M_2} - \left(\eta_{M_1} \cdot \eta_{M_2}\right)$$

where η are the collection efficiency of the mechanisms M_1 and M_2. M_1 and M_2 could be for example : diffusion and interception or electrostatic effect and sedimentation.

Under normal conditions fine aerosols follow the same path as the stream lines of the carrier gas. However, in the presence of acoustic wave they could leave these stream lines by oscillation, by the drift forces and by the acoustic turbulence diffusion. In this way they can get closer to the collection surface and can be deposited there. This explains the importance of these acoustic phenomena, which could have strong effects on particles of any sizes, including the size range corresponding to the minimum in the overall collection efficiency.

The details of theoretical studies of the acoustic waves effects in the granular bed have been undertaken by us and could be reviewed in the following publications (Tavossi et al. 1983a ; Tavossi et al. 1983b).

In order to verify the theoretical predictions, we have carried out a program of experimental investigation.

2. EXPERIMENTAL

2.1. Experimental set up

The main part of our system consist of a glass column of two meters high and 15 cm diameter in the middle of which we have placed our granular bed medium with the following characteristics : glass beads diameter 0.2 cm, depth of the bed 22 cm and porosity 0.35.

Experimental conditions and working parameters :

    - The flow rate through the bed is in the range of 3 to 40 m³/h. The superficial velocity is in the range of 4 to 50 cm/s under standard ambient atmosphere.
    - The maximum acoustic intensity attainable upstream of the bed is about 153 dB in the frequency range 400 Hz to 3000 Hz.
    - Monodisperse oil aerosol (DOP) could be generated in the size range 0.05 μm to 15 μm.

    The particle concentration is analyzed by condensation nuclei counter for submicronic aerosol and by an optical counter for the larger particles.

2.2. Experimental results

    Initially we have determined the collection efficiency of our granular bed, under given experimental conditions, using the theoretical overall collection efficiency reviewed by us (Tavossi and al. 1983b).

* The minimum of collection efficiency, "which is our main interest in this
  study", was calculated and found to be 16% for the size range 0.1 μm to
  0.3 μm and for the superficial velocity of about 5 cm/s was 16%.

    Our experiments were performed with aerosol in the size range mentioned above. The range of particles concentration was from 50 000 to 100 000 particles per cc.

* Our main interest in this study is the minimum of collection efficiency.
  For the size range corresponding to this minimum (0.1 μm to 0.3 μm) and
  for the superficial velocity of about 5 cm/s the calculated value was 16%.

    Preliminary experiments were aimed at measuring the collection efficiency without acoustic waves. The results of these investigations are in good agreement with the calculated values. For example for an aerosol diameter of 0.3 μm the mean value of the experimental collection efficiency for forty runs was 16.5% with a standard deviation of 3.8%.

    The above experimental investigations were repeated this time with the presence of acoustic waves. The first results demonstrate that they were significant increase in the collection efficiency for the size range corresponding to the minimum in efficiency. But this effect was not observed below an acoustic intensity level of about 145 dB which suggest the existence of a threshold in efficient intensity. For example, for a diameter of 0.3 μm, (the most penetrating size in this case), an acoustic frequency of 400 Hz and an acoustic intensity level upstream of about 153 dB, the collection efficiency was about 60%, under the experimental conditions described above, which is more than three times the value obtained in the absence of acoustic waves.

    These significant results have encouraged us to undertake further detail experimental and theoretical investigations in order to find the precise relationships between, collection efficiency, acoustic parameters (intensity, frequency, propagation direction, etc...), particle size and flow characteristics of the system.

References

Mednikow, E.P (1965) "Acoustic Coagulation and Precipitation of Aerosol" Trans from Russian by Larrick, L.V. Consultants Bureau, New York.

Patterson, H.S. and Cawood, W. (1931) "Phenomena in a Sounding Tube", Nature, 127, 667.

Tavossi, H., Renoux, A., Madelaine, G., Boulaud, D. (1983a) "Influence d'une onde acoustique sur l'efficacité de captation des aérosols par un lit granulaire", Pollution atmosphérique, 98, 123.

Tavossi, H., Renoux, A., Madelaine, G., Boulaud, D. (1983b) "Influence of an Acoustic wave on the Aerosol Collection efficiency of a Granular bed", Idöjaras, 87, n° 6, 327.

Published 1984 by Elsevier Science Publishing Co., Inc.
Aerosols, Liu, Pui, and Fissan, editors

The Filtration of Soot Particles by
Different Types of Filter Systems

T. Glück, S. Jordan, W. Lindner

Kernforschungszentrum Karlsruhe GmbH
Laboratorium für Aerosolphysik und Filtertechnik I
7500 Karlsruhe, W.-Germany

Extended Abstract

Introduction

Despite several precautions, incidents in reprocessing plants cannot be
excluded. One hypothetical accident scenario might be that solvent leaking
out from a reprocessing column spreads over the ground and starts burning.
Solvent fires are a burden on structures and components by pressure and heat
development. There is also a potential risk from the release of fuel and
fission product particles in the fire. It is assumed that these particles
are attached to solvent fire aerosols. Solvents used in reprocessing plants
are mixtures of Kerosene and tributyl phosphate (TBP). The formation of
aerosols during kerosene fires is due to incomplete burning. The aerosols
are composed mainly of soot but contain HDBP and phosphoric acid in
kerosene/TBP mixtures fires.

Soot Particles

The aerosol formation is for kerosene fires about 2 % of the burned
solvent, for kerosene-TBP mixtures (70/30) about 10 % with maximum of 25 %
shortly before the extinguishment of the fire /1/.

Soot aerosols are composed of chain-like agglomerated small primary
particles. The evaluation of electron microscopic photographs yields a
geometric diameter of primary particles of d = 0.05 µm. These particles
agglomerate already in the flame; single airborne particles have not been
identified.

The size and shape of particles from kerosene fires depend on solvent
composition, type of burning (pool fire, spray fire) and composition of the
atmosphere (relative humidity). Particles released from fires involving
kerosene TBP mixtures have a chain structure at the beginning of the fire;
at the end of the fire (high TBP concentration) particles are more resemble
droplets /2/. The smallest aerodynamic mass equivalent mean diameter was
found for particles from pool fires in the free atmosphere: d = 0.22 µm.
Particles in closed containments and spray fire aerosols were slightly
larger, attaining up to d = 0.45 µm.

Filtration of solvent fire particles

Depending on the volume of the cell and on the process cycle, requirements for the filters have to be defined. The resistance to shock overpressure should be up to 10 bar. The temperature resistance should be as high as
800 °C.

622

On account of these extreme operation data sand-bed filters were optimized for the removal of soot particles. Besides, the performance of fibreglass filters during loading with aerosols from kerosene fires was investigated. The different filter types were tested in the test loop "AEPOS". In
this loop the performance of filtration systems can be investigated under
different atmospheric conditions. Flow rates up to 2000 m^3/h are possible.

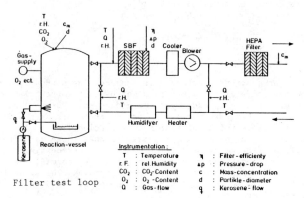

<u>Fig. 1</u>   Filter test loop

Using model calculations a basic multi-layer sand-bed filter for soot
particles was designed. This basic filter was tested experimentally and optimized with respect to particle load, pressure drop and removal efficiency.
The filtration material was crushed basalt gravel. To increase the loading
capacity of the filter at moderate efficiency, the granulation of the first
layer as well as the depth of the filter were increased. This increased the
loading capacity to about 1000 g per m^2. The intergral efficiency was
increased from 90 % to 99 % by mixing finely granulated sand (1-3 mm) into
the last layer of the filter.

The loading capacity and efficiency were improved by using crushed
lava. This sand has a high porosity offering a large surface area for particle retention. Another feature of this material is a much lower specific
weight. The removal efficiency of the filter was increased to an integral
value of 99.9 % and an initial efficiency of 90 % by using a combination of
basalt and lava gravel. The loading capacity was optimized to 1500 g per m^2
at a pressure drop lower than 30 mbar. The composition of the basic and the
optimized filter are placed together in table 1.

The optimized filter was tested under normal and accident conditions with
respect to temperature and humidity. Figure 2 compares for the optimized
filter the increasing efficiency with increasing load at 20 °C and 300 °C
and at 90 % rel. humidity and at water condensation in the filter material.

Furthermore the performance of the optimized filter was tested by using
artificial gasconcrete. From Fig. 3 can be concluded that the efficiencies
of the tested filter materials are quiet similar. The increasing pressure
drop with increasing load for the different filter materials can be seen
from Figure 4. Gasconcrete has during the whole period of load of the filter

Basis Filter	Granulation /mm/	5/11	5/11 + 2/5	2/5	1/3		
	Mixing factor	–	1 : 1	–	–		
	Material	Basalt	Basalt	Basalt	Basalt		
	Layer depth /mm/	200	200	150	150		
Optimized Filter	Granulation /mm/	5/11	5/11 + 2/5	5/11 + 2/5	2/5	0,8/3	0,8/3 + 0,4/0,8
	Mixing factor	–	2 : 1	1 : 1	–	–	3 : 1
	Material	Basalt	Basalt	Basalt	Lava	Lava	Lava
	Layer depth /mm/	200	100	200	150	150	50

Table 1: Composition of sand-bed filters

the lowest pressure drop which is at higher loads about 25 % lower than basalt or lava sand. From Figure 3 and 4 can be concluded that the optimized filter (combination Basalt-LAVA) has similar performance than the gasconcrete filter. The substantial advantage of the gasconcrete filter is that the weight of the filter material is by a factor of 3 lower than the Basalt-Lava filter (400 kg and 1350 kg per m^2 filter area).

SB-filter might be used as safety filters in reprocessing plants to reduce pressure, temperature and particle concentration in case of an accident. They might be used as prefilters followed by high-efficiency fibreglass filters.

Different types of fibreglass filters were tested with respect to their loading capacity, pressure drop and failure limitation. The pressure increase with increasing challenge by soot particles is included in Figure 4 for a glassfibre filter. The loading capacity per m^2 ventilation cross section is higher by a factor of 5 for SBF's than for fibreglass filters. Usually, soot loaded fibre filters with metal frames fail at a pressure drop of 100 mbar while filters with wooden frames fail at 150 mbar, while SB-filters have high resistance to pressure, temperature and chemicals.

References

/1/ S. Jordan, W. Lindner: The Behaviour of Burning Kerosene, Aerosol Formation and Consequences, CSNI Specialist Meeting, 1983, Los Alamos

/2/ S. Jordan, W. Lindner: Aerosolentstehung bei Kerosinbränden, 9. Conference Aerosols in Science, Medicine and Technology, 1981, Duisburg/FRG

624

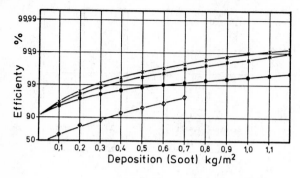

Fig. 2: SBF-performance under idfferent conditions

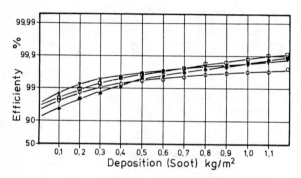

Fig. 3: SBF-performance with different filter materials

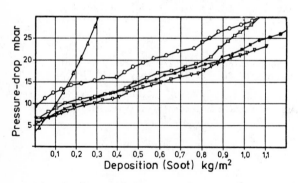

Fig. 4: Comparison SBF - Fiber glas filter

Published 1984 by Elsevier Science Publishing Co., Inc.

Aerosols, Liu, Pui, and Fissan, editors

MEASUREMENTS OF THE PENETRATION OF HIGH EFFICIENCY
AIR FILTERS USING PLUTONIUM OXIDE AEROSOLS
M.I. Tillery, M. Gonzales, J. Elder,
G. Royer and H.J. Ettinger
LOS ALAMOS NATIONAL LABORATORY

High efficiency air filters are frequently used to limit the
release of airborne particulates containing toxic materials. The goal
of limiting the release of radioactive materials to as low as
practical has resulted in the use of several stages of high efficiency
filters and the requirement for a method to insure the integrity of
the filters prior to installation. The Department of Energy (DOE)
requires that all filters used to control the release of radioactive
materials pass a quality assurance test prior to acceptance or
installation in a DOE facility. The quality assurance tests are
conducted at one of three DOE filter test facilities (FTF) using an
aerosol of di(2-ethylhexyl)phthalate (DEHP) or
di(2-ethylhexyl)sebacate (DEHS). The test aerosol is generated by a
controlled vaporization condensation process to produce a monodisperse
aerosol (all particles being approximately the same size) with a
desired diameter of 0.3 microns. The filters are challenged with a
concentration of about 100 mg/m^3 with the upstream and downstream
concentrations measured with a forward light-scattering photometer.
Acceptance requires that the collection efficiency determined by the
ratio of the two photometer readings must be at least 99.97%. The
material used to generate the aerosol has a density of approximately 1
and will form a spherical liquid particle at ambient temperatures.

The aerosols encountered in DOE facilities will frequently include
irregularly shaped, solid particles with a density greater than 1.
Filtration efficiency measurements have been carried out with nuclear
grade high efficiency air (HEPA) filters using aerosols of plutonium
oxide (density approximately 10 mg/m^3) to correlate the results of
the quality assurance tests with performance against the type of
aerosols encountered in field operations.

The aerosols used in this study were produced by grinding a
suspension of PuO$_2$ powder in a ball mill or a centrifugal mill to
provide small particles. The milled material was then diluted in water
to a concentration of 2.5 to 8.0 mg/ml and ultrasonically agitated to
break up agglomerates. An anionic surfactant was added to maintain
suspension and dispersion. The water suspension of small PuO$_2$
particles was placed in modified Retec[1] compressed air nebulizers to
produce the aerosol. Six nebulizers were used in parallel to provide
sufficient concentration to test three, 25 cubic feet per minute, HEPA
filters connected in series. The nebulizers were modified by doubling
the diameter of the jet and jet cap and replacing the plastic
reservoirs with Teflon-coated brass reservoirs with greater capacity.
During operation the reservoirs received mild ultrasonic agitation to
maintain suspension of the PuO$_2$. This system provided aerosol

626

concentrations ranging from 8 X 10^9 to 3.3 X 10^{10} disintegrations per minute per cubic meter. These concentrations include the addition of sufficient room air to insure dry particles.

Filtration efficiency was determined by the isokinetic collection of filter samples and cascade impactor samples upstream of each of the three HEPA filters and filtration of all of the air downstream of the third filter. The concentration at each location was determined by alpha counting with gas flow proportional counters or alpha scintillation counters. Corrections were made for dilution of high concentration samples, resolving time, efficiency of sampling filters, and self-adsorption in the samples and sample holders. The low concentration samples were counted with scintillator counters after holding the sample to allow for the decay of radon and thoron daughters.

The HEPA filters were separated into two groups according to the results of the quality assurance tests. The high penetration group were filters for which the quality assurance test efficiencies were greater than 99.97% but less than 99.985%. The low penetration group had quality assurance test efficiencies of greater than 99.985%. The overall efficiency for plutonium oxide aerosols for each of these groups is shown in Table I.

TABLE I

PERCENT COLLECTION EFFICIENCY (E) FOR NUCLEAR GRADE HEPA FILTERS

QUALITY ASSURANCE TEST	PuO_2 EFFICIENCY		SAMPLE NUMBER
99.970 < E < 99.985	E = 99.9939	2σ = .00690	38
99.985 $\leq$ E	E = 99.9973	2σ = .00149	48

These results indicate that the quality assurance test provides a conservative estimate of these filters performance against the PuO_2 aerosol. These measurements of efficiency are not equivalent as the QA test measures the total light scattered from a collection of particles having approximately the same diameter. This would be a measure of the number efficiency. The use of radioactive counts result in the PuO_2 efficiency measurements being measurements of the mass collection efficiency. The mass collection efficiency is biased by the large particles in the distribution. The PuO_2 aerosols used for these measurements had an activity median aerodynamic diameter (amad) of 0.58 $\pm$ 0.29 microns with a geometric standard deviation (σ_g) of 2.39 $\pm$ 0.63. The range in amad was from 0.22 to 1.61 and in σ_g from 1.50 to 3.90.

The number collection efficiencies for PuO_2 aerosols can be determined from the cascade impactor samples collected for all positions except for the aerosol penetrating the third HEPA filter. The results of looking at the average of at least 21 samples for each

particle size are shown in Figure 1. These efficiencies were
determined by comparing the collection on the same stages of upstream
and downstream impactor samples. Collection efficiencies are plotted
with respect to the average particle size collected on each stage.

These results also indicate that for all particle sizes observed
the collection efficiencies are considerably better than the minimum
value for the quality assurance test (99.97%). Collection efficiencies
as a function of particle size for DEHP particles are also shown in
figure 1. The single measurements of collection efficiency for DEHP
particles are slightly higher than the average efficiency measurements
for PuO_2 particles of equal aerodynamic diameter. The difference in
efficiency is small and probably not significant as the standard
deviation of the PuO_2 efficiencies encompasses the DEHP
measurements. The PuO_2 efficiency measurements should be more
accurate through the use of radioactive material to determine the
efficiencies and the measurement of a number of filters.
These results indicate no measurable difference between the performance of
HEPA filters for these two types of particles over the aerodynamic size range
from 0.06 to 0.30 microns. Collection efficiency differences associated with
particle density seem to be smaller than the collection efficiency variability
between different filters.
The efficiency curve may indicate a minimum in efficiency in the particle
size range from 0.3 to 0.4 microns for the PuO_2 aerosol. However, the
decrease in collection efficiency is slight and the variation in performance
by different filters is relatively large in this size range. The high
efficiency for all particle sizes indicate overlap between the different
collection mechanisms for these relatively dense (~10 gm/cm^3) particles in
high efficiency (HEPA) filters.

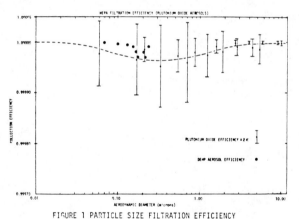

FIGURE 1 PARTICLE SIZE FILTRATION EFFICIENCY

1. M. Gonzales, J.C. Elder and H.J. Ettinger Performance of Multiple HEPA
Filters Against Plutonium Aerosols, Proceedings of the 13th Air Cleaning
Conference (AEC), CONF-740807 (1974)

CHAPTER 14.

CONTROL TECHNOLOGY

Published 1984 by Elsevier Science Publishing Co., Inc.

Aerosols, Liu, Pui, and Fissan, editors

WALL FLOWS IN A CYCLONE SEPARATOR; A DESCRIPTION OF INTERNAL PHENOMENA

William E. Ranz

Department of Chemical Engineering and Materials Science
University of Minnesota
Minneapolis, MN 55455

It is remarkable that particles of size 10^{-5} m can be separated in a device of scale 1 m with inlet gas velocities of order 10 m/s and mechanical energy expenditures equivalent to about eight times the inlet kinetic energy. In this study various types of experiments were performed and analyses developed to discover how cyclones really work and how particles, concentrated near the wall by centrifugation, flow out of a cyclone in a fluidized state. Because of its high frequency of use as a unit operation, the cyclone needs critical assessment of its design and operational limits for separating small particles at high loadings. Furthermore, a detailed picture of mechanisms is desired if the cyclone is to be applied to chemical treatment of particles and if heat and mass transfer, such as drying, occur simultaneously.

Inside a Test Cyclone

A cyclone vortex should be pictured primarily as a highly turbulent region of uniform particle concentration, that is, a uniform stirred tank. Particles are centrifuged into a wall flow layer at the outside boundaries of this central vortex. Here they are swept up into a concentrated, fluidized spiral dune of nearly stationary form. Flow inside the dune is along the spiral; and an axial component velocity is maintained on the cone wall surface by an overriding secondary gas flow. Generated by an interaction between wall shear and radial pressure gradient of the vortex, this secondary motion is in the form of a boundary layer with velocities in the direction of the cone apex. Gas and particle motion is interactive and inseparable. At the same inlet flow velocity the mechanical energy loss is decreased by particle loading. Without forced underflow there is an axial gas reflux from the dust bin where the dune is deposited.

The fluidized state near the wall can be explained by a kinetic model that predicts a condensed phase when $2\pi\rho_p(\rho_p - \rho_g)g_L'D_p/27C_D g_g \rho_g p < 1$. A self-preserving wave-top form for the dune can be explained by imaginary waves whose tops have a slip boundary condition at the wall and whose wavelength is equal to the distance between branches of the spiral. While the secondary gas flow sweeps it toward the dust bin, the dune tries to climb the cone wall because of a centrifugal acceleration.

The present point of view is still too narrow to give a model which serves all of the many purposes of cyclone application, but it does give a model with some of the necessary internal detail. Separation efficiency now becomes less than an ideal value because of reentrainment from complex wall flows and because the dune's motion is limited in its vertical flow of particle material.

Central Flow Field

The central flow field is dominated by a turbulence modified free vortext outside a central rotating vortex and inside a complex wall boundary layer flow. In the free vortex $v_\theta(r)/v_\theta(r=R_c) = (r/R_c)^{-n}$ where $n = 0.60$ for gas flow alone and $r \geq R_c/2$. $v_\theta(r=R_c)/V_c$ is very nearly unity for the standard geometry of the test cyclone. For other tangentail inlet designs this ratio would have to be established from calculations of conservation of angular momentum, or measured experimentally. Fortuitously, v_θ appears to be nearly independent of z for axial positions below the outlet tube extension to the cone outlet, with the modified free vortex located at r less than the inside radius of the cone wall and greater than that for a line between $r = R_c/2$ at the junction of cylinder and cone and the cone apex. The value of n can be related to turbulence intensity and a turbulent dispersion coefficient of random-walk type.

Axial velocities are set by an annular downflow confined by the outside wall of the outlet tube extension into the cyclone and the inside diameter of the cyclone itself. This continues downward with one free side into the cone section where the resistance of the wall is substituted by a helping shear from the secondary flow on the cone wall. Superimposed on this axial flow is a negative radial flow v_r, and reverse axial flow at small r, as though there was a uniform sink flow to the axis of rotation from the bottom of the outlet tube extension to the cone apex.

When particles were present, angular velocity profiles were similar in shape, but v_θ was somewhat decreased in value near the wall and much decreased at the radius of maximum v_θ for pure gas flow. Since v_θ dominates the central flow field, and its character near the wall boundary layers is critical to separation, static pressure measurements on the wall and static pressure traverses near the wall appear to be the only practical way to diagnose the central flow field of a cyclone while separating particles.

Mechanical Energy Losses and Turbulent Mixing

The mechanical energy loss per unit mass of fluid flowing, as measured by static pressure on the walls of the approach section and exit pipe is seven times $V_c^2/2g_c$ and with no dust load. Since swirl energy is the exit pipe is not recovered, the effective energy loss is 7.5 to 8 $V_c^2/2g_c$ per unit mass of fluid processed. The static pressure is found to be constant along the axis of the cyclone, into the dust bin and along the axis of the outlet pipe, even around bends. Axial flow in such a region is small, perhaps reversed. Downstream disturbances, such as fires and explosions, are easily propagated upstream.

With dust loading mechanical energy loss is immediately less than the loss when there is no dust loading. The decrease is as much as one-quarter and appears to grow larger for larger particles and larger loadings. Since the tangential velocities are also smaller near the wall, it is apparent that the fluid mechanics of the core is remarkably affected by the fluid mechanics of the wall layer. The dune cannot be treated as a "roughness" because a roughness would cause an increase in mechanical energy loss at the same V_c.

Wall shear contributes little to energy loss, which is due primarily to turbulent dissipation in the main vortex flow. Since all flow passes through a state of maximum v_θ^2, it is that kinetic energy which is lost during expansion in the rotating core.

Intensity of turbulence in the central flow field is at least 10^{-1}. The turbulence scale is of order $D_c/4$; dispersion is of order $10^{-2}V_cD_c$. The parameter $(2\pi N_{ep}\rho_pV_c/18\mu_c)^{1/2}D_p \leq 2.5$ for mixed conditions, and an appropriate model for collection efficiency is centrifugation, at tangential velocity V_c, of a particle concentration equal to the concentration of the outlet flow, into the wall layer.

Dynamics of a Fluidized Particle Phase on the Cyclone Wall

For a cyclone to remove dust from a gas flow the particulates first form a fluidized layer on the wall of the cyclone. The liquid-like layer built up on the cyclone wall must then flow from the cyclone through the cone outlet. The layer possesses properties of the emulsion phase of an axial fluidized particle bed. In a fluidized bed the remarkable feature is bubble formation; on the wall of a cyclone the complication is a spiral dune moving lengthwise toward the cone outlet. A useful equation of state, for volume fraction ϵ_p of particles in the dense phase, can be developed from simple kinetic theory of dense gases. Centrifugal forces, and gravity, are responsible for continuous input of random kinetic energy and maintenance of the fluidized state.

Suspended particles not yet centrifuged into the wall layer or resuspended from the dune are centrifuged into the wall layer and swept into the side of a dynamically stationary dune by the overriding boundary layer gas flow. The dune is rotating in cross section, with slip conditions on the wall. It continues into the dust bin as a twisting "rope". Here it is supposed to deflate and deposit. However some particles will be reentrained in the gas reflux from the dust bin and be recycled into the cyclone.

Particle residence time, for $10~\mu m$ $CaCO_3$, in the dune was found to be two to three times the superficial gas holdup time in the cyclone. Gas flow carried into the dust bin by the secondary flow and recycled back into the cyclone is found to be 5% of the total inlet flow. For marginal separation of fine particles, a controlled underflow with its own separator is recommended.

Particle Loading Limits

Theoretical treatment of the dune as the top portion of ideal liquid waves, with total slip on the wall, gives advice on what governs dune cross section. Motion along the spiral, which can also be predicted and measured experimentally, then limits outflow of separated particles. In terms of inlet concentrations, $\epsilon_{po} \lesssim 10^{-2}$ for the cyclone to function as described.

When liquid particulates were separated, a few experiments showed the wall flow to be streaked with broad spiral bands rather than a single spiral dune. At high loadings liquid covered the cone wall completely and a stationary collar or dam appeared high on the cone wall, held up by centrifugal forces.

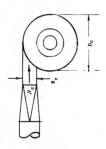
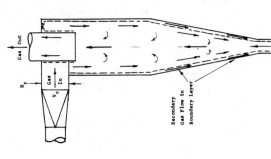

Separation Efficiency

In a turbulently mixed model the mass ratio Y_c of dust to air in the outlet flow is the same as that in the gas flow over the wall layer. Flux of dust to the wall layer is assumed everywhere to be $\rho_g Y_c$ times the terminal radial velocity of the particle at the outer edge of the central flow, and the wall area is taken to be the same as in a trajectory model ($N_e \pi D_c H_c$ for the standard design). The mass of particles of size D_p entering the wall layer to the mass entering the cyclone gives

$$N_I(D_p) = 1 - (1/(1 + 2\pi N_e(\rho_p V_c D_c^2/18 \mu_g B_c)))$$

N_e can be interpreted as "number of turns" in a trajectory model or "measure of wall collection area" in the turbulently mixed model. For practical purposes N_e is a single number to be established by experiment for each geometry. In the case of the geometry tested, $N_e \approx 5$ gives a fair fit of experimental data.

Particle Density in Fluidized States

Simplified kinetic theory for a dense array of bouncing balls gives an equation of state for the volume fraction of particles ϵ_p,

$$\frac{2\pi}{27} \frac{\rho_p(\rho_p - \rho_g)D_p^3 g_L'}{g_c \rho_g \rho C_D} = (\pi/6\epsilon_p) - (\pi/6\epsilon_p)^{2/3} \qquad \epsilon_{p,max} = \pi/6$$

where g_L' is the effective acceleration acting on the particle to maintain its agitated state and C_D, a drag coefficient for relative motion in the gas volume fraction, will be of order unity.

Test Cyclone Configuration

With high-efficiency geometry and showing regions of axial and secondary flow:

References:

Test cyclone, Page 20-82, "Chemical Engineers' Handbook", 5th Ed., McGraw-Hill (1973)

Details of experiments and analysis, "Experimental and Theoretical Studies on Gas Cyclone Separators Operating at High Efficiencies", Thinh Van Tran, PhD thesis, University of Minnesota (1981)

D_c = diameter of cyclone

V_c = entrance velocity (tangential)

ρ_p = true particle density

D_p = particle size (as sphere)

ρ_g = gas density

μ_g = gas viscosity

N_e = effective number of flow turns in cyclone (≈ 5)

Published 1984 by Elsevier Science Publishing Co., Inc.

Aerosols, Liu, Pui, and Fissan, editors

THE PERFORMANCE OF THE COOLED-ELECTRODE PRECHARGER
ON HIGH-RESISTIVITY FLY ASH

Michael Durham, George Rinard, Donald Rugg, and Ted Carney
Denver Research Institute
P.O. Box 10127
Denver, Colorado 80210

Introduction

This paper describes the field evaluation of the Cooled-Electrode Precharger (Rinard and Durham, 1984) that was developed as part of an EPA funded program to provide cost effective particulate control technology for sources of high-resistivity dusts. Results of laboratory tests and small scale field tests indicated that this concept warranted further evaluation (Durham, et al., 1982). A near full scale evaluation of the Cooled-Electrode Precharger was performed at the 15,000 acfm EPA/DRI Valmont ESP Test Facility (Rinard, et al., 1981). The results of this evaluation are presented here to demonstrate the tremendous potential of this concept. It has the capability to greatly reduce the size of an ESP required for high-resistivity dusts, and because of its small size, it can easily be retrofit into existing ESP's.

Cooled-Electrode Precharger

Figure 1 is a schematic of the Cooled-Electrode Precharger design used in the pilot scale evaluation. As can be seen it is a very simple system consisting of a manifold of welded 2.375 inch pipes, with water entering the bottom header rising through the 12 foot high vertical pipes and exiting the precipitator through the upper header. The vertical pipes are spaced 9 inches center to center. Corona is produced on 1/8 inch wires which are suspended between adjacent pipes by a hanger system. Rapping is performed by means of a single impact pneumatic rapper connected to the bottom header.

It should be noted that the purpose of cooling the anode is not to cool the gas stream and reduce the resistivity of the particles entrained in the gas, but to cool only the dust collected on the surface of the anode. It is the dust on the anodes that causes the detrimental effects in an ESP. Therefore, by cooling the anodes, the resistivity of the collected dust is reduced which allows corona current to flow freely through the dust layer eliminating the effects of back ionization.

Four Cooled-Electrode Prechargers were installed in the first four precharging sections of the ESP and were each followed by parallel plate collectors. Grounded plates, 1.83 m (6 ft) x 3.67 m (12 ft), were spaced 22.9 cm (9 inches) apart forming 6 passages. The high voltage electrodes also were plates of the same size placed midway between the grounded plates and parallel to them. The parallel plate collectors were used because they avoid high-resistivity problems by operating at an electric field below corona onset.

636

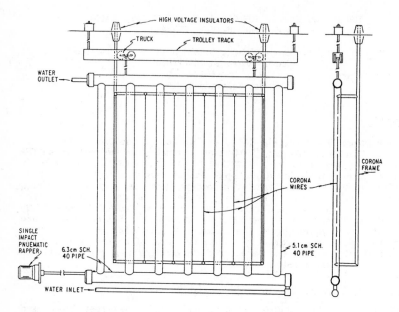

Figure 1.   Schematic of Cooled-Electrode Precharger.

Results

The tests were run on a 15,000 acfm test unit operating on a slipstream of the 180 MW unit 5 of the Valmont Station in Boulder, Colorado. The boilers fires a 0.6% sulfur pulverized coal which produces a fly ash with a resistivity of 4.2×10^{12} ohm-cm. The size distribution of the fly ash at the inlet had a 11.4 micrometer MMD aerodynamic with a geometric standard deviation of 2.7. The flue gas had a moisture content of 6.8% with 150 ppm SO_2 with no measurable SO_3.

During most tests, the current in each precharger was 7 mA and the current density on the cooled-pipe surface was 160 nA/cm^2. The voltage for the first precharger was about 45 kV for a field strength of 5.3 kV/cm, and 39 kV or 4.6 kV/cm for the other three prechargers. The parallel plate collectors operated at about 33 kV and the field strength was 2.9 kV/cm.

Table 1 shows a summary of the results of the performance of the Cooled-Electrode Prechargers. All of the results in Table 1 represent averages of 3 or 4 runs. Several combinations of chargers and collectors were tested and represented by the test numbers. The "A" designation after a test number indicates that the test were run after baffles were added to the sides of the prechargers to prevent sneakage. The "B" designation indicates that the tests were run at a lower flow rate to determine the effect of velocity on non-rapping reentrainment.

TABLE 1. TEST RESULTS WITH THE COOLED-ELECTRODE PRECHARGERS
WITH PARALLEL PLATE COLLECTORS

TEST NO.	1	2	3	4	2A	3A	4A	2B	4B
PRECHARGERS	1	2	2	4	2	2	4	2	4
COLLECTORS	1	2	4	4	2	4	4	2	4
BAFFLES	no	no	no	no	yes	yes	yes	yes	yes
VELOCITY (ft/s)	4.6	4.6	4.6	4.6	4.5	4.6	4.6	2.2	3.0
SCA (sqft/kacf)	60	115	235	230	120	235	230	235	350
INLET (g/dscf)	2.00	2.06	2.09	2.35	1.68	1.78	2.04	1.56	1.86
OUTLET (g/dscf)	0.249	0.057	0.049	0.044	0.054	0.037	0.024	0.010	0.008
PENETRATION (%)	12.4	2.9	2.5	1.9	3.2	2.5	1.2	0.6	0.5
EFFICIENCY (%)	87.6	97.1	97.5	98.1	96.8	97.5	98.8	99.4	99.5

Discussion

The test results provide a considerable amount of information on two-stage precipitators. The VI curves for the cooled-pipe prechargers showed that back corona was eliminated when the cooling water temperature into the cooled-pipe manifold was 43°C (110°F) or less. The temperature of the water was increased about 2.7°C (4.8°F) and the temperature decrease of the flue gas was about 2°C (3.6°F). Without back corona, field strengths of 5 kV/cm could be maintained in the charging region of the precharger and the current density on the cooled surfaces was 160 nA/cm^2. Q/M ratio measurements for particles in the 1 to 2 micron diameter size range showed that the particles were charged to about 85% or 90% of the theoretical unipolar ion charging level. Again, this showed the effectiveness of the cooled-pipe precharger in eliminating back corona when charging dust which had resistivity of 4 x 10^{12} ohm-cm.

Performance measurements on several two-stage ESP configurations revealed the limitations of the parallel plate collector sections. The problem appeared to be caused by some non-ideal effect such as rapping or non-rapping reentrainment or sneakage. The frequency of rapping prechargers and collectors were varied and the resulting change in penetration was not measureable. Baffles were added to eliminate sneakage around the prechargers but this did not have much effect.

638

Non-rapping reentrainment from the parallel plate collector sections appeared to be the main factor limiting performmance. This was indicated by charge measurements which showed that particles lost charge as they passed through a collector section. Cascade impactor measurements which showed a considerable amount of large particles at the ESP outlet also indicated reentrainment.

Reentrainment is also indicated by analyzing the effects of velocity by comparing test 2b with test 3. Both configurations contain two prechargers and the identical plate area per volume of flow. The only difference is that in test 2b, the velocity is reduced resulting in a decrease in emissions by a factor of 4.

Conclusions

Although the performance was limited by non-rapping reentrainment from the parallel plate collectors, the test results showed that two-stage precipitators can meet the 43 ng/J (0.1 lb/MBTU) emission level or the New Source Performance Standard of 13 ng/J (0.03 lb/MBTU). The coal analysis showed that a penetration of 1.9% would meet the 43 ng/J (0.1 lb/MBTU) emission level and 0.6% penetration would meet New Source Performance Standards.

Even lower penetration could be achieved if reentrainment from the collector sections is reduced. This can be achieved by interfacing the Cooled-Electrode Precharger with a conventional wire plate ESP which would be operated at current levels below that required to produce back ionization but sufficient to hold the collect dust on the plate. If this is not possible with dc energization then pulsed energization could be applied.

The Cooled-Electrode Precharger has tremendous potential as a retrofit device on an existing ESP that is not performing as designed or at an installation that has undergone a process change that resulted in higher resistivity particles than the ESP was designed for. Examples of applications of the latter case would include changes resulting from acid rain mitigation strategies such as coal switching, coal washing, Limestone Injection Multistage Boilers, and dry injection. The precharger can be installed in less than a foot of space and provides high collection efficiencies for even extremely high resistivity dust. An economic analysis has shown that it can be installed a half the cost of an SO_3 conditioning system.

References

Durham, M.D., G.A. Rinard, D.E. Rugg, and L.E. Sparks (1982). "The Development of a Charging/Collecting Device for High Resistivity Dust Using Cooled Electrodes"; J. Air Poll. Cont. Assn., Vol. 32, No. 11, November.

Rinard, G.A., M.D. Durham, J.A. Armstrong, D.E. Rugg, and L.E. Sparks (1981b). "The EPA/DRI Valmont Electrostatic Precipitator Test Facility"; J. Air Poll. Cont. Assn., Vol. 31, No. 8, pp. 908-911; August.

Rinard, G.A. and M.D. Durham (1984). "Electrostatic Precipitator using a Temperature Controlled Electrode Collector," U.S. Patent 4,431,434, February 14.

Published 1984 by Elsevier Science Publishing Co., Inc.

Aerosols, Liu, Pui, and Fissan, editors

SIZE DEPENDENT PENETRATIONS FROM 0.01 TO 10 MICROMETERS THROUGH
SEVERAL MODERN PARTICULATE CONTROL DEVICES ON UTILITY BOILERS

Gregory R. Markowski
Consulting Aerosol Scientist
2009 N. Madison Ave.
Altadena, Ca. 91001

David S. Ensor
Research Triangle Institute
P. O. Box 12194
Research Triangle Park, N. C. 27709

EXTENDED ABSTRACT

Introduction

Control of fine particles from pulverized coal combustion
has received increasing attention during the past two decades.
Since the early 1970's, federal regulations have increased particle mass removal requirements by a factor of about 30 for new
utility boilers. These regulations have been needed to reduce
adverse enviromental effects, such as visibility degradation,
from the large amounts of fly ash generated during coal combustion (McElroy et al., 1982). Although current regulations are
relatively size independent, environmental effects, control
device particle collection, and boiler emissions are strongly
dependent on particle size. Thus, particle collection as a
function of particle size is needed to predict overall control
device performance and enviromental impact.

The data presented here were gathered as part of a program
funded by the Electric Power Research Institute to measure
emission control performance of the major types of state-of-the-
art particulate control devices. Particular attention was given
to submicrometer particle measurement since little quantitative
data in this size range was available. Some results of this
program, in particular those dealing with fine particles generated during pulverized coal combustion, have been previously
published by Markowski et al. (1980) and McElroy et. al. (1982).

Experimental

Particle size distributions from about 0.01 to 10 micrometers were measured simultaneously at the inlet and outlet of
modern particle control devices on 5 pulverized coal utility
boilers. The control devices included a baghouse, a combination
venturi particle and absorbtion tower SO_2 scrubber, a cold and a
hot side electrostatic precipitator (ESP) collecting high resistivity ash from Western low-sulfur coal, and an ESP collecting
low resistivity ash from Eastern high-sulfur coal. Particle
size distributions from about 0.01 to 1 micrometers (actual
diameter) were measured using TSI Model 3030 Electrical Aerosol

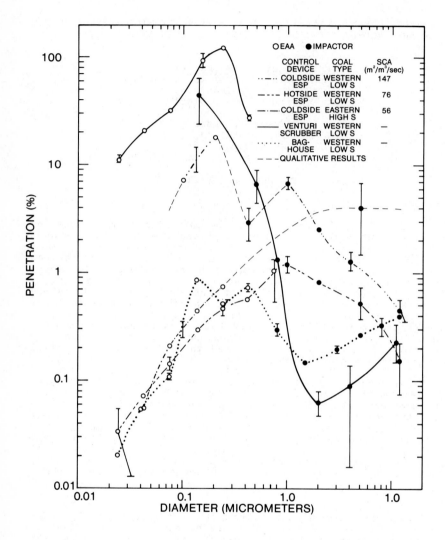

Figure 1. Average particle penetration through 5 particle control devices on pulverized coal utility boilers. Error bars show one standard deviation of the average values. Sampling periods were typically 2 to 8 hours.

Analyzers (EAAs) and dilution systems. Particle size distributions from about 0.3 to 10 micrometers (actual diameter) were
measured with MRI cascade impactors and, in the case of the
scrubber, University of Washington Mark 10 and Mark 20 low
pressure cascade impactors. Measurement techniques are discussed in more detail in McElroy et al. (1982) and Smith et al.
(1983). Particle penetration through the control devices as a
function of particle size was determined by dividing the size
distribution at the control device outlet by the size distribution at the control device inlet.

Results and Discussion

Particle penetration (1 minus the collection efficiency) as
a function of particle size is shown in Figure 1 for the 5
control devices we studied. The most obvious feature of these
results is the wide variation in collection efficiency among the
various devices, even among devices of a similar type, i. e.
ESPs. Note that the venturi scrubber shows high penetration
(low collection efficiency) for particles between about 0.02 and
0.5 micrometers and the baghouse shows low penetration relatively independent of particle size. The results for these
devices appear to be reasonably representative for these types
of control devices.

The cold-side ESP collecting ash from Western low-sulfur
coal showed anomolously high penetration in the region of the
submicrometer mode (0.1 to 0.2 micrometers). This occurred even
though the ESP plate area per unit gas flow (SCA) was much
larger than the other ESPs. This high penetration was likely
related to the high resistivity ash and consequent backcorona in
the ESP. This effect was likely also responsible for the higher
penetration at larger particle sizes.

The cold-side ESP collecting ash from high-sulfur coal
showed the opposite effect: low penetration in the region of
the submicrometer mode compared to higher penetration for larger
particles. The low submicrometer penetration indicated that the
ESP was functioning well electrically. The higher penetration of
larger particles appeared to result from higher than normal ash
reentrainment from the ash hoppers and plate rapping. These
processes are unlilely to resuspend the submicrometer ash.

Note that all types of devices tested appeared capable of
high collection of larger particles. Since most of the boiler
fly ash emissions on a mass basis was in larger (>4 micrometers)
particles, on this basis all types of devices were capable of
high overall collection of particles.

References

Markowski, G., D. Ensor, R. Hooper, and R. Carr, (1980) "A Submicron Aerosol Mode in Flue Gas from a Pulverized Coal Utility Boiler", Environ. Sci. Technol. 14:1400.

McElroy, M., R. Carr, D. S. Ensor, and G. R. Markowski (1982) "Size Distribution of Fine Particles from Coal Combustion", Science 215:13.

Smith, E., W. Morgan, A. Furguson, Principal Investigators, (1982) "Full Scale Scrubber Characterization of Colstrip Unit 2", EPRI Report CS-2764, Research Project 1410-3, R. G. Rhudy, EPRI Project Manager.

Published 1984 by Elsevier Science Publishing Co., Inc.

Aerosols, Liu, Pui, and Fissan, editors

Dust Pilling on Pulse-Cleaned Fabric Filters

by

Michael J. Pilat and Paul G. Burnet
Department of Civil Engineering
University of Washington
Seattle, Washington 98195 USA

I. Introduction

Dust pilling into nodules on fabric filters were causing increased
gas pressure drops at a primary aluminum reduction plant. The filter
bags also required periodic removal from the bag house to be physically
cleaned. Research was conducted to determine the dust nodule composition,
the effects and magnitude of the gas cooling caused by the baghouse pulse-
jet cleaning system, and the effects of the SO_3 and H_2O vapor concentrations
on nodule formation.

A review of the literature showed very little information on dust pilling.
The Enviromental Consultant Company (1982) analyzed filter bags from an
aluminum plant and found nodules to contain 10 to 18% by weight sulfate.
Bag fibers on which dust nodules had formed showed evidence of chemical
degradation while no evidence of chemical attack was found on the bag
interior (clean gas side). Dethloff (1982) mentioned scale formation on
fabric filters at a Norwegian aluminum plant. Soot particles were theorized
to bind the alumina particles together. Miller (1975) reported on bag scale
formation at the same Norwegian aluminum plant and proposed that moisture
and pot impurities were possible factors in scale formation.

II. Description of Filter Baghouse

The emissions from the prebake anode pots are ducted to alumina dry scrubbers
and then onto filter baghouses for collection of both the gaseous and
particulate flourides. The injected alumina is carred in the gas stream to
the baghouses. The alumina collected in the baghouses is returned to the
hoppers which feed the reduction pots. The filter baghouse is cleaned using
an air pulse-jet system which operates by rapidly expanding air through
a venturi nozzle, creating the "pulse" that flexes the fabric bag outward,
thus removing the dust cake collected on the outside of the cylindrical bags.
The dust nodules are not removed with this cleaning process. The pulse is
formed by compressed air (125 psig) which is released through a diaphram
valve which is open for 0.1 second. The compressed air "blast" is
transported through an approximately 1" diameter pipe to 0.25" diameter
holes located directly above the cylindrical filter bags of 4.5" diameter.
Air exiting the 0.25" diameter blow holes acts like a jet pump, entraining
surrounding air and carrying it down into venturi nozzles located at the
top of the bags. The compressed air blast entrains baghouse air equal to
about 3 times its mass. The pulse results in a sudden pressure increase
(about 0.5 psig) inside the bag. Time required for one air pulse cleaning
of one bag is less than one second. The bags are cleaned very 8 minutes.
The bags were felted 16 ounces/sq. yard type 54 Dacron polyester, 4.5"
diameter and 8 ft. long.

644

III. Composition of Dust Nodules

Nodules were present down the entire exterior (dirty gas side) of the
bags. Nodule size and concentration were heaviest at the top 12 inches
of the bag. Nodule diameter ranged from about 2 mm at the top of the
bag to 0.1 mm at the bottom. Inspection of the dust nodules under an
optical microscope at 300X showed an agglomeration of light-colored
crystalline particles (presumably alumina) with occasional small black
particles (presumably carbon). Four samples were viewed with a scanning
electron microscope at about 2000X magnification. Dust nodules analyzed
by EDAX were found to contain about 2% by weight sulfur (elements below
atomic number 9 including oxygen and carbon are not detectable with this
technique).

Dust nodules were analyzed for soluble sulfates by leaching the samples
with boiling water and analyzing the leachates for SO_4. Dust nodules
were found to average 4.6% water-soluble sulfate by weight. The average
soluble weight fraction of the nodules was 29.3%. Recycled ore averaged
0.9% water soluble weight fraction and 0.5% water-soluble sulfate.

Fresh ore, recycled ore, and nodules were analyzed by x-ray diffraction.
The dust nodules showed a significantly higher sulfate levels than either
the recycled or fresh alumina ore. Sulfate was found to average 4.6% in
nodules, 0.7% in recycled ore, and 0.15% in fresh ore.

IV. SO_3 Measurements

Measurements of the SO_3 or H_2SO_4 concentration in the stack gases was
conducted using and EPA Method 8 type sampling train. Due to the presence
of fluorides and the low SO_3 concentrations expected, the analysis of the
isopropyl alcohol and water solution were performed using ion
chromatography (Dionex 2010i). The SO_3 concentrations for 10 source tests
45 minutes of sampling ranged between 0.34 and 0.76 by gaseous volume.

V. Cooling of Pulse Air

The reduction in pressure from about 125 psig to about 0.5 psig in the
cleaning air pulse results in a reduction of the air temperature. With
the pulse air at 125 psig, 50 to 80°F, and a water dewpoint of 50°F and
the baghouse air at a pressure of -0.3 psig, 180-200°F, and water dewpoint
of 95°F, calculations of the pulse air temperature after expansion and
mixing with the baghouse air were made. The results of these calculations
are shown in Fig. 1. Assuming an initial compressed air temperature of
50°F, the air temperature upon complete expansion from 125 psig (without
mixing with the baghouse air) will be -191°F. With mixing with baghouse
air in the 180 to 230°F range, the expanded air pulse has calculated
temperatures in the 90 to 130°F range. In Fig. 1, the shaded area
represents the range in which the possibility of moisture condensation
exists, assuming a 95°F moisture dewpoint in the baghouse air.

VI. Dewpoint of Gases

The gas dewpoint temperature will vary as a function of the concentrations
of H_2SO_4 and H_2O vapors. Dewpoints, calculated using an equation reported
by Verhoff and Banchero (1974) are presented in Fig. 2 for 4 concentrations
of water vapor (0.5, 1, 2, and 5%). Using the SO_3 source test data, the
gas dewpoints average at about 183°F. The variability in the dewpoints is

shown in Fig. 3 and shows the dewpoints ranged from about 178 to 192°F.

VII.   Discussion

The calculated acid dewpoints in the 178 to 192°F range are very close to the baghouse operating temperatures of about 180 to 200°F. No evidence of acid corrosion was reported in the ducts upstream or downstream of the alumina injection or in the baghouse. However, the cool cleaning pulse air, if back-flushed through the dirty side of the bags, amy sufficiently cool the baghouse gases in the local bag environment to cause acid condensation. During a cleaning pulse, the dust cake is knocked off the bag exterior, exposing the bag surface. At this moment, acid condensation is possible at or near the bag surface. Sulfuric acid could bond the alumina particles to the bag fibers and allow subsequent bonding of alumina particles to each other.

References

Dethloff, F.H. (1982). "Dry Scrubbing of Pot Gas from Aluminum Electrolysis," Ardal og Sunndal Verk, A.S. (ASV), Oslo, Norway, p. 5.

Environmental Consultant Company (1982). Report TLN 2464, Oct. 1, 1982, to Intalco Aluminum Corp., p. 3.

Lundgren, D.A. and T.C. Gunderson (1976). "Filtration Characteristics of Glass Fiber Filter Media at Elevated Temperatures", EPA-600/2-76/192, p. 66.

Miller, J. (1975). "Environmental Conditions Inside and Outside a Modern Aluminum Smelter at A/S Ardal og Sunndal Verk (ASV), Norway,", in Engineering Aspects of Pollution Control in Metals Industries, proceedings of a conference by Activities Group Committee III of The Metals Society, London, p   165-172.

Pilat, M. J. and J. M. Wilder (1983) "Opacity of Monodisperse Sulfuric Acid Aerosols", Atmospheric Environment 17 1825-1835.

Burnet, Paul G. (1983) Dust Pilling on Fabric Filters at Intalco Masters Thesis, Department of Civil Engineering, University of Washington, Seattle, Washington.

Verhoff, F. H. and J. T. Banchero (1974). "Predicting Dew Points of Flue Gases," Chemical Engineering Progress 70 p. 71-72.

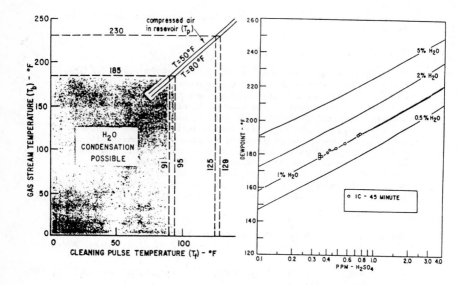

Fig. 1 Temperature Effects   Fig. 2   Acid Dewpoints

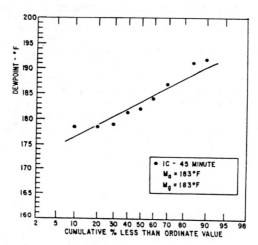

Fig. 3 Acid Dewpoint Distribution

Published 1984 by Elsevier Science Publishing Co., Inc.
Aerosols, Liu, Pui, and Fissan, editors

APPROXIMATION FORMULAS DESCRIBING AEROSOL COLLECTION IN CYCLONES

A. Bürkholz

Bayer AG
5o9o Leverkusen
W.-Germany

EXTENDED ABSTRACT

Cylones habe been used for many years as dust and droplet separators in industrial practice. Recently they have also acquired increasing importance as measuring instruments in aerosol technology.[1]

Although cyclones have been studied thoroughly both theoretically and experimentally, there seems to be no agreement in literature about a reliable way to calculate the pressure loss and the particle cut-cize.

During the past years the author has made extensive investigations with 14 different cyclones. The outer dimensions of the cyclones ranged from 3o mm to 6oo mm, test aerosols were oil and acid mists. The fractional separation efficiencies were measured with cascade impactors.[2]

From these studies it was found that the experimenteal results were in good agreement with calculations based on the work of Barth, Muschelknautz and co-workers [3]. In addition, it was possible to formulate uncomplicated approximation formulas without empirical constants and which describe the experimental results with sufficient accuracy. The formulas were also verified by applying them to other authors.

As an upper value for the pressure loss factor we found

$$\zeta = 0.9 \cdot R \cdot (F^2 + D_a/h_i) \cdot (\log Re - 1). \tag{1}$$

For Re from 10^4 to 10^5 we get approximately

$$\zeta = 3 \cdot R \cdot (F^2 + D_a/h_i). \tag{2}$$

This leads to a pressure loss of

$$\Delta p = \zeta \cdot 0.5 \cdot \rho \cdot v_e^2 \tag{3}$$

and with

$$v_e = Q/(\pi \cdot r_e^2)$$

$$r_e = F \cdot r_i^2$$

$$\Delta p = 0.15 \cdot Z \cdot \rho \cdot Q^2 \cdot r_i^{-4} \tag{4}$$

where

$$Z = R \cdot (1 + D_a/(h_i \cdot F^2)) \tag{5}$$

is a dimensionsloss geometrical factor which only depends on the type of cyclone.

In the figure we can see the relation between the pressure loss, the cyclone exit tube radius and the particle cut-cize.

The open points and $(\boxtimes, \oplus)$ are calculated values according to the exact cyclone theory, the closed values and (x, +) are results obtained by our experiments. It can be see that theoretical and experimental values can be approximated by the simple formula

$$d_{50} \; (\mu m) = 1.7 \cdot \left[r_i (mm) / \Delta p (mbar) \right]^{1/3} . \tag{6}$$

This latter formula can also be obtained as part of a dimensionless number ψ_A, which itself is a product of three dimensionless numbers $(\psi = \frac{1}{2} St)/4/$:

$$\psi_A^{1/2} = 1.74 \cdot \psi^{1/2} \cdot Re^{1/6} \cdot \gamma^{1/3} \tag{7}$$

For the particle cut-sizes d_{50}, we get from (6):

$$\psi_A^{1/2} = 1.7; \tag{8}$$

explicitly, we thus obtain

$$1.7 = 0.5 \cdot (\rho_T^{1/2} \cdot \rho^{1/6} \cdot \mu^{-2/3})(\Delta p^{1/3} \cdot r_i^{-1/3} \cdot d_{50}) . \tag{9}$$

From (9) we get

$$\Delta p = 39.3 \; (\rho_T^{-2/3} \cdot \rho^{1/2} \; \mu^2) \cdot (r_i \cdot d_{50}^{-3}) . \tag{10}$$

By equating formulas (4) and (10), we obtain the general approximation formulas for the particle cut-size of cyclones:

$$r_i = 0.33 \; Z^{1/5} \cdot S^{1/5} \cdot Q^{2/5} \cdot d_{50}^{3/5}$$

$$d_{50} = 6.35 \; Z^{-1/3} \cdot S^{-1/3} \cdot r_i^{5/3} \cdot Q^{-2/3}$$

$$Q = 16 \cdot Z^{-1/2} \cdot S^{-1/2} \cdot r_i^{5/2} \cdot d_{50}^{-3/2} \tag{11}$$

where

$$S = \rho_T^{3/2} \cdot \rho^{1/2} \cdot \mu^{-2} \tag{12}$$

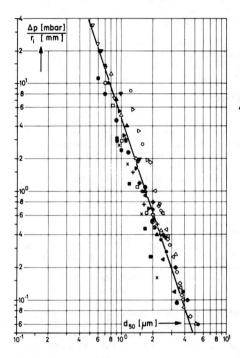

Nomenclature

D_a diameter of cyclone

d   particle diameter

d_{50} particle cut size

$F = r_e^2 / r_i^2$

h_i effective cyclone length

Δp pressure loss

Q   gas flow

$R = r_a / r_i$

Re Reynold's number

r_a radius of cyclone

r_e radius of entry tube

r_i radius of exit tube

S   physical properties factor

v_e gas entry velocity

Z   cyclone geometric factor

ζ pressure loss factor

η_F fractional separation efficiency

μ gas viscosity

ρ gas density

ρ_T particle density

ψ inertial parameter

ψ_A separation parameter

References

Smith, W.B., D.L. Iozia and D.B. Harris (1983) " Performance of small Cyclones for Aerosol Sampling ", J. Aerosol Sci. 3:402

Bürkholz, A. (1978) " Zyklone als Abscheider für Nebeltröpfchen ", Staub-Reinhaltung der Luft,9:449

Muschelknautz, E. (197o) " Auslegung von Zyklonabscheidern in der technischen Praxis", Chemie-Ing.-Technik 5:187

Bürkholz, A. (1976) " Charakterisierung von Trägheitsabscheidern durch praktische Kennzahlen", Verfahrenstechnik 1:29

Published 1984 by Elsevier Science Publishing Co., Inc.

Aerosols, Liu, Pui, and Fissan, editors

650

'ABSOLUTE' FINE PARTICLE CONTROL USING AN
ACOUSTIC AGGLOMERATOR-ELECTROSTATIC
PRECIPITATOR COLLECTION TRAIN

Sushil Patel and D. T. Shaw
Laboratory for Power and Environmental Studies
State University of New York at Buffalo
Buffalo, New York

EXTENDED ABSTRACT

Electrostatic precipitators are used widely in the removal of particulate
from industrial flue gas, and operate in excess of ninety percent on a mass
basis. However, there exists a window in its efficiency curve allowing
submicron particles of 0.1 to 1 micron diameter to penetrate relatively easily.
Acoustic preconditioning of the particulate is used to move particles from
the submicron size range (< 1 µm) to a coarser size range (>5 µm) where
particles are more readily removed by the electrostatic precipitator.

Experimental results of this concept of fine particle control is presented
using the acoustic agglomerator-electrostatic precipitator (AA-ESP) collection
train. A schematic of the experimental set-up is shown in Figure 1. Acoustic
preconditioning of ammonium chloride aerosol was performed for sound pressure
levels ranging from 145 db to 160 db, frequency from 0.5 kHz to 3 kHz under
travelling wave conditions and acoustic exposure times of 1 to 5 seconds. The
size distribution of the aerosol after acoustic exposure was found to be
bimodal. An example of the shift in size distribution is shown in Figure 2.
The ESP is of the wire-cylinder type. The applied voltage and gas velocity
were used as control parameters to study changes in the penetration of the
AA-ESP collection train. The fractional penetration of the ESP and AA-ESP
train was calculated from measurements of the size distribution using the
Berner low pressure impactor and from electromicrographs of particles deposited
on the collecting electrode of the ESP. An example of the increase in collection
efficiency is shown in Figure 3. Results also indicate that, for inlet mass
loading of 10 to 50 gm/m^3 studied, the penetration of submicron particles
(< 1 µm) through the AA-ESP collection train decreased a hundred fold following acoustic agglomeration. This decrease in penetration increases with mass
loading, acoustic intensity, frequency and exposure time. The decrease in penetration of large particles following acoustic preconditioning was substantially
lower. Furthermore, a decrease in corona quenching within the ESP is observed
by current measurements during the collection of acoustically agglomerated
aerosols. These results indicate the potential of using acoustic agglomeration
as preconditioner to conventional electrostatic precipitation for achieving
'absolute' fine particle control.

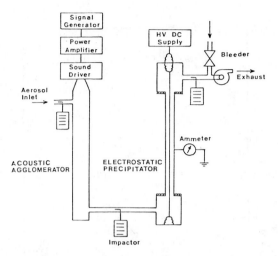

Figure 1. Experimental set-up of the Acoustic Agglomerator-
Electrostatic Precipitator Collection Train.

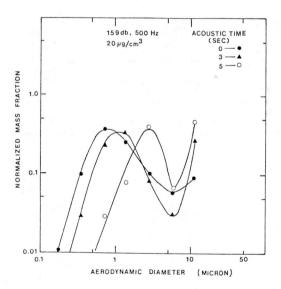

Figure 2. Change in particle size distribution of ammonium
chloride aerosol following acoustic agglomeration.

652

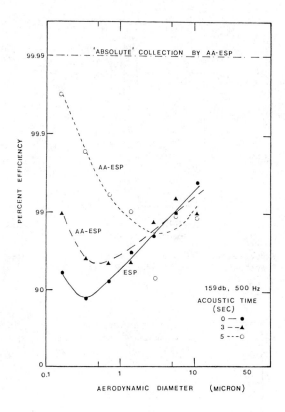

Fugure 3. Collection efficiency for the ESP and AA-ESP
collection train using ammonium chloride aerosol.

Published 1984 by Elsevier Science Publishing Co., Inc.

Aerosols, Liu, Pui, and Fissan, editors

Performance of an Electrostatic Precipitator for Tar Removal
in a low BTU Coal Gasification Facility

B.Y.H. Liu, D.B. Kittelson, D.Y.H. Pui, T. Kotz and W. Szymanski
Particle Technology Laboratory
Mechanical Engineering Department
University of Minnesota
Minneapolis, MN 55455-0111

D. Thimsen, D. Lonick, D. Pooler and R. Maurer
Black, Sivalls and Bryson, Inc.
7324 Southwest Freeway
Houston, TX   77074

Introduction

A research program to study the conversion of coal to low BTU gas has been untaken since May, 1982 at the Twin Cities Research Center, U.S. Department of Interior, Bureau of Mines. This is a joint U.S. Department of Energy, Bureau of Mines and industry funded program (U.S. Bureau of Mines Contract No. H0222001 and Bureau of Mines-Dept. of Energy Interagency Agreement DE-AI21-77ET10205). The test facility includes a 6-foot, 6-inch Wellman Galusha, single stage, fixed-bed atmospheric gas producer with a nominal rated capacity of 32 x 10^6 kJ/hr (30 x 10^6 Btu/hr).

Part of the study was concerned with evaluating an industrial electrostatic precipitator for tar removal. The ESP, with a design efficiency of 99.14% at 1400 ACFM, was installed on a slip stream to treat about 1/3 of the total gas from the gasifier. The raw coal gas contained a very high tar concentration (up to 50 g/m^3) at output temperatures ranging from 250 to 600^0C. The gas consisted mainly of H_2, CO, CO_2, N_2, small amounts of CH_4, C_2H_4, C_2H_6, water vapor and the condensable organic material, here referred to as "tar". The gas entering the ESP was first quenched by a water spray and the tar droplets formed were then removed by the ESP. This paper describes the experiments performed to determine the efficiency of this Quench/ESP unit.

Concentration and Size Distribution of Tar Droplets

The upstream tar concentration was obtained by means of a special gas sampling and conditioning system described by Liu et al (1983). The system consists of a indirect contact heat exchanger/condenser for cooling the gas, a high efficiency (>99.99%) electrostatic precipitator for collecting the nucleated tar droplets and a refrigeration unit for removing the water vapor from the gas stream. The collected tar was drained periodically and its mass and moisture content determined. The gas volume between the drain periods was measured by a gas meter. For the test described below on the Kentucky Elkhorn bituminous coal, the tar concentration was found to be between 2.5 to 3.3 lb/10^3 SCF (40 to 53 g/m^3).

The size distribution of the tar droplets was measured by a Sierra Model 266 Marple cascade impactor at a point immediately downstream of the condenser. The condenser was air-cooled to about 90-100^{0}F and the tar vapor was completely condensed. The measured droplet size distribution should be the same or similar to that for tar droplets entering the Quench/ ESP. Figure 1 shows the measured size distribution. The mass median diameter is seen to be 2.6 um and the geometrical standard deviation, 2.0.

Efficiency of the Electrostatic Precipitator

To determine the efficiency of the Quech/ESP, a sampling system was developed to measure the mass concentration of the tar droplets at the exit of the Quench/ESP. Figure 2 is a schematic diagram of the sampling system which was installed at a point immediately downstream of the Quench/ESP. The system consisted of an isokinetic sampling probe, a high efficiency filter, a condenser trap and a flow measuring system. The sampling system upstream of the filter was heated to prevent moisture condensation. The filtered gas was reinjected into the gas stream.

Figure 3 shows the Quench/ESP efficiency as a function of the specific area of the ESP. The data can be fitted by an equation of the classical Deutsch type giving the penetration, P, as a function of the specific plate area, A/Q (surface area/volumetric flowrate), and the migration velocity V. The empirical fit of data resulted in an experimental migration velocity of 3.4 cm/sec. Included in the graph is the design point given by the manufacturer showing the relative poor performance of the ESP when compared to the design value. The cause of the poor performance is not completely clear, but may be due to ion depletion caused by high tar droplet concentration, non-uniformity in gas flow in the precipitator, the improper electrical operation of the ESP power supply or excessive sparking in the ESP due to the properties of the coal gas. Using the combined field-diffusion charging theory of Liu and Kapadia (1978), a theoretical operating point was found that was in reasonable agreement with the design values given by the manufacturer.

In an attempt to understand the poor performance of the Quench/ESP for tar collection, an experiment was performed to see whether there was any difference in the characteristics of corona discharge in air and in the coal gas. The experiment was performed using a small wire-in-tube (0.001" diameter wire in a 0.4" diameter tube) precipitator. Figure 4 shows the current-voltage curves for positive and negative corona in air and in the coal gas. The steep slope of the current-voltage curve for negative corona in the coal gas suggests that there are probably considerably more spark-over in the coal gas, resulting in the lowered efficiency of the ESP. However, this conclusion is tentative and needs further verification before it can be accepted.

References

1. Liu, B.Y.H., D.B. Kittelson, D.Y.H. Pui, J.A. Armstrong, D. Thimsen, D. Pooler and R. Maurer (1983) "Aerosol Problems in Coal Gasification Research," Presented at the Annual Meeting of Gesellschaft fuer Aerosolforschung (Munich, West Germany; September 14-16, 1983) and for publication in J. Aerosol Sci.

2. Liu, B.Y.H. and A. Kapadia (1978) "Combined Field and Diffusion Charging of Aerosol Particles in the Continuum Regime," J. Aerosol Sci. 9:227-242.

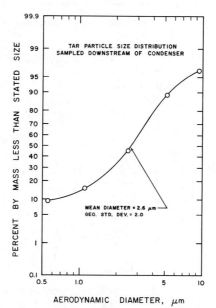

Figure 1. Tar particle size distribution measured by the cascade impactor.

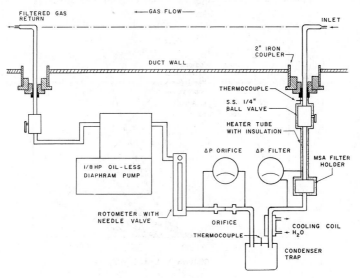

Figure 2. Schematic diagram of the tar particle sampling system downstream of the ESP.

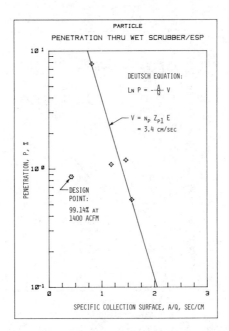

Figure 3. Quench/ESP efficiency as a function of the specific area.

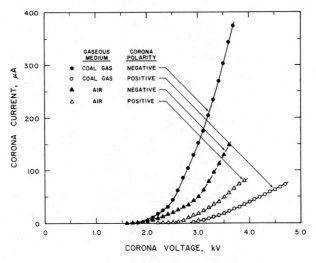

Figure 4. Corona Characteristics for discharges in air and in coal gas.

Published 1984 by Elsevier Science Publishing Co., Inc.
Aerosols, Liu, Pui, and Fissan, editors

TEST RESULTS OF AN ELECTROSTATIC PRECIPITATOR
AT SIMULATED PFBC CONDITIONS

James Armstrong, Michael Durham, George Rinard and Donald Rugg
Denver Research Institute
P. O. Box 10127
Denver, Colorado 80210

EXTENDED ABSTRACT

Introduction

This paper describes the evaluation of a subpilot scale electrostatic precipitator (ESP) operating at pressurized fluidized bed combustor (PFBC) conditions. Tests were conducted on the cylindrical tube ESP at temperatures ranging from 745° to 975°C (1370° to 1790°F) and a pressure of 640kPa (6.4 atm). Electrical operating conditions are reported for three different electrodes; mass collection efficiencies are reported for one electrode.

ESP System and Test Conditions

The High Temperature High Pressure (HTHP) ESP was designed to simulate a wide range of PFBC operating conditions (Rinard, et al., 1981). Figure 1 shows a schematic of the HTHP-ESP. Pressurized hot gas to the ESP which is housed in a pressure vessel is provided by a 93 kW (125 hp) compressor and a 264 kW (900,000 Btu/hr) methanol burner. Flow through the ESP is controlled by a throttle valve and is monitored by an orifice meter. The unit is capable of operating at temperatures up to 980°C (1800°F) and pressures up to 100 kPa (10 atm) with a flowrate of 0.078 m³/sec (165 acfm) at these conditions. The ESP collector tube is 30.5 cm (12 in) in diameter. A tube section 2.1 m (6.9 ft) long between the inlet and outlet is electrically isolated so that corona current to it can be measured. At both the ESP inlet and outlet, representative fly ash samples are withdrawn isokinetically through nozzles and collected in total mass samplers maintained at stack pressure.

To simulate PFBC operations, fly ash was used which was from the second cyclone of the Curtiss-Wright (C-W) Small Gas Turbine Test Rig. A pressurized fluidized bed aerosol generator delivered the resuspended ash to the gas through a port located just downstream of the system burner.

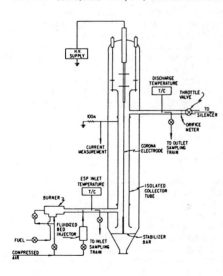

Figure 1. HTHP-ESP Schematic

Sketches of the three corona electrodes tested are shown in Figure 2. The first corona electrode was a smooth wire; the second was a scalloped design by Research Cottrell (R-C) for use at the C-W PFBC facility (Feldman, 1977); and the third was a vane electrode also designed by R-C.

Test Results

Stable corona was obtained for all operating conditions both with and without reentrained ash. In all cases corona current, for a given voltage, increased with temperature and decreased with pressure. With positive corona the ESP was spark limited and with negative corona the maximum voltage was generally limited by the power supply.

Corona current density as a function of average field strength is shown in Figure 3 for each of the three electrodes under clean and dirty conditions. While no signs of thermal ionization were apparent at even the highest temperature test, the addition of fly ash in the gas stream did cause an unexpected increase in corona current. The voltage-current characteristics, could be controlled by corona electrode design. The highest field strengths were achieved with the vane electrode. The wire and scalloped electrodes produced the same field strengths under dirty conditions.

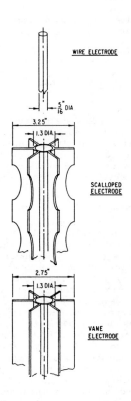

Figure 2. Corona Electrodes

Mass efficiency measurements were made for negative corona for temperatures at 743°, 821° and 916°C (1370°, 1510°, and 1680°F). The gas velocity through the collector tube was maintained at 1.07 m/sec (3.5 ft/sec) which produced a specific collection area (SCA) of 25.6 sec/m (130 ft²/kacfm) and a volume flow rate of 0.078 m³/sec (165 acfm). The scalloped corona electrode, shown in Figure 2, was selected for the tests since it was being used in tests at the C-W PFBC facility. The C-W PFBC dust was redispersed into the HTHP-ESP at an average rate of about 1.3 g/Nm³ (0.6 gr/scf). At the start of each run, the collector tube was rapped about 10 times to remove any previously collected dust.

A summary of the results is given in Table 1. The best ESP efficiency, 99.4%, was obtained for Test B where the temperature was 743°C (1370°F) and the ESP was operating at the maximum possible voltage of -89 kV (here the unit was spark limited). At the same temperature and pressure conditions, i.e. Test A, the operating voltage was reduced to -46 kV resulting in a collection efficiency of 90.9%. This indicates that the efficiency is a strong function of voltage.

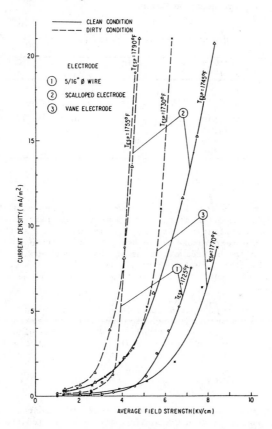

Figure 3. Field Strength vs Current Density for Negative Corna at 6.4 atm

Furthermore, for Test D where the operating voltage was comparable to Test A, but the temperature was 916°C (1680°F), the current was 40 times greater and the collection efficiency only increased to 93.1%. This indicates that efficiency is not a strong function of current.

The lower efficiencies at the higher temperatures (negative corona) seen in Tests C and D were caused by reduced operating voltages. For these tests the ESP was power supply current limited.

The efficiencies shown in Table 1 for the positive corona cases (Tests E and F) are considerably lower than those for negative corona. Also at higher operating voltages, the collection efficiency for positive corona did not increase as was the case for negative corona.

Table 1. HTHP-ESP Efficiency Measurements

Test	Pressure (atm)	Temperature ($^\circ$F)	Voltage (kv)	Current (mA)	Efficiency (%)
A	6.4	1370	−46	1.0	90.9
B	6.4	1370	−89	18.8	99.4
C	6.4	1510	−58	40.0	95.3
D	6.4	1680	−47	40.0	93.1
E	6.4	1680	+46	8.3	87.6
F	6.4	1680	+72	18.9	86.8

Concluding Remarks

Tests conducted to date confirm that higher efficiencies at HTHP conditions are obtained with negative corona than positive corona. For negative corona operations, at constant density and voltage conditions, the corona current increases as the temperature increases. It is believed that this is due to a greater percentage of free electrons as charge carriers in the inner electrode space at the higher temperatures. Also for negative corona, the collection efficiency is seen to be a strong function of voltage and therefore field strength.

A major question to be answered concerning ESP operations at extended PFBC conditions is how to obtain higher negative voltage levels at higher operating temperatures. Possible solutions include: a) changing electrode geometries [see Figure 3] and b) operating at higher pressures which gives higher voltages. These possibilities are presently evaluated in a continuing test program.

References

Feldman, P. L. (1977) "High Temperature, High Pressure Electrostatic Precipitator", EPA/ERDA Symposium on High Temperature/Pressure Particulate Control, Washington, D.C.

Rinard, G., M.Durham, J.Armstrong, and R.Gyepes (1981) "The DRI High Temperature/High Pressure Electrostatic Precipitator Test Facility", Proceedings: High Temperature, High Pressure Particulate and Alkali Control in Coal Combustion Process Streams, DOE/METC Contractor's Meeting, Morgantown, W.V.

CHAPTER 15.

CLEAN ROOM

Published 1984 by Elsevier Science Publishing Co., Inc.

Aerosols, Liu, Pui, and Fissan, editors

AEROSOL SAMPLE MULTIPLEXING FOR CLEAN ROOM CHARACTERIZATION

A. Lieberman
Particle Measuring Systems, Inc.
1855 S. 57th Ct.
Boulder, CO 80301
U. S. A.

EXTENDED ABSTRACT

In large multifunction clean rooms, aerosol monitoring may be required at many locations at frequent intervals. In some cases, more than 100 sample locations may have to be monitored, each at intervals of one hour or less. Particle count data are used as a production control input. Because of the cost and reliability considerations in procuring and operating many optical particle counters simultaneously, sample multiplexing has become a necessity. Two means of sample multiplexing can be considered. In one, an optical sensor assembly is located at each of the points to be measured and each sensor's output is multiplexed electronically to a single data processing system. The data are then procured simultaneously from all of the locations where there is a sensor and analyzed in the data processing system. In the other multiplexing mode, air samples are transported sequentially to a single optical sensor through an aerosol manifold system and a number of sample locations are monitored sequentially by a single sensor. In some cases, combined systems are used with computer control of sample sequencing or selection.

In the case of multiplexing by use of simultaneous sensor operation, the advantages of the ability to procure sample data simultaneously for many sample points is obvious. At the same time, the problems of sensor-to-sensor variability and the effects that this situation has on the reliability of the data are pointed out. In addition, the cost and maintenance requirements are mentioned. In the case of multiplexing by aerosol sample manifolding, the advantages in terms of cost and reliability are pointed out. The problems of aerosol loss in the tubing during transport are considered and the losses are defined as they are affected by particle size. Generally, it is found that the losses are negligible for particles smaller than one or two micrometres in diameter. Data are shown in Figure 1 for particle penetration at a flow rate of 85 liters per minute through polyester lined vinyl tubes of 3/8 inch inside diameter for tubes up to 38 meters long.

The operation of both systems is described, with some discussion of their advantages and disadvantages. Cost, reliability, particle sizing and counting discrepancies along with sample transport line losses are considered. Since sample line multiplexing is normally dictated by cost and reliability considerations, some of the requirements for transport line installation and maintenance are discussed. Some empirical data are presented to show particle loss for several line lengths (up to 45 meters) for the particle sizes of concern in clean rooms. These data also show the effects of two sample line materials and two flow rates in the favored line material.

664

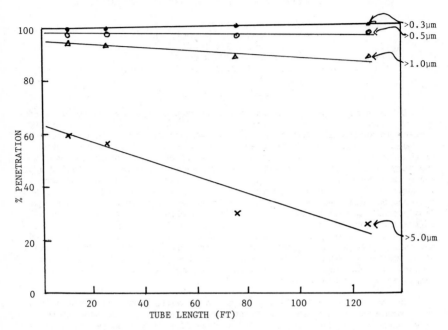

Figure 1. Particle penetration through polyester lined tubing, 3/8 "I.D., at 3 CFM aerosol flow rate.

Published 1984 by Elsevier Science Publishing Co., Inc.
Aerosols, Liu, Pui, and Fissan, editors

ONE-DIMENSIONAL MODEL OF COUNTER-DIFFUSION IN A LEAK FROM A CLEAN ROOM

Douglas W. Cooper
IBM T.J. Watson Research Center
Yorktown Heights, NY 10598
U.S.A.

EXTENDED ABSTRACT

Introduction

Clean rooms are widely used in the electronics, computer, and pharma-ceutical industries. Often clean rooms provide concentrations of contam-inants that are many orders of magnitude smaller than the environments in which they are placed. Typically, clean rooms are kept at pressures somewhat higher than that of the areas next to them so that any leakage (or flow through doors, pass-through windows, etc.) would be out of the clean area into the dirty area. It is generally assumed that such flow is adequate to prevent the transport of contaminants through the leakage path back into the room, against the leakage flow. We present a simple model to help determine what flow velocities are adequate to prevent con-tamination of the room.

Transport Equations

Figure 1 shows an idealization of the leakage path, modeled as a chan-nel of parallel sides that has a length L. It joins two perfectly mixed regions that have concentrations a and b respectively of some contaminant. The gas flows through the leak at a velocity u in the x direction. The flow is assumed to be one-dimensional, constant, incompressible (plug flow).

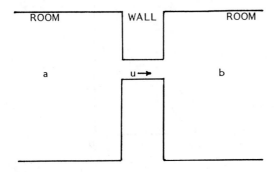

Figure 1. Leak Geometry: Room with Concentration a, Wall of Thickness L, and Room of Concentration b. Leak Velocity Is u in x Direction.

1

666

The differential equation governing the concentration in the flow path is

$$\partial c/\partial t + u \, \partial c/\partial x = D \, \partial^2 c/\partial x^2$$

where $\partial c/\partial t$ is the partial derivative of the concentration with respect to time; $u\partial c/\partial x$ is the convective component of contaminant transport; $D\partial^2 c/\partial x^2$ is the diffusive component.

The approach to steady-state is over a period of time having the same magnitude as the larger of the characteristic times: for convection, L/u, and for diffusion, L^2/D. We can show that the steady-state concentration is given by the following expression:

$$c = a + (\exp(ux/D)-1) \, (b-a)/(\exp(uL/D)-1)$$

Figure 2 shows concentration versus normalized x coordinate for two Peclet numbers, $Pe=uL/D=1$ and $=10$, for $a=1$ and $b=10$. Near $x=0$ the gradient for $uL/D=1$ is obviously much greater than it is for $uL/D=10$. Often, a clean room is vented to a much less clean environment, so that ratios of b/a may be 1000 or more.

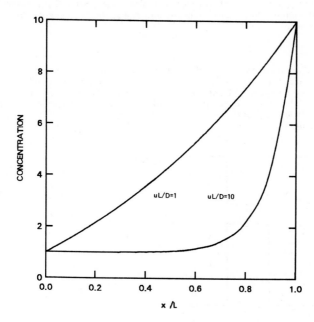

Figure 2. Concentration versus Normalized Distance Along Leak Path

For values of uL/D much smaller than 1, this expression for the concentration becomes

$$c = a + (b-a)x/L,$$

which is the linear profile expected where diffusion dominates convection. The diffusive flux would become $-D(b-a)/L$ operating against a convective flux of ua, giving a net flux of $ua-D(b-a)/L$, which flux decreases the room contamination if the flux is positive, but increases the room contamination if the flux is negative (thus back into the room, against the flow). For uL/D of any magnitude, the more exact expression gives a net flux at x=0 of

$$ua - ua(b/a-1)/(exp(uL/D)-1)$$

and this flux is either positive (cleaning the room) or negative (contaminating the room) depending on whether $exp(uL/D)$ is greater than b/a or smaller. Since clean rooms are typically much cleaner than their surroundings, b/a is generally much much greater than 1, so uL/D must be much greater than 1 to have the leak provide a net cleansing effect.

Examples

For example, if the leakage path is a centimeter long and the contaminant is a gas with diffusivity about 0.1 cm^2/s and the flow is 1000 cm/s (potential flow, pressure drop of a fraction of an inch of water), then uL/D is about 10,000 and the counter-diffusional flux is usually negligible. If, however, the leakage path is 0.1 cm and the flow is more nearly 1 cm/s (viscous flow), then uL/D is about 1, the counter-diffusional flux is the same magnitude as the convective flux and perhaps even greater, giving a net flow of contaminant into the clean room, rather than a net flow out. An open window with a thickness of a few cm and having a flow turbulent diffusivity of about 10^3cm^2/s would have appreciable counter-diffusion even if the outflow were about 100 cm/s. Thus, having the clean room under positive pressure is not a panacea for assuring no transport of contaminant through leakage paths.

Discussion

The model is a major simplification of reality. Both the room and its environment are assumed to be perfectly mixed so as to produce constant concentrations a and b at the beginning and end of the leakage path. No losses are allowed for in the flow, nor does the flow have a velocity profile, and the momentum of the exiting flow is not taken into account (producing a concentration profile in the receiving volume). Still, it does identify major elements in determining the net flow of contaminants and allows for a rapid estimation of whether counter-diffusion is likely to be appreciable.

Published 1984 by Elsevier Science Publishing Co., Inc.

Aerosols, Liu, Pui, and Fissan, editors

668

CRITICALITY OF AIR FLOW BALANCE AND THE IMPACT OF OBSTRUCTION TO AIR FLOW

George H. Cadwell, Jr.
Flanders Filters, Inc.
Washington, North Carolina 27889
U.S.A.

With current technologies it is possible to manufacture and install a hepa filter ceiling that emits air with virtually no particles at nominally 0.1 micrometers and larger. Uniform velocity begins about three to six inches downstream of the hepa filter. In order to utilize that uniform velocity to remove particles that are generated in the clean room, it is essential to have control devices at the air exits so that balanced air flow can be maintained from the filter ceiling to some point below the level of the process or product activity. If the air flow is not balanced, there will be lateral and/or cross currents to carry internally generated particles to the surface of the product or process.

Obstruction to parallel air flow often create traps for particles as well as "cork screw" conduits in which particles can migrate. There are, however, a few situations in which obstruction can be used to enhance air flow and thereby reduce particle deposition on the process or product.

"Air flow patterns observed under conditions of imbalance in a wafer fab tunnel are compared with patterns that were observed in a partially balanced tunnel. Also presented are air flow patterns and pump down characteristics that were observed for various arrangements of objects in an experimental clean room."

Published 1984 by Elsevier Science Publishing Co., Inc.
Aerosols, Liu, Pui, and Fissan, editors

SEMICONDUCTOR CLEAN ROOM PARTICLE SPECTRA

B. R. Locke,* R. P. Donovan,* D. S. Ensor* and C. M. Osburn[†]

*Research Triangle Institute
P.O. Box 12194
Research Triangle Park, NC 27709

[†]Microelectronics Center of North Carolina
P.O. Box 12889
Research Triangle Park, NC 27709

EXTENDED ABSTRACT

Introduction

Aerosol size distributions in semiconductor clean rooms have been measured under several different conditions. Measurements were made at two state-of-the-art class 100 manufacturing/research facilities and one class 10,000 university laboratory facility. The aerosol size measurements spanned the range from 0.01 to 5.0 μm. No large concentrations of sub-0.1 μm aerosol were found in any of the facilities. Indeed, the aerosol concentration below 0.1 μm was near the lower limits of the instrument sensitivity and detectability. Further refinement of low concentration measurement capability is needed.

The instruments used for aerosol measurement included the TSI differential mobility particle spectrometer (DMPS), the TSI condensation nuclei counter (CNC), the PMS optical particle counter (LPC), and two PMS laser aerosol spectrometers (both LAS-X). Prior to clean room measurements several of the instruments were compared with monodisperse polystyrene latex aerosol (PSL) and with polydisperse room air aerosol. From the instrument comparisons with the test aerosols, it can be concluded that the DMPS and the LAS-X agree fairly well in the regions of overlapping size. Both number concentration and peak resolution were comparable at the higher aerosol concentrations used in the calibration runs (> 1 cm^{-3}).

Instrument Comparison

The DMPS and one of the LAS-X spectrometers were compared using high concentrations of monodisperse PSL aerosols and lower concentrations of polydisperse room air. The operation of the DMPS is discussed by Keady, Quant, and Sim (1983). The DMPS has a size range capability of 0.01 to 1 μm, although its accuracy is greater for the size range below 0.5 μm. The background theory of the laser spectrometer is discussed by Knollenberg (1984). The LAS-X used for these tests had a size range of 0.09 to greater than 3.0 μm. The total concentration comparisons for these two instruments are presented for the overlapping size range.

Figure 1 illustrates the number concentration comparison between the LAS-X and the DMPS for a series of 25 experiments with PSL particle sizes of 0.1 to 0.5 μm. Figure 2 illustrates a comparison between the DMPS and the LAS-X for 0.312 μm PSL particle size. The LAS-X was run in two different

size ranges. In the Probe 1 setting, the 16 channels of the spectrometer
span the size range from 0.22 to 0.70 μm with a resolution of 0.1 μm; in
Probe 3, the 16 channels of the spectrometer span the entire range of the
instrument (0.09 to 3.0 μm) but with a reduced resolution. The number
concentration in Probe 1 and the DMPS agree well. For all data the posi-
tioning of the peak is remarkably close. Further comparison with other size
PSL particles showed good agreement for the location of the center peak.

Several tests with polydisperse room air filtered through a low efficiency prefilter were performed to compare the two instruments at low particle
concentrations. Figure 3 gives the distribution for one such case. The
region of instrument overlap, from 0.09 to about 0.3 μm, shows very good
agreement. However, other runs showed a larger deviation near the upper
size limit of the DMPS. This deviation at the larger size limit may be due
to impactor cutoff and possible generation of particles by the center rod of
the electrostatic classifier in the DMPS. (Keady 1984)

It was concluded from this work that comparative measurements of aerosol
number and size agree quite well between the DMPS and the LAS-X when aerosol
concentration exceeds 1 cm^{-3}.

Clean Room Measurements

Three clean room facilities were monitored with an instrument array
consisting of a TSI DMPS, a PMS LPC-555, and two PMS LAS-X spectrometers.
Additionally, the TSI CNC was run for one test. The CNC measures particles
from 0.01 to greater than 3 to 5 μm with a sensitivity of less than 1×10^{-5}
particles/cc (Rubow and Liu 1984); however, just concentration (no size
information) is obtained.

The major objective of these tests was to determine the particle size
distribution in the clean rooms. The DMPS was used to measure the size
distribution of particles below 0.1 μm; the PMS instruments were used for
comparisons in the regions of overlap and for determination of the size
distribution of the larger particles.

No significant concentrations of condensation nuclei were found with
the DMPS. Indeed, in the two class 100 facilities the DMPS was unable to
adequately determine the size spectra due to the low number of particles
counted (total CNC count < 0.01 cm^{-3}). Particles below 0.1 μm were detected
at the class 10,000 university facility (total CNC count $\lesssim$ 1 cm^{-3}); however,
even this concentration did not allow a highly accurate determination of the
aerosol spectrum. Figure 4 illustrates this spectrum. Note the variability
in the DMPS readings; this is probably due to the low number of counts.

Figure 4 also shows a comparison between the three PMS instruments.
The overlap is quite good considering the different flow rates of the instruments. The two LAS-X instruments correspond well in the region of 0.15 to
0.60 μm. However, operation of one of the LAS-X instruments at a different
probe range showed a greater deviation.

Figure 5 compares the particle spectra measured by the various optical
counters in one of the class 100 facilities. The error bars on the figure

correspond to one standard deviation for a Poisson distribution. The points not marked with error bars had standard deviations larger than their means and thus the probability for any count from the mean to zero is within one standard deviation.

In conclusion it appears that particle spectra in class 100 clean rooms are difficult to acquire with current instrumentation. An instrument array consisting of a CNC counter, an LPC, and an LAS-X generated the most useful data but could not provide size information below 0.1 μm at aerosol concentrations below about 1 cm^{-3}.

References

Keady, P., F. Quant, and G. Sem (1983) "Differential Mobility Particle Size: A New Instrument for High Resolution Aerosol Size Distribution Measurement Below 1 μm," TSI Quarterly, 9(2) (April-June).

Keady, P. (1984) Development engineer, TSI, Incorporated. Minneapolis, MN, personal communication (May).

Knollenberg, R. (1984) "The Measurement of Particle Sizes Below 0.1 Micrometers," Proceedings of 30th Annual Technical Meeting of the Institute of Environmental Sciences, Orlando, Florida, p. 46 (May 1-3).

Rubow, K., and B. Liu (1984) "Evaluation of Ultra-High Efficiency Membrane Filters," Proceedings of 30th Annual Technical Meeting of the Institute of Environmental Sciences, Orlando, Florida, p. 64 (May 1-3).

Acknowledgment

This research is jointly sponsored by the Microelectronics Center of North Carolina (MCNC) and the Semiconductor Research Corporation (SRC), both located in Research Triangle Park, NC.

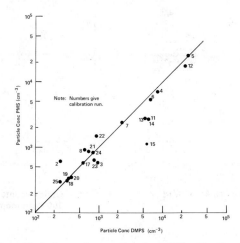

Figure 1. Instrument comparison from PLS calibration runs

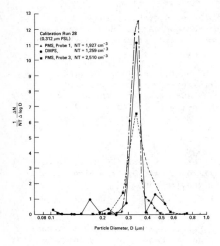

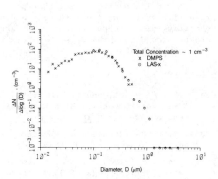

Figure 2. PMS vs. DMPS comparison with PSL aerosol

Figure 3. Prefiltered room air

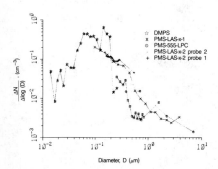

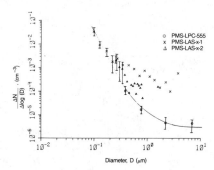

Figure 4. University Lab 4/16–4/18

Figure 5. Optical counter data in a class 100 clean room

Published 1984 by Elsevier Science Publishing Co., Inc.
Aerosols, Liu, Pui, and Fissan, editors

AEROSOL CONTROL IN SEMICONDUCTOR MANUFACTURING

Carlton M. Osburn
Microelectronics Center of North Carolina
P.O. Box 12889
Research Triangle Park, NC 27709
and
Department of Electrical and Computer Engineering
North Carolina State University

EXTENDED ABSTRACT

Trends in Semiconductor Technology

Semiconductor manufacturing technology has made tremendous progress over the last two decades. The number of transistor devices on a chip of silicon has doubled every 1-2 years. Since the cost of manufacturing a chip has been relatively constant, this compounding of integration levels has resulted in dramatically lowered prices. Affordable electronic systems have in turn been the heart of the "high technology" revolution. Consumers of electronics systems have come to expect 20-50% price reductions every year--a dramatic contrast to the inflation expectations of other consumer goods. This year total semiconductor component sales will be over $10 billion, and the value-added by incorporating these components into systems increases this amount by 1-2 orders of magnitude.

These dramatically falling prices exemplify the fierce world-wide competition in the semiconductor industry. To remain a viable competitor, every manufacturer must be able to continue lowering prices every year. Since most of the "fat" has already been taken out of the cost of manufacturing a wafer, the approach has been to increase there number of devices on a wafer by increasing the wafer size and by shrinking the device.

Shrinking device dimensions pose a serious problem in aerosol control. The size of a critical defect, i.e. one that is large enough to cause the yield loss of a chip, is believed to decrease with the scaling factor. The density of particles increases dramatically as the particle size decreases; Figure 1 shows typical data for the MCNC clean room showing that this increase in particle density varies between the reciprocal square and cube of the particle size. Thus a dimensional scaling by a factor of two could result in 4-8 times more critical particulates with that much more yield loss. If the yield loss due to particulates is 25% on a base process, a 1/2 X scaling would not give any yield at all under these conditions. It is no mystery then why improved aerosol control is a fundamental requirement in advancing the state-of-the-art semiconductor technology.

Submicron device technology can be used to illustrate the problem. The dimensional requirements for this technology are given in Table I. Typically ten levels of lithography are required on a chip that is one centimeter on a side. If the wafer is exposed to an airstream at 100 FPM (typical of HEPA outputs) for a 30 minute period at each mask level, then each chip is washed with about 30 cu ft. If the yield objective is 90%, then only 0.1 critical particles can be allowed in this 30 cu. ft. Now the critical defect size for particulates to cause a lithography loss would be about 0.15 μm (1/4 to 1/2 of the linewidth). Figure 1 shows about ten times more 0.15 μm particles than 0.5 μm particles. Bottom line this means that a class 3×10^{-4} room would be required to achieve this yield. An even more stringent requirement comes from the gate oxidation level where a 5 nm particle would cause a failure in a 10 nm oxide. Here a 3×10^{-6} class room would be necessary.

This extreme example illustrates how very few yield-killing particles can be actually tolerated in submicron technology. The assumptions used can be seriously challenged: air flow should be parallel to wafers not perpendicular; not all defects of critical size hit a critical area or even stick to the surface; and recent measurements (1) suggest that the particle density remains relatively con-

674

stant and does not increase as the reciprocal square of size below 0.1 μm. Nevertheless, the allowable concentration of critical particles is orders of magnitude smaller than the 100 particles per cubic foot over 0.5 μm in size that are allowed in a Class 100 clean room. Strictly speaking, there is no official definition of a room cleaner than Class 100, although there has been widespread use of characterizing a clean room by the number of particles over 0.5 μm per cu. ft. Thus people refer to "Class 1" areas as having no more than 1 particle per cu ft. Such areas represent about the best state-of-the-art in semiconductor manufacturing areas. Under such conditions, i.e., when not used, these areas may only have 0.01-0.1 particles/CF. However, measurement of such areas is very difficult since the best particle counters typically only sample 1 CFM. Measurement times of an hour or more may be needed to reduce the statistical fluctuations in the counts.

Control of Aerosol Sources

To achieve the high cleanliness levels required in semiconductor manufacturing, increasing attention is being paid to clean room design. HEPAs having higher efficiency and designed to remove smaller particles are one solution to the problem. Higher air flows and increased attention to achieving laminar flow perpendicular to wafers are another. Ceiling coverage of HEPAs is approaching 100% in state-of-the-art rooms. Raised floors to achieve VLF patterns and ionization systems have also raised a lot of attention. Increased care is going into the control of particulate emmissions from people: antistatic coats, air showers, etc. Many of these later techniques are still undergoing evaluation to determine their effectiveness.

There is widespread agreement, however, that in most cases the clean room itself is not the major cause of aerosol contamination on wafers. Most contamination can be traced back to the processing tools themselves. Deposition processes, for instance, are notorious sources since thick films are ultimately deposited on the wafer fixturing in such systems. Sooner or later these thick films peel or flake particles onto wafers. The turbulence in vacuum systems during pumpdown leads to particulate contamination; new systems are often equipped with anti-turbulence pumping. Wet chemical operations are increasingly using particulate-free chemicals and in addition they recirculate and filter the chemicals during use. Outgassing of plasticizers used in wafer carriers has been a serious source of aerosol contamination on wafers. Gases also have been major sources of contamination in the past. Many conventional facilities use soldered copper pipe. After exposure to moisture over a period of time, oxide scale is released into the lines. Purification systems that are effective in removing chemical impurities in gases rely on high surface area catalysists or collectors and, as a result, often shed copious numbers of particles into the system; conversely, filters which may be efficient in removing particles can add chemical contaminants back into lines.

The strong trend to automated wafer handling provides another serious challenge to aerosol control. A machine is not necessarily cleaner than a person, and horror stories abound on the severe contamination introduced by some new wafer transport system. In fact today many customers place specifications on the maximum number of particles that can be introduced on a wafer in any new tool.

Considerable progress in aerosol control in the semiconductor industry can be expected as people systematically track down the specific generator sources and clean them up. The focus of attention has already shifted from air or liquid borne particulate detection to surface detection on wafers. Fortunately, today we see signs of a good, cooperative effort between semiconductor manufacturers, tool makers and gas/chemical suppliers to reduce wafer contamination levels so that future advances in semiconductor technology will be possible.

References

B.R. Locke, R.P. Donovan, D.S. Ensor, and C.M. Osburn, "Semiconductor Clean Room Particle Spectra", First International Conference of the American Association for Aerosol Research, Sept. (1984).

Table I. Dimensional Requirements for a Half-Micrometer Technology

Lateral Dimensions:
Pattern size	0.5 μm
Pattern Tolerance	0.15 μm
Level-level registration	0.15 μm

Vertical Dimensions:
Gate oxide thickness	10 nm
Field Oxide thickness	200 nm
Film Thicknesses	0.25-0.5 μm
Junction Depth	0.05-0.15 μm

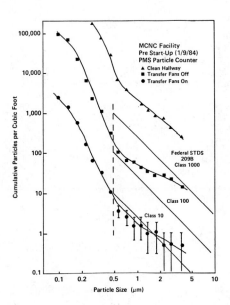

Figure 1. Particulate Distributions in the MCNC Clean Room

CHAPTER 16.

INDUSTRIAL FINE PARTICLE PRODUCTION

Published 1984 by Elsevier Science Publishing Co., Inc.

Aerosols, Liu, Pui, and Fissan, editors

SILICON PRODUCTION IN AN AEROSOL REACTOR

J.J.Wu, M.K.Alam[*], B.E.Johnson, and R.C.Flagan
California Institute of Technology 138-78
Pasadena, CA 91125
U. S. A.

EXTENDED ABSTRACT

Introduction

The formation of silicon powder by thermal decomposition of silane gas in an aerosol reactor is a promising method for the production of high purity bulk silicon. In this process, silicon particles are generated by homogeneous nucleation of the products of gas phase pyrolysis reactions. The particles are then grown by condensation of reaction products and by gas-solid reactions. In order to facilitate the separation of the particles from the gas stream and the subsequent processing of the collected powder, particles should be grown to diameters larger than 10 microns and, preferably, larger than 50 microns. The number concentration of seed particles must be reduced substantially below that resulting from homogeneous nucleation if such large particles are to be grown by vapor deposition. A two-stage aerosol reactor has been developed in which seed particles generated by homogeneous nucleation in the first stage are grown by vapor deposition in a second separate stage.(Alam, 1984) Growth of submicron seed particles to the desired size requires that the rate of reaction in the primary growth stage not exceed the ability of the seed particles to scavenge vapors and prevent secondary nucleation.

Theory

Vapor deposition on the surface of a growing particle depletes a region surrounding the particle of the vapor species. The reduced vapor concentration in turn reduces the nucleation rate in the vicinity of the particle. The net reduction in the rate of new particle formation due to a single growing particle may be evaluated by integrating the rate of particle formation per unit volume from the particle surface to infinity. It is convenient to describe the reduction in the net nucleation rate using a clearance volume which is defined such that, if the nucleation rate inside that volume is taken to be zero and outside that volume to be equal to the free stream value, J , the total nucleation rate equals that determined by exact integration. The dimensionless clearance volume is proportional to the Knudsen number in the free molecular limit and asymptotically approaches a constant value in the continuum limit. It also increases with decreasing partial pressure (Alam and Flagan, 1984). An asymptotic analysis of the dimensionless clearance volume has been carried out to simplify the evaluation of the conditions under which nucleation can be controlled.

[*]Present address: Department of Mechanical Engineering
Ohio University
Athens, Ohio 45701

The collective influence of aerosol particles on the global rate of nucleation can be estimated by computing the fraction of the total volume of gas which is contained in clearance volumes. Assuming that the domains of influence of individual particles do not overlap, the total clearance volume fraction becomes

$$\Omega = \int_0^\infty \frac{4\pi}{3} \rho^3 \, a^3 \, n(a,t) \, da$$

where ρ^3 is the dimensionless clearance volume, a is the particle radius, and $n(a,t)=dN/da$ is the particle size distribution function at time t. This condition has been used to determine operating conditions of the aerosol reactor under which particle growth could be achieved without nucleation.

Experimental

Figure 1 illustrates the two-stage aerosol reactor which has been developed for silicon processing. Seed particles are produced in the first stage by nucleation of the products of homogeneous gas phase pyrolysis of silane. The seed aerosol is diluted with a large excess of silane, thoroughly mixed, and then grown in a second stage. The temperature profile in the second stage has been designed using the aforementioned theory such that vapor deposition on the seed particles depletes the vapor and prevents secondary nucleation. Initially the reaction rate must be severely limited due to the small size and clearance volumes of the seed particles. As the particles grow, their clearance volumes increase so higher reaction rates can be tolerated. The reaction rate is, therefore, accelerated by increasing the reactor wall temperature.

Results

Early experiments were carried out using a dilute mixture of silane in nitrogen in the second reactor stage. A constant, slow temperature increase rate, 330 K/sec from 770 K to 1100 K, achieves complete reaction of the silane within a 1 second residence time without inducing secondary nucleation at the design seed aerosol concentration of 10^{11} particles/Nm^3. 0.2 micron radius seed particles were grown to a mass median radius of 3.1 microns when a mixture of 1 percent silane in nitrogen was reacted. 4.5 micron particles were produced when 2 percent silane was used. The measured particle size distribution for the latter case is shown in Fig. 2. The collected aerosol accounted for 73 percent of the input silicon in the former case and 68 percent in the latter case. X-ray diffraction analysis of the powder indicates that the product is amorphous silicon.

Reactor design studies indicate that, by increasing the residence time from 1 to 10 seconds, pure silane can be used to grow particles up to 25 microns radius. A larger reactor has been constructed to examine aerosol reactor operation under these more severe conditions. Results for a range of silane and seed concentrations will be discussed.

681

References

Alam, M.K. (1984) <u>Nucleation and Condensational Growth of Aerosols: Application to Silicon Production</u>, Ph.D. Thesis, California Institute of Technology.

Alam, M.K., and Flagan, R.C. (1984) Simultaneous Nucleation and Aerosol Growth, <u>J. Colloid Interface Sci. 97</u>, 232-245.

Acknowledgements

This work was supported by the Jet Propulsion Laboratory of the California Institute of Technology and the Department of Energy under the Flat Plate Solar Array Project.

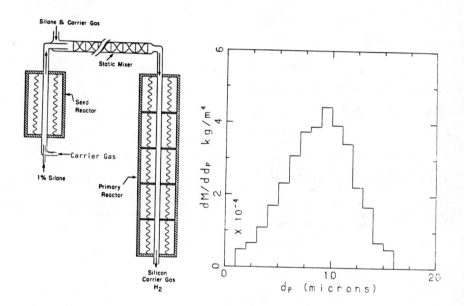

Figure 1. Aerosol reactor

Figure 2. Mass distribution of aerosol after reacting 2% silane in presence of seed aerosol (total seed concentration = $1.02 \times 10^{11}/m^3$).

Published 1984 by Elsevier Science Publishing Co., Inc.
Aerosols, Liu, Pui, and Fissan, editors

THE EFFECT OF IRON ON SOOT GROWTH RATES IN PREMIXED FLAT ETHYLENE FLAMES

K.E. Ritrievi, J.P. Longwell, and A.F. Sarofim
Department of Chemical Engineering
Massachusetts Institute of Technology
Cambridge, Massachusetts 02139

Submicron inorganic particulates formed from vaporized mineral matter from coal have been observed in coal combustion and gasification systems. This organic aerosol may have a major effect on the resulting soot aerosol formed. The interactions between an iron oxide and the carbonaceous aerosols formed in a premixed flat (one dimensional) ethylene flame doped with dicyclopentadienyl iron (Ferrocene) were studied. Inorganic particulates of iron oxide are of interest since iron is a significant component of submicron particulates generated during pulverized coal combustion. The physical characteristics of the aerosols - particle size, number density, and loading - were examined through the use of laser light scattering and extinction measurements. Comparisons will be made among the diameters measured by these optical techniques, the aerodynamic diameters provided by electrical size analysis and the primary particle sizes from transmission electron micrographs of particles collected with an electrostatic precipitator.

Published 1984 by Elsevier Science Publishing Co., Inc.
Aerosols, Liu, Pui, and Fissan, editors

COMPUTER MODELING OF A SOLAR GRADE SILICON PRODUCTION PROCESS

C.H. Berman
AeroChem Research Laboratories, Inc.
P.O. Box 12
Princeton, NJ 08542
U.S.A.

EXTENDED ABSTRACT

Introduction

A computer model is used to indicate how silicon collection can be optimized in turbulent flow tube reactors similar to that tested by Westinghouse (Fey, 1978). Here Si is produced via a chemical reaction between $SiCl_4$ and Na, carried out in an arc-heated hydrogen stream as shown schematically in Figure 1. The present paper is based on the work of Gould and Srivastava (1979) carried out at AeroChem.

Computer Model

This paper focuses on CHEMPART, an extension of the well-known LAPP (Mikatarian, Kau, and Pergament, 1972) code for rocket plume exhaust studies, which solves the turbulent jet mixing problem with detailed gas-phase chemistry and particle formation while treating mass, momentum, and energy flux to the walls.

The gas phase is computed using the principles of conservation of mass, species, momentum, and energy. Adjustments to the resultant equations are made to treat interactions with particles. The governing set of parabolic partial differential equations is rewritten in finite difference form and then solved using standard forward-marching techniques in the axial direction with stream function as the radial coordinate. The program provides five possible reaction types involving atomic, simple molecular, and polymeric association and decomposition, seven different temperature dependencies for rate coefficients, and six eddy viscosity models. Thermodynamic properties data (specific heat, Gibbs free energy, and enthalpy) for each species are input to the program in tabular form as a function of temperature.

The code computes the flux of condensible species, particulate matter, momentum, and energy across the outermost stream tube to obtain rates of pressure drop and mass and energy deposition on the reactor walls. For particulate transport, thermophoretic diffusion becomes dominant close to the wall. Silicon vapor deposition on the walls is treated via analogy to a mixing length heat transfer model by assuming that the Lewis number is unity.

Very small or large silicon droplet deposition on the reactor walls is modeled respectively by Brownian and turbulent diffusion. However, the often ignored transport mechanism of Soret diffusion (or thermophoresis) due to large temperature gradients controls the deposition of silicon droplets in the diameter range 0.01 μm to 1.0 μm, within a viscous sublayer region close to the wall.

The particle nucleation model postulates a limited series of simple gas phase addition reactions resulting in a "critical nucleus" size above which growth to particle species with bulk properties is rapid (Bauer and Frurip, 1977). The critical size is determined by finding that n-mer resulting from a series of Si n-mer reactions for which the rates of reactions creating it are most nearly equal to the rates destroying it. This model finds a critical size larger than the liquid drop theory (e.g., Abraham, 1974), which predicts critical radii the size of one Si atom. The model is thus physically more realistic than the liquid drop theory for cases where very large supersaturations are encountered.

The particle growth is described by means of Brownian, turbulence-enhanced and particle-particle impaction agglomeration models. The effects of condensation from the gas phase and surface evaporation on particle size are modeled as a function of Sherwood number over a wide range of Knudsen numbers. The latent heat liberated or absorbed by the phase change is assumed to be transformed immediately to the gas while the particle temperature remains constant.

Small particles with Knudsen number greater than 2 are considered to follow the gas flow and adjust instantaneously to its temperature. Such particles are treated as gases themselves and contribute to the total gas density, enthalpy, heat capacity, etc. Radiation from the particles is assumed to cool the gas. The model treats the momentum and energy transfer (including radiation) between larger particles ($Kn \leq 2$) and the gas flow.

CHEMPART is time consuming to run and so the code is limited to a maximum of 30 gridpoints with points near the wall being much closer together than in the core to better define the transport processes near the wall. Except for the first few steps, the axial step size is determined by the computed mixing parameters.

To make more efficient computations downstream of the initial reaction region, a special code called MPDEU, which handles mass transfer to the walls, is used with a modified version of the well-known GENMIX program (Spalding, 1977). GENMIX first solves the gas-phase momentum and energy equations, and these solutions are then stored and read by MPDEU, which treats the mass transfer rate of silicon droplets by convection, Fick and eddy diffusion, and Soret effects.

Results

There is no clear-cut indication from this work as to what constitutes an optimum silicon particle size distribution. Experimental evidence at AeroChem and Westinghouse tends to indicate that the actual particle size range is 0.1 to 1 μm. This range is intermediate between the optimum sizes for impact collection of larger particles and Brownian deposition of much smaller particles. Thus, thermophoresis is the dominant particle transport/deposition mechanism in the process.

For the case of the turbulent flow reactor (8 m long, 0.15 m i.d.), a flow velocity of < 50 m s^{-1} is required to obtain more than 50% deposition of Si particles on the walls when the temperature of the gas is above the dew point of Si.

Large input enthalpies, supplied by arc-heated hydrogen (H atoms) are needed to suppress the formation of large Si(ℓ) droplets. The large droplets' radiation of heat to the reactor wall decreases the gas phase temperature gradient needed for effective collection by thermophoresis. The majority of the particle sizes computed from the nucleation/growth models fell into the range of 0.1 to 1.0 μm, consistent with available experimental evidence. One must also be careful not to increase the flow of arc-heated hydrogen (i.e., the input enthalpy supply) to the point where collection efficiency is reduced simply by reducing reactor residence time.

This example shows the utility of this computer model in optimizing the performance of such complex devices as the turbulent flow silicon reactor.

References

Abraham, F.F. (1974) Homogeneous Nucleation Theory, Academic Press, New York.

Bauer, S.H. and Frurip, D.J. (1977) "Homogeneous Nucleation in Metal Vapors. A Self-Consistent Kinetic Model," J. Phys. Chem. 81:1015.

Fey, M.B. (1978) "Development of a Process for High Capacity Arc Heater Production of Silicon for Solar Arrays," Westinghouse Electric Corp., Quarterly Report, DOE/JPL 954589-78/6.

Gould, R.K. and Srivastava, R. (1979) "Development of a Model and Computer Code to Describe Solar Grade Silicon Production Processes," Final Report, AeroChem TP-392, DOE/JPL 954862-79/8.

Mikatarian, R.R., Kau, C.J., and Pergament, H.S. (1972) "A Fast Computer Program for Nonequilibrium Rocket Plume Predictions," Final Report, AeroChem TP-282, AFRPL-TR-72-94, NTIS AD 751 984.

Spalding, D.B. (1977) GENMIX: A General Computer Program for Two-Dimensional Parabolic Phenomena, HMT Series, Vol. 1, Pergamon Press, New York.

686

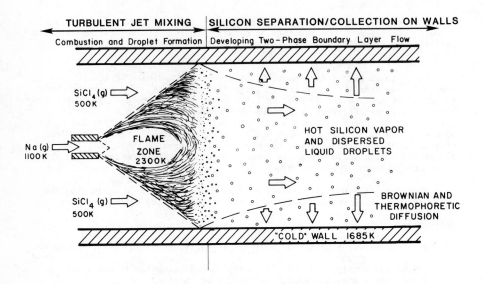

FIGURE 1   SCHEMATIC OF THE SILICON REACTOR

Published 1984 by Elsevier Science Publishing Co., Inc.
Aerosols, Liu, Pui, and Fissan, editors

PLASMA GENERATED METALLIC AEROSOL

J. W. George, T. N. Meyer, C. B. Wolf
Westinghouse Electric Corp.
Applied Products Division
400 Penn Center Blvd.
Pittsburgh, Pa. 15235

EXTENDED ABSTRACT

Introduction

A large scale experiment in Sweden is underway to study the behavior of aerosols which could be generated in a nuclear reactor core meltdown. Fume deposition sites and rates within the reactor containment building and the potential for escape of fume to the atmosphere is to be assessed in a full scale experiment. The project is directed by Studsvik Energiteknik AB (Sweden) in cooperation with the sponsors of the Marviken Aerosol Transport Test Project. To satisfy the aerosol simulation, it was considered necessary to vaporize 450 kg of metal in a 5000 second period. Further, the aerosol must approach a concentration of 1.96 kg/Nm^3. To vaporize metals consisting of iron, copper*, and manganese at the above flow rates, various plasma methods were investigated. The method described herein is the only method known to satisfy concentration and generation rate requirements. Using iron only, concentrations of 0.6 to 0.850 kg/Nm^3 were obtained in early feasibility tests. (Ciliberti, et al) These results along with theoretical calculations were reported.[1]

A plasma system was designed, built and tested to vaporize metal and subsequently form a highly concentrated aerosol. Metals consisting of iron, copper, manganese were vaporized at a rate of 6.7 kg/min via a plasma heated nitrogen gas stream. The vaporized metal (mixture of iron, copper and manganese) condensed to produce a fume of aerosol having a density of 1.3 kg/Nm^3 (kg per normal cubic meter). To accomplish this vaporization, two plasma torches operated at a power level of 2000 kilowatts. With these results, the large aerosol generator described below was built and tested.

Aerosol Generator

The vaporization system is shown in Figure 1. Three 1500 kW Westinghouse plasma torches are connected to the vaporization chamber (V/C). The plasma torches electrically heat a nitrogen gas stream to temperatures of 6000 to 7000°C. This gas exits the torch and enters a nozzle connecting the torch to the V/C. The nozzle is modified to permit

* Copper is to represent silver in the nuclear reactor meltdown simulation.

the metal powder transported in an argon gas stream to enter the plasma heated gas stream. The powder is then heated and ultimately vaporized in the V/C. A plasma torch, feed nozzle, and vaporization chamber are shown in Figure 2. As the metal vapor cools and exits the V/C, condensation occurs to form the desired metallic aerosol fume.

The water-cooled elbow downstream of the V/C provides area for a significant quantity of aerosol to be collected for analysis. Just prior to the venturi scrubber, samples from the stream can be taken through an access port.

The mixture of material to be vaporized contained copper, manganese, iron and zirconium dioxide in proportions (see Table 1) to thermally represent the aerosol elements expected in a reactor meltdown.

Test Results

A number of shakedown tests were conducted and the results of two full scale operational tests are reported in Table 1. For the two data sets, concentrations of 0.8 and 1.300 kg/Nm^3 were achieved. Relative to the nuclear composition containing silver in place of copper, these are thermally equivalent to 1.08 and 1.8 kg/Nm^3, respectively.

Since all analyses of material deposited on the elbow wall showed particles of submicron size (0.05 micron) to as large as two micron, it was concluded that all material leaving the V/C must have been vaporized at some point. The distribution of particles need not be the same but essentially no large particles were found. Since no large particles were found in the elbow or sampling area, the size distribution to determine fraction of vaporization is unnecessary.

Conclusion

The system described represents the only known system capable of providing the concentration and vaporization rate levels indicated for such high temperature elements. When the minimum energy required for vaporization is compared with the electrical input to the torches the overall efficiency is found to be quite high at 33%.

References:

1. Ciliberti, D.F., Meyer, T.N., Wolf, C.B., Taylor, R.F. (1983) "Metal Vaporization Using a Plasma Torch", 6th International Symposium on Plasma Chemistry, McGill University, Montreal Quebec, July, 1983

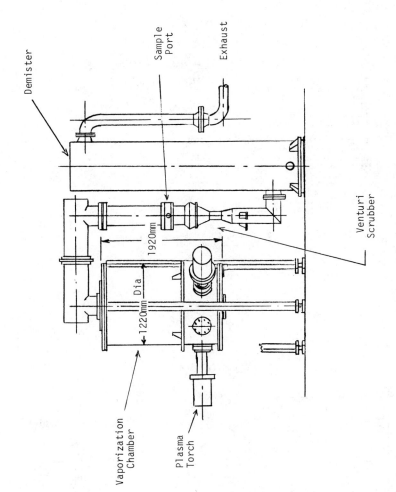

Figure 1. Aerosol Generation System

TABLE 1   AEROSOL GENERATION TESTS

	Low Concentration	High Concentration	
Plasma Torch A Power	1172	1188	kW
Plasma Torch A Gas Flow (N$_2$)	155	153	Nm^3/H
Plasma Torch A Efficiency	83	83	%
Plasma Torch B Power	788	840	kW
Plasma Torch B Gas Flow (N$_2$)	95	94	Nm^3/H
Plasma Torch B Efficiency	73	70	%
Graphite Liner Wall Temperature	2280	2155°C	
Vaporization Chamber and Feed Nozzle Thermal Loss	703	920	kW
*Feed Rate Thru Nozzle	4.0	6.7	kg/min
Feed Transport Gas (Argon)	22.0	25.0	Nm^3/H
Total Powder Feed	100	77.9	kg
Material Remaining in Vaporization Chamber	5	3	kg
Calculated Aerosol Concentration	0.8	1.3	kg/Nm^3

a) Feed Composition: 38% Copper, 41% Manganese, 14% Iron,
                       7% Zirconium Dioxide
b) Size: -325 Mesh

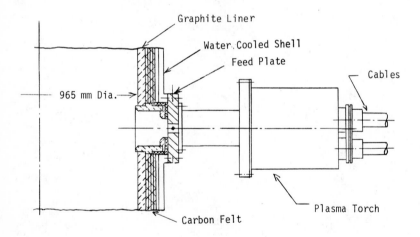

Figure 2. Plasma Torch, Feed Nozzle, Vaporization Chamber.

Published 1984 by Elsevier Science Publishing Co., Inc.
Aerosols, Liu, Pui, and Fissan, editors

691

THERMODYNAMICS OF THE SILICON-CHLORINE-HYDROGEN SYSTEM:
APPLICATIONS TO POLYSILICON PRODUCTION BY AEROSOL GROWTH

Kenneth D. Allen and Herbert H. Sawin
Department of Chemical Engineering
Massachusetts Institute of Technology, Cambridge, MA 02139

EXTENDED ABSTRACT

MOTIVATION

Polycrystalline silicon (polysilicon) is the basic source material from which many photovoltaic and microelectronic devices are fabricated. The purity requirements are stringent, particularly in the microelectronic industry, where ppb contaminants can result in device failure. Polysilicon production is a multistep process in which the raw material (quartz) is reduced, volatilized as a mixture of chlorosilanes, purified, and deposited. The Siemens process, in which polysilicon is deposited by chemical vapor deposition (CVD), reacts trichlorosilane ($SiHCl_3$) and hydrogen over a hot (1000 °C) silicon wire. Polysilicon produced by this process is expensive due to low conversion and thermal inefficiencies [Yaws et al., 1981].

Recently, several alternatives to the Siemens process have been proposed. In each of these processes, the homogeneous nucleation of silicon from the gas is important. The fluidized bed reactor (FBR) process [McHale, 1981] utilizes small (.1 μm) silicon particles to seed the fluidized bed. Silane (SiH_4) and hydrogen are passed through the bed, depositing polysilicon on the particles. Free-space reactors (FSR) are also being studied [Lay and Iya, 1981; Alam and Flagan, 1983]. In the FSR, seed particles are injected into the reactor (at a lower concentration than in the FBR) and grow by CVD as they proceed through the reactor. Particle contact with the wall (and hence contamination) is minimal. In both cases, the large surface area and reduced heat losses lead to lower costs.

It was the objective of this work to examine the thermodynamic properties of the Si-H and Si-Cl-H systems relevant to FSR operation to determine the feasibility of chlorosilane deposition. The equilibrium species distribution, the chemical potential driving force for silicon deposition ($\Delta\mu$), and the equilibrium deposition efficiency (ξ) were calculated over a wide range of temperature and composition. The cited processes utilize silane as a reactant due to higher deposition rates and deposition efficiency. However, the use of silane has several disadvantages. It is considerably more expensive than $SiHCl_3$, which is produced as an intermediate in the existing process. It is unstable above 600 °C and is pyrophoric. Homogeneous nucleation of silane to form a fine silicon powder is a major concern, and has been observed even at low silane concentrations [Eversteijn, 1971]. The resulting fines are highly susceptible to contamination due to their high surface area. The addition of chlorine to the system has been reported to suppress homogeneous nucleation [Murthy et al., 1975].

THEORY

The rate of homogeneous nucleation is determined by the chemical potential driving force of the system. In physical condensation, (e.g. water droplets from water vapor), $\Delta\mu$ can be written in terms of the supersaturation S:

$$\Delta\mu_i = RT \cdot \ln[S] = RT \cdot \ln[p_i/p_i^{\,0}] \qquad [1]$$

where p_i is the partial pressure of species i with no solid present and $p_i^{\,0}$ is the vapor pressure of species i. In a chemically reacting system, equation [1] does not hold, and the rigorous definition [Modell and Reid, 1983] must be used:

$$\Delta\mu_{Si} = \mu_{Si(s)} - \mu_{Si(g)} \equiv \left[\frac{\partial G}{\partial N_{Si}} \right]_{T,P,N_j [Si]} \qquad [2]$$

where $\underline{G}$ is the total Gibbs energy of the system. This distinction is not well recognized in the nucleation literature. The values of $\Delta\mu$ obtained from equations [1] and [2] differ by a factor of two for the Si-Cl-H system due to the formation of chlorinated silicon species. The distinction is an important one since in classical nucleation theory the rate depends on exp{$-\Delta\mu$}.

A theory of homogeneous nucleation in the FBR/FSR system must account for the effect of the CVD reaction [Katz and Donohue, 1982] as well as the influence of the growing aerosol on the formation of new particles [Pesthy et al., 1981]. Both processes tend to suppress homogeneous nucleation by scavenging silicon from the gas phase.

RESULTS AND DISCUSSION

Fifteen species were accounted for in the equilibrium calculations. Examination of the species distribution results yields that the system moves to form light chlorinated silicon species (e.g. $SiCl_2$) and hydrogen as temperature is increased. Heavier and more hydrogenated silicon species are favored at lower temperatures and high atomic Si/Cl ratios.

The calculated values of $\Delta\mu$ are shown in Figures 1-4. The driving force increases with temperature for a given composition. This implies that the rate of homogeneous nucleation will be very sensitive to temperature. For large Si/Cl ratios (≥ 0.5), $\Delta\mu$ is fairly insensitive to Cl/H ratio (mole fraction silicon-bearing gas in H_2). The driving force drops sharply as the atomic Si/Cl is decreased (from SiH_4 to SiH_2Cl_2 to $SiHCl_3$). Consequently, the addition of chlorine to the Si-H system greatly reduces the nucleation rate by making the deposition reaction reversible. The suppression of the nucleation rate relative to that in the Si-H system offers tremendous advantages in polysilicon processing. Higher silicon loading can be utilized in the Si-Cl-H system without inducing homogeneous nucleation. Calculations based on the theory of Alam [1983] show that a factor of 10^5 fewer seed particles are required in the Si–Cl–H system to prevent nucleation. Consequently, large polysilicon particles can be obtained, resulting in a higher quality, more easily processed product.

The deposition efficiencies are shown in Figures 5–6. As expected, ξ decreases with decreasing Si/Cl ratio (SiH$_4$ deposition is essentially 100% efficient). Surprisingly, ξ increases with temperature. The most important result is that ξ is essentially independent of Cl/H ratio above 20% silicon-bearing gas in H$_2$. Previous studies have not included the high reactant concentration range. This previously unrecognized fact implies that polysilicon throughput in a FSR monotonically increases with the reactant concentration in the feed, without affecting the efficiency of the process. Operation at high reactant concentrations offers significant reduction in capital and operating costs.

A priori prediction of homogeneous nucleation rates is not presently possible, even for simple physical condensation systems. Consequently, critical conditions for nucleation (rates if possible) must be determined experimentally for the Si–Cl–H system using techniques such as pulsed-laser heating [Haggerty and Cannon, 1981]. Such experiments would also provide vital data on polysilicon growth kinetics at high reactant concentrations.

REFERENCES

M.K. Alam, PhD Thesis, California Institute of Technology, 1983

M.K. Alam and R.C. Flagan, Aerosol Science and Technology 2 (1983), pp. 27

F.C. Eversteijn, Phillips Research Reports 26 (1971), pp. 134

J.S. Haggerty and W.R. Cannon, MIT Energy Lab Report MIT–EL–81–003, 1981

J.L. Katz and M.D. Donohue, J. Colloid and Interface Sci. 82 (1982), pp. 267

J.R. Lay and S.K. Iya, 15th IEEE Photovoltaics Conf., 1981, pp. 565

E.J. McHale, JPL Invention Report SC–917/NPO–14912, 1981

M. Modell and R.C. Reid, "Thermodynamics and its Applications", 2nd edition, Prentice–Hall, Englewood Cliffs, NJ, 1983

T.U.M.S. Murthy, N. Miyamoto, M. Shimbo, and J. Nishizawa, J. Crystal Growth 33 (1975), pp. 1

A.J. Pesthy, R.C. Flagan, and J.H. Seinfeld, J. Colloid and Interface Sci. 82 (1981), pp.465

C.L. Yaws, K–Y Li, J.R. Hopper, Y.S. Fang, and K.C. Hansen, Final Report, NASA–CR–164009, N81–19570, 1981

694

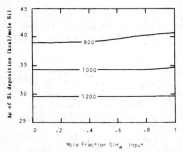

Figure 1: Chemical Potential Driving Force for Nucleation vs.
Mole Fraction SiH_4 in H_2 at 800, 1000, and 1200 °C.

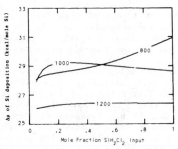

Figure 2: Chemical Potential Driving Force for Nucleation vs.
Mole Fraction SiH_2Cl_2 in H_2 at 800, 1000, and 1200 °C.

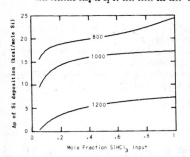

Figure 3: Chemical Potential Driving Force for Nucleation vs.
Mole Fraction $SiHCl_3$ in H_2 at 800, 1000, and 1200 °C.

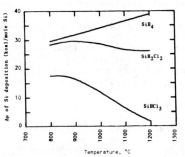

Figure 4: Chemical Potential Driving Force for Nucleation vs.
Temperature for 0.10 Mole Fraction SiH_4, SiH_2Cl_2, and
$SiHCl_3$ in H_2.

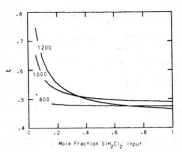

Figure 5: Equilibrium Deposition Efficiency vs. Mole Fraction
SiH_2Cl_2 in H_2 at 800, 1000, and 1200 °C.

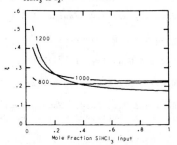

Figure 6: Equilibrium Deposition Efficiency vs. Mole Fraction
$SiHCl_3$ in H_2 at 800, 1000, and 1200 °C.

Published 1984 by Elsevier Science Publishing Co., Inc.

Aerosols, Liu, Pui, and Fissan, editors

FINE PARTICLES FORMED IN A SILANE - FLUIDIZED BED REACTOR

G. Hsu, R. Lutwack, and N. Rohatgi
Jet Propulsion Laboratory
Pasadena, CA
U. S. A.

EXTENDED ABSTRACT

Silane(SiH_4)-fluidized bed technology is being developed for a low cost process for producing semiconductor grade Si for terrestrial photovoltaic solar cells. The advantages of SiH_4 for the preparation of very pure Si are: (1) SiH_4 is readily purified to electronic grade or better levels; (2) the thermal decomposition to Si occurs at comparatively low temperatures; and (3) the decomposition rection gives an essentially 100% conversion yield to Si in an irreversible reaction with H_2 as an easily handled non-corrosive by-product. The advantages of using a fluidized bed reactor are the high deposition rate and low energy-use characteristics. A disadvantage of using a thermal decomposition reaction to convert SiH_4 to Si is the tendency for a homogeneous gas phase reaction and the formation of fine particles.

The research with the 6" I.D. reactor at the Jet Propulsion Laboratory has been directed to studying deposition on seed particles in the bed and the concurrent problem of the elutriation of fine particles as the range of SiH_4 concentrations in the experiments was extended from 10% by volume in H_2 to 100% SiH_4. The operating conditions of the reactor have been controlled to maintain a bubbling bed, to obtain dense deposition, to minimize elutriation, and to prevent plugging of the distributor. The extent of the elutriation of fine particles under varying experimental conditions was determined by measurements of the amounts collected by filters of 0.1µm porosity in the exit gas and calculations of the mass balance. The primary variables were the SiH_4 concentration, the temperature, the bed height, and the gas velocity.

The lower limit of the temperature range was set by considerations of the quality of the deposition on the growing seed particles in the bed. At 550°C the deposit is loosely adhering and becomes increasingly dense above 600°C; at 700°C it is very coherent and dense. Since very high levels of fine particles were found at 800°C, the reaction temperature zone of the fluidized bed was kept between 650° and 700°C. Although the amounts of fines collected by the filters increased with temperature, comparisons of the experimental data with values calculated using specific rate constants indicated that the predominant amounts of fines were incorporated on the seed particles in the fluidized bed. A scavenging mechanism can be invoked to account for this result.

The capability of operating at high SiH_4 concentrations in a fluid-ized bed without losing unacceptable quantities of Si as elutriated fine particles was demonstrated. The results were contrary to what would be expected if a simple comparison is made with previous observations using other reactors. Although the fines collected in the filters increased by ten fold to ca 11% of the total Si over the concentration range of 10 to 80% by volume, the fluidized bed enabled the incorporation of nearly all of the fine particles formed in the reaction zone of the bed by SiH_4 decomposition and subsequent nucleation, condensation, and coagu-lation. Recent experiments indicate further reductions of elutriated fines are obtained by changes in the reactor operating conditions.

The capability for the incorporation of elutriable fine particles on the seed particles is also dependent on the residence time of these particles in the bed. Residence time in turn is directly affected by the bed height and the gas velocity. There are trade-offs in optimizing the velocity, since the effect on the bubbling bed needs to be considered also. For example, the lower limit is set by the requirement to keep the velocity high enough to prevent exceedingly high densities of fine particles which lead to agglomeration and then quickly to reactor shutdown.

The distributor plate through which the gas flows forming the bubbles to fluidize the bed is an additional factor affecting elutriation. It would be expected that large bubbles not only favor the formation of fine particles but also inhibit fine particle-seed interactions and the consequent incorporation of the fine particles on the seed surfaces. Conversely, small bubbles allow much greater interaction between the fine particles and the seed. This simple modeling scheme was supported by the data obtained for various distributor designs.

The problem of fine particle elutriation will continue to be studied in our program to characterize and develop the conditions for steady-state operation of a fluidized bed reactor for the conversion of silane to silicon.

Published 1984 by Elsevier Science Publishing Co., Inc.

Aerosols, Liu, Pui, and Fissan, editors

NUCLEATION AND GROWTH OF SILICON POWDERS FROM
LASER HEATED GAS PHASE REACTANTS

Robert A. Marra
Alumina, Chemicals and Ceramics Division
Alcoa Technical Center
Alcoa Center, PA 15069

John S. Haggerty and John H. Flint
Department of Materials Science and Engineering
Massachusetts Institute of Technology
Cambridge, MA 02139

EXTENDED ABSTRACT

The need for ceramic bodies with improved properties has suggested that uniform diameter constituent powders should be substituted for the traditionally used powders with broad particle size distributions. Achieving the desired product reliability requires highly controlled powder purity, stoichiometry, crystallinity, agglomeration, shape, and size. Since comminution will not produce submicron powders exhibiting all of these characteristics, they must be synthesized in a manner that achieves these properties simultaneously. In turn, this requires that the nucleation growth and agglomeration processes be understood and controlled.

We have investigated laser-driven gas phase synthesis because this process achieves the required attributes. In this process, an optical energy source, a CO_2 laser, is used to initiate and sustain a vapor phase chemical reaction. The gaseous molecules absorb the laser energy, which is thermalized through molecular collisions. The establishment of high supersaturation levels results in small, equiaxed particles because nucleation rates are high and growth is isotropic. The reaction vessel walls remain cold, avoiding contamination and heterogeneous nucleation sites. The reaction zone is well defined by the intersection of the laser beam and reactant gas stream. A wide range of uniform and precise heating rates with controlled maximum temperatures can be achieved by varying the laser intensity as well as the gas mixture, pressure, and flow rates. Agglomeration can be controlled with dilution and minimizing exposure to high temperatures.

This work describes the nucleation and growth mechanisms of silicon powders synthesized from laser-heated silane gas. The synthesis of Si_3N_4 and SiC from silane-ammonia and silane-ethylene gas mixtures, respectively, has also been studied.

The brightness temperature of the particulate cloud was measured with a micro-optical pyrometer. The emissivity was determined from the scattering and extinction of HeNe laser light incident on the particulate cloud. The scattering and transmission measurement also allowed the determination of the size and number density of the particles as a function of position in the reaction zone. With heating rates between 10^4-10^8 °C/s, silicon particles nucleate at rates in excess of 10^{14}/$cm^3 \cdot$s. These rates are in reasonable agreement with the Volmer-Weber theory. The nucleation rate is controlled primarily by the concentration of silicon atoms in the vapor phase; the high level of supersaturation results in almost a negligible thermodynamic barrier. Nucleation ceased after the reaction proceeded only a few percent. The cessation is a result of the depletion of silicon vapor reducing the supersaturation as well as the kinetic favoring of particle growth by the accretion of vapor molecules over further nucleation. The growth is controlled by the vapor phase transport of the reactants to the particle surface. The nature of the growth process results in a narrow distribution of particle sizes. The growth process is terminated by the depletion of reactants; the final particle size is established by the initial SiH_4 concentration and the number of nuclei formed.

The number density measurements suggested that the particles form as amorphous silicon and then crystallize in the hotter regions of the reaction zone. This is consistent with the final particle microstructure and possible particle formation mechanisms. The particles do not agglomerate within the observed region of the reaction zone.

CHAPTER 17.

AEROSOL GENERATION

Published 1984 by Elsevier Science Publishing Co., Inc.
Aerosols, Liu, Pui, and Fissan, editors

701

A DUST GENERATOR FOR CHRONIC INHALATION TOXICITY STUDIES

I. Tanaka and T. Akiyama
Department of Environmental Health Engineering
School of Medicine
University of Occupational and Environmental Health, Japan
Kitakyushu, 807
Japan

EXTENDED ABSTRACT

Introduction

As a result of improved dust control in work and atmospheric environments, investigation of chronic health effects of inhaled dust have become much more important than the acute disease. In animal exposure experiments, therefore, the dust concentration and size distribution in the exposure chamber must be kept constant over a long period.

This paper describes a fluidized bed dust generator (Tanaka and Akiyama, 1984) with continuous screw feeder and overflow pipe, developed for long-term chronic inhalation toxicity studies.

Experimental

Apparatus

The dust exposure system is shown in Fig.1. The dust generator consists of a screw feeder (8) supplying a fluidized bed (11) which has an overflow pipe (13). Small glass beads of dia. 210-297 μm and density 2.5 g/cm^3 are used as the fluidizing particles. Coal fly ash is used as dust. The fly ash is dust for industrial testing, No.10 in JIS-Z8901 and is obtained from the Association of Powder Process Industry and Engineering,Japan. The fluidized bed is contained in an acrylic resin column 67 mm dia. $\times$ 60cm high, supported on a distributor (7) comprising a perforated plate (hole dia. 1 mm,pitch 4 mm)

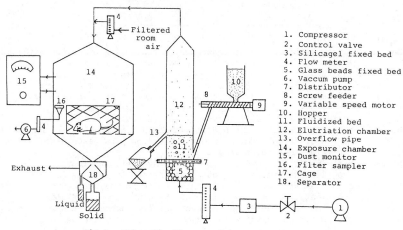

1. Compressor
2. Control valve
3. Silicagel fixed bed
4. Flow meter
5. Glass beads fixed bed
6. Vaccum pump
7. Distributor
8. Screw feeder
9. Variable speed motor
10. Hopper
11. Fluidized bed
12. Elutriation chamber
13. Overflow pipe
14. Exposure chamber
15. Dust monitor
16. Filter sampler
17. Cage
18. Separator

Fig.1   Schematic diagram of dust exposure system.

on a 250 mesh bronze screen. Air from the compressor (1) is supplied to a bead-filled chamber (5) beneath the distributor via the control valve (2),silica-gel air drier (3) and flow meter (4). The dry air then passes upwards through the fluidizing bed, in which the powder is dispersed, and into the elutriation chamber. The air velocity is high enough to transport the dust particles but not the glass beads, which fall back into the bed and maintain fluidization until discharged via (13) into the sampler for overflow solids.

Mixed dust and fluidizing particles (beads) are fed into the hopper (10) and transported to the fluidized bed by the screw feeder, which is regulated by a variable speed motor (9).

The air flow rate for the fluidization was 25 L/min. The corresponding maximum particle sizes which can pass through the elutriator is calculated to be 63 μm aerodynamic dia. The coal fly ash used were well below these limits.

The flow rate for incipient fludization was found experimentally to be 14 L/min. The mass concentration of dust was controlled primarily by measurement of the rate of overflow of beads from the bed and then adjusted by the speed of the screw feeder. Dusts from the generator was introduced into the exposure chamber and then was with filtered room air. The dust concentration in the exposure chamber was always monitored by a light scattering method (15) (Digital dust indicator, Shibata AP 632L),and was adjusted automatically by the on-off control of screw feeder. The air flow was downward. The volume of the exposure chamber was $0.1m^3$. For low and high concentrations, We used two dust generators and two exposure chambers.

The exposure conditions are shown in Table 1. The coal fly ash aerosol was exposed to rats in two exposure chambers for 23 hours per day, 5 days per week for twelve months.

Table 1   Exposure conditions

Volume (m^3)	0.1
Air flow rate (m^3/min)	0.05
Diferential pressure from atomospheric pressure (Pa)	+10 ~ +50
Temperature (°C)	27 ± 2
Relative humidity (%)	52 ± 6

Dust feeding

The special feature of this dust generator is the mixing of powder with the fluidizing particles before feeding. If powder only is fed into the fluidized bed, it is very difficult to keep a constant rate over a long period. The reason for this is that the properties of fine particles are adhesive and agglomerative. In our generator, this defect was overcome by mixing the powder with the fluidizing particles. The mixed particles were smoothly fed via the screw feeder into the fluidized bed.

The dispersal of powder from the bed into the elutriation chamber and the discharge of the fluidizing particles from the overflow pipe reach a steady state that can be maintained for a long time.

The uniform mixing of the powder with fluidizing particles is very important to keep the dust generation rate constant. If the mixing is poor, there will be large variations in the dust generation in the dust generation rate. In our experiments, the mixing ratios of the powder to the fluidizing particles were kept at 0.01 and 0.001 in high- and low- concentration chambers, respectively. The powder was mixed with the fluidizing particles by hand. One or 2 kg of the fluidizing particles was put in a shallow dish and then the weighed dust was added into the fluidizing particles. The mixing was continued until the colour of the mixture was judged to be uniform by visual observation. The mixing time was about 10 min. The fluidizing particles were used repeatedly without washing.

The feed rate, F, and the dust generation rate, V, can be estimated from the mixing ratio, C, and the rate of collection, R, of fluidizing particles in the overflow receiver by use of the relationships:

$$F = R/(1 - C)$$ (1)

and

$$V = CF = RC/(1 - C)$$ (2)

assuming that all the powder fed into a fluidized bed is aerosolized and eluted. R is used as the indicator for the constant feed rate.

Result and Discussion

One example of the daily dust concentration monitored by the light scattering method is shown in Fig.2. The upper line in the figure shows the results in the lower exposure concentration chamber and the lower line shows the concentration in the higher one. The concentration of fly ash aerosol in the exposure chamber seems to be kept at constant. These monitors were continued during the exposure.

The daily average mass concentration for twelve months by the glass fiber

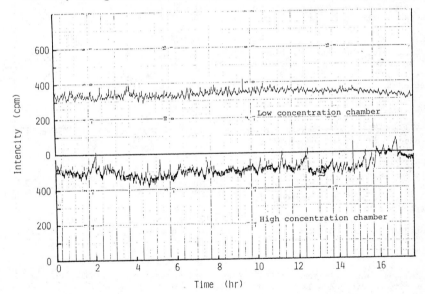

Fig.2 Dust concentrations in exposure chamber monitored by digital dust indicator (Shibata AP 632L).

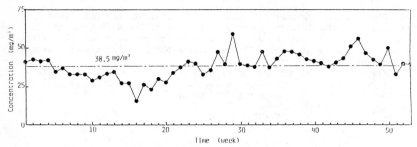

Fig.3 Weekly average mass concentration in high exposure chamber for twelve months.

filter in the high and low concentration
exposure chambers were 38.5±11.1 mg/m³
and 5.0±2.3 mg/m³, respectively. In the
higher concentration exposure chamber,
the weekly average mass concentration
for twelve months is shown in Fig.3. The
mass concentration shows to be kept at
constant.

The size distribution of fly ash in
the exposure chamber is shown in Fig.4.
The distribution was measured by the
Andersen sampler at the biginning and
the end of twelve months exposure
period. As the difference between them
were slight, the size distributions were
assumed to be constant during the
experiments. The mass median aerodynamic
diameter and the geometric standard
deviation was 2.7 Аum and 1.9,
respectively.

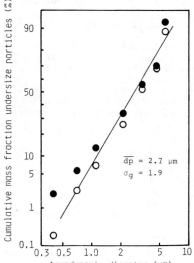

Fig.4 Cumulative size distribution of
coal fly ash in exposure chamber.
 O : Beginning of exposure
 ● : End of exposure

Conclusion

A dust generator was developed to
produce dust aerosol from powderous
materials for animal exposure
experiments and was tested by the coal
fly ash aerosol. The fly ash was exposed
to rats in two exposure chambers for 23
hours per day, 5 days per week for
twelve months.

It was found that the generator developed has the advantage that is the
retention of the constant mass concentration and constant size distribution.

References

Tanaka, I. and T. Akiyama (1984) " A New Dust Generator for Inhalation
Toxicity Studies" Annals of Occupational Hygiene, 28:(in press).

Published 1984 by Elsevier Science Publishing Co., Inc.
Aerosols, Liu, Pui, and Fissan, editors

GENERATION OF HIGH CONCENTRATION MONODISPERSE WATER AEROSOLS

F.Prodi

Istituto FISBAT-CNR, Reparto Nubi e Precipitazioni, 40126 Bologna,
and Osservatorio Geofisico dell'Università, 41100 Modena, Italy

and

V.Prodi

Dept. of Physics, University of Bologna, and Lavoro e Ambiente scrl,
40137 Bologna, Italy

EXTENDED ABSTRACT

There are many instances where is is highly desirable to generate an
aerosol of monodisperse water droplets: this would allow more controlled
experiments to be performed. The techniques more commonly used include
vibrating orifice generation, spinning disk or top atomizers, vibrating
needle: nevertheless these have in common low concentrations, generally high
electrostatic charge, and a lower size limit too high for some specific
applications.

The condensation generator has been previously described by V.Prodi
(1972) and has been operated only with low vapor pressure materials. It has
been recently modified with the replacement of the silicon oil thermostatic
bath with a forced air circulation thermostat which allows a wider range of
temperatures and a faster temperature rise (Monodisperse Aerosols Generator,
LA, Bologna).

A schematic diagram is shown in Fig.1. As an additional modification for
the generation of water droplets an aluminum reheater with a longer heated
section was introduced. In this type of generators the condensation nuclei
source is critical; for this reason it was kept strictly constant making use
of a Collison generator (British Standard Institution, 1955) operated at
constant pressure. In this case the drying of the droplets was not necessary
and the dessicator was omitted. The stream of gas (filtered air in this case)
is then let to the thermostated bubbler containing water. The exiting stream
contains at this stage a constant concentration of nuclei and, at a given
temperature, of water vapor. The size can be adjusted by acting on the
temperature of the bubbler: the higher is the temperature the larger is the
amount of water available to each nucleus. The stream is again vaporized in
the reheather and recondensed downstream in a controlled way, thus producing a
cloud of droplets. These exhibit vivid higher order Tyndall spectra and thus
are monodisperse. Due to their rapid changes, as a characterization technique
the coated slide method was choosen, both with magnesium oxide (May, 1950) and
carbon black (Squires, 1958). In addition, for sampling the droplets generated

in the smaller size ranges the gelatine naphtol green coating (Liddle and Wootten, 1957) has been used.

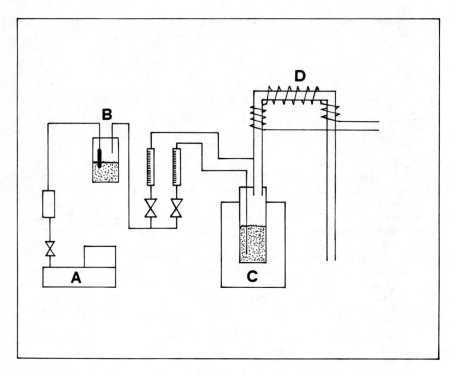

Fig.1. Scheme of the monodisperse water aerosol generator. A: Pressurized air supply; B: Collison generator; C: Thermostated bubbler; D: Reheater

In Fig.2 the optical microscope picture shows the spots produced by droplets generated with a bubble temperature of 76°C and reheater temperature of 68°C (Test c).

The glass slide was 5 mm wide, in this case coated with carbon black, and was put across the stream at high velocity.

The Table summarizes the earlier results obtained. At this stage the calibration factors, the ratio of the droplet diameter to the pit or spot diameter, given in the literature of 0.86 for the carbon black and 0.4 for the gelatine method have been used.

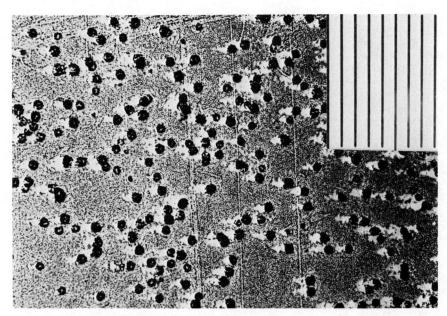

Fig.2. Optical microscope picture of the spots produced on the sampling slide covered with carbon black, by droplets generated during Test c. Separation of the lines in the scale is 10 μm.

Table
Size of the monodisperse water droplets at the indicated conditions

Test	T_B,°C	T_R,°C	F_BNl/h	F_PNl/h	Size(μm)	s
a	61	41	220	140	1.6	0.26
b	67	56	225	150	3.8	0.24
c	76	68	250	50	5.4	0.24

Where T_B, T_R are the bubbler and reheater temperatures, F_B and F_P are the flow rates trough the bubbler and in by-pass repectively, and s is the relative standard deviation. The size is the average of the values.

The data pertaining to the smallest size is probably adversely affected by the small relative resolution of the technique.

The concentration is of the order of 10^5 cm^{-3} for the largest droplets and of the order of 10^6 cm^{-3} for the smaller sizes.

The particles carry very little electric charge: they are statistically neutral and the average absolute charge is of the order of 2 electronic units.

These particles are therefore very suitable for experiments of cloud microphysics, optical visibility calibration, gas scavenging and reactions in the liquid phase.

Aknowledgments

The research has been supported by the Centro Ricerca Termica Nucleare (CRTN) of the Ente Nazionale per l'Energia Elettrica (ENEL) under grant to the Osservatorio Geofisico dell'Università di Modena. Partial support of the Piano Finalizzato Energetica 2, Sottoprogetto "Ambiente e Salute", is also gratefully knowledged.

References

British Standard Institution (1955). British Standard 2577, 2 Park Street, London W.I..

Liddel,H.F. and Wootten, N.W. (1957) "The detection and measurement of water droplets".Q.J.Roy.Met.Soc., 83, 1263.

May,K.R. (1950) "The measurement of airborne droplets by magnesium oxide method" J.scient. Instrum., 27, 128.

Prodi, V. (1072) "A condensation aerosol generator for solid monodisperse particles" Assessment of airborne particles (Edited by T.T.Mercer, P.E.Morrow, W.Stöber), C.C.Thomas Publ., Springfield, Illinois. p.169.

Published 1984 by Elsevier Science Publishing Co., Inc.

Aerosols, Liu, Pui, and Fissan, editors

CHAIN AGGREGATE AEROSOLS PRODUCED BY LIQUID PHASE REACTION

S. G. Kim and J. R. Brock
Chemical Engineering Department
University of Texas
Austin, Texas 78712

EXTENDED ABSTRACT

Introduction

There is current interest in the formation and growth of particulate aggregates. We have investigated the formation of metal aggregates , including linear chains, in the liquid phase and their subsequent dispersion as an aerosol. In the literature, various methods have bee used to produce metallic particles and aggregates in the liquid phase. These include electro-deposition, electrocondensation, and reduction by chemical reaction. Owing to its simplicity, we have used the reduction of metal salts by borohydride for production of metal aggregates.   We describe that portion of our work dealing with the formation of ferromagnetic aggregates and chains in the liquid phase.

Procedures and Results

The following studies have been carried out on formation of ferromagnetic particles by liquid phase reduction of iron, cobalt and nickel salts by boro-hydride with the view of production of chain aggregate aerosols.

(1) A stopped flow-type reactor has been used to study the effects of reactant concentrations, surface active agents, solvent and external mag-netic field on primary particle size, chain length (and fractal dimension) of random aggregates. The liquid phase dispersion from this reactor has been nebulized to produce chain aggregate aerosols. Aggregates have been studied using transmission electron microscopy (TEM) and for this purpose have been sampled both from the liquid phase and from the aerosol. The growth process has also been studied in situ at low reactant concentration using quasi-elastic light scattering. The following observations are found.

(a) Above reactant concentrations of 0.02 M, primary particle size remains approximately constant and only average chain length increases with increasing concentration.

(b) Chain length increases from ~1 -2 μm to 40 μm as externally imposed magnetic field is increased from 0.65 to 12.5 KG. At higher fields, straight chains are produced with little or no branching.

710

(c) Certain surface active agents were effective in inhibiting chain formation.

(d) Alloy particles produced by solutions of iron, cobalt and nickel salts showed increase of primary particle size as iron content was increased.

(e) Non-aqueous solvents as the reaction medium resulted in smaller primary particle sizes than found with aqueous solvent.

(f) All particles, except for very small nickel primary particles, were magnetic. Also, all such particles, except for those produced by a chelation process, failed to exhibit any crystalline pattern upon examination by TEM.

(2) Ferromagnetic chains and aggregates have also been produced by the collision and reaction of aerosol particles containing metal salts and borohydride. From this process, very small primary particle sizes resulted (less than 100 A). In the presence of external magnetic fields, linear chains were formed in individual droplets with diameters in the range $0.2 \sim 0.5$ μm.

Published 1984 by Elsevier Science Publishing Co., Inc.
Aerosols, Liu, Pui, and Fissan, editors

GENERATION OF LARGE, SOLID, CALIBRATION AEROSOLS

R.W. Vanderpool, Dept. of Environmental Engineering Sciences, University of Florida, Gainesville, FL 32611, and K.L. Rubow, Particle Technology Laboratory, University of Minnesota, Minneapolis, MN 55455.

Extended Abstract

During calibration of four, large-particle impactors developed at the University of Florida, a technique was developed for the generation of large, solid calibration aerosols. Slight modifications to a vibrating orifice aerosol generator (Model 3050, TSI Inc., St. Paul, Minn.) enabled the successful generation of solid ammonium fluorescein particles up to 70 um aerodynamic diameter. When generated under the proper test conditions, the particles were found to be spherical and of uniform size.

The developed generation technique does not require the somewhat awkward inversion of the aerosol generator. The dilution flowrate of the generator, however, is insufficient to suspend droplets significantly larger than 70 um. As a result, large droplets will normally settle out before having sufficient time to dry to the desired particle size. The successful generation of large particles, therefore, requires the use of liquid solutions of high volume concentrations which allows production of droplets of suspendable size which dry to form particles of the desired diameter. By dissolving fluorescein powder in aqueous ammonia, volume concentrations as high as 30% were produced. Although the use of smaller orifices and stronger liquid solutions requires more patience to start and maintain the liquid jet through the orifice, the overall generation process is more convenient than inversion of the generator.

Unlike the generation of smaller particles, the generation of particles larger than 20 um requires careful optimization of the operating parameters of the aerosol generator including orifice diameter, liquid feedrate, vibrational frequency, and dilution flowrate. Guidelines for the proper selection of these parameters are outlined. In addition, the properties of the fluorescein generation material are discussed.

Although the described generation method requires considerable operator technique, it is felt that the method will be useful in situations where the use of liquid calibration aerosols may be inappropriate.

Published 1984 by Elsevier Science Publishing Co., Inc.
Aerosols, Liu, Pui, and Fissan, editors

712

EFFECT OF INTERPARTICLE FORCES ON DISPERSION OF
FINE POWDERS IN AIR

Madhav B. Ranade and Michael E. Mullins
Research Triangle Institute
P.O. Box 12194
Research Triangle Park, NC 27709

Abstract

Dispersion of powders into single particles in an airborne state is
determined by interparticle forces of attraction and repulsion. The
energy input and the rate of energy application are both important in
achieving good dispersion. Parameters, such as the powder tensile
strength, which characterize the interparticle forces and dispersion
energy, will be discussed along with experimental data on several
powders ranging from cohesive to granular in characteristics.

Published 1984 by Elsevier Science Publishing Co., Inc.

Aerosols, Liu, Pui, and Fissan, editors

THE STABILITY OF BES AEROSOL GENERATED BY CONDENSATION METHOD

I. K. T. Lau, F. C. Hiller, M. K. Mazumder, and J. D. Wilson
University of Arkansas
Graduate Institute of Technology
P. O. Box 3017
Little Rock, Arkansas 72203

EXTENDED ABSTRACT

Introduction

The measurement of evaporation rate of oily droplets with ultralow vapor pressure has been of great interest in different applications. For use in human inhalation studies (Renninger, et. al., 1981), it is important that the aerosol droplets under test be inhaled before the aerosol evaporation changes the size distribution. In military applications it is useful to predict the residence-time of oil screening smokes. Also, in testing the efficiency of HEPA-filter (Kirsch and Zhulano, 1978), it is critical to assure that the aerosol does not evaporate completely instead of being filtered out.

Although many reports on the evaporation rate of non-volatile liquids have been published, the evaporative behavior of ultrafine droplets smaller than 0.1 μm in diameter has not been sufficiently studied experimentally because of measurement difficulties with ultrafine particles. Also, according to the Kelvin effect, pure liquid droplets smaller than 0.1 μm in diameter are unstable and are predicted to evaporate completely even if the saturation ratio is greater than one. The experimental measurements of this study, however, show that the evaporation of ultrafine BES particles did not proceed as fast as theory predicts (Renninger, 1981).

The purpose of this work was to study the evaporation of ultralow vapor pressure droplets smaller than 0.1 μm in diameter. A heterodisperse aerosol was generated by using a condensation aerosol generator. Particle concentration and decay were measured by using a microcomputer-modified electrical aerosol analyzer (EAA). The composition of ultrafine aerosol particles were then studied analytically using nuclear magnetic resonance spectroscopy (NMR), mass spectrometery (MS), pyrolysis gas chromatography (GC), and mass spectrometer (MS) techniques.

Theoretical Consideration

Reported studies on the kinetics of submicron size low vapor pressure droplets due to the combined effects of evaporation, coagulation and diffusion (wall losses) inside a closed chamber are limited in number. Most investigations on evaporation were separated from those on coagulation or wall losses. Since in most practical cases, these three phenomena occur simultaneously, their effects were considered together here in order to find out the stability of ultrafine BES aerosol.

On the assumption that quasi-steady state conditions exist, the expression for the rate of evaporation of liquid droplets having low vapor pressure can be shown (Renninger et al., 1981), that the rate of change of a particle of mass m_p can be written as

$$\frac{dm_p}{dt} = 2\pi d\overline{D}(C-C_s), \qquad (1)$$

where d is the diameter of the droplet. C_s and C are the vapor mass concentrations at the surface and far from the droplet. $\overline{D} = D/Cm$, D is the vapor diffusion coefficient and Cm is a dimensionless correction factor. This equation can be solved by numerical method.

Therefore, in the total mass loss of aerosols by evaporation, equation (1) becomes

$$\frac{dM_T}{dt} = 2\pi dn\overline{D}(C - C_s), \qquad (2)$$

where $M_T = nm_p$, M_T is the total mass of the aerosols and n is the total number of particles per unit volume.

A combined equation for the rate of decrease of concentration by number for coagulation of polydisperse aerosols (Lee and Chen, 1984) and wall losses (Rocker and Davis, 1978) in gas-slip regime is given by

$$\frac{dN}{dt} = -(\beta N^2 + LN), \qquad (3)$$

where N is the number of particles per unit volume and β is the polydisperse coagulation rate constant. L = SD/Vh, S is the surface of the vessel, V is the volume of the vessel, and h is a layer of thickness adjacent to the surface where the air viscosity damps out convection.

Also, the rate of decrease of total mass of aerosols by coagulation and wall losses can be derived by writing the total mass of the droplets within the containment vessels as

$$M_T = WN,$$

where W is a conversion mass constant and substitutes into equation (3)

$$\frac{dM_T}{dt} = \frac{(\beta M_T + LM_T)}{W}. \qquad (4)$$

Since the total mass of the aerosol is conserved (Green and Lane, 1964, Lee, 1979) during coagulation, the term of $\beta M_T/W$ is zero. Therefore, the rate of reduction of total mass of the aerosols is

$$\frac{dM_T}{dt} = -LM_T. \qquad (5)$$

715

Integration of equation (5) by neglecting any time dependence of L, the result is

$$\frac{M_T(t)}{M_T(0)} = \exp(-tL), \qquad (6)$$

where $M_T(t)$, $M_T(0)$ are the loss mass of aerosols at time t and zero. Thus, the total mass loss of aerosols over a period of time by wall losses can be predicted from equation (6).

Finally, using equations (2) and (5), the combined effects of evaporation and wall losses give

$$\frac{dM_T}{dt} = 2\pi dn\overline{D}(C - C_s) - LM_T. \qquad (7)$$

In an experimental situation where an aerosol is contained within a closed volume, the mass loss due to evaporation of particles will change the vapor mass concentration far from the droplet. Therefore, the solution for equation (7) becomes very complex when C varies with time. Equation (7) can be solved by numerical methods if it is assumed that the vapor mass concentration far from the droplet inside the box is independent of the change of mass loss of the droplet.

References

Green, H. L. and Lane, W. R., (1964) Particulate Clouds: Dust, Smokes, and Mists, Universities Press, Belfast.

Kirsch, A. A., and Zhulano, U. V., (1978) "Measurement of Aerosol Penetration Through High Efficency Filters", J. Aerosol Sci., 9:291.

Lee, K. W., and Chen, H., (1984) "Coagulation Rate Polydisperse Particle", Aerosol Science of Technology, Vol. 3.

Lee, K. W., (1979) "Change of Particle Size Distribution During Brownian Coagulation", J. of Colloid and Interface Sci., 92:315.

Renninger, R. G., Hiller, F. C., and Bone, R. C., (1981) "The Evaporation and Growth of Droplets Having More Than One Volatile Constituent", J. Aerosol Sci., 6:505.

Rocker, S. J., and Davis, C. N. (1979) "Measurement of the Coagulation Rate of a High Knudson Number Aerosol with Allowance for Wall Losses", J. Aerosol Sci., 10:139.

Published 1984 by Elsevier Science Publishing Co., Inc.

Aerosols, Liu, Pui, and Fissan, editors

716

EXPERIMENTAL INVESTIGATION OF POWDER DISPERSIBILITY

Sushil Patel, M. T. Cheng and D. T. Shaw
Laboratory for Power and Environmental Studies
State University of New York at Buffalo
Buffalo, New York

EXTENDED ABSTRACT

A versatile compressed air powder redispersing ejector (CAPRE) has been designed and tested for the generation of high density aerosols from powder materials. Three criteria were emphasized during the development of the CAPRE. First, the complete dispersion of the powder batch. Partial redispersion results in undesirable aggregates which rapidly settle out of the aerosol cloud. Second, the generation of high density aerosol to up to 50 gm/m^3 based on unit bulk density of the powder. Third, the generation of a steady airborne concentration of the powder for appropriate lengths of time depending upon the application.

The CAPRE is a pneumatic type powder ejector. It operates by aspirating powder into a high velocity airstream flowing through a fine nozzle. The high velocity gradients provide the shear force necessary to overcome the adhesion forces binding particles of the aggregate. Several models have been tested for air flowrates of 2 scfm using a single nozzle to 200 scfm using a compact multinozzle ejector. The powder feedrates range from a controlled 0 to 0.04 lbs. per minute for a single nozzle ejector to up to 5 lbs. per minute for a multi-nozzle ejector. Microscopic analysis of dispersed powders show complete dispersion for supply pressures as low as 20 psig to 70 psig operable pressure. A vacuum is sufficient to fluidize powders and draw it into the high velocity airstream making external forced powder feed redundant.

Various paint pigments have been tested for dispersibility and size distribution using a Scanning Electron Microscope and the 11-stage low pressure impactor. These results will be presented.

Published 1984 by Elsevier Science Publishing Co., Inc.
Aerosols, Liu, Pui, and Fissan, editors

AEROSOL GENERATION FROM A VISCOUS LIQUID:
CHARACTERIZATION OF A MEDICATION NEBULIZER

Rolf Bretz*, Robert Hess*, and David M. Bernstein**
* CIBA-GEIGY LTD., Experimental Toxicology, CH-4332 Stein/Switzerland
**BATTELLE Geneva Research Centres, CH-1227 Carouge/Switzerland

EXTENDED ABSTRACT

Introduction

Compressed air nebulizers are widely used for medical purposes, and a number of commercially available units have been characterized in terms of their output concentration and particle size (Raabe, 1976; Crider et al., 1980; DeFord et al., 1981; Marks et al., 1983), however, only for aqueous media. In solutions of low viscosity containing a relatively volatile solvent, the nebulization energy supplied by the compressed air breaks up the liquid into fine droplets, which then undergo a further size reduction through solvent evaporation.

The situation is often different in inhalation toxicity testing: Here, aerosols which are respirable for small laboratory rodents (<3 μm mass median aerodynamic diameter (MMAD)) must be generated from pure liquid test substances of low volatility and in some cases high viscosity. Such liquids require more energy to produce small particles, and they have little possibility to reduce their droplet size by surface evaporation. Three commercial medication nebulizers were used by Miller et al. (1981) to generate the high aerosol concentrations required for acute toxicity testing (OECD, 1981). However, an organic liquid of relatively low viscosity (2.5 cP) was used.

This paper describes a reliable, low-cost device of sufficient throughput, good resistance to chemicals, and versatility for toxicological assays, the HOSPITAK medication nebulizer, which generates high concentrations of respirable aerosols from aqueous as well as viscous nonaqueous media.

Materials and Methods

HOSPITAK nebulizers (Cat. No. 950, Hospitak, Farmingdale, N.Y. 11757) use the Wright jet baffle principle (Raabe, 1976), with a double orifice system (diameter 0.6 mm) and a hemispheric baffle. They were operated with filtered compressed air at subsonic velocities. The square of the total airflow was found to be proportional to the pressure supplied to the nebulizer (up to 0.7 bar), as is the case for nozzles and orifices.

Aerosols were generated from an aqueous solution of glycerol (10 % w/w, viscosity 1.3 cP) containing a tracer dye (0.9 % Neutral Red), a fluid similar to that used by Drew et al. (1978). Additionally, undiluted olive oil was used, which has a viscosity of 84 cP at 20° C (Handbook of Chemistry and Physics, 1982).

The total aerosol output was collected on an in-line filter following the nebulizer. Cellulose nitrate membrane filters (0.2 μm) and glass fiber depth filters with and without Teflon coating proved to be equally effective, except for the tendency of the membrane filters to clog at higher load. For the aqueous aerosols, the filters were extracted with ethanol. The dye was quantitated spectrophotometrically (at 537 nm), and glycerol was determined enzymatically

(Triglyceride Rapid Test ROCHE). Both values, together with the respective concentrations (and the density) measured in the nebulizer reservoir, were used to calculate the usable aerosol concentration (in mg/l of air). The concentrations of the olive oil aerosols were determined gravimetrically.

To measure the particle size distribution of the aerosols, the nebulizer output was diluted with filtered air and passed into a 100 l mixing chamber. Samples were taken from the mixing chamber with an eight-stage Andersen Mark I Sampler and an eight-stage Sierra Series 290 Marple cascade impactor (Rubow et al., 1984). The amount of material on each stage was determined as described above. The particle size measurements were analyzed assuming a log-normal distribution (Raabe, 1971). Comparable results were obtained from both instruments.

Aerosols Generated from Glycerol/Water

When the nebulizer was operated with a single filling of the test solution (10 to 20 ml), the volatile solvent (water) evaporated from the refluxing solution. After 4 hours of operation at a rather small airflow (2 l/min), both the glycerol and the dye concentrations approximately doubled (fig. 1). The amount of the two nonvolatile components in the aerosol increased linearly with their actual concentrations in the nebulizer reservoir (fig. 2). This enrichment may severely influence the toxicity testing of substance mixtures containing volatile solvents.

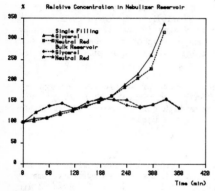

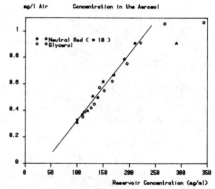

Figure 1. Nebulizers with single filling and with attached bulk reservoir: Reservoir concentration as a function of time (starting conc. = 100 %)

Figure 2. Nebulizer with single filling (glycerol/dye solution, airflow 2 l/min): Aerosol output (nonvolatile solutes) with increasing reservoir concentration

The concentration increase also induced a shift of the mean particle size (i.e. a smaller size reduction by solvent evaporation). Within 290 min of operation at 2 l/min, the MMAD increased by 24 % (4.9 μm to 6.1 μm). From the rise in glycerol concentration (about twofold), an increase of 26 % would be predicted. Within the detection limits, the liquid (glycerol) and solid (dye) nonvolatile components showed the same distribution over the particle size.

Following the technique of DeFord et al. (1981), the nebulizer was attached to a bulk reservoir by means of a siphon, so that the concentrating effect was confined to a small aspirating volume of about 1 ml. This rendered the aerosol output much more constant after an equilibration period of about 60 min. However, a concentration increase in the small reservoir of about 50 % (fig. 1) and a corresponding change in the output were observed for both nonvolatile solutes during the first hour of operation.

Olive oil aerosols

Olive oil was chosen as a test medium to avoid the enrichment effects observed above and the particle size changes caused by evaporation. Additionally, its use demonstrated the good performance of the nebulizer at higher viscosity.

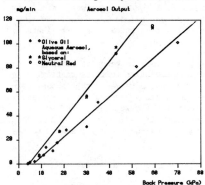

Figure 3. Olive oil and aqueous aerosols: Usable output as a function of back pressure

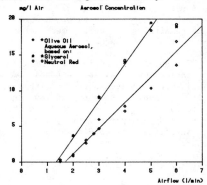

Figure 4. Olive oil and aqueous aerosols: Concentration as a function of airflow

Both the aqueous and the oil aerosol outputs increased linearly with the back pressure (fig. 3). At any given pressure, the oil output (0.1 mg/min at 0.04 bar to 100 mg/min at 0.7 bar) was about 60 % of the yield of aqueous aerosol (produced from a freshly filled reservoir), indicating only a moderate influence of viscosity on the nebulizer performance. A minimum threshold pressure of 0.04 bar was required for aerosol production with both media.

The corresponding airflows ranged from 1.5 to 6 l/min, producing olive oil aerosol concentrations of 0.1 to 13-16 mg/l. The aerosol concentrations were found to vary linearly with the airflow through the nebulizer (fig. 4), i.e. with the square root of the back pressure.

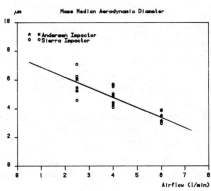

Figure 5. Olive oil aerosols: MMAD

In addition to generating higher aerosol concentrations, an increase in airflow also produced a smaller droplet size (fig. 5). At low flow rates (2.5 1/min), the MMAD of the oil droplets was found to be between 5 and 7 μm (with a geometric standard deviation $\sigma_g \simeq 2$), depending upon the nebulizer tested. However, with a flow rate of 6 1/min aerosols of 3.1 – 3.9 μm MMAD ($\sigma_g \simeq 2.4$) were produced, a size range which is more useful for inhalation experiments.

Conclusion

Olive oil as a testing medium for aerosol generation equipment proved to be easier to handle and analyze than the commonly used aqueous mixtures, and it is not subject to changes by evaporation. Its high viscosity and low volatility simulate conditions frequently met in toxicity testing, thus putting the nebulizers to a more realistic test.

The HOSPITAK nebulizer was found to be a useful tool to generate the high concentrations of respirable aerosols necessary for inhalation toxicity testing. High yields were achieved not only with aqueous solutions, but even with viscous, non-volatile liquids. However, if low aerosol concentrations are required (especially to rodents), it may be necessary either to remove the resulting larger particles by means of an appropriate cyclone, or to operate the nebulizer at higher back pressure (and consequently higher output), and dilute the aerosol produced to the desired concentration.

We wish to thank Mr. D. Senften for his expert technical assistance.

References

Crider, W.L., Landis, D.A., and Weaver, F.H. (1980) "Computer-Controlled Human Exposure Facilities Utilizing Water-Soluble Aerosols", Generation of Aerosols and Facilities for Exposure Experiments (Edited by K. Willeke), Ann Arbor Science Publishers, Ann Arbor.

DeFord, H.S., Clark, M.L., and Moss, O.R. (1981), Am. Ind. Hyg. Assoc. J. 42:602.

Drew, R.T., Bernstein, D.M., and Laskin, S. (1978), J. Toxicol. Environm. Health 4:661.

Handbook of Chemistry and Physics (1982), (Edited by R.C. Weast and M.J. Astle), 62nd edition, p. F-47, CRC Press, Boca Raton, Florida.

Marks, L.B., Oberdoerster, G., and Notter, R.H. (1983), J. Aerosol Sci. 14:683.

Miller, J.L., Stuart, B.O., DeFord, H.S., and Moss, O.R. (1981), "Liquid Aerosol Generation for Inhalation Toxicology Studies", Inhalation Toxicology and Technology (Edited by B.K.J. Leong), Ann Arbor Science Publishers, Ann Arbor.

OECD (1981) Guidelines for Testing of Chemicals, Method 403: "Acute Inhalation Toxicity", Organisation for Economic Co-operation and Development, Paris.

Raabe, O.G. (1971), Aerosol Science 2:289.

Raabe, O.G. (1976) "The Generation of Aerosols of Fine Particles", Fine Particles: Aerosol Generation, Measurement, Sampling, and Analysis (Edited by B.Y.H. Liu), Academic Press, New York.

Rubow, K.L., Marple, V.A, and Olin, J.G. (1984) "A New Personal Cascade Impactor", paper in preparation. Personal communication of the calibration data by Dr. Rubow.

Published 1984 by Elsevier Science Publishing Co., Inc.

Aerosols, Liu, Pui, and Fissan, editors

A BRUSH FEED MICRONISING
JET MILL POWDER AEROSOL GENERATOR
FOR PRODUCING A WIDE RANGE OF CONCENTRATIONS
OF RESPIRABLE PARTICLES

David M. Bernstein*, Ph.D., Owen Moss**, Ph.D.
Hermut Fleissner* and Rolf Bretz[+], Ph.D.

* Battelle-Geneva, 1227 Carouge, Geneva, Switzerland
** Battelle Northwest Lab., Richland Washington 99352
+ Ciba-Geigy AG, 4332 Stein, Switzerland

[EXTENDED ABSTRACT]

For many inhalation toxicology studies aerosol concentrations ranging from several µg/l up to 5 mg/l are often necessary. While methods have been available to produce these concentrations, they often suffer from either not working with "sticky" compressible powders, not producing the concentrations desired or producing aerosols which are of too large a diameter to be inhaled by the test animals. With nose breathing, the relative distribution of particles in the various regions of the respiratory tract in rodent and man have been found to be similar only for particles less than 3 µm MMAD[1].

To overcome these difficulties an aerosol generation system has been developed by Battelle which uses a specially designed brush mechanism to feed material to a micronising jet mill (Figure 1). The processes of feeding the powder and micronising it have been separated and are performed by two functionally independent devices.

The brush feed mechanism is based on a dual flexible brush design by Milliman et al.[2]. The Milliman design employs two brushes, one a larger hopper brush which keeps the powder in motion, and a smaller tangential feed brush. Material is transferred from the hopper to the feed brush gravitationally through a circular channel of 1 to 2 mm diameter. This design, however, was not capable of feeding "sticky" compressible powders as the powders would clog in the transfer channel.

The mode of operation was conceptually modified to allow the continuous fluid dynamic motion of the powders created by the movement of the larger hopper brush to uniformly pass material onto the feed brush. As shown in Figure 2, instead of a circular channel a very thin slit was machined which precludes the possibility of clogging at this point as the brushes themselves keep the passage clear. In addition, slits of different lengths and widths are used depending upon the nature of the powder and the concentration to be generated.

Both brushes are driven by electronic stepping motors which create a slight snapping action which further helps to minimise the possibility of clogging.

The powder is fed by the smaller spiral brush to the aspiration tube from the jet mill where it is aerosolized. An aspiration flow of approximately 10 l/min is created by a venturi following the left air jet of the jet mill.

The jet mill is of a Trost (Garlock, Inc.) design. The commercial Trost unit however requires a total flow rate of over 100 l/min which for the Battelle flow-past nose exposure systems[3] used in this laboratory was too great. The jet mill was therefore redesigned to operate with a total flow of 30 to 50 l/min.

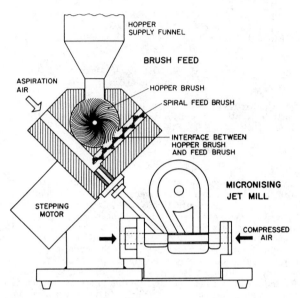

Figure 1: Schematic Representation of the Brush Feed Micronising Jet Mill Aerosol Generator.

The jet mill micronises particles by impaction of particle against particle in two opposing air streams. An integral part of the microniser is a cyclone type device which ensures that only particles of desired diameter leave the jet mill while the rest are returned for further micronisation.

Materials and Methods

The operation of the aerosol generator was tested using a range of powders (supplied by Ciba-Geigy AG) that might be used for inhalation toxicology tests and with talc. Each of the powders was aerosolized for approximately one hour. During this period 3 filter samples were taken and between 2 and 3 impactor measurements. A Gellman in-line stainless steel filter holder was used with Pallflex high efficiency teflon coated glass fiber filters for concentration measurements. A Mercer seven stage cascade impactor (IN-TOX Products, 02-100) with ECD of 0.33 μm to 4.6 μm was used for particle size measurements. Each stage was coated with Apiezon L grease. In addition, to monitor the real time variation of the concentration, a RAM-1 (GCA Corp.) light scattering aerosol monitor was used with a venturi dilution system.

Upon leaving the generator the aerosol passed by a KR-85 line neutralizer to bring the particles to Boltzmann equilibrium and was then fed into the Battelle flow-past exposure chamber[3]. All sampling was performed from one exposure port, with all the aerosol emitted from the port collected by the samplers, thus minimizing sampling bias.

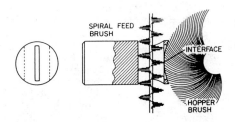

Figure 2: Schematic Representation of the Interface Between the Hopper and Spiral Feed Brush.

Results and Discussion

With each powder a low and a high concentration were generated using a narrow and wide interface slit respectively between the brushes. For each of these concentration ranges the same interface slit was used for all the powders.

The results from these tests are shown in Table I. The high and low concentration variations between powders are a result of the physical characteristics of the powders. These range from the BD40A which was very soft and compressible to CG 8300631 which was very hard. The same concentrations could have been obtained for all the powders by using different interface slits and adjusting the brush speeds.

The particle size measurements show that an aerosol of less than 3.0 μm MMAD was produced for all powders except the high concentration talc.

References

[1] Raabe, O.G., H.C. Yeh, G.J. Newton, R.F. Phalen and D.J. Velasquez (1977), "Deposition of Inhaled Monodisperse Aerosols in Small Rodents". Inhaled Particles IV. (W.H. Walton, ed.), Pergamon Press, New York, p. 3-32.

[2] Milliman, E.M., D.P.Y. Chang and O.R. Moss (1981), "A dual flexible-brush dust-feed mechanism". Am. Ind. Hyg. Assoc. J. 42: 747-751.

[3] Cannon, W.C., E.F. Blanton, and K.E. McDonald (1982) "The flow-past chamber: an improved nose-only exposure system for rodents". Am. Ind. Hyg. Assoc. J. 44(12): 923-928.

724

Table I: Aerosol Concentration and Particle Size Produced from a Selection of
Test Powders with the Brush Feed-Jet Mill Aerosol Generator

Powder	Aerosol concentration (mg/l, n=3) (mean ± S.D.)	Particle size Mass Median Aerodynamic Diameter (Geometric standard deviation)		
CG 836610	0.22 + 0.03	2.0 (1.8)	1.1 (2.4)	1.3 (2.2)
CG 836610	4.37 ± 0.11	1.5 (2.1)	2.0 (1.9)	2.2 (1.9)
CG 830589	0.17 + 0.02	1.3 (2.2)	1.4 (2.1)	1.6 (1.9)
CG 830589	1.75 ± 0.29	1.4 (1.8)	1.5 (1.7)	**
CG 830631	0.42 + 0.05	1.9 (2.1)	1.4 (1.3)	2.2 (1.9)
CG 830631	7.48 ± 2.63	2.0 (1.7)	2.2 (1.2)	**
CG 830617	0.57 + 0.18	1.9 (1.9)	2.2 (1.6)	**
CG 830617	2.18 ± 0.35	*	2.4 (1.6)	**
BD 40A	0.30 + 0.13	1.3 (1.8)	1.2 (1.8)	**
BD 40A	5.70 ± 0.76	1.6 (2.0)	1.0 (1.7)	**
TALC	0.49 + 0.03	2.1 (1.9)	2.4 (1.8)	2.5 (1.6)
TALC	4.46 ± 0.05	> 4.6	3.3 (1.5)	> 4.6

* lost sample due to fractionation of impactor substrate
** Only two measurements were taken

CG 830631 = 3,4-Dichloroaniline
CG 836610 = 4,5-Dichloro-2-nitroacetanilide
CG 830589 = 5-Chloro-6-(2,3-dichlorophenoxy)-benzimidazole-2-thion
CG 730617 = 4-Chloro-5(2,3-dichlorophenoxy)-2-nitroacetanilide
BD 40A =   Inhalation pharmaceutical product

Published 1984 by Elsevier Science Publishing Co., Inc.
Aerosols, Liu, Pui, and Fissan, editors

A NEW AEROSOL GENERATOR FOR STUDIES OF ULTRAFINE PARTICLES IN THE
SIZE RANGE BELOW 0.01 μm

H. Bartz, H. Fissan, B.Y.H. Liu[+]
Process- and Aerosol Measurement Technology
University of Duisburg, FRG
+University of Minnesota, Minneapolis, Minnesota, USA

EXTENDED ABSTRACT

Introduction

Aerosol particles in the size range below 0.01 μm are of special interest in the area of gas-to-particle conversion and play a major role as condensation nuclei in certain technical processes. Here, as well as in instrument calibration, a stable aerosol source with known particle size and concentration is required.

Several different techniques have been used in the past to produce submicron aerosol particles. A common method is to heat either a coated or a pure metal wire. The disadvantage of this method, however, is that the particle size is a function of time and the aerosol thus produced is fairly polydisperse. This is true also for the so-called exploding wire technique, where the particles tend to form chain aggregates. To achieve fairly monodisperse aerosols, Sinclair and LaMer (1949), generated particles by condensing the vapor of a high boiling point liquid on condensation nuclei. This method in which the vapor is slowly and uniformly cooled, is suitable for producing good monodisperse aerosols in the size range between 0.01 and 1.0 μm diameter. For smaller particles difficulties arise in obtaining sufficiently high supersaturations, which are necessary to get not only condensation but also a high nucleation rate.

For ultrafine aerosols a method has been proposed in which the aerosol material is evaporated from a boat in a tube furnace (Scheibel et al., 1981). The disadvantage of this technique is that the precise control of the amount of evaporated material is difficult, which in turn affects the particle size being produced. Further, due to the low cooling rate, the aerosol is also rather polydisperse. In our review, a very promising way to generate a nearly monodisperse ultrafine aerosol is to cool a vapor of the aerosol material far below the saturation temperature. There are several possibilities to achieve a rapid cooling. In this study, the rapid mixing of vapor with a cold gas stream is used.

Description of the system

In figure 1 the system used in this study is shown schematically. The aerosol material (salt) together with a solvent (water) is dispersed with a common atomizer. Here a

constant output atomizer from TSI was used, producing a mass flow
of aerosol solution of about 75 mg/h at a gas flow rate of
2 l/min. This spray is then dried in a diffusion dryer and
vaporized in the tube furnace.

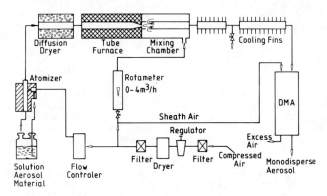

Figure 1: System for generating monodisperse ultrafine aerosols

To achieve the complete evaporation of the salt the furnace
temperature is held at 900°C. The vapour thus produced is then
accelerated in a nozzle and expands into the mixing chamber where
it is cooled by mixing with cold air entering through six radial
jets at an angle of 45° towards the center axis. Thus a very
intense and rapid cooling is achieved along with a dilution of
the generated aerosol. To cool this aerosol down to ambient
temperature, cooling fins are added to one section of the tube. A
sample of the aerosol at 2 l/min is then extracted and fed into a
differential mobility analyzer (TSI Model 3071) to obtain a
monodisperse aerosol. The same instrument together with an
electrometer as a detector can also be used to determine the
particle size distribution of the original aerosol obtained by
mixing the vapor with the cold air.

Theoretical considerations

Because the cooling in this case takes place very rapidly and
the original particles are generated nearly instantaneously by
nucleation, the most important process in the aging of the
aerosol is coagulation. Thus the concept of the self preserving
size distribution for Brownian coagulation in the free molecule
regime can be applied in this case (Lai et al., 1972). According
to this theory the resulting particle size distribution can be
described in terms of a dimensionless distribution curve, from
which the particle volume and the corresponding distribution
function can be numerically obtained.

From this theory it is concluded that the key parameters
which influence particle size as well as particle number
concentration in this generation process are the residence time

for the particles and the volume flow of salt together with the
volume flow of cooling air, combined to determine the dimensionless volume fraction of the aerosol material. It is concluded
that the furnace temperature is not important as long as the
complete evaporation of the salts can be guaranteed. This
could also be shown experimentally.

As a result of calculation from this theory figure 2 shows
the dependance of the mean particle size on these two key parameters.

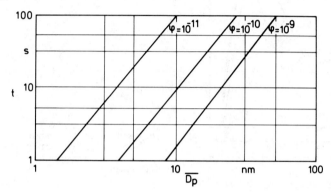

Figure 2: Dependance of mean particle size on volume fraction and
residence time (theory)

A value of $\varphi = 1$ E-11 is roughly equivalent to a 0.05% NaCl
solution and a mixing air flow of 30 l/min. For residence times
in the order of about 5 to 10 seconds, which correspond to the
experimental setup, a mean particle size in the range of 4 to
5 nm should be obtained. The total number concentration in this
case clearly exceeds 1 E9 cm-3, so that, assuming a Boltzmann
charge distribution and a fraction of 10% of the original
polydisperse distribution, at least several thousand monodisperse
particles/cm3 should leave the mobility analyzer.

Experimental results

In figure 3 the electrometer current is plotted against the
collector rod voltage of the mobility analyzer for a typical set
of parameters and for three different dimensionless volume
fractions. The experiment gives mean sizes of 7.5, 6.5 and 5 nm,
whereas according to theory they should be smaller by about 20%.
This is however still a fairly good agreement. One can also see
that the distribution is very narrow and that the number
concentration is clearly above 1 E4 cm-3. The mean particle size
changes in the way predicted by theory. Besides this result,
another objective of this project, to produce a stable aerosol,
was reached satisfactorily. The electrometer current from the
monodisperse aerosol leaving the mobility analyzer was recorded
over a time period of four hours and showed that the number

concentration was stable to within 5% of the mean during this time period. The absolute value is high enough to be used for several applications as mentioned in the beginning.

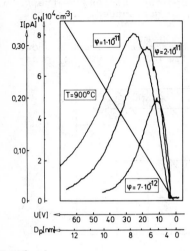

Figure 3: Electrometer current vs collector rod voltage for a typical set of parameters

Conclusions

An aerosol generator based on the rapid mixing of vapor and cold air to produce a continuous and stable stream of ultrafine particles was developed and investigated. It was shown that the particle size and in turn the particle number concentration can be influenced by the residence time of the particles and, more efficiently by the dimensionless volume fraction of the aerosol material. To achieve a sufficiently high concentration of monodisperse particles it is possible, based on theoretical calculations, to feed the mobility analyzer with a narrow distribution having a mean size near the size of the monodisperse aerosol desired.

References

Lai, F. S.; Friedlander, S. K.; Pich, J.; Hidy, G. M.: The self preserving particle size distribution for Brownian coagulation in the free-molecule regime. Journal of Colloid and Interface Science, 39, (1972), pp. 395-405.

Sinclair, D.; LaMer, V. K.: Light scattering as a measure of particle size in aerosols. The production of monodisperse aerosols. Chemical Review, 44, (1949), pp. 245-267.

Scheibel, H. G.; Porstendörfer, J.; Becker, K.-H.; Hussin, A.: An aerosol generator for monodisperse Ag- and NaCl aerosols in the particle size range between 2 nm and 300 nm. Association for Aerosol Research, 9 Conference (1981), Duisburg, FRG.

CHAPTER 18.

COMBUSTION AEROSOLS

Published 1984 by Elsevier Science Publishing Co., Inc.
Aerosols, Liu, Pui, and Fissan, editors

PARTICLE DISTRIBUTIONS FOR SMOLDERING AND FLAMING FIRE
SITUATIONS

B.C. Krafthefer and G.E. Lee II
1. Honeywell, Bloomington, MN 55420
2. Southwest Research Institute, San Antonio, TX 78284

EXTENDED ABSTRACT

Introduction

Each year more than 700,000 fires occur in residences in
which 6600 people lose their lives and $800 million of property
are destroyed. Of these household fires, some 62% (U.S. Dept. of
Commerce,]976) involved appliances of some type and over half of
these fires (34%) involved the ignition of grease or other food-
stuffs. In 33% of all reported fires, grease and food were the
first items to ignite, and cooking was involved in nearly three-
quarters of all fires in which human activity was involved.

To gain a beter understanding on the combustion process and
effluents, experiments were conducted under actual fire conditions
using materials involved in residential fires. The materials used
in the tests were: 1) polyurethane bolsters with cigarette ignition, 2) a Christmas tree with burning newspaper ignition, 3)
cooking oil ignited by a hot-plate, and 4) cooking oil ignited by
a gas flame. The ignition sources for the combustion modes were
chosen to parallel the major causes of home fires in the United
States. The primary locations of these fires were in the living
room and the kitchen.

Experimental Procedures

For the test a full scale multi-room burn facility consisting
of a kitchen, living room, and both upstairs and downstairs
bedrooms was used. Measurements of particles were made in the upstairs corridor.

Two locations were chosen as source locations for the fires;
the living room area and the kitchen. In the living room, experiments were conducted using a cigarette to initiate a smoldering
front on an array of ten polyurethane bolsters consisting of 77 gm
flexible polyurethane foam (]3x47x6 cm), 98 gm cotton batting and
49 gm cotton fabric.

A flash fire was created by placement of a commercially
purchased "fresh cut" Scotch pine Christmas tree in the same location as the smoldering bolsters. Newspapers covered the base of
the tree in an open single sheet fashion and the fire was initiated
using a match.

For the kitchen smoldering fires an iron skillet was filled

half full with a major brand of corn oil and placed over an electric hotplate in the kitchen at high heat. A thermocouple was placed in the oil to monitor the temperature. This same skillet and oil was also placed over a gas stove top burner at full flame. The gas flame remained on until all the oil had been consumed.

To measure the smoke density in the upstairs corridor, a one-meter pathlength densitometer was used. Along with this the particle size distributions in the size range of 0.01 to 10 um were measured using a light-scattering Royco particle counter and an Electrical Aerosol Size Analyzer manufactured by TSI, Inc.

Results

During the fire tests, smoke was measured at the upstairs corridor in concentrations up to 10^6 particles/cc. As the smoke reached this position agglomeration has already shifted the mean diameter. This shift in mean diameter appeared to be more rapid for the flaming fires than for the smoldering fires, see Figure 1 1, though this could be accounted for by the flaming fires

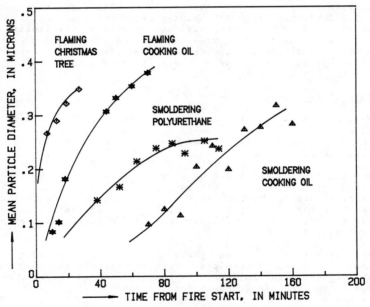

FIG. 1. CHANGE IN THE MEAN PARTICLE DIAMETER WITH TIME

exhausting their fuel into the room more quickly. One interesting observation is that the mean particle diameter reached during the smoldering fires did not appear as large as the mean diameters attained for the flaming fires, even though the smoke was in the corridor for a longer period of time. This can be seen in Figure 1 where, for example, the mean diameter reached for the smoldering cooking oil was about 0.30 microns, yet for the flaming fires diameters of above 0.35 microns were reached.

The light transmission during the tests was found to decrease smoothly during the tests with the increase in smoke concentration. However, transmission values were found to be lower for smoke produced during the flaming tests than for smoke from the smoldering tests.

References

U.S. Dept. of Commerce (1976), <u>Highlights of the National Household Fire Survey</u>, National Fire Prevention and Control Administration, Washington D.C. 20230

Published 1984 by Elsevier Science Publishing Co., Inc.
Aerosols, Liu, Pui, and Fissan, editors

734

PARTICLE CHARGE IN COMBUSTION AEROSOLS

H. Burtscher and A. Schmidt-Ott
Lab. for Solid State Physics, ETH-Hönggerberg
CH-8093 Zürich, Switzerland

and

A. Reis
Cerberus AG, CH-8708 Männedorf, Switzerland

EXTENDED ABSTRACT

Introduction

In the equilibrium state aerosol particles are charged according to the Boltzmann distribution at ambient temperature with deviations for particles below 20 nm. When the aerosol is produced in a combustion process, the particle charge is usually not in the equilibrium state immediately after formation, but depends on the characteristics of the combustion such as ion and smoke concentration or temperature.

Then, during aging, the particle charge approaches the equilibrium. In the following, some results obtained by measuring the fraction of charged particles as a function of size for different types of fires are presented.

Experiment

Fig. 1 shows the arrangement used for these measurements. The combustion takes place in a 0,7 m^3-"fire-box", where a small fan mixes the aerosol to obtain a homogeneous

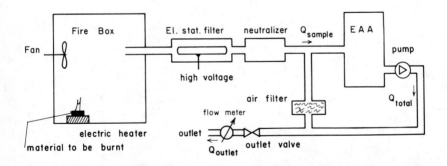

Figure 1. System for measurement of the particle charge

distribution in the box. After the combustion, while the aerosol is aging, a
small amount of aerosol is continuously taken from the fire-box and passed
through an electrostatic filter, followed by a Neutralizer.

Finally, the remaining size distribution is determined with an Electrostatic Aerosol Analyzer (EAA).

By feedback of the air outlet of the EAA, the aerosol is diluted before
measuring. Since the flow Q_{sample} (see Fig. 1) is equal to Q_{outlet}, the dilution ratio is

$$\frac{Q_{sample}}{Q_{total} - Q_{sample}}$$

and can be varied by the outlet valve

By measuring with the electrostatic filter on and off, the fraction of
charged particles can be determined. Measurements have been performed with
cotton-wick, smoldering and open wood fires and Heptan-alcohol mixtures. For
approximately one hour the EAA-spectrum is registrated with the electrostatic
filter alternatingly switched on and off, to monitor the aging of the aerosol.

Results

As an example Fig. 2 shows the fraction of charged particles versus time
for a smoldering wood fire. In this case the initial charge on the particles is
very low and rises during aging. Performing this experiment with different
amounts of burnt material or different smoke concentrations, respectively,
shows that

- higher smoke concentration leads to a lower initial total
  charge and to a slower rise of the charge with time

- all aerosols are initially charged far below the Boltzmann
  equilibrium and it takes very long time (>> 1h) to reach
  equilibrium.

For an open flame fire, however, the particles already are charged as
expected for equilibrium at room temperature from the beginning. During
aging, the charge remains almost constant.

These results and results from experiments with cotton-wick and Alcohol-
Heptan fires show that

- when the fire is smoldering or burning with a small flame
  and high smoke-production, the particles carry a low
  initial charge that rises during aging

- when there is an open flame, the initial charge is much
  higher, usually not far from equilibrium.

736

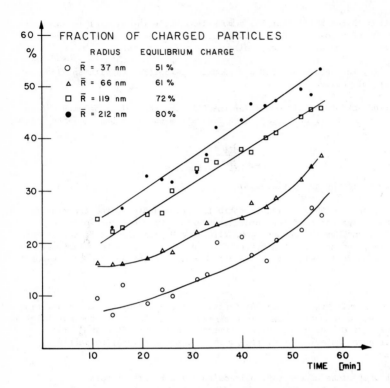

Figure 2. Fraction of charged particles versus time
for a smoldering wood fire

This leads to the following conclusions:
In a smoldering fire the smoke production is very high compared to the ion production and the resulting lack of ions leads to the low initial charge. As the smoke concentration is high it takes a long time until enough ions are produced by cosmic rays etc. to establish equilibrium. On the other hand, the high ion density in an open flame allows to reach the equilibrium almost immediately after combustion. Since the initially measured charge corresponds to the equilibrium at room temprature, it has to be assumed that this charge distribution is established at temperatures not far from room temperature.

Other evaluations concerned the development of the size spectrum during aging. There it has been observed that in several cases the size distribution was unimadal in the beginning and turned to a bimodal shope later.

Published 1984 by Elsevier Science Publishing Co., Inc.
Aerosols, Liu, Pui, and Fissan, editors

PARTICLE SIZE DISTRIBUTION OF MAINSTREAM CIGARETTE SMOKE UNDERGOING DILUTION

Pao-T. Chang, Leonard K. Peters, and Yasuo Ueno
Department of Chemical Engineering
University of Kentucky
Lexington, Kentucky 40506
U.S.A.

EXTENDED ABSTRACT

Introduction

Information on particle size distribution and number concentration of cigarette smoke aerosol is necessary to evaluate its effects on smokers and non-smokers. This paper describes research to study these properties of mainstream cigarette smoke undergoing different dilution conditions. Both composition and flowrate of the dilution stream were varied to determine the effects on the size characteristics of cigarette smoke.

Previous Research

The uncommonly high concentration and correspondingly rapid coagulation of cigarette smoke have made it difficult to measure the size distribution and number concentration. Investigators have used a variety of techniques for such experiments. However, one factor frequently discussed was the amount of air used to dilute the aerosol before the concentration and size distribution measurements. Several of the more relevant previous investigations are summarized in Figure 1. The data reported on mass mean diameter and number concentration have been used to estimate the equivalent total particulate matter (TPM) generated by each cigarette, assuming a particle density of 1 g/cm3. The calculated TPM results are compared to a TPM value of 38.8 mg/cigarette, which corresponds to the TPM of the Kentucky 2R1 Reference cigarette. It should be noted that the 2R1 was not used in all studies, but variations in the experimental measurements generally override the calculated TPM variations among the cigarettes.

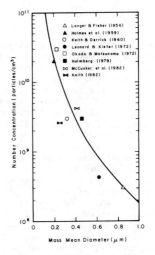

FIGURE 1. Comparison of the number concentration and mass mean diameter of previous studies to the corresponding values for a TPM of 38.8 mg/cigarette

Experimental

A Smoke Exposure Machine (SEM-II), which generates a continuous stream of smoke with reverse puffing, and a flow dividing dilution system were the major components of the apparatus used in this study. In some runs, two SEM-II's were

738

employed. The first, or primary SEM-II, provided cigarette smoke for analysis. The second SEM-II was used to supply a stream of filtered cigarette smoke for diluting the smoke aerosol from the primary SEM-II. For both SEM-II's, the puff pressure was held constant and the cigarette smoke was immediately diluted with laboratory air, dry air, or nitrogen. This primary dilution ratio was either 6:1, 10:1, or 18:1.

After the initial dilution, the mainstream smoke from the primary SEM-II passed through a TSI Model 3054 Neutralizer to bring the particles to charge equilibrium. Then, about 90% of the smoke passed through an Andersen Cascade Impactor and backup filters. A stream of either nitrogen or filtered cigarette smoke was introduced to dilute the remaining 10% of the cigarette smoke in the flow dividing dilution system. The overall dilution ratio was typically 50,000 but varied from 13,600 to 79,000. Following dilution, the smoke aerosol was directed to an Electrical Aerosol Size Analyzer (EAA) and Environment One Rich 100 Condensation Nuclei Monitor (CNC).

To calibrate the aerosol analyzing instruments, a DeVilbiss Nebulizer was used to generate aerosols by spraying aqueous NaCl solutions of concentration 0.1%, 1% and 10%. The generated aerosol passed through a flask partially filled with silica gel to promote evaporation of the water. The diluted aerosol passed through another flask and then analyzed with the EAA, CNC, and Andersen Cascade Impactor. The size distribution curves obtained from the EAA and the Andersen Cascade Impactor agreed reasonably well.

Results

A series of experiments was made by using filtered, dry air as the first dilution stream and the vapor phase from the secondary SEM-II as the second dilution stream. The number and volume distribution curves obtained with the EAA and the Andersen Cascade Impactor agreed with each other only approximately well. A typical volume distribution is shown in Figure 2. The volume distribution was distinctly bi-modal, which has not been reported before for mainstream cigarette smoke. For the small-diameter mode which was principally established by the EAA results, the average $D_g^{\overline{1}}$ was 0.239 μm, average $D_g^{\overline{3}}$ was 0.259 μm, and average σ_g was 1.18. The total number concentration obtained from the EAA was 3.6×10^9 particles/cm^3, and the calculated equivalent TPM was 11.2 mg/cigarette. The corresponding CNC result was 2.1×10^9 particles/cm^3. The second mode was considerably smaller (5 to 30% of the total volume) and centered around a particle diameter of 5.5 μm.

In another series of experiments, the primary dilution ratio was changed from 10

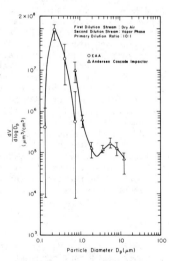

FIGURE 2. Average volume distribution of mainstream cigarette smoke for a primary dilution of 10:1

to 6 or 18 by adjusting the flowrate of the first dilution stream. Figure 3 shows that the equivalent TPM from the EAA results decreased when the primary dilution ratio increased. Both $\overline{D^1_g}$ and $\overline{D^3_g}$ behaved similarly. Comparison of the trends of the sub-micron TPM from the EAA and the Impactor suggested that the cigarette smoke particles decreased in size from the super-micron to the sub-micron range due to greater evaporation at the higher primary dilution.

From the number and volume distribution curves for all the cases studied, the system showed considerable self-consistency. When the primary dilution ratio was held constant (10:1), the TPM values determined using the Andersen Cascade Impactor were about 20 mg/cigarette and agreed reasonably well with Cambridge filter TPM measurements made by Schillinger (1983). However, the equivalent TPM calculated from the EAA data were only about half of the Andersen Cascade Impactor results. Several reasons can be advanced to explain these differences. Two possible reasons are (a) water and volatile organic particulate matter evaporated during the primary dilution process; and/or (b) the concentration of the cigarette smoke increased when

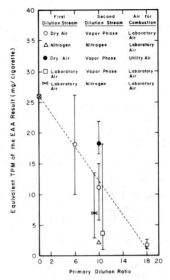

FIGURE 3. Effects of primary dilution ratio on the equivalent TPM from EAA results

the cigarette butt became shorter, and the Impactor sampled for the entire cigarette burn while the EAA sampling took only about 2 puff-cycles. Neither number concentration nor size distribution appeared to be substantially affected by the composition of the secondary dilution medium.

When the primary dilution ratio was increased from 6 to 18, the total TPM results obtained with the Andersen Cascade Impactor remained nearly the same, the super-micron TPM decreased, and the sub-micron and backup filter TPM increased. The equivalent TPM calculated from the EAA results (basically sub-micron TPM also) decreased when the dilution ratio increased. The mean values of $\overline{D^1_g}$, $\overline{D^3_g}$, and total number concentration obtained with the EAA showed trends similar to the EAA TPM results. In addition, the size distribution curves obtained from both the EAA and the Andersen Cascade Impactor shifted toward smaller particle diameters. In other words, the peak particle diameter of both modes decreased when the primary dilution ratio increased. This indicates that evaporation caused by higher primary dilution decreased the cigarette smoke particle size from super-micron to sub-micron. At the same time, the number and mean size of the particles contained in the small-size mode decreased. The EAA results showed trends similar to the data of Holmes et al. (1959) and Hinds (1978).

In summary, increased primary dilution from 6 to 10 to 18-fold caused evaporation of the particles. From these data, the calculated effective vapor pressure of the volatile species in cigarette smoke was 0.3 to 4 mm

Hg. For the small-size mode, complete evaporation of some particles occurred while partial evaporation of others was indicated. The large-size mode also shifted to a smaller size overall, indicating particle evaporation. The secondary dilution, typically 50,000 fold, did not further affect the size distribution.

Acknowledgements

This research was supported by the Kentucky Tobacco and Health Research Institute.

References

Hinds, W. C. (1978) "Size Characteristics of Cigarette Smoke", J., Am. Ind. Hyg. Assoc. 39:48-54.

Holmberg, R. W. (1978) "Determination of Particle Size in Tobacco Smoke Inhalation Exposure Devices Using Methylcyanoacrylate Fixation and Scanning Microscopy", Tobacco Smoke Inhalation Bioassay Chemistry (M. R. Guerin, J. R. Stokely, C. E. Higgins, and G. W. Griffith, ed.), ORNL-5524 Report, Oak Ridge National Laboratory, Oak Ridge, Tennessee, 103-118.

Holmes, J. C., Hardcastle, J. E. and Mitchell, R. I. (1959) "The Determination of Particle Size and Electric Charge Distribution in Cigarette Smoke", Tobacco Sci. 3:148-153.

Keith, C. H. (1982) "Particle Size Studies on Tobacco Smoke", Beit. zur Tabakfor. Int. 11:123-131.

Keith, C. H. and Derrick, J. C. (1960) "Measurement of the Particle Size Distribution and Concentration of Cigarette Smoke by the 'Conifuge'", J. Coll. Sci. 15:340-356.

Langer, G. and Fisher, M. A. (1956) "Concentration and Particle Size of Cigarette-Smoke Particles", Arch. Ind. Health 13:372-378.

Leonard, R. E. and Kiefer, J. E. (1972) "Coagulation Rate of Undiluted Cigarette Smoke", Tobacco Sci. 16:35-37.

McCusker, K., Hiller, F. C., Wilson, J. D., McLeod, R., Sims, R., and Bone, R. C. (1982) "Dilution of Cigarette Smoke for Real Time Aerodynamic Sizing with a SPART Analyzer", J. Aerosol Sci. 13:103-110.

Okada, T. and Matsunuma, K. (1974) "Determination of Particle Size Distribution and Concentration of Cigarette Smoke by a Light Scattering Method", J. Coll. Interface Sci. 48:461-469.

Schillinger, R. E. (1983) "Size Characterization of Mainstream Cigarette Smoke", M.S. Thesis, Department of Chemical Engineering, University of Kentucky, Lexington, Kentucky.

Published 1984 by Elsevier Science Publishing Co., Inc.
Aerosols, Liu, Pui, and Fissan, editors

MODELING OF SOOT GROWTH

G. W. Mulholland and R. D. Mountain
National Bureau of Standards
Washington, DC 20234
USA

EXTENDED ABSTRACT

Introduction

There are many situations where it is desirable to control the formation and emission of soot. One would like to minimize the amount of soot produced by internal combustion engines, gas-turbine combustors, and large fires, for which the dominant mode of fire spread results from radiation arising from hot soot particles. There are also situations where it is desirable to increase the soot production including the production of carbon black and the efficient operation of a furnace.

The stages of soot particle growth include particle inception (sometimes called nucleation), surface growth, coagulation, and burnout (Haynes and Wagner, 1981). A global model including these processes has been developed for describing the formation and emission of soot for a homogeneous system. The focus here is on integrating the various stages of growth, each of which is a major research area in its own right.

The growth of aggregate structures typical of soot is studied separately by a computer simulation technique. In this case, the primary interest is in the structure of the aggregate and in the growth kinetics.

Global Soot Growth Model

The growth of soot is described by a free radical growth model, which is really more a framework for studying soot growth than being a specific model. The particle inception corresponds to the pyrolysis of the fuel, F, to form free radicals, R_1 and R_A. One of the free radicals, R_1, acts as an initiator for growth through the successive addition of fuel molecules. Each of the growing radicals is considered to be a soot particle and growth by coagulation is also included in the model. The three steps are given below in their simplest form.

$$F \xrightarrow{k_1} R_1 + R_A \qquad \text{fuel pyrolysis} \qquad (1)$$

$$F + R_1 \xrightarrow{k_2} R_2 + H_2$$

$$F + R_2 \xrightarrow{k_2} R_3 + H_2 \qquad \text{free radical growth} \qquad (2)$$

$$F + R_n \xrightarrow{k_2} R_{n+1} + H_2$$

$$R_i + R_j \overset{\Gamma}{\rightarrow} R_{i+j} \qquad\qquad \text{coagulation} \qquad\qquad (3)$$

The amount of soot surviving the burnout region is determined by using the Nagle and Strickland-Constable (1962) kinetics for the oxidation of carbon.

$$C_s + O_2 \rightarrow CO \qquad\qquad \text{burnout} \qquad\qquad (4)$$

The rate of mass loss is proportional to the particle surface area, is weakly dependent on O_2 concentration below 1800 K, and has an Arrhenius temperature dependence with an actuation energy of 34 kcal.

Solving this model in the limit of small fuel depletion, one finds that the soot concentration has a quadratic dependence on both time and the fuel concentration. One also finds that the characteristic time for surface growth and coagulation are similar. The basic model has been extended to include a growth rate proportional to the surface area rather than constant and to include the additional pyrolysis of the fuel to form a growth species which is taken to be acetylene.

Soot Aggregation

The aggregation process is a consequence of chance encounters of particles undergoing Brownian motion. The kinetics of this process is modeled in the following way. Initially the system consists of N individual, isolated particles of unit diameter σ and mass m_0 undergoing diffusive motion described by the Langevin equation for the velocity,

$$m_0 \frac{d\vec{v}}{dt} = - m_0 \beta \vec{v} + \vec{F} \qquad\qquad (5)$$

where $\vec{F}$ represents the random forces provided by collisions with the host gas molecules. Aggregation occurs whenever the separation of two particles becomes unit distance or less. In that case the two particles "stick" and then diffuse as a rigid entity. As particles continue to diffuse, successively larger aggregates are formed.

The primary quantities of interest in this simulation are the radius of gyration, R_g, of the aggregate, which gives a measure of the overall size, and the time dependence of the number concentration of clusters. We find that over the cluster size range from 2 to 50, R_g has a power law dependence on the number of spheres in the aggregate, k, given by

$$R_g \sim k^\alpha , \qquad\qquad (6)$$

where $\alpha = 0.56$. A value of α of 1/3 would correspond to a close packed aggregate; whereas, a value of α of 0.56 corresponds to an open structure. We find that the coagulation rate for aggregates is much greater than for coalescing spheres. This is a consequence of the much larger collision cross section of an aggregate compared to a sphere with the same mass.

References

Haynes, B.S. and Wagner, H.G. (1981) "Soot Formation", Prog. Energy Combust. Sci. 7:229.

Nagle, J. and Strickland-Constable (1982) "Oxidation of Carbon Between 1000-2000°C, Proceed. of Fifth Conf. on Carbon, 154.

Published 1984 by Elsevier Science Publishing Co., Inc.
Aerosols, Liu, Pui, and Fissan, editors

IN-CYLINDER MEASUREMENTS OF DIESEL PARTICLE SIZE DISTRIBUTIONS

C.J. Du* and D.B. Kittelson
Department of Mechanical Engineering
University of Minnesota
111 Church Street S.E.
Minneapolis, MN 55455, U.S.A.

EXTENDED ABSTRACT

Introduction

The diesel engine is more fuel efficient than the spark ignition engine in many applications. As fuel costs have risen, its use has spread from heavy-duty vehicles to other applications, including light-duty trucks and passenger cars. This has led to concern about diesel engine emissions (1). Although, it is relatively easy to control diesel exhaust hydrocarbon and carbon monoxide emissions, control of emissions of oxides of nitrogen and particulate matter is difficult. In order to understand the formation and removal of particulate matter during diesel combustion, both physical and chemical processes taking place must be examined. This paper presents preliminary measurements of an important physical characteristic of diesel particles, i.e., their size distribution, as they are formed and consumed within the combustion chamber.

Experimental Apparatus

The heart of the experimental apparatus is a unique single-cylinder version of a 1979 5.7-liter Oldsmobile V-8 diesel engine. This engine is equipped with a sampling system which is capable of removing, quenching, and diluting essentially the entire contents of the cylinder during actual engine combustion. The system has been described in detail in several previous publications (2,3,4). Figure 1 shows the general arrangement of the single-cylinder engine and sampling system. The cylinder contents are dumped at a predetermined point in the engine cycle by cutting a 14-mm diameter diaphragm which is mounted flush with the surface of the pre-chamber. The characteristic time for blowdown and primary quenching is about 0.5 ms. Samples undergo an essentially adiabatic expansion into the sampling apparatus, where they are diluted with nitrogen in a ratio of about 10. The diluted combustion-chamber contents then flow into a poly-ethylene sampling bag which has been filled with enough nitrogen to give an overall dilution ratio of approximately 330. The contents of the sample bag are analyzed using a TSI Model 3030 Electrical Aerosol Analyzer (EAA). The sample is further diluted with clean air prior to entering the EAA. The secondary ratio is about 16, which gives an overall dilution ratio at the EAA inlet of about 5000. The EAA measures particle concentration and size distribution in the 0.01 to 1.0 um diameter range. It is described in further detail in (3).

*Department of Research and Development, Shanghai Diesel Engine Works, Shanghai, People's Republic of China.

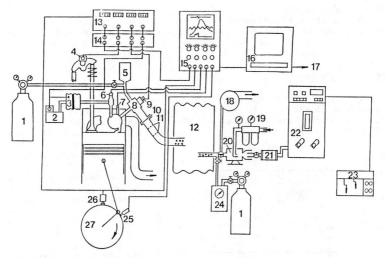

Figure 1 - General Arrangement of the Single-Cylinder Engine and Sampling System

1. nitrogen cylinder; 2. fuel injection pump; 3. magnetic fuel-line break device; 4. magnetic valve deactuator; 5. dilution nitrogen bottle; 6. injector needle-lift pickup; 7. combustion-chamber pressure transducer; 8. precombustion chamber dumping apparatus; 9. squib; 10. sampling valve; 11. shock-wave silencer; 12. sampling bag; 13. four-channel, phase-locked loop electronic timer; 14. four-channel high-current driver; 15, digital oscilloscope; 16. computer terminal; 17. computer center; 18. vacuum pump; 19. filter and dryer; 20. low dilution system; 21. krypton-85 neutralizer; 22. electrical aerosol analyzer; 23. strip chart recorder; 24. gas meter; 25. crankangle marker pick-up; 26. magnetic TDC pick-up; 27. flywheel of the engine

Procedure

The engine was operated at a single speed-load condition, 1000 RPM, and an equivalence ratio of 0.32. Number No. 2 diesel fuel was used.

The EAA measures a number weighted size distributions which are converted to volume weighted form by assuming solid spherical particles. The aerosol in the sampling bag changes fairly rapidly with time as a result of coagulation and wall losses. In order to determine size distributions at the instant of blowdown, a series of EAA measurements were made and the results were extrapolated back to blowdown time. This extrapolation was facilitated by fitting a polynomial to the variation of current in each EAA channel with time.

The total cylinder sampling system used for these experiments suffers from a significant flaw when used for particle sampling: the sampling process itself reentrains particulate matter from walls of the combustion chamber. The mass of reentrained particles produced by the blowdown process is similar to the mass of suspended particles present in the cylinder. Thus, it is necessary to correct our results for this reentrained mass. This is done by conducting fired and non-fired blowdown experiments under essentially identical conditions and assuming that the difference between fired and non-fired size distributions is due to suspended particles present in the combustion chamber at the instant of blowdown. The details of this procedure are given by Du and Kittelson (2).

Results and Discussion

Figure 2 shows fired and non-fired particle volume concentrations plotted against blowdown angle. Non-fired concentration levels are of the same order as fired levels. Thus, the reentrainment correction is very important. The difference between the fired and non-fired curve is taken

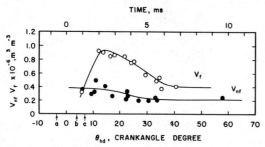

Figure 2 - Fired and Non-Fired Volume Concentrations in the Cylinder Plotted Against Blowdown Angle. The arrows indicate: (a) start of fuel injection, (b) start of combustion, and (c) end of fuel injection. All concentrations are refired to 300K and 1 atm.

as the combustion generated particle concentration. Combustion generated particles appear shortly after the start of combustion and reach a peak at 20 CA (CA denotes crankangle degrees after top dead center). This peak corresponds to roughly three times the concentration observed at 40 CA. After 40 CA, the difference between fired and non-fired remains approximately constant and is essentially equal to the particle concentration in the exhaust.

Figure 3 shows the variation of the combustion generated particle size distribution with crankangle. These combustion-generated size distributions were obtained by taking the difference between the fired and non-fired distributions. The exhaust size distribution is also shown. The last distribution shown, i.e., 40 CA is very similar to that of the exhaust, except for the largest size range, i.e., 0.75 um diameter parti-

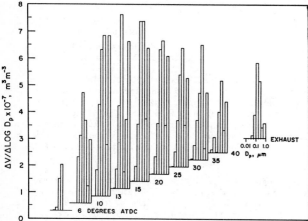

Figure 3 - Combustion Generation Particle Dise Distributions at Different Crankangles. The diameter scale is shown on the exhaust distributions. All concentrations are refired to 300K and 1 atm.

cles. The relatively large concentration of these particles in the cylinder sample may be due to sampling reentrainment which is not adequately corrected for by taking the difference between fired and non-fired distributions. Figure 4 shows plots of the geometric volume mean diameters versus crankangle for the fired, non-fired, and combustion generated size distributions. The particles produced by non-fired blowdowns and are relatively large which would be expected for reentrained particles. The fired particles are a mix of smaller particles produced by combustion and larger particles produced by reentrainment and thus show an intermediate size. The combustion generated particles start very small early in the combustion process, reach a maximum diameter of about 0.24 um at about 20 CA, and then slowly decrease in size with increasing crankangle toward the exhaust diameter of 0.17 um.

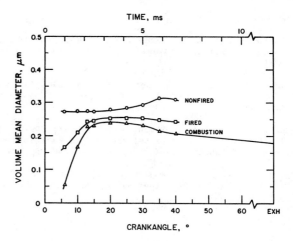

Figure 4 - Geometric Volume Weighted Mean Particle Diameters Plotted Against Blowdown Angle and Time. Results for fired, non-fired, and combustion-generated (the difference between fired and non-fired) size distributions are shown.

We have not attempted to apply detailed growth models to these results because sampling artifacts as evidenced by the large concentrations of particles in the last EAA channel are evidently present and may bias the results. However, we believe that the general trends of the particle size distribution development shown in this paper are correct. Particles are presumably formed initially by nucleation, grow by coagulation and surface growth, and finally decrease somewhat in diameter as a result of oxidation. Since these results were obtained, a new system which greatly reduces the concentration of reentrained particles has been developed and will be used to make more detailed size distribution measurements.

Acknowledgements

The authors wish to thank General Motors Research Laboratories for their financial support of this project. We wish to thank Mr. Michael J. Pipho for his assistance in conducting these experiments and Mr. Jeffrey L. Ambs and Ms. Betsy A. Antinozzi for their assistance in preparing this manuscript.

748

References

1. Amann, C.A., D.L. Stivender, S.L. Plee, and J.S. MacDonald, (1980) "Some Rudiments of Diesel Particulate Emissions," SAE Paper No. 800251.

2. Du., C.J. and D.B. Kittelson, (1983) "Total Cylinder Sampling from a Diesel Engine: Part III - Particle Measurements," SAE Paper No. 830243.

3. Lipkea, W.H., J.H. Johnson, and C.T. Vuk, (1978) "The Physical and Chemical Character of Diesel Particulate Emissions--Measurement Techniques and Fundamental Considerations," SAE Paper No. 780108.

4. Liu, X. and D.B. Kittelson, (1982) "Total Sampling from a Diesel Engine (Part II), SAE Paper No. 820360.

Published 1984 by Elsevier Science Publishing Co., Inc.
Aerosols, Liu, Pui, and Fissan, editors

SHAPE AND SIZE DISTRIBUTION OF COMBUSTION GENERATED AEROSOLS

J. J. Helble, M. Neville, and A. F. Sarofim
Department of Chemical Engineering
Massachusetts Institute of Technology
Cambridge, MA   02139
U.S.A.

EXTENDED ABSTRACT

Introduction

Ultrafine particles (0.01 to 0.10 um diameter) are produced during pulverized coal combustion by vaporization and subsequent condensation of a small fraction of the mineral matter present. Homogeneous nucleation occurs when mineral matter vaporized in a suboxide or elemental form is oxidized in the particle boundary layer, leading to supersaturation followed by nucleation (Flagan 1979, Quann 1982). Particles then grow by a coagulation and condensation mechanism until the temperature of the surroundings decreases. Aggregates of distinct spherical particles (primary particles) are then observed (Neville, 1982). The factors governing the transition from coalescence to aggregation are examined in this paper. Particle size distributions for the primary particles within the aggregates are also discussed.

Results of Coal Studies

The influence of mineral matter content on aerosol particle size was examined by burning four coals of varying composition (Table I) in a laminar flow drop tube furnace. Under identical combustion conditions of gas temperature, oxygen content (30%) and residence time, the volume mean primary particle diameter (as determined by automated image analysis of TEM micrographs) varied with coal type, as shown in Table II. At both temperatures considered, the Montana lignite produced the largest mean particle size, while the Alabama Rosa bituminous produced the smallest.

Table I. Coal Properties (Quann, 1982)

	Montana Lignite	Wyoming Commanche sub-bituminous	Pittsburgh #8 (bituminous)	Alabama Rosa #18 (bituminous)
1) Raw Coal (wt%)				
Si	0.87	0.63	1.31	1.65
Al	0.64	0.51	0.90	0.70
Fe	0.30	0.34	0.81	0.56
Ca	1.44	0.99	0.56	0.70
Mg	0.58	0.27	0.08	0.06
2) Ash Oxide Composition (20% O_2) - (wt%)				
SiO_2	3.7	3.2	41.2	19.6
Al_2O_3	0.6	0.8	1.3	1.9
Fe_2O_3	11.5	35.2	18.4	47.4
CaO	7.8	6.9	1.4	1.1
MgO	64.4	31.3	2.9	3.7

750

Table II. Mean Aerosol Diameter - 30% O_2

	*	dp	*	dp
Montana Lignite	8.4%	409 A	6.7%	394 A
Wyo Com sub-bit	10.5	328	7.0	273
Pitts #8	5.8	226	2.6	193
Ala Rosa #18 bitum	4.1	185	2.8	131
Mg/SC	3.0	237	-	-
Fe/SC	35.9	296	26.9	233

* Percentage of ash recovered in aerosol

Aggregates of spherical particles were observed to form for all of the coals studied (Figure 1). Post-combustion cooling effects are directly responsible for this. Temperatures decrease rapidly in the post combustion zone because of the influence of a water-cooled collection probe. Coalescence times are no longer instantaneous (< 1 msec), and growth by solid state diffusion becomes limiting, causing aggregates to form. Because diffusion effects vary between compounds and hence between aerosol particles from different parent coals, it is expected that coalescence will become growth limiting for different aerosols at different temperatures.

For the Montana lignite studied, MgO constitutes more than 60% of the submicron ash. Modelling the ash as pure MgO, one can predict coalescence times controlled by solid-state diffusion from

$$t = (\Delta L/Lo)^{5/2} \frac{\sqrt{2kT}}{(20D^*\gamma a^3)} r^3 \dots\dots\dots\dots\dots (1)$$

where $\Delta L/L_0$ is the fractional shrinkage in diameter of 2 coalescing spheres, k is Boltzmann's constant, D* is the self diffusion coefficient for the mobile species, γ is the oxide surface tension, and a^3 is the atomic volume of a diffusing vacancy (Kingery 1965). These times become appreciable (> 10 msec) at 1500 K, for dp = 100 A, and approach 100 msec at 1400 K for the lignite.

The residence time prior to quenching in the collection probe is of order 300 msec. It is expected that coalescence times become limiting for the lignite as the gas temperature falls below 1400 K. This will occur as the collection probe is approached due to the cooling effects of the water cooled surfaces of the probe. As expected, the time required to reach this transition temperature increases as gas temperature is raised from 1500 K to 1750 K. A corresponding increase in primary particle diameter is expected, and is experimentally observed (Table II).

Ultrafine particles produced by vaporization and condensation are expected to obey a self-preserving primary particle size distribution (Figure 2) (Friedlander, 1977). A model based on this principle was developed to predict mean particle diameter as a function of aerosol volume fraction and growth time (Neville et. al., 1981).

$$\overline{dp} = (6/\pi)^{1/3} (k_{theory} Vt)^{2/5} \dots\dots\dots\dots\dots (2)$$

Equating growth times with residence times, as suggested by equation (1), one can predict the mean primary particle diameter expected. For the Montana lignite studied, equation (2) accurately predicted mean diameter up to the time at which aggregates were observed. At longer residence times the theory gave an approximate measure of the diameter, determined from projected areas, of the aggregates.

High iron content ashes, such as those produced from the Wyoming Commanche or Alabama Rosa, are predicted by equation (1) to coalesce at lower temperatures than the MgO dominated lignite. For equal volume fractions, the high iron ash is therefore expected to produce larger particles. This was not observed, as the results in Table 2 indicate. Multicomponent effects may be responsible for this, as compounds such as SiO_2, which coat the primaries (Neville 1982), form aggregates at much higher temperatures. Coalescence of SiO_2 by viscous flow (Ulrich et. al. 1977) reaches 100 msec for 100 A diameter particles at 1850 K.

Results of Model Coals Studied

To completely eliminate multicomponent effects, a high surface area pure carbon product ("spherocarb") was immersed in either a $Mg(NO_3)_26H_2O$/water solution, or a $Fe(NO_3)_39H_2O$/water solution. Particles were then evaporated to dryness and analyzed for metal content. In each case the metal concentration in the particles was approximately 5% by weight.

Spherocarbs were burned at the same residence time and oxygen content as the coals. Chain aggregates of primary particles were produced when the iron impregnated spherocarbs (Fe/SC) were burned (Figure 1). Primary particle sizes are as shown in Table 2. A comparison of mean primary particle diameters indicates that Fe/SC produced particles were smaller than those obtained from a MgO rich Montana lignite fume. This suggests that particle burning times (and therefore growth time prior to coalescence-controlled growth) are much shorter than those for the Montana lignite.

Magnesium impregnated spherocarbs (Mg/SC) also produced aerosol particles of a smaller mean diameter than the Montana lignite. This is likely due to the low degree of vaporization of Mg observed (Table 2). Energy dispersive x-ray analysis (EDXA) indicated Mg had surface coated the spherocarbs, leaving virtually no detectable Mg in the particle interiors. (Fe was found to be well dispersed by EDXA.) Large surface inclusions such as these tend to melt and fuse together, forming large (> 1 um) residual ash particles (Quann, 1982). Most of the mass of the Mg was in fact recovered as larger particles of this type.

A low degree of Mg vaporization, as compared with extensive Fe vaporization, suggests that iron may be catalyzing the combustion of spherocarb, producing higher temperatures and shorter burning times. Additional study is being conducted to elucidate these findings.

Primary particles produced during spherocarb combustion were found to obey the self preserving size distribution function (Figure 2). Determination of burning times should therefore permit accurate prediction of mean primary particle size via equation (2).

Concluding Comments

The primary particle sizes of aerosols produced during combustion vary widely, by a factor of two for the conditions of this study. These variations are a function of combustion conditions which determine both the amount and composition of the aerosol produced and the temperature history which determines coalescence times.

752

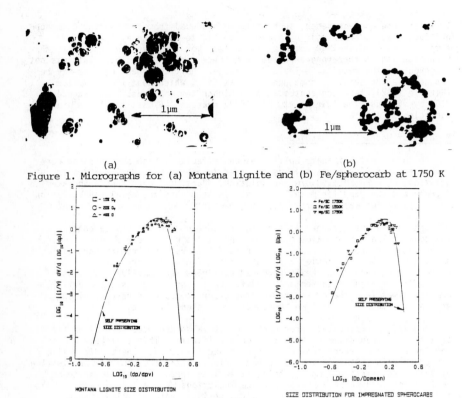

(a)   (b)

Figure 1. Micrographs for (a) Montana lignite and (b) Fe/spherocarb at 1750 K

Figure 2. Self preserving size distributions

References

Flagan, R.C. (1979), Seventeenth Symposium (International) on Combustion, p.97, The Combustion Institute.

Friedlander, S.K. (1977), Smoke, Dust, and Haze, J. Wiley and Sons, New York.

Kingery, W.D. (1965), Introduction to Ceramics, J. Wiley and Sons, New York.

Neville, M., R.J. Quann, B.S. Haynes and A.F. Sarofim (1981), Eighteenth Symposium (International) on Combustion, p.1267, The Combustion Institute.

Neville, M (1982), Sc.D. Thesis, Department of Chemical Engineering, M.I.T.

Quann, R. (1982), Sc.D. Thesis, Department of Chemical Engineering, M.I.T.

Ulrich, G.D., and N.S. Subramentan (1977), Combustion Science and Technology, 17:119.

Published 1984 by Elsevier Science Publishing Co., Inc.

Aerosols, Liu, Pui, and Fissan, editors

SIZE DISTRIBUTION OF FINE PARTICLE EMISSIONS FROM A
FLUIDIZED BED COAL COMBUSTOR AT A STEAM PLANT*

R. L. Carpenter, Y. S. Cheng, and E. B. Barr
Inhalation Toxicology Research Institute
Lovelace Biomedical and Environmental Research Institute
P. O. Box 5890
Albuquerque, NM 87185
U. S. A.

EXTENDED ABSTRACT

Introduction

Fluidized bed coal combustion (FBC) offers possible improvements over conventional coal combustion in the complexity of the equipment needed to clean stack emissions to environmentally acceptable standards. Potential applications of this technology include augmentation of existing coal fired plants for central steam generation, electrical power generation, and combined cycle electrical power generation using gas turbines in the combustor flue gases in addition to conventional steam turbines.

A fluidized bed coal combustor contains a bed of material containing approximately 1% coal and 99% coal ash and limestone. Combustion in this bed is uniform throughout the bed at temperatures between 700-900°C. This low temperature operation with a limestone rich bed results in low emissions of nitrogen and sulfur oxides. If FBC is to be a viable alternative combustion technique in the U.S., FBC flue gas emissions should meet the 1978 new source performance standards (NSPS) rquired by the EPA for sulfur dioxide (1.2 lb/million Btu) and particulates (0.1 lb/million Btu).

We wish to report the physical characterization of fly ash from a 20 meter square FBC as part of an effort to provide emission source data on a commercial scale FBC. Fly ash particle size distribution of flue gas aerosol before and after the plant bag house filter are reported for both micron- and submicron-sized fly ash particles. The formation of submicron fly ash particles has been reported for pulverized coal power plants but little information is available for the production of these particles from FBC (McElroy et al. 1982; Schmidt et al. 1976; Desrosiers et al. 1979; Neville et al. 1983). These particles are of interest due to their potential for haze formation as well as the potential for toxic effects from trace element enrichment on their surfaces.

Methods

The sampling technique used for collecting materials during this study has been described previously (Newton et al. 1980). A probe was inserted into the FBC flue to extract a portion of the flue gas aerosol. This aerosol was diluted and cooled immediately with filtered air to quench any further chemical reactions and delivered to a sampling chamber. Several

*Research performed under U. S. Department of Energy Contract Number DE-AC04-76EV01013.

754

aerosol sampling instruments could be operated in parallel to collect
aerosol material from the chamber. Cascade impactors were used to measure
fly ash aerodynamic diameter, filter samples were collected to measure
aerosol concentration and a point-to-plane electrostatic precipitator was
used to collect particles for electron microscopy. A train of cyclones was
used to collect size-classified fly ash and particles penetrating the
cyclones were sized by an electrical aerosol analyzer (EAA). Samples were
collected before and after the baghouse filter in order to assess its
collection efficiency.

Results

Fly ash mass concentration before the baghouse was about 1.5 g/cu.M but
was reduced to about 0.35 mg/cu.M after the baghouse. Fly ash size
distribution was bimodal. The mass fraction of the submicron particles was
4.3% of the total aerosol distribution before the baghouse and 18% after the
baghouse. Figures 1 and 2 show the aerosol size distribution before and
after the baghouse respectively. Figure 3 shows the number size
distribution of the fly ash after the baghouse. The solid line is a
theoretical model (Flagan and Friedlander 1978). The empty circles (EAA)
and triangles (impactor) are our results and the dots are data from a
pulverized coal combustor (McCain et al. 1975).

Conclusions

A bimodal size distribution of the fly ash emitted from the Georgetown
AFBC was observed. The mass fraction of the submicron mode was enhanced in
the stack emission because the baghouse had high efficiency for larger
particles. The high collection efficiency (99.98%) of the newly installed
baghouse removed most of fly ash; therefore the emission factor of particles
was below the EPA guideline.

References

Desrosiers, R. E., J. W. Riehl, G. D. Ulrich, and A. S. Chiu (1979)
"Submicron Fly-Ash Formation in Coal-Fired Boilers", in Proceeding 7th
Symposium of Combustion, pp. 1395-1403.

Flagan, R. C. and S. K. Friedlander (1978) "Particle Formation in Pulverized
Coal Combustion-A Review", in Recent Developments in Aerosol Science (D. T.
Shaw, ed.), Wiley, New York, NY, pp. 25-60.

McElroy, M. W., R. C. Carr, D. S. Ensor, and G. R. Markowski (1982) "Size
Distribution of Fine Particles from Coal Combustion", Science 215:13-19.

Neville, M., J. F. McCarthy, and A. F. Sarofim (1983) "Size Fractionation of
Submicrometer Coal Combustion Aerosol for Chemical Analysis", Atmos.
Environ. 17:2599-2604.

Newton, G. J., R. L. Carpenter, H. C. Yeh, and E. R. Peele (1980) "Respirable Aerosols from Fluidized Bed Coal Combustion. 1. Sampling Methodology for an 18-inch Experimental Fluidized Bed Coal Combustor", _Environ. Sci. Technol._ 14:849-853.

Schmidt, E. W., J. A. Gieseke, and J. M. Allen (1976) "Size Distribution of Fine Particulate Emissions from a Coal-Fired Power Plant", _Atmos. Environ._ 10:1065-1069.

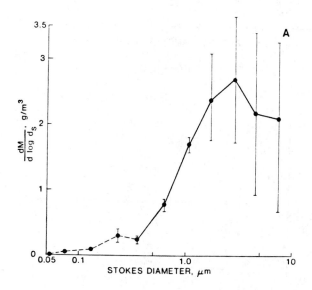

FIGURE 1

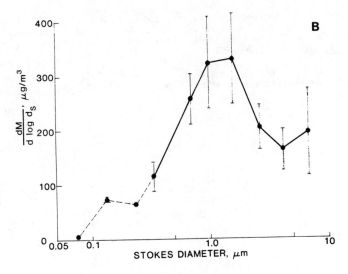

FIGURE 2

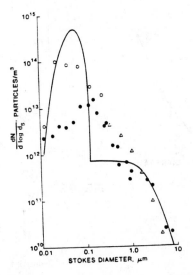

FIGURE 3

Published 1984 by Elsevier Science Publishing Co., Inc.

Aerosols, Liu, Pui, and Fissan, editors

SUBMICROMETER AEROSOL SIZE DISTRIBUTIONS FROM AN OIL-FIRED BOILER

Gregory R. Markowski
Consulting Aerosol Scientist
2009 N. Madison Ave.
Altadena, Ca. 91001

Jerry L. Downs
Consultant
8341 Grenoble Street, #18
Sunland, Ca. 91040

James L. Reese
Blair A. Folsom
Energy and Environmental Research Corp.
18 Mason St.
Irvine, Ca. 92714

EXTENDED ABSTRACT

We measured submicrometer aerosol size distributions using
a dilution system and TSI Model 3030 Electrical Aerosol Analyzer
(EAA) at the outlet of a 45 MW utility boiler. Measurements were
made under normal and "low NO_x" firing modes for each of two
fuels: No. 4 fuel oil (the normal fuel), and a synthetic liquid
fuel derived from coal using the Exxon Donor Solvent (EDS)
process. Measurement techniques are described in more detail in
McElroy et al.(1982). Data inversion was done using Twomey's
algorithm (1975). This work was part of a larger study funded
by the Electric Power Research Institute to evaluate the suit-
ablity of EDS fuel as a replacement fuel for utility boilers.
Details of the entire study can be found in Reese et al.
(1984).

The volume distributions from the No. 4 oil were characterized by two distinct modes, at about 0.0105 and 0.20 micrometers. Both modes were approximately log normal with peak
concentrations (dV/d(logD)) of about 7×10^{-5} cc/normal cubic
meter and geometric standard deviations of 1.4. Little difference was seen between the normal and low NO_x firing modes. The
small mode was fairly stable in concentration and was probably
formed, at least in part, by ash vaporized during combustion.
The large mode was highly variable over periods as short as a
few seconds and probably was soot.

The volume distributions from the EDS liquid synthetic fuel
were less well defined in shape and had only one obvious mode.
Figure 1 shows the results for the EDS fuel and includes a comparison to the No. 4 fuel oil results. The smaller mode between
0.02 and 0.1 micrometer present during both types of firing
seemed too broad to have been formed by a simple vaporization-
condensation process alone. The large mode at about 0.3 micro-

758

meters increased by a factor of about 10 during the 5 hour test
period (hence the large error bars) and was likely soot. The
number concentration of this larger mode was too small to indicate   formation by a simple homogeneous nucleation process
followed by coagulation. Probable factors contributing to the
increase in the large mode were the higher aromatic content of
the EDS fuel and a 0.5 percent decrease in excess oxygen during
the sampling period.

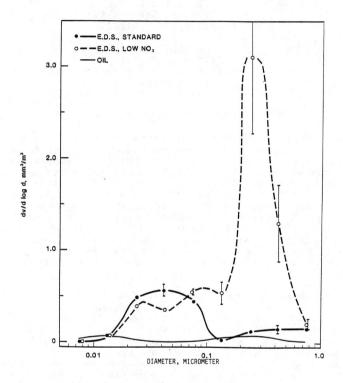

Figure 1. Volume distribution for EDS fuel. Fuel oil results
          are shown for comparison. Error bars reflect the
          relative standard error of the mean of the raw data
          for each point. For points without error bars, the
          variability was smaller than the size of the dot or
          circle. Variation of the 0.3 micrometer peak is dis-
          cussed above.

References

McElroy, M., R. Carr, D. Ensor, and G. Markowski (1982) "Size Distribution of Fine Particles from Coal Combustion", Science 215:13.

Reese, J. L., B. Folsom, and F. Jones, Principal Investigators (1984) "Evaluation of Exxon Donner Solvent as a Utility Boiler Fuel", EPRI Research Project RP 2112-4, Henry Schreiber, EPRI Project Officer.

Twomey, S. (1975) "Comparison of Constrained Linear Inversion and an Iterative Nonlinear Algorithm Applied to the Indirect Estimation of Particle Size Distributions", J. Comp. Physics 18:188.

Published 1984 by Elsevier Science Publishing Co., Inc.

Aerosols, Liu, Pui, and Fissan, editors

AEROSOL DYNAMICS IN THE CONVECTION SECTION
OF A
LABORATORY SCALE COAL COMBUSTOR

William P. Linak
Thomas W. Peterson
Department of Chemical Engineering
University of Arizona
Tucson, Arizona 85721

Simulation results of the particle dynamics in the convection section of a
laboratory-scale pulverized coal combustor were compared to data previously
reported (Linak & Peterson, 1984). A multi-component aerosol simulation code,
MAEROS (Gelbard, 1982), augmented by a stiff differential equation package,
EPISODE (Hindmarsh & Byrne, 1976) was used to simulate the aerosol dynamics.
Volume distributions measured at the entrance to the convection section were
used as initial conditions to the simulation, and predicted distributions
were compared to data at the appropriate aerosol residence times. Data from
three different coals including a Utah bituminous coal, a Beulah (North
Dakota) lignite and a Texas lignite were studied. These simulations, and the
subsequent data comparison, were useful in determining which aerosol dynamics
mechanisms most influence the evolution of the aerosol size distribution.

In addition, the MAEROS code, as modified to handle the fragmentation process,
was used to study the formation of ash particles during the combustion of
pulverized coal, taking into account fragmentation, nucleation, condensation,
coagulation and a variable temperature profile. The MAEROS results from this
implementation pointed out several characteristic differences in the resultant
ash distribution, depending on what the chosen mechanism for fragmentation was.

Gelbard, F. (1982). Sandia Natl. Lab. Reprt. No. SAND80-0822.

Hindmarsh, A.C. & Byrne, G.D. (1976). In "Numerical Methods for Differential
Systems", L. Lapidus & W.E. Schiesser, Eds. Academic Press, New York, NY.

Linak, W.P. & Peterson, T.W. (1984). Aerosol Science and Tech. (3),77-96.

Published 1984 by Elsevier Science Publishing Co., Inc.

Aerosols, Liu, Pui, and Fissan, editors

CHAMBER INVESTIGATIONS OF THE PHYSICAL, CHEMICAL AND BIOLOGICAL
FATE OF DIESEL EXHAUST EMISSIONS IN THE ATMOSPHERE

by

Thomas M. Albrechcinski, John G. Michalovic, Bruce J. Wattle,
Elizabeth P. Wilkinson
Environmental Sciences Department, Calspan Advanced Technology Center

and

David B. Kittelson
University of Minnesota, Department of Mechanical Engineering

Introduction

There have been many investigations into the potential effects of diesel exhaust emission on air quality due to the copious quantities of particulate matter emitted from diesel powered vehicles. Some of these studies have addressed purely the physical characteristics of the emissions, focusing on aerosol characteristics such as mass emissions, size distributions, coagulation rates, aerosol formation, interactions, and visibility. Numerous other studies have focused on the chemical composition of the soluble organic extracts derived from diesel particulate matter, especially since these extracts have shown to be mutagenic in the Ames assay.

The vast majority of these studies were conducted using standard laboratory dilution tunnel techniques where particulate samples were collected directly from the emission source. In such cases, the aerosol is non-aged with respect to residence in the atmosphere and, therefore, does not reflect modifications that may occur with atmospheric aging and exposure. The study described in this paper was initiated in response to the need to extend the characterization of diesel exhaust particulate emissions to longer residence times than one is capable of by conventional means.

In 1980 Calspan was awarded a contract by the Coordinating Research Council to conduct a two-task laboratory investigation of the fate of diesel emissions in the atmosphere. Experiments were conducted utilizing Calspan's 600 m^3 chamber to quantify the chemical, physical and biological changes that diesel particulate emissions undergo upon their initial dilution and during their lifetime in simulated atmospheres.

Program Discussion and Results

The principal objective of the first phase effort was to validate the use of the chamber for vehicle emissions studies. The basis for validation was a comparison of data from a series of parametrically similar chamber and over-the-road highway experiments using a 1978 Volkswagen Rabbit as the source of diesel exhaust.

To simulate the exhaust mixing and dynamic aerosol sampling processes that occurred in the highway experiments, the chamber/dynamometer experiments were formulated to reproduce these dynamic exhaust injection/mixing processes. This was achieved by continually flushing the chamber with non-filtered clean rural air during the introduction of prediluted diesel exhaust. Upon achieving an equilibrium exhaust/air mixture in the chamber, aerosol samples and related measurements were obtained with the continued introduction of exhaust and chamber flushing. The filtered mixture was discharged from the chamber. Vehicle operation on the dynamometer was matched to reproduce the speed and load of highway tests. During the initial task, experiments were conducted at nominal exhaust dilution ratios of 25:1 and 80:1. During highway experiments this ratio was determined from the raw exhaust mass concentration and the mass concentration in the exhaust plume at the sampling station. During chamber experiments, exhaust dilution ratios were established on the basis of the NO_x ratio between the raw exhaust and chamber mixture.

Figure 1 is a schematic representation of the exhaust injection/dilution scheme employed during the chamber experiments and also notes the test parameters and defines the variety of measurements made.

Comparisons were made of these non-aged aerosol systems on the basis of mass loading, size distribution, the percentage of the organic soluble fraction derived from aerosol samples, the chemical signature of organic aerosol extracts based on the EPA-SWRI fractionation procedure and the mutagenic activity of the extracts determined by Ames Salmonella assay.

Paired filter samples were obtained from several experiments which permitted duplicate analyses to be conducted at two laboratories. These results yielded useful information relating to the reliability and variability of data attributable to differences in analytical techniques and permitted quantitative assessment of the consistency of the results.

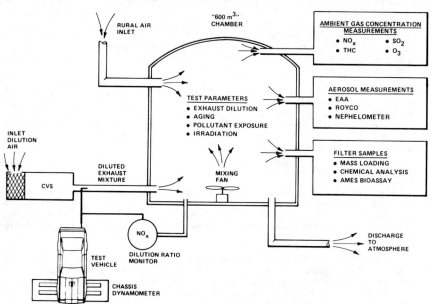

Figure 1 SCHEMATIC ILLUSTRATION DEPICTING THE CONFIGURATION OF CALSPAN'S 600 m^3 ATMOSPHERIC SIMULATION CHAMBER FOR EXPERIMENTAL INVESTIGATIONS OF THE FATE OF DIESEL EMISSIONS IN THE ATMOSPHERE AND A SUMMARY OF THE ARRAY OF MEASUREMENTS PERFORMED

The comparison of aerosol data obtained from these experiments yielded consistent results. Aerosol mass loadings averaged 0.74 mg/m^3 for both highway and chamber experiments at high exhaust dilution and 2.8 and 2.9 mg/m^3 for highway and chamber experiments respectively at low dilutions. Peak particle sizes based on volume distributions ranged between 0.1 and 0.18μm diameter. These EAA data are consistent with data previously reported for non-aged diesel exhaust aerosols.

Chemical analysis of extracts obtained by solvent extraction of the aerosol samples yielded comparable data between highway and chamber experiments. The soluble organic fraction of the aerosols generally ranged between 16 and 22 percent. Mass data obtained from class fractionation, using column chromatography, of the diesel extracts into paraffins, aromatics, transition and polar compounds resulted in similar distributions for highway and chamber experiments.

Ames bioassay data obtained for low (25:1) and high (80:1) exhaust dilutions similarly demonstrated good agreement between highway and chamber experiments. Ames activities ranged between 1.7 and 2.4 revertants per microgram of extract for strain TA-100 (-S9) and between 1.6 and 1.9 revertants per microgram of extract for TA-98 (-S9) at low dilutions. Activities ranged between 0.9 and 1.2 revertants per microgram for TA-100 (-S9) and between 1.0 and 1.2 revertants per microgram of extract for TA-98 (-S9) for high exhaust dilution experiments.

The results of this task demonstrated the validity of chamber experiments for the investigation of vehicle emission systems and provided baseline data for non-aged diesel aerosol systems.

In the second phase of the program, the chamber experiments focused on aged diesel aerosol systems using the same vehicle as the source of exhaust. Experiments were conducted at exhaust dilution ratios ranging between 80:1 and 300:1. However, the majority of experiments were conducted at a nominal exhaust dilution of 150:1 due to the finite volume of the chamber and the aerosol mass requirements for various analyses. This task included a much more rigorous analysis of the diesel aerosol systems. In a series of controlled experiments, the effects of atmospheric dilution, aging, and irradiation were quantified for unmodified diesel exhaust mixtures and also for diesel emission systems where excess concentrations of oxides of nitrogen ($\sim$2ppm NO$_x$) and ozone ($\sim$.25ppm O$_3$) were added to the chamber.

Diesel aerosol samples were collected at various intervals during these experiments for the determination of aerosol mass loading and total carbon (organic and elemental). Aerosol samples were also collected for Ames assay and for chemical characterization including the extractable fraction and HPLC analyses of the organic extracts for various PNAs (BaP and pyrene) and their nitro-derivatives. HPLC techniques were also implemented to obtain separation of the organic extracts based on polarity. In addition, real time measurements were made of aerosol size distributions and ambient gas concentrations for NO$_x$, O$_3$, SO$_2$ and THC.

Aerosol size distributions and number densities were similar to data obtained during the first task study. Peak aerosol diameters based on a volume size distribution were between 0.1 and 0.18 μm for non-aged aerosol systems. The peak diameter shifted towards larger sizes with aging due to coagulation and aerosol growth. A reduction in the total particle concentration was also observed with

764

aging and was attributed to coagulation, fall out and wall losses in the chamber. Aerosol mass decay curves showed, on the average, a 20% reduction in aerosol concentration in the chamber over an 8-hour interval and a 60% reduction in aerosol mass over a 24-hour aging interval. These data were supported by similar curves generated from elemental carbon analysis of aerosol filter samples.

Chemical and bioassay data obtained from these experiments identified trends with respect to the mutagenic acitivity and the chemical composition of extracts derived from the aerosol samples. The chemical composition and mutagenic activity of samples obtained from non-aged (baseline) tests closely resembled values reported for non-aged aerosol samples collected in the first task highway experiments. The results from samples derived from eight-hour dark aging experiments were also similar to those obtained in non-aged tests. Measured increases in mutagenic activity were observed in aerosol samples derived from experiments when the diluted diesel exhaust was aged for eight hours with UV irradiation and when aged for twenty-four hours with twelve hours of irradiation followed by twelve hours of dark. In both cases changes in the chemical composition of the particulate extract was observed. Concentrations of BaP and pyrene decreased sharply while the fraction of the extract found in the more polar δ_2 peak increased. These data indicate that photochemical processes are involved in the modification of the chemical composition of diesel exhaust and result in increases in mutagenicity.

In separate experiments, excess concentrations of NO_2 were added to the chamber diluted diesel exhaust mixture. Again increases in mutagenic activity were observed, especially with irradiation. Meanwhile, increases in the concentration of the nitro-PAHs such as 1-nitropyrene, 1,6 dinitropyrene, 1,8 dinitropyrene and 6-nitroBaP were observed. The increased concentrations of these nitro-compounds could only account for a small percentage of the observed increase in mutagenic activity. Again, decreases in pyrene and BaP were noted. These results were more pronounced during irradiation. Additional experiments conducted with excess ozone added to the chamber similarly produced increases in observed mutagenic activity and corresponding changes in chemical composition of the organic extract.

Summary

These results were obtained from a limited number of experiments, however, they consistently demonstrated that under certain experimental conditions, photo-oxidation processes and nitration reactions are important in the modification of the chemical composition and mutagenic activity of diesel exhaust.

A third phase of the program is currently underway, to improve our understanding of the atmospheric processes and interactions involved in the modification of diesel emissions in the atmosphere. The experiments will concentrate on those parameters, which, on the basis of the previous program, were shown to have an effect on the chemistry and mutagenic activity of the emissions. These experiments will include photochemically reactive diesel aerosol systems which will require the addition of reactive hydrocarbons, to simulate urban environments. Such systems were not specifically addressed in the previous studies. Extensive aerosol, gas chemistry and analytical chemical analyses will be conducted to trace the complex chemistry and evolution of classes of chemical species which may contribute to changes in mutagenic activity of diesel aerosols.

Published 1984 by Elsevier Science Publishing Co., Inc.

Aerosols, Liu, Pui, and Fissan, editors

765

PARTICLE EMISSIONS AND CONCENTRATIONS OF TRACE ELEMENTS

BY A COAL FIRED POWER PLANT

I. Kalman, Eisenforschungsinstitut Budapest,Ungarn
and F. Sporenberg, Universität Essen, Institut für
Umweltverfahrenstechnik, Deutschland

EXTENDED ABSTRACT

Introduction

The emissions of anorganic substances of coal fired
power plants are not extensively studied, there are up to
not relatively limitated information about the emissions
and enrichment. The aim of this investigations is to
determine the characteristics of size distributions and
concentration of trace elements in particulate emissions.
The elemental size dependences are due to volatilization of
certain elements during combustion. Small particles have
greater spezific surface are as than large particles, this
volatilization deposition mechanismen leads to higher
spezific concentrations of trace elements in small particles.

Experiments

The charakteristics of element concentration (lead,
zinc and cadmium) are depicted in Table 1.

Table 1. Specific Concentrations And Particle
Enrichment In The Stack Emissions From
A Coal Fired Power Plant

Element	Size cut off 18,5 μm ppm	Size cut off 6 μm ppm	Size cut off 3,7 μm ppm	Size cut off 2,4 μm ppm
Cd	0,4	1,6	2,8	4,6
Pb	73	169	226	278
Zn	68	189	301	590
Enrichment				
Cd	0,4	1,6	2,8	4,7
Pb	1,4	3,1	4,1	5,2
Zn	1,0	2,6	4,0	8,1

The measurements of specific concentrations of trace elements in coal fly ash shows that the enrichment depends both on particle density and physical size. The observed element concentration and enrichment for Cd, Pb and Zn indicate that the highest concentrations of anorganic compounds are found in small particles. Small particles have a greater specific surface than large particles, this deposition mechanism leads to higher concentrations in small particles (Natusch, 1982). Beside the experiments of particle enrichment the particle size distribution and the separation efficiency of fly ash from coal fired power plants are measured with inertial impactors.

References

Natusch, D.F.S (1982) "Size Distributions and Concentrations of Trace Elements in Particulate Emissions from Industrial Sources", VDI-Berichte 429, Schwebstoffe und Stäube, Analytische und technische Aspekte, VDI-Verlag GmbH, Düsseldorf

Published 1984 by Elsevier Science Publishing Co., Inc.
Aerosols, Liu, Pui, and Fissan, editors

SIZE DISTRIBUTION AND CHEMICAL COMPOSITION OF
THE SUBMICRON CONDENSATION AEROSOL FROM
A COAL FIRED POWER PLANT

J L Wolf, NSP Co, Minneapolis, Minnesota
D A Lundgren, University of Florida, Gainsville
V A Marple, University of Minnesota, Minneapolis
USA

EXTENDED ABSTRACT

Introduction

A large power plant operating in Minnesota burns two types of low-sulfur
western coal. The two coals are supplied by different companies and mined less
than fifty miles apart in Montana. In spite of this geographical proximity,
the coals exhibit markedly different combustion characteristics. The most
apparent differences are in outlet mass emission rate and stack opacity.

The power plant employs a venturi scrubber for particulate and sulfur
control. Measurements were made before and after the scrubber for both coals,
and particle size and chemical characteristics were determined and compared.

Sampling Method

Samples of combustion ash from both coals were collected during two
periods, August and November 1983. A University of Washington in-stack
impactor was used to size separate and collect ash particles larger than 0.38μm
aerodynamic diameter. A Lundgren in-stack diffusion classifier (Lundgren,
1982) was used to sample particles less than 0.38μm aerodynamic diameter.

These devices provided total aerosol size distributions and collected
samples on 37mm teflon filters for subsequent chemical analysis.

Results

Aerosol mass distributions for the two coals, designated "A" and "B", are
shown in Figures 1 and 2. Mass distributions measured at the scrubber inlet
and outlet (shaded) are shown.

Chemical Analysis

Chemical analysis was conducted on samples collected by the diffusion
classifier (less than 0.38μm) using x-ray flourescence. Two runs for each coal
at both the scrubber inlet and outlet were averaged for comparison. Data for
elements of interest are presented in Tables 1 and 2.

Table 1.   X-RAY FLOURESCENCE ANALYSIS
"A" COAL < 0.38 μm, μg/cm^2

Element	Inlet	Outlet	Ratio
NA	.29	.10	.34
MG	2.95	1.54	.52
AL	4.53	1.65	.36
SI	12.58	6.50	.52
S	5.06	5.06	1.00
CA	25.81	13.24	.51
FE	1.94	.84	.43

Table 2.   X-RAY FLOURESCENCE ANALYSIS
"B" COAL < 0.38μm, (μg/cm^2)

Element	Inlet	Outlet	Ratio
NA	.51	.41	.80
MG	3.10	2.00	.65
AL	4.05	2.47	.61
SI	21.10	12.47	.61
S	6.97	12.72	1.82
CA	42.18	24.50	.58
FE	2.88	1.48	.51

Discussion

Submicron condensation aerosols produced from the volatile material of the two coals are substantially different. The "B" coal has 30% more mass below 0.38μm diameter at the inlet than does the "A" coal. At the outlet however, "B" coal has 120% more mass. Figures 1 and 2 show that aerosol is generated in the 0.14 to 0.38μm range for both coals.

X-ray flourescence data in Tables 1 and 2 show a reduction in all elemental concentrations except for sulfur. "A" coal outlet concentrations are about 50% of inlet, and "B" coal outlet concentrations are about 60% of inlet. However, there was no reduction in sulfur quantity for the "A" coal and an 80% increase for "B" coal. As most elements decreased through the scrubber the percentage of sulfur in the outlet aerosol increased dramatically. Sulfur is an important part of condensation aerosol formation in the venturi scrubber. The formation process is significantly affected by the subtle differences in coal combustion chemistry.

References

Lundgren, D A and Rangaraj, C N (1982) "Diffusion Classification of Submicron Aerosols", Environmental Progress (Vol 1, No. 2).

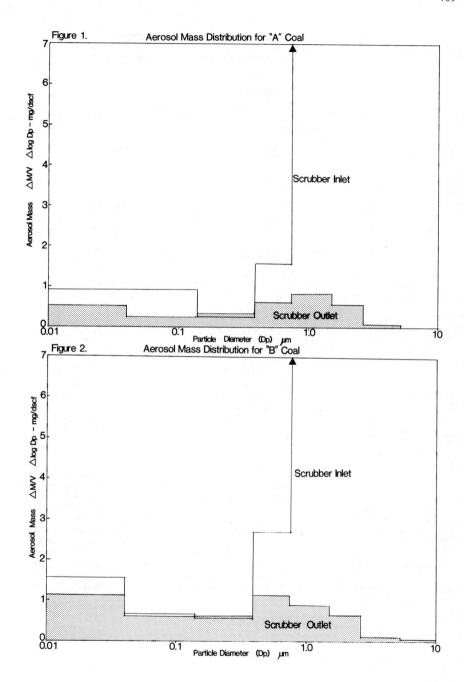

Figure 1. Aerosol Mass Distribution for "A" Coal

Figure 2. Aerosol Mass Distribution for "B" Coal

Published 1984 by Elsevier Science Publishing Co., Inc.
Aerosols, Liu, Pui, and Fissan, editors

THE INFLUENCE OF ADSORPTION/DESORPTION PROCESSES

ON THE NATURE OF DIESEL PARTICLES IN THE ATMOSPHERE

D.B. Kittelson and M.A. Barris
Department of Mechanical Engineering
University of Minnesota
Minneapolis, MN 55455, U.S.A.

EXTENDED ABSTRACT

Introduction

The soluble organic fraction (SOF), which comprises typically 5 to 50% of the mass of diesel particles (1,2,3), has been shown to contain a variety of hydrocarbons and related compounds including potentially harmful agents such as polynuclear aromatic compounds (PACs). It is widely accepted that PACs originate in the vapor phase. However, when laboratory or atmospheric measurements are conducted, almost all of the heavier PACs (5 rings and larger) are found to be associated with particulate matter. The partitioning of these compounds between the gas and particulate phase depends upon the process of dilution and cooling of the sample as it passes from the tailpipe to a filter or other particle collection device. It is unclear whether laboratory experiments truly simulate atmospheric dilution (1,4). Experiments to determine the relationship between SOF and dilution have given contradictory results. Some researchers (4,5) have found an influence of the dilution ratio on SOF while others (2) report no significant dependence.

The principal gas-to-particle conversion mechanism believed to be important during dilution and sampling of diesel particles is adsorption (6). This paper describes a simple kinetic model of adsorption which may be used to determine the influence of the dilution and sampling process on the SOF.

Model Development

This model is described in detail by Barris (7). Diesel exhaust is assumed to contain carbon particles (the adsorbent) and high molecular weight hydrocarbons (the adsorbate). As the exhaust is diluted and cooled, the partitioning of the adsorbate between the gas and adsorbed phase is determined by adsorption kinetics. The model divides the adsorption process into two steps: diffusion of the adsorbate through the boundary layer surrounding the particle, and adsorption at the particle surface. Although the SOF consists of many compounds, the model considers only single component adsorption. Pyrene, a PAC is used as the single adsorbate. The rate of diffusion through the boundary layer, $\overset{.}{m}$, has been determined using the interpolation formula of Fuchs and Sutugin (8) for particles in the transition regime.

$$\dot{m} = \frac{2\ D\pi d_p\ (\rho_\infty - \rho_s)}{1 + Kn\ \dfrac{1.33 + .71/Kn}{1 + 1/Kn}} = h_m\ (\rho_\infty - \rho_s) \tag{1}$$

Where, D is the diffusion coefficient of the adsorbate, d_p is the particle diameter, ρ_∞ and ρ_s are the gas phase adsorbate densities far from the surface and at the surface of the particle, respectively, Kn is the Knudsen number, and h_m is the overall mass transfer coefficient.

Adsorption at the particle surface is described using a simple Langmuir model (9)

$$\dot{m} = k_1\ (1 - \theta)\ \rho_s - k_2\ \theta\ M_m \tag{2}$$

Here, θ is the fraction of monolayer coverage, k_1 and k_2 are the forward and reverse rate constants, and M_m is a mass of the surface monolayer. In equilibrium, $\dot{m}$ is zero, and ρ_s is equal to ρ_∞. The mass adsorbed is directly proportional to θ. Under these conditions Equation may be solved for the equilibrium fraction coverage, θ_e.

$$\theta_e = (1 + k_2\ M_m/k_1\ \rho_\infty)^{-1} \tag{3}$$

The forward and reverse rate constants are given by

$$k_1 = c\ \pi d_p^2 (RT/2\pi M_p P)^{1/2}\ \exp\ (-E_a/RT)$$

and

$$k_2 = kT/h\ \exp\ (-E_d/RT)$$

Here, c is the sticking coefficient, T is temperature, M_p is the molecular weight of pyrene, E_a is the adsorption activation energy, and E_d is the activation energy for desorption. The adsorption energy, E, is given by

$$E = E_d - E_a$$

Equation 3 may be used directly to find the equilibrium coverage, θ_e. The kinematically limited coverage, θ_k, may be determined by numerically solving Equations 1 and 2 using the steady state assumption to find ρ_s. Both the equilibrium and kinetic solutions depend upon ρ_∞ and T. These, in turn, depend upon the tailpipe conditions and the dilution schedule. Two dilution schedules were investigated, the first is intended to simulate dilution in a standard laboratory dilution tunnel, while the second is intended to simulate dilution over a roadway. Both are shown in Figure 1. The variation of ρ_∞ and temperature with time was determined from the dilution schedule using a simple adiabatic dilution model. Calculations were based on the following conditions: tailpipe exit temperature, 480K; ambient temperature, 300K; 1 ppm pyrene in the exhaust; 100 mg m^{-3} carbon particles in the exhaust with specific surface area of 100 m^2 gm^{-1} and a diameter of 0.033 um; sticking coefficient, 1; and on adsorption activation energy of 2 kcal mole^{-1}. These values will tend to make adsorption relatively fast and probably represent an upper limit on adsorption rate. One additional parameter is needed to do these calculations, either the adsorp-

772

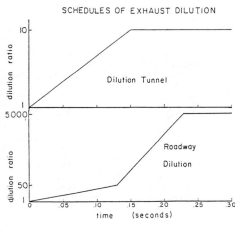

Figure 1 - Dilution Schedules

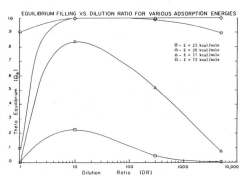

Figure 2 - Surface Coverage in Equilibrium Plotted Against Dilution Ratio

tion energy, E, or the desorption activation energy, E_d. The equilibrium coverage does not depend upon kinetic parameters, but depends strongly upon E. We have calculated the variation of θ_e with dilution ratio for several different adsorption energies. These results are plotted in Figure 2. The results show that an adsorption energy of 17 kcal mole^{-1}, θ_e strongly depends on dilution ratio while at 25 kcal/mole^{-1}, it is almost independent of dilution ratio. At 20 kcal/mole^{-1} θ_e depends mildly upon dilution ratio, which is qualitatively the same as the dependence of SOF upon dilution ratio which has been reported by several investigators (4,5,10). Consequently, an adsorption energy of 20 kcal/mole^{-1} has been used for these calculations.

Results

Figure 3 shows values of θ_e and θ_k which were calculated for the laboratory tunnel type dilution schedule plotted against time. Initially, the kinetic results lag behind the equilibrium results, but only slightly more than 1 s is required to achieve equilibrium. Thus, under steady state laboratory conditions, it is probably reasonable to assume that the SOF has been adsorbed to an equilibrium level. Figure 4 shows the variation of θ_e and θ_k with time for roadway dilution. Again, the kinetic results lag equilibrium, but here a considerably longer time, approximately 1000 s is required to reach equilibrium. In experiments where particle samples are collected near or on the roadway, collection times of several hours are usually necessary to obtain a sufficient quantity of particulate matter for analysis. Thus, if the roadway source is relatively steady, much of the particulate matter collected on a filter might be expected to be in adsorption equilibrium. On the other hand, the typical residence time of particles over or near a roadway is likely to be much shorter than the 1000 s adsorption equilibration time. Thus, these calculations suggest that there might be significant differences between the SOF of particles to which people on or near

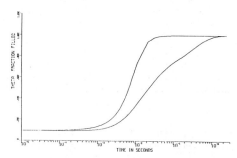

Figure 3 – Surface Coverage Calculated for Equilibrium and Kinetically Limited Conditions with Laboratory Dilution

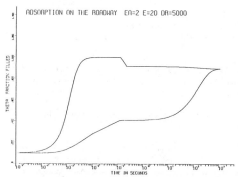

Figure 4 – Surface Coverage Calculated for Equilibrium and Kinetically Limited Conditions with Roadway Dilution

the roadway are exposed and the SOF of particles collected on a filter under similar conditions. Consequently, if the health effects of particle bound organics are significantly different from those of the same organics in the gas phase, adsorption kinetics might influence these health effects.

The model used here was a very simplified one. Another model and set of assumptions might give different results. However, the kinetics used were relatively fast ones and it is unlikely that a more complex model would give higher adsorption/desorption rates. Thus, departures from adsorption equilibrium might be even worse than suggested by this model. There is clearly a need for more work in this area, both in the development of suitable adsorption/desorption models and in the determination of the influence of partitioning of hydrocarbons between the gas and particle phase on the health impacts of diesel particles.

Acknowledgements

This work was supported by the Coordinating Research Council uner the direction of the CAPA-13-76 Project Group. We wish to thank Mr. Jeffrey L. Ambs and Ms. Betsy A. Antinozzi for their assistance in preparing this manuscript.

References

1. Kittelson, D.B. and D.F. Dolan (April 4, 1979) "Diesel Exhaust Aerosols," Symposium on Aerosol Generation and Exposure Facilities, ACS/CSI Chemical Congress, Honolulu, HI.

2. Black, F. and L. High (1979) "Methodology for Determining Particulate and Gaseous Diesel Hydrocarbon Emissions," SAE Paper No. 790422.

3. Funkenbusch, E.F., D.G. Leddy, and J.H. Johnson (1978) "The Characterization of the Suluble Organic Fraction of Diesel Particulate Matter," SAE Paper No. 790418.

4. Frisch, L.E., J.H. Johnson, and D.G. Leddy (1979) "Effect of Fuels and Dilution Ratio on Diesel Particulate Emissions," SAE Paper No. 790412.

5. MacDonald, J.S., S.L. Plee, J.B. Darcy, and R.M. Schneck (1980) "Experimental Measurement of the Independent Effects of Dilution Ratio and Filter Temperature on Diesel Exhaust Particulate Samples," SAE Paper No. 800185.

6. Plee, S.L. and J.S. MacDonald (1980) "Some Mechanisms Affecting the Mass of Diesel Exhaust Particulate Collected Following a Dilution Process," SAE Paper No. 800186.

7. Barris, M.A. (1981) "Mass Transfer Processes Affecting the Nature of Diesel Exhaust Aerosols," M.S. Thesis, submitted to the Graduate School, University of Minnesota.

8. Shaw, D.T. (1978) Recent Development in Aerosol Science, John Wiley & Sons, p. 12.

9. D.F.S. Natusch and B.A. Tomkins (1978) "Theoretical Considerations of the Adsorption of Polynuclear Aromtic Hydrocarbon Vapor onto Fly Ash in the Coal-Fired Plant," Carcinogenesis, Vol. 3, Polynuclear Aromatic Hydrocarbons, Eds.: P.W. Jones and R.I. Freudenthal, Raven Press, NY, pp. 145-153.

10. Kittelson, D.B. (Sept. 17-21, 1984) "Physical Characteristics of Fresh Diesel Aerosols in the Atmosphere," for presentation at First International Aerosol Conference, Mpls, MN, Paper No. 261.

Published 1984 by Elsevier Science Publishing Co., Inc.
Aerosols, Liu, Pui, and Fissan, editors

SOURCE CONTRIBUTIONS TO ATMOSPHERIC CARBON PARTICLE CONCENTRATIONS

H. Andrew Gray and Glen R. Cass
Environmental Engineering Science Department and
Environmental Quality Laboratory
California Institute of Technology
Pasadena, California 91125 USA

EXTENDED ABSTRACT

Carbonaceous aerosols released from combustion sources are one of the largest contributors to fine particle concentrations in the atmosphere of cities. Recent experiments in Los Angeles, for example, show that circa 40% of the fine aerosol mass in particles of diameter below 2.1 μm is composed of organics plus elemental carbon.[1] The elemental carbon fraction of this aerosol carbon loading is an effective light absorber and contributes substantially to visibility reduction, while polynuclear aromatic hydrocarbons adsorbed onto soots are a matter of current public health concern. As a result, there is considerable interest in determining how primary aerosol carbon concentrations might best be controlled. Methods for identifying the relative contribution of different source types to primary aerosol carbon air quality must be developed and tested if efficient solutions to the aerosol carbon air quality problem are to be identified.

A mathematical model for the emission and atmospheric transport of primary aerosol carbon particles has been developed. Pollutant concentrations are computed using a Langrangian particle-in-cell technique that incorporates atmospheric transport and diffusion, a vertically stratified atmosphere, and ground level dry deposition. The model provides an accounting at each monitoring site of the air quality increment contributed by each major class of aerosol carbon emission source. Elemental carbon and primary organic aerosol concentrations are computed separately.

The air quality model is evaluated by application to the Los Angeles Basin during the year 1982. A spatially resolved primary aerosol emission inventory is constructed for more than 50 classes of mobile, stationary and fugitive sources. Data on the size distribution and chemical composition of the emissions from these sources are used to convert the total emissions inventory into separate inventories for primary fine particle total carbon and elemental carbon emissions, following a procedure like that of Cass et al. (1980).[2] The emission inventory is matched to the air quality model. Elemental carbon emissions are treated as an inert form of primary carbonaceous aerosol. The model's performance is assessed by comparison of predicted elemental carbon concentrations to the 1982 elemental carbon concentrations measured in the Los Angeles area by Gray et al. (1984a,b)[1,3]

The completed air quality model will be used in the future to evaluate alternative strategies for abatement of primary organic and elemental carbon particle emissions.

References

Gray, H.A., Cass, G.R., Huntzicker, J.J., Heyerdahl, E.K. and Rau, J.A. (1984) "Elemental and Organic Carbon Particle Concentrations: A Long-Term Perspective," <u>Sci</u>. <u>Tot</u>. <u>Environ</u>., (in press).

Cass, G.R., Boone, P.M. and Macias, E.S. (1982) "Emissions and Air Quality Relationships for Atmospheric Carbon Particles in Los Angeles," in <u>Particulate Carbon</u>: <u>Atmospheric Life Cycle</u>, (Edited by G.T. Wolff and R.L. Klimisch), Plenum Press, New York.

Gray, H.A., Cass, G.R., Huntzicker, J.J., Heyerdahl, E.K., Rau, J.A. (1984) "Characteristics of Atmospheric Organic and Elemental Carbon Concentrations in Los Angeles," to be submitted for publication.

Published 1984 by Elsevier Science Publishing Co., Inc.
Aerosols, Liu, Pui, and Fissan, editors

COAL GENERATED INORGANIC AEROSOLS: A REVIEW

R. C. Flagan
California Institute of Technology
Pasadena, CA 91125
U.S.A.

and

A. F. Sarofim
Massachusetts Institute of Technology
Cambridge, MA 02139
U.S.A.

EXTENDED ABSTRACT

Recent studies of coal combustion effluents have shown that the fly-ash is bimodally distributed with respect to size. The larger ash particles have diameters in the range of 1 to 50 μm and correspond to the ash particles freed in the coal grinding process or present as inclusions in the coal matrix and released on combustion. Not all the ash is generated in the same way, however, as several studies have shown that coal ash vaporizes during combustion. It is apparent that, as the combustion gases are cooled, the vaporized mineral matter may either homogeneously nucleate or condense on the surface of the existing ash particles. The nuclei grow by coagulation to form a distinct submicron mode. The paper reveiws the processes that govern the size distribution and chemical composition of both the residual fly ash and the submicron condensation aerosol. The results of field studies will be presented first, followed by a discussion of laboratory studies which provide mechanistic insights into the processes occuring in pulverized coal-fired boilers.

Field Studies

Analyses of fly ash collected from coal-fired power plants have shown (Davison et al., 1974; Klein et al., 1975, Kaakinen et al., 1975; Gladney at al., 1976; Ondov et al., 1976) that the smaller ash particles are enriched in trace volatile elements such as So, As, Sb, Cd, Zn. Furthermore, these trace elements are concentrated in the surface layers of these particles (Linton et al., 1976). These observations have been explained (Davidson et al., 1975; Flagan and Friedlander, 1978) by a model based on the vaporization and recondensation of the mineral matter. As a consequence of transport limited condensation on particles the size of which varies from much smaller to much larger than the mean free path of the gas, the variation of concentration volatile of trace elements with size (d) varies from a $1/d$ dependence for submicron particles to $1/d^2$ for the supramicron size model. Biermann and Ondov (1980) have applied surface-deposition models to correlate the size dependence of concentration in coal fly ash. The vaporized mineral matter, in addition to depositing on the surface of residual fly ash, homogeneously nucleates to form a submicron aerosl in a sharply distinct size region from that of the residual ash, as shown by the results of McElroy et al. in Fig. 1. Tests on six power plants showed a range of 0.2 to 2.2 percent of the fly ash entering an electrostatic precipitator was in this submicron mode, but as much as 20 percent of the ash leaving the precipitator.

Laboratory Studies

Detailed information on the processes governing the formation of fly ash have been obtained in well-defined laboratory units. Combustion of size and density-graded coal particles and measurement of the size distribution of the ash has been used to determine the number of ash particles produced during combustion of a coal (Quann and Sarofim, 1984). The results shown in Table 1 indicate that most of the mass of ash results from the fusion and coalescence of the ash in the coal to yield of the order of six particles in the 10-20 μm size range. These particles result from the fragmentation of coal as it burns. In addition micron size ash particles are shed during combustion to produce a large number of smaller ash fragments, which were collected on the different stages of a Anderson impactor. The condensation aerosol is collected on the final filter and dominates the number of particles produced while accounting for of the order of one percent of the mass.

Because the fine particles are formed entirely from volatilized ash, their composition may differ substantially from that of the bulk ash. This is illustrated in Fig. 2, where the concentrations of the major ash constituents in size-fractionated ash samples are plotted versus particle size (Taylor and Flagan, 1982). For this relatively low-temperature laboratory combustion experiment, the fume consists primarily of iron oxides. The concentrations of Si and Al in the fume are virtually undetectable.

The quantity and composition of the fine particles is highly dependent on coal composition and combustion operating conditions. Finely dispersed elements, such as those organically bound in the coal, will vaporize to a larger extent than those found in micron-size mineral inclusions (Quann and Sarofim, 1983). Alkali metals and other high vapor pressure compounds will vaporize preferentially. Elements such as Si, Ca, Mg, and Fe, the oxides of which have a low vapor pressure, will vaporize by reduction of the oxide to a suboxide of metal by reactions of the type

$$MO_n + CO \rightleftharpoons MO_{n-1} + CO_2$$

The CO is provided in the locally reducing atmosphere of the particle. A model based on this mechanism predicts adequately the dependence of the vapor pressure of the metal or metal suboxide, and its dependence on temperature and size of mineral inclusion, as shown for magnesium in Fig. 3. The three lines about which the data are clustered correspond to the theoretical lines for vaporization of atomically dispersed mineral matter, such as encountered in the Montana lignite, or in 2 or 5 μm inclusions as encountered in the Illinois No. 6 and Alabama Rossa coal.

The reoxidation of the vaporized suboxide or metal and the subsequent homogeneous nucleation have been modeled by Senior and Flagan (1983). The results, illustrated for SiO_2 nucleation in Fig. 4, show that high nucleation rates are encountered near the particle surface.

The particle size distribution of the particles after a few milliseconds should become self preserving. The mean particle diameter calculated from coagulation theory has been found to be in good agreement with values measured in both laboratory (Neville and Sarofim, 1983) and field studies (Flagan, 1979), as predicted by theory the mean particle diameter varies as 2/5 power of the amount of ash vaporized and the mean residence time in a furnace.

Concluding Comments

The topics touched lightly in the abstract will be used to illustrate progress that has been made in characterizing coal generated inorganic aerosols and to draw attention to some of the problems remaining unresolved.

References

1. Bierman, A.H., and Ondov, J.M. (1980) Atmospheric Environment 14:289-295.

2. Davidson, R.L., Natusch, D.F.S., Wallace, J.R., and Evans, C.A. Jr. (1974) Environ. Sci. Tech. 12:447-451.

3. Flagan, R.C., and Friedlander, S.K. (1978) in Recent Development in Aerosol, D.T. Shaw, Ed,, Wiley-Interscience.

4. Flagan, R.C. (1979) Seventeenth Symposium (International) on Combustion, The Combustion Institute, Pittsburgh, Pa.

5. Gladney, E.S., Small, J.A., Gordon, G.E., and Zoller, W.H. (1976) Atmospheric Environ. 10:1071-1077.

6. Kaakinen, J.W., Jorden, R.M., Lawasan, M.H., and West, R.E. (1976) Environ. Sci. Tech. 9:862-869.

7. Klein, D.H., Andren, A.W., Carter, J.A., Emergy, J.F., Feldman, C., Fulkerson, W., Lyon, W.S., Ogle, J.C., Talm, Y., Van Hook, R.I., and Bolton, W. (1975) Environ. Sci. Tech. 9:973-979.

8. Linton, R.W., Loh, A., Natusch, D.F.S., and Evans, C.A. Jr. (1976) Science 191:852-854.

9. McElroy, M.W., Carr, R.C., and Markowski, G.R. (1982) Science 215:18.

10. Neville, M., and Sarofim, A.F. (1983) Nineteenth Symposium (International) on Combustion, The Combustion Institute, Pittsburgh, Pa., pp. 1441-1449.

11. Ondov, J.M., Ragaini, R.C., and Biermann, A.H. (1977) Atmospheric Environ. 12:1175-1185.

12. Quann, R.J., and Sarofim, A.F. (1983) Nineteenth Symposium (International) on Combustion, The Combustion Institute, Pittsburgh, Pa., pp. 1421-1440.

13. Quann, R.J., and Sarofim, A.F. (1984) "An Electron Microscopic Study of the Transformation of Organically Bound Metals During Lignite Combustion," submitted to Fuel.

14. Senior, C.L., and Flagan, R.C. (1982) "Laboratory Studies of Submicron Aerosol Formation in Pulverized Coal Combustion Using Model Compounds," Aerosol Science and Technology 1:71-383.

15. Taylor, D.D., and Flagan, R.C. (1981) ACS Symposium Series, No. 167, pp. 157-172.

Table 1. The Distribution of Ash Particles on Cascade Impactor
Stages After Complete Combustion of the Lignite

Stage	Approximate Diameter (μm)	Weight (g/g coal burned)	Calculated No. of Ash Particles Produced per Coal Particle Burned
0	20	.0388	6
1	7.9	.0079	21
2	5.4	.0043	36
3	5.7	.0033	87
4	2.4	.0024	230
5	1.2	.0013	1000
6	0.77	.0006	2×10^3
7	0.54	.0004	4×10^3
Filter	0.02	.0032	5×10^8

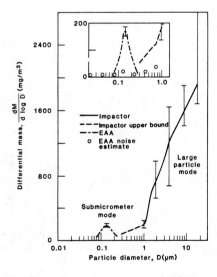

Figure 1. Typical bimodal differential size distribution at boiler outlet
(electrostatic precipitator inlet). The submicrometer mode
contains 1½ percent of the total mass. The large-particle mode
reaches a peak and then tails off at particle sizes greater than
10 μm.

782

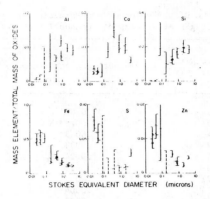

Figure 2. Variation of elemental composition with particle size in ash
produced by low temperature combustion of Utah subbituminous coal.
(From Ref. 15.)

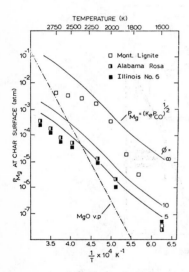

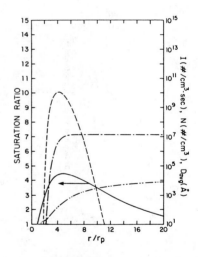

Figure 3. Comparison of experimentally determined average
partial pressures of Mg
at the char surface with
model predictions.
(From Ref. 12.)

Figure 4. Calculated profiles of
saturation ratio (——),
nucleation rate (I,---),
number concentration
(N, — — —), and average
diameter (— — —) for
silica vaporized from
a burning char particle.
(From Ref. 14.)

Published 1984 by Elsevier Science Publishing Co., Inc.
Aerosols, Liu, Pui, and Fissan, editors

FRESH DIESEL AEROSOLS IN THE ATMOSPHERE

D.B. Kittelson
Department of Mechanical Engineering
University of Minnesota
Minneapolis, MN 55455, U.S.A.

EXTENDED ABSTRACT

This paper describes some results of a study of the influence of atmospheric dilution and aging on the nature of diesel aerosols. Samples were obtained from diesel exhaust plumes from both stationary and mobile sources. Atmospheric residence times ranged from about 0.3 s to 120 s. Dilution ratios ranged from about 2 to 11,000. The University of Minnesota Mobile Laboratory was used to monitor the plumes. Continuous measurements of oxides of nitrogen, light scattering, and condensation nuclei were made. A bag sampler was used for intermittent size distribution determinations using an electrical aerosol analyzer (EAA) and an optical particle counter (OPC). Filter samples were collected for chemical analysis and Ames testing using an ultra-high volume sampler system (UHVS) which sampled at a rate of 8 m^3 min^{-1}. Low volume filter samples were collected to determine light adsorption cross sections using the integrating plate method (1,2). All filter samples were collected with a 0.9 µm aerodynamic diameter cut-size impactor upstream of the filter. Thus the filters only collected sub-0.9 µm particles.

Figure 1 shows size distributions obtained upstream and downsteam of a large diesel engine test facility and ahead and behind a heavy-duty highway tractor operating on a rural freeway. In both cases, the dilution ratios were high so that the difference between the background aerosol and the plume was modest. In the freeway experiments, the particles contributed to the plume by the diesel engine were relatively small with a volume mean diameters in the 0.15 µm diameter range. This is quite comparable to diameters reported in laboratory studies of diesel exhaust particle size distributions (3,4,5). This is not unexpected because the rapid mixing and short residence times (< 5 s) between the tailpipe and sampling inlet give little opportunity for aerosol growth. In the engine test facility study, the particles contributed to the plume by the diesel engines have a some-what larger volume mean diameter, about 0.5 µm. This increase is probably due to coagulation and reentrainment. Several diesel engines feed the stack through a common manifold at the diesel engine test facility. The stack and sampling manifold provide a longer time at high concentration than would be encountered in the normal engine exhaust system. This will enhance particle coagulation in the stack. Coagulation is further enhanced by the fact that once the particles are emitted into the atmosphere, the time scale of dilution is considerably slower than over a roadway. This slower mixing results from a larger plume diameter and lower wind veloci-ties than in the roadway study. Reentrainment from the walls of the stack and exhaust system will further increase the diameter of particles emitted from the stack of the test facility. Many of the engines were operating under transient conditions. Such conditions favor deposition and reen-trainment of particles from the surface of the engine exhaust system and

784

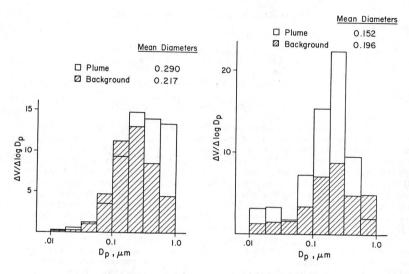

Figure 1 — EAA Measurements of Particle Size Distribution.

I — Measurements made upwind and
    downwind of diesel test facil-
    ity. Dilution ratio approxi-
    mately 10,000.

II — Measurements made ahead and
    behind diesel powered highway
    tractor operating on rural
    freeway at 50 mph. Dilution
    ratio approximately 5000.

stack. These reentrained particles are significantly larger than particles formed by primary combustion processes and increase the mean diameter
of emitted particles (6).

The soluble organic fraction (SOF) of the particles collected with the
ultra-high volume filter sampler were extracted using standard Soxhlet
techniques. The extracts were analyzed using a high pressure liquid chro-
matograph and were also Ames tested. The results of the analyses of these
abstracts will be described in a forthcoming paper. Only the influence of
sampling conditions upon the SOF will be considered here. Figure 2 shows
the variation of SOF with dilution ratio for particles collected in the
plume and from the stack of a four-stroke highway tractor operating at 50-
55 mph on a rural freeway. The SOF for the plume samples are corrected for
background particles already present over the roadway. Stack samples were
obtained using a mini-dilution tunnel mounted on the cab of the tractor.
The results obtained with the mini-tunnel are indistinguishable from the
results obtained by plume sampling at the same dilution ratio. The SOF
appears to depend upon dilution ratio in a manner which is consistent with
adsorption/desorption equilibria controlling the SOF (7,8).

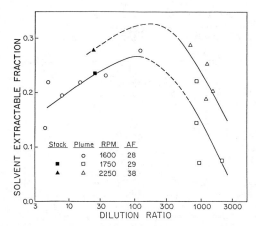

Figure 2 - Solvent Extractable Fractions Versus Dilution Ratio.

These measurements were made with a four-stroke highway tractor operated at 50-55 mph on a rural freeway. Plume samples were obtained using the UMML. Stack-to-filter residence times ranged fro 0.3 to 5 s. Stack samples were collected using a mini-dilution tunnel mounted on top of the tractor cab.

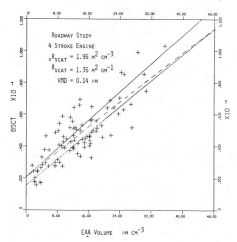

Figure 3 - Scattering Coefficient Plotted Against EAA Volume.

Particles measured in the plume of a four-stroke diesel tractor operating on a rural freeway at 50-55 mph.

Both light scattering and adsorption measurements were made in this study. Scattering cross sections were determined from simultaneous measurements of scattering coefficient with an integrating nephelometer, m^{-1}, total volume with an EAA, μm^3 cm^{-3}, and submicron aerosol mass with the UHVS, $g\ m^{-3}$. Figure 3 shows a typical plot of scattering coefficient against EAA volume. These measurements were made in the roadway study with a four-stroke diesel engine powered tractor providing the plume. The slope of the curve gives the volumetric scattering cross section of $1.96\ m^2\ cm^{-3}$. An apparent particle density of $1.44\ g\ cm^{-3}$ was determined in the same experiments from the ratio of particle mass to EAA particle volume. Dividing the volumetric scattering cross section by density gives a mass scattering cross section of $1.36\ m^2\ g^{-1}$. Table 1 summarizes both light scattering and absorption measurements. Particles from a two-stroke diesel powered highway tractor showed the smallest scattering cross section, $1.00\ m^2\ g^{-1}$ while the particles from the engine test center showed the largest cross section, $4.05\ m^2\ g^{-1}$. This difference may be at least partly attributable to the fact that particles from the test center were considerably larger in diameter than those from the two-stroke, i.e., $0.5\ \mu m$ versus $0.12\ \mu m$ volume mean diameter. This is roughly

Table 1 - Optical Properties of Diesel Aerosols

	Light Absorption		Light Scattering		
	R_{abs}	R_{absC}	R_{scatV}	R_{scat}	ρ_p
Roadway Studies					
2 Stroke Diesel			1.44	1.00	1.44
4 Stroke Diesel	5.7 + 0.8	6.9 + 0.6	1.96	1.36	1.44
Background	1.4				
Stationary Source					
4 Stroke Diesel	6.3 + 1.5	6.4 + 1.5	3.66	4.05	0.90
Background/Ambient	2.1			2-4	
Laboratory Studies					
4 Stroke Diesel	6.6 + 1.5	9.1 + 0.3		1.2 + 0.3	

R_{abs} = Light absorption cross section, $m^2\ g^{-1}$
R_{absC} = Light absorption cross section, m^2 per gram of non-extractable material
R_{scatV} = Light scattering cross section, $m^2\ cm^{-3}$. Volumes measured with EAA
R_{scat} = Light absorption cross section, $m^2\ g^{-1}$
ρ_p = Apparent particle density, ratio of submicron filter mass to EAA volume, $g\ cm^{-3}$

consistent with Mie theory which suggests that light scattering should increase with increasing particle diameter in this range.

Light absorption measurements were also made and are summarized in Table 1. The light absorption cross sections are expressed both in terms of m^2 per gram of particulate matter and m^2 per gram of equivalent carbon. The carbon mass is taken as the total particle mass multiplied by (1-SOF). Since diesel particles may contain small concentrations of oxygen and other materials, we may be overestimating the carbon content. However, the error is not likely to be more than a few percent. Comparison of laboratory and roadway experiments show that the absorption cross sections measured over the roadway are slightly lower than those obtained in the laboratory. On the other hand, both roadway and laboratory measurements show that diesel particles absorb light much more strongly than they scatter light. Consequently, any visibility reduction associated with diesel particles would be expected to be mainly due to light absorption rather than scattering.

The results presented above show that, in general, the physical properties of freshly emitted diesel particles are similar to those measured in laboratory dilution tunnel studies. There are, however, some differences. Under conditions such as those encountered in the engine test facility study, relatively slow dilution may allow significant particle growth by coagulation. This not only increases the particle size, but may also increase the light scattering cross section of the particles. The residence time of freshly emitted diesel particles over a roadway is probably in the range of 1-100 s. On the other hand, the collection time for our filter samples was always greater than 10,000 s. Consequently, although the behavior of the SOF with respect to dilution ratio measured in roadway

studies was similar to that observed in dilution tunnels, the SOF of the
freshly emitted particles as they exist before they have been collected and
aged upon a filter could be different from both laboratory and roadway
experiments. This could arise from incomplete equilibration of the particle bound SOF with the corresponding gas phase constituents. This is
discussed in more detail in a companion paper (8).

Acknowledgements

This work was supported by the Coordinating Research Council under the
direction of the CAPA-13-76 Project Group. We wish to thank Mr. Jeffrey
Ambs, Michael J. Pipho, and Ms. Betsy Antinozzi for their assistance in
preparing this manuscript.

References

1. Scherrer, H.C. (1982) "Optical Properties of Diesel Exhaust Aerosol,
    MS Thesis, Mechanical Engineering Department, University of Minnesota.

2. Lin, C.F., M. Baker, and R.J. Charlson (June 1973) "Adsorption Coef-
    ficient of Atmospheric Aerosol: A Method for Measurement," Applied
    Optics, 12(6):1356-1363.

3. Kittelson, D.B. and D.F. Dolan (1979) "Diesel Exhaust Aerosol,"
    ACS/CSJ Chemical Congress, Honolulu, Hawaii.

4. Dolan, D.F., D.B. Kittelson, and D.Y.H. Pui (1980) "Diesel Exhaust
    Particle Size Distribution Measurement Technique," SAE Paper No.
    800187.

5. Groblicki, P.J. and C.R. Begeman (1979) "Particle Size Variation in
    Diesel Car Exhaust," SAE Paper No. 790421.

6. Fang, C.P. and D.B. Kittelson (1984) "The Influence of a Fibrous
    Diesel Particulate Trap on the Size Distribution of Emitted Parti-
    cles," SAE Paper No. 840362.

7. Plee, S.L. and J.S. MacDonald (1980) "Some Mechanisms Affecting the
    Mass of Diesel Exhaust Particulates Collected Following a Dilution
    Process," SAE Paper No. 800186.

8. Kittelson, D.B. and M.A. Barris (1984) "The Influences of Adsorp-
    tion/Desorption Processes on the Nature of Diesel Particles in the
    Atmosphere," to be presented at First International Aerosol Confer-
    ence, September 17-21, 1984, University of Minnesota, Minneapolis, MN.

PART IV.

BASIC PROPERTIES AND BEHAVIOR;
HEALTH EFFECTS

CHAPTER 19.

BASIC PROPERTIES AND BEHAVIOR

Published 1984 by Elsevier Science Publishing Co., Inc.

Aerosols, Liu, Pui, and Fissan, editors

TRANSPORT MECHANICS OF NONSPHERICAL AEROSOL PARTICLES: PAST PRESENT AND FUTURE RESEARCH

Isaiah Gallily
Department of Atmospheric Sciences
The Hebrew University of Jerusalem
Jerusalem,91904 Israel

EXTENDED ABSTRACT

Introduction

The study of the transport mechanics of nonspherical aerosol particles is complicated due to their general lack of isotropic symmetry. Usually, the motion of these particles is significantly affected by their translation and rotation which endows it with additional degrees of freedom. In a real situation, one encounters within a cloud of such particles an extremely great number of shapes and so questions of a suitable strategy of research as well as of parameterization arise. These questions are unsolved as yet.

Past Studies

Past studies of the motion of nonspherical aerosol particles were mostly concerned with their gravitation-like settling as expressed by their so-called Aerodynamic Diameter (Stoeber,1972; Gallily and Cohen, 1976). Investigations were dedicated, however, to the creeping motion of regular small bodies in a continuous fluid with the result of an analytic equation for the fluiddynamic force (Oberbeck,1876; Brenner,1964) and torque (Jeffery,1923; Brenner, 1964), and the Slender Body Theory (Batchelor,1970; Cox,1970 and others) as well as a formal general treatment for the irregular shapes (Brenner,1964). Also, related reological studies were performed on the orientation distribution function of some types of particles in a continuous media (Peterlin,1938; Brenner and Condiff,1974 ; Hinch and Leal,1976,1979; and Rallison 1978). The special problems of aerosol science were not considered then.

Present Studies

Present studies on the subject have been concerned with typical aerosol phenomena such as inertial impaction (Gallily and Cohen,1979; Cohen et al. and Hollaender et al.,1980), and Brownian (aerosol) diffusion (Eisner and Gallily,1981,1982,1984). From these studies, one can already note that nonspherical particle transport is strongly influenced by particle rotation and, in many situations but not all (Gallily and Eisner,1979), by the initial orientation of the particles, which necessitates the introduction of a collision efficiency based on the (initial) orientation-averaged value. Likewise, one can already observe that the rotational diffusion tensor of very small cylindrical and discoidal particles in a rarified gas flow is almost two orders of magnitude smaller than the value calculated on classical grounds. The last result has been reflected in the ensemble (orientation-averaged) translational diffusion tensor in a shear flow, the value of which is essentially identical with that for a quiet air (Eisner and Gallily, 1984).

The so called wall effect representing the collector-particle fluiddynamic interaction has been also studied for situations of cylindrical particles in a creeping motion near a rigid plane (Russel et al.,1977; Schiby and Gallily,1980).

A different aspect of the problem which is related to the mechanical behavior of nonspherical aerosol particles as a cloud, namely the orientation distribution function, has been studied anew for the case of a general laminar flow and spheroidal bodies, the latter being a model for fibers and platelets respectively (Krushkal and Gallily,1984).

Future Research

With the usual danger embedded in predictions, one may delineate unsolved future problems which seem important to the author.

The first and major one relates to the strategy of further studies, which involevs from the multitude of particle forms encountered in a real aerosol cloud. A second one concerns the problem of the calculation of the fluiddynamic force and torque operating on irregularly shaped particles.

A no less important problem is related to the field of the motion of nonspherical particles in a turbulent field which is unexplored as yet but for a study by Kagermann and Koeler (1982).

One should have considered here also the experimental determination of those particle properties which matter in their transport, but this is a subject by itself.

References

Batchelor, G.K. (1970). J. Fluid Mech. 44:419.

Brenner, H. (1964). Chem. Eng. Sci. 19:703.

Brenner, H. and Condiff, D.M. (1974). J. Colloid Interface Sci. 47:199.

Cohen, A.H., Schiby,D., Gallily, I, Hollaender, W., Mause,P. Schless, D. and Stoeber, W. (1980). Proc. 8th Conference, GAF, FR Germany, pp. 299-309.

Cox, R.G. (1970), J. Fluid Mech. 44:791:

Eisner, A.D. and Gallily, I. (1981). J. Colloid Interface Sci. 81:214.

Eisner, A.D. and Gallily, I. (1982). ibid. 88:185.

Eisner, A.D., and Gallily, I. (1984). ibid. (in press).

Gallily, I. and Cohen, A.H. (1976). ibid. 56:443.

Gallily, I. and Eisner, A.D. (1979). ibid 68:320.

Gallily, I. and Cohen, A.H. (1979). ibid.68:338,

Hinch, E.J. and Leal, L.G. (1976). J. Fluid Mech. 76:187.

Hinch, E.J. and Leal, L.G. (1979) ibid.92:591

Jeffery, G.B. (1923). Proc. Roy. Soc. London A 102: 161.

Kragermann, H. and Koehler, W.E. (1982). Physica 116A :178.

Krushkal, E.M., and Gallily, I. (1984) J. Colloid Interface Sci. (in press).

Oberbeck, H.A. (1876). C relles, J. 81:79.

Peterlin, A. (1938). Z. Phys. 111:282.

Rallison, J.M. (1978)J. Fluid Mech. 84:237.

Russel, W.B., Hinch, E.J., Leal,Leal, L.G. and Tiefenbruck,G. (1977). ibid.83:213.

Schiby, D. and Gallily, I. (1980). J. Colloid Interface Sci. 77:328.

Stoeber, W. (1972), in "Assessment of Airborne Particles"(T.T. Mercer, P.E. Morrow, W. Stoeber, eds.), Thomas, Springfield, Ill., Chap. 14.

Published 1984 by Elsevier Science Publishing Co., Inc.
Aerosols, Liu, Pui, and Fissan, editors

796

APPLICATION OF THE MONTE CARLO METHOD TO SIMULATION OF PHOTOPHORESIS OF COATED SUBMICRON AEROSOL PARTICLES

M. Sitarski and M. Kerker
Department of Chemistry
Clarkson University
Potsdam, NY   13676
U. S. A.

EXTENDED ABSTRACT

A numerical method, including dual Monte Carlo sampling, has been developed for simulation of the photophoretic force acting on an illuminated coated aerosol particle. At low light intensity, of the order of the solar constant, the response of the system (photophoretic force, photophoretic velocity) to the beam is linear. However, at high intensity the strong nonlinear dependence of evaporation rate and reradiation on temperature in boundary conditions and in the expression for the photophoretic force can cause significant deviation from linear response. In the simulated theoretical model of photophoresis, the mass, momentum and energy fluxes at the particle surface were obtained in the "free molecule" regime and the distribution of heat sources within the particle was calculated according to the extension of the electromagnetic theory of light scattering from concentric spheres (Aden and Kerker, 1951). In addition to the dual Monte Carlo sampling developed before (Sitarski and Kerker, 1984) we applied external Newton iteration for the nonlinear boundary condition at the surface of the volatile aerosol particle. Test computations were carried out for two-layered particles at the accumulation mode (0.16 µm radius) of atmospheric aerosols for conditions corresponding to the altitude of 10 km. The particles consist of spherical soot cores coated with concentric shells of dirty ice. The effect of ice sublimation is taken into account in these calculations. The two assumed radii: 0.09 µm and 0.15 µm of the internal sphere correspond to the minimum and maximum of the photophoretic force (Sitarski and Kerker, 1984), respectively. The present computer study shows significant deviation from the linear response model. The calculated values of the photophoretic force at luminous intensities of the order of 10^5 mW/cm^2 are several times lower than the estimations based on the linear extrapolation of the former results for low intensities.

References

Aden, A.L., and M. Kerker (1951) "Scattering of Electromagnetic Waves from Two Concentric Spheres", J. Appl. Phys. 22:1242.

Sitarski, M., and M. Kerker (1984) "Monte Carlo Simulation of Photophoresis of Submicron Aerosol Particles", J. Atmos. Sci. submitted for publication.

Published 1984 by Elsevier Science Publishing Co., Inc.

Aerosols, Liu, Pui, and Fissan, editors

PARTICLE SHAPE EFFECTS IN THERMOPHORESIS AND DIFFUSIOPHORESIS

K. H. Leong
Department of Civil Engineering
University of Illinois
Urbana, IL 61801

EXTENDED ABSTRACT

Introduction

Thermophoresis and diffusiophoresis of aerosol particles have been researched extensively (e.g., Waldmann and Schmitt, 1966; Brock, 1962; Derjaguin and Yalamov, 1972), but theoretical formulations for thermophoresis and diffusiophoresis have been developed mostly for spherical aerosol particles. There have been a limited number of publications on cylindrical shapes only. Reed (1970) investigated the thermophoresis of cylindrical particles in slip flow and found substantial differences compared to a spherical particle. However, Yalamov and Afanas'ev (1977) found no difference in the diffusiophoretic velocity between a cylindrical and a spherical particle. The intention of this work is to examine the shape dependency of thermophoresis and diffusiophoresis of large aerosol particles.

Theory

Four different particle shapes (sphere, cylinder, oblate spheroid and prolate spheroid) will be considered to provide a representative variation in shape (Davies, 1979). Only the thermophoretic and diffusiophoretic velocities for an oblate spheroid and a prolate spheroid will be derived since the equation for a sphere and a cylinder are available elsewhere (e.g., see Reed, 1970). The coordinate systems used are standard oblate and prolate spheroidal coordinates and the gradients in temperature or concentration are aligned along the axis of symmetry (z-axis) of the oblate or prolate spheroid. The alignment of the gradient is chosen such that the particle moves in the positive z direction through the gas at rest at infinity. The method of solution used for the particle velocity follows that given by Derjaguin and Yalamov (1972).

The velocity of thermophoresis obtained for an oblate spheroid with its axis of symmetry parallel to the uniform temperature gradient ∇T is

$$U_{TO} = -\frac{b}{a} k_T \, \nu\gamma \, \frac{1-(\lambda+1/\lambda)\cot^{-1}\lambda}{T\sqrt{1+\lambda^2}\,[(1-\gamma)\cot^{-1}\lambda+\gamma\lambda/(1+\lambda^2)-1/\lambda]} \, \nabla T \qquad (1)$$

a and b are the major and minor semiaxes, k_T is the thermal creep coefficient, ν is the kinematic viscosity, γ is the ratio of the thermal conductivity of the gas to the particle, T is the temperature and λ is the ratio of b to c where

$$c = (a^2 - b^2)^{1/2} \qquad (2)$$

For the corresponding case of a prolate spheroid, the thermophoretic velocity is

$$U_{TP} = - \frac{a}{b} k_T \ v\gamma \ \frac{1+(1/\tau-\tau)\coth^{-1}\tau}{\tau\sqrt{\tau^2-1}[(1-\gamma)\coth^{-1}\tau+\gamma\tau/(\tau^2-1)-1/\tau]} \ \nabla T \qquad (3)$$

where τ is the ratio of a to c

The equations derived above are based on boundary conditions that neglect momentum slip and temperature jump and are hence strictly valid only for large particles. However available results for spheres and cylinders indicate that momentum slip and temperature jump do not alter the trend of the dependency of thermophoresis on particle shape, the problem of interest.

The same method of solution was applied to the case of diffusiophoresis for oblate and prolate spheroids. However, the velocities obtained are identical to that of a sphere. This indicates that diffusiophoresis is independent of particle shape in contrast to thermophoresis. This is surprising considering that thermophoresis and diffusiophoresis are analogous phenomena. However, the boundary conditions are substantially different and may lead to different dependencies.

Results and Discussion

The variation of the thermophoretic velocities with the aspect ratios of oblate and prolate spheroids are shown in Figure 1 for a value of γ of 0.2, The velocities are monotonic functions of (b/a) with the prolate spheroids having high velocities. The thermophoretic velocities obtained have shown large variations in magnitude depending on the shape of the particle. In a uniform gas flow, if the Reynolds number of the particle is less than 0.1, there is no preferred orientation of a particle if the particle has three mutually perpendicular planes of symmetry (Davies, 1979). However, the problem of interest often involves the dynamics of an aerosol particle near an object where a shear flow exists. In such instances, the particle will tend to align with a minimum surface area in the direction of flow (Fuchs, 1964). In the case of a prolate spheroidal particle with the thermal gradient normal to the flow, the thermophoretic velocity would approximate that of a cylindrical particle and hence will be lower than that of a sphere. For oblate spheroidal particles with the semi-minor axis aligned with the thermal gradient, the thermophoretic velocity derived earlier would be applicable. However, a disk-shaped particle with its edge aligned with the thermal gradient will have substantially higher thermophoresis. Hence, thermophoresis of nonspherical particles is a strong function of their shape and alignment with the thermal gradient

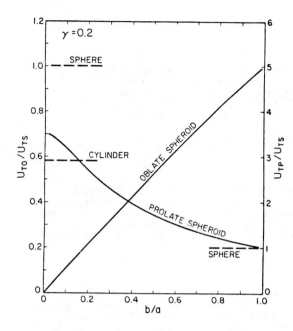

Figure 1: The thermo-
phoretic velocities of
oblate and prolate
spheroidal particles.
The abscissa is the
ratio of the minor
to the major semiaxis.
The left ordinate is
for oblate spheroidal
particles and the right
ordinate is for prolate
spheroidal particles.

References

Brock, J. R. (1962) J. Colloid Sci., 17:768.

Davies, C. N. (1979) J. Aerosol Sci., 10:477.

Derjaguin, B. V. and Yalamov, Yu. I. (1972) International Reviews in Aerosol
Physics and Chemistry, Vol. 3 (Topics in Current Aerosol Research, Part 2)
edited by G. M. Hidy and J. R. Brock, Pergamon Press, London.

Fuchs, N. A. (1964) The Mechanics of Aerosols, Pergamon Press, New York.

Reed, L. D. (1971) A Continuum Slip Flow Analysis of Steady and Transient
Thermophoresis. M.S. thesis, University of Illinois, Urbana–Champaign.

Waldmann, L. and Schmitt, K. H. (1966) Aerosol Science (Edited by C. N.
Davies) Chapter 6; Academic Press, New York.

Yalamov, Yu. I. and Afanas'ev, A. M. (1977) Sov. Phys. Tech. Phys., 22:1165.

Published 1984 by Elsevier Science Publishing Co., Inc.

Aerosols, Liu, Pui, and Fissan, editors

800

ADHESION OF AEROSOL PARTICLES TO SURFACES

S. Pickering

Commission of the European Communities
Joint Research Centre, Karlsruhe,
Institute for Transuranium Elements,
Postfach 2266, D-7500 Karlsruhe,
Federal Republic of Germany

EXTENDED ABSTRACT

Introduction

Experimental results are presented on the adhesion of aerosol particles of the nuclear fuel material $(U,Pu)O_2$ to various surfaces. The surfaces investigated were those of materials frequently used in the construction of the gloveboxes in which nuclear fuels are manufactured. Measurements were made of the centrifugal acceleration needed to detach particles from a substrate.

Experimental

Aerosol particles of $(U,Pu)O_2$ were generated from sintered fuel pellets by mechanical abrasion. This was achieved by vibrating a number of pellets together in a perforated container. The resulting poly-disperse aerosol of irregular particles was separated according to the aerodynamic diameter of the particles using an inertial spectrometer (INSPEC). Particles of known aerodynamic diameter were transfered to the chosen substrate simply by pressing the filter against the substrate with a pressure of 1.3×10^{-6} Pa. Tests showed that the transfer pressure had no measurable effect on the degree of adhesion.

Rectangular specimens of aluminium, plexiglas, steel and brass were made with the dimensions 20 x 5 x 3 mm to serve as the test substrates. The aerosol particles were loaded onto the 5 x 3 mm faces of these specimens which were then mounted in a centrifuge rotor and taken up to a pre-deter-mined speed for 1 minute. The alpha activity on the ends of the specimen before and after this treatment was measured to give the percentage of particles retained on the surface.

The centrifuge consisted of a May spinning top aerosol generator with the rotor modified to accomodate the specimens. This device, essentially an air turbine, was capable of speeds of up to 2350 rps at 6 bar air pressure. A centrifugal acceleration of up to 225000g could be applied to the end faces of the specimens by this means. The rotor was magnetized and the speed of rotation was measured using an electromagnetic tachometer.

The experiments were performed in a glove box in an atmosphere either of air with a relative humidity of 40-50% or of nitrogen with RH = 10%.

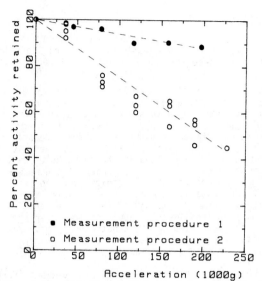

Fig.1: Adhesion of 7μm (U,Pu)O$_2$ particles on aluminium in air at 50% R.H.
Procedure 1: All data from a single deposit of particles
Procedure 2: A fresh deposit of particles used for each data point.

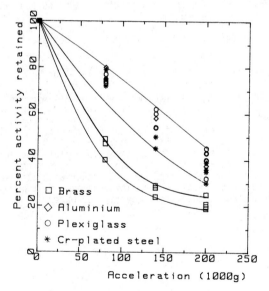

Fig.2: Adhesion of irregular 7μm (U,Pu)O$_2$ particles in N$_2$ at 10% R.H.

Results and Discussion

 The adhesion of particles was expected to depend on a) the chemical
nature and surface roughness of the substrate and on b) the relative humidity of the atmosphere. These are the parameters that were investigated.

 A further parameter was discovered unexpectedly when two different
experimental procedures were tried. The first of these consisted of
repeatedly taking the same sample to successively higher accelerations to
provide all the data points on the plot of retained activity against
acceleration. The second procedure was to use a fresh deposit of particles
on the specimen for each data point. The results obtained with these two
procedures for the retention of irregular shaped $(U,Pu)O_2$ particles on
machined aluminium are shown in Fig.1 from which it can be seen that the
retention is strongly dependent on the technique used to measure it.
It would appear that repeatedly stressing the particles at less than the
force needed to remove them in some way increases the force of adhesion.
It should be noted, that these results apply to irregular particles so that
a reorientation of the particles on the substrate may have been possible.
The technique adopted for all subsequent measurements was to use a fresh
deposit of particles for each data point.

 The results obtained in air and nitrogen at the two different humidities were the same within the experimental scatter. Similarly there
was no measurable difference in retention for machined or polished surfaces. All the results for Al, plexiglas and Cr-plated steel fell within
the same scatter band and only the results with the brass substrate differed
significantly. These results are shown in Fig.2.

 Thus, with the exception of brass all data can be described by the
linear relationship,

$$\% \text{ retention} = 100 - 2.6 \times 10^4 \text{ acceleration}$$

To calculate the force of adhesion the mass of the particle must be
known. For the irregular particles used in the this work, however, the
mass was not known. If nevertheless it is assumed that the particles are
spherical, then the force of adhesion, F, for 50% removal, at 160000g,
is 8×10^{-8} N. This value is only 8% of the value of 10^{-6} N calculated
for the force of adhesion of the same particle to a plate due to capillary
water using the formular: $F = 4\pi r\sigma$. The much lower experimental force of
adhesion compared with the calculated value together with the fact that
adhesion did not depend on relative humidity indicates that adhesion in
these experiments was not due to a surface film of water.

Published 1984 by Elsevier Science Publishing Co., Inc.
Aerosols, Liu, Pui, and Fissan, editors

SMOKE PARTICLE DEPOSITION ON CHARGED COLLECTORS

J. Podzimek, V. Wojnar, A. Keshavarz, and G. Stowell
Department of Mechanical and Aerospace Engineering
and Graduate Center for Cloud Physics Research
University of Missouri-Rolla
Rolla, MO 65401

Introduction

The study of aerosol particle or droplet deposition on collectors of different shapes has many practical applications. One of the most important is the investigation of the particulate scavenging by natural and manmade scavengers. According to the particle size range, and to the shape and physico-chemical properties of the scavenger as well as the particle dynamic interactions, one defines the following different regimes of particle deposition on scavengers: Brownian diffusion, diffusio- and thermophoresis, deposition in turbulent shear flow, inertial deposition and deposition under the influence of electrical forces. Theoretical and electrostatic forces can enhance the particle deposition in the very important particle size range (0.1 $< d_p < 1.0$ µm) named "Greenfield gap" where particle diffusion, gravitational coagulation and inertial deposition are marginally effective. There are, however, different types of particle charging during particle formation and disintegration, due to ion diffusion, charging through induction in an outer electric field etc. The potential combination of several mechanisms may explain the anomalous pattern of deposited particulates on the collector's surface which even a sophisticated model is unable to describe.

This contribution focuses on simple experiments with the deposition of spherical smoke particulates on charged nonspherical collectors and on the explanation of the observed deviations from a numerical solution of particle deposition on planar collectors.

Experimental Arrangement

The original free fall experiments in a large tank filled with smoke had been abandoned for low reproducibility of counted aerosol particles on the collector's surface. The electric charge effect on particle deposition is now studied exclusively in a laboratory wind tunnel (Turbofan-Scott Engineering Science, Model 9005) with the measuring section diameter of 13.2 cm. Most of the experiments have been performed at airflow velocities between 0.25 m s^{-1} and several m s^{-1}. The flowrate is permanently monitored by a calibrated electronic pressure gage and all velocity profiles were measured for each test velocity by a hot-wire anemometer.

All experiments with charged collectors have been performed with Titanium Chloride (TiCl$_4$) aerosol produced by the passage of dry nitrogen gas through a humidifier into a generating flask where TiCl$_4$ reacted with water vapor. After removing the excess water from the aerosol-gas mixture the aerosol passed through two furnaces operated at temperatures around 700°C. The aerosol of a narrow size distribution passed afterwards through a coarse filter and through a Kr-85 (10 mCi) aerosol neutralizer into the wind tunnel (Podzimek and Martin, 1984). Aerosol particles had usually a mean diameter around 0.5 µm and their geometric standard deviation varied between 0.1 and 0.15. Their concentration and size distribution was monitored by a Laser Active Scattering Aerosol Spectrometer, Model ASAS-300A. In most of the measurements the aerosol generation and size distribution remained similar for 30 minutes.

The models (collectors) for wind tunnel measurements--such as disks, cylinders and grids--were made of conductive (metals) and dielectric (glass, plexiglass, PVC) material as well. Their main dimension was smaller than 1.0 cm. The models exposed in the center of the tunnel's test section were mounted on a special stage which enabled to scan them with a simple manipulation in a scanning electron microscope. Deposited smoke particulates were evaluated from the electron microphotographs under the assumption that the particle deposition pattern on symmetrical models is symmetrical with respect to the center on the front and back side as well. Models were charged by connecting them with a generator keeping them at a potential difference (with reference to the grounded tunnel wall) between 0.5 kV and 8 kV. For instance to the potential difference of 1 kV corresponds a gradient of 3.3 esu cm^{-1} at a distance of 0.25 cm from the model with diameter of 0.45 cm. Currently, experiments with field charged collector and particles are performed.

Results of Measurements

Conclusions from the general picture of collection efficiencies, E, of a charged glass disk for smoke particle diameters d_p < 0.7 μm, < 1.4 μm, < 2.1 μm and Re between 70 and 320 (Fig. 1) are the following: There is a large scatter of measured E with a general trend that smoke particles with d_p < 0.7 μm are removed at higher electrostatic charges more effectively. However, there seems

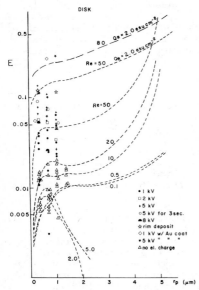

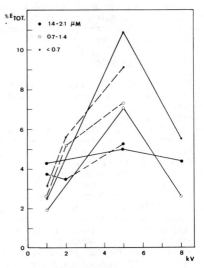

Figure 1. Calculated (dashed lines) and measured E for a disk (R_C=0.225 cm and V=1 ms^{-1}).

Figure 2. Measured E for a charged disk and different particle sizes.

to be a maximum deposition at 5 kV (Fig. 2). Increasing further the potential difference up to 8.0 kV is not paralleled by a higher particle deposition (see also Podzimek and Martin, 1984). There is an indication that at 8.0 kV the deposited aerosol forms more chain-like aggregates at the collector's rim. This might cause possibly the undercounting of particles in larger aggregates. There was also an interesting dependence of the number of deposited particles on the surface and physico-chemical properties of the collector. Half of the glass disk coated by gold showed high particle deposit (N_{un}/N_{co}) = 0.594 on the front and back collector's surface while the other (uncoated) half was featured by high deposit along the rim (N_{un}/N_{co} = 1.133). At the collector charge 5 or 8 kV was several times observed that almost 50% of all particulates are found along the edge. This might contradict the calculated collision efficiencies by Martin et al (1981) in Fig. 1. Also, the numerical model based on "creeping" flow equations underestimates the particle deposition on the reverse side of a disk especially for particles with d_p < 0.7 μm if only simple Coulomb forces are considered. This can explain why the measured total E-curves do not follow the calculated collection efficiencies with a typical "saddle" domain in particle diameter range 0.7 < d_p < 5.0 μm.

Many experiments with smoke particle deposition on dielectric and metallic cylinders at Re ranging from 3.4 to 44.0 have been performed with the following results: The collection efficiency of brass wires and cylinders increases almost linearly with the potential difference (E = 0.003 for 1 kV; E = 0.02 for 5 kV and E = 0.03 for 8 kV). The pattern of deposited particles reminds one strongly of the theoretically calculated values (high deposit of largest particulates around the front stagnation point and the high deposit of small particulates around the rear stagnation point. There is, however, an indication that at a potential difference of 8 kV the deposition of particles with d_p > 1.40 μm on the reverse side of a cylinder is increasing. The number of deposited particles on glass and PVC cylinders at 5 KV overrides that on dielectric materials which combined with uneven surface plays an important role in particle deposition.

Several pilot experiments have been performed with charged ELM metallic grids (Mesh 300 and 50 with the diameters of 0.3 cm). Compared to the collection efficiency of a solid disk was found a higher particle deposition rate especially close to the grid's rim. Many particles form long chain aggregates. For relatively low "charge" (1 kV) the collection efficiency for particles with d_p < 0.7 μm was high (5.21%).

Preliminary results of the particle deposition on the fibers (nylon or "dust magnet") indicate also a high potential of using these mesh type scavengers for particle deposition (Fig. 3).

Conclusion

Electric charge changes dramatically the collection efficiency of both metallic and dielectric materials. Dielectric materials seem to enhance particle deposition along the rim, which can make more than 50% of the total particle deposition. There is an optimum electric charge of a disk (5 kV) beyond which the collection efficiency is decreasing. Grid type scavengers are very effective at relatively low electrostatic charge (1 kV).

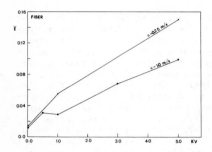

Figure 3. Collection efficiency
of a charged "dust
magnet" fiber of
R_c = 15 μm.

References

Martin, J. J., P. K. Wang, and H. R. Pruppacher (1981) "A Numerical Study of
the Effect of Electric Charges on the Efficiency with Which Planar Ice Crystals
Collect Supercooled Drops", J. Atmos. Sci., 38:2462.

Podzimek, J. and J. J. Martin (1984) "Deposition of Particulates on Planar
Scavengers, Preprint VII-th Int. Conf. Atmos. Electricity, Albany, AMS, 57.

Published 1984 by Elsevier Science Publishing Co., Inc.

Aerosols, Liu, Pui, and Fissan, editors

SLIP CORRECTION MEASUREMENTS OF POLYSTYRENE AND POLYVINYLTOLUENE AEROSOL PARTICLES IN AN IMPROVED MILLIKAN APPARATUS

M. D. Allen* and O. G. Raabe
Laboratory for Energy-related Health Research
University of California
Davis, CA 95616
U. S. A.

EXTENDED ABSTRACT

Introduction

A slip correction factor is used to correct Stokes' law for the fact that the no-slip boundary condition is violated for small aerosol particles moving with respect to the gaseous medium. The Knudsen-Weber form of the slip correction is given by $C(Kn) = 1 + Kn [\alpha + \beta \exp (\gamma/Kn)]$. The parameters α, β, and λ are customarily those based upon the experiments of Millikan (1917; 1923) for aerosol droplets of oil. Because of differences between molecular interactions with the surfaces of solid particles and oil drops, different parameters should be used for solid particles.

There is a significant difference between the reported values of α for polystyrene spheres (PSL), other solid particles, and watch oil droplets. The values of β and γ are much more difficult to measure than the value of α and do not appear in the literature for PSL spheres. Since PSL spheres are a commonly used laboratory aerosol, the values of α, β, and γ need to be precisely and accurately measured. Also, these hard plastic spheres should be representative of other solid aerosol particles.

In this research, an improved version of the Millikan apparatus was designed and constructed to precisely and reproducibly measure the slip correction factor for solid, spherical particles in air. The slip corrections for PSL and PVT spheres were measured for particle Knudsen numbers ranging between 0.03 and 7.2, and the most appropriate values of α, β, and γ were determined by fitting the Knudsen-Weber formula to these experimental results.

Methods

Figure 1 is a schematic drawing of the improved Millikan oil-drop apparatus used in these experiments. Major differences in this apparatus and the one used by Millikan were: (1) the capacitor plates were immersed directly in the oil bath, (2) particle illumination was accomplished with a vertical helium-neon laser beam, (3) solid particles of known monodispersity were used, and (4) particles were introduced into the interelectrode space using a glass settling tube. Most of these improvements were aimed at either suppressing or detecting natural convection in the interelectrode space.

---

*Present address: Lovelace Inhalation Toxicology Research Institute, P.O. Box 5890, Albuquerque, NM 87185.

808

Results and Discussion

The physical density of the PSL and PVT particles used in this study were determined utilizing density gradient ultracentrifugation with solutions of sucrose. The density of the PSL particles with the surfactant removed was found to be 1.0468 ± 0.0006 g/cm^3, and of the PVT particles was 1.0227 ± 0.0006 g/cm^3. Particle size and slip correction measurements were made on eleven 4.6-μm PSL-DVB spheres, twenty-five 2.08-μm PVT spheres, forty-eight 1.18-μm PSL spheres, and six 0.79-μm PSL spheres at particle Knudsen numbers between 0.03 and 7.2. Sizes and pressures were varied to change the particle Knudsen number.

In Figure 2, the particle Knudsen number (Kn) is plotted as the independent variable and the slip correction parameter (A) is plotted as the dependent variable. The fitted nonlinear least squares values of α, β, and γ in the Knudsen-Weber form of the slip correction equation are 1.142, 0.558, and 0.999, respectively, for the assumed mean free path of 0.0673 μm for air at 760 mm Hg and 23°C with a viscosity of 183.245 micropoise. These values of α, β, and γ are in good agreement with Millikan's (1923) result (adjusted to the

Figure 1. Improved Millikan Apparatus Used in These Experiments

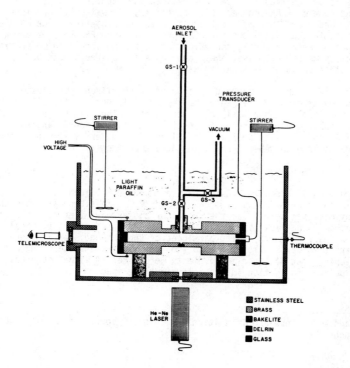

same mean free path). The sum of $\alpha + \beta = 1.70$ was 4.45 percent higher than Millikan's value for oil droplets; this observation is consistent with the expectation that a higher percentage of molecules undergo specular reflections from the surface of a solid particle (23.7 percent) than from the surface of an oil droplet (8.6 percent).

References

Allen, M. D. and O. G. Raabe (Submitted) "Slip Correction Measurements of Spherical Solid Aerosol Particles in an Improved Millikan Apparatus." Aerosol Sci. Technol.

Millikan, R. A. (1917) "A New Determination of e,N, and Related Constants." Phil. Mag. 34, 1-30.

Millikan, R. A. (1923) "The General Law of Fall of a Small Spherical Body Through a Gas, and Its Bearing Upon the Nature of Molecular Reflection from Surfaces." Phys. Rev. 22, 1-23.

Figure 2. Slip Correction Parameter (A) versus Particle Knudsen Number for Solid Spherical Particles

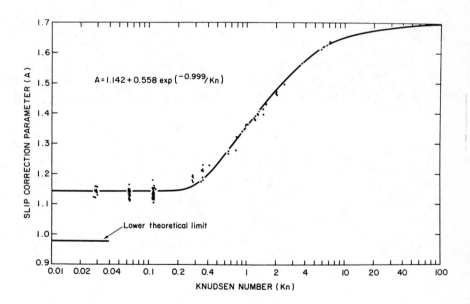

Theoretically Computed Optical Constants and Scattering Properties of Nonspherical Particles

by

O. I. Sindoni[(o)], F. Borghese[(+Δ)], P. Denti[(+Δ)], R. Saija[(+)], and G. Toscano[(+)]

(*)   Based on work supported by the U. S. Army European Research Office through Contract DAJ437-81-C-0895 and Grant DAJ445-84-C-0005

(+)   Universita' di Messina, Istituto di Struttura della Materia, 98100 Messina, Italy

(o)   Chemical Research and Development Center, Aberdeen Proving Ground, Maryland, USA

(Δ)   Gruppo Nazionale Struttura della Materia, Sezione di Messina

Recently we have published several papers (Borghese, 84) (Borghese, 84); (Sindoni, 84) in which we show our theoretical model, its symmetization and through the use of Group Theory its simplification to compute the optical properties of molecular or molecular-like systems. The scattering particles may be of any size or shape. The model describes the particle as a hierarchical sequence of clusters of spherical domains down to the level of resolution of individual atoms. The expressions for the cross sections are:

$$\sigma_\eta^{(s)} = \frac{1}{2k^2} \sum_{L'M'} (|\bar{A}_{\eta L'M'}|^2 + |\bar{B}_{\eta L'M'}|^2),$$

$$\sigma_\eta^{(abs)} = \frac{1}{8k^2} \sum_{L'M'} (2|W_{\eta L'M'}|^2 - |2\bar{A}_{\eta L'M'} + W_{\eta L'M'}|^2 - |2\bar{B}_{\eta L'M'} + \eta W_{\eta L'M'}|^2),$$

$$\sigma_\eta^{(tot)} = \frac{-1}{2k^2} \sum_{L'M'} \mathrm{Re}[W_{\eta L'M'}{}^*(\bar{A}_{\eta L'M'} + \eta \bar{B}_{\eta L'M'})].$$

The solutive equations for A and B, which describe the optical properties of the cluster, were written in matrix form as:

$$
\begin{vmatrix} \mathfrak{R}^{-1} + \mathfrak{H}^r(m) & \mathscr{K}^r(m,e) \\ \mathscr{K}^r(e,m) & \mathfrak{S}^{-1} + \mathfrak{H}^r(e) \end{vmatrix} \begin{vmatrix} \mathscr{A}^{rp} \\ \mathscr{B}^{rp} \end{vmatrix} = - \begin{vmatrix} \mathscr{P}^{rp} \\ \mathscr{Q}^{rp} \end{vmatrix}.
$$

Where the components of the left-hand tensor and the right-hand side are known from the cluster structure and the incomming wave.

The definitions of all terms in these equations and details of their computation from the physical description of the scatterers and EM wave can be found in the papers cited above. The computer codes developed on the basis of this theoretical model have yielded interesting results such as the discovery of the existence of an isobestic point (with respect to change in shape or orientation), characteristic of each material at a specific set of values for the wavevector, k, and domain radius, b_H. At this point, the scattering and absorption properties are independent of the shape and orientation, i.e, the system behaves as a sphere. Figures 1 and 2 illustrate this phenomenon, where $\sigma^{(s)}$ is the scattering cross section for a cluster of domains and σ_0 is that for a single domain.

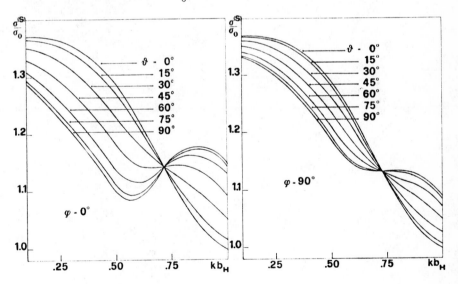

Figures 1 and 2. Both figures show the ratio $\sigma^{(s)}/\sigma_0$ for a cluster of three spheres with the structure of water as a function of kb_H for several values of Θ. Where Θ is the scattering angle. ϕ is the azitumal (incidence angle), σ_0 is the sum if the cross section of the constituent spheres, b_H is the radius of the smallest sphere of the cluster.
Indexes of refraction: $n_0 = 1.5$

$$n_H = 1.3$$

In Figure 1 $\phi = 0$, In Figure 2 $\phi = 90$

We also show that when studying the optical properties (index of refraction and absorption coefficient) of a system of multiple clusters having arbitrary orientation, we are always able to find such an isobestic point. The only difference is a shifting of the wavelength with respect to the single cluster system.

References

Borghese, F., et. al., JOSA, Vol 1, No. 2, 1984.

Borghese, F., et. al., ATS, Vol 3, No. 2, 1984.

Sindoni, O. I., et. al., ATS, Vol 3, No. 2, 1984.

Published 1984 by Elsevier Science Publishing Co., Inc.
Aerosols, Liu, Pui, and Fissan, editors

Direct Calculation of Attachment Coefficients for the
Unipolar Charging of Ultrafine Particles

Stowell W. Davison
Engineering Department, University of Maryland Baltimore County
Catonsville, Md. 21228

S. Y. Hwang and J. W. Gentry
Chemical and Nuclear Engineering Department
University of Maryland, College Park, Md. 20742

Introduction

   The development over the last ten years of the electrical aerosol analyzer and the electrical aerosol classifier has generated increased interest in
the charging behavior of ultrafine particles (under 100 nm in diameter).
These particles acquire charge by diffusion of ions, a mechanism distinct from
field charging, which becomes significant for particles over 100 nm and controls for particles over one micron in diameter (Friedlander, 1977, pg. 41).
Initially it was believed that particles at bipolar equilibrium were charged
according to the Boltzmann distribution (Liu and Pui, 1974), and this distribution was used to interpret data from the electrical aerosol classifier. It
rapidly became apparent, however, that this distribution was seriously in
error for particles of less than about 40 nm diameter. This discovery led to
the formulation of several empirical modifications to the Boltzmann distribution which were applied with some success (Kojima, 1978; Haaf, 1980).

   These modifications did not, however, attempt to relate charging behavior
to the physical parameters of the charging phenomenon such as particle dielectric constant, ion number density, and ion species. These last two may be
responsible for the recently observed asymmetry in the bipolar equilibrium
charge distribution. A further shortcoming of the modified Boltzmann disribu-
tions is that they do not address the problem of unipolar charging, which is
central to the interpretation of results from the electrical aerosol analyzer
and from devices that count particles by measuring total charge. At present
these results are interpreted entirely by means of calibration data obtained
from test aerosols.

   Several models have been developed to fill this need. Calculations based
on four of them are employed in the simulations discussed here. Solutions of
these models indicate that the experimental data necessary to refine them and
discriminate among them must come from experiments with extremely small particles (under 15 nm), from transient bipolar charging experiments, or from
unipolar charging experiments. It is the design of experiments to measure the
unipolar charging coefficient that is the subject of this presentation.

Charging Theories

The four charging theories have been discussed in detail elsewhere (Davison and Gentry, 1983). They differ in how they handle the problems of the image potential and the transition regime.

As an ion approaches a conductive particle a local charge is induced in the particle opposite in polarity to the ion charge. This is true regardless of the polarity or magnitude of the total charge on the particle. Because of this induced charge an additional term (called the image potential) is added to the potential energy expression. The image potential becomes infinite at the surface of the particle and thus causes some computational difficulties. Fuchs, Gentry, and Hoppel deal with this problem by defining a capture sphere concentric with the particle. The potential is evaluated at the capture sphere, not the particle surface.

Fuchs and Hoppel use similar classical mechanical arguments to define the capture sphere. This results in the sphere radius being a function of ionic kinetic energy. Approximate averaging techniques are used to eliminate this dependence. Gentry defines the capture sphere to be at the radius where the current of ions in toward the particle reaches an apparent minimum. This is called the minimum flux radius (MFR). Laframboise and Chang neglect the image force. Their model can be precise, then, only for the charging of perfectly dielectric particles.

To deal with the problem of the transition regime Fuchs and Hoppel define limit spheres about the particle. Inside the limit sphere ion behavior is assumed to be governed by kinetic gas theory. Outside the limit sphere the gas is assumed to be a contnuum with the ions moving through it by diffusion. The ion number densities are matched at the limit sphere. The two authors use slightly different definitions of the sphere.

Gentry derives an approximate solution to the BGK version of the Boltzmann equation to provide a solution for the charging rate in the transition regime.

Laframboise and Chang use a single, closed-form expression to calculate charging rates in all regimes. It reduces in the extremes (free molecule and continuum regimes) to the expected forms, and in the transition regime interpolates between them.

Bipolar Data

Data on the charging of ultrafine particles are sparse, and confined to the bipolar equilibrium case. Figures 1 and 2 show some recently reported measurements for positively and negatively charged particles respectively. Also shown on the figures are some theoretical curves. Hussin, who obtained some of the data shown here, decided that the Fuchs model represented them well with positive and negative ion masses of 140 and 101 respectively and positive and negative ion diffusivities of 0.029 and 0.036 square centimeters per second respectively. These parameters were not measured, but were instead treated as being adjustable. The other curves shown were obtained from the MFR theory with the indicated parameters. The ion masses and diffusivities are important and unless they can be independently determined discrimination among the models based on bipolar measurements is difficult.

Unipolar Charging

The equilibrium bipolar charge distribution depends on ratios of the unipolar and bipolar charging coefficients (Davison, 1984). As a result the differences between the theories which would affect predictions of transient behavior might be masked in steady state solutions. Unipolar charging calculations and measurements, on the other hand, are likely to be more revealing of distinctions. Figure 3 shows that solutions of the four models for unipolar charging of 5 nm particles vary greatly. Figure 4 shows a more fundamental value: the unipolar charging coefficient. Experimental measurement of these coefficients would discriminate among the theories and elucidate the fundamental processes involved in aerosol charging.

References

Davison, S. (1984) Evaluation of Kinetic Models for the Charging of Ultrafine Particles by Diffusion of Ions, Ph. D. Thesis, University of Maryland, College Park, Md.

Davison, S.; Gentry, J. (1983) "Fundamental Hopotheses on the Charging of Ultrafine Particles by Diffusion of Ions", Pacific Region Meeting, Fine Particle Society, August 1 - 5, 1983, Honolulu, Hawaii.

Friedlander, S. (1977) Smoke, Dust and Haze , John Wiley & Sons, Inc. New York, N.Y.

Haaf, W. (1980) "Accurate Measurements of Aerosol Size Distribution I: Theory of a Plate Condenser for Bipolar Mobility Analysis", Journal of Aerosol Science, v. 11, pg. 189.

Kojima, H. (1978) "Measurements of Equilibrium Charge Distribution on Aerosols in Bipolar Ionic Atmosphere", Atmospheric Environment, v. 12, pg. 2363

Liu, B.; Pui, D. (1974) "Equilibrium Bipolar Charge Distribution of Aerosols" Journal of Colloid and Interface Science, v. 49, pg. 305.

816

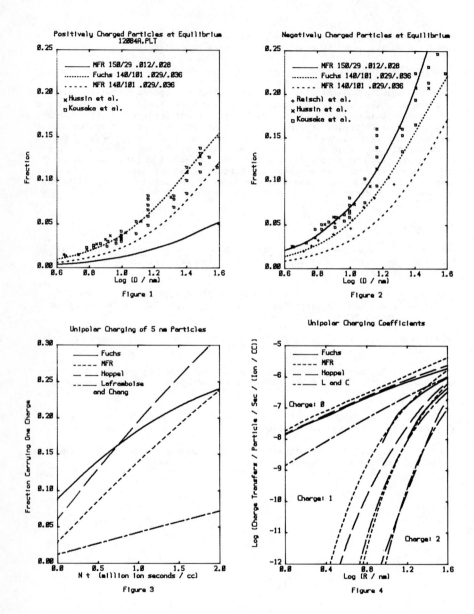

Positively Charged Particles at Equilibrium
12084A.PLT

MFR 150/29 .012/.028
Fuchs 140/101 .029/.036
MFR 140/101 .029/.036
x Hussin et al.
□ Kousaka et al.

Fraction

Log (D / nm)

Figure 1

Negatively Charged Particles at Equilibrium

MFR 150/29 .012/.028
Fuchs 140/101 .029/.036
MFR 140/101 .029/.036
+ Reischl et al.
x Hussin et al.
□ Kousaka et al.

Fraction

Log (D / nm)

Figure 2

Unipolar Charging of 5 nm Particles

Fuchs
MFR
Hoppel
Laframboise
and Chang

Fraction Carrying One Charge

N t (million ion seconds / cc)

Figure 3

Unipolar Charging Coefficients

Fuchs
MFR
Hoppel
L and C

Log (Charge Transfers / Particle / Sec / (Ion / CC))

Charge: 0

Charge: 1

Charge: 2

Log (R / nm)

Figure 4

Effects of the "Chemical Reactions" on the Macroscopic Optical Constants of a Model Aerosol[*]

by

R. Saija[+], O. I. Sindoni[o], G. Toscano[+], P. Denti [+Δ],
and F. Borghese [+Δ]

(*)   Based on work supported by the U.S. Army European
        Research Office through Contract DAJ437-81-C-0895 and
        Grant DAJ445-84-C-0005

(+)   Universita' di Messina, Istituto di Struttura della
        Materia, 98100 Messina, Italy

(o)   Chemical Research and Development Center, Aberdeen
        Proving Ground, Maryland, USA

(Δ)   Gruppo Nazionale Struttura della Materia, Sezione di
        Messina

It is well known that the refractive index of a low-density dispersion of scatterers is a matrix whose (complex) elements are given by (Newton, 1966).

$$N_{\eta'\eta} = \delta_{\eta'\eta} + \frac{2\pi}{Vk^2} \sum_{\nu} f_{\nu, \eta'\eta} \tag{1}$$

In eq. (1) k is the magnitude of the wavevector of the incident light and ν numbers the scatterers within the volume V. Furthermore, if $\underset{\sim}{u}_{\eta'}$ is the polarization vector of the incident light, $f_{\nu, \eta'\eta}$ is related to the normalized forward-scattering amplitude for polarization η, of the ν-th scatterer, through

$$f_{\nu, \eta'\eta} = \underset{\sim}{u}^*_{\eta'} \cdot \underset{\sim}{f}_{\nu,\eta} \tag{2}$$

The elements $N_{\eta'\eta}$ are related to the macroscopic refractive index, n_{η}, and to the absorption coefficient, γ_{η}, for polarization η, through

$$n_{\eta} = \text{Re} (N_{\eta\eta}), \quad \gamma_{\eta} = 2 k \, \text{Im} (N_{\eta\eta})$$

respectively. We recall that if all scatterers are identical and spherically symmetric, $f_{\nu,n'n}$ does not depend on ν, while if the scatterers are identical but nonspherical, $f_{\nu,n'n}$ depends in their orientation with respect to the incident field.

In a preceding paper (Borghese, 1984) we proposed to account for the lack of spherical symmetry of some kind of scatterers by modeling them as clusters of spheres, and, making full use of the features of the model, succeeded in factorizing $\tilde{f}_{\nu,n'n}$ into a part depending only on the structure and a part depending only on the orientation of the clusters. For a dispersion of identical clusters we were then able to run over the orientations and to express $\mathcal{N}_{n'n}$ in terms of the multipolar amplitudes scattered by a cluster whatever. The result is

$$\mathcal{N}_{n'n} = \delta_{n'n} + \frac{N}{4k^3 i} \sum_{LM} \frac{1}{2L+1} \; W^*_{n'LM} \; W_{nLM} \sum_{M'} (\overline{U}^{(A)}_{nLMLM'} +$$

$$n' \; \overline{U}^{(B)}_{nLMLM'}) \tag{3}$$

where the $\overline{U}$'s depend only on the structure of the clusters, whose number density is N, and W_{nLM} is proportional to the multipolar amplitudes of the incident plane wave field. Eq. (3) can be immediately extended to dispersions of more than one kind of clusters. If $N_\alpha = N c_\alpha$ is the number density of the α-th kind of cluster, we can write

$$\mathcal{N}_{n'n} = \delta_{n'n} + N \sum_{\alpha,L} c_\alpha \; \xi_{n'nL} \; S_{\alpha,n'nL} \tag{4}$$

where we define

$$\xi_{n'nL} = \frac{1}{4\pi i} \frac{1}{2L+1} \sum_M \; W^*_{n'LM} \; W_{nLM}$$

$$S_{\alpha n'nL} = \frac{1}{k^3} \sum_{M'} (\overline{U}^{(A)}_{\alpha nLMLM'} + n' \; \overline{U}^{(B)}_{\alpha nLMLM'}) \; .$$

The index α added to the $\overline{U}$'s individuate the quantities appropriate to the α-th kind of cluster.

The cluster model and eq. (4) lend themselves to the calculation of the changes induced on the optical constants of the dispersion by structural rearrangements of the clusters or by redistribution of the spheres among different clusters; these rearrangements should account for the occurrence of "chemical reactions" among the scatterers of the dispersion. In fact, eq. (4) can be used for different values of the relative concentrations, c_α, of the reagents and of the reaction products, subject to the restraint

$$\sum_\alpha C_\alpha = 1$$

of course. The results of the calculations are well described by the quantity

$$\Gamma = \lfloor p\,\gamma^{(f)} + (1 - p)\,\gamma^{(i)} \approx /\gamma^{(i)}$$

where $\gamma^{(f)}$ and $\gamma^{(i)}$ are the absorption coefficients of the dispersion of the reaction products and of the reagents, respectively, and p is the completion parameter of the reaction. The parameter Γ has been computed for a few "reactions" chosen so as to evidence the effect of the rearrangement of the spheres. In all cases considered the changes in the absorption coefficient are more visible when dramatic changes of the structure of the clusters occur. Of course, the results we present here are to be considered as preliminary ones and do not pretend, at this stage, to fit any actual experimental data. In any case our work shows that the anystropy of the scatterers plays a relevant role in the process of coherent electromagnetic propagation and that, neglecting the possible structural changes of the scatterers may prevent any realistic interpretation of the experimental data.

References

Newton, R. G., "Scattering Theory of Waves and Particles", McGraw-Hill, New York, 1966.

Borghese, F., Denti, P., R. Saija, G. Toscano, and O. I. Sindoni, "Macroscopic Optical Constants of a Cloud of Randomly Oriented Nonspherical Scatterers, Nuovo Cim., to be published, 1984).

Published 1984 by Elsevier Science Publishing Co., Inc.

Aerosols, Liu, Pui, and Fissan, editors

820

STUDIES OF THE PHYSICAL DENSITY OF FINE PARTICLES*

Otto G. Raabe
Paul E. Brunemeier
Amiram Rasolt
Stephen V. Teague

Laboratory for Energy-related Health Research
University of California
Davis, California 95616 USA

ABSTRACT

The physical density of aerosol particles has a direct influence on their inertial and aerodynamic properties . Particles in a given aerosol may have a wide distribution of physical densities which may not be apparent from either aerodynamic size measurements or from physical size determinations. In some cases the aerosol may consist of essentially monodense particles that have physical densities that vary little such that the coefficient of variation of the densities is less than twenty percent.

In this study the density of polystyrene latex aerosol particles, polyvinyltoluene particles , fused aluminosilicate particles formed by heating aerosols of montmorillonite clay, and respirable coal fly ash were measured with collected samples by liquid displacement and density gradient centrifugation techniques. These methods provide information on the physical density of the bulk materials and the density distribution of the particles, respectively. By study of size-classified monodisperse samples it is possible to evaluate the distribution of densities at given particle sizes using density gradient centrifugation. The results show that the polystyrene latex, polyvinyltoluene particles , and fused aluminisilicate particles made from montmorillonite clay are monodense with densities of 1.0468 g/cm^3 , 1.0227 g/cm^3, and 2.457 g/cm^3 respectively. The error estimate for these values is 0.1%, 0.1%, and 0.2%, respectively. In contrast the respirable coal fly ash was polydense with some particles with densities less than 2.0 g/cm^3 and some heavier than 3.0 g/cm^3, while most of the ash had a median density of 2.47 g/cm^3. The use of polydisperse fused aluminosilicate particles formed from montmorillonite clay for calibrations of sampling instruments is recommended because the particles are polydisperse to allow a broad calibration spectrum but provide a accurate calibration baseline because the particles are spherical and monodense. Their use in calibration of a spiral duct centrifuge is demonstrated and compared to calibrations performed with monodisperse polystyrene latex particles.

* Research supported by the Office of Health and Environmental Research of the U. S. Department of Energy under Contract No. DE-AM03-76SF00472 with the University of California, Davis.

MOTION OF A SPHERE IN A RAREFIED GAS:
ROLE OF GAS-SURFACE INTERACTIONS

K.A. Hickey and S.K. Loyalka
Nuclear Engineering Program
University of Missouri-Columbia
Columbia, MO   65211
U.S.A.

EXTENDED ABSTRACT

The motion of a sphere in a non-uniform rarefied gas is a
fundamental problem that has applications in a variety of systems
(Loyalka, 1983). Previous studies have examined this problem
using the linearized time-independent Boltzmann equation with the
Bhatanager-Gross-Krook collisional model (Cercignani and
Pagani,1968, and Lea and Loyalka, 1982). Cercignani, et al calculated drag forces for all rarefactions using a variational method.
Lea and Loyalka obtained velocity profiles as well as the drag
force by numerical methods. For both situations, the boundary
conditions used were diffuse reflection on the surface of the
sphere.

A more general and realistic model of the boundary conditions
would include the possibility of specular reflection on the surface of the sphere. Sone and Aoki (1977) have considered specular
reflection in the restricted near continuum region by the use of
asymptotic expansions. Wang Chang (1950) obtained expressions for
the drag force on a sphere for an arbitrary diffuse-specular
reflection boundary condition in the free molecular regime.

Following the methods of Wang Chang, we obtained expressions
for velocity and density profiles for arbitrary diffuse-specular
accommodation in the free molecular regime. These results are
reported in Table I.

For an arbitrary Knudsen number, it is necessary, however, to
solve the boundary value problem using the Boltzmann equation. A
computer program was constructed to solve the linearized Boltzmann
equation with the BGK collision model and arbitrary diffuse-
specular boundary conditions. This is accomplished by transforming the initial integro-differential equation into a system of
integral equations which are then regularized and solved by numerical methods. The results are verified by comparing them with the
exact results in the limit of the free molecular and continuum
regimes.

The full development of the equations and the results will be
reported, and their implications discussed.

References

Loyalka, S.K. (1983) "Mechanics of Aerosols in Nuclear Reactor
Safety: A Review", Progress in Nuclear Energy, 12, 1-56.

Cercignani,C. and C.D. Pagani (1968) "Flow of a Rarefied Gas Past an Axisymmetric Body", Physics of Fluids, 117, 1395-1403.

Lea, K.C. and S.K. Loyalka (1982) "Motion of a Sphere in a Rarefied Gas", Physics of Fluids, 25(9), 1550-1557.

Sone, Y. and K. Aoki (1977) "Slightly Rarefied Gas Flow over a Specularly Reflecting Body", Physics of Fluids, 20(4), 571-576.

Wang Chang, C.S. (1950) "Transport Phenomena in Very Dilute Gases, II", University of Michigan CM-654.

TABLE 1: Free Molecular $(x = R/r)$

Density Profiles

Diffuse: $n = n_\infty [1 - \dfrac{u_\infty \cos\theta_o}{\sqrt{\pi}} \{ x^2 (1 + \dfrac{\pi}{6}) + \dfrac{\pi}{2}\dfrac{1}{x} (1 - \sqrt{1 - x^2})$

$- \dfrac{\pi}{6}\dfrac{1}{x} (1 - (1 - x^2)^{3/2}) \}]$

Specular: $n = n_\infty [1 - \dfrac{u_\infty \cos\theta_o}{\sqrt{\pi}} \{ \dfrac{3}{2}\dfrac{1}{x} - \dfrac{1}{2}x + x^2 + \dfrac{1}{x^2} (-\dfrac{3}{2} + x^2 + \dfrac{1}{2}x^4)$

$\ln \dfrac{(1 + x)}{(1 - x^2)^{1/2}} \}]$

Velocity Profiles

$$c_{||} = u_\infty (\tfrac{2kT}{m})^{1/2} (b + a \sin^2\theta_o)$$

$$c_\perp = -u_\infty (\tfrac{2kT}{m})^{1/2} (a \sin\theta_o \cos\theta_o)$$

where, for diffuse reflection:

$a = -\dfrac{1}{16} + \dfrac{3}{16}x^2 + \dfrac{3}{8}x^3 + \dfrac{1}{4x} (1 - x^2) \ln \dfrac{1 + x}{\sqrt{1 - x^2}}$

$- \dfrac{3}{16x} (1 - x^2)^2 \ln \dfrac{1 + x}{\sqrt{1 - x^2}} + \dfrac{3}{4}x^2 \sqrt{1 - x^2}$

$b = \dfrac{3}{8} - \dfrac{x^2}{8} - \dfrac{x^3}{4} + \dfrac{1}{8x} (1 - x^2)^2 \ln \dfrac{1 + x}{\sqrt{1 - x^2}} + \dfrac{1}{2} (1 - x^2)^{1/2}$

and for specular reflection:

$a = \dfrac{2}{5}\dfrac{1}{x^2} + \dfrac{3}{5}x^3 + \dfrac{3}{5}\dfrac{1}{x^2} (1 - x^2)^{5/2} - \dfrac{1}{x^2} (1 - x^2)^{3/2}$

$b = \dfrac{2}{5}\dfrac{1}{x^2} - \dfrac{2}{5}x^3 - \dfrac{2}{5}\dfrac{1}{x^2} (1 - x^2)^{5/2}$

Drag (from Wang Chang, 1950)

$F = \dfrac{16}{3} \sqrt{\pi} \rho_\infty R^2 u_\infty (1 + \dfrac{\pi}{8})$ diffuse

$\} F = 4\sqrt{\pi} \rho_\infty R^2 u_\infty (\dfrac{\alpha\pi}{6} + \dfrac{4}{3})$

$F = \dfrac{16}{3} \sqrt{\pi} \rho_\infty R^2 u_\infty$ specular

Published 1984 by Elsevier Science Publishing Co., Inc.
Aerosols, Liu, Pui, and Fissan, editors

824

Near-Forward Scattering from Nonspherical Particles

C. D. Capps, R. L. Adams, G. M. Hess
Boeing Aerospace Company
P.O. Box 3999
Seattle, WA 98124

Introduction

Most aerosols, except for liquids, consist of nonspherical particles. A major problem in designing instrumentation to characterize these aerosols or designing aerosols to exploit particular properties of nonspherical scatterers, such as increased absorption or scattering directionality, is the lack of solutions to the electromagnetic scattering equations for arbitrary particle geometries. The general problem of scattering of electromagnetic radiation from particles with a size on the order of a wavelength has been solved only for a few geometric shapes: spheres (Mie, 1908), infinite cylinders (Wait, 1955), and spheroids (Asano and Yamamoto, 1975). If the aerosols of interest are not closely approximated by one of these shapes, then one must adopt some form of empirical approach to determining how light is scattered by one of these particles or, for the inverse problem, what the particle size and shape is if the scattering pattern is known.

This paper will present results of a phenomenological investigation into the shape and size information contained in the near-forward scattering pattern of some nonspherical particles. The choice of the near-forward region was made on the basis of a previous study (Capps, et al., 1983) which suggested that specific features in this region could be directly related to size. The following section describes the experimental apparatus and the final section the analysis technique.

Experimental Apparatus

The heart of the experimental system is an electrodynamic balance (Davis and Ray, 1980) which will stably suspend a single micron-size particle for long periods of time. A laser beam is used to illuminate the suspended particle. As shown in Figure 1, a beam stop blocks the unscattered radiation while the scattered rays pass through a collimating lens and into a beamsplitting cube. The two beams of scattered light then strike two 1024 element linear arrays arranged so that a cruciform pattern of detectors is formed, as shown in the figure. In this configuration the arrays cover an angular range of $\pm 20^{\circ}$ from the incident beam direction in two orthogonal scattering planes with the scattering from -3° to $+3^{\circ}$ being occluded by the beam stop. A microscope is incorporated into the system so that the in situ orientation of the particles can be recorded with a photomicrograph, at least to the limits allowed by optical resolution. The arrays are connected through a digital storage oscilloscope to an HP 9825 computer where data is stored for later analysis.

Data Analysis

Scattering data were collected from four geometric classes of particle shape; spheres, fibers, flakes, and cubes. These data consist of relative

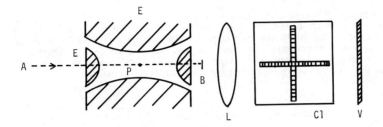

Side View

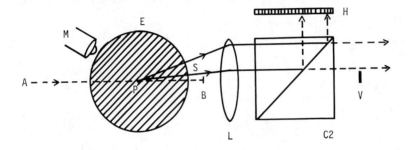

Top View

Figure 1. Schematic view of the system showing: A - incident laser beam,
E - electrodes of the electrodynamic balance, P - suspended particle,
B - beam stop, L - collimating lens, C1 - beamsplitter cube with
images of horizontal and vertical arrays, V - vertical linear array,
H - horizontal linear array, S - scattered rays, C2 - top view of
beamsplitter cube showing paths of scattered rays in the horizontal
plane, M - microscope objective.

826

intensity versus scattering angle in two orthogonal scattering planes plus the information on particle size, shape, and orientation available from the photomicrographs. The intensity curves were Fourier analyzed and features of the Fourier spectra compared with known particle sizes. In addition, correlation values for scattering patterns of particles of the same class were computed as well as cross correlations between different classes.

References

Asano, S. and Yamamoto, G. (1975) Appl. Opt. 14, 29.

Capps, C. D., Adams, R. L., Tokuda, A. R., Hess, G. M. (1983)
Proc. of the 1983 CSL Conference on Obscuration and Aerosol Research,
CRDC, Aberdeen, Maryland.

Davis, E. J. and Ray, A. K. (1980) J. Colloid and Interface Sc. 75, 566.

Mie, G. (1908) Ann. Physik 25, 377.

Wait, J. R. (1955) Can. J. Phys. 33, 189.

Published 1984 by Elsevier Science Publishing Co., Inc.
Aerosols, Liu, Pui, and Fissan, editors

LIGHT EXTINCTION SIGNATURES BY PREFERENTIALLY/RANDOMLY ORIENTED NONSPHERICAL PARTICLES

R. T. Wang
Space Astronomy Laboratory, Univ. of Florida
1810 NW 6th St., Gainesville, FL 32601
U. S. A.

EXTENDED ABSTRACT

Introduction

Obscuration (or extinction) of incident light by a particulate should not be construed as the mere blocking of light by the particle, but should be interpreted as a subtle interference phenomenon between the scattered wave in the beam direction ($\theta=0$) and the incident wave (van de Hulst, 1957). This interpretation leads to the well-known *Optical Theorem*:

$$C_{ext} = \frac{4\pi}{k^2} \left| S(0) \right| \sin \phi(0) = \frac{4\pi}{k^2} \text{Re} \left\{ S(0) \right\} \qquad (1),$$

which relates the extinction cross section C_{ext} of an arbitrarily oriented particle to the scattering amplitude $\left| S(0) \right|$ and the phase shift $\phi(0)$ of the $\theta=0$ scattered wave, measured along the incident polarization direction. To determine the relationship between single-particle extinction and particle orientation, $\left| S(0) \right|$ and $\phi(0)$ must be measured simultaneously. Presently, this is feasible only through the use of the microwave analog technique (Lind, et al., 1965; Schuerman, et al., 1981). This technique allows the precise control of pertinent parameters such as particle size, shape, refractive index and orientation, and also permits us to discriminate $S(0)$ against the large, yet coherently arriving background incident wave.

Extinction Signatures of Oriented Particles

We routinely display a single-particle extinction measurement result in the form of a P,Q plot, the cartesian display of complex scattering amplitude $S(0)$ versus particle orientation. Such a P,Q plot can be called the signature of a particle's extinction process. We show examples, Figs. 1A–1D, for four differently shaped but similarly sized particles with the same complex refractive index $m=m'-im''=1.61-i0.004$, corresponding to silicate at visible wavelengths (Wang, Ref. 7, 1983). A vector drawn from the origin to each orientation mark along our 2 experimental curves in each figure is the $S(0)$ vector at that particle orientation; i.e., the tilt angle of the $S(0)$ vector from the P-axis gives the $\phi(0)$, while the length of this vector is the scattering amplitude $\left| S(0) \right|$. The projection of this $S(0)$ vector onto the calibrated Q-axis gives the volume-equivalent extinction efficiency $Q_{ext,v}=C_{ext,v}/\pi a_v^2$, where a_v is the radius of a sphere equal in volume to the scattering particle. The higher the $Q_{ext,v}$, the more efficiently the incident light is obscured *per unit volume* at that particle orientation. At $\theta=0$, the $S(0)$ vector of an axisymmetric particle, arbitrarily oriented in space, is a linear superposition of 2 $S(0)$ vectors obtained respectively in 2 mutually orthogonal planes (Wang and Schuerman, 1981). These 2 planes are the k-E and the k-H planes, which contain the incident $(\vec{k},\vec{E})$ and $(\vec{k},\vec{H})$ vectors, respectively. This remarkable superposition property is independent of the target material, and it considerably reduces the number of measurements required to evaluate the averaged efficiency over random particle orientations, $Q_{ext,v}$.

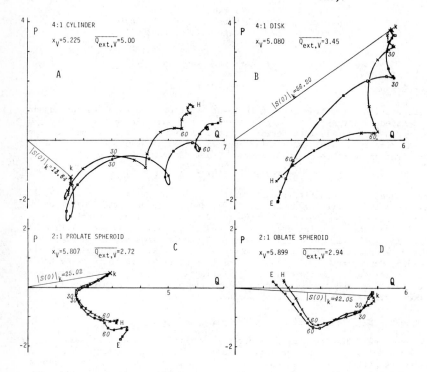

Figures 1A-1D.   Four experimental P,Q plots for four target shapes

Extinction Signatures of Randomly Oriented Particles

In Figs. 2A-2C the laboratory measurements of extinction efficiency averaged over random particle orientations, $\overline{Q}_{ext,v}$, are plotted versus the volume-equivalent phase-shift parameter $\rho_v = 4\pi a_v(m'-1)/\lambda = 2x_v(m'-1)$ for 49 axisymmetric particles possessing a range of refractive index m characteristic of natural aerosols/hydrosols (Wang and Schuerman, 1981; Wang, Refs.7 & 8, 1983). These figures are distinguished according to the particles' shape and aspect ratios: 2:1 prolate spheroids are shown in Fig. 2A, 2:1 oblate spheroids in Fig. 2B, and 4:1 prolate/oblate spheroids, cylinders and disks in Fig. 2C. The abscissa ρ_v gives the measure of phase-lag suffered by a centrally transmitted ray with respect to that of a free-space ray across the diameter of the equal-volume sphere. In each figure a theoretical extinction curve obtained by the Eikonal approximation (Wang, Refs. 6 & 7, 1983) for randomly oriented spheroids of the same aspect ratio but with arbitrary *nonabsorbing* refractive index (m"=0), is also shown to demonstrate its power of qualitatively predicting the salient resonance extinction phenomena exhibited by these randomly oriented nonspheres. The approximation, often known as the anomalous diffraction approximation (van de Hulst, 1957), ascribes the resonance extinction phenomenon to the resultant interference between the unperturbed incident rays and the undeviated but phase-shifted rays transmitted through various parts of a given particle. Although the Eikonal approximation underestimates the magnitude near extinction peaks (the larger the m' value, the larger the error), it predicts rather well the extinction profile versus ρ_v and the value of ρ_v at which the extinction peaks occur (except for the 4:1 flattened particles shown in Fig. 2C). Also shown in Figs. 2A-2B are the extinction curves for randomly oriented 2:1 spheroids of m=1.33-i0.0, as deduced from exact theoretical results (Asano and Sato, 1980; Wang, Ref. 7, 1983). The observations agree with the exact theory even more closely than the correlation with the Eikonal approximation, but these two theories are seen to predict remarkably similar extinction profiles.

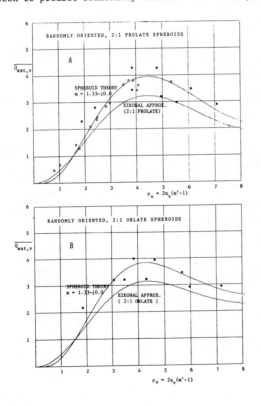

Figures 2A-2B. $\overline{Q_{ext,v}}$ vs. ρ_v plots

Refractive index m is identified by:
◐ m=1.61-i0.004;
● m=1.33-i0.05 ;
□ 1.1 ≲ m' ≲ 1.36 ,
m" ≲ 0.005 .

830

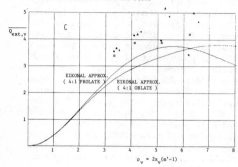

RANDOMLY ORIENTED,
4:1 SPHEROIDS, CYLINDERS AND DISKS

$\rho_v = 2x_v(m'-1)$

Figure 2C. $\overline{Q_{ext,v}}$ vs. ρ_v plots

$m=1.61-i0.004$ for all
particles. Target shape
is identified by:
+ 4:1 prolate spheroids;
□ 4:1 oblate spheroids;
▲ 4:1 cylinder;
◇ 4:1 disk.

References

1. Asano, S. and Sato, M. (1980) "Light Scattering by Randomly Oriented Spheroidal Particles", Appl. Opt. 19:962.

2. Lind, A.C., Wang, R.T. and Greenberg, J.M. (1965) "Microwave Scattering by Nonspherical Particles", Appl. Opt. 4:1555.

3. Schuerman, D.W., Wang, R.T., Gustafson, B.Å.S. and Schaefer, R.W. (1981) "Systematic Studies of Light Scattering. 1:Particle Shape", Appl. Opt. 20:4039.

4. van de Hulst, H.C. (1957) "Light Scattering by Small Particles", Wiley, New York.

5. Wang, R.T. and Schuerman, D.W. (in press) "Extinction by Randomly Oriented, Axisymmetric Particles", Proc., 1981 CSL Scientific Conference on Obscuration and Aerosol Research (Edited by R. Kohl), Army CSL, Aberdeen, MD.

6. Wang, R.T. (1983) "Similarities and Differences Between Light-Wave and Scalar-Wave Extinctions by Spheres", Proc., 1982 CSL Scientific Conference on Obscuration and Aerosol Research (Edited by R. Kohl), ARCSL-SP-83011, Army CSL, Aberdeen, MD. p.187.

7. Wang, R.T. (1983) "Extinction Signatures by Randomly Oriented, Axisymmetric Particles", in the same "Proceedings" as above. p.223.

8. Wang, R.T. (1983) "Obscuration of Light by Randomly Oriented Nonspherical Particles", Aerosol Sci. and Tech. 2:187.

Published 1984 by Elsevier Science Publishing Co., Inc.

Aerosols, Liu, Pui, and Fissan, editors

EXPERIMENTAL MEASUREMENTS OF UNIPOLAR CHARGING OF ULTRAFINE AEROSOLS

Shyh-Yuan Hwang and James W. Gentry
Department of Chemical and Nuclear Engineering
University of Maryland, College Park, MD 20740
U. S. A.

EXTENDED ABSTRACT

Introduction

The unipolar charging of ultrafine aerosol particles has been studied extensively on a theoretical basis. In this work, the charge distribution on aerosol particles of diameters 18, 24, 32, and 42 nm is measured as a function of the product of N_o (ion number density) and t (aerosol residence time). The experimental data are compared with four existing charging theories proposed by : Fuchs (1963), Hoppel (1977), Laframboise and Chang (1977), and Gentry (1972). The charge attachment coefficients calculated by Davison (1984) are used in this work.

Experimental

As shown on Figure 1, the carrier gas (air) is first dried in a drierite dryer where water vapour is partially removed. After the airbone particles have been removed in an absolute

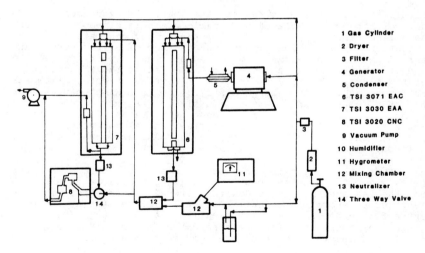

1 Gas Cylinder
2 Dryer
3 Filter
4 Generator
5 Condenser
6 TSI 3071 EAC
7 TSI 3030 EAA
8 TSI 3020 CNC
9 Vacuum Pump
10 Humidifier
11 Hygrometer
12 Mixing Chamber
13 Neutralizer
14 Three Way Valve

Figure 1

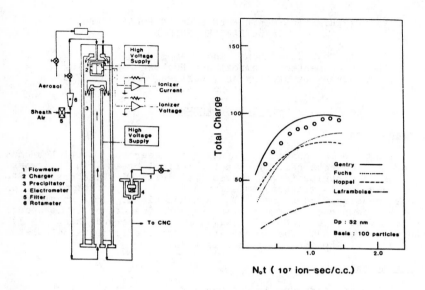

Figure 2   Figure 3

filter, the air stream is fed to the particle generator where
silver filters are placed in ceramic boats in a combustion tube
and heated above 550°C. Silver particles generated in the
combustion tube by homogenous condensation of silver vapour are
carried away by the air stream. Aerosol flow is then cooled to
room temperature in a condenser and fed to the Electrostatic
Aerosol Classifier (EAC). In the EAC, the aerosol flow is
brought to equilibrium charge distribution in a Kr-85 neutralizer
and then classified in the classifier. Highly monodisperse
aerosols of the desired size can be obtained by adjusting
flowrates in the EAC and the classifier central rod voltage. The
size of the aerosol particles leaving EAC has been checked for
uniformity.

After the positive charges carried by particles are
neutralized in another Kr-85 neutralizer, monodisperse aerosols
from the EAC are then diluted with filtered air and fed to the
Electrical Aerosol Size Analyzer (EAA). In the EAA (as shown on
Figure 2), aerosols are electrically charged in a diffusion
charger by unipolar ions generated by a corona discharge. A
fraction of charged particles are removed from aerosol flow due

to the electric field in the charger. Finally, the charge and number distributions of the particles in the aerosol stream are measured with the analyzer in the EAA and a condensation nuclei counter (CNC). The ion density N_o and the particle residence time t in the charger can be adjusted independently.

Results

Table 1 compares the experimental data with simulation for 32 nm particles with $N_o t = 1.20 \times 10^7$ ion-sec/cm^3. For the total current carried by particles (last column), The theory of Laframboise and Chang underpredicts by more than 60% while the theories of Gentry and Fuchs give close estimations to the experimation data. All the four theories predict the fraction of particles lost in EAA fairly well. The theories of Gentry and Fuchs work well in predicting the fractions of particles that carry one and two charges.

Figure 3 shows the total number of charges carried by 100 particles as a function of $N_o t$ for 32 nm particles. Experimental data fall between the theories of Gentry and Fuchs. Figure 4 shows the fraction of particles lost in the EAA. All the four theories (not shown) give close estimation to the experimental data. Figure 5 shows the fraction of particles remaining uncharged in the corona charger. All the four theories and the experimental data show linear dependence on semi-log paper.

Conclusion

By measuring the charge distribution of aerosol particles between 18 and 42 nm undergoing unipolar charging, five conclusions were obtained:
(1) The model incorporates a parameter L_{c3} to estimate the particle loss outside the charging section. Good agreement between the simulation and the experimental data were obtained.

Table 1   Comparison of Simulation and Experimental Data

$D_p = 32$ nm, $N_o t = 1.20 \times 10^7$ ion-sec/cm^3

	Particle Distribution (%)				Total*
	0	1	2+	Loss	Charge
Exptl.	4.9	61.1	15.9	18.1	95.3
Gentry	1.4	57.4	21.3	19.9	100.2
Fuchs	10.7	53.8	15.7	19.8	85.5
Hoppel	6.5	66.9	6.6	20.0	80.1
Laframboise	47.0	33.0	0.3	19.7	33.7

Note* : In elementary charge per 100 particles

833

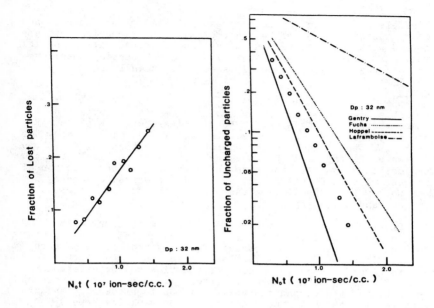

Figure 4   Figure 5

(2) The fraction of particles remaining uncharged in the corona charger shows a straight line dependence with respect to the charging parameter $N_o t$ on semi-log paper. This agrees with the simulation.

(3) The Laframboise and Chang theory, which neglects the image force contribution, underpredicts the total charge by more than 50%.

(4) The Hoppel theory fits the data for the smallest particle but underpredicts the total charge for larger particles.

(5) The theories of Gentry and Fuchs follow the trend of the experimental data well and are within the experimental error.

References

Davison, S. (1984) "Evaluation of Kinetic Models for Charging of Ultrafine Particles by Diffusion of Ions", Ph.D. Thesis, Univ. of Maryland

Fuchs, N. (1963) Geofisica Pura e Applicada, 56:185

Gentry, J.W. (1972) J. of Aerosol Sci., 3:56

Hoppel, W. (1977) Proc. of 5th Int. Conf. Atom. Elect.

Laframboise, J.G. and Chang (1977) J. of Aerosol Sci., 8:331

Published 1984 by Elsevier Science Publishing Co., Inc.

Aerosols, Liu, Pui, and Fissan, editors

MULTIPLE LIGHT SCATTERING OF HIGHLY CONCENTRATED AEROSOL PARTICLES

S. Okuda, H. Takano and T. Yamamura
Department of Chemical Engineering,
Doshisha University, Kyoto 602
Japan

EXTENDED ABSTRACT

Introduction

The validity of a physical model based upon Hartel's multiple scattering theory was experimentally confirmed in terms of particle diameter and number concentration of aerosols. In the experiments, a specially designed optical system connected with high speed poto-amplifier, large scale data memory, and a 16-bit microcomputer was employed for the data treatments, and the various sizes of monodispersed polystyrene latex were also used as a sample aerosol.

Physical model based upon multiple light scattering theory

The angular intensity of multiple light scattering, $I(\theta)$, was obtained by functions related to the particle size, the scattering angle, θ, the particle concentration, and the wavelength of light defined as follows:

$$I(\theta) = \sum_{k=1}^{\infty} Q_k \, f_k(\theta) \tag{1}$$

where, Q_k and f_k were the scattering intensity and the angular distribution functions of k-th order of multiple scattering respectively. These functions were consequently expressed as next equations:

$$Q_k = \frac{exp(-NV)}{k!} \, (NV)^k \tag{2}$$

where, V is the scattering volume and N is the particle number concentration. For the angular distribution function (Chu and Churchill, 1955),

$$f_k = \frac{1}{4\pi} \sum_{n=0}^{\infty} \frac{An^k}{(2n+1)^{k-1}} \, P_n(cos\theta) \tag{3}$$

where, P_n is general Legendre polynomials, and A_n is the angular coefficients obtained by Hartel's fromula (Hartel, 1940). As shown in Fig. 1, a numerical result of this phase function indicated that the angular dependence upon the scattering intensity was decreased with increasing the order of multiplicity.

Experimental apparatus and procedure

Schematic diagram of the optical device and its detail were illustrated in Fig. 2 and Fig. 3. The scattered light of aerosol particles gathered by a parabolic mirror in the resonanced light was condensed into a glass fiber connected with a selfoc micro-lens, and then transfered to the high speed photoelectric amplifier. After that, the signals for this light intensity through the A/D converter were stored into the data memory of 256 KB in order to analyze the peak height for measuring the diameter and the concentration of aerosol particles from those signals.

836

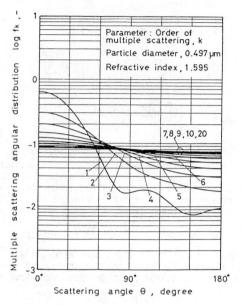

Fig. 1 Numerical results of phase function related to the multiple scattering order

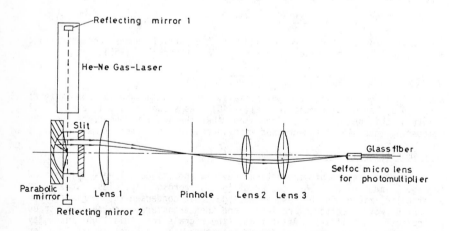

Fig. 2 Schematic diagram of optical device in the measuring system

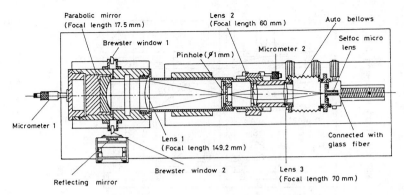

Fig. 3 Top view of the optical device

Results and discussion

Figure 4 shows the correspondence of the particle concentrations obtained by this system in comparison with the calculated results from other measuring device such as mass monitor. In this figure, it was obviously noticed that even in relatively higher concentrations, the particle number concentration can be obtained.

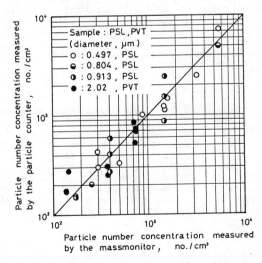

Fig. 4 Comparison of particle number concentrations
between this system and the mass monitor

838

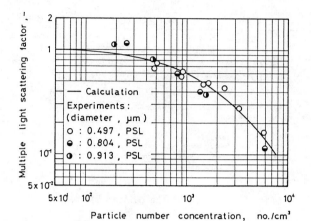

Fig. 5   Comparison of experimental data with numerical
results on multiple light scattering factor

In order to evaluate the multiple light scattering effects on the highly
concentrated aerosol particles, a special factor was defined as the ratio of
multiple scattering intensity to the single scattering intensity.   As shown
in Fig. 5, in the case where the wide range of collecting angle of 0°-180°
was employed for the optical geometry, this multiple light scattering factor
indicated a good agreement between the numerical results and the experimental
data related to the properties of phase function (Okuda and Takano, 1983).
This figure also showed that the scattered intensity on a single particle
was decreased as the particle number concentration increased, although this
factor is shifted corresponding to the scattering volume.

Conclusion

The characteristics of multiple light scattering of monodispersed aerosol
particles were analyzed in terms of particle sizes and number concentrations.
From the numerical results, it was noticed that the phase function became
smooth as the multiple scattering order increased, and that the wide renge of
collecting angle was sufficiently applicable to the size measurements.
In the experiments, it was found that the multiple light scattering was
occured even in relatively lower particle concentrations.   Furthermore, the
measured values of scattering intensity were obviously in good agreement with
the numerical results, especially in the relationship between the particle
number concentration and the multiple scattering factor.

References

Chu, C. M. and S. W. Churchill (1955): J. Opt. Soc. Amer., 45, 958.
Hartel, W. (1940): Das Licht, 40, 141.
Okuda, S. and H. Takano (1983): Proc. of Paci. Reg. Meeting of FPS, Hawaii.

Published 1984 by Elsevier Science Publishing Co., Inc.
Aerosols, Liu, Pui, and Fissan, editors

INTERMOLECULAR FORCES IN THE COLLISION RATES OF SMALL MOLECULAR CLUSTERS

William H. Marlow and Leonard A. Slatest
Brookhaven National Laboratory
Upton, New York 11973

EXTENDED ABSTRACT

For a number of years, experimental results have indicated that inter-molecular forces are important in the collision rate densities of free-molecu-lar regime aerosols.[1-2] More recently, limited experimental evidence[3] has pointed to the possibility[4] that these forces also play a small role in the collisions of aerosol particles as large as 200 nm. Very recently, new methods for the measurement of near-free molecular and transition regime aerosols have given the most detailed experimental data on ultrafine aerosol coagulation yet obtained.[5] Those experimental results explicitly show the following: (1) aerosol collision rates require enhancements over their hard-sphere (i.e., no long-range interactions) values; (2) those enhancements are dependent upon the compositions of the colliding aerosol particles; and (3) the enhancements are dependent upon particle sizes. In all cases, these results are in agreement with the theoretical predictions of aerosol coagula-tion published during 1980-1982.[4,6,7] In those papers, the Lifshitz-van der Waals attraction provides the basic collision enhancement with retardation of their long-range force and transport (i.e., Knudsen number dependence) lessening the enhancement with increasing particle size.

The successes of this theory for particles in the 1-100 nm radius range and the increasing importance of the intermolecular force with decreasing particle size suggest that molecular cluster growth may also be dependent upon the inter-cluster long-range forces. Evidence for the validity of this conjecture is found in measurements of carbon dioxide cluster size distribu-tions published in 1982.[8] In the original paper, those results were modelled with a simple hard-sphere collision rate giving relatively poor agreement with the data. Besides improving the description of cluster collision rates, advancing the understanding of the roles of long-range forces in the physics of clusters should also contribute to the understanding of molecule-to-condensed-phase transformation processes such as nucleation and growth.

The theory of aerosol coagulation given in 1980-1982 includes a formula-tion[9] of free-molecular collisions that exploits the mathematical singularity of the contact potential which is characteristic of macroscopic theories of long-range interactions of condensed media. Since these descriptions treat the interacting bodies as though they were continuous, they are inapplicable to clusters. However, clusters do consist of numerous, interacting molecules requiring a many-body description of their long-range potentials. To treat these interactions in a manner consistent with both the discrete nature of the particle and with the requirements of collective interactions, this paper uses the point-dipolar interaction model. Following Langbein's derivation of the continuum Lifshitz theory according to this model,[10] intermolecular interac-tions are evaluated to all orders (i.e., including pair, triplet, etc. summa-tions). Unlike Langbein, we do not take the limit of a continuous body, requiring the introduction of the bulk material's dielectric function, but use the individual molecular polarizabilities.

The calculations we have done are for $(CO_2)_n$ clusters assuming the CO_2 molecule is spherically symmetric. For the sake of definiteness, the cluster structures for n = 1-9, 13 given by Hoare and Pal[11] as most stable for pair interactions are used. Since the clusters are not spheres, the intercluster potentials are not spherically symmetric. The correct computation of collision rates therefore should include a complete trajectory analysis which is not feasible for the number of combinations of colliding cluster pairs needed to model the data.[8] Consequently, an averaging procedure is used to estimate the effective collision rate.

Computational results will be presented at the time of the conference.

References

1. N.A. Fuchs and A.G. Sutugin, J. Colloid Interface Sci. 20, 492 (1965).

2. S.C. Graham and J.B. Homer, Discuss. Faraday Soc. 7, 85 (1973).

3. Paul E. Wagner and Milton Kerker, J. Chem. Phys. 66, 638 (1977).

4. William H. Marlow, J. Chem. Phys. 73, 6288 (1980).

5. K. Okuyama, Y. Kousaka, and K. Hayashi, J. Colloid Interface Sci. 100 (1984).

6. William H. Marlow, Surf. Sci. 106, 529 (1981).

7. William H. Marlow, J. Colloid Interface Sci. 87, 209 (1982).

8. J.M. Soler, N. Garcia, O. Echt, K. Sattler, and E. Recknagel, Phy. Rev. Lett. 49, 1857 (1982).

9. William H. Marlow, J. Chem. Phys. 73, 6284 (1980).

10. D. Langbein, Theory of Van der Waals Attraction (Springer-Verlag, Berlin, Heidelberg, New York, 1974).

11. M.R. Hoare and P. Pal, Adv. Phys. 20, 161 (1971).

This work was performed under the auspices of the United States Department of Energy under Contract No. DE-AC02-76CH00016.

Published 1984 by Elsevier Science Publishing Co., Inc.

Aerosols, Liu, Pui, and Fissan, editors

841

ON THE INFLUENCE OF LOW FREQUENCY ELECTRIC FIELDS ON CHARGED AEROSOLS

H.O. Luck and N.G. Bernigau
Department of Communication Technology
Duisburg University, Fed. Rep. Germany

EXTENDED ABSTRACT

Introduction

The ionization chamber is a well known tool in aerosol measurement techniques. The underlying theory has intensively been studied in the literature and many applications are reported. Applications of practical importance are the use as a dust monitor /1/, the ionization type smoke detector for automatic fire detection /2-6/ and the use of ionization chambers in connection with optical methods for aerosol parameter measurements /8/.

The classical ionization chamber

The ionization chamber consists essentially of an airfilled chamber-volume under normal pressure with two electrodes to which a dc-voltage is supplied. The air is made conductive by a small radioactive source (e.g. AM 241 of some μCi) and a dc of the order of 10^{-9} A normally results in the aerosol-free case. The current is reduced by the presence of aerosol particles because of the capture of clusters by aerosol particles which have a reduced electrical mobility of the order of 5 compared with the clusters. The associated theory considers the ionization chamber mainly in two different geometrical arrangements, the parallel-plate and the cylindrical form. The parameters to be taken into account, are the volumetric ion generation rate (strength of radioactive source) g, ion recombination coefficient α, electrical mobility of the charge carrying clusters $b^{\pm}$ and the chamber voltage supply U. The aerosol-free ionization chamber is described by its I-U characteristic. Several methods to calculate this characteristic are given in the literature /2-7/. In the case of an aerosol being present in the chamber its current reading I in addition depends on the aerosol parameters (e.g. size-distribution, mean particle diameter $\bar{d}$, particle number concentration N etc.) Some authors give their results in the form of a characteristic ionization-chamber equitation /2-4,6,8/ others present it in form of numerical calculations by curves /5, 7/. A most simple and very clearly arranged formula can be drawn from /2/.

Defining $\qquad x = \dfrac{I_o - I}{I_o} \qquad$ and $\qquad y = x \cdot \dfrac{2-x}{1-x}$

the characteristic ionization chamber equation working below the current saturation region reads

$y = \dfrac{2}{3} \dfrac{c_B}{\sqrt{\alpha \cdot g}} \cdot \bar{d} \cdot N \qquad$ where c_B = BRICARD attachment coefficent.

In /2/ a cylindric geometry is used and the result indicates that such an ionization chamber measures the product of aerosol number concentration N and the mean particle diameter $\bar{d}$. In any case all the work on ionization chamber aerosol measurement available to the authors considers the dc electrical behavior of this device.

Dynamic behavior of ionization chambers

Consider an ionization chamber with a parallel-plate geometry where the plates serve as electrical electrodes. This chamber filled with a charged aerosol can be described as an electrical admittance Y_2 if the following conditions are met with suitable accuray /9/:

- The distance D between the plates is small compared with the other chamber dimensions, so that the electrical field is a function of one coordinate only.
- The aerosol particles have a spherical shape and the specific particle mass γ is known.
- The particle number concentration N is known and in an order of magnitude that makes the space charge effect on the electric field negligible.
- No diffusion or other effects (e.g. forced flow) have a considerable influence on the particle movement.

Using the normal Laplace-transform notations for electrical quantities the chamber admittance is $Y_2(s) = Y_c(s) + Y(s)$. $Y_c(s)$ represents the normal admittance without any aerosol and $Y(s)$ depends on the aerosol parameters in the following way

$$Y(s) = \frac{A \cdot N}{D} \int_{d_{min}}^{d_{max}} \frac{q^2(d)}{\lambda(d)} \cdot \frac{p(d)}{1+s\tau(d)} \, dd \qquad \text{where}$$

A is the chamber-plate surface, $\lambda(d)$ is the coefficient of friction between particles and the surrounding air and $\tau(d) = \pi\gamma d^3/6\lambda(d)$ the relaxation time constant. As $Y_c(s) >> Y(s)$ it is only possible to measure $Y(s)$ by using a Wheatstone-bridge circuit as shown in Fig. 1

With $u(t) = U_o \cos\omega t$ and $Y(j\omega) = G(\omega)-jB(\omega)$ and the condition

$$R|Y_c(j\omega)| \ll \frac{3}{4} \qquad \text{it can be shown that}$$

$$u_B(t) = \frac{1}{3} RU_o(G(\omega)\cos\omega t + B(\omega)\sin\omega t))$$

if two identical ionization chambers are used in the Wheatstone-bridge circuit arrangement one with [Y_2] and the other without the aerosol [Y_c]. In order to reduce the inherent measurement noise and to seperate the two monofrequent components in $u_B(t)$ an arrangement can be used that is indicated in Fig. 1. The result is

$$m(\omega_i, \varphi) = \begin{cases} m(\omega_i, \varphi_v) \approx K(\omega_i) \cdot G(\omega_i) \\ m(\omega_i, \varphi_v + \frac{\pi}{2}) \approx K(\omega_i) \cdot B(\omega_i) \end{cases} \text{with suitable accuracy}$$

$$K(\omega_i) = \frac{1}{6} RU_o^2 |H_v(\omega_i)| \qquad \text{and}$$

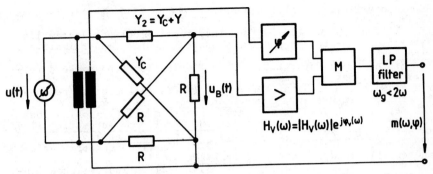

Fig. 1 Wheatstone-bridge circuit for the measurement of the aerosol-dependent admittance $Y(j\omega)$

$$(1) \quad G(\omega_i) = \frac{A \cdot N}{D} \cdot \int_{d_{min}}^{d_{max}} \frac{q^2(d)}{\lambda(d)} \cdot \frac{p(d)}{1+\omega_i^2\tau^2(d)} \, dd$$

$$(2) \quad B(\omega_i) = \frac{A \cdot N}{D} \int_{d_{min}}^{d_{max}} \frac{q^2(d)}{\lambda(d)} \cdot \frac{\omega_i\tau(d)}{1+\omega_i^2\tau^2(d)} p(d) \, dd$$

The only unknown quantity in (1) and/or (2) is the distribution function of the particle diameter $p(d)$. (1) and (2) are integral equations of the linear Fredholm type. If it is possible to measure a series of values $m(\omega_i, \varphi)$ with $i \in [1, M]$ the attempt can be made to solve eq. (1) or (2) by using the approximation

$$p(d) \approx p_A(d) = \sum_{\nu=1}^{M} a_\nu \text{rect} \left(\frac{d-\frac{d_{\nu+1}d_\nu}{2}}{d_{\nu+1}-d_\nu}\right)$$

A system of M linear simultaneous equations follows, e.g.

$$m(\omega_i, \varphi_\nu) \rightarrow G(\omega_i) = \frac{A \cdot N}{D} \cdot \sum_{\nu=1}^{M} a_\nu \int_{d_\nu}^{d_{\nu+1}} \frac{q^2(d)}{\lambda(d)} \frac{dd}{1+\omega_i^2\tau^2(d)}$$

which can be solved for the M unknown coefficients a_ν /9/. The method briefly mentioned in this section has been investigated in a first step in a theoretical study. A series of simulations have been made. Experimental work has been started. Fig. 2 shows the result of an simulation experiment where $p(d)$ has been prefixed and parameters as indicated in Fig. 2 have been used.

References
/1/ Hasenclever, D.; Siegmann, H. Chr. (1960) "Neue Methode zur Staubmessung mittels Kleinionenanlagerung", Staub 20, Nr. 7
/2/ Gilson, H.H.; Hosemann, J.P. (1973) "Meßkammer nach dem Kleinionenanlagerungsprinzip zum quantitativen Nachweis von

844

Aerosolpartikeln", Forschungsbericht NRW Nr.2336, Westdeutscher
Verlag, Köln
/3/ Litton, C.D. (1977) "A Mathematical Model for Ionization-
Type Smoke Detectors and the Reduced Source Approximation",
Fire Techn. 13/4
/4/ Simon, F.N.; Rork, G.D. (1976) "Ionization type smoke detectors", Rev. Sci. Instrum. 47/1
/5/ Mcclure, B.T.; Krafthefer, B.C (1978) "Ion chamber response
to particulate matter", Journal of Electrostatics 5
/6/ Meyer, G. (1982) "Ein mathematisches Modell für Ionisations-
rauchmelder", Proc. of 8th intern. Conf. on Automatic Fire
Detection AUBE'82 (H.O. Luck ed.), Duisburg University
/7/ Hoppel, W.A.; Gathman, S.G. (1970) "Charge Transport Through
an Aerosol Cloud", Journal o. Appl. Phys. 41/15
/8/ Kraus, F.J.; Luck, H.O. (1978) "Zeitkontinuierliches Meß-
system zur Charakterisierung von Aerosolen", Forschungsbericht
NRW Nr. 2728, Westdeutscher Verlag, Opladen
/9/ Bernigau, N.G. (1984) "Ein Modell zur Wechselwirkung eines
monodispersen Aerosols mit einem elektrischen und/oder magneti-
schen Wechselfeld", Diplomarbeit, Universität Duisburg
/10/ Bademosi, F.; Lui, B.Y.H. (1971) "Diffusion Charging of
Knudsen-Aerosols", Part I, Particle Techn. Lab., University of
Minnesota

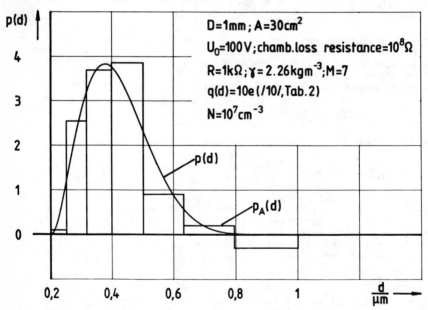

Fig. 2 Aerosol particle distribution p(d) estimated from
simulated ionization chamber measurements

Published 1984 by Elsevier Science Publishing Co., Inc.

Aerosols, Liu, Pui, and Fissan, editors

Direct Measurement of Diffusiophoretic Deposition of
Particles at Elevated Temperatures

H. Bunz, W. Schöck

Laboratorium für Aerosolphysik und Filtertechnik I
Projekt Nukleare Sicherheit
Kernforschungszentrum Karlsruhe GmbH
7500 Karlsruhe, W.-Germany

Extended Abstract

Introduction

Diffusiophoretic deposition of aerosol particles, compared to other
deposition mechanisms, is a rather weak effect under normal atmospheric
conditions. On the other hand it may become dominating in some extreme
environments, when the condensing component of the atmospheric has a vapor
pressure comparable to the non-condensing components and when the condensation rates are high enough. E.g. a saturated steam-air-particle mixture at
temperatures above 100 °C in a closed vessel with cold internal surfaces
will exhibit significant diffusiophoretic deposition of particles. Such a
situation is of interest for certain accident evaluation investigations for
light water reactor power plants, where diffusiophoresis may play an important role in the natural cleaning process of the atmosphere in the reactor
building.

This paper describes a model for diffusiophoretic deposition in such
extreme environments and experimental investigations which were carried out
to validate the model.

Theory

A detailed description of the mechanism of diffusiophoretic deposition
is given in /1/ and will not be discussed here. For our purpose it is only
necessary to relate the diffusiophoretic deposition velocity v_p of a
particle to measurable thermodynamic quantities. In the basic equation /2/
v_p is related to the molar gradient of the diffusing vapor in the boundary
layer at the cold surface which is not very well known in many technical
applications. Therefore, this gradient is substituted by an expression containing the diffusion velocity v_s of the vapor which leads to a simple
relation of v_p to v_s

$$v_p/v_s = (\gamma_s \sqrt{M_s}) / (\gamma_s \sqrt{M_s} + \gamma_a \sqrt{M_a}) \tag{1}$$

where M_s and M_a are the molar weights of the vapor and of the air, and
γ_s and γ_a are the respective molar fractions. So, when the composition
of the gas phase is known, the ratio R of the two velocities v_p and v_s
can be calculated.

Both species, particles and condensing vapor, are deposited onto the same
surface F and the change in airborne mass in the volume V follows the
general law

$$dm/dt = - m \cdot v \cdot F/V \qquad (2)$$

which leads to the well known exponential decrease of airborne mass in a stirred closed system. In our case, substituting $C_m = m_p/V$, the mass concentration of the particles, and $\rho_s = m_s/V$, the vapor density, we obtain

$$dm_p/dt = - C_m \cdot v_p \cdot F$$

$$dm_s/dt = - \rho_s \cdot v_s \cdot F \qquad (3)$$

Eqs. (1) and (3) can be combined to eliminate v_s and v_p

$$dm_p/dt = dm_s/dt \cdot C_m \cdot R/\rho_s$$

Here the particle removal rate is correlated to the vapor condensation rate dm_s/dt which is known in many cases. So eq. (4) was implemented in our aerosol behavior model, and experiments were performed to validate this formulation.

Experimental methods

For several reasons, which will not be discussed here, a direct mass balance of airborne particles could not be used to check eq (4).

Therefore, a method had to be developed to measure directly the deposit of particles due to diffusiophoresis on a samll surface and likewise the corresponding amount of condensate. In the case of small amounts of deposited masses Δm_p and Δm_s, compared to the total masses $C_m \cdot V$ and $\rho_s \cdot V$, eq (4) may be simplified to

$$\Delta m_p = \Delta m_s \cdot C_m \cdot R/\rho_s \qquad (5)$$

An experimental check of our model is possible by evaluating and comparing the quantity R as given in both eqs (1) and (5).

Fig. 1. shows an axial section of the probe which was developed for the measurements. The cooled brass surface has a diameter of 14 mm, and is oriented with its face vertical. The cooling medium is either ambient air or water, transported through coaxial tubes. The water which is condensed runs down the cooled surface and is soaked and retained by an absorbing medium in a ring shaped gap in the outer wall of the probe. To separate the diffusiophoretic deposit from the much larger amount of condensed water a diaphragm

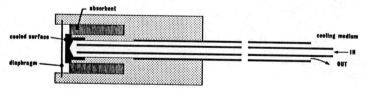

Fig.1: Diffusiophoretic sampling probe, schematic

is mounted in front of the cooled surface at a distance of 2 mm. It consists of a nuclepore membrane with 0.1 μm pore sizes and retains all particles while letting pass the steam. By this arrangement we obtained separated deposits of particles and water.

The probe can be inserted into a 3 m³ vessel containing the aerosol via a thermostated lock. After sampling the probe is removed from the vessel and immediately weighed. By this the mass of condensed steam is obtained which is typically in the order of 100 to 500 mg. This represents a partial volume of steam of a few tenths of a liter depending on the temperature and composition of the atmosphere. Since aerosol mass concentrations in the vessel are in the order of 10 to 100 mg/m² the aerosol mass collected on the filter in the probe could be less than 1 μg. In order to achieve a sufficiently good sensitivity in evaluating the samples we used uranium dioxide as aerosol material. The samples can then be evaluated by fluorometric techniques with a sensitivity of $\cong$ 10 nanograms. The same technique was used to evaluate the mass concentration of the aerosol from filter samples.

The accuracy of measuring the aerosol deposit is much higher than that of determining the mass of the condensate. As the main problem in measuring the condensate mass we detected the possible re-evaporation of the hot condensate in the time interval between extracting the probe and weighing it. Different methods of preventing re-evaporation have been tried, a combination of using absorbing materials in the probe which are still working at the elevated temperatures and capping the probe after extraction from the vessel gave the best results. Nevertheless, the reproducibility of single measurements was not very good and therefore multiple measurements were done and the results averaged to obtain a higher accuracy for a given set of thermodynamic parameters.

Results

The thermodynamic regime of interest was saturated steam in air in a rather limited temperature range around 110 °C which leads to a value of $R \cong 0.45$. Measurements were done in this area but also at lower vapor (steam) fractions. In Table 1 the results of the measurements are given. The first column gives the temperatures T, the second and third column the partial pressures of steam p_s and air p_a which are proportional to the molar fractions γ_s and γ_a in eq (1). The degree of saturation was about 80 % in all cases. Column four contains the mass concentration C_m of the particles, column five and six the sampled masses Δm_p of particles and Δm_s of condensate. From these data R is calculated from eq (5) giving the values listed in column seven. They are results of single measurements with large errors as mentioned above. In the eighth column the averaged values of those in column seven are given which have been measured under identical steam air ratios, and these averaged values may be compared to the theoretical values of R in the last column as obtained from eq (1).

The first group of measurements at 115 °C represents the case which is important for our application. A good agreement between $\bar{R}(exp)$ and R(theor) can be noted, the difference being less than 10 %. The second group at 116 °C was done with decreased air fraction, here the differences are 13 %. The last group at temperatures below 80 °C was an attempt to measure

at the lower and of the R-range. Here the difference between measured and
theoretical R-values are rather large. This is certainly due to inaccuracies
in measuring Δm_s. The sampled condensate mass is much less than in the
other cases due to the low steam content, and subsequently any re-evaporation of condensate before weighing the sample leads to systematically
higher R(exp.)-values. So we do not believe that a disagreement with the
model exists, but rather that the experimental technique contains a source
of systematic errors. We regard the experimental results as a sufficient
validation of the model for our application, but we will continue the
measurements with improved methods to obtain a better accuracy.

References

/1/ J. F. v. d. Vate, Investigations into the Dynamics of Aerosols...,
     ECN-86, Petten (1980)

/2/ K. H. Schmitt, Z. Naturf. 16 a, p. 144 (1961)

T	P_s	P_a	C_m	Δm_p	Δm_s	R	$\bar{R}$	R
°C	Bar	Bar	mg/m³	µg	mg	exp	exp	theor.
115	1.37	1.61	65	5.2	230	.23		
			34	3.8	170	.51		
			21	1.9	180	.39		
			18	1.3	220	.25		
			15	3.0	310	.50	.376	.401
116	1.38	1.00	32	5.0	250	.50		
			31	4.6	230	.51		
			29	2.8	110	.70		
			26	4.6	230	.61		
			18	2.4	170	.63	.590	.520
76	0.24	1.83	85	1.4	36	.07		
77	0.26	1.83	60	3.5	51	.19		
78.5	0.30	1.83	46	3.9	59	.26		
79.5	0.32	1.83	37	3.2	63	.27	.197	.11

Table 1: Results of the diffusiophoretic deposition measurements

Published 1984 by Elsevier Science Publishing Co., Inc.
Aerosols, Liu, Pui, and Fissan, editors

RESIDENCE TIME AND DEPOSITION VELOCITIES OF PARTICULATE SUBSTANCES

Jürgen Müller

Umweltbundesamt, Pilotstation Frankfurt
45 Feldbergstr., 6000 Frankfurt (M)

EXTENDED ABSTRACT

Introduction

Airborne particulate matter (PM) either is directly injected into the atmosphere or is produced by gas-to-particle conversions of reactive trace gases. Tropospheric PM by steady state of sources and sinks has a bimodal mass size distribution (1). Total PM is an agglomeration built up of elements and compounds existing on the earth. Every element or compound incorporated in PM has a characteristic mass size distribution which like total PM in steady state is constant within certain limits. Measurements of mass size distributions revealed characteristic differences between the different elements (2). Volatile elements like Cd or Pb have their mass predominantly in the fine mode of PM ($<$ 2 µm particle diameter) whereas less volatile elements like Si or Fe are mainly found in the coarse mode.

The atmospheric residence time of a particle is a function of its size. Accumulation mode particles (diameter range between 0,1 and 2 µm) have the longest residence time whereas coarse particles ($>$ 2 µm) due to enlarged sedimentation are quickly deposited. Very fine particles belonging to the so-called nucleus mode ($<$ 0,1 µm) which are generated by gas-to-particle-conversions are quickly transformed into the accumulation-mode-size by coagulation.

Theoretical

The residence time of a particle in dependence of its diameter (d) and the meteorological parameters approximately can be quantitied (3). The atmospheric sinks can be divided into wet and dry removal processes. In a first approach the residence time of an airborne particle can be calculated by the expression:

$$① \quad \frac{1}{\tau} = \frac{v_0}{2H} \left[\left(\frac{d_0}{d}\right)^2 + \left(\frac{d}{d_0}\right)^2 \right] + \frac{v_I}{H} + \frac{1}{\tau_{wet}}$$

The first and second term are responsible for dry deposition. The first term is subdivided into removal by coagulation due to Brownian diffusion and Stokes' sedimentation. The diameter $d_0 \approx 0,6$ µm maximum of the accumulation mode represents the particle size with the longest residence time due to a minimum in the effects of coagulation and sedimentation. $v_0 = v(d_0) \approx 0.01$ cm/s empirically found by wind tunnel experiments (4) represents the respective deposition velocity at the size d_0.

$v_I \approx 0.01$ cm/s empirically assessed (4) describes the impaction of particles induced by interaction of turbulent air streams with the ground. This value is dependent of wind speed, atmospheric mixture and surface vegetation. H is the height of the tropospheric mixing layer which in the yearly average amounts to about 1 000 m.

τwet describes the atmospheric water cycle. The global water atmosphere is turned over about 45 times in the course of a year resulting in an average atmospheric residence time of about 8 days. For Middle Europe in the influence of the west wind drift a τwet-value of seven days is assumed. This means that the atmosphere is cleaned one time by rain-out and wash-out processes during this period. τwet represents the limiting parameter for the long-living accumulation mode particles.

Measurements

By aid of impactor measurements (2) equation ① can be applied in order to determine the residence time of total PM. Each impactor stage corresponds to a certain size interval to which a mean diameter can be assigned. For every mass fraction x_i collected on the different stages a mean residence time τ_i can be calculated. By summation over all stages the total residence time of PM can be determined:

$$\textcircled{2} \quad \tau = \frac{1}{\sum \frac{x_i}{\tau_i}} \qquad \sum x_i = 1$$

This method can be applied to any substance in PM for which the mass size distribution has been determined.

In Fig. 1 the determined residence time of several particulate substances are represented. Furtheron, the percentage of the wet deposited fractions under mean European conditions are given.

The wet deposited fraction of a substance increases the more mass is bound in the accumulation mode. This means that the removal of a particulate substance is determined by physics and its chemical properties are "lost" due to incorporation in the particulates.

The deposition velocity (v) of a substance can be determined by measuring the deposited mass per area and time (F) as well as the concentration (c) in the atmosphere:

$$\textcircled{3} \quad v = \frac{F}{c} \qquad\qquad \textcircled{4} \quad \tau \sim \frac{1}{v}$$

In Fig. 2 the results of deposition measurements by aid of equation $\textcircled{3}$ are represented (5).

On account of the fact that τ is inversely proportional to v the results of the different methods represented in Fig. 1 and 2 should be comparable. The relative sequence for the elements Ca, Mg, Fe, Cu, Na and Cd is the same. The position for Mn is slightly different. Pb has a lower measured v than Cd. This can be explained by the higher gaseous mass of Cd (≈ 30 %) in the atmosphere compared to that of Pb (≈ 4 %).

The relative sequence of the substances is a function of their volatilities which in approximation can be quantified by their boiling and melting points.

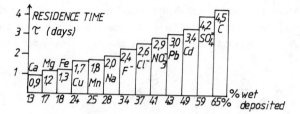

Fig.1: Residence time of particulate substances and their wet deposited fraction. Calculated from measured mass size distributions.

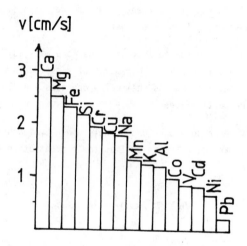

Fig.2: Measured deposition velocities for particulate substances (Rohbock 1984).

References

1. Whitby, K. T., EPA Report R 800971 (1975), c/o EPA, Raleigh (NC)
2. Müller, J.: VDI-Kolloquium "Schwebstoffe" (1979)
        c/o VDI-Luftreinhaltung, Düsseldorf
3. Jaenicke, R.: Ber. Bunsenges. Phys. Chemie 82, 1198-1202 (1978)
4. Chamberlain, A. C.: Proc. Royal Soc. A299, 45-70 (1966)
5. Rohbock, E.: Thesis c/o Meteorological Inst., Univ. Frankfurt (1984)

Published 1984 by Elsevier Science Publishing Co., Inc.

Aerosols, Liu, Pui, and Fissan, editors

EFFECT OF THE BASSET TERM ON PARTICLE RELAXATION
BEHIND NORMAL SHOCK WAVES

L. J. Forney and A. E. Walker

School of Chemical Engineering
Georgia Institute of Technology
Atlanta, Georgia   30332-0100

W. K. McGregor
Sverdrup Technology, Inc.
Arnold AFS, TN 37389

EXTENDED ABSTRACT

Small particles and droplets encounter normal shocks in a variety of applications. The particle-shock interaction subjects the particles to large unsteady drag forces behind the shock front. In the present paper, an analysis has been made of the relative importance of the Basset history integral for particle displacement and velocity behind a normal shock wave. The effect of the Basset integral has been related to gas stagnation conditions and the local gas Mach number.

Figures 1 and 2 illustrate the change in the particle Reynolds number and relaxation distance, with and without the Basset term, for a 10 μm particle moving behind a Mach 2 shock front in air . The stagnation conditions for this case are a particle Knudsen number Kn_o = 9.8 x 10^{-4} and the ratio of gas- to-particle density ρ_o/ρ_p = 8 x 10^{-3}. These reservoir conditions correspond to P_o = 100^{P} psi, T_o = 300°K, d_p = 10 μm and ρ_p = 1 gm/cc.

Figures 1 and 2 illustrate that the particle displacement relative to the gas back of the shock is unaffected by the inclusion of the Basset term until the latter stages of particle relaxation. The effect of the Basset history integral which results from diffusion of vorticity from the particle is shown to increase with increasing stagnation pressures and particle diameters but with decreasing gas stagnation temperatures.

Acknowledgment

This study was supported by grant No. AFOSR-83-0182 from the Air Force Office of Scientific Research.

854

References

Forney, L. J., McGregor, W. K., and Girata, P. T., Jr., "Sampling with Supersonic Probes: Similitude and Particle Breakup", AEDC-TR-83-26, Arnold Air Force Station, Tennessee (1983).

Forney, L. J. and McGregor, W. K., "Scaling Laws for Particle Breakup in Nozzle Generated Shocks", Particulate Science and Technology 4, (1984).

855

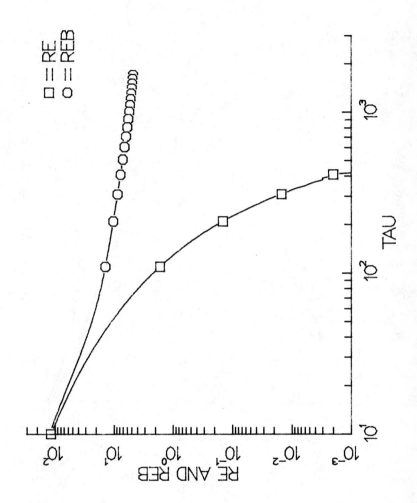

FIGURE 1 — RE vs TAU

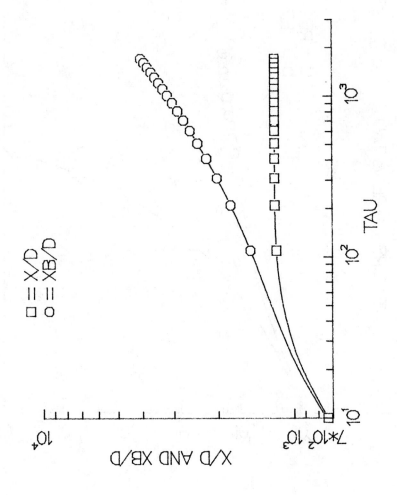

FIGURE 2 – X/D vs TAU

□ = X/D
O = XB/D

Published 1984 by Elsevier Science Publishing Co., Inc.

Aerosols, Liu, Pui, and Fissan, editors

TURBULENT AND BROWNIAN DIFFUSIVE DEPOSITION AND GRAVITATIONAL SEDIMENTATION OF AEROSOL PARTICLES IN A STIRRED TANK

K. Okuyama, Y. Kousaka and T. Hosokawa
Department of Chemical Engineering
University of Osaka Prefecture
Sakai 591 Japan

EXTENDED ABSTRACT

Introduction

The loss of aerosol particles in a stirred tank, which is of great importance for the investigation of the production of fine powders in chemical reactors, has been studied experimentally using monodisperse uncharged particles with diameters between 0.006μm and 2μm.

Experimental apparatus and method

A schematic diagram of the experimental apparatus is shown in Fig. 1. The stirred tank used was made of acrylic resin, and equipped with four vertical baffles each of which had a width of one tenth of the tank diameter. The stirrer was six-bladed turbine impeller and three bladed propeller. The aerosol particles used were monodisperse uncharged NaCl, DEHS and polystyrene latex particles with diameters between 0.006μm and 2μ m. NaCl and DEHS particles were generated by an evaporation-condensation type generator and were electrically classified by means of the differential mobility analyzer. The number concentrations of aerosol particles were controlled below 10^3 particles/cm^3 to avoid the effect of coagulation. Aerosols thus generated were introduced continuously throughout the chamber and was mechanically stirred for a short period with extremely low revolution to make the aerosol uniform. Then the revolution of the stirrer was raised to the desired speed. The stirrer speed was checked by a tacho meter. Aerosol sampled at any given residence time was introduced into the particle size magnifier and light scattering particle counter to measure its particle number concentration, and the decreases in particle number concentration with time were observed under various conditions.

Figure 1. Experimental apparatus

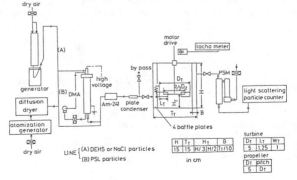

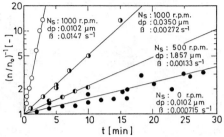

Figure 2. Decrease in particle
number concentration
with the stirring time

Experimental results and discussion

Figure 2 shows experimental relations between the ratio of particle number concentrations at initial time to those at lapsed time and stirring time in a tank. It is seen that the decreases in particle number concentration can be arranged by the next equation.

$$dn/dt = -\beta n, \quad n = n_0 \exp(-\beta t) \tag{1}$$

where, t is the stirring time, n_0 the initial particle number concentration, β the deposition rate constant(decay constant).

Figure 3 shows the dependence of the deposition rate constant β experimentally obtained on the stirred speed and particle size. It is seen that there is no difference due to the particle materials, and that the values of β for the same particle size increase with the stirrer speed due to turbulent diffusion. The values of β have distinct minima, where decreasing parts are the diffusion-predominant region while increasing parts are the gravity-predominant region.

Generally, the deposition flux of particles onto a wall surface can be described by;

$$N = -(D+D_e)dn/dy - nu_t \cos\theta \quad \text{for } 0<y<\delta \tag{2}$$
$$\text{B.C.: } n=0 \text{ at } y=0 \text{ and } n=n_c \text{ at } y=\delta$$

where, D and D_e are Brownian and turbulent diffusion coefficients, δ the boundary layer thickness, u_t the gravitational settling velocity, y the distance from the surface and θ the angle between the opposite direction of y and the direction of the gravity.

In Eq.(2), Crump and Seinfeld (1981) gave the next equation for the deposition velocity K(=deposition flux/particle number concentration above the surface) approximating the turbulent diffusion coefficient D_e to $K_e y^m$.

$$K = \frac{u_t \cos\theta}{\exp\left[\frac{\pi u_t \cos\theta}{(m\sin\frac{\pi}{m})^m \sqrt{K_e D^{m-1}}}\right]} \tag{3}$$

Here, K_e is approximated to k(du/dx) and du/dx is evaluated from the energy dissipation rate per unit mass of fluid ε_0 as follows.

$$du/dx = (2\varepsilon_0/15\nu)^{1/2} \tag{4}$$

In the case of stirred tank, the average value of the energy dissipation rate ε_0 is equal to the power consumption rate per unit mass of mixing fluid which is given as follows

$$\varepsilon_0 = (4/\pi)N_p(N_s^3 D_T^5 T_T^2 H) \tag{5}$$

where, Np is the power number, N_s the stirrer speed, D_T the stirrer

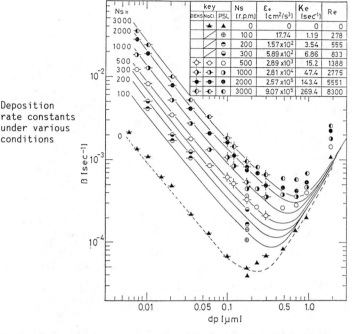

key			Ns	ε₀	Ke	Re
DEHS	NaCl	PSL	[r.p.m]	[cm²/s³]	[sec⁻¹]	
▲		▲	0	0	0	0
		⊕	100	17.74	1.19	278
		⊖	200	1.57×10^2	3.54	555
		⊜	300	5.89×10^2	6.86	833
◇	◇	○	500	2.89×10^3	15.2	1388
◈	◈	◑	1000	2.81×10^4	47.4	2775
●	●	●	2000	2.57×10^5	143.4	5551
◆	◆	⬤	3000	9.07×10^5	269.4	8300

Figure 3. Deposition rate constants under various conditions

diameter, T_T the diameter of stirred tank and H the height of a stirred tank.

When the surface areas of upper, lower and side walls of the stirred tank are S_1, S_2 and S_3, respectively, the deposition rate of the stirred tank can be given as

$$\beta = (K_1 S_1 + K_2 S_2 + K_3 S_3)/V \qquad (6)$$

K_1, K_2 and K_3 are deposition velocities for upper, lower and side walls given by Eq.(3) as $\theta = \pi$, $\theta = 0$ and $\theta = \pi/2$, respectively.

In evaluating the β, however, the values of k and m in turbulent diffusion coefficient must be determined experimentally. Figures 4 and 5 show the dependences of the deposition rate β obtained experimentally in the turbulent and the Brownian diffusion-predominant region on the energy dissipation rate ε_0 and Brownian diffusion coefficient D. From the comparisons of the experimental data of β with the calculated results for various values of k and m, k=0.3 and m=2.7 are found to explain experimental results well as shown by the solid lines in Figs. 4 and 5. The solid lines in Fig. 2 also shows the calculated results for k=0.3 and m=2.7. It is seen that the experimental results agree well with the calculation results when particle diameter is smaller than about $0.2\mu m$. However, with the increasing of particle size and ε_0, experimental data are seen to deviate from the theory, which is considered to be caused by the enhancement of particle loss due to particle inertia.

860

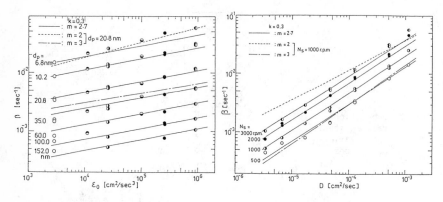

Figure 4. Dependence of the
deposition rate constants
on the energy dissipation
rate

Figure 5. Dependence of the
deposition rate constants
on the Brownian diffusion
constant

　　　In the case of three-bladed propeller, the experimental data agreed
well with those by six-bladed turbine impeller when the value of were ε_0
equal.
　　　In the case of no stirred condition, the deposition rate constant can
be evaluated by the next equation as indicated by broken lines in Fig. 3.

$$\beta = 0.347D^{2/3}\left(\frac{S}{V}\right) + \frac{u_t}{H}　　　　　　(7)$$

And the diffusion boundary layer thickness is found to be given by the next
equation.

$$\sigma = 2.884D^{1/3}　　　　　　(8)$$

Conclusion

　　　The loss of aerosol particles in a stirred tank has been studied
experimentally, and the dependence of the deposition rate on particle size
and turbulent intensity of stirring has been clearly evaluated. And the
experimental results are found to be explained by the theory assuming that
the turbulent diffusivity near the wall is proportional to 2.7th power of
the distance from the wall.

Reference

Crump, J. G. and J. H. Seinfeld (1981) "Turbulent Deposition and Gravitational Sedimentation of an Aerosol in a Vessel of a Arbitrary Shape"
J. Aerosol Sci. 12:405

Published 1984 by Elsevier Science Publishing Co., Inc.
Aerosols, Liu, Pui, and Fissan, editors

861

BROWNIAN DYNAMICS OF AEROSOL
DEPOSITION ON SPHERICAL COLLECTORS

M. H. Peters and D. Gupta
Department of Chemical Engineering
Rensselaer Polytechnic Institute
Troy, NY 12181

EXTENDED ABSTRACT

Introduction

When aerosol particles are removed from a fluid phase by impingement
on collecting elements the physicochemical surface characteristics of the
collecting bodies may be changed significantly. A simple example would be
the mere increase in the effective size of the collecting element due to
the presence of deposited particles (Tien et al., 1974). More complicated
effects include, for example, changes in the electrical properties of the
collector, such as the net static charge of the surface (Peters et al.,
1984), and changes due to external electric field interactions (Auzerais et
al., 1983). Since the properties of the collecting bodies change as particles deposit, the aerosol removal process is inherently time-dependent.

An important theoretical advancement in the area of transient particle
deposition behavior was prescribed by Tien et al. (1974), in which it was
shown that considerations of the spatial randomness of the particles on a
"microscopic" scale[1] can lead to the formation of particle chains, or
dendrites, on the collecting surface. The analysis, in this case, is based
on a stochastic simulation procedure, which has been recently reviewed and
generalized by Peters et al. (1984a). In another paper, the inclusion of
the Brownian motion of the particles, through the solution of the so-called
ordinary Langevin equation was accomplished (Peters and Gupta, 1984b). This
method is also followed in the present study.

The results of the Brownian dynamic computations alluded to above, may
be expressed in terms of a time-dependent, single collector efficiency.
Unfortunately, there have been few studies on the integration of such
results with the particle mass conservation equations associated with a
swarm of collecting bodies, such as in fixed-bed filtration or fluidized-
bed coating processes. The scope of the present work is to address this
inadequacy.

---

[1]By "microscopic" scale, it is meant a scale on the size of an individual
aerosol particle.

862

Brownian dynamics simulation

The analysis of the single, spherical collector efficiencies are based on a dynamical simulation of the movements of a large number of aerosol particles within a control volume which contains a collecting sphere at its center. With the inclusion of Brownian motion, the displacement of a single aerosol particle may be accomplished through integration of the ordinary Langevin equation[2] (Chandrasekhar, 1943)

$$\frac{d\underset{\sim}{v}_i}{dt} = -\beta_i \underset{\sim}{v}_i + \frac{\underset{\sim}{F}_i}{m_i} + \frac{\underset{\sim}{A}_i(t)}{m_i} + \sum_{\substack{j \\ (i \neq j)}} \frac{\underset{\sim}{F}_{int_{ij}}}{m_i} \qquad [1]$$

The first term on the right-hand side of Eqn. [1] represents the drag force due to the motion of the ith particle in a quiescent fluid, where β_i is assumed to be given by Stokes' law as, $\beta_i = 6\pi\mu r_{pi}/C_s m_i$. The force F_i includes external forces due to gravity and Coulombic interactions between the particle and collector, as well as the drag force due to the motion of the fluid over a stationary particle. Finally, $A(t)$ is a random force due to particle collisions with the surrounding fluid molecules (sometimes called the "Brownian force"), and $\sum_{\substack{j \\ (i \neq j)}} F_{int_{ij}}$ represents a Coulombic interaction force between particle i and all of the surrounding particles. Note that the assumption of pairwise additivity has been invoked.

As discussed in detail by Chandrasekhar (1943) the random force $A(t)$ follows a white-noise or Gaussian process, with the following statistical distribution

$$w[\underset{\sim}{A}_i(t)] = (\frac{1}{4\pi q_i \Delta t})^{3/2} \exp\{- \frac{[\underset{\sim}{A}(t) \cdot \underset{\sim}{A}(t)]}{4q_i \Delta t}\} \qquad [2]$$

where $q_i = kT\beta_i/m_i$

---

[2]The descriptor "ordinary" is used to distinguish Eqn. [1] from generalized Langevin equations (e.g., Albers et al., 1971).

The integrations of Eqn. [1], for the particle velocities and displacements, may be carried out over time steps small enough such that the external and particle-particle interaction forces may be treated as constants. Further details may be found in Chandrasekhar (1943).

The time-dependent motion, and deposition behavior, of the particles within the control volume surrounding the collector is determined by the use of a control volume subdivision method given previously (Peters et al. 1984a). Modifications for the case of Brownian motion, which include, for example, the use of a Maxwellian distribution in the initial particle velocities and an associated Poisson distribution in "arrival" times, are given in Peters and Gupta (1984b).

Our interests in the present study are in the application of the results of the dynamic calculations to the multicollector situation that exists in fixed-bed processes.

Application to fixed-bed processes

In reporting the single collector efficiencies from the simulation technique briefly presented above, the results can be expressed by the following expansion in terms of the total number of deposited particles, M, as

$$\eta = \eta_0 + \eta_1 M + \eta_2 M^2 + \ldots \tag{3}$$

When Eqn. [3] above is combined with the particle mass conservation equation for a fixed-bed process, the resulting equations are of the nonlinear, integrodifferential variety. For example, in the case of a simple linear form of Eqn. [3], the resulting particle mass conservation equation may be written (Peters and Gupta, 1984b)

$$\frac{\partial c}{\partial t} + u_0 \frac{\partial c}{\partial z} = c u_0 \{A_1 + A_2 \int_0^t [\frac{\partial c(z,\xi)}{\partial z}] d\xi\} \tag{4}$$

where

$$A_1 = -\frac{(1-\varepsilon)}{\varepsilon} \eta_0 \frac{3}{4R} \tag{5}$$

and

$$A_2 = \eta_1 \pi R^2 u_0 \tag{6}$$

864

Equation [4] above may be solved by a method of successive approximations. For example, the $(n+1)^{th}$ approximation is written as

$$c_{n+1} = \frac{\dfrac{\partial c_n}{\partial t} + u_0 \dfrac{\partial c_n}{\partial z}}{u_0 \{A_1 + A_2 \int_0^t [\dfrac{\partial c_n(z,\xi)}{\partial z}] d\xi\}} \quad ; n = 1, 2, \ldots \quad [7]$$

The solution for c_1 is obtained from Eqn. [4] by setting $A_2 = 0$.

In general, higher approximations are required at larger times, although the solution given by Eqn. [7] always converges.

Notation (abbreviated list)

$\underset{\sim}{A}(t)$	Brownian force
c	particle number concentration
$\underset{\sim}{F}, \underset{\sim}{F}_{int}$	external and particle-particle interaction forces, respectively
m	particle mass
M	total number of deposited particles at a given time
r_p	particle radius
u_0	superficial fluid velocity
$\underset{\sim}{v}$	particle velocity vector

References

Albers, J., Deutch, J.M. and Oppenheim, I. (1971), J. Chem. Phys. 54:3541.

Auzerais, R., Payatakes, A.C. and Okuyama, K. (1983), Chem. Eng. Sci. 38:447.

Chandrasekhar, S. (1943), Rev. Mod. Phys. 15:1.

Peters, M.H., Jalan, R.K. and Gupta, D. (1984a, in press) "A Dynamic Simulation of Particle Deposition on Spherical Collectors", Chem. Eng. Sci.

Peters, M.H. and Gupta, D. (1984b), A.I.Ch.E. Symp. Ser. No. 234, 80:98

Tien, C., Wang, C. and Barot, D. T. (1977), Science, 196:983.

Published 1984 by Elsevier Science Publishing Co., Inc.
Aerosols, Liu, Pui, and Fissan, editors

DEPOSITION OF SUBMICRON PARTICLES ON ROUGH WALLS
FROM TURBULENT AIR STREAMS

L.A. Hahn, J.J. Stukel, K.H. Leong, and P.K. Hopke
Department of Civil Engineering
and
Institute for Environmental Studies
University of Illinois
Urbana, Illinois 61801

EXTENDED ABSTRACT

Introduction

Exposure to respirable radioactivity is the major health threat faced by uranium miners. The radiation dose received by their respiratory tract is due primarily to the short lived daughter products of radon, ^{222}Rn, a chemically inert intermediate in the ^{238}U decay chain. The radon daughters, once formed, readily attach to airborne particles and are then transported with the particles. Therefore in dust-laden uranium mines, the transport of radioactivity is directly related to the transport of airborne particulate matter. Submicron particles in particular can carry the radioactivity deep into the tracheobronchial and pulmonary regions of the respiratory tract. In order to reduce the radiation dose it is necessary to obtain a better understanding of the mechanisms governing the transport and removal of submicron airborne particles.

The transport and deposition of entrained particles from a turbulent gas stream onto pipe or channel walls has been investigated in the past. Most of the early work involved the study of deposition onto smooth walls and in general the investigations have been limited to either particles greater than 1 μm in diameter or molecular-size particles. The purpose of the present work is to present results of a rough wall deposition study for particles having diameters ranging from 0.04 to 0.2 μm and to review the available theoretical literature to correlate the deposition data.

Experimental Design

The experimental system is given in Figure 1. A monodisperse mobility-classified submicron aerosol (uranine particles) is generated and flows through a rough walled pipe. The pipe centerline concentration is continuously monitored by a condensation nuclei counter as well as with the collection of an integrated filter sample. The deposited particles are washed from the walls of the pipe test section and the resulting solution is analyzed using fluorescence spectroscopy for the mass of particles deposited.

The roughness configuration chosen to conduct the deposition measurements was a repeated rib geometry to represent the wall roughness found in uranium mines. Two distinct types of wall roughness were found to occur in uranium mines including a closely spaced and a more widely spaced configuration. To allow for a flexible spacing to height ratio in a repeated rib configuration, roughness elements were machined from thick walled (.95 cm) aluminium tubing. The outer diameter of the tubing was just less than the inner diameter (9.8 cm) of the stainless steel pipe so that the tolerance was less than 0.05. The roughness elements, 0.64 cm wide, were separated using spacer rings, 0.64 cm and 5.1 cm wide, cut from thin walled (.32 cm) tubing. A schematic diagram of the roughness rings is shown in Figure 2.

Results and Discussion

The experimental deposition velocities, non-dimensionalized by the friction velocity, from this study are plotted in the lower left side of Figure 3. The curve in Figure 3 is computed from a semi-empirical equation given by Kader and Yaglom (1977). The nondimensional deposition velocity is

$$\frac{V_d}{u_s} = \frac{1}{\alpha \ln(L/K) + b_1''(K^+)^q (Sc^{2/3} + b_2'') + C' + \beta_1} \tag{1}$$

u_s is the friction velocity, α and β_1 are dependent on the flow in the core of the pipe, L is the characteristic length for the flow, K and K^+ are the dimensional and nondimensional roughness height, b_1'', b_2'', q and C' depend on the shape and distribution of the roughness elements and Sc is the Schmidt number for the particles. For the present case, equation (1) reduces to

$$\frac{V_d}{u_s} = \frac{1}{b_1''(K^+)^{1/4} Sc^{2/3}} \tag{2}$$

The results, within experimental error, are consistent with equation (2) over the wide range of roughness geometries studied. Also plotted in Figure 3 are the data from a wind tunnel study by Chamberlain (1968). Chamberlain measured the concentrations of water vapor and thorium B vapor over rough surfaces. Collectively, the results show that Kader and Yaglom's equation can be used to predict rough wall particle deposition accurately from molecular sizes to 0.2 μm diameter.

References

Chamberlain, A.C. (1968). Q. J. Roy. Meteor. Soc. 94:314-332.

Kader, B.A. and Yaglom, A.M. (1977). Int. J. Heat Mass Transfer 20:345-357.

Detail of Roughness Elements

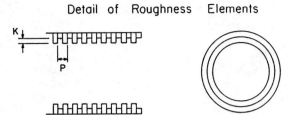

Figure 1.   Laboratory model of rough surfaces
using a repeated rib geometry.

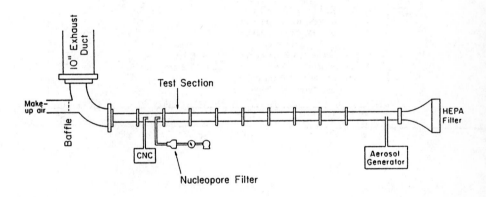

Figure 2.   Experimental set-up used to determine particle deposition.

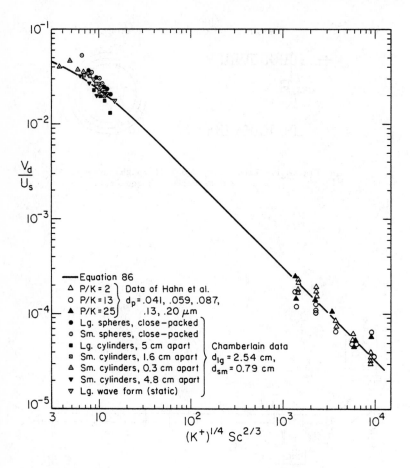

Figure 3. Deposition data compared with the correlation by Kader and Yaglom.

Published 1984 by Elsevier Science Publishing Co., Inc.

Aerosols, Liu, Pui, and Fissan, editors

STUDIES ON THE DENSITY, POROSITY AND WATER CONTENT
OF HYGROSCOPIC AEROSOLS BY A DIFFERENTIAL SETTLING
VELOCITY ANALYSIS AND THE COMPARISON OF COAGULATION
RATE CONSTANTS CALCULATED FROM THREE KINDS OF
DENSITIES

Yasuo Ueno
Department of Chemical Engineering, University of Kentucky
Lexington, KY 40506
U. S. A.

EXTENDED ABSTRACT

Introduction

One of the problems in aerosol studies is with the densities of
different kinds of aerosols. Patterson et al.(1927) found that the densities
of metal oxide particles were far smaller than the true densities of the
materials in bulk. Usually we use aerodynamic diameter instead of true
density. This is a very convenient and efficient procedure when we
deal with atmospheric aerosols or water mists whose densities are
approximately unity. On the other hand, true density of aerosol material
cannot be always used since sometime most of aerosols have porous or
chainlike structure. Without the knowledge of the density of particle,
aerodynamic diameter cannot be converted to Stokes' diameter. We
cannot fully evaluate aerosol behaviors in terms with mobility, diffusion
coefficient, collision radius or Cunningham's correction factor. This
paper presents a differential settling velocity method to estimate hygroscopic solid aerosol of sodium sulfate. Comparison has also been made
between the coagulation rate obtained by estimated density and that
calculated by aerodynamic diameter or the true density of sodium sulfate.
Further, the water content and porosity (the ratio of radius/thickness) of
hygroscopic aerosol has been investigated.

Theoretical

The settling velocity of a particle of a monodisperse aerosol
confined in a chamber under agitation may be expressed as follows:

$$- dn_t/dt = n_t \cdot V_t /H \dots\dots\dots\dots\dots\dots (1)$$

Here n is number concentration at time t, V Stokes' velocity of settling
particle at time t and H height of chamber. The integration of Equation
(1) between time t to t+Δt yields

$$\ln n_{t+\Delta t}/ n_t = - V_t/H \cdot \Delta t \dots\dots\dots\dots\dots (2)$$

Vt being assumed to be constant for a short interval of time, Δt, it
follows that

$$- \ln [m_{t+\Delta t}/m_t \cdot (D_t/D_{t+\Delta t})^3]$$

$$= V_t \cdot \Delta t /H - \ln[\rho_{t+\Delta t}/\rho_t] \quad \ldots\ldots\ldots(3)$$

Here, m is mass concentration, D diameter of particle and density of particle. It can be expected that the plots of the left-hand side of Equation (3) against Δt will pass through the point of origin, and the slope at the origin will give the settling velocity at time, t, with the height of chamber H. It can also be considered that the slope of the line will get steeper as the time becomes longer. By equating the effective weight of particle at time t, to the Stokes' drag, we obtain

$$\pi/6 \cdot D^3 (\rho - \rho') g = 3\pi \eta D V \ldots\ldots\ldots\ldots(4)$$

where D is the diameter of particle, ρ and V, its density and settling velocity, η and ρ' the viscosity and density of the medium (air), and g, the gravitational acceleration. By neglecting ρ' in comparison with ρ, we obtain the following equation

$$\rho = 18 \eta V/g D^2 \ldots\ldots\ldots\ldots\ldots\ldots\ldots\ldots(4)$$

From Equation (4), the value of particle density at time t can be estimated if D and V at time t are known.

The water content of a dried particle can be determined as follows: It is assumed that the droplets generated have passed through a furnace to be dried particles without encountering mutual collision. Since the solid content of a droplet will remain unaltered, there holds a mass-balance relation between the diameter (D_l) of a droplet produced by atomization and the diameter (d_s) of a particle formed by the dehydration of the droplet. We obtain

$$d_s / D_l = [\rho_l (1 + W_s) / \rho_s (1 + W_l)]^{1/3}$$

ρ_l: density of droplet, ρ_s: density of dried particle, W_l: water content of liquid droplet (grams of water per gram of solid particle), W_s: water content of dried particle (grams of water per gram of solid content).

Experimental Procedure

Test aerosol of sodium sulfate was generated with a generator passed through an electric furnace with a constant temperature controller and introduced into an aerosol chamber for 10 minutes. The relative humidity in the chamber was controlled. The aerosol was subjected to aging with a constant rate agitation for more than an hour. Portions of the aerosol were sampled and collected by an electrostatic precipitator and a thermal precipitator for aerosol mass and size analysis (Ueno, 1976). Estimates of particle size were made on the thermal precipitator samples with an electron microscope. The particles were formed to be either isolated and rounded ones or chainlike aggregates. As the size distribution of the former, the number-averaged diameter was taken, and while as that of the latter, the number average of the volume-equivalent diameters was taken.

Results and Discussion

Experimental data could be well analyzed with theory. It was found that as the humidity is increased, the half-decay period gets shorter, the particle size larger, and the particle density smaller. Spray-dried sodium sulfate particles are hollow and crystallized, and they show holes with edges rounded inward and have thin, compact shells with radius/thickness ratios of the order of 10. The formation of hollow particles due to the evaporation rate of water exceeding the diffusion rate salt back into the inside, thereby creating internal voids after dryness is reached. The fact observed by Duffie et al.(1953) agreed partially with the present results. with aerosol region. Hygroscopic aerosols such as sodium sulfate aerosol look growing in size and being collected easily from their system (Ueno, 1984). Results obtained can be summerized as follows: 1) the values of density estimated by a differential settling velocity analysis change with the concentration of the dispersing solution and range from 0.93 to 0.62 in dried atmosphere. They are also decreasing with the increase of relative humidity and range from 0.62 to 0.43 in terms with initial aerosols (without aging). 2) the ratios of radius/thickness calculated range approximately from 7 to 13 in terms with initial aerosols. 3) the diameter ratios of solid particle/droplet change with the concentration of the dispersing solution and range approximately from 0.2 to 0.5 in terms with initial aerosols. 4) the number of H_2O molecules per sodium sulfate molecule under various relative humidities change approximately from 0 to 5. 5) the coagulation rate constants calculated by these estimated values of density in the atmosphere with different relative humidity show quite reasonable values in comparison with the results calculated from aerodynamic diameter or true density.

References

Duffie, J.A. and Marshall, Jr.,W.R. (1953) "Factors Influencing the Properties of Spray-Dried Materials", Chem. Engr. Progress 49:417.

Patterson, H.S. and Whytlaw-Gray, R. (1927) "A density of Smoke Particles", Proc. Roy. Soc.(London), A113:302.

Ueno, Y.(1976) "Effect of Surface-active Substances on the Stability of Aqueous Salt Solution Aerosols", Atmos.Environ. 10:409.

Ueno, Y. the proceedings of the 3rd International Conference on Indoor Air Quality and Climate, Stockholm, Sweden, August 1984.

Published 1984 by Elsevier Science Publishing Co., Inc.
Aerosols, Liu, Pui, and Fissan, editors

872

A THEORETICAL STUDY OF THE SCAVENGING AND REDISTRIBUTION OF AEROSOL
PARTICLES DURING THE CONDENSATION AND COLLISION-COALESCENCE GROWTH
OF CLOUD DROPS

by

A. Flossmann and H. R. Pruppacher
Meteorologisches Institut, Johannes Gutenberg Universität, Mainz, FRG

and

Wm. D. Hall
National Center for Atmospheric Research (NCAR), Boulder, Colorado, USA

Abstract

A theoretical model has been formulated with which the scavenging of
aerosol particles and the re-distribution of the scavenged aerosol particle
mass can be followed during the process of condensation and the subsequent
collision-coalescence growth of the formed drops. For this purpose a parcel-
model with entrainment was formulated for describing the condensation growth
and the nucleation-scavenging of aerosol particles. Added to it was a stochastic collision-coalescence model which was used to describe the formation
and evolution of the drop size spectrum formed from a given aerosol particle
size spectrum. Coupled with these models was a stochastic model which
allowed to follow the aerosol particle mass which entered the drops as a
result of nucleation scavenging, impaction scavenging, and as a result of
collision between the drops. The combined models also allowed to study the
effects of particle scavenging on the size distribution of the interstitial
aerosol particles, i.e. the aerosol particles which remained between the
drops during the drop growth process.

Our study shows: (1) the manner in which nucleation scavenging depends
on the size distribution of the aerosol particles and their composition, and
the manner in which impaction scavenging depends on the aerosol particle and
drop size distributions; (2) that inside a cloud the main aerosol mass enters
the cloud drops via nucleation scavenging rather than via impaction scavenging; (3) that the aerosol outside a cloud suffers a significant reduction in
mass and number while undergoing transformation through scavenging to the
cloud-interstitial aerosol, both (1) and (2) in agreement with most recent
observations; (4) that the main aerosol mass scavenged is bound to the main
water mass in a cloud, indicating that through precipitation the main
airborne aerosol mass indeed may become returned to the ground. This intuitively expected result (4) crucially hinges on correctly including in the
model the process which delivers scavenged aerosol mass to other drops by
colliding and coalescing with them. The crucial nature of this requirement
is shown by evaluating our models for the case that the redistribution of
the aerosol mass during the stochastic collision and coalescence process
is not taken into account, i.e. for the case that each drop is considered
to obtain aerosol material only via aerosol particles it captures by itself
rather than what it also obtains by collision and coalescence with other

drops. For this case our computation shows that the main aerosol mass would remain bound to the small drops which do not precipitate.

The present model is presently formulated to be built into the convective cloud model of Clark and Hall (NCAR) to describe the scavenging potential of a given convective cloud-precipitation-cycle.

CHAPTER 20.

COAGULATION, CONDENSATION AND NUCLEATION

Published 1984 by Elsevier Science Publishing Co., Inc.

Aerosols, Liu, Pui, and Fissan, editors

CONDENSATION OF ARGON BY HOMOGENEOUS NUCLEATION IN THE NOZZLE FLOW OF A CRYOGENIC LUDWIEG-TUBE

J. Steinwandel[*] and Th. Buchholz[**]

Department of Mechanical Engineering, Fluid Dynamics Laboratory
Yale University, New Haven CT 06520, USA

EXTENDED ABSTRACT

1. Introduction

The classical theory of homogeneous nucleation (CNT) predicts reasonably well the observed onset of condensation in many vapors. However, there are problems connected with the free enthalpy of formation of condensation nuclei, where the macroscopic surface tension enters the theory quite substantially. This quantity is often poorly known even for the bulk liquid (or solid) phase. Moreover, the surface tension of small droplets certainly differs from that of the bulk, a problem that has often been discussed. Therefore, a recent modification of the CNT has become of particular interest, because it avoids the use of the macroscopic surface tension. Here, the kinetic formalism of CNT is retained, but the formation free enthalpies of small clusters containing 2-100 monomers are calculated ab initio from the interatomic (molecular) potentials by the methods of statistical mechanics. Because of the relatively simple dispersion interaction potential, argon clustering is often studied in atomistic theories (e.g. McGinty (1973); Hoare et al. (1980); Garcia and Torroja (1981)). The statistical mechanics calculations provide the standard free enthalpy change $\Delta G^o(i, p^o, T)$ at a reference pressure p^o for clusters containing i monomers. From that, the free enthalpy change can be calculated for any thermodynamic state by the standard relation:

$$\Delta G(i, p, T) = \Delta G^o(i, p^o, T) + (1-i)k_B T \ln(p/p^o)$$

The nucleation rate is calculated by:

$$J = \frac{\alpha p}{(2\pi m_o k_B T)^{1/2}} \Big/ \sum_{i=2}^{i_{max}} \left\{ 4\pi r_i^2 \left(p/k_B T\right) e^{-\Delta G/k_B T} \right\}^{-1}$$

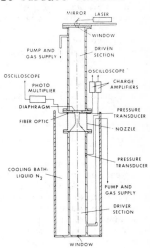

Fig. 1   Schematic experimental arrangement

r_i is the radius of a cluster containing i monomers of individual mass m_o. α is the condensation coefficient. Due to the
low condensation temperatures of argon (Fig. 3), experiments
on the subject must be performed in rapid gasdynamic expansions (steady free jet, hypersonic or supersonic nozzle expansions, unsteady shock tube expansions). Such experiments
have been performed and are discussed briefly in section 3.
However, different values of argon condensation onset were observed in the steady expansions and the unsteady shock tube expansion. Therefore, the present Ludwieg tube study was undertaken in order to provide additional experimental information.

2. Experimental Method

The Ludwieg tube arrangement is shown schematically in
Fig. 1. The tube was operated in vertical position and only
the high pressure section including the Mach 1.62 nozzle was
precooled to 77.4 K. The precooling was necessary in order to
expand the test gas (argon at various mole fractions in helium
carrier gas) beyond the argon phase equilibrium to supersaturated states. The high pressure tube was operated at total
initial pressures 3.3 atm $\leq p \leq 5$ atm at initial argon partial
pressures 2.8 torr $\leq p \leq 102$ torr. Condensation at the nozzle
exit was observed by light scattered from the growing particles
in an axially centered beam of a He/Cd-Laser operated at 442 nm.
The scattered light was detected by an optical fiber system in
conjunction with a photomultiplier tube (RCA 7265). The static
pressure at the observation station was measured simultaneously
by a piezoelectric transducer (Kistler 606 L). From the measured
pressure in the nozzle expansion, the corresponding temperature
was calculated assuming isentropic flow. A typical oscilloscope
recording is shown in Fig. 2. (Trace (1) is the pressure transducer signal, numbers indicate the steady flow periods. Trace
(2) shows the scattered light inten-
sity. Condensation onset is indicated
by a steep rise in scattering intensity.

3. Results and Discussion

Argon condensation onset conditions were determined in the range
0.6 torr $\leq p \leq 57$ torr at corresponding temperatures 41 K $\leq T \leq 62$ K.
The average adiabatic supercooling on
the isentropes with respect to phase
equilibrium was about 10 K. The results are shown in an argon phase diagram in Fig. 3. In addition, all previous experiments on the subject known
to the authors are shown. Curves of
constant nucleation rates calculated

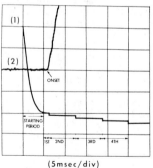

Fig. 2 Typical experi-
ment recording

from atomistic theories ((a):
Hoare et al. (1980); (b): Garcia
and Torroja (1981); (c): McGinty
(1973)) and finally the predictions of the CNT (curve (d)) are
also shown in Fig. 3. (Curves (a)-
(d) refer to a nucleation rate of
$J = 10^0$ cm^{-3} taken as the boundary
for detectable condensation). The
atomistic nucleation theories agree
among themselves while the CNT predicts condensation of argon much
closer to phase equilibrium.
Presently, there is no explanation for this discrepancy. Moreover, not even the various sets
of experiments on the subject lead
to consistent results, which is
obvious from Fig. 3. The results
of Lewis and Williams (◊ , 1976),
showing the largest values of
supercooling (or supersaturation),
were performed in free expansion

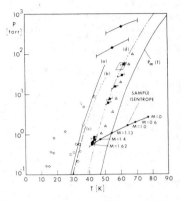

Fig. 3 Argon condensation
onset. Comparison of experiments and theory

of argon starting from room temperature. Molecular beam type
of experiments generally occur under non-equilibrium conditions
and are characterized by extremely high cooling rates (dT/dt ≈
-5×10^6 K/sec), which in turn excludes such experiments from a
comparison with the steady-state theories presented here. The
experiments of Pierce et al. (♦ , 1971) were performed in a
blow-down wind tunnel employing a hypersonic Mach 5 nozzle,
again starting from room temperature (dT/dt ≈-5×10^5 K/sec).
Due to the experimental uncertainty of ± 10 K at condensation
onset, a correct interpretation of these experiments seems
difficult. Another set of experiments was performed by Wu et al.
(o , 1978). Here, a Mach 3 nozzle mounted in a continuously
operated wind tunnel was used to expand mixtures of argon in
helium from supply temperatures 104 K ≤ T ≤ 144 K (dT/dt ≈-
7×10^5 K/sec). Finally, Matthew and Steinwandel (Δ , 1983) took
data on the onset of condensation of argon highly diluted in
helium using the expansion waves of cryogenic shock tubes operated at a starting temperature of 77.4 K. These experiments had
the lowest cooling rate (dT/dt ≈-2×10^4 K/sec) of all previous
work. These experiments show condensation close to phase equilibrium in contrast to previous work. Maximum adiabatic supercooling on the isentropes was about 5 K.
The present experiments were performed in order to clarify the
differences. Because higher cooling rates cannot be obtained at
controlled conditions in shock tube expansions, a Ludwieg-tube
nozzle expansion was used. The Mach number was 1.62 and the
cooling rate was adjusted to a value of -9×10^6 K/sec comparable
to that of Wu et al. (1978). Otherwise, the experimental conditions of Matthew and Steinwandel (1983) were duplicated almost
exactly. Considering the well known effect of the cooling rate

880

on the observed condensation onset, the present experiments (■)
are in general agreement with the cryogenic shock tube experiments. Therefore, the differences in condensation onset of Wu et
al. (1978) and Matthew and Steinwandel (1983) cannot be related
to the different cooling rates alone.

A comparison of the last three sets of experiments can be
based on the sample isentrope shown in Fig. 3. Initial temperature for the shock- and Ludwieg-tube experiments was 77.4 K in
contrast to the steady nozzle flow experiments having initial
temperatures $104 \text{ K} \leq T \leq 144 \text{ K}$. In addition, the helium pressure
at the intersection of the isentrope with the saturation line
$p_\infty(T)$ is about a factor of 100 higher for the shock- and Ludwieg-tube experiments. This difference in absolute carrier gas
pressure is probably much more important in order to explain
the differences between the experiments. The carrier gas has to
remove reaction energy from the smallest clusters in a three-
body collision process with a cluster and a monomer. Despite
the fact, that helium is one of the less efficient collision
partners, a substantially higher carrier gas concentration leads
to a higher probability of reaction energy removal and thus increases the rate of production of clusters in the initial stages
of nucleation. The application of classical nucleation kinetics
depends on the efficiency of energy removal. However, rotational
and vibrational deactivation of molecules (or clusters) are
serious problems in modern molecular reaction dynamics. Therefore, only a rough estimate about the rate constants involved
is possible. Such calculations suggest that classical nucleation kinetics might not be applicable a priori for all experiments reported here.

4. References

Garcia, N.G. and Torroja, J.M.S. (1981) Phys.Rev.Lett. 47, 186
Hoare, M.R. et al. (1981) J.Colloid Int.Sci. 22, 73
Lewis, J.W.L. and Williams, W.D. (1974) AEDC-TR-74-32
Matthew, M.W. and Steinwandel, J. (1983) J.Aerosol Sci. 14, 755
McGinty, D.J. (1973) J.Chem.Phys. 58, 4733
Pierce, T. er al. (1972) Astronautica Acta 16, 1
Wu, B.J.C. et al. (1978) J.Chem.Phys. 69, 1776

This work was supported by the National Science Foundation
under Grant CPE-7909683 and by a Feodor Lynen Fellowship of
the Alexander-von-Humboldt Foundation for J. St.

Present Addresses

*) Department of Physical Chemistry, University of Stuttgart,
   FRG
**) Department of Mechanical Engineering, University of Karls-
   ruhe, FRG

Published 1984 by Elsevier Science Publishing Co., Inc.
Aerosols, Liu, Pui, and Fissan, editors

COAGULATION RATE OF POLYDISPERSE PARTICLES

K. W. Lee and H. Chen
BATTELLE
Columbus Laboratories
505 King Avenue
Columbus, Ohio 43201
U. S. A.

EXTENDED ABSTRACT

Introduction

The rate of coagulation is conveniently represented by the following simple equation for the particle number decay:

$$\frac{dN}{dt} = -\gamma N^2, \tag{1}$$

where N is the total particle number concentration, t is the time, and γ
is the coagulation rate. The coagulation rate based on a model of monodisperse particles in the continuum and gas-slip regimes is

$$\gamma = \frac{4kT}{3\mu} \left[1 + 1.246Kn + 0.42Kn \exp(-0.87Kn)\right] \tag{2}$$

where k is Boltzmann's constant, Kn is the Knudsen number, T is the absolute
temperature, and μ is the gas viscosity. For the free-molecule regime, the
rate is written as below from the kinetic theory:

$$\gamma = 6\sqrt{2} \left(\frac{\mu r}{\rho} K\right)^{\frac{1}{2}} \tag{3}$$

where r is the particle radius.

The rates given by Equations (2) and (3) are very useful for many experimental and analytical applications. It should be recognized however that
all of these results are based on a model of monodisperse particles. In
general, the coagulation rate depends not only upon the particle size but
upon dispersity of the size distribution. Therefore, the rates given by
Equations (2) and (3) are subjected to considerable error when a polydisperse
aerosol is to be dealt with (Lee, 1983; Lee et al, 1984).

In the present work, new coagulation rates are presented accounting for
polydispersity of the size distribution. The present paper has been accepted
and details can be referred to Lee and Chen (1984).

Model Feature and Results

The complete size distribution of an aerosol or hydrosol undergoing
coagulation is governed by the following nonlinear integro-differential
equation if particle size is taken to be continuous:

$$\frac{\partial n(v,t)}{\partial t} = \frac{1}{2} \int_o^v \beta(v-\bar{v},\bar{v})n(v-\bar{v},t)n(\bar{v},t)d\bar{v} - n(v,t) \int_o^\infty \beta(v,\bar{v})n(\bar{v},t)d\bar{v} \tag{4}$$

where $n(v,t)$ is the particle size distribution function, v and $\bar{v}$ are the particle volumes, $\rho(v,\bar{v})$ is the collision kernel for two particles of volumes v and $\bar{v}$.

In order to represent a polydisperse aerosol size distribution, the lognormal function which is one of the most commonly used mathematical forms is used here. After a rearrangement, Equation (4) becomes:

$$\frac{dN}{dt} = \frac{1}{2} \int_o^\infty \int_o^\infty \beta(v,\bar{v})n(\bar{v},t)n(v,t)dvd\bar{v} \tag{5}$$

For the case of coagulation due to Brownian Motion in the continuum regime, the collision kernel is:

$$\beta(v,\bar{v}) = K(v^{1/3} + \bar{v}^{-1/3}) \left(\frac{C(v)}{v^{1/3}} + \frac{C(\bar{v})}{\bar{v}^{-1/3}} \right) \tag{6}$$

where C is the slip correction factor ($= 1 + A\ Kn$) and $A = 1.246$.

After substituting Equation (6) into Equation (5) and using moment functions of lognormal distribution, we obtain:

$$\frac{dN}{dt} = -K \left\{ (N^2 + M_{1/3}M_{-1/3}) + A\ \lambda(\frac{4}{3}\pi)^{1/3}(N\ M_{-1/3} + M_{1/3}M_{-2/3}) \right\} \tag{7}$$

where M_k is the kth moment,

$$M_k = (\frac{4}{3}\pi r_g^3)^k\ N \exp(\frac{9}{2}k^2 \ln^2 \sigma) \tag{8}$$

Utilizing Equation (8), we have

$$\frac{dN}{dt} = -K[\left\{ 1 + \exp(\ln^2 \sigma) \right\} + \left\{ A\ Kn\ \exp(\frac{1}{2}\ln^2 \sigma) \right\} \left\{ 1 + \exp(2\ln^2 \sigma) \right\}]N^2 \tag{9}$$

where Kn is now defined based on r_g. Equation (9) indicates that in the gas-slip regime the coagulation rate depends not only on spread of the size distribution but on the mean particle radius. Figure 1 demonstrates that Hinds' numerical calulation results (Hinds, 1982) can fall on the single correlation curve which is directly available from Equation (9).

In the free-molecule regime where particle sizes are much smaller than the mean free paths of gas molecules, λ, the collision kernel has the following form:

$$\beta(v,\bar{v}) = \overset{\circ}{K}(v^{1/3} + \bar{v}^{-1/3})^2(\frac{1}{v} + \frac{1}{\bar{v}})1/2, \tag{10}$$

where $\overset{\circ}{K}$ is the collision coefficient for the free-molecule regime

$[= 3\left(\frac{3}{4\pi}\right)^{1/6}\left(\frac{\mu}{\rho}K\right)^{1/2}]$, and ρ is the particle density. Following a procedure similar to the above derivation combined with some simplifications, we obtained:

$$\frac{dN}{dt} = -\left[\frac{3\sqrt{3}}{\sqrt{2}}\left(\frac{\mu^2 r_g}{\rho kT}\right)^{1/2} b\ K\left\{\exp(\frac{25}{8}\ln^2\sigma) + 2\exp(\frac{5}{8}\ln^2\sigma) + \exp(\frac{1}{8}\ln^2\sigma)\right\}\right]N^2 \tag{11}$$

Again, the effects of the geometric standard deviation on the coagulation rate is substantial.

Having obtained analytic expressions for γ that provide the size distribution dependent coagulation rates for various Knudsen number regimes, the existing diagram showing the Knudsen number dependent rate can now be expanded to include the polydispersity effect. By defining the Knudsen number based on the geometric number median particle radius, r_g, Figure 2 has been prepared. Figure 2 compares this theory with existing experimental data.

Conclusions

The coagulation rate of particles in various Knudsen number regimes have been studied analytically utilizing the lognormal function as the size distribution. The obtained expressions take into account the effects of both the particle size and the particle size spread. It is concluded from the results that the coagulation rate for polydisperse aerosols is substantially higher than that for monodisperse aerosols. The classical diagram showing the coagulation rate as a function of the Knudsen number has been expanded as a result of the present study. The results indicate that large discrepencies noted in past comparisons of the monodisperse aerosol model with experimental data are partially due to the polydispersity effects and suggest strongly that the coagulation rates measured experimentally have to be correlated not only with the mean particle size but with the size spread.

884

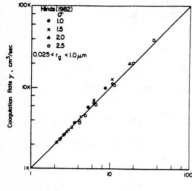

Figure 1.   Comparison of Present Theory
With Hinds (1982) Numerical
Calculations.

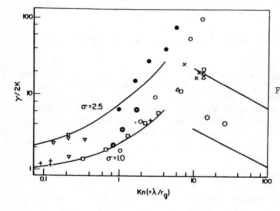

Figure 2.   Comparison of the
Present Prediction
With Various
Experimental Data.

References

(1) Hinds, W. C., (1982), Aerosol Technology, Chapter 12, p. 233, Wiley,
New York.

(2) Lee, K. W., (1983), J. Colloid Interface Sci., 92, (2), 315-326.

(3) Lee, K. W. and Chen H., (1984), "Coagulation Rate of Polydisperse Particles",
in press, Aerosol Science and Technology.

(4) Lee, K. W., Chen, H., and Gieseke, J. A., (1984), Aerosol Science and
Technology, 3, 53-62.

Published 1984 by Elsevier Science Publishing Co., Inc.
Aerosols, Liu, Pui, and Fissan, editors

ON OSTWALD RIPENING

T.H. Tsang and J.R. Brock
Chemical Engineering Department
University of Texas
Austin, Texas 78712

EXTENDED ABSTRACT

Introduction

Subsequent to their appearance by homogeneous nucleation from a monomer, particles grow by coagulation and condensation/evaporation processes.
Such growth has been titled "Ostwald ripening" in the literaturature. Numerous studies have been concerned with the existence of asymptotic limit
distributions arising from Ostwald ripening by condensation/evaporation
and by coagulation . We have obtained results by numerical simulation
of these aspects of Ostwald ripening.

Results

From our numerical studies of Ostwald ripening by the condensation/
evaporation process we conclude the following:

(a) Starting with some arbitrary initial distribution, for long times
an asymptotic limit distribution is approached which agrees, for the linearized Kelvin term, with the analytical similarity solutions for continuum and
kinetic (free molecule) growth laws. At this asymptotic limit, details of
the initial distribution are completely "forgotten."

(b) The time evolution starting with some arbitrary initial distribution
involves two epochs. In the first, the total number concentration remains
sensibly constant. In the second, particles begin to be lost from the
distribution and the rate of decrease of total particle concentration approaches
that predicted by the asymptotic similarity theory.

(c) The time necessary to achieve the asympotic similarity solutions
for continuum and kinetic (free molecule) growth laws increases with
increasing dispersion in initial particle size distribution.

(d) For long times , the differences in the distributions resulting from
use of nonlinear and linear Kelvin terms (eqs. (2) and (8)) in the growth
law are confined to a boundary region near $x*$ outside of which the distributions are very nearly identical. $x*$ is mass of smallest particle.

(e) As is implicit in the similarity theory development, no unique asymptotic limit distribution occurs in the transition region of Knudsen number.
This is borne out by our numerical simulations using transtion region growth
laws.

(f) For conditions where coagulation may be neglected, it does not appear
necessary to include competitive effects in describing growth by condensation/
evaporation.

From our numerical studies of Ostwald ripening by simultaneous condensation/ evaporation and Brownian coagulation in near-continuum and continuum regimes of Knudsen number we conclude the following:

(a) Starting with some arbitrary initial distribution, for long times an asymptotic limit distribution is approached.

(b) For long times the total number concentration of particles, Mo, decays with time, t, as: $M_0 \sim 1/t$. For the same Knudsen regime, the separate processes exhibit the same decay of M_0 with time. Therefore in experimental investigation, in the same Knudsen regime, the decay of M_0 with time is not a reliable guide to identification of the relevant rate process.

(c) The contributions of the condensation/ evaporation and coagulation processes to the asymptotic evolution of the particle size distributions depend on a dimensionless ratio A. For $A \gg 1$ coagulation is the dominant process. For intermediate and small values of A, the contribution of coagulation is a monotonic decreasing function of time.

Published 1984 by Elsevier Science Publishing Co., Inc.

Aerosols, Liu, Pui, and Fissan, editors

HIGH TEMPERATURE AEROSOL EVAPORATION IN THE TRANSITION REGIME

R. Fischer and P. Roth
Fachgebiet Strömungstechnik
Universität Duisburg
4100 Duisburg, West Germany

Introduction

The process of aerosol particle evaporation is of great importance in the field of aerosol physics and technology. If the ratio of the mean free path of the gas phase and the particle radius is of the order of unity the process is in the transition regime between continuum and free molecular conditions. In this case the theoretical description of the transfer processes between the aerosol particles and the surrounding gas is quite complex. Because of the known mathematical difficulties in solving the Boltzmann equation, only approximate solutions and semi-theoretical expressions are available for calculating the mass flux of an evaporating aerosol droplet.

The purpose of this paper is to study the evaporation process in this transition regime at high temperatures, both experimentally and theoretically. The experiments are done in a conventional diaphragm shock tube. The decrease in particle size during the evaporation process is measured by laser light scattering under fixed angles using MIE's theory for data interpretation. The shock wave technique seems to be a successful method to study particle evaporation and related processes at high temperatures.

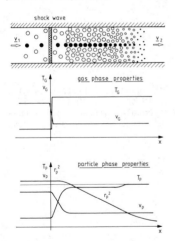

In Fig. 1 the modelized situation for stationary conditions and low particle volume concentrations behind a shock wave is schematically shown (Fischer and Roth, 1983; Roth and Fischer, 1983). In this model the particle behaviour is determined by three relaxing quantities, that is particle velocity v_p, particle temperature T_p and particle radius r_p, which have different characteristic relaxation times. Now the experimental conditions are chosen in such a manner, that the relaxation time for particle evaporation is much longer than for momentum and energy transfer. In this case the process of mass transfer is isolated from the other exchange processes.

Figure 1:   Schematic of droplet evaporation behind a shock wave

Experimental

The experimental setup is discussed in some detail in Roth and Fischer (1983). The mean particle radius of the aerosol, generated with an evaporation/recondensation generator (Liu and Lee, 1975), can be varied in the range $0.2\ \mu m < r_p < 0.5\ \mu m$ with a small geometric standard deviation $\sigma < 1.2$. The aerosol is very rapidly heated by a shock wave and the initiated evaporation process is measured by laser light scattering.

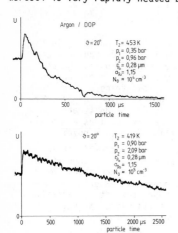

Fig. 2 shows typical results of directly measured scattered light intensities of two individual shock wave experiments. The equilibrium properties and the time scales are different in both cases. The first fast increase in the intensity results from the increase in particle number density because of the shock compression. It is also influenced by other particle transfer processes. The second more extended intensity decrease must be related to the decrease in particle size. Using the MIE theory, a theoretical relation between the measured scattered light flux and the particle size can be derived (Roth and Fischer, 1984)

Figure 2:   Measured scattered light intensities

Results and Discussion

An analytical description of shock waves in two phase aerosol systems is given by Roth and Fischer (1983). This analysis leads in the case of low particle volume concentration to the following system of coupled differential equations for the relaxating quantities (conservation of mass, momentum and energy):

$$\frac{d\,r_p^{\,2}}{dt} = -\frac{2}{3\,k\,r_p}\,j \qquad\qquad \frac{d\,v_p}{dt} = -\frac{1}{k\,r_p^{\,3}}\,f_d$$

$$\frac{d\,T_p}{dt} = -\frac{1}{k\,c_p\,r_p^{\,3}}\,q - j\,r\,(T_p) \qquad k = 4/3\ \pi\rho_p^{\bullet}$$

In these equations j, f_d and q are the fluxes of mass, momentum and heat from the particle to the surrounding gas, $\rho_p^{\bullet}$, c_p are the density and the specific heat capacity of the droplet material and $r(T_p)$ is the specific enthalpy of evaporation.

Apart from the very short relaxation region immediately behind the shock wave, the dominant process is the mass exchange between gas and particle phase. Therefore the only cruxial term is the mass flux j. In the transition regime it can be expressed in the following way:

$$j = j_T = j_c \beta_T (Kn_V, \alpha_j, \ldots)$$

$$j_c = 2 r_p D_{VG} Sh \left[\rho_V (T_p) - \rho_{V\infty} \right]$$

In these equations j_T, j_c are the mass fluxes in the transition and the continuum regime, respectively. D_{VG} is the binary diffusion coefficient, Sh is the Sherwood number which is 2 for neglecting relative motion between the gas and the particle, $\rho_V(T_p)$, $\rho_{V\infty}$ are the vapor densities at the droplet surface and at infinity, β_j is the mass flux correction term in the transition regime and α_j is the accomodation coefficient, usually chosen to be 1.

There are several correction expressions β_j published, based on the "flux-matching" model (Fuchs, 1959), the "flux-resistance" model (Bademosi and Liu, 1971). Sahni's solution of the Boltzmann equation uses the BGK lineraziation (Smirnov, 1964; Fuchs and Sutugin, 1970) and the solution of the Boltzmann equation with the method of Grad was done by Sitarski and Nowakowski (1979).

For testing these expressions it is important to recognize that all the authors use different definitions for the binary diffusion coefficient and the mean free path, depending on the used approximation of the kinetic theory of gases. Therefore it is necessary to test the equations on the same basis as pointed out by Davis and Ray (1978). We use the result of Jeans (1954) for calculating the effective mean free path for the vapor molecules.

$$l_V = (\sqrt{1 + \frac{m_V}{m_G}} \cdot \pi \cdot \sigma_{VG}^2 \cdot n_G)^{-1}$$

The equation of the binary diffusion coefficient for Lennard-Jones molecules was given by Hirschfelder et al. (1954).

$$D_{VG} = \frac{3}{8 \sqrt{\pi}} \cdot \frac{\left[kT \cdot \frac{m_V + m_G}{2 \cdot m_V \cdot m_G} \right]^{1/2}}{\sigma_{VG}^2 \cdot n_G \cdot \Omega_{VG}^{(1,1)}}$$

In these equations m_V, m_G are the mass of a vapor and gas molecule, respectively, n_G is the number of gas molecules per unit volume, k is the Boltzmann constant, σ_{VG} is the mean collision diameter and $\Omega_{VG}^{1,1}$ is the collision integral tabulated by Hirschfelder et al. (1954).

Fig. 3 shows the normalized transitional correction factors as a function of the Knudsen number Kn_V.

All experimental results of more than 100 individual droplet evaporation experiments can be fitted by numerical simulations of the given differential equations. The used fitting parameter is β_j. As demonstrated in Fig. 3, it is between the theoretical solutions of Smirnov (1970) and Sitarski and Nowakowski (1979). In this way our experiments confirm these theoretically derived equations of the transitional correction factor for mass exchange processes.

890

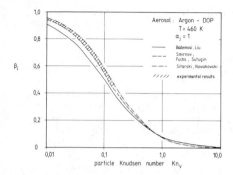

Figure 3: Transitional correction factors of mass transfer

References

Bademosi, F. and Liu, B. Y. H. (1971) "A Universal Law for Transfer Processes in Knudsen Aerosols", Publication No. 157, Particle Technology Laboratory, University of Minnesota.

Davis, E. J. and Ray, A. K. (1978) "Submicron Droplet Evaporation in the Continuum and Non-Continuum Regimes", J. Aerosol Sci. 9:411.

Fischer, R. and Roth, P. (1983) "Evaporation Rates of Submicron Aerosol Droplets at High Temperatures", J. Aerosol Sci. 14:376.

Fuchs, N. A. (1959) "Evaporation and Droplet Growth in Gaseous Media", Pergamon, London.

Fuchs, N. A. and Sutugin, A. G. (1970) "Highly Dispersed Aerosols", Ann Arbor, London.

Hirschfelder, J. O., Curtiss, C. F. and Bird, R. (1954) "Molecular Theory of Gases and Liquids", John Wiley & Sons, New York.

Jeans, J. H. (1954) "The Dynamical Theory of Gases", Dover, New York.

Liu, B. Y. H. and Lee, K. W. (1975) "An Aerosol Generator of High Stability", Am. Ind. Hyg. Ass. J. 36:861.

Roth, P. and Fischer, R. (1983) "Shock Tube Measurements of Submicron Droplet Evaporation", 14. Symp. on Shock Tubes and Waves, Sydney, to be published.

Sitarski, M. and Nowakowski, B. (1979) "Condensation Rate of Trace Vapor on Knudsen Aerosols from the Solution of the Boltzmann equation", J. Colloid Interface Sci. 72:113.

Smirnov, V. I. (1971) "The Rate of Quasi-Steady Growth and Evaporation of Small Drops in a Gaseous Medium", Pure and Applied Geophysics 86:184.

THEORY OF HOMOGENEOUS NUCLEATION—A CHEMICAL KINETIC VIEW

C. H. Yang and H. Qui
Mechanical Eng. Department, S.U.N.Y. Stony Brook
Stony Brook, N.Y. 11794

EXTENDED ABSTRACT

The basic kinetic mechanism of the theory of homogeneous nucleation considers a sequence of bimolecular reactions between a vapor molecule m_1 and a cluster of i vapor molecules m_i:

$$m_1 + m_i \underset{k_{i+1}^v}{\overset{k_i^c}{\rightleftharpoons}} m_{i+1}$$

The steady state flux of nucleation for a supersaturated vapor with a supersaturation ratio S and temperature T is obtained by Becker and Doring as follows:

$$I = \frac{\alpha k^c C_\infty^2 A_1 S_2}{1 + \dfrac{k_2^v}{S\alpha k^c C_\infty} + \dfrac{k_2^v k_3^v}{S^2 (\alpha k^c C_\infty)^2} + \ldots + \dfrac{k_2^v \cdots k_n^v}{S^{n-1}(\alpha k^c C_\infty)^{n-1}}}$$

where α is the accommodation factor, k^c is the condensation rate constant, C_∞ is the concentration of vapor molecules at equalibrium pressure, A_1 surface area of the vapor molecule and k_∞^v is the evaporization rate constant under equilibrium vapor pressure. Write the evaporization rate constant in the Arrhenius form for an i-cluster, E_i, and note that $k_\infty^v = \alpha k^c C_\infty$ at the

$$k_i^v = A e^{-\frac{E_i}{RT}}$$

liquid and vapor interface we have

$$\frac{k_i^v}{\alpha k^c C_\infty} = \frac{k_i^v}{k_\infty^v} = e^{-\frac{E_i - E_\infty}{RT}}$$

Let $\Delta E_i = E_\infty - E_i$ the flux becomes

$$I = \frac{\alpha k^c C_\infty^2 S^2 A_1}{1 + \dfrac{1}{S} e^{\frac{\Delta E_2}{RT}} + \ldots + \dfrac{1}{S^{n-1}} e^{\frac{\sum \Delta E_i}{RT}}}$$

Since the condensation rate may be taken to be the collision rate of clusters with vapor molecules the nucleation flux is determined if the activation energies of the evaporization reaction which varies with the cluster size are known.

892

The Kelvin's relation of cappillerity is used in the classical theory for all size clusters to determine ΔE_i.

$$\log \frac{P_i}{P_\infty} = \frac{2\sigma m}{\rho r_i RT} \qquad i = 2,3\ldots n$$

where P_i = Equilibrium vapor pressure for an i-cluster, m is the molecular weight of vapor molecule and ρ is the density of the liquid phase.

$$RT \log \frac{P_i}{P_\infty} = RT \log \frac{k^c C_{P_i}}{k^c C_\infty} = RT \log \frac{k^v_i}{k^v_\infty} = \Delta E_i = \frac{2\sigma m}{\rho r_i}$$

The assumption that the Kelvin's relation is valid for clusters with only a few molecules cannot be justified. Furthermore, it rigidly ties the rate constants to surface tension which leaves little flexibility for modification of the theory to accommodate all the nucleation data and all the substances tested.

The present work proposes the following form of parameterizing of the activation energies as a function of cluster size:

$$E_i = E_\infty (1 - \frac{\beta}{i^\Gamma})$$

where β and Γ are parameters that may be determined from nucleation data. The activation energy for an infinite size cluster turns out to be the latent heat for vapor-liquid transition prescribed by the Clausius-Clapeyron equation. When

$$\beta = \frac{\alpha m \sigma}{\rho E_\infty} (4\pi\rho/3m)^{\frac{1}{3}} \text{ and } \Gamma = \frac{1}{3}$$

the fitted parameter formalism becomes identical with the classical theory.

Best values of β and Γ for water, benzene and methanol are presented in Figs. 1-3. Critical vapor pressures for different levels of flux rate in a range of temperatures have been calculated with both the classical theory and the fitted parameters approach. They are compared with available data in these figs.

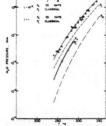

Figure 1. Critical pressure of water vapor for homogeneous nucleation in supersonic nozzles and expansion chamber
  • Yellot[3], ▢ Binnie & Wood[4], △Gyarmathy & Meyer[5],
  O Jaeger et al[6] and ◯ Volmer and Flood[7]

Both the classical theory and fitted parameters predicted critical water vapor pressures close to experimental values of the expansion chamber where the flux I is in the order of 1 molecules, cc^{-1}, S^{-1}. The classical theory predicted much lower critical vapor pressure for the nozzle experiment where the flux is of the order of 10^{15}, molecules, cc^{-1}, S^{-1}.

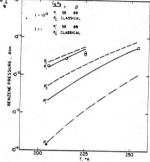

Figure 2. Critical pressure of benzene vapor for homogeneous nucleation in supersonic nozzle and expansion chamber. □ Scharrer[8], △ Dawson[9] and 0 Jaeger[6] et al.

The classical theory predicted higher critical vapor pressure than values obtained from both types of experimental apparatus.

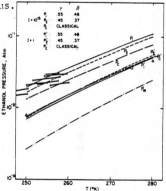

Figure 3. Critical pressure of C_2H_5OH vapor for homogeneous nucleation in supersonic nozzle and expansion chamber
△ Scharrer[8], 0 Volmer & Flood[7] and ⊢—⊣ Jaeger[6] et al.

References

Becker, R. and Doring, W., (1935), Ann. Physik, 24, 719.

Thomson, Sir William (Lord Kelvin), (1870), Proc. Roy. Soc. Edinb. 7, 63.

Yellot, J. L., (1934), Trans. ASME 56, 411.

Bonnie, A., M. and Woods, M. W., (1942), Proc. Inst. Mech. Eng. (London) 181A; 134.

Gyarmathy, G. and Meyer, H., (1965), Spontane Kondensation, Verlag Düseldorf, Germany, 508.

Jaeger, H. L., Wilson, E. J., Hills, P. G. and Russell, K. C. (1969), J. Chem. Phys. $\underline{51}$, 5380 and 5389.

Volmer, M. and Flood, H. Z. (1934), Physik, Chem. (Leipzig), $\underline{170}$, 273.

Scharrer, I. (1939), Am. Phys. $\underline{35}$, 619.

Dawson, D. B. (1967), Report No. 90, MIT, Gas Turbine Lab.

Published 1984 by Elsevier Science Publishing Co., Inc.
Aerosols, Liu, Pui, and Fissan, editors

CALCULATION OF DROPLET EVAPORATION USING ESTIMATES

A. E. Seaver
3M, Central Research Laboratories
Bldg. 208-1-01
St. Paul, MN 55144
USA

EXTENDED ABSTRACT

To calculate droplet evaporation some physical properties of the background gas and vapor must be known. The physical properties of any pure substance depend directly on the nature of the molecules of which that substance consists. Hence a complete understanding of the molecular behavior is needed for accurate calculation of physical properties. Since the understanding of molecular behavior is not yet complete the engineer is forced to either measure physical properties or make reasonable estimates of their values.

Although measured data is sometimes available the engineer often is faced with a situation where very little data on the physical properties may be known. In such situations the estimation of physical properties becomes a necessity.

One purpose of this paper is to describe a method to estimate the properties needed in droplet evaporation calculations. A second purpose is to examine the use of these practical estimates via the analytic equations developed for the Knudsen regime by Fuchs (1959) and by Seaver (1984).

An excellent review on methods of physical properties estimation can be found in the book by Reid et. al. (1977). As they point out the Lennard-Jones 6-12 potential gives one of the more tractable relations for the intermolecular energy between two molecules. The Lennard-Jones 6-12 potential is only valid for non-polar molecules and even then it is based upon rather tenuous theoretical grounds. However, it is the most often used potential and is still usefull for determining the diffusion coefficient provided either the droplet vapor or the background gas is non-polar [Reid, p552]. Droplet evaporation estimates in this paper are based on the Lennard-Jones 6-12 potential. In this potential the two parameters which must be determined are the characteristic size d and characterestic energy E of the molecule. For many background gases these two parameters are already tabulated [Reid et. al. (1977),Hirschfelder et. al. (1954)].

When a chemist synthesizes a liquid the most common information known is the molecular structure. Furthermore, the molecular weight M and mass density p are two properties which are relatively easy to measure. In this paper besides the above propeties it is assumed that the normal boiling point T_b will also be measured. If a boiling point at other than standard pressure is measured the method of Hass and Newton (1982) is suggested to correct the value to T_b.

The method of Lydersen (1955) is used in this paper to estimate the critical point (V_c, P_c, T_c) using T_b, M and a knowledge of the chemical structure of the liquid. Of specific interest are the critical pressure P_c and critical temperature T_c, since the acentric factor w is then estimated [Reid et. al. (1978)] using T_b and T_c. Next the characteristic size d and characteristic energy E are estimated using P_c, T_c, and w with the method of Tee et. al. (1966). Finally, the vapor pressure P_{vp} is estimated using w, P_c, T_c and the method of Pitzer et. al. (1955) with the functions of Lee and Kesler (1975).

In Table 1 estimated values of d, E, and P_{vp} for DBP, DOP and DBS are compared to measured values taken from Davis and his collegues. The characteristic size d has agreement within 5% while the characteristic energy E is within 10% except DOP which is 21% high. Fortunately, the error in estimating E contributes only a small error in droplet evaporation calculations. The estimated values of the vapor pressure P_{vp} are quite poor due to the decreased accuracy of Lydersen's method at higher molecular wieght.

TABLE 1   COMPARISON TABLE OF MEASURED AND CALCULATED PARAMETERS FOR DIBUTYL PHTHALATE (DBP), DIOCTYL PHTHALATE (DOP) AND DIBUTYL SEBACATE (DBS). Measured values from Ray et. al. (1979) for vapor pressure, and from Davis et. al. (1981) for characteristic length, d, and energy,E/k parameters. All vapor pressure values at 20 degrees C.

COMPOUND	PARAMETER	MEASURED	CALCULATED	ERROR
DBP	d (angs)	8.84	8.77	1%
	E/k (K)	672	739	-10%
	P_{vp} (kPa)	2×10^{-6}	5.9×10^{-7}	71%
DOP	d (angs)	9.95	9.81	1%
	E/k (K)	689	832	-21%
	P_{vp} (kPa)	9.9×10^{-9}	8.4×10^{-10}	92%
DBS	d (angs)	9.97	9.5	5%
	E/k (K)	688	741	-8%
	P_{vp} (kPa)	2.4×10^{-7}	3×10^{-7}	-26%

In Table 2 the estimated parameters of Table 1 are used to calculate the evaporation time of a DBS droplet evaporating in the Knudsen regime (see No. 6). The procedure is straightforward

and is discussed elsewhere [Davis et. al. (1981)]. Calculations
are given using both the equation of Fuchs (1959) and the
equation of Seaver (1984). Fuchs' delta is set equal to the mean
free path of the DBS vapor molecule. Also shown in Table 2 are
calculations using the measured values of d, E and P_{vp} of Davis
and his collegues (see No. 1, 2) for two specific d, E values of
the nitrogen background gas. Finally, other evaporation times are
calculated (see No. 3, 4, 5) using various combinations of the
estimated d, E, and P_{vp} with values found in the literature.

TABLE 2   TABLE OF EVAPORATION TIMES AND PRECENT ERROR FOR
           A DBS DROPLET WITH AN INITIAL RADIUS OF 1.05
           MICROMETERS TO EVAPORATE TO 0.502 MICROMETERS.
           A measured time of 880 seconds was found by
           Davis and Ray (1978) for this droplet
           evaporating in nitrogen.

NUMBER	FUCHS TIME (sec)	FUCHS ERROR	SEAVER TIME (sec)	SEAVER ERROR
1	786	10.7%	869	1.3%
2	786	10.7%	869	1.3%
3	784	10.9%	870	1.1%
4	708	19.5%	781	11.3%
5	458	48.0%	505	42.6%
6	611	30.6%	678	23.0%

For N_2
  1 d, E/k values from Davis
  2, 3, 4, 5, 6 d, E/k values from Hirshcfelder

For DBS
  1 d, E/k and P_{vp} values from Davis
  2 d, E/k and P_{vp} values from Davis
  3 d, E/k values from present work; P_{vp} from Davis
  4 d, E/k values from Reid; P_{vp} from Davis
  5 d, E/k values from Reid; P_{vp} from Small
  6 d, E/k and P_{vp} values from present work

For the Seaver equation the error in the completely
estimated time (No. 6) is found to be comparable to the error in
the estimation of the vapor pressure. Very good agreement is
found using the Seaver equation whenever the measured vapor
pressure is used (No 1, 2, 3). This suggests the electroballance
method developed by Ray et. al. (1979) is an excellent method for
measuring the vapor pressure and allows the Clapeyron equation to
be used to compute vapor pressure near the temperature of the
measurement. The Clapeyron equation is known to give a poor
estimate away from the measured temperature, [Reed. et. al.
(1977)].

REFERENCES:

Davis, E. J. and Ray, A. K. (1978) J. Aerosol Sci. 9: 411

Davis, E. J., Ravindran, P. and Ray, A. K. (1981) Adv. Colloid Interface Sci. 15: 1

Fuchs, N. A. (1959) "Evaporation and Droplet Growth in Gaseous Media", Pergamon Press, New York.

Hass, H. B. and Newton, R. F. (1982) CRC Handbook of Chem. and Phys., CRC Press, Ohio [see boiling point, correction].

Hirschfelder, J. O., Curtiss, C. F. and Bird, R. B. (1954) "Molecular Theory of Gases and Liquids", Wiley, New York.

Lee, B. I. and Kesler, M. G. (1975) AIChE J. 21: 510

Lydersen, A. L. (1955) Estimation of Critical Properties of Organic Compounds, Univ. Wisconsin Coll. Eng., Eng. Exp. Stn. Rep. 3, Madison, Wis.

Pitzer, K. S., Lippmann, D. Z., Curl, R. F., Huggins, C. M. and Petersen, D. E. (1955) J. Am. Chem. Soc. 77: 3433

Ray, A. K., Davis, E. J. and Ravindran, P. (1979) J. Chem. Phys. 71: 582

Reid, R. C., Prausnitz, J. M. and Sherwood, T. K. (1977) "The Properties of Gases and Liquids", McGraw-Hill, New York.

Seaver, A. E. (1984) Aerosol Sci. & Tech. 3: No. 2 (in press).

Small P. A., Small, K. W. and Cowley, A. (1948) Trans. Faraday Soc. 44: 810 [see also Ray et. al. (1979)]

Tee, L. S., Gotoh, S. and Stewart, W. E. (1966) Ind. Eng. Chem. Fundam. 5: 356, 363

Published 1984 by Elsevier Science Publishing Co., Inc.
Aerosols, Liu, Pui, and Fissan, editors

TURBULENT COAGULATION OF SMALL AEROSOLS: AN EXPERIMENTAL STUDY

D. L. Swift and W. H. Wells
Aerosol Research Laboratory
The Johns Hopkins University
School of Hygiene and Public Health
Baltimore, Maryland 21205 U.S.A.

EXTENDED ABSTRACT

Introduction

Coagulation of aerosol particles is often an important mechanism in the dynamics of size distributions. There are several specific coagulation processes including Brownian motion, laminar shear, electrical charge attraction, differential setting, and turbulent fluid motion. Most experimental studies of coagulation have been carried out either in still air or in laminar flow, examining the first two processes. Very few experimental studies of coagulation in turbulent flow have been reported, so that there is little basis for predicting turbulent coagulation rates in the atmosphere or in technical aerosol applications. In the study reported herein, turbulent flow was achieved by shear motion in a chamber and results were compared to pipe tubulent studies reported by Yoder and Silverman (1967).

Experimental

Monodisperse wax aerosols (σ=1.2) were produced in a LaMer-Sinclair generator (Swift, 1967) and their aerodynamic diameter determined in a sedimentation cell. Their concentration was from 2×10^5–4×10^5 particles/cm^3. The particles were subjected to turbulent shear flow in a rectangular chamber (1.8m high, .28m deep, and .15m wide) in which 2 counter rotating rubber belts were mounted with 1.25 cm space between belt and wall and 2.5 cm space between belts. In this fashion a uniform shear rate was provided throughout the chamber. The aerosol concentration was determined by CNC before and after a two minute period of coagulation as well as the percentage of doublet particles. The coagulation constant and wall loss constant was calculated from these data according to the method of Yoder and Silverman. Two shear rates were employed in the studies, 84 sec^{-1} and 180 sec^{-1} and particle diameters from 0.50 to 1.4 μm were studied.

Results

Values of the coagulation coefficient, K ($cm^3 \cdot part^{-1} sec^{-1}$) and the wall deposition factor, β (sec^{-1}) were calculated for 28 experimental runs. The coagulation coefficient was plotted against particle diameter and found to be in good agreement with theory for shear or turbulent coagulation, i.e., $k \propto d^3$. The coagulation coefficients ranged from 1×10^{-10} to 8.6×10^{-9}; over the limited range of shear rates, coagulation rates increased linearly both shear rate, also in accord with theory. Considerably greater experimental scatter was observed for the wall deposition coefficients, but the trend with size and shear rate was also in accord with physical principles, i.e. increasing wall deposition with particle size and shear rate.

As overall equation for coagulation coefficient was obtained from the data:

$$k = 24Gd^3$$

A pseudo-Reynolds number, based on belt velocity and spacing was calculated for each shear rate; these values were 950 and 2000 for the low and high shear rate respectively. The average coagulation coefficients based on Reynolds number were compared with the experimental results of Yoder and Silverman (1967), obtained in pipe turbulent studies, and found to be on the same line when plotting log K vs. log Re. Yoder and Silverman's studies were carried out at significantly higher pipe Reynolds numbers.

Conclusions

1. Coagulation rates of small aerosol particles in turbulent shear flow increase linearly with shear rate and with the third power of particle diameter over the range of variable studied experimentally. This is in agreement with theory.

2. Experimental results of coagulation constant, K, when plotted in terms of a pseudo Reynolds number are in reasonable agreement with the experimental values of Yoder and Silverman. The concept of such a Reynolds number for prediction of a coagulation rate seems justified.

3. Studies at larger particle sizes should be performed to confirm size and shear rate dependencies at these particle sizes.

References

Yoder, J. D. and Silverman, L. (June 1967) "Influence of Turbulence on Aerosol Agglomeration and Deposition in a Pipe", Paper 67-33, APCA Meeting.

Swift, D. L. (1967) "A Study of the Size and Monodisparity of Aerosols Produced in a Sinclair-LaMer Generator", Ann. Occ. Hygiene 10:337.

Published 1984 by Elsevier Science Publishing Co., Inc.
Aerosols, Liu, Pui, and Fissan, editors

THE EFFECT OF VAN DER WAALS' AND
VISCOUS FORCES ON AEROSOL COAGULATION

M. K. ALAM
Dept. of Mechanical Engineering
Ohio University
Athens, OH 45701

EXTENDED ABSTRACT

INTRODUCTION

An analysis is carried out to determine the combined effect of Van der Waals' and viscous fluid forces on coagulation of aerosols in the free-molecular, transition and continuum regime. Van der Waals' forces, which result from the interaction of fluctuating dipoles in the particles, increase the rate of aerosol coagulation.

The collision frequency is also modified by viscous interactions between the particles. A particle in motion in a fluid generates velocity gradients around it. This velocity gradient influences the motion of other particles in the fluid. This effect, like the Van der Waals' forces, becomes increasingly important as the particles come closer to each other, and influences the collision rates of the particles.

The analysis considers the effect of the viscous forces by modifying the particle diffusion coefficient (Spielman, 1970) of particles in the continuum size range. An interpolation formula is then used to extend the results into the transition and free-molecular size ranges.

THEORY

The coagulation enhancement factor due to van der Waals forces has been calculated by a number of authors, including Friedlander (1977), Twomey (1977), and Schmidt-Ott, et al. (1982). The collision enhancement factor E_b due to van der Waals forces is given by

$$E_b^{-1} = (a+b) \int_{a+b}^{\infty} \exp\left[\frac{V(x)}{KT}\right] \frac{dx}{x^2} \qquad (1)$$

where a, b are the radii of the particles, and V is the van der Waals interaction potential. E_b is the inverse of the stability coefficient. The above equation does not take into account the effect of viscous interactions between the particles. The motion of two particles become interdependent as they approach each other; velocity gradients generated by one particle influences the motion of the second particle. Such viscous interactions retard the coagulation rate. The forces in these interactions can be calculated from the solutions obtained by Stimson and Jeffery (1926). Spielman (1970) utilized these solutions to incorporate viscous effects in particle coagulation. This was done by modifying the diffusion coefficient in the coagulation equation. This introduces a diffusion ratio D_{12}^{∞}/D_{12} in Eq.(1):

$$E_b^{-1} = (a+b) \int_{a+b}^{\infty} \exp \left[\frac{V(x)}{KT}\right] \frac{D_{12}^{\infty}}{D_{12}} \frac{dx}{x^2} \qquad (2)$$

Spielman's analysis applies to particles in the continuum size range.

In this paper, the above analysis is extended to include particles in the transition and free molecular size ranges. This is achieved by modifying the force on the particles by Sherman's (1963) interpolation formula. The resistance force on a moving particle is then given by

$$\frac{F}{F_c} = \left(1 + \frac{F_c}{F_k}\right)^{-1} \qquad (3)$$

where F_c and F_k are the forces on particles in the continuum and free molecular size ranges, respectively. By incorporating Eq.(3) in the analysis of particle motion, the diffusion coefficients in Eq.(2) become a function of the particle Knudsen numbers.

RESULTS

Calculations were carried out for the following van der Waals interaction potential:

$$V(x) = -\frac{\pi^2 Q}{6} \left[\frac{2ab}{x^2-(a+b)^2} + \frac{2ab}{x^2-(a-b)^2} + \ln \frac{x^2-(a+b)^2}{x^2-(a-b)^2}\right] \qquad (4)$$

The collision enhancement factors obtained are shown in Fig.(1) and Fig.(2). It can be seen that viscous forces reduce the effectiveness of the van der Waals forces in coagulation. This reduction in collision efficiency is more pronounced for smaller Knudsen numbers, and for lower values of van der Waals interaction energy. As an approximation to van der waals interaction potential, the formula in Eq.(4) is not always valid.

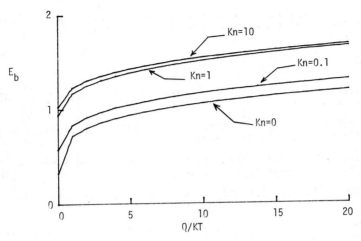

Fig. 1. Collision Enhancement Factor
$Kn=\lambda/a=\lambda/b$. $\lambda=6.5\times10^{-7}m$

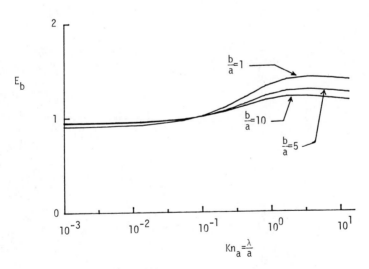

Fig. 2. Collision Enhancement Factor (Q/KT=4)

Results from calculations with more accurate expressions of van der Waals interaction potential are also presented.

REFERENCES

1) Friedlander, S. K. (1977) "Smoke, Dust and Haze," p-184, Wiley, New York.

2) Schmidt-ott, A., and Burtscher, H. (1982), J. of Colloid and Interface Sci., Vol. 89, p-353.

3) Sherman, P. (1963) "Rarified Gas Dynamics, 3rd Symposium," Vol. 2, p-228,Academic Press, New York.

4) Spielman, A. L. (1970), J. of Colloid and Interface Sci., Vol. 33, p-562.

5) Stimson, M., and Jeffery, G. B. (1926) Proc. Roy. Soc. London Ser. A, Vol. 111, p-110.

6) Twomey, S. (1977) "Atmospheric Aerosols," Elsevier, Amsterdam/New york.

Published 1984 by Elsevier Science Publishing Co., Inc.

Aerosols, Liu, Pui, and Fissan, editors

DROPLET COALESCENCE IN TURBULENT PIPE FLOW

F.Ebert, Lehrstuhl für Mechanische Verfahrenstechnik und Strömungsmechanik
Fachbereich Maschinenwesen
Universität Kaiserslautern
Kaiserslautern, FRG

EXTENDED ABSTRACT

Experimental:

The paper concerns turbulent pipe flow of air in which small droplets are suspended. The droplets result from condensation in the mixing zone of two coaxial, confined jets consisting both of nearly water vapor saturated air at temperatures of approx 15^{0}C resp. 65^{0}C (ambient pressure conditions). The mixing zone the length of which amounts to ca. 0,5 m passes into a glass pipe of 80 mm diameter and 8 m length. The Reynoldsnumber of the pipe flow, calculated for saturated mixed air of 50^{0}C, varied between $1,6 \cdot 10^{4}$ and $2,22 \cdot 10^{4}$. As condensation nuclei a narrow fraction of glass ballotinis with mass mean diameter of about 3,5 μm was added to the colder air stream. The number concentration amounted to $5 \cdot 10^{5}$cm^{-3}.

The development of the droplet size distribution was observed by isokinetic sampling at different positions along the pipe and measurement of the droplets in the partial stream by a light scattering analyser. A system with 90^{0} scattering angle was used. The calibration method which is suitable for particles of different materials and shape has been reported elsewhere [1]. The result of experiments shows figure 1. Curve 1 gives the distribution of the condensation nuclei, curve 2 gives the theoretical droplet size distribution based on the assumption that droplet growth is due to condensation exclusively. The curves I, II give the experimental distributions measured at a fixed position but for increasing velocity and consequently lowering residence time. It can be seen that increasing residence time shifts the distributions to greater droplet sizes. From the situation of the curves it can be concluded that the residence time exerts the main influence of droplet growth whereas the turbulence level, depending on the Reynoldsnumber, is of minor importance, at least for the experimental conditions.

Theoretical:

The marked difference between the distribution, calculated for condensation to equilibrium, and the experimental curves suggests that a mechanism, coupled with the flow conditions, must be responsible for the droplet growth. The obvious presumption that turbulence caused coalescence would be the dominant mechanism has been checked by theoretical estimation as follows: One starts with the basic equation for the time dependent droplet distribution, the terms accounting for turbulent particle diffusion included. Wall deposition of droplets is allowed for by the boundary conditions at the pipe wall. Numerical information for the wall deposition is taken from Wood [2]. The turbulence induced droplet diffusion is considered in radial direction only. The particle diffusion coefficient is estimated adopting the proposal of Meek and Jones [3].

906

The kinetic terms in the basic equation can be specialized by different relationships for turbulence - caused coalescence. It can be shown that the models of Saffman and Turner [4] give a lower limit, the model of Abrahamson

[5] an upper limit for the coalescence outcome: With regard to the type of droplet motion the models of Saffman and Turner resp. Abrahamson represent two limiting cases namely. For two droplets being candidates for coalescence the stochastic motion of the gas in their immediate neighbourhood must be highly correlated in the model of Saffman and Turner but uncorrelated in the model of Abrahamson. The first model is valid for very small particles suspended in a flow of low turbulence intensity, the second for more massive particles in flow of high intensity turbulence. The large gap existing between the validity domains of the two models is claimed to be filled by a theory published recently by Williams and Crane [6]. The point is that these authors succeeded to calculate the relative velocity of different particles coming from partial correlated fluid regions.

Accounting for the different models the basic equation was solved by means of a finite difference method. The results are plotted in figure 1 as slightly S-shaped curve covering the experimental distributions. It is to note, notwithstanding the simplifications implied, the theory gives the correct order of magnitude. Further remarks: A computation, neglecting the particle diffusion, gives practically the same result. For the system presented here, the Saffman and Turner model underestimates the coalescence outcome by far. The difference between droplet distribution calculated with the models of Abrahamson respectively Williams and Crane is unexpectedly small.

[1] Büttner, H. (1984), "Messung von Partikelgrößenverteilungen an bewegten dispersen Systemen mit einem optischen Partikelzähler unter Zuhilfenahme von Impaktoren und Meßzyklonen", Preprints, S. 287-303, 3. Europäisches Symposium Partikelmeßtechnik 1984, Nürnberg

[2] Wood, N.B. (1981), "A simple method for the calculation of turbulent depositions to an rough surfaces", J.Aerosol Science 12 : 275

[3] Meek, C.C.; Jones, B.G. (1973), "Studies of the behaviour of heavy particles in a turbulent fluid flow", J.Atmospheric Science 30 : 239

[4] Saffman, P.G.; Turner, J.S. (1956), "On the collision of drops in turbulent clouds", J.Fluid Mechs. 1 : 16

[5] Abrahamson, J. (1975), "Collision rates of small particles in a rigorously turbulent fluid", Chem.Engng. Science 30 : 1371

[6] Williams, J.J.F.; Crane, R.J. (1983), "Particle collision rate in turbulent flow", Int.J.Multiphase Flow 9 : 421

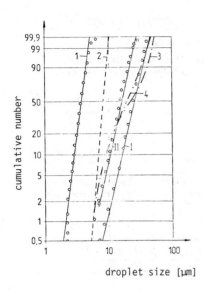

droplet size [µm]

1: Condensation nuclei(glass spheres)
2: Theory, condensation only
3: Theory, turbulent coalescence(Abra-
          hamson)
4: Theory, turbulent coalescence(Willi-
          ams, Crane)

Experimental :

I: Re = $1{,}6 \cdot 10^4$, t_{res}= 2,5 sec.
II: Re = $2{,}22 \cdot 10^4$, t_{res}= 1,86 sec.

Published 1984 by Elsevier Science Publishing Co., Inc.
Aerosols, Liu, Pui, and Fissan, editors

908

ACOUSTIC DEPOSITION AND AGGLOMERATION OF MONODISPERSE AEROSOL

D. Boulaud, C. Malherbe and G. Madelaine

Laboratoire de Physique et Métrologie des Aérosols, IPSN/DPT/SPIN,
Commissariat à l'Energie Atomique, B.P. n° 6, 92260 Fontenay aux Roses,
France

EXTENDED ABSTRACT

Introduction

The acoustic agglomeration of the fine fraction of an aerosol is a
process wherely the average diameter of this aerosol is increased and the
numerical concentration reduced on exposure to an acoustic field. This
change in the particle size distribution improves the collecting efficiency
of downstream cleaning systems since these larger particles are more easily
separated by conventional methods (cyclones, sedimentation chambers,
granular beds, ...).

Acoustic agglomeration has already been used successfully to size
condition an aerosol. Most of these experiments are summed up in EP
Mednikov"s book (1965). The subject has lately been revised and a model
developed by D.T. Shaw (1978) and K.H. Chou and al. (1981) to predict the
efficiency of acoustic agglomerators.

These latter studies incorporate recent findings on the strong
renforcement of agglomeration mechanisms at acoustic intensities of about
160 dB (1 W/cm^2), duc to increased turbulence in the fluid.

On the basis of these investigations our laboratory is now engaged in
an experimental study of acoustic agglomeration in order to determine the
experimental sensitivity of these processes towards different parameters
(sound pressure level, acoustic frequency, median diameter and standard
deviation of the aerosol size distribution, residence time in the
agglomeration chamber).

1. PREVIOUS RESULTS

The results previously obtained on the acoustic agglomeration of a
polydisperse aerosol of mass concentration between 0.5 and 1 g/m^3 can be
summarized in the following way (Boulaud and al. (1983)) :

- Little variation in the size distribution is observed between 100 and
140 dB.

- The mass median diameter increases from 1.5 to 4.5 µm between 140 and
155 dB reaches a maximum between 155 and 158 dB then drops suddenly when the
acoustic intensity is raised further.

- The influence of acoustic frequency on the variation of the size
distribution is insignificant in the 500 to 1000 Hz range.

During these experiments, large aerosol deposits were observed on the agglomerator walls and led us to undertake a specific study of particle precipitation in a acoustic field. A preliminary description of these experiments has been given during the GAeF meeting held in Munchen (1983) (Boulaud and al. 1983) here a more extended description is given.

2. ACOUSTIC PRECIPITATION

The aim here is to determine the relations between the establishement of an acoustic field and the precipitation of an aerosol on the walls of a pipe. For these experiments the aerosol is monodispersed and has a low numerical concentration (under 50 particles cm^{-3}) to reduce the collision frequency when the acoustic field is applied.

The experimental sep up and the experimental procedure have already been described (Boulaud and al. 1983).

From the fraction of particles deposited on the wall (an acoustically induced deposition velocity (Ka) may be deduced as follows :

$$K_a = - \frac{UD}{4L} \, Ln(C/C_o)$$

where U is the flow rate of the fluid in the tale, D the tube diameter L its length, C and C_o the concentration respectively downstream and upstream.

More generally it is useful to relate acoustic deposition velocity to particles diameter through dimensionless quantities. For this purpose a dimensionless acoustic deposition velocity is introduced, defined by :

$$K_a^+ = K_a / (\epsilon . \nu)^{1/4}$$

ν being the cinematic viscosity and ϵ the energy dissipated per unit time and fluid mass owing to turbulences induced by the acoustic field.

To describe the inertial behaviour of aerosols caused by acoustically induced turbulences, dimensionless relaxation time is introduced, defined by :

$$\tau^+ = \tau . (\epsilon/\nu)^{1/2}$$

where τ is the particle relaxation time.

All the experimental results being calculated in dimensionless form K_a^+ can be plotted against τ^+ if the different values of are known.

If we choose for ϵ the relation given by Mednikov (1965)

$$\epsilon = \left(I^{3/2} f \right) / \left(\rho_g^{3/2} c_g^{5/2} \right)$$

where I is the acoustic intensity, f the frequency, ρ_g the fluid mass per unit volume and C_g the acoustic wave propagation velocity in the fluid, we obtain the following expression relating K_a^+ to τ^+ by a power function :

$$K_a^+ = 0,031 \; \tau_+^{0,471}$$

This dimensionless relation obtained with a correlation coefficient of 0.834 offers a first empirical approcach to acoustic deposition phenomena.

The quantitative understanding of these mechanisms request the exact knowledge of ε, so to attain this end an experiment is now undertake to perform a direct determination of ε.

References

Boulaud, D., Frambourt, C., Madelaine, G., Malherbe, C., (1983) "Experimental study on the acoustic agglomeration and precipation of an aerosol", GAeF meeting, Munchen, september, 1983.

Chou, K.H., Lee, P.S., Shaw, D.T. (1981) J. Colloid and Interface Sci., 83, 335.

Mednikov, E.P. (1965) "Acoustic coagulation and precipitation of aerosols", Translated from Russian by Larrick, C.V. Consultants Bureau, New York.

Shaw, D.T. (1978) "Recent developments in Aerosol Science" (Edited by Shaw, D.T.) Wiley, New York.

Published 1984 by Elsevier Science Publishing Co., Inc.

Aerosols, Liu, Pui, and Fissan, editors

911

CHANGE IN SIZE DISTRIBUTION OF TWO-COMPONENT
ULTRAFINE AEROSOLS UNDERGOING BROWNIAN COAGULATION

K. Okuyama, Y. Kousaka and K. Hayashi
Department of Chemical Engineering
University of Osaka Prefecture
Sakai 591 Japan

EXTENDED ABSTRACT

Introduction

Brownian coagulation of ultrafine aerosol particles consisting of two different species has been experimentally studied in a laminar pipe flow. The obtained experimental results have been compared with the solution of population balance equation, and the influence of van der Waals force on coagulation rate between different species has been evaluated.

Experimental apparatus and method

Figure 1 shows the experimental apparatus used in the coagulation experiment. High concentration $ZnCl_2$, NaCl and Ag aerosol particles having geometric mean diameters of 7-70nm and geometric standard deviations around 1.4 were generated separately, and two of these three aerosols were mixed to obtain two-component aerosols. The aerosols having one or two peaks thus obtained were continuously flowed through the brass pipe of 11.24m in length and of 2cm in diameter in a laminar state. The particle densities of $ZnCl_2$, NaCl and Ag particles are 2.91, 2.16 and 10.5 g/cm^3, respectively. The particle size distributions at the inlet and the outlet of the coagulation pipe were measured by means of a differential mobility analyzer(DMA) combined with a mixing type condensation nucleus counter(CNC). The particle number concentrations at the inlet and the outlet of the pipe were counted independently by an ultramicroscopic system after enlarging their sizes in a particle size magnifier(PSM).

Figure 1. Experimental
apparatus

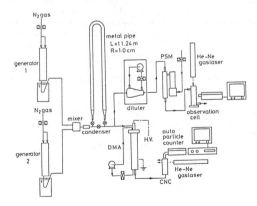

Experimental results and discussion

Figure 2(a) shows the experimental size distributions of $ZnCl_2$-NaCl aerosol system at the inlet and the outlet of the pipe by the frequency against particle diameter. It is seen that the particle size distribution shifts toward larger particle diameter accompanying the decrease in particle number concentration, but keeping two peaks. And in the case of $ZnCl_2$ aerosol having smaller particle sizes, the coagulation is found to proceed faster than the larger NaCl aerosol. Figure 2(b) shows the change in size distribution of $ZnCl_2$ aerosol which was obtained by supressing the generation of NaCl aerosol. Comparing the smaller tail of the size distribution of $ZnCl_2$ aerosol in Fig. 2(b) with that of $ZnCl_2$-NaCl aerosol system in Fig. 2(a) at the outlet of the pipe, considerable numbers of smaller $ZnCl_2$ particles in the latter are found to coagulate with larger NaCl particles. Figure 3 shows the experimental results of Ag-NaCl aerosol particles whose particle densities are largely different. Similar tendency as Fig. 2 can be seen in this experimental results.

The dashed lines in the figures indicate the calculated results for the outlet of the pipe obtained by numerically solving the population balance equation for simultaneous Brownian coagulation and diffusive deposition of aerosol in a laminar flow. In this calculation, the solid curves were used as the inlet particle size distributions, and the Brownian coagulation rate function proposed by Fuchs(1964) was used. Since the Fuchs' coagulation rate function depends on particle density, in the case of two-component aerosol, its particle density is assumed to be that of smaller particles($ZnCl_2$ particles in $ZnCl_2$-NaCl aerosol system and Ag particles in Ag-NaCl aerosol system). As seen from the figures, coagulation is found to proceed faster than the calculated result. In the case of single component aerosols, this enhancement of coagulation rate is considered to be caused by the van der Waals force between two colliding particles, which was qualitatively explained using the Marlow's coagulation rate function(Marlow, 1980) and Hamaker's formula as the van der Waals force in the previous study(Okuyama et al., 1984). On the other hand, in the case of two-component aerosols, since the particle densities are not same, the deviation of the experimental results from the dashed lines is considered to be caused by the both effects of van der Waals force and particle density on coagulation rate function. In order to account for these effects, a correction factor β was multiplied with Fuchs' coagulation rate function and the fit value of β was determined by comparing the calculated results obtained under different values of β with the experimental data. The values of β for single component aerosol agreed well with those obtained in the previous study(Okuyama et al., 1984).

Figure 4 shows the values of β obtained for two-component aerosols. The values of the abscissa show the ratios of geometric mean diameter of lager aerosol, d_{pL}, to that of smaller aerosol d_{ps}. It is seen that the factors β of Ag-NaCl aerosol system are larger than those of $ZnCl_2$-NaCl aerosol system, and they tend to become smaller with the ratio d_{pL}/d_{ps}. The solid lines in the figure indicate the calculated factors by the next equation.

$$\beta = (\ R_{ss}^{M} + R_{sL}^{M} + R_{LL}^{M}\ )/(\ R_{ss}^{F} + R_{sL}^{F} + R_{LL}^{F}\ )$$

Figure 2. Change in size
distribution of
$ZnCl_2$–NaCl
aerosol system

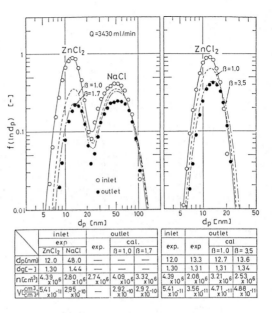

| | inlet | | outlet | | | inlet | outlet | | |
| | exp. | | exp. | cal. | | exp. | exp. | cal. | |
	$ZnCl_2$	NaCl		$\beta=1.0$	$\beta=1.7$			$\beta=1.0$	$\beta=3.5$
d_p[nm]	12.0	48.0	—	—	—	12.0	13.3	12.7	13.6
σ_g[—]	1.30	1.44	—	—	—	1.30	1.31	1.31	1.34
n[cm³]	4.39×10^6	2.80×10^6	2.74×10^6	4.09×10^6	3.32×10^6	4.39×10^6	2.08×10^6	3.21×10^6	2.53×10^6
$V[\frac{cm^3}{cm^3}]$	5.41×10^{-11}	2.95×10^{-10}	—	2.92×10^{-10}	2.92×10^{-10}	5.41×10^{-11}	3.56×10^{-11}	4.71×10^{-11}	4.88×10^{-11}

Figure 3. Change in size
distribution of
Ag–NaCl
aerosol system

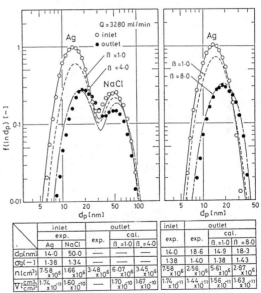

| | inlet | | outlet | | | inlet | outlet | | |
| | exp. | | exp. | cal. | | exp. | exp. | cal. | |
	Ag	NaCl		$\beta=1.0$	$\beta=4.0$			$\beta=1.0$	$\beta=8.0$
d_p[nm]	14.0	50.0	—	—	—	14.0	18.6	14.9	18.3
σ_g[—]	1.38	1.34	—	—	—	1.38	1.40	1.38	1.43
n[cm³]	7.58×10^6	1.66×10^6	3.48×10^6	6.07×10^6	3.45×10^6	7.58×10^6	2.56×10^6	5.61×10^6	2.97×10^6
$V[\frac{cm^3}{cm^3}]$	1.74×10^{-11}	1.60×10^{-10}	—	1.70×10^{-10}	1.67×10^{-10}	1.74×10^{-11}	1.44×10^{-11}	1.56×10^{-11}	1.63×10^{-11}

$$R_{ss}^{X} = \frac{1}{2} \int_{r_{smin}}^{r_{smax}} \int_{r_{smin}}^{r_{smax}} K_B^X(r_s, r_s; \rho_{ps}, \rho_{ps}) n(r_s) n(r_s) dr_s dr_s$$

$$R_{sL}^{X} = \int_{r_{smin}}^{r_{smax}} \int_{r_{Lmax}}^{r_{Lmax}} K_B^X(r_s, r_L; \rho_{ps}, \rho_{pL}) n(r_s) n(r_L) dr_s dr_L$$

$$R_{LL}^{X} = \frac{1}{2} \int_{r_{Lmin}}^{r_{Lmax}} \int_{r_{Lmin}}^{r_{Lmax}} K_B^X(r_L, r_L; \rho_{pL}, \rho_{pL}) n(r_L) n(r_L) dr_L dr_L$$

X = M or F

$n(r_s)$ is the particle number distribution function of the smaller aerosol at the inlet of the pipe, $K_B^F(r_s, r_L; \rho_{ps}, \rho_{pL})$ indicates the Fuchs' coagulation rate function between the particle of radius r_s with the density ρ_{ps} and the particle of radius r_L with the density ρ_{pL}, and $K_B^M(r_s, r_L; \rho_{ps}, \rho_{pL})$ the corresponding Marlow's coagulation rate function where Hamaker's formula was used as the van der Waals force. As the value of Hamaker's constant, A=8.98 x 10^{-13} erg for NaCl and ZnCl2 particles and A=4.00 x 10^{-11} erg for Ag were used. In the case of two-component aerosol, the Hamaker's constant was calculated as the geometrical mean value. It is seen that the calculated results can qualitatively explain the experimental data.

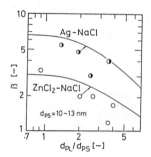

Figure 4. Coagulation rate correction factor of two-component aerosol system

Conclusion

Brownian coagulation of two-component ultrafine aerosols in transition regime has been experimentally studied in a pipe flow, and the effects of particle density, particle size distribution and van der Waals force on coagulation rate are found to be qualitatively explained using the Marlow's coagulation rate function and Hamaker's van der Waals force.

References

Fuchs, N. A. (1964) "The Mechanicas of Aerosols" Pergamon, New York.

Marlow, W. H. (1980) "Derivation of Aerosol Collision Rate for Singular Attractive Contact Potentials", J. Chem. Phys. 73:6284.

Okuyama, K., Y. Kousaka and K. Hayashi (1984) "Change in Size Distribution of Ultrafine Aerosol Particles Undergoing Brownian Coagulation", J. Colloid Interface Sci. in press.

Published 1984 by Elsevier Science Publishing Co., Inc.
Aerosols, Liu, Pui, and Fissan, editors

BROWNIAN COAGULATION OF CHAINLIKE PARTICLES

M. T. Cheng and D. T. Shaw
Laboratory for Power and Environmental Studies
State University of New York at Buffalo
Buffalo, New York

EXTENDED ABSTRACT

Introduction

Chainlike aggregates are frequently encountered in combustion processes. They have received an increasing interest in the programs related to industrial emission control. Dynamically, chainlike particles behave differently from sperical particles. Being asymmetrical in shape, fine chainlike aggregates diffuse with translational and rotational motion. Little work has been done on Brownian coagulation of such aggregates.

The objective of this work is to experimentally study Brownian coagulation of chainlike aerosol particles. A theoretical model (Lee, 1984) employing the concept of an effective collision diameter and assuming equal a priori probability for all orientations of an aggregate will be discussed and compared with the experiment.

Experimental

The coagulation experiment has been conducted in a flowing system. Chain aggregates of Iron Oxide (Fe_2O_3) are generated by introducing Iron Pentacarbonyl ($Fe(CO)_5$) vapor into a laminar carbon monoxide flame. By this technique, the primary particle diameter can be controlled in the range of 0.02 μm to 0.07 μm. The aggregates consist of up to 100 of sperical primary particles.

The generated aggregates pass through a Kr^{85} neutralizer into a Differential Mobility Analyzer (DMA) and then neutralized again before entering a coagulation tube. For a fixed primary particle diameter, the length of the aggregates are controlled by varying the voltage of DMA from 2 KV to 10 KV. Size distribution of aerosols are determined using a thermal precipitator with the aid of a transmission electronmicroscope (TEM). Measurements of changes in particle number concentration after flowing through the coagulation tube are performed with TSI Model 3020 Condensation Nucleus Counters (CNC). The particle number concentration ranges from 1.0×10^5 to 4×10^5 cm^{-3}. The coagulation time varies from 2 to 4 minutes.

Results

Experimental results of coagulation constants of Fe_2O_3 chainlike aggregates in the range of mean primary particle numbers between 27 and 75 are summarized in Table 1. The primary particle size distribution and the mean number of primary particles per aggregate were determined from TEM micrographs. The measured coagulation rates were about an order of magnitude higher than those of volume equivalent sperical particles. The coagulation constants show an increasing trend with mean number of primary particles. This is expected since the effective collision diameter increases with the increasing number of primary particles. The difference between experimental and theoretical prediction for the coagulation constants (Table 1) are minimized when the polydispersity and slip effect are considered in the model.

Table 1. Experimental Data for Coagulation Constants of Fe_2O_3 Chainlike
          Aggregates.

d_g [a] (µm)	n [b]	N_o [c] ($\times 10^5$ cm^{-3})	N_o/N_t	t [d] (sec)	k_{exp} [e] ($\times 10^{-9}$ cm^{3}sec^{-1})	K_{exp}/K_s [f]
0.018	30	3.93	1.09	162	1.47	4.97
0.018	64	2.10	1.11	162	3.16	10.68
0.026	50	3.15	1.11	162	2.10	7.09
0.026	75	2.37	1.09	162	2.32	7.84
0.031	27	3.33	1.21	218	2.85	9.62
0.031	45	2.40	1.09	218	1.74	5.87
0.031	75	1.04	1.11	218	4.69	15.85
0.033	33	3.25	1.12	162	2.34	7.91
0.033	45	2.90	1.15	162	3.28	11.08
0.033	56	2.05	1.14	162	4.28	14.46
0.039	50	2.30	1.13	197	2.93	9.90
0.064	38	3.37	1.10	197	1.47	4.96
0.064	45	1.88	1.13	197	3.39	11.45

(a) d_g = Primary particle geometric mean diameter
(b) n = Mean number of primary particles per aggregate
(c) N_o = Initial particle number concentration
(d) t = Coagulation time
(e) K_{exp} = Experimental coagulation constant
(f) K_s = Coagulation constant for spherical particles in continuum regime
              = 2.96×10^{-10} cm^3 sec^{-1}

References

Lee, P.S. and Shaw, D.T. (1984) "Dynamics of Fibrous-Type Particles: Brownian Coagulation and the Charge Effect," Aerosol Science and Technology 3:9-16.

Published 1984 by Elsevier Science Publishing Co., Inc.
Aerosols, Liu, Pui, and Fissan, editors

AEROSOL GROWTH FROM VAPOUR CONDENSATION IN VAPOUR-GAS MIXTURES

C. F. Clement
Theoretical Physics Division
A.E.R.E. Harwell
Didcot, Oxon OX11 ORA
England

EXTENDED ABSTRACT

The formation and growth of aerosols by condensation is important in many contexts, including those where a dangerous condensible vapour could be emitted such as the possible overheating of a reactor core. The coupling of aerosol growth to heat and mass transfer processes in vapour-gas mixtures is being investigated and many results have already been obtained (Clement, 1982 and 1984; Clement and Taylor 1984). We describe here how a growth in a saturated vapour-gas mixture at constant pressure is characterised by dimensionless numbers. The theory applies when radiative heat transfer is relatively unimportant.

Mass diffusion and heat conduction control evaporation and condensation on the aerosols considered so that a crucial role is played by the Lewis number:

$$Le = \frac{k}{D\rho\bar{c}_p} = \frac{\text{rate of heat transport by conduction}}{\text{rate of mass transport by diffusion}}$$

where
k = thermal conductivity of the mixture
D = vapour-gas diffusivity
ρ = total density
$\bar{c}_p$ = mean density-weighted specific heat of the mixture.

A second dimensionless number specifying the growth rate may be termed the 'condensation number' (Clement 1982 and 1984):

$$Cn(p,T) = \frac{k}{LD\rho c_E'(T)} = \frac{\text{rate of latent heat removal by conduction}}{\text{rate of mass transport}} \quad ,$$

L = latent heat of vaporisation
$c_E(T) = \rho_{VE}(T)/\rho(T)$ = equilibrium vapour concentration
$c_E'(T) = dc_E/dT$
p = total pressure.

This number is approximately proportional to p and varies rapidly with temperature through the rapid variation in the equilibrium vapour density, $\rho_{VE}(T)$.

At small concentrations with the neglect of temperature variations of all quantities other than c_E, the aerosol growth rate per unit volume is

$$\dot{m}_v = \frac{1}{L} \frac{Cn}{(Le+Cn)(1+Cn)} \left[- (1-Le) \; \nabla.\underset{\sim}{q}_t - \frac{Cn+Le}{Cn+1} \frac{c_E''(T)}{c_E'(T)} \; \underset{\sim}{q}_t.\nabla T \right] \quad ,$$

where
$$\underset{\sim}{q}_t = -k \; \nabla T - LD\rho \; \nabla c_E$$
is the total heat transfer current which would be observed at a wall.

The final term in $\dot{m}_v$ is always positive and dominates for sufficiently large temperature differences, ΔT. With $Le = 1$ the first term is larger for small ΔT, and the sign of $\dot{m}_v$ is then determined by the signs of $1-Le$ and $\nabla.\underset{\sim}{q}_t$. For a fixed q_t, maximum values of $\dot{m}_v$ occur in the temperature region where $Cn \approx 1$.

The sensitivity to Le implies that accurate transport data for k and D are needed to specify $\dot{m}_v$, and even whether an aerosol will grow or evaporate in some circumstances. We discuss the cases of water vapour in air and sodium vapour in argon, whose values of Le and Cn are shown in table 1. In many cases, values of D are not known accurately enough to be able to predict aerosol formation rates in vapour-gas mixtures.

Table 1. Lewis and Condensation Numbers for Water Vapour-Air
and Sodium Vapour-Argon Mixtures at a Total Pressure,
$p = 1$ Atm

H_2O - air		Na - Ar	
$Le \approx 0.85$	$0 - 60°C$	$Le \approx 1$	$300 - 600°C$
$Le \approx 0.54$	$100°C$	$Le > 1$	$< 300°C$
		$Le < 1(?)$	$> 600°C$
$Cn \approx 1.25$	$0°C$	$Cn \approx 10^4$	$200°C$
$Cn \approx 1$	$4°C$	$Cn \approx 1$	$520°C$
$Cn \approx 0.1$	$50°C$	$Cn \approx 0.1$	$700°C$
$Cn \approx 0.01$	$100°C$		

References

Clement, C.F. (1982) "Aerosol Growth in Vapour-Gas Mixtures Cooled Through Surfaces", AERE-TP.897.

Clement, C.F. (1984) "Aerosol Formation from Heat and Mass Transfer in Vapour-Gas Mixtures", AERE-TP.1003.

Clement, C.F. and Taylor, A.J. (1984) "Aerosol Growth by Steam Condensation in a PWR Containment", AERE-TP.1048.

CLOUD FORMATION IN LOW GRAVITY UNDER
THERMAL/PRESSURE WAVE FORCING

D. A. Bowdle and M. T. Reischel
(Universities Space Research Association)
B. J. Anderson, V. W. Keller, and O. H. Vaughan
(National Aeronautics and Space Administration)

Atmospheric Sciences Division
Systems Dynamics Laboratory
NASA/Marshall Space Flight Center
Huntsville, Al 35812
U.S.A.

EXTENDED ABSTRACT

Introduction

Several important cloud/aerosol microphysical processes – such as droplet growth and evaporation and gas-to-particle conversion in growing and dissipating clouds, fogs, and hazes – occur in the atmosphere over time scales of minutes to days or more. Such phenomena are therefore difficult to study in earth-based cloud chambers, where sedimentation and convection usually limit the usable observation time for experiments with free-floating droplets. On the other hand, mechanical, acoustical, and electrostatic droplet supports allow long observation times, but they also introduce perturbations which may be large enough to mask the phenomena under study.

The thermal wave experiment is designed to evaluate a completely new low gravity technique for detailed study of these processes. The initial investigations involve cloud microphysical processes with time scales of several seconds. These proof-of-concept experiments are being conducted in low gravity conditions (usually less than $\pm$ 10 mG for about 20 sec) during ballistic parabola maneuvers on the National Aeronautics and Space Administration (NASA) KC-135 aircraft. Follow-on experiments with durations up to 10 days are possible on NASA's Space Shuttle.

Experiment

In the basic thermal wave experiment, controlled cooling of a dry surface generates a cloud of water droplets on polystyrene latex (PSL) microsphere nuclei in an initially isothermal air parcel at water saturation. Controlled heating then evaporates the droplets. Since the experiments are performed in low gravity and in a sealed chamber, this technique allows extended observations of cloud microphysical processes in a stationary, free-floating droplet field, in an isolated, stagnant air parcel. Under these conditions, droplet growth and evaporation are regulated only by thermal and moisture boundary conditions, by molecular diffusion of heat and water vapor, and by bulk pressure fluctuations and airflows arising from temperature and concentration gradients. Time-registered measurements of the chamber air pressure, chamber wall temperature, thermal wave surface temperature, and the concentration and critical supersaturation of the PSL nuclei are input to XNADIA, a one-dimensional, time dependent numerical model of the relevant physical processes. Predictions of cloud density and the position of the cloud/clear air interface from XNADIA are compared with microdensitometric traces from time-registered 35 mm photographs of the experimental clouds. Model parameters are then adjusted to optimize the agreement between model predictions and experimental results.

The basic thermal wave experiment can be modified for investigations of longer duration microphysical processes, comparisons with more traditional ground-based cloud microphysical studies, and performance evaluation of the thermal wave chamber or the XNADIA model. Controllable variables include the moisture boundary condition, geometry, and dimensions of the thermal wave surface; the inclusion of controlled expansions or compressions: and the polarities, ramp rates, shapes, durations, and timing of the thermal waves and pressure waves. Various combinations of carrier gases, condensible gases, and other trace gases are also possible.

The chamber is flushed with filtered dry air and then filled with monodisperse PSL nuclei (usually $0.234~\mu m$ diameter) from a Collison atomizer. Nucleus concentrations usually range from about 300-1000 cm^{-3} at 0.1% supersaturation. The nuclei-laden air is allowed to still for several minutes to achieve thermal, moisture, and momentum equilibrium. Temperature gradients are minimized by passive thermal control. Chamber air is brought to water saturation by a wetted filter paper lining. Two types of thermal wave chambers have been successfully flown on the KC-135 (Table 1).

Results

Photographs of cloud dissipation in low gravity by the thermal waves in each chamber are shown in Fig. 1. Note the sharp, regular cloud/clear air interfaces, indicative of minimal motion. These results demonstrate the feasibility of the thermal wave experimental method.

Figure 2 shows the results of an XNADIA simulation for a sinusoidal thermal wave with initial cooling on a partially wetted thermoelectric module (TEM) surface. The wave was centered around 17.65 C, with an amplitude of 5.65 C, a period of 10 sec, and an accompanying expansion of -0.3 mb/sec. Fogging of the TEM surface was simulated by introducing "dummy" droplets (100,000 nuclei cm^{-3} active at 5% supersatuation) in the thermal wave generating surface. The growing edge of the simulated cloud lasts considerably longer than the actual cloud, particularly near the TEM surface. Model outputs of auxiliary data show that this mismatch is produced in the simulation by a weak supersaturation pulse which propagates into the chamber after the dummy droplets evaporate during the thermal wave heating cycle. On the other hand, if the TEM is treated as completely dry, an even larger supersaturation pulse is produced by the evaporation of the droplets in the chamber near the TEM. Simulations with a wide range of surface moisture parameters show that only a rather narrow range of parameters is capable of matching both the growth and evaporation lines simultaneously.

To minimize the dramatic effects of uncertainties in the moisture boundary conditions on the thermal wave surface, subsequent experiments have used either a fully wetted TEM surface or a fine hot wire (Table 1). Simulations of the wet TEM showed improved agreement with measurements. Analysis of the recent hot-wire experiments is in progress.

Conclusions

The primary result of this study to date is the identification of strong non-linear interactions between adjacent thermal, water vapor and water droplet fields. Interactions of this type are likely to generate significant spatial and temporal inhomogeneities in the droplet fields of natural clouds, and possibly even conventional laboratory cloud chambers.

Acknowledgements

This work was supported under National Aeronautics and Space Administration Contract NAS8-33882. The invaluable assistance of Robert Shurney and of the KC-135 flight crew is gratefully acknowledged. Myron Plooster developed the original XNADIA model. Rhonda Blocker performed the model runs.

Table 1. Thermal Wave Cloud Chamber Designs

Chamber Parameter	Original Chamber	New Chamber
a. Chamber Body		
Thermal Control	Thermal Inertia	Multiple Lamina, Thermal Conductors/Insulators
Cavity Diameter	5.0 cm	14.0 cm
Cavity Length	25.0 cm	28.0 cm
b. Thermal Wave		
Wave Generator	Thermoelectric Module (TEM) on Cavity Side-wall (2 cm x 2 cm Borg-Warner 940-31)	Axial Hot Wire, Metal Oxide Insulation, 1/16" Outer Diameter, 28 cm long Stainless Steel Sheath
Wave Geometry	Three-dimensional Cartesian (One-dimensional near TEM)	One-dimensional Cylindrical
Wave Control	Sinusoidal Heating/Cooling	Rapid Heating, Passive Cooling
Wave Temperature (Copper/constantan Thermocouple)	Fine Bead on TEM Surface	Fine Butt-welded Junction in Hot-wire Sheath
Moisture Boundary Condition	Dry/Fogged, or Wet	Dry
c. Experimental Clouds		
Cloud Formation	Sealed Chamber Cooling of Dry TEM Heating of Wet TEM	Controlled Expansion
Cloud Photography (one/sec)	30° Forward Scattering	90° Scattering

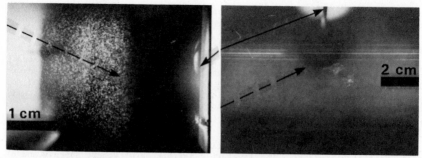

Figure 1. Photographs of water droplet cloud dissipation by a thermal wave in
low gravity. Bars show spatial scales; bright arcs are background
lighting. a) Original chamber, with thermoelectric module (solid
arrow) and plane wave front (dashed arrow). b) New chamber with
hot wire (solid arrow) and cylindrical wave front (dashed arrow).
Horizontal streaks are film scratches.

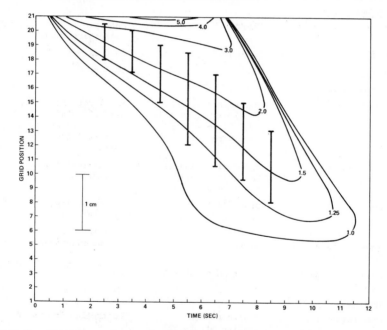

Figure 2. Comparison of experimental measurements (heavy bars) and XNADIA
model predictions (contours of droplet radius in μm) of cloud development in space and time during low gravity thermal wave experiment. Spatial scale is shown by light bar. Optical detection
limit was between 1 μm and 1.5 μm radius.

Published 1984 by Elsevier Science Publishing Co., Inc.

Aerosols, Liu, Pui, and Fissan, editors

EXPERIMENTAL STUDIES OF COAGULATION OF FREE MOLECULE AEROSOLS

Chong S. Kim[1] and Benjamin Y.H. Liu[2]

[1]Aerosol Research Laboratory
Division of Pulmonary Disease
Mount Sinai Medical Center
Miami Beach, FL 33140
[2]Particle Technology Laboratory
Mechanical Engineering Department
University of Minnesota
Minneapolis, MN 55455
U.S.A.

EXTENDED ABSTRACT

For a highly dispersed aerosol, the coagulation is probably the most important process governingthe dynamic behavior of the aerosol system. The rate of coagulation for monodisperse arosols is given by the classical Smoluchowski's equation as

$$\frac{dN}{dt} = -KN^2$$

where K is the coagulation coefficient and N is the total number concentration of aerosol. The value of K for a free molecule aerosol may be obtained from the kinetic theory of dilute gases. However, validity of this has not been thoroughly tested.

In the present study, coagulation coefficients of free molecule aerosols have been measured with sodium chloride aerosols ranging in size between 0.0023 and 0.0063 um in diameter (Kn = 19 to 57). The test aerosols were generated by rapidly mixing a heated NaCl vapor laden jet with clean room air concentrically. Particle size was varied by changing concentration of NaCl vapor in the jet and jet-air mixing ratio. Size distribution of the aerosols were measured by diffusion battery method utilizing three different size diffusion tubes (0.47 cm i.d.), viz 50, 150 and 250 cm in length in conjunction with a GE condensation nuclei counter (CNC). The CNC was overhauled thoroughly prior to use and counting efficiency was established for particles as small as 0.002 um diameter. The coagulation tube was in the form of a 3.91 cm i.d. aluminum tube with a total length of 207 cm. Sampling probes were positioned along the axis of the tube and located at 25 and 202 cm from the tube inlet, making a coagulation distance of 177 cm. Test aerosols were introduced into the coagulation tube at 5 and 10 Lpm and coagulation times for these flow rates were found 9.5 and 20 sec experimentally. During this time period, diffusion loss of particles in the coagulation tube was negligible and diffusion layer remained very near the tube surface without affecting axial region of the tube. Therefore, concentration difference between the two probes could solely be related to coagulation. Sample aerosols from the inlet and outlet probes were diluted up to 1000 times, and maintained at concentration levels below 10^5 particles/cm^3 before being measured by CNC. After integrating and rearranging the Smoluchowski's equation, the coagulation coefficient could be calculated by the following equation

$$K = \frac{1-N_2/N_1}{N_2/N_1} - \frac{1}{N_1 t}$$

where N_1 and N_2 are the aerosol concentration at the inlet and outlet of the coagulation tube. N_1 and N_2 were then obtained from the following equations,

$$N_1 = N_1' \, z/s_1 \, P_1$$

$$N_2 = N_2' \, z/s_2 \, P_2$$

where N_1' and N_2' are the measured aerosol concentrations, z is the dilution ratio, s is the counting efficiency of the CNC and P is the aerosol loss factor by diffusion between the sampling probe and CNC.

Results of 22 experiments showed that coagulation coefficients for particles in the Knudson number range of 19 to 57 ranged between 5×10^{-10} and 9.5×10^{-10} cm³/sec (Figure 1). These values were an order of magnitude lower than those predicted by the continuum theory with Cunningham slip correction which was supported by some recent studies (Davies, 1979; Rooker and Davies, 1979), but were in alignment with the data of Fuchs and Sutugin (1965) and 20 to 200% higher than those predicted by the kinetic theory of diluted gases.

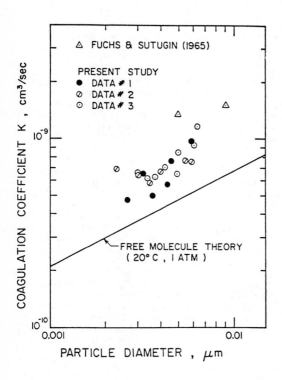

Figure 1: Experimental K values of free molecule aerosols compared with the predicted values from the kinetic theory of dilute gases

References

Davies, C.N. (1979) "Coagulation of aerosols by Brownian motion", J. Aerosol Sci. 10: 151.

Rooker, S.J. and Davies, C.N. (1979) "Measurement of the coagulation rate of a high Knudsen number aerosol with allowance for wall losses", J. Aerosol Sci. 10: 139.

Fuchs, N.A. and Sutugin, A.G. (1965) "Coagulation rate of highly dispersed aerosols", J. Colloid Sci. 20: 492.

CHAPTER 21.

GAS-PARTICLE INTERACTION

Published 1984 by Elsevier Science Publishing Co., Inc.

Aerosols, Liu, Pui, and Fissan, editors

INVESTIGATION OF THE REACTION BETWEEN SINGLE AEROSOL
ACID DROPLETS AND AMMONIA GAS

Glenn O. Rubel
US Army Armament, Munitions and Chemical Command
Chemical Research and Development Center
Aberdeen Proving Ground, MD 21010
U. S. A.

and

James W. Gentry
Department of Chemical Engineering
University of Maryland
College Park, MD 20783
U. S. A.

EXTENDED ABSTRACT

Introduction

One of the earlier investigations of the reaction between acid droplets
and ammonia gas was conducted by Robbins and Cadle (1958). Determining the
aerosol ammonium ion concentration as a function of downstream distance in an
aerosol flow reactor, the extent of reaction of the aerosol was determined as
a function of particle size for discrete times. The extent of reaction, for a
fixed time, increased with decreasing particle size and the initial reaction
rate was governed by a second order surface phase reaction. However surface
phase reaction models, with constant velocity coefficients, could not predict
the entire reaction history of the aerosol. In general the experimental
reaction rate was significantly smaller than that predicted by a surface phase
reaction model which employed the aforementioned velocity constant. Changing
the aerosol carrier gas from pure nitrogen to a mixture of helium and nitrogen,
no measurable difference in the aerosol reaction rates was observed. It was
concluded the later stages of reaction was not gas phase diffusion but internal
particle diffusion controlled.

Also using an aerosol flow reactor, Huntzicker, Cary and Ling (1980)
studied the reaction between H_2SO_4 droplets and NH_3 gas for relative humidities
between 8 and 80%. The ambient NH_3 gas pressure was roughly two orders of magnitude lower than those values used by Robbins and Cadle. Furthermore as
opposed to the conclusion of Robbins and Cadle, the reaction dynamics was
controlled by gas phase diffusion and internal particle diffusion. No evidence
of surface phase reaction was found. It is the intent of this study to investigate the reaction dynamics between acid aerosol droplets and ammonia gas.
Understanding the role that surface phase reaction, gas phase diffusion and
internal particle diffusion play in the aerosol reaction dynamics will be of
central importance.

Experiment

A new experimental method is employed to study the aerosol reaction dynamics continuously. The continuous monitoring of the aerosol reaction dynamics permits the identification of transistion points in the reaction history which are associated with catastrophic events as particle crystallization.

The central component of the experimental system is an electrodynamic balance chamber which stabilizes single droplets at the null point of an alternating current driven electric field. Continuously mixing NH_3 laden gas into the suspension chamber, the single droplet reaction dynamics is monitored by balancing the droplet weight against a direct current electric field. The rate at which NH_3 molecules are transported to the droplet is determined from the rate at which the balancing voltage increases. Employing a 35 mm. objective, the system has the additional capability to monitor the time dependent morphology of the particle and so permit correlating specific morphology changes to specific rate controlling processes.

Figure 1 shows the reaction dynamics of phosphoric acid droplets for NH_3 partial pressures varying from 115 to 1000 ergs/c.c. For such a range

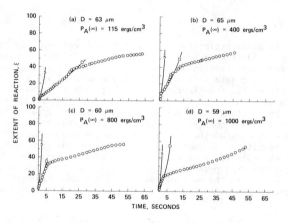

Figure 1. Reaction dynamics of phosphoric acid droplets in the presence of ammonia gas

of conditions the characteristic time for surface phase reaction is intermediate to those values corresponding to the experiments of Robbins and Cadle and Huntzicker, Cary and Ling. It is found the reaction dynamics are sequentially controlled by surface phase reaction, gas phase diffusion and internal particle diffusion. As the NH_3 partial pressure increases the time of particle crystallization decreases. If the NH_3 partial pressure is high enough, the reaction dyanamics are controlled by an initial surface phase reaction and a subsequent internal particle diffusion controlled reaction. As the NH_3 partial pressure is reduced, the gas phase diffusion process plays a more dominant role.

Being able to suspend the particle indefinitely in the electrodynamic chamber, the final extent of reaction of the acid droplet is measured as a function of particle size and ammonia gas pressure. Figure 2 shows the final extent of reaction increases with increasing particle size and decreasing ammonia gas pressure. It appears the final extent of reaction is dependent on the previous reaction dynamics of the acid droplets.

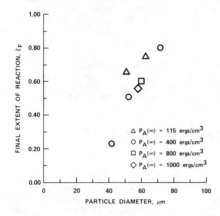

Figure 2. Final extent of reaction of acid droplets as a function of particle size and NH_3 pressure

References

Robbins, R.C. and Cadle, R.D., J. Phys. Chemistry, 62, 469 (1958)

Huntzicker, J.J., Cary, R.A., and Ling, C., Env. Sci. and Tech., 14, 819(1980)

Published 1984 by Elsevier Science Publishing Co., Inc.

Aerosols, Liu, Pui, and Fissan, editors

MEASUREMENT OF THE KINETICS OF SOLUTION DROPLETS IN THE
PRESENCE OF ADSORBED MONOLAYERS: DETERMINATION OF WATER
ACCOMMODATION COEFFICIENTS

Glenn O. Rubel
US Army Armament, Munitions and Chemical Command
Chemical Research and Development Center
Aberdeen Proving Ground, MD 21010
U. S. A.

and

James W. Gentry
Department of Chemical Engineering
University of Maryland
College Park, MD 20783
U. S. A.

EXTENDED ABSTRACT

Introduction

The evaporation rate of water through insoluble monolayers has been of
considerable interest for the past fifty years. Langmuir and Schaefer(1943)
developed an experimental technique for the simultaneous measurement of the
evaporation resistance and surface pressure of plane surfaces of water covered
with compressed monolayers. The researchers demonstrated the evaporation resistance increased with surface pressure and with the chain length of fatty
acid monolayers. Furthermore it was found rapid changes in the evaporation resistance with increasing surface pressure correlated well with the rapid
changes in the surface pressure with decreasing molecular area. Performing
a series of compression/expansion cycles, the same researchers demonstrated a
hysteresis loop in the evaporation resistance-surface pressure curves for the
long chain fatty acid monolayers.

Using a similar experimental technique, Archer and La Mer(1954) studied
the evaporation rate of water through fatty acid monolayers directing particular attention to the employment of specific spreading solvents. It was concluded for the fatty acids ranging in carbon number from 17 to 20, the evaporation resistance is independent of surface pressure and subphase PH in the
liquid condensed state. In disagreement with the work of Langmuir and Schaefer,
Archer and La Mer detected no sharp rise in the evaporation resistance with
surface pressure and suggested impurities as the cause for the pressure dependent resistance in the liquid condensed state. In the present work a new
experimental technique is established for the measurement of the kinetics of
solution droplets in the presence of adsorbed monolayers with the objective
to identify discontinuities in droplet mass transport rates associated with
variations in the monolayer surface coverage.

Experiment

The experimental system was employed by Rubel and Gentry(1983) to measure
the evaporation and growth kinetics of phosphoric acid droplets in continuous-

ly mixed humidified/dehumidified chambers. By stabilizing charged solution
droplets in an electrodynamic chambers, it was demonstrated the droplet's
kinetics deviate from isothermal continuum mass transfer theory by less than
5%. The ability to study the kinetics of solution droplets under isothermal
conditions permits the accurate determination of the water accommodation
coefficient without the complicating effects of temperature dependent vapor
pressures and thermal accommodation coefficients being introduced.

Introducing a surfactant gas, in this case hexadecanol, into the humidified air stream, evaporation and growth of the solution droplets occur with a
concomitant compression and expansion of the monolayers. The water accommodation coefficient is determined as a function of time by fitting experimental
mass histories to those theoretically predicted from a flux matching theory.
Figure 1 shows a representative evaporation run for a phosphoric acid droplet
in the presence of hexadecanol gas. At a critical coverage, corresponding to

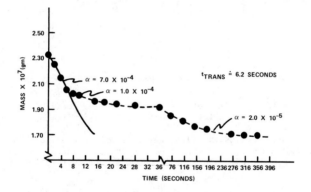

Figure 1. Evaporation of phosphoric acid droplet in the
presence of hexadecanol gas

a phase transistion in the monolayer from the liquid condensed to the solid
state, the water accommodation coefficient sharply decreases from a value of
7.0×10^{-4} to a value of 1.0×10^{-4} and subsequently decreases as evaporation
continues to compress the monolayer. Initiating the growth cycle, the water
accommodation coefficient increases smoothly as the solid monolayer is expanded and then abruptly increases as the surface coverage attains its
critical value. The arguement that the sharp change in the evaporation resistance is due to the presence of impurities is contradicted by these findings.

Employing mass transfer theory for both the volatile solvent(water) and
for the two dimensional surface phase (hexadecanol), the surface coverage is
predicted as a function of time. Consequently the surface coverage is correlated with the water accommmodation coefficient for surface coverages between 0.80 and 1.00. Figure 2 shows a comparison between the accommodation
coefficient-surface coverage dependence determined from droplet mass transfer

934

rates(R) and from bulk liquid mass transfer rates(L). Data sets are in agree-

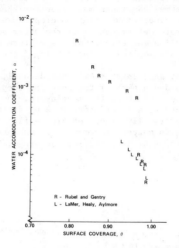

Figure 2. Comparison of water accommodation coefficient determined
from droplet and bulk mass transfer rates

ment for surface coverages exceeding the critical coverage of o.976, however
for smaller surface coverages the accommodation coefficient in the present
study greatly exceeds the values reported by La Mer et. al.(1964).

References

Langmuir,I.L. and Schaefer,V.J., J. Franklin Institute, 235,119 (1943)

Archer,B.J. and La Mer,V.K., J. Physical Chemistry, 59,200 (1954)

Rubel,G.O. and Gentry,J.W., Particulate Science and Technology (In press)

La Mer,V.K., Healy,T.H. and Aylmore,L.A.G., J.Colloid Science, 19, 673 (1964)

Published 1984 by Elsevier Science Publishing Co., Inc.

Aerosols, Liu, Pui, and Fissan, editors

PREDICTION OF AEROSOL CONCENTRATIONS RESULTING FROM A BURST OF NUCLEATION

D. R. Warren and J. H. Seinfeld
Chemical Engineering Department
California Institute of Technology
Pasadena, CA 91125
U. S. A.

EXTENDED ABSTRACT

Introduction

Homogeneous nucleation theory attempts to predict the instantaneous rate at which new particles will be formed. For many applications, including urban atmospheric modeling and smog chamber experiments, the more important question is how many particles ultimately will result from a burst of nucleation, as well as when and at what size they will be formed. This study will attempt to answer these questions by combining classical nucleation theory and particle growth theory to predict the number of particles produced in a spatially uniform system.

We shall consider a system consisting of a condensable vapor and aerosol. The system state is given by the temperature T, total pressure P, the vapor concentration (either number concentration N_v or mass concentration $M_v = m_1 N_v$), and the aerosol size distribution. For simplicity, the aerosol size distribution may be roughly described by its first two moments, the total aerosol number concentration N_p, and the total aerosol mass concentration M_p. We assume that the rates of three processes are of interest: vapor generation R_G, homogeneous nucleation R_J, and condensation R_C, which may be expressed in terms of number of molecules per unit volume per unit time. Mass and number balances for the system yield the following:

$$dN_v/dt = R_G - g_s R_J - R_C \qquad (1)$$

$$dN_p/dt = R_G \qquad (2)$$

$$dM_p/dt = m_1 g_s R_J + m_1 R_C \qquad (3)$$

The somewhat arbitrary diameter at which nucleated clusters are considered to be particles will be denoted by d_s, corresponding to g_s molecules. The exact value of g_s is not crucial, so long as it is a bit above the largest critical number at which significant nucleation can occur. A total mass balance for this system, which may be described as a continuously-reinforced batch reactor, gives $d(M_v + M_p)/dt = m_1 R_G$. The value of the vapor source rate R_G is assumed to be known, and, for simplicity, constant.

Nucleation and Growth Rates

Classical homogeneous nucleation theory expresses nucleation rate $(cm^{-1}sec^{-1})$ as

$$R_J = \left[\frac{\varsigma \pi d_c{}^2 N_v}{\sqrt{2\pi m_1/kT}} \right] \left[\frac{2v_1}{\pi d_c{}^2} \sqrt{\frac{\sigma}{kT}} \right] \left[N_v \exp\left(-\frac{\pi d_c{}^2 \sigma}{3kT} \right) \right] \qquad , \qquad (4)$$

where the critical diameter $d_c = 4\sigma v_1/kT \, lnS$, σ is the surface tension, and v_1 is the molecular volume of the nucleating species. The accomodation coefficient ς will be assumed to be unity.

The rate of condensation is a sum of the rate of condensation onto all particles contained within the system volume. For a particle of diameter d_p, the rate at which molecules of vapor condense onto it is given by

$$R_{C_p} = N_s \frac{\bar{c}_1}{4} \pi d_p{}^2 \left(S - e^{d_K/d_p} \right) f(Kn) \quad , \tag{5}$$

where the saturation ratio $S = N_v/N_s$ and the mean molecular speed $\bar{c}_1 = \sqrt{8kT/\pi m_1}$. The characteristic Kelvin diameter d_K equals $2\pi d_1{}^2 \sigma/3kT$, where d_1 is the apparent molecular diameter (for a sphere of volume v_1). The size regime interpolation function $f(Kn)$ is defined to go to unity in the kinetic limit where $Kn \to \infty$. The Knudsen number Kn is defined by $Kn = (2\lambda_1/d_p)(3D/\lambda_1\bar{c}_1) = 6D/\bar{c}_1 d_p$. This definition of the Knudsen number allows the use of a simple size regime interpolation formula $f(Kn)$, such as the well-known Fuchs and Sutugin expression, which adequately approximates the behavior of more rigorous transition regime formulae. Integrating over the size distribution, where $n(d_p)$ is the number density function, provides the total condensation rate R_C.

It is now convenient to nondimensionalize mass and number concentrations, as well as the rates, by scaling with respect to the saturated vapor. The vapor concentration will simply be expressed by the saturation ratio S. The aerosol is described by $M = M_p/m_1 N_s$ and $N = N_p/N_s$. A dimensionless surface tension σ^* may be defined by $\sigma^* = 2\pi d_1{}^2 \sigma/3kT$, so $d_K = \sigma^* d_1$. A characteristic monomer-monomer collision rate R_β is given by $R_\beta = N_s{}^2 \pi d_1{}^2 \bar{c}_1/4$. This allows the rate expressions to be written as

$$R_J = R_\beta S^2 \sqrt{\sigma^*/6\pi}\, e^{-\sigma^{*3}/2ln^2 S} \tag{6}$$

$$R_C = \frac{R_\beta}{N_s} \int_{d_s}^{\infty} \frac{d_p{}^2}{d_1{}^2} \left(S - e^{\sigma^* d_1/d_p} \right) f(6D/\bar{c}_1 d_p) n(d_p)\, \mathrm{d}d_p \quad . \tag{7}$$

The equation for R_C may be expressed in terms of the number mean diameter $\bar{d}_p$ or the dimensionless number mean diameter $\bar{d}_r = \bar{d}_p/d_1 = (M/N)^{1/3}$, provided that a correction factor α, somewhat less than unity, is introduced to account for the polydispersity. Defining the dimensionless (molecular) Knudsen number as $Kn^* = 6D/\bar{c}_1 d_1$, the condensation rate may be expressed as

$$R_C = \alpha R_\beta \left(S - e^{\sigma^*/\bar{d}_r} \right) \bar{d}_r{}^2 f(Kn^*/\bar{d}_r) N \quad , \tag{8}$$

Assuming that the vapor generation rate R_G is constant, the differential equations may expressed more simply in dimensionless time $\tau = t/\tau_G$ by introducing the time scale $\tau_G = N_s/R_G$, the source regeneration time for the saturated state. Eqs. (1)-(3) now may be written in dimensionless form as

$$dS/d\tau = 1 - g_s J/R^* - C/R^* \tag{9}$$

$$dN/d\tau = J/R^* \tag{10}$$

$$dM/d\tau = g_s J/R^* + C/R^* \quad , \tag{11}$$

where C and J are the nondimensional forms for the rates of condensation and nucleation, respectively, and are given by

$$J = \frac{R_J}{R_\beta} = S^2 \sqrt{\frac{\sigma^\star}{6\pi}} e^{-\sigma^{\star 3}/2ln^2 S} \qquad (12)$$

$$C = \frac{R_C}{R_\beta} = \alpha \left(S - e^{\sigma^\star/\bar{d}_r} \right) \bar{d}_r^{\,2} f(Kn^\star/\bar{d}_r)N \qquad (13)$$

The dimensionless source rate $R^\star$ is defined by $R^\star = R_G/R_\beta$. As mentioned by Warren and Seinfeld (1984), classical homogeneous nucleation theory will break down for $R^\star$ values greater than approximately unity, as the vapor concentration will change too rapidly for a steady state cluster profile to develop, and cluster-cluster or cluster-aerosol collisions also may become significant.

The set of three simultaneous ordinary differential equations given by Eqs. (9)-(11) are soluble numerically for any given initial conditions. The relevant physical parameters of the system reduce to three nondimensional groups, $R^\star$, $\sigma^\star$, and $Kn^\star$. Of the three, $R^\star$ may be varied by changing the source rate, and $Kn^\star$ may be varied by changing the pressure of the system, while $\sigma^\star$ is intrinsic to the compound of interest (at a given temperature).

To solve the set of differential equations as posed, it is necessary to use a sensible value of g_s, which has been taken as 200 for the simulations here. If g_s is too small (less than the critical number), the freshly nucleated aerosol will not grow as it should; a larger than necessary value of g_s does not significantly affect the results, except for $R^\star$ approaching unity (or larger), where nucleation ceases to be steady state and the droplet current begins to show a cluster size dependence. An approximation for α is also required. The simplest is $\alpha = 1$, which corresponds to a monodisperse aerosol distribution, and will be used in these simulations. It can be shown that, for any given N and M, a monodisperse aerosol maximizes C, and thus $\alpha < 1$ for a polydisperse distribution. But even for very broad aerosol size distributions, $\alpha > 0.6$; only when the bulk of aerosol mass and number occur in different, well-separated modes can α become substantially smaller than one. In such cases where the number and mass concentrations are dominated by two different modes, a pair of differential equations corresponding to Eqs. (10) and (11), may be set up for each mode. (The homogeneous nucleation term appears only in the smallest size mode.)

Simulation Of Aerosol Evolution

The nondimensional model has been used to simulate the evolution of aerosol in an initially aerosol-free system for varying values of the dimensionless parameters $R^\star$, $\sigma^\star$ and $Kn^\star$. The model will not be applicable when $R^\star$ is greater than unity or when $Kn^\star$ is less than unity, since classical nucleation theory requires that the clusters be in steady state with the vapor concentration and that cluster growth occur in the free molecule regime.

Figure 1 shows the dimensionless aerosol number N resulting as a function of dimensionless source rate $R^\star$ for values of $\sigma^\star$ ranging from 4 to 12, with $Kn^\star = 100$. A larger $R^\star$ requires a larger C to halt the increase in S, so N becomes larger. A larger $\sigma^\star$ drastically decreases J for a given S, without changing C, so understandably the resulting N is also smaller. It is significant to note that N shows a dependence close to $R^\star$ to the 1.4 power over most of the plot, which is apparently characteristic for the case of homogeneous

938

nucleation competing with condensational growth in the continuum regime. Since peak nucleation rates typically occur with S and M around 10^1, the mean Knudsen number should be unity (transition regime) for N equal to 10^{-8}, and this is about where the Figure 1 curves become noticeably nonlinear. Other simulations for very large Kn^* where the peak nucleation rate competes with free molecule regime condensational growth reveal that N then goes with R^* to about the 2.5 power. Regardless of the growth regime of the freshly nucleated aerosol, if $R^* < 1$, the vast majority of vapor which goes to the aerosol phase condenses onto supercritical nuclei rather than homogenously nucleates. Simulations also show that the resulting N will go inversely with Kn^* raised to approximately the 1.4 power when nucleation competes with continuum condensation, while the resulting N is, of course, independent of Kn^* when nucleation competes with free molecule aerosol growth.

A sufficiently large quantity of initial aerosol will lead to a rapid rate of condensation, preventing the saturation ratio from rising high enough to allow homogeneous nucleation. Simulations show that if the initial N is about one-tenth or more of the resulting N for an initially aerosol-free system with identical dimensionless parameters, nucleation is almost totally suppressed. Figure 2 illustrates this dependence of resulting dimensionless aerosol number on initial dimensionless aerosol number for various R^*, where the initial aerosol is taken to be monodisperse with a dimensionless diameter $\bar{d}_r = 100$. Thus the proper amount of initial seed aerosol will suppress homogenous nucleation and minimize the number of particles resulting in the system.

References

Warren, D.R., and Seinfeld, J.H. (1984) "Nucleation and Growth of Aerosol from a Continuously Reinforced Reactor", *Aerosol Sci. Technol.* 3:xxx.

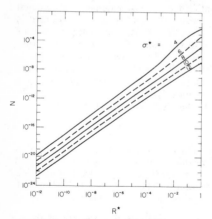

Figure 1. Predicted dimensionless aerosol number concentration as a function of dimensionless source rate for various dimensionless surface tensions and a dimensionless Knudsen number of 100.

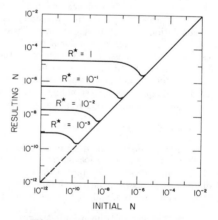

Figure 2. Predicted dimensionless aerosol number concentration as a function of initial dimensionless number concentration for various dimensionless source rates with $\sigma^* = 8$ and $Kn^* = 100$.

Published 1984 by Elsevier Science Publishing Co., Inc.

Aerosols, Liu, Pui, and Fissan, editors

PHOTOCHEMICAL FORMATION OF ORGANIC AEROSOLS: GROWTH LAWS AND MECHANISMS

P.H.McMurry
Particle Technology Laboratory
University of Minnesota
111 Church Street Southeast
Minneapolis, MN 55455-01111

and

D. Grosjean
Daniel Grosjean and Associates, Inc.
Suite 645, Paseo Camarillo
350 North Lantana Street
Camarillo, CA 93010

EXTENDED ABSTRACT

Experiments were conducted to study secondary aerosol formation by chemical reactions involving organics, oxides of nitrogen, and ammonia. The organics that were studied included O-cresol, 1-heptene, 1-butene, and dimethyl sulfide. Aerosol size distributions were measured as a function of time with an electrical aerosol analyzer (0.01 < particle diameter < 0.56 µm) and with a single particle optical counter (0.48 < particle diameter < 3.0 µm).

Aerosol size distribution data were analyzed with the growth law technique. With this approach, measured changes in aerosol size distributions with time are used to determine particle growth rates. The diameter dependence of these growth rates is compared with expected behavior for several different conversion mechanisms. It is concluded that aerosol formation is due to gas phase chemical reactions (rather than reactions that take place on or within aerosol droplets), and that the data are qualitatively consistent with condensational growth provided that the effect of particle curvature on vapor pressure (i.e. the Kelvin effect) is accounted for. A comparison between measured and theoretical particle diameter growth rates is shown in Figure 1.

In fitting theory for condensational growth to the data, it is possible to infer saturation ratios, S, and saturation vapor concentrations, N_S, for the condensing species. A summary of these values for all experiments that were performed is presented in Table I. Note that saturation vapor concentrations varied from about 10^8 to 10^{12} molecules cm^{-3} (3×10^{-9} to 3×10^{-5} torr).

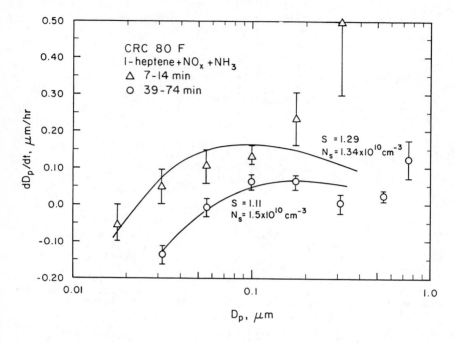

Figure 1. Comparison of measured and theoretical particle diameter growth rates at two times during one experiment. Note that growth rates of small particles are negative. This shows that small particles were evaporating due to the Kelvin effect.

Table 1

SUMMARY OF KELVIN EFFECT GROWTH LAW ANALYSES

Chemical System	Run No.	T_0, min$^+$	T_f, min$^+$	Time at Start$^\beta$	Ambient Temp, °C	N_s, cm^{-3}*	S^Δ
1-heptene + NO_x	36	24	68	11:44	25	$(4.2+3.1) \times 10^8$	$1.64+1.55$
	41 port2	195	231	14:35	21	$(1.7 \pm 0.4) \times 10^{10}$	1.13 ± 0.33
		6	33	11:06	21	$(1.3 \pm 0.9) \times 10^9$	$1.46 + 1.41$
		101	204	12:51	23	$(9.6 \pm 2.4) \times 10^8$	1.57 ± 0.61
		204	428	14:24	22	$(1.8 \pm 0.3) \times 10^9$	1.32 ± 0.27
	42 port2	67	94	11:56	22	$(1.4 \pm 0.5) \times 10^9$	$1.43 + 0.74$
		171	204	13:40	24	$(3.7 \pm 0.8) \times 10^9$	1.34 ± 0.39
	43 port1	28	61	11:19	21	$(7.1 + 4.4) \times 10^8$	$1.57 + 1.25$
		159	192	13:30	23	$(7.1 \pm 1.1) \times 10^9$	1.19 ± 0.42
1-heptene + O_3	30	32	71	11:57		$(2.9 \pm 0.6) \times 10^{11}$	1.04 ± 0.30
O-cresol + NO_x	38	14	21	11:07		$(2.7 \pm 0.9) \times 10^{10}$	1.24 ± 0.54
		148	176	13:21		$(1.4 \pm 0.3) \times 10^{11}$	1.02 ± 0.32
	48	39	61	12:24		$(4.2 + 3.5) \times 10^{10}$	1.08 ± 1.27
	117	77	99	13:54		$(4.5 \pm 1.9) \times 10^{11}$	$1.03 + 0.61$
	118	74	110	12:56		$(1.0 \pm 0.2) \times 10^{12}$	1.03 ± 0.25
1-heptene + NO_x + NH_3	80	3	37	10:17		$(3.1 \pm 0.5) \times 10^{10}$	$1.10 + 0.25$
		37	112	10:53		$(2.8 \pm 0.3) \times 10^{10}$	$1.09 + 0.18$
		66	152	11:19	30	$(6.5 \pm 1.2) \times 10^{10}$	1.06 ± 0.27
	80F	7	14	13:13		$(1.3 \pm 0.3) \times 10^{10}$	$1.29 + 0.39$
		17	39	13:23		$(1.9 \pm 0.3) \times 10^{10}$	1.15 ± 0.38
		35	49	13:41		$(2.5 \pm 0.5) \times 10^{10}$	1.12 ± 0.32
		39	74	13:45		$(1.5 \pm 0.6) \times 10^{10}$	1.11 ± 0.60
	81	2	64	10:02		$(5.2 \pm 1.1) \times 10^9$	2.23 ± 0.54
		64	108	11:04		$(7.5 \pm 1.2) \times 10^{10}$	1.10 ± 0.25
		71	112	11:13	30	$(2.3 \pm 0.4) \times 10^{11}$	1.06 ± 0.24
	82	-35	-1	9:38		$(4.8 \pm 0.6) \times 10^{10}$	1.08 ± 0.20

SUMMARY OF KELVIN EFFECT GROWTH LAW ANALYSES
- CONTINUED -

Chemical System	Run No.	T_0, min	T_f min	β Time at Start	Ambient Temp, °C	N_s, cm^{-3} *	Δ S
	82F	34	65	12:37		$(8.7\pm1.4)\times10^9$	1.17 ± 0.27
		34	118	12:37		$(5.3\pm0.6)\times10^9$	1.16 ± 0.18
	95	0	21	11:15		$(2.9\pm1.1)\times10^{10}$	1.11 ± 0.60
		0	32	11:15		$(4.2\pm0.9)\times10^{10}$	1.11 ± 0.40
		10	43	11:37		$(4.2\pm0.8)\times10^{10}$	1.10 ± 0.31
DMS + NO$_x$ (dry)	39						
		60	104	12:19		$(4.5\pm0.8)\times10^{11}$	1.04 ± 0.25
		154	213	13:53		$(7.0\pm1.2)\times10^{11}$	1.02 ± 0.24
	65	53	85	12:10	25	$(6.0\pm1.0)\times10^{10}$	1.07 ± 0.25
	67	-2	3	10:52		$(4.7\pm0.5)\times10^{10}$	1.43 ± 0.25
		-2	31	10:52		$(1.1\pm0.1)\times10^{10}$	1.51 ± 0.18
	68	-2	22	10:28	26	$(2.7\pm0.4)\times10^{10}$	1.27 ± 0.28
	77	16	44	10:41		$(1.3\pm1.1)\times10^{10}$	1.28 ± 1.47
		69	86	11:33		$(3.6\pm0.8)\times10^{10}$	1.22 ± 0.39
	77F	16	67	15:26		$(2.0\pm0.5)\times10^{10}$	1.09 ± 0.39
DMS + NO$_x$ (50 RH)	74	50	83	12:35		$(4.3\pm0.8)\times10^9$	1.39 ± 0.35
		67	83	12:50		$(7.7\pm1.3)\times10^9$	1.29 ± 0.29

[+]T_0 and T_f are measurement times for the initial and final size distributions that were used in the growth law analyses. For photochemical experiments they represent time after the cover was removed. For dark reactions (1-heptene + ozone) they represent time after reactants were first mixed. Where the average of several size distributions was used for a characteristic size distributions, these are average times.

[β]Time of day at which the initial distribution was measured. If several distributions were averaged to obtain a characteristic initial distribution, this is the time of the first distribution.

[*]Saturation vapor concentration of condensing molecules from least squares analysis.

[Δ]Saturation ratios for condensing vapor from least squares analysis.

Published 1984 by Elsevier Science Publishing Co., Inc.

Aerosols, Liu, Pui, and Fissan, editors

PHOTOLYTIC AND RADIOLYTIC AEROSOL FORMATION IN A CONTINUOUSLY STIRRED TANK
REACTOR ; EXPERIMENTS AND MODEL CALCULATIONS (*)

F.RAES and A.JANSSENS
Nuclear Physics Laboratory RUG,
Proeftuinstraat 86, 9000 GENT (Belgium)

EXTENDED ABSTRACT

The question whether ions and ionizing radiation may play a role in the
formation of aerosol particles in the atmosphere is examined.

A SO_2- NO_2- synthetic air mixture is irradiated with U.V. light (300-
450 nm) or both U.V. light and (^{60}Co) radiation. The gas mixture is contained
in a flow reactor of the completely stirred type (V = 3l.), its mean residence (and irradiation) time $\bar{t}$ can be varied between 0.5 and 20 min. The gases
are supplied from gas cylinders (L'Air Liquide : N2 > 99.9992%, O2 > 99.995%,
SO_2 and NO_2 > 99%).
    The steady state concentration of the aerosols formed in the reactor is
measured with a continuous condensation nucleus counter (TSI model 3020). The
counter is coupled to what we call our diffusion carrousel (Raes et al, 1983);
this is a new design for a screen type diffusion battery which allows the determination of particle diameters between 2 and 30 nm.
    The photolyticaly formed aerosol particles are thought to be H_2SO_4- H_2O
droplets. H_2SO_4 being formed by the reaction of SO_2 with OH radicals which
themselves are produced by the photodissociation of HNO_2. Experiments and chemical model calculations suggest that HNO_2 is formed mainly by heterogeneous
reactions with the reactor wall.

    The experiments described in this paper were performed at a fixed U.V.
intensity (17 Wm^{-2}), a fixed γ dose rate (32 $\mu Gy.s^{-1}$) and as a function of the
mean residence time. The relative humidity was set at 75%. The results of
these experiments are shown in Figure 1. It should be mentioned that these results are not corrected for the decrease in counting efficiency of the CCNC
for particle diameters smaller than 20 nm. It is observed that :
1) the measured increase in particle concentration due to ionizing radiation
at a mean residence time > 5 min. is a true effect and not an artefact due to
the counting efficiency of the CCNC, since the mean particle size is identical
in both radiation conditions.
2) at $\bar{t}$ < 5 min. part of the increase in $N_{U.V} + \gamma/N_{U.V}$ must be attributed to
the difference in particle size in both radiation conditions. If we consider
the counting efficiency for the photometric mode of the CCNC determined by
Brockmann (1981), an actual increase in $N_{U.V} + \gamma/N_{U.V}$ does still occur with
decreasing mean residence time.
3) The particle number and volume concentrations become independent of $\bar{t}$, for
$\bar{t}$ > 5 min. This can partly be explained by the fact that the dilution loss
rate (= $1/\bar{t}$) becomes negligible in comparison to the wall deposition loss rate.
The other necessary condition for the constant behaviour of the parameters at
$\bar{t}$ > 5 min. is that the formation of H_2SO_4 molecules is also independent of $\bar{t}$,
for $\bar{t}$ > 5 min.

(*) This work is supported by the C.E.C., contract nr BIO-B-327-B(RS)

944

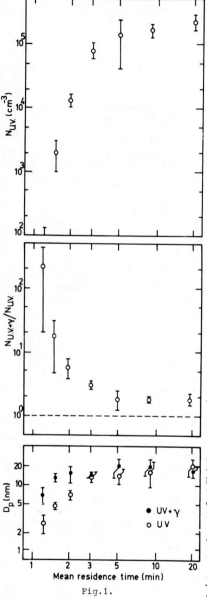

Fig.1.

For a more thorough interpretation of these observations a numerical model describing the formation and growth of H_2O-H_2SO_4 aerosols is described by the classical thermodynamic theory of nucleation (Kiang et al 1973, see also Hamill et all 1982 for a review of different nucleation mechanisms). According to the classical theory, nucleation rates can only be calculated in environmental conditions that allow an equilibrium between the gas phase and the droplet. In the presence of ions however, it is possible that this situation does not occur, in which case nucleation becomes kinetically rather than thermodynamically controled. For an estimation of the nucleation rate we then take the ion formation rate which is the upper limit for the ion induced nucleation. In the case of an experiment with U.V. radiation alone, only homogeneous nucleation is considered in the model, in the case of simultaneous U.V. and γ radiation, the sum of homogeneous and heterogeneous nucleation is taken as the total nucleation rate (see Figure 2) concentration,

Condensational growth is described as a collision process of H_2SO_4 molecules with the droplet, whereafter the droplet immediately acquires enough water molecules to become in equilibrium with the water vapour again (Hamill, 1975).

Coagulation was not included for the calculations in this paper.

The particle size distribution is approximated by 50 discrete diameter classes, logarithmically evenly spaced between 0.4 and 200 nm. The time evolution of the number concentration of each class is given by a set of coupled first order equations of the form :

$$\frac{dN_{i^*}}{dt} = I(H) - \frac{F_{i^*} N_{i^*}}{m_{i^*+1} - m_{i^*}} - \left(\lambda_{dep}^{i^*} - \lambda_{dil}\right)N_{i^*}$$

$$\frac{dN_i}{dt} = \frac{F_{i-1} N_{i-1}}{m_i - m_{i-1}} - \frac{F_i N_i}{m_{i+1} - m_i} - \left(\lambda_{dep}^i - \lambda_{dil}\right)N_i$$

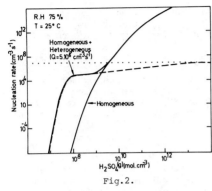

Fig.2.

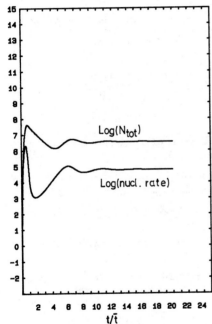

Log(N_{tot})

Log(nucl. rate)

t/t̄

Fig.3.

with i* indicating the class containing the critical particle, Ni the
number concentration of particles of
size i, n_i the number of H_2SO_4 molecules in a particle of size i, I(H) the
nucleation rate and F_i the flux of
H_2SO_4 molecules colliding with a particle of size i :

$$F_i = 4\pi r_i^2 \cdot H \sqrt{\frac{kT}{2\pi m_a}}$$

where H is the H_2SO_4 concentration in
the gas phase, r_i the particle radius
and m_a the molecular weight of H_2SO_4.
The variation of the H_2SO_4 molecules
in the gas phase is described by :

$$\frac{dH}{dt} = S - \sum_i F_i N_i - \left(\lambda''_{dep} + \lambda_{d,l}\right)H$$

with S the production rate of H_2SO_4
molecules. This set of equations is
solved by a variable step fourth order
Runge-Kutta algorithm.

Figure 3 shows the evolution of
the nucleation rate and total number
concentration in the C.S.T.Reactor in
the case of pure photolytical aerosol
production. It is shown that at a given
mean residence time, a high H_2SO_4 production rate leads to oscillations in
the particle concentration. This behaviour is also predicted at a fixed
H_2SO_4 production rate and high mean
residence times. This behaviour is explained by the cycle : nucleation- depletion of H_2SO_4(gas) by nucleation and
condensation and subsequent inhibition
of nucleation - dilution - nucleation
... .

Simulations of the experiments gave the following results :
1) constant behaviour of the parameters
is found with a diffusion boundary layer smaller than $0.48D^{0.274}$ cm, with D

the diffusion coefficient. This corresponds with a mean deposition rate
> 0.14 min^{-1} for the particles and > 10 min^{-1} for H_2SO_4 molecules.
2) the formation rate of H_2SO_4 molecules must be larger than 5 10^8 cm^{-3}s^{-1}
to obtain a detectable amount of aerosol particles in the reactor. This however should correspond with a SO_2 transformation rate of more than 50% per
hour in our experimental conditions !
3) under some conditions, an increase in aerosol number concentration due to
ionizing radiation may be explained by ion induced nucleation alone. It is
not clear as yet whether an enhancement of the SO_2 transformation rate should
also be considered in the model in order to fully explain the effects induced
by ionizing radiation.

FIGURE CAPTIONS

Fig.1. Results of irradiation experiments in the C.S.T.Reactor. Shown are
the parameters at steady state conditions. SO_2 = 0.5 ppmV, NO_2 = 0.015 ppmV,
T = 22°C, R.H. = 75%. U.V. intensity = 17Wm^{-2}, ionisation rate = 5 10^6 ion
pairs cm^{-3}s^{-1}. Error bars are 90% confidence limits on the mean.

Fig.2. Homogeneous nucleation rate and heterogeneous nucleation rate around
ions as a function of the H_2SO_4 concentration in the gas phase for T = 25°C
and R.H. = 75% according to the classical theory of nucleation.

Fig.3. Simulation of the nucleation rate and total particle concentration as
a function of time in a C.S.T.R., in the case of pure photolytic aerosol
formation. T = 22°C, R.H. = 75%, H_2SO_4 production rate = 1 10^8 cm^{-3}s^{-1}, S/V
of the reactor = 0.5, t = 100 s.

REFERENCES

Brockmann J.E. (1981) "Coagulation and deposition of ultra-fine aerosols in
turbulant pipe flow.", Ph.D.Thesis Minnesota

Hamill P. et al (1982) "An Analysis of Various Nucleation Mechanisms for
Sulfate Particles in the Stratosphere", J.Aerosol Sci. 13, 561

Hamill P., (1975) "The time dependant growth of H_2O-H_2SO_4 aerosols by
heteromolecular condensation", J.Aerosol Sci., 6, 475

Kiang, C.S. et al (1973) "Chemical nucleation theory for various humidities
and pollutants", Faraday Symp. Chem. Soc. 7, 26

Raes F. et al (1984) "A new diffusion battery design : The diffusion carrousel" to be published as a technical note in Atmospheric Environment.

Published 1984 by Elsevier Science Publishing Co., Inc.

Aerosols, Liu, Pui, and Fissan, editors

FORMATION OF NH_4NO_3 AEROSOLS BY NH_3 + NO_2 REACTION

Tuan T. Tran and Mayis Seapan
School of Chemical Engineering
Oklahoma State University
Stillwater, Oklahoma 74078
(405) 624-5280

EXTENDED ABSTRACT

Introduction

The presence of ammonium nitrate in atmospheric aerosols has been documented by many researchers (1,2,3,4). Since the level of nitrogen oxides has increased in the urban regions of the U.S. over the years, the atmospheric burden of particulate nitrates has become increasingly important. The study of the formation, its controlling factors, and concentrations of ammonium nitrate have become of substantial interest. Ammonium nitrate is usually considered to form from the reaction of ammonia and nitric acid, primarily because the atmospheric concentrations of these compounds follow the chemical equilibria for the ammonia/nitric acid reaction (2,5). However, ammonium nitrate aerosols may also form from direct reaction of ammonia and nitrogen dioxide. This reaction is also important in plumes and stack gases where ammonia injection is used as a means of reducing nitrogen oxides emissions.

At high temperatures, ammonia and nitrogen dioxide react to form nitrogen gas and water. But at ambient temperatures they result in ammonium nitrate aerosols.

According to Falk (6), below 100°C the reaction between nitrogen dioxide and ammonia can be represented as

$$2NO_2 + 2NH_3 \longrightarrow NH_4 NO_3 + N_2 + H_2O$$

The present project was undertaken to study the formation of ammonium nitrate particles, their size distribution and stoichiometry of the reaction.

Experimental Apparatus and Results

A Pyrex tubular reactor with 70 mm ID and 830 mm length was used to react ammonia and nitrogen dioxide gases in air at ambient conditions. A movable sampling probe at one end of the reactor allowed the residence time to be varied for a given flow rate. A condensation nuclei counter/diffusion battery system was used to measure aerosol number density and particle size distribution. An electrostatic aerosol sampler was also used to collect samples for electron microscopy study. A chemiluminescent NO_x analyzer was used for analysis of gaseous streams.

The solid product of the reactor was analyzed and was shown to be mainly ammonium nitrate. Traces of ammonium nitrite was also identified. The ratio of nitrate to nitrite varied from 99:1 to 99.9:0.1. These results are consistent with those reported by Falk (6). Preliminary study on the reaction stoichiometry, however, revealed that the ratio of NO_2/NH_3 consumed in the reaction is not 1:1 as reported previously (6).

During the electron microscopy study, the ammonium nitrate particles appeared to be round and of very fine structure which sublimated quickly under low intensity electron beam.

Flow dynamics study showed that the reaction zone is mainly the boundary layer of the NO_2 jet stream where mixing took place. This is consistent with calculations showing that only ten percent of reactants were consumed in the reaction.

Particle size distributions in the range of 0.030 - 1 µm were measured where more than half of the total number of particles were larger than 0.4 µm. Residence time was varied from 2 to 16 seconds by changing reactor length. As the residence time was decreased, an increase in the number of particles having diameters below the median value was measured while the total number of particles remained fairly constant. The main growth mechanism is concluded to be coagulation. A typical particle size distribution is shown in Figure 1.

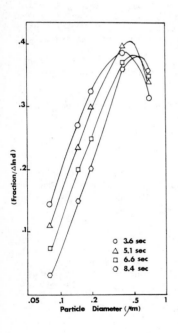

Both ammonia and nitrogen dioxide concentrations were varied from 15 to 60 ppm in dry air. Particle size distributions did not show any significant change. Increase in the total number of particles was measured only when an appreciable increase in one or both reactants concentrations was introduced.

Figure 1. Effect of residence time on particle size distribution.

References

Lundgren, D. A. (1970) "Atmosphere Aerosol Composition and Concentration as a Function of Particle Size and of Time," J. Air Pollution Control Ass., 20:603.

Cadle, S. H., Countess, R. J. and Kelly, N. A. (1982) "Nitric Acid and Ammonia in Urban and Rural Locations," Atmospheric Environment, 16:2501.

Gordon, R. J. and Bryan, R. J. (1973) "Ammonium Nitrate in Airborne Particles in Los Angeles," Envir. Sci. Technol., 7:645.

Mamane, Y. and Pueschel, R. F. (1980) "A Method For the Detection of Individual Nitrate Particles," Atmospheric Environment, 14:629.

Stelson, A. W., Friedlander, S .K. and Seinfeld, J. H. (1979) "A Note on the Equilibrium Relationship Between Ammonia and Nitric Acid and Particulate Ammonium Nitrate," Atmospheric Environment, 13:369.

Falk, F., "Stoichiometry and Kinetics of the Gas Phase Reaction of Nitrogen Dioxide and Ammonia," Ph.D. Dissertation, Princeton University (1955).

Published 1984 by Elsevier Science Publishing Co., Inc.
Aerosols, Liu, Pui, and Fissan, editors

950

AN ESTIMATE OF GAS-SOLID AEROSOL INTERFACE REACTION RATE BY ICE NUCLEATION METHOD

Yasuo Ueno

Department of Chemical Engineering, University of Kentucky
Lexington, KY 40506
U. S. A.

EXTENDED ABSTRACT

Introduction

It has been well-known that a crystal with hexagonal shape such as silver iodide or lead iodide has ice nucleation capability at high ice nucleation temperature.   In previous papers (Ueno, 1982), physico-chemical studies were made on the mechanism of lead aerosol formation and formation conditions for the monodisperse aerosol.   The ice nucleation temperatures of these typical aerosols (lead, lead iodide, iodide) could be individually marked using ice nucleation chamber.   In this paper, an experiment of the reactivity of lead aerosol toward iodine vapor was carried out using lead aerosol and iodine vapor under different temperatures.   The reactivity between lead aerosol and iodine vapor could be estimated from the number of ice crystal in the ice nucleation chamber.

Experimental Procedures

Test aerosols of lead, lead iodide and iodide were generated by a condensation method.   The experimental apparatus and operating procedure were almost similar with previous reports (Ueno, 1982). Particle sizes of lead aerosol were determined with an electronmicroscope and those of lead iodide and iodide aerosols were calculated from their settling velocities by a ultramicroscope.   Their particle number concentration was calculated from mass concentration and particle size.

"Lead iodide aerosols" were obtained from the reaction between lead aerosol and iodine vapor.   A small amount of lead aerosol was introduced into a test chamber in which the atmosphere of iodine vapor had already been prepared.   Then, small portions of the aerosol reacted were sampled from the chamber to be further diluted with clean nitrogen gas in another large chamber, whose atmosphere was moderately agitated.   The final dilution ratio in each experiment was 1:12000 or more.   A small amount (usually 20cc) of the aerosol thus diluted was slowly injected onto the ice nucleation chamber previously reported (Ueno, 1982).   The number of ice crystal which was observed in the ice nucleation chamber gave the number of "lead iodide aerosol".

Results

Generation of lead aerosol and ice nucleation

It was found that the mass concentration and the particle size alter with the flow rate of nitrogen stream and the temperature of lead vapor: both of them reach a peak at a certain rate of flow under any of the temperatures examined (950°C to 1150°C) making a contrast to the number concentration passing through a trough. There is a linear relation between the logarithm of mean particle size and reciprocal of vapor temperature for all the rates of flow tested (1.0 to 4.5 l/min). Aerosol of high monodispersity might be obtained by vaporizing metal at a temperature in the range of 1000°C to 1150°C and by flowing nitrogen at a rate of 1.5 l/min. The number concentrations were of the order of magnitude from 10^4 to 10^8 particles/cc. The ice nucleation temperature of lead aerosol was examined and seemed to be low than -30°C.

Generation of lead iodide aerosol and ice nucleation

Both the aerosol mass concentration as well as the particle sizes goes through a peak with the flow rate of nitrogen gas stream. The heat of vaporization of lead iodide was estimated as nearly as 23 kcal/ mole in the range from 360° to 490°C. It was also estimated that the operating conditions for highly monodisperse aerosol are approximately 3 l/min of the flow rate of nitrogen gas stream at the temperature of 400°C. The number concentrations were of the order of magnitude of 10^7 particles/cc. The ice nucleation temperature of lead iodide was examined and approximately -7°C.

Generation of iodide aerosol and ice nucleation

Iodide aerosol generated seemed unstable and very adsorptive on the wall of a test aerosol chamber. Particle size ranged from 0.2 to 1 um upwards. Usually, the aerosol generated had a high monodispersity in the small size range. The ice nucleation temperature of iodide aerosol found to be -10.5°C.

Reactivity ratio of lead aerosol to iodine vapor

Lead aerosol with the average sizes of 0.11 μm and 0.78 μm was used for gas-solid interface reaction experiment. At the temperature of 30°C and for the reaction time of 15 minutes, the former gave the yield of 1.5% and the latter 2.8%. At the temperature of 50°C and for the reaction time of 15 minutes, the former gave the yield of 16% and the latter 31%. It seems that the coarser particles of lead react with iodine vapor to produce lead iodide more readily than finer ones. The reactivity is affected by the temperature of reaction.

References

Ueno, Y. (1982) "Generation of Lead Aerosols in Nitrogen, Argon or Carbon Dioxide", the proceedings of the 75th Annual Meeting of the Air Pollution Control Association #82-33.1, New Orleans, LA, June 20-25.

Ueno, Y. (1982) "Formation of Monodisperse Lead Aerosols and Identification of Particle Number Concentration by Ice Nucleation",

the Proceedings of the 15th International Colloquium of Polluted
Atmosphere, Paris, France, May, 1982.   Atmospheric Pollution 1982
(edited by M. M. Benarie), Elsevier Scientific Publ. Co., Amsterdam
1982, 251pp.

Published 1984 by Elsevier Science Publishing Co., Inc.

Aerosols, Liu, Pui, and Fissan, editors

SO₂ CONVERSION RATE INFERRED FROM IN-SITU MEASUREMENTS OF
AEROSOLS AND SO₂ IN THE STRATOSPHERIC CLOUD FORMED BY THE
ERUPTION OF THE EL CHICHON VOLCANO

by

James Charles Wilson
Particle Technology Laboratory
University of Minnesota
Minneapolis, Minnesota 55455

EXTENDED ABSTRACT

INTRODUCTION

The eruptions of El Chichon between March 25 and April 5, 1982 injected
large quantities of gases and particles into the stratosphere. The resulting
enhancement of the stratospheric aerosol layer increased optical depths
significantly and may have perturbed the earth's climate. The primary
component of the stratospheric aerosol cloud formed by El Chichon is generally
understood to be sulfuric acid formed from sulfur dioxide. Formation rates
and mechanisms effect the size distribution of the aerosol and hence its
impact on radiation and climate. This paper reports aerosol formation rates
and mechanisms inferred from measurements made in the lower portions of the
lower stratospheric cloud formed by El Chichon.

NASA U2 aircraft flew several missions in 1982 and 1983 in the Aerosol
Climatic Effects Study. The data used in this paper are from the University of
Minnesota Condensation Nucleus Counter (UMCNC) (Wilson et al.,1983), Particle
Measuring Systems ASAS-X and FSSP optical particle counters (OPC) (Knollenberg
and Huffman, 1983), and a cryogenic gas sampler (Vedder, et al.,1983). The
UMCNC measures the number concentration of particles larger than about 0.01
micron in diameter. The OPC's measure size distributions in the 0.1 to 30
micron range and the gas sampler was used in the measurement of sulfur dioxide
concentration . The U2 data set also inlcudes impactor and filter samples
which permit determination of chemical composition and provide independent
measures of size and concentration.

INFERENCE OF CONVERSION RATE

The calculation of gas-to-particle conversion rates involves determining
the concentration of sulfur dioxide immediately after the eruption, $[SO_2]_i$,
and $[SO_2]_t$, the concentration at a time t later. The average rate conversion
rate is then given by

$$R_{av} = ([SO_2]_i - [SO_2]_t)/t. \quad (1)$$

Since measurements were not made immediately after the eruption, it is
necessary to infer $[SO_2]_i$ from the measurements made at time t. In order to do
this, it was assumed that both gas and particles were injected at the time of
the eruption, and that they were mixed into the background aerosol and sulfur
dioxide. From analysis of the background aerosol and that measured in the
cloud at time t, it was determined that the evolution of the aerosol was
consistent with the formation of sulfuric acid in gas phase reactions and

condensation of the acid on the background aerosol. The condensing acid
molecules also formed new particles. Particles were observed in the 1 to 20
micron range which could not have been created in this way and these were
assumed to be primary aerosol which settled from higher in the cloud. This
assumption is consistent with the observations of Oberbeck et al. (1983). The
volume of secondary aerosol formed in these processes was estimated then from
comparisons of the background aerosol and that measured in the cloud. Then the
amount of sulfur dioxide required to produce the secondary aerosol volume was
calculated and added to $[SO_2]_t$ to obtain $[SO_2]_i$.

ANALYZED CASES AND RESULTS

Figures 1 and 2 show the number concentration measured by the UMCNC
(upper trace) and that measured by the PMS OPC's (lower trace) during flights
on April 19 and May 5, 1982. Sulfur dioxide and size distribution measurement
intervals are indicated on the plots. The displayed flight portions were at
nearly 20 km altitude. The intervals over which sulfur dioxide was measured
were broken into segments during which the aerosol was nearly uniform. Size
distributions were analyzed for each segment. The concentration of small
particles was strongly enhanced during two of the analyzed episodes inidcating
in-situ gas-to-particle conversion (Wilson et al., 1983). (The pre-eruption
CNC concentration was typically between 6 and 10 particles-cm^{-3}.)

Analysis of the aerosol growth was done using the techniques of McMurry
and Wilson (1981). The bacground aerosol was assumed to evolve to the cloud
aerosol and the diameter growth rates were calculated. The results for two of
the episodes are shown on Figure 3. Growth rates which are nearly constant as
a function of diameter are expected for free-molecule condensation. The
observed rates are nearly constant up to about 0.6 microns which supports the
assumption of condensational growth for that part of the size distribution.
The sharp jump in growth rates at .6 probably shows the effects of injected
primary aerosol rather than growth. Figure 4 illustrates the effect of
condensation on the background distribution of 5 May. The (*) symbol
indicates the measured volume distribution for SD 6 and the (+)'s indicate
volume distribution that would be expected if the background distribution
experienced a diameter change of .17 microns due to condensational growth. The
value of .17 resulted from the growth analysis. Most of the aerosol added to
the distribution below 1 micron can be explained by this process. Above one
micron,   however, the measured volume far exceeds that projected by
condensation. These particles may be large silicates settling out of the
upper cloud and they are excluded from the tally of secondary aerosol. Also
excluded are the sub-.1 micron particles which appear in abundance in the
number distribution. Since their size is not known, their volume cannot be
estimated. However, they are clearly products of gas-to-particle conversion
and could easily contribute a significant amount of secondary aerosol in this
case.

Table 1 summarizes the results for the analyzed episodes.
V_c is the aerosol volume concentration added by condensation. V_t is the total
aerosol volume measured at time t. The units of both V_c and V_t are
$micron^3/cm^3$. The values of $[SO_2]_t$ given in the table are higher than those
reported by Vedder et al., because all the sulfur dioxide was assumed to be in
the segment being analyzed. The time constant is calculated assuming an
exponential decay of sulfur dioxide consistent with first order reaction.

It is calculated from the percent of the sulfur dioxide which has been removed. R_{av} is given in molecules/sec.

Table 1. Analysis of Sulfur Dioxide Conversion

Episode	V_t	V_c	$[SO_2]_t$ pptv	$[SO_2]_i$ pptv	R_{av} s^{-1}	SO_2 Removed %	Time Const. weeks
April 19							
Background	0.024						
SD 8	0.58	.132	122	590	440	80	2
May 5							
Background	0.021						
SD3	0.67	.14	40	610	340	93	
SD4	0.91	.07	40	320	170	87	2
SD5	1.37	.56	40	2330	1370	98	
SD6	0.35	.07	40	340	180	88	

The rates presented here are, as can be seen from the restrictive assumptions, likely to be underestimates. An additional source of systematic error must be considered. There is a suspected error in ASAS-X flowrates at high altitude. Comparison with impactor data suggest that rates and volumes tabulated here may be a factor of 3 too small. Even if that is the case, the shape of the size distributions and the growth analysis will remain unchanged.

CONCLUSIONS

        The analysis of aerosol growth strongly suggests that secondary aerosol was formed in gas phase reactions and condensed on the particles in the background aerosol. This process appears to explain the observed submicron cloud size distribution. The suspected flow errors in the ASAS-X at 20 km altitude does not effect this conclusion. Larger particles in the cloud may have settled from higher in the cloud. The time constant for SO_2 removal inferred from this analysis is about one half the value generally accepted for the cloud as a whole. This suggests that either rates were faster on the lower edges of the cloud or that a significant precursor gas was not measured. Correcting for the suspected flow problem would increase the rates even more.

REFERENCES

Knollenberg, R.G., D. Huffman,"Measurement of Aerosol Size Distributions in the El Chichon Cloud,",Geophys. Res. Lett. 10:1025

McMurry, P.H., J.C.Wilson,"Growth Laws for the Formation of Secondary Ambient Aerosols:Implications for Chemical Conversion Mechanisms,"Atmos. Envir. 16:121.

Oberbeck, V.R.,E.F. Danielsen,K.G. Snetsinger, G.V. Ferry, W. Fong,D.M. Hays,"Effect of the Eruption of El Chichon on Stratospheric Aerosol Size and Composition,"Geophys. Res. Lett.10:1021.

Vedder,J.F.,E.P.Condon,E.C.Inn,K.D.Tabor,M.A.Kritz,"Measurements of Stratospheric SO2 after the El Chichon eruptions," Geophys. Res. Lett. 10:1045.

956

Wilson,J.C.,E.D. Blackshear,J.H.Hyun,"Changes in the Sub-2.5 Micron Diameter
Aerosol Observed at 20KM Altitude after the Eruption of El Chichon" Geophys.
Res. Lett. 10:1029.

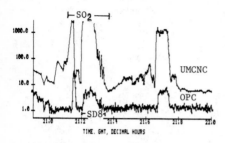

Fig. 1. UMCNC and OPC Concentration
as a function of time. April 19.
Measurement intervals indicated.

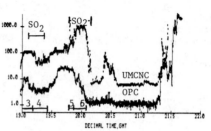

Fig. 2. UMCNC and OPC Concentrations
as a function of time. May 5.
SO2 and size distribution
measurement intervals indicated.

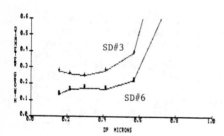

Fig. 3.Growth of background aerosol
as a function of particle size
for two size distributions.
Nearly ocnstant growth rates
indicate condensation growth.

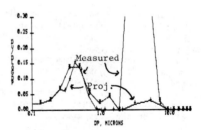

Fig. 4. Measured volume distribution
and that projected by condensation
growth hypothesis. SD 6, May 5.

Published 1984 by Elsevier Science Publishing Co., Inc.
Aerosols, Liu, Pui, and Fissan, editors

PRIMARY OXIDANTS AND THE SO_2 OXIDATION PROCESS*

T. Novakov

Applied Science Division
Lawrence Berkeley Laboratory
University of California
Berkeley, California 94720

Deposition of acidic forms of rain, fog, and particulate sulfate is one of the outstanding contemporary environmental problems. Because specific catalysts and/or oxidants are required in aqueous SO_2 oxidation pathways, the amount of SO_2 oxidized will depend not only on the SO_2 concentration but, under many circumstances, may also depend critically on the availability and origin of the oxidants and catalysts. Greater knowledge of the types and sources of SO_2 oxidants is crucial to solving this problem.

Certain catalytic materials such as soot and transition metals are co-emitted into the atmosphere with SO_2 from the same combustion sources. The atmospheric concentrations of these catalysts are directly related to source emissions and are therefore, in principle, controllable at the source. However, one of the most important oxidants for SO_2 oxidation is hydrogen peroxide (H_2O_2), which is believed to be produced entirely in the atmosphere by a series of photochemical reactions that involve many reactants not emitted directly from sources. Thus, the occurrence of H_2O_2 in the atmosphere is essentially unrelated to source emissions and is therefore uncontrollable at the source.

In this paper, we present evidence that incomplete combustion, besides being a source of particulate catalysts, is also a source of primary oxidants, which in the aqueous phase are mostly present as H_2O_2 and, to a lesser extent, organic peroxides.[1]

For a particulate catalyst, soot, we will demonstrate that the oxidant involved in aqueous oxidation of sulfite is chemisorbed O_2. The amount of chemisorbed oxygen depends on the type of carbon, i.e., its surface chemical properties.[2] Experiments will be described that demonstrate that incomplete combustion is a source of reactive species. These species readily react with SO_2 in the gas phase, forming a compound not detected by a pulsed fluorescence SO_2 detector. Reactive gas-phase combustion products, when dissolved in water, are manifested predominantly as H_2O_2 and, to a lesser extent, as organic peroxides.

These primary oxidants oxidize SO_2 to sulfate or sulfuric acid in either bulk or dispersed (fog) aqueous systems. For a constant SO_2 concentration and amount of fuel combusted, the amount of SO_2 oxidized depends critically on the combustion conditions, being highest for incomplete combustion and negligible for stoichiometric flames.

This evidence suggests that primary oxidants may play a role in atmospheric SO_2 oxidation, particularly in formation of acid fog by involvement of

*This work was supported by the Director, Office of Energy Research, Office of Health and Environmental Research, Physical and Technological Research Division of the U.S. Department of Energy, under contract DE-AC03-76SF00098 and by the National Science Foundation under contract ATM 83-15442.

958

automobile-emitted oxidants and in the formation of particulate sulfate in environments characterized by poor combustion technology.

The oxidant involved in the surface oxidation of S(IV) is adsorbed oxygen, $C_x \cdot O_2^*$. The more $C_x O_2^*$ is available, the more S(VI) can be formed. The size and microstructure of the carbon surface also influence the accessibility of S(IV) to $C_x \cdot O_2$. The wide variation in measured oxidation rates for different carbons is due to variability in their capacities for chemisorbing oxygen. Removal of unreactive surface functional groups by heat treatment with subsequent re-exposure to air at room temperature allows for enhanced chemisorption because more sites are available, so the rates of both types of S(IV) oxidation increase. Conversely, fresh combustion particles are more reactive than aged particles, whose surfaces are covered with unreactive species.

References

1. W. H. Benner and T. Novakov, "Oxidation of SO_2 in Fog Droplets by Primary Oxidants," Lawrence Berkeley Laboratory Report LBL-17703 (1984).
2. L.A. Gundel, W. H. Benner, and T. Novakov, "The Role of Surface Oxygen in S(IV) Oxidation by Different Carbons," Lawrence Berkeley Laboratory Report LBL-17909 (1984).

Published 1984 by Elsevier Science Publishing Co., Inc.
Aerosols, Liu, Pui, and Fissan, editors

DESIGN AND PERFORMANCE OF AN AEROSOL REACTOR
FOR PHOTOCHEMICAL STUDIES

W. HOLLÄNDER, W. BEHNKE, W. KOCH and G. POHLMANN
Fraunhofer-Institute for Toxicology and Aerosol Research
Nottulner Landweg 102, D 4400 Münster-Roxel

1. INTRODUCTION

In a study on the photochemical degradation of environmental chemicals with special emphasis on the effect of airborne particles we face the problem of aerosol decay in the reaction chamber. The atmospheric aerosol residence time is - depending on particle size, emission height etc. - of the order of a few days. Hence, one cannot expect to correctly measure slow photochemical aerosol reactions of tropospheric relevance, if the aerosol chamber residence time is significantly shorter than the reaction time.

The intensive illumination causes additional problems, because it heats the reactor volume and consequently enhances the turbulent deposition of particles as well as their thermophoretic motion to the walls. In addition, depending on the optical properties of the paricles, photophoretic motion may be of importance.

The aerosol in a containment may be stagnant or in ordered or in turbulent motion. The simplest behavior should be found under the stagnant or ordered motion condition. From hydrodynamic stability considerations it is clear, however, that turbulent motion prevails in our reactor.

In a recent theoretical work Crump and Seinfeld (1981) obtained an analytical solution for the aerosol decay in a containment of arbitrary shape.

2. DESIGN OF THE PHOTOREACTOR

As the residence time of particles increases with increasing chamber dimensions, a large chamber is favorable, but financial constraints forced us to use commercially available 1 m diameter Pyrex glass rings. The volume should not be smaller than approximately 2 m^3 because of the need for enough material for subsequent chemical analysis. So, three of the Pyrex rings were considered sufficient.

All the model experiments showed that high temperature differences are detrimental to residence time and hence, climatization of the top, where the radiation enters, and of the bottom, where most of it is absorbed, is essential.

Therefore, both the bottom plate and the top plate of the reaction chamber are water cooled. The top plate is a Pyrex glass plate of 3 mm thickness which is inserted in a slotted aluminum ring. This configuration is mounted on the topmost of the three glass rings. A laminar and vortex free stream of water is flowing over the glass plate absorbing a part of the infrared radiation of the lamps and carrying away the heat of absorption produced in the glass plate. The flow rate is of the order of 1 m^3/h so that the temperature gradient of the water in flow direction is less than 1 K/m.

The bottom of the chamber is a hollow aluminum plate with

960

a black oxide layer on its outer surface. This reduces diffuse reflection of the light which would lead to an undesirable heating of the reactor walls. The inner surface of the plate which is in contact with the water is riffled. In combination with the high flow rate of the water (3,6 m^3/h) a good exchange of the large amount of radiation energy (7 kW) absorbed at the black outer surface is achieved. For climatization we use two separate water circuits, one for the top plate and another for the bottom plate. One top circuit is connected to a heat exchanger with a large exchange area which is ventilated with room air. This leads to a temperature of the top glass plate which is slightly (3°C) above room temperature. The desired temperature difference between the top and the bottom plate is adjusted by controlling the temperature in the bottom water circuit by means of a computer aided control circuit. The design is shown in Fig. 1.

3. PERFORMANCE OF THE PHOTOREACTOR

The decay constant β for different monodisperse latex particle sizes and a temperature difference of 5 K is shown in Fig. 2. The evaluation was done by electron microscopy and/or an optical particle counter (Royco 226). The best fit constants for the dark experiment are

k_e (dark)   = 0,018 sec^{-1} and
k_e (illumination) = 0,165 sec^{-1}.

It is obvious, that Crump's and Seinfeld's theory (1981) can correctly describe the general size dependence of the deposition factor, because it has an adjustable parameter. With the limited data available so far, it cannot be decided yet, whether the larger k_e value in the illumination experiment should be attributed to a photophoretic effect or simply an increased turbulent dissipation caused by the illumination.

The influence of temperature difference on the residence time is shown in Fig. 3 for concentrated, polydisperse aerosols under investigation (TiO$_2$, SiO$_2$, SiO$_2$ coated with 5 % DOP-Dioctylphthalate). The large scatter is probably caused by the different mass concentrations which result in different sizes due to coagulation.

4. ACKNOWLEDGEMENT

The present study is supported by a joint grant of the Union of Chemical Industries (Verband der Chemischen Industrie) and the Federal Ministery for Research and Technology (BMFT). This support is gratefully acknowledged.

5. REFERENCE

Crump, J.G. and J.H. Seinfeld (1981) "Turbulent deposition and gravitational sedimentation of an aerosol in a vessel of arbitrary shape", J. Aer. Sci 12, 405

FIGURE CAPTIONS
Fig. 1: Scheme of the reactor temperature control circuit
Fig. 2: Decay constant for Latex particles of various sizes (ΔT=5K)
Fig. 3: Decay constant for different aerosols (c ≈ 2 mg/m^3) as a function of temperature difference ΔT

961

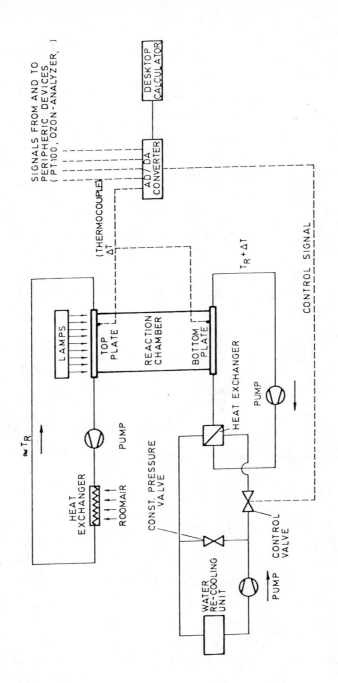

Fig. 1

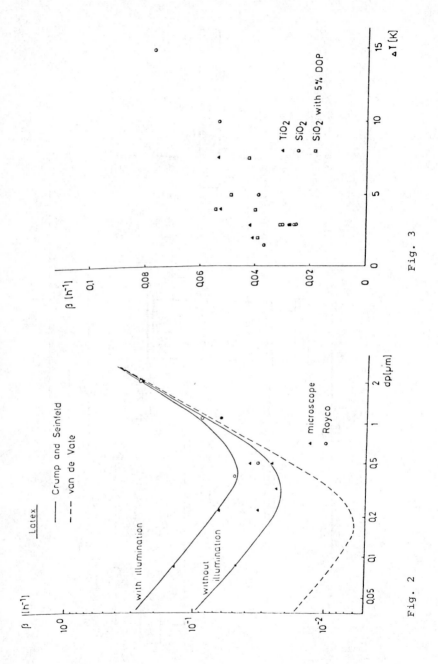

Fig. 3

Fig. 2

Published 1984 by Elsevier Science Publishing Co., Inc.
Aerosols, Liu, Pui, and Fissan, editors

Studies on Aerosol Behaviour in Smog Chambers
A Comparison between Experimental Results and Model Calculations

H. Bunz and R. Dlugi

Laboratorium für Aerosolphysik und Filtertechnik I
Kernforschungszentrum Karlsruhe GmbH
7500 Karlsruhe, W.-Germany

Extended Abstract

Introduction

Accompanying smog chamber experiments by theoretical calculations have been recognized to be necessary as it is not possible in many experimental situations to distinguish between the different physical processes and to analyze the experimental results. Therefore, a numerical model was developed (Bunz, Dlugi; 1983) including most aerosol-physical processes, being regarded as important in the experiments. A process still not treated in the model was the condensation of sulfuric acid on pre-existing particles which might be important in many situations. The model was extended, therefore, by this process, and a number of parametric calculations and comparisons to experiments will be presented in this paper.

Description of the Numerical Model

Since the basis of the model has already been described elsewhere (Bunz, Dlugi; 1983), only the extension made - the treatment of the condensation process - will be presented.

The transport of the gaseous sulfuric acid and the water vapor are controlled by diffusion. Therefore, the diffusion equation has to be solved with the appropriate boundary conditions to obtain the rate of mass transport to the particle. In the case of particles being comparable in size to the mean free path of the gas molecules, the kinetic boundary conditions have to be applied to account for the fact that the continuous diffusion equation is not valid up to the surface of the particles. In addition, the increase of the vapor pressure with increasing curvature of the surface has to be taken into account. These two effects can be combined (Dahneke; 1983) to get a formulation for the total mass flow I to the particle (radius r)

$$I = 4 \pi r D \beta (n_\infty - n_s (T, r)) \tag{1}$$

(D: diffusivity ; n : bulk concentration of vapor molecules ; $n_s (T, r)$: saturation concentration over the particle at temperature T).
The coefficient β takes the kinetic boundary conditions into account and approximates 1 in the case of $r \to \infty$, giving the expresssion for the Maxwell flow.

$$\beta = (Kn_D + 1) / (2 Kn_D(Kn_D+1) / \delta + 1) \tag{2}$$

where $Kn_D = 2D/c \cdot r$; C: average thermal velocity and δ: the evaporation/condensation accommodation coefficient.

In the parameter study presented in the next section of this paper it can be seen that the results are very sensitive to the not very well known accommodation coefficient. Multiplying by the molecular mass, equation (1) can easily be transferred into a differential equation for the gaseous sulfuric acid concentration C_s:

$$C_s = -\sum_i \alpha_i (C_s - C_{SO}(r_i)) + R_s \qquad (3)$$

where $C_{SO}(r)$: saturation mass concentration over the particle, $\alpha_i = 4\pi r_i D \cdot \beta$ and $R_s = R \cdot C_{SO_2}$ = production rate of s.a.

Since the α_i are only material and particle size dependent, they change very slowly in time as well as R_s (Eq.(3) and can, therefore, very easily analytically be solved for short time periods as e.g. the time steps of the numerical aerosol model. This solution can be used to integrate eq. (1) and to calculate the mass increment of the particle within one time step.

The formulation of the correction factor β (eq. (2)) used in this paper, agrees in principle with formulations given by other authors (e.g. (Peterson and Seinfeld; 1980).

Parameter Studies and Comparisons with Experimental Results

As in the paper in 1982, the calculations are mainly based on the experiments of Perrin (1980). The values of the diffusional boundary to be found to achieve the best agreement with the experiments are used throughout all calculations. To find out the parameters of greatest influence, e.g. the relative humidity, the SO_2-conversion rate, the initial mass of preexisting particles and the accommodation coefficient as the parameter of greatest uncertainty in the model were varied over a realistic range.

The range of the first three parameters is defined by the experimental conditions and the error in measuring these parameters. The humidity was varied between 30% - 70%, the conversion rate between 0.2 - 0.5%/h and the initial mass between 10^{-9} - 10^{-7} g. The variation of the accommodation coefficient was found in the literature to be between 0.001 to 1. But the calculation showed that a value greater than 0.1 leads to unrealistic results which are in total disagreement with the experimental findings (see Fig. 1). On the other hand the initial mass influences the results only in the case of such high accommodation coefficients and is, therefore, unimportant (see Fig. 1). Calculations using values for δ between 0.01 to 0.03, agree very well with the experiments. Variations concerning the relative humidity, the SO_2-conversion rate, and the accommodation coefficient compensate each other as can be seen in Fig. 2. The independent measurement of humidity and conversion is, therefore, necessary to interpret the experiments and to obtain information concerning the accommodation coefficient.

Conclusion

The numerical model to recalculate smog chamber experiments is now able to treat nucleation and condensation of sulfuric acid simultaneously. In the parameter study the strong influence of the condensation/evaporation accommodation coefficient could be shown as well as the necessity to measure the other parameters of influence independently. The influence of a small amount of preexisting particles was found to be cancelled shortly after the beginning of the experiment.

References

Bunz, H.; Dlugi, R. (1983): J. Aer. Sci. 14, pp. 288 - 292

Dahneke, B. (1983) in: "Theory of Dispersed Multiphase Flow", Academic Press

Peterson, T.W.; Seinfeld, I.M. (1980) in:"Adv. in Env. Sci. and Technol.", John Wiley & Sons

Perrin, M.-L. (1980): Thesis, Univiversité Pierre et Marie Curie, Paris

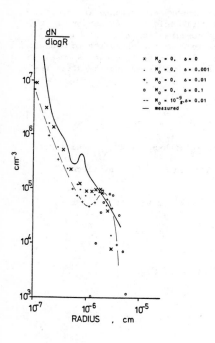

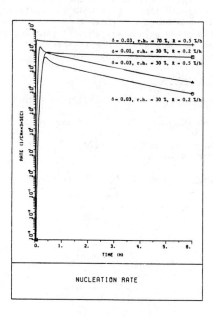

Figure 1. Particle size distribution in the experiments of Perrin after 3 h, measured and calculated as a function of different parameters

Figure 2. Calculated nucleation rate of $H_2SO_4 - H_2O$-vapor as a function of the time, depending on different parameters

Published 1984 by Elsevier Science Publishing Co., Inc.
Aerosols, Liu, Pui, and Fissan, editors

SIMULATION OF AEROSOL FORMATION AND GROWTH IN MULTI-COMPONENT SYSTEM

M. Kasahara, K. Takahashi and Y. Moriguchi
Institute of Atomic Energy
Kyoto University
Uji, Kyoto 611
Japan

EXTENDED ABSTRACT

1 Introduction

During last 15 years, the gas-to-particle conversion mechanism has been extensively investigated both theoretically and experimentally. Theoretical studies were mostly performed using simulation model based on both nucleation theory and aerosol dynamic theory.

In this study, a new combined simulation model was led to investigate the photochemical particle formation and growth mechanisms in multi-component system. The model is a combination of two sub-models of a gas-phase reaction model and an aerosol formation-growth model, and was modified after comparison of calculated results with experimental values. The present simulation model can elucidate well the general tendency of aerosol formation and growth processes in multi-component system.

2 Modeling of Aerosol Formation and Growth in Multi-component System

It is well understood that one of the main processes of the photochemical aerosol formation in polluted atmosphere is the conversion of sulfur dioxide gas to particulate sulfate. In our previous experimental studies (Kasahara and Takahashi, 1978), it was clarified that the most dominant reactant initiating photochemical aerosol formation is SO_2 in multi-component system containing SO_2, NO_2 and hydrocarbon (HC), especially at the early stage of reaction, and that the presence of NOx and/or HC effects on the aerosol formation in very complicated way depending on both the kind of gaseous system and concentration of reactants. Accordingly, it is reasonably assumed in modeling of aerosol formation in multi-component system that the nucleation from sulfuric acid-water vapor following the oxidation of SO_2 is the most dominant process in the formation of new particle.

A kinetic model was proposed (Takahashi, et al., 1975) to simulate the sulfuric aicd formation and its growth in SO_2/air system, and it was based on the nucleation theory and aerosol dynamic theory. On the other hand, a large number of models have been reported to simulate photochemical gas-phase reaction in polluted atmospheric air.

In this study, the photochemical gas-phase reaction model and the kinetic model is combined together to examine aerosol formation and growth mechanisms in multi-component system containing representative air pollutants such as SO_2, NOx and HC. In this model, eighteen principal reactions were adopted for the members of gas-phase reaction model and are shown in Table 1 with their reaction rate constants. The 11th through 13th and 15th reactions contribute to the production of sulfuric acid vapor continuing to formation of sulfuric acid mist. The 16 - 18th reactions are quasi-quenching reactions on the oxidation of SO_2 which were discussed in detail by Sidebottom et al. (1972), and their reaction rate constants were determined from our experimental results.

Table 1 Simplified gaseous reaction model

Reactions		Model value#	
		Ps(mmHg) 1.0-6	Ps(mmHg) 3.6-4
(1)	$NO_2+h\nu$ —— $NO+O$	9.4-2	
(2)	$O+(O_2)+(M)$ —— $O_3+(M)$	4.0+6	
(3)	$NO+O_3$ —— NO_2+O_2	3.0+1	
(4)	$HC+O$ —— $axRO_2$		
(5)	$HC+O_3$ —— $bxRO_2+O_2$		
(6)	RO_2+NO_2 —— XX	5.0-1	
(7)	RO_2+NO —— $NO_2+cxOH+YY$	1.0+2	
(8)	$HC+OH$ —— $dxRO_2$		
(9)	RO_2+RO_2 —— ZZ	1.0+4	
(10)	$NO_2+OH+(M)$ —— $HNO_3+(M)$	1.2+4	
(11)	$SO_2+O+(M)$ —— $SO_3+(M)$	2.0+1	
(12)	SO_2+RO_2 —— SO_3+RO		
(13)	$SO_2+OH+(M)$ —— $HOSO_2+(M)$	7.8+2	
(14)	$SO_2+h\nu$ —— SO_2^*	8.0-4	5.0-3
(15)	SO_2^* —— SO_3	2.0+3	2.0+4
(16)	SO_2^*+HC —— XXX		
(17)	$SO_2^*+NO_2$ —— YYY	6.0+5	2.0+5
(18)	$SO_2^*+(M)$ —— $SO_2+(M)$	9.8+4	8.0+4

unit = ppm,min; 9.4-2 means 9.4x10-2
a, b, c and d are constants depending on HC

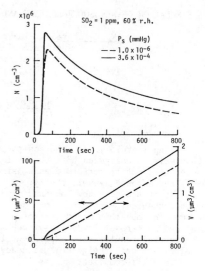

Fig.1 Aerosol formation and growth in SO_2/air system (Simulation): at $Ps=10^{-6}mmHg$, $k=1.6x10^{-5}min^{-1}$, $AF=1.0$, $CGC=10^{-7}cm^3/s$; at $Ps=3.6x10^{-4}mmHg$, $k=10^{-3}min^{-1}$, $AF=1.0$, $CGC=10^{-9}cm^3/s$.

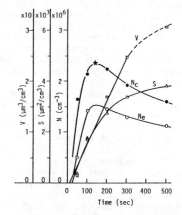

Fig.2 Aerosol formation and growth in SO_2/air system (Experiment): Number(Nc, Ne), surface area(S), volume(V) concentration changes.

3 Simulated Results

Two types of calculation were performed. The one is sensitivity analysis, which was carried out for the most simple system of SO_2/air, to determine the magnitude of parameters relating to aerosol formation and growth. Another is to evaluate the photochemical aerosol forming potentialin (PAFP) of various gaseous systems.

3.1 Sensitivity analysis : Time change of particles in SO_2/air system

Number concentration, N and average size, r of particles were calculated as a function of the reaction time, t with changing the value of each parameter. The formation (nucleation) of sulfuric acid particles depends strongly on the saturated vapor pressure, Ps of sulfuric acid. The values of 10^{-6} and $3.6x10^{-4}$ mmHg were used as Ps. The production rate of SO_3, k is a dominant factor on the particle number concentration. The coagulation constant and the accomodation factor (AF) influenced definitely on the shape of $N-t$ curve after the peak concentration. Fig.1 shows the fitting curves for experimental results illustrated in Fig.2. At 1.0×10^{-6} mmHg of P_s, the particle size was always smaller than the experimental value, and so the volume formation rate of 0.0021 $\mu m^3 cm^{-3} s^{-1}$ was fairly less than the experimental value of 0.081 $\mu m^3 cm^{-3} s^{-1}$. And in this case, considerably large coagulation constant of 10^{-7} $cm^3 s^{-1}$ (at AF=1.0) were requested to fit better. On the contrary, at 3.6×10^{-4} mmHg of P_s, the particle size was slightly larger than the experimental results. Consequently, the volume formation rate of 0.143 $\mu m^3 cm^{-3} s^{-1}$ was greater than the experimental results. It is concluded that the better fitting condition may be obtained by adopting an intermediate value between 1.0×10^{-6} and 3.6×10^{-4} mmHg, as reported by Ayers, et al. (1980).

3.2 Effect of reactants on photochemical aerosol formation

PAFP was evaluated for various $SO_2/NOx/HC$ systems. In this work, it was expressed by the number concentration, N of particles formed during a fixed reaction time (50 sec here) in which new particle formation is dominant rather than growth.

The reaction rate constants in gas–phase reactions were determined by referencing previous experimental reports and are shown in Table 1. However, the reaction rate constants of 4, 5, 8, 12 and 16th reactions were changed in accordance with the photochemical reactivity of each HC. HC was ranked in 3 groups of low, middle and high based on the photochemical reactivity.

Effect of reactants on PAFP was examined for various $SO_2/NOx/HC$ systems. HCs taken up in our study were ethane, propane and benzene (low photochemical reactivity), propylene, 1–butene and toluene (middle), and trans–2–butene and m–xylene (high).

Calculated results for propylene and m–xylene are illustrated in three dimensional way in Figs.3 and 4, respectively, and compared with experimental results. PAFP is raised with the increase of SO_2 concentration regardless of both HC and NO_2 concentration though it is not shown here. The independent existence of either NO_2 or HC in SO_2/air mixture results in a hindrance effect

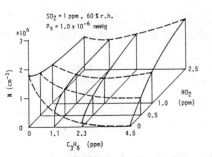
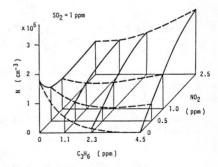

(a) Simulation; Ps=1.0x10⁻⁶mmHg   *(b) Experiment*

Fig.3 Three dimensional illustrations of PAFP in $SO_2/NOx/propylene$ system; SO_2=1ppm, 60% r.h., t=50 sec.

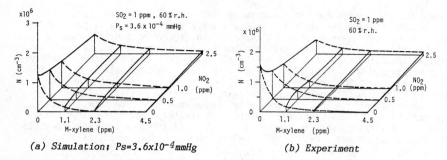

(a) Simulation; Ps=3.6x10⁻⁴mmHg   (b) Experiment

Fig.4   Three dimensional illustrations of PAFP in SO_2/NOx/m-xylene system;
SO_2=1ppm, 60% r.h., t=50 sec.

on PAFP through 16th and 17th reactions. The simultaneous presence of both HC
and NO_2 influences on PAFP in very complicated way by reactions of NOx, HC and
their intermediate products.
   As shown in Figs.3 and 4, experimental results can be simulated pretty
well by determining appropriately the reaction rate constants and the values
of aerosol dynamic parameters. And the new model can be utilized to study the
aerosol formation processes in the atmosphere and to evaluate the PAFP.

4 Conclusion

   New simulation model was led to examine the aerosol formation and growth
in multi-component system. The model supports experimental results and may
be helpful in search of photochemical aerosol formation processes. Further
investigations are expected on gas-phase reaction model and on size divided
calculation.

References

M. Kasahara and K. Takahashi (1978) "Photochemical aerosol formation from
gaseous air pollutants", Proc. International Clean Air Conference (Edited by
E.T. White), Ann Arbor Sci., 687

K. Takahashi, M. Kasahara and M. Itoh (1975) "A kinetic model of sulfuric
acid aerosol formation from photochemical oxidation of sulfur dioxide vapor",
J. Aerosol Sci., 6:45

H.W. Sidebottom, C.C. Badcock, G.E. Jackson, J.G. Calvert, G.W. Reinhardt and
E.K. Damon (1972) "Photooxidation of sulfur dioxide", Environ. Sci. Tech.,
6:72

G.P. Ayers, R.W. Gillett and J.L. Gras (1980) "On the vapor pressure of
sulfuric acid", Geopys. Res. Letters, 7:433

Published 1984 by Elsevier Science Publishing Co., Inc.

Aerosols, Liu, Pui, and Fissan, editors

A Discussion on the Heterogeneous Oxidation
of Trace Gases in Aqueous Aerosols

R. Dlugi and S. Jordan

Laboratorium für Aerosolphysik und Filtertechnik I
Kernforschungszentrum Karlsruhe GmbH
7500 Karlsruhe, W.-Germany

Extended Abstract

Introduction

A number of experiments on the catalytic trace gas removal by artificial and anthropogeneous aerosol particles and liquid systems (bulk solutions, droplets) are performed in the past with a main interest in the SO_2- and NO_x-oxidation process. Different analytical methods, chemical reactors and theories were used to get reaction rates for the systems studied. Beside the problem of rate determination the total removal of a gas by a certain aerosol mass must be known (for example the capacity of an aerosol system: the mass of products per mass of aerosols). Only with some restrictions the results being published are applicable for problems of atmospheric gas oxidation by real aerosol systems.

Rate laws:

In general the experimental data on the catalyzed SO_2-oxidation (by Mn(II), Fe(III), Fe(II)) are fitting rate laws of the form

$$d/dt \ [S(VI)] = k \ (I) \ [Me(n+)]^{\alpha_1} \ [S(IV)]^{\alpha_2} \ [H^+]^{\alpha_3} \ [O_2]^{\alpha_4}$$

with k = rate constant, I = ionic strength; α_i = const.; Me(n+) = catalyst; S(IV) = SO_2-species in solution; $[O_2]$, $[H^+]$ = oxygen- and hydrogen concentration. The sulfate formation is a function of catalyst -, S(IV)-, O_2- and H^+ -concentration; for Mn(II)-ions the coefficients are calculated to be: $0 \leq \alpha_1 \leq 2$, $0 \leq \alpha_2 \leq 1$, $\alpha_3 \approx -1$, $0 \leq \alpha_4 \leq 1$ and for Fe-ions $1 \leq \alpha_1 \leq 2$, $0.5 \leq \alpha_2 \leq 2$, $-1 \leq \alpha_3 < 0$, $\alpha_4 = 0$. It is partly an open question which type of chemical reactor should be used to minimize the errors in the determination of the reaction rates.

Results of comparison

As an example the results of the published data for the SO_2-oxidation by Mn(II)-ions are shown in table 1 and figure 1 for the conditions $[Mn(II)] \geq 10^{-3}$ mole 1^{-1}, $[S(IV)] \geq 10^{-4}$ mole 1^{-1}. The line in figure 1 fits the relation: Rate = $2.6 \cdot 10^{-9} / [H^+]$ mole $1^{-2} \ s^{-1}$ and can probably be related to a more general mechanism (see for example Crump et al., 1983; Dlugi, 1983 and literature therein).

One would expect that experiments with artificial aerosols give correct rates and capacities which can be used to estimate these quantities for real atmospheric aerosols. The catalyst concentration is large for a low water content of particles (low relative humidity) and decreases with increasing

Table 1: The results for the SO_2-oxidation by Mn(II)ions in aerosol particles for $[$ Mn (II) $] > 10^{-3}$ mole l^{-1} (according to Crump et al. (1983) and Dlugi (1983), letters see figure 3)

		SO_2- C. (ppm)	pH_o	SO_4^{2-}formed (M/kg_{H_2O})	Time min	Rate moles s^{-1}	r.H. %
1	Cheng et al., 1981 (G)	$MnCl_2$	18	$\approx$ 3.4	$\sim$ 0.34	60	$9.5 \cdot 10^{-5}$	$\approx$90 - 95
2	Dlugi et al., 1981	NaCl	0.8	$\approx$ 4.0	$\sim$ 0.09	125	$1.5 \cdot 10^{-5}$	95
3	Matteson et al., 1969	$MnSO_4$	67	$\approx$ 3,0	0.69	10	$1.15 \cdot 10^{-3}$	$\geq$ 95
4	Haury et al., 1978	$MnSO_4$	1 - 2	$\approx$ 3.8	0.71	480	$2.66 \cdot 10^{-5}$	$\approx$ 92
5	Haury et al., 1978 (H) (neu berechnet gemäß Dlugi, 1982)	$MnSO_4$	1 - 2	$\approx$ 3.8	0.71	$\sim$200	$6.66 \cdot 10^{-5}$	$\approx$ 92
6	Crump et al., 1983 (C)	$MnSO_4$	14 - 50	$\approx$3.5-3.1	$10^{-3}-10^1$	$\sim$60	$8.33 \cdot 10^{-5}$	$\approx$95 - 97
7	Kaplan et al., 1981 (E)	$(NH_4)_2SO_4$ + $MnSO_4$	0.7	$\approx$4.0-4.5	0.0025	0.75	$5.5 \cdot 10^{-5}$	$\approx$92 - 95
8	Dlugi, 1982 (D)	$MnCl_2$	0.1 - 2	$\approx$4.0-3.8	0.65	140	$8.33 \cdot 10^{-5}$	< 95
9	Clarke et al., 1981	$MgCl_2$, NaCl, $(NH_4)_2SO_4$	1 - 20	$\approx$3.9-3.35	0.16	140	$1.83 \cdot 10^{-6}$	$\approx$ 85
10	Runca-Köberich, 1979 (A)	A: Mn(II)-Ac.	0.1 - 1		$\sim$0.0005	$\sim$160	$2.08 \cdot 10^{-6}$	$\approx$ 95
	(B)	B: NH_4NO_3+NaCl+$MnCl_2$	0.1 - 1		$\sim$0.0001	$\sim$160	$1.05 \cdot 10^{-6}$	$\approx$ 95

*) estimated using published reaction rates, $3 \leq pH_o \leq 4$, t = 1 h

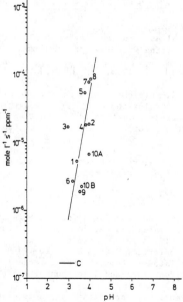

Figure 1: Mean reaction rates from Table 1, normalized for mean SO_2-gas concentrations (C = Coughanowr and Krause, 1965; see table 1, No. 6; numbers from table 1)

humidity when droplets are formed (e.q. Dlugi, 1983). For low water content the reaction rate depends only on the Mn(II)-concentration. When the Mn-concentration decreases below 10^{-3} mole 1^{-1}, the kinetics change to $\alpha_1 = 2$. As shown for special types of aerosols (fly ash from coal combustion; cement dust) in figure 2 and figure 3 both the rates and the capacities of such artificial and real systems differ largely for about the same concentrations of metal ions.

Synergistic effects of different catalysts (especially Mn-Fe) and the enhancement of reaction rate by Cl^- and NO_x^-, as reported in the literature might be a possible explanation for these differences of the reaction rates (see for example L.R. Martin, 1984). An interesting effect is shown in Fig.3 where the capacity is constant with decreasing Mn-concentrations for fly ash particles (particles containing a water mass up to 10^2 times the aerosol mass). The sulfate concentration at the end of the reaction is consistent with measured pH-values of $\text{pH} \lesssim 1$. From own experiments it is shown that the influence of NO_2 on the reaction rate was to increase the rate by a factor of 2-3 for NO_2-concentrations of 1-2 ppm (SO_2 = 1-2 ppm) and pH_o = 4.5 - 4.9.

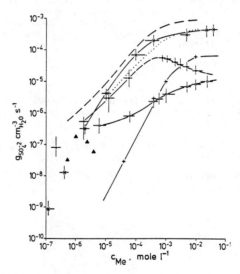

Figure 2: Reaction rates as a function of catalyst concentration for coal fly ash (x, .), cement dust (o); - + - $MnSO_4$-solution (C, Fig. 1); Δ Cohen et al. (1981) see Dlugi, 1983),calculated for: 1 % (- -), 0.1 % (...) of the available Fe-content is a catalyst.

974

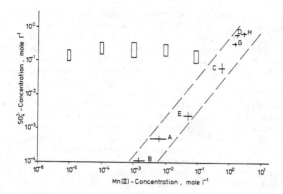

Figure 3: The total amount of SO_4^- being formed at the end of the reaction for artificial aerosols (table 1) and coal fly ash $(3,8 < pH_o < 5.0)$ (◻) (the letters are according to table 1).

These results show that for real atmospheric aerosol systems the catalytic reactions on paricles and in small droplets obviously proceed in a different way than in synthetic particles or droplets. Consequently, the gas oxidation by these real systems (fly ashes, dusts, soots) have to be studied too for atmospheric applications.

References:

Crump J. G., R. C. Flagan, J. H. Seinfeld (1983). "An experimental study of the oxidation of sulfur dioxide in aqueous mangnese sulfate aerosols", Atmos. Environ. 17, 1277

Dlugi, R. (1983): "SO_2-oxidaton in aerosol particles ad droplets", J. Aerosol. Sc. 14, 292

Martin, L.R. (1984): "Kinetic studies of sulfite oxidation in aqueous solution", Acid Precipitation Series Vol. 3, 63, Butterworth Publ., Boston

CHAPTER 22.

LUNG DEPOSITION AND HEALTH EFFECTS

Published 1984 by Elsevier Science Publishing Co., Inc.

Aerosols, Liu, Pui, and Fissan, editors

RECENT ADVANCES IN EXPERIMENTAL STUDIES OF PARTICLE DEPOSITION
IN THE HUMAN RESPIRATORY TRACT

Joachim Heyder
Gesellschaft für Strahlen- und Umweltforschung m.b.H.
Abteilung für Biophysikalische Strahlenforschung
Paul-Ehrlich-Str.20, D-6000 Frankfurt am Main 70
Federal Republic of Germany

EXTENDED ABSTRACT

An inspired particle can be transported towards airway surfaces by a number of mechanisms where it will be deposited upon
contact. Even if all particles inspired in one breath were identical, the probability of being deposited would be different for
each paticle. This is because the air adjacent to each particle
remains for a different period of time in the respiratory tract
and penetrates to a different depth in the respiratory tract,
and because of the stochastic nature of particle transport.

Therefore, the term "deposition" refers to the mean probability of an inspired particle being deposited in the respiratory
tract by collection on airway surfaces. Total deposition refers
to particle collection in the whole respiratory tract and regional deposition to particle collection in a region of the respiratory tract. Although this definition of deposition is generally accepted, several conclusions inherent in this definition
have often been ignored:

- The determination of deposition requires the use of aerosols of uniform particle number concentration.
- The concept of the mass median aerodynamic diameter is
irrelevant for the estimation of deposition.

Deposition of particles in the human respiratory tract has
been extensively documented since its first observation in 1881
by Tyndall and a considerable amount of reliable data have been
accumulated in recent years. It should however be emphasized
that caution must be executed in interpreting and comparing deposition data found in the literature since they are often based
on a different definition of deposition than that used here. Deposition data are now available for a wide range of particle sizes (0.005 - 15 um) and for various patterns of oral and nasal
breathing. Recent advances cover the following topics:

- deposition of hydrophobic and -philic particles smaller
than 0.1 um in diameter due to mechanical transport
- deposition of hydrophobic particles larger than 10 um in
aerodynamic diameter due to mechanical transport
- deposition of hydrophobic particles due to electrical
transport
- interaction of diffusional and gravitational particle
transport onto airway surfaces
- biological variability of particle deposition

Concepts of human deposition studies

Total deposition can be determined during steady breathing
of well-defined monodisperse aerosols by either counting the
number of inspired and expired particles per breath in situ or
measuring the mean particle number concentration in samples taken from inspired and expired aerosols over periods of several
breaths and taking particle losses due to the sampling procedure
into account. The latter method can be used in inhalation studies with particles of any size whereas the counting method is
only applicable in inhalation studies with particles larger than
about 0.1 um. Particle labelling techniques employing tracer
elements can be used to partition total deposition into extrathoracic and thoracic components by external detection of the
amount of tracer element in head and thorax immediately after
short-term aerosol administration.

Deposited matter however will not remain permanently in the
respiratory tract, but will be removed from it by a number of
clearance mechanisms. For matter inspired through the mouth and
which is relatively insoluble in body fluids, several phases of
removal from the thorax can be observed; a fast phase due to
particle removal from the ciliated region and a slow phase due
to removal of matter deposited in the nonciliated region. These
kinetics can be used to calculate the amount of tracer element
initially deposited in these regions which are usually refered
to as bronchial and alveolar regions. Therefore, thoracic deposition can be partitioned into bronchial and alveolar deposition. However, it should be emphasized that partitioning the
lungs into bronchial and alveolar region is based upon the behavior of matter deposited in the lungs and not on anatomical or
physiological considerations. Nasal deposition can be studied by
determining total deposition for mouth-, nose-, mouth-in-nose-
out-and nose-in-mouth-out-breathing.

Deposition depends not only on the properties of aerosols
particles and the characteristics of breathing but also on the
morphology of the respiratory tract and the distribution of the
inspired aerosols in the lungs. The influence of these factors
upon deposition can be studied in subjects breathing the same
aerosol under the same breathing conditions.

Current knowledge on deposition of compact particles

The size range of particles which can be inhaled and which
are therefore subjected to deposition can be partitioned into
several domains.

Thermodynamic domain:

In this domain, particle transport onto human airway surfaces is due to diffusion. Total deposition decreases with particle diameter, is independent of particle density and increases
with the mean residence time of the inspired aerosol in the va-

rious regions of the respiratory tract. For unit density spheres, the upper limit of this domain is 0.16 um, for 3 g cm-3
density spheres it is 0.09 um. The time dependence of total deposition of spheres larger than 0.05 um in diameter indicates
that these particles are only deposited in the alveolar region
so that total and alveolar deposition are identical. Regional
deposition data for mouth- and nose-breathing are lacking for
particles smaller than 0.05 um. For nonspherical particles, the
thermodynamic diameter is a suitable equivalent diameter to
describe deposition.

Aerodynamic domain for mouth-breathing:

In this domain, particle transport onto airway surfaces is
due to gravitational sedimentation and inertial impaction. Total
deposition increases with particle diameter, particle density,
flow rate and mean residence time of the aerosols in the respiratory tract. For unit density spheres, the lower limit of this
domain is 1 um, for spheres of 3 g cm-3 density it is 0.58 um.

When an aerosol with larger particles enters the respiratory tract its particles first experience inertial transport.
The further it penetrates into the respiratory tract the efficacy of this transport diminishes and the efficacy of gravitational transport is enhanced. The upper respiratory tract can be
considered as an impactor and the lower respiratory tract as an
elutriator. This is due to the fluid dynamics of the respiratory
tract. An inspired aerosol flows through the upper respiratory
tract at high velocity and remains only for a short period in
this region, so that particle deposition is governed by inertial
impaction. On the other hand, the long residence time of an aerosol in the lower respiratory tract is associated with a low
velocity, so that particle deposition is governed by gravitational sedimentation. Consequently, with increasing particle size
the site of deposition is shifted from the lower to the upper
respiratory tract, so that alveolar deposition of unit density
spheres decreases with particle size in the range above 3 um and
bronchial deposition in the size range above 8 um.

For mouth-breathing of unit density spheres, the aerodynamic
domain can be partitioned into three subdomains:

- a sedimentation domain where particles smaller than 2 um
are deposited solely by gravitational sedimentation, this
taking place in the nonciliated thoracic airspaces.
- a sedimentation-impaction domain where particles in the
size range 2 - 15 um are deposited by inertial impaction
in the extrathoracic and upper ciliated thoracic airways
and by gravitational sedimentation in the lower ciliated
thoracic airways and nonciliated thoracic airspaces.
- an impaction domain where particles larger than 15 um
are deposited solely by inertial impaction, this taking
place in the extrathoracic and upper ciliated thoracic
airways during inspiration. These particles are unable to
penetrate to the nonciliated portion of the lungs.

The large intersubject variability of deposition observed in the aerodynamic domain is due to the intersubject variability of airway dimensions. However, the intersubject variability of deposition rate is not determined by the intersubject variability of deposition but by the intersubject variability of ventilation rate.

For nonsperical particles, the aerodynamic diameter is a suitable equivalent diameter to describe deposition.

Aerodynamic domain for nose-breathing:

For nose-breathing of unit density spheres, the aerodynamic domain can be partitioned into two subdomains:

- an impaction-sedimentation domain where particles in the size range 0.3 - 9 um are deposited by inertial impaction in extrathoracic and upper ciliated thoracic airways and by gravitational sedimentation in lower ciliated thoracic airways and nonciliated thoracic airspaces.
- an impaction domain where particles larger than 9 um are solely deposited by inertial impaction, this taking place in extrathoracic and upper ciliated thoracic airways during inspiration. These particles are unable to penetrate to the nonciliated portion of the lungs.

Intermediate domain:

For nose-breathing the aerodynamic domain is contiguous with the thermodynamic domain. For mouth-breathing however, these domains are seperated by an intermediate domain. This domain is characterized by simultaneous diffusional and gravitational particle transport towards airway surfaces.

Deposition assumes a minimum when diffusional and gravitational particle losses are equal. This minimum is independent of the breathing pattern and only dependent on the particle properties. For particles of unit density the lowest deposition is observed for spheres of 0.36 um diameter and for particles of 3 g cm-3 density, for 0.26 um spheres.

While thermodynamic and aerodynamic diameters are useful equivalent diameters for describing pure diffusional and pure gravitational particle transport, no equivalent diameter can be defined which accounts for simultaneous diffusional and gravitational particle transport. This refers not only to nonspherical particles but also to spheres of unknown density.

While diffusional and gravitational particle transport are least effective in the intermediate domain, electrical transport is most effective. It is however unknown whether or not electrical and mechanical particle transport are independent processes.

Published 1984 by Elsevier Science Publishing Co., Inc.

Aerosols, Liu, Pui, and Fissan, editors

THEORETICAL LUNG DEPOSITION OF HYGROSCOPIC AEROSOLS

G.B. Xu and C.P. Yu
Department of Mechanical and Aerospace Engineering
State University of New York at Buffalo
Amherst, NY 14260
USA

Introduction

It has been well established that the amount of deposition of inhaled aerosol particles in the lung depends strongly upon particle size. When a hygroscopic aerosol is inhaled into the lung, the particles will grow in size as they travel along the humid airways. The deposition pattern of this aerosol may therefore differ considerably from that of a nonhygroscopic aerosol.

Particle growth in a humid environment is a process involving simultaneous heat and mass transfers. The growth rate of the particle depends not only upon its physical properties and water content, but also upon the local water vapor concentration, temperature and flow field of the surrounding air. Ferron (1977) was the first one to summarize the mathematical equations for calculating particle growth during respiration. These equations were later used in a study (Ferron and Hornik, 1983) to predict the deposition of dry NaCl particles in the lung for several initial particle diameters. Recently, deposition data of monodisperse NaCl particles in the lung under controlled conditions have become available (Tu and Knutson, 1983; Blanchard and Willeke, 1983). The purpose of this paper is to provide a further theoretical study of the problem and to obtain results which can be compared directly with the experiment.

Equation for Particle Growth

The mathematical equations given by Ferron (1977) for determining particle growth in the airways are quite complex. To facilitate calculation, we have derived a simple empirical equation for the particle diameter d as a function of time t in the following form:

$$\frac{d-d_o}{d_\infty-d_o} = 1-\exp(-2.3 \ \frac{t^{0.62}}{d_o^{1.12}}) \tag{1}$$

where d_o and d_∞ are, respectively, the initial and final equilibrium diameters of the particle. The value of d_∞ is determined from the equation (Ferron, 1977)

$$\frac{d_\infty}{d_o} = \left[\frac{\rho_o}{\rho_\infty} \ (1 + \frac{M_w}{M_o} \ \frac{iH}{1-H})\right]^{1/3} \tag{2}$$

in which ρ_0 and d_0 are the initial mass density and diameter of the dry particle, M_w and M_0 are molecular weights of water and dry particle respectively, H is the relative humidity, and i is the number of ions into which a salt molecule dissociates in water.

Deposition Model and Results

Using Equations (1) and (2), we have calculated deposition in a Weibel's lung for dry NaCl particles using a deposition model previously developed by us (Yu and Diu, 1982). Figure 1 presents total and regional deposition results with mouth breathing at a respiratory condition of 500 ml tidal volume and 13.7 respirations per minute with no pause. It is shown that the minimum total deposition for NaCl particles occurs at an initial particle diameter of 0.08 µm. This is smaller by a factor of six to seven than the usual value of 0.5 µm found for nonhygroscopic aerosols. Consequently, we may expect that a 0.08 µm NaCl particle to grow six to seven times of its initial size in the airways during a respiratory cycle. The result is consistent with experimental observation (Tu and Knutson, 1983). It is also shown in Figure 1 that for initial particle diameters larger than 0.2 µm, NaCl particles have larger total deposition than their nonhygroscopic counterpart; whereas total deposition is lower for NaCl particles if the initial particle diameter is less than 0.2 µm.

Figure 1 also shows the changes in regional deposition due to particle growth. Although the head deposition is larger for NaCl particles at all initial particle sizes, deposition in the tracheobronchial and alveolar regions for NaCl particles can be larger or smaller than nonhygroscopic particles depending upon the initial particle size.

In Figure 2, we present a comparison between the calculated total deposition results and experimental data for NaCl particles. The recent data of Tu and Knutson (1983) and Blanchard and Willeke (1983) were used for this comparison. The agreement between the theoretical deposition and Tu's data is reasonably good for all particle sizes. The data of Blanchard and Willeke (1983) however are consistently higher, due possibly to the effect of intersubject variability.

Tu and Knutson (1983) have also made measurements on particle recovery from the exhaled air after inhalation of NaCl particles and followed with a period of breath-holding. Such an experiment is useful from the point of view of medical application since most medical aerosols are hygroscopic and the breathing-holding exercise will help increase the dosage of these aerosols on the lung tissues. Figure 3 shows the experimental and theoretical particle recovery results for inhalation of both NaCl and nonhygroscopic particles of 550 ml. The theoretical results were obtained following the previous work by Yu and Thiagarajan (1979). It is seen that the theoretical recovery agrees fairly well with the data, and as expected, it decreases with increasing breathing-holding time. At an initial particle diameter of 0.05 µm, the recovery fraction is much larger for NaCl particles, due to the decrease in diffusional deposition as the particles grow in size. However, for an initial particle diameter of 0.15 µm, it is observed that the recovery fractions for both NaCl and nonhygroscopic particles are about equal, but NaCl particles have a faster deposition rate during breath-holding. This is caused by the presence of sedimentational deposition for NaCl particles.

Acknowledgment

This work was supported by Grant No. OH-00923 from the National Institute for Occupational Safety and Health and Grant No. ES-02565 from the National Institute of Environmental Health Sciences.

References

Blanchard, J.D. and Willeke, K. (1983), "An Inhalation System for Characterizing Total Lung Deposition of Ultrafine Particles", Am. Ind. Hyg. Assoc. Conf., Philadelphia.

Ferron, G.A. (1977), "The Size of Soluble Aerosol Particles as a Function of Humidity of the Air. Application to the Human Respiratory Tract", J. Aerosol Sci. 8:251.

Ferron, G.A. and Hornik, S. (1983), "Influence of Different Humidity Profiles on the Deposition Probability of Soluble Particles in the Human Lung", GAeF Conf., Neuherberg.

Tu, K.W. and Knutson, E.O. (1983), "Total Deposition of Ultrafine Hydrophobic and Hygroscopic Aerosols in the Human Respiratory System", AAAR Conf., Maryland.

Yu, C.P. and Thiagarajan, V. (1979), "Decay of Aerosols in the Lung During Breath Holding", J. Aerosol Sci. 10:19.

Yu, C.P. and Diu, C.K. (1983), "Total and Regional Deposition of Inhaled Aerosols in Humans", J. Aerosol Sci. 14:599.

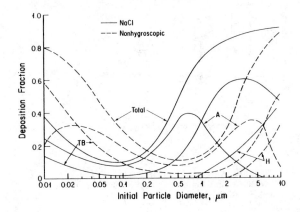

Figure 1. Calculated Total and Regional Deposition of NaCl and Nonhygroscopic Particles. H: Head Deposition, TB: Tracheobronchial Deposition, A: Alveolar Deposition.

984

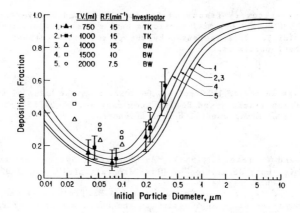

Figure 2.   Comparison of Calculated Total Deposition Results
for NaCl Particles with Experimental Data.

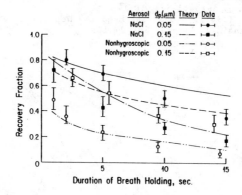

Figure 3.   Theoretical and Experimental Recovery Fractions
for NaCl and Nonhygroscopic Particles versus
Breath Holding Time.

Published 1984 by Elsevier Science Publishing Co., Inc.

Aerosols, Liu, Pui, and Fissan, editors

PHYSICAL AND CHEMICAL ANALYSIS OF COBALT OXIDE AEROSOL PARTICLES USED FOR INHALATION STUDIES

W.G. Kreyling, G.A. Ferron

Institut für Strahlenschutz, Gesellschaft für Strahlen- und Umweltforschung mbH München, Ingolstädter Landstr. 1, D-8042 Neuherberg, F.R.Germany

EXTENDED ABSTRACT

Introduction

Cobalt containing industrial aerosols will frequently be produced by high temperature processes. The resulting aerosol particles originate either by thermal degradation of cobalt compound particles or by condensation of cobalt oxide which had been decomposed from vaporized cobalt compounds. In any of these cases the particles may contain di- and threevalent cobalt. The corresponding cobalt oxides are: divalent cobaltous oxide CoO, threevalent cobaltic oxide Co_2O_3, and di- and threevalent Co_3O_4 cobaltosic oxide.

According to the literature of inorganic chemistry it is impossible to compose Co_2O_3 by high temperature processes in the ambient air. Co_3O_4 is obtained at temperatures above 300°C. Above 1200°C CoO is yielded in a reducing atmosphere. This composition is not stable but oxidizes to Co_3O_4. Depending on the generation process the particles might be porous with varying density and might consist of cristalline and amorphous structures.

Recently we have studied the long term lung retention depending on some physicochemical parameters of cobalt oxide aerosols which had been inhaled by dogs, Kreyling et al. (1984). They are produced under controlled conditions including a high temperature process described earlier by Kreyling and Ferron (1984). In this contribution the physical and chemical properties of the cobalt oxide particles produced under various conditions are discussed.

Experimental

The aerosols are produced by thermal degradation of monodisperse cobalt nitrate aerosols at degradation temperatures between 800°C and 1200°C resulting in monodisperse cobalt oxide particles as demonstrated in the micrograph of fig. 1a. As shown previously (Kreyling and Ferron, 1984) the solubility of the particles studied by a long term dissolution test varies considerably depending on their size and the degradation temperature. The solubility increases with decreasing particle size. Particles treated at 800°C show minimal solubility during two months, whereas particles treated at 1000°C and 1150°C are nearly completely dissolved during the first 10 days. The particle solubility decreases slightly with increasing reaction time in the tube furnace.

The density ρ of the particles is calculated from the mean geometric diameter DE measured by a scanning electron microscope (SEM) and the median aerodynamic diameter DA measured by the Stöber centrifuge (Ferron and Kreyling, 1979) using the equation:

$$\rho \ / \ \rho_0 = (\ DA^2 \ C(DA) \ ) \ / \ (\ DE^2 \ C(DE) \ ) \tag{1}$$

where ρ_0 is the unit density and C is the Cunningham slip correction.

The density of bulk CoO is 6.5 g/cm^3 and that of bulk Co$_3$O$_4$ is 6.1 g/cm^3. In case of monodisperse cobalt oxide particles decomposed from cobalt nitrate particles the density is about half of the bulk cobalt oxide and decreases with increasing particle diameter. In a series of eight experiments monodisperse aerosols with varying sizes between 1.5 μm and 5.1 μm aerodynamic diameter are produced under similar generation conditions. The density ρ can be described by a linear regression function:

$$\rho\,(DA) = 4.1 - 0.28\,DA \qquad\qquad (2)$$

with a correlation factor of 0.83. The density of 4.9 g/cm^3 for particles with a mean geometric diameter of 0.3 μm produced under slightly different conditions supports equation 2.

Starting with monodisperse cobalt chloride aerosols a bimodal distributed cobalt oxide aerosol is obtained. Since the cobalt chloride particles partially evaporate before being thermally degraded, this vaporised amount decomposes and condenses to a small sized fraction besides the original particle fraction. The latter is still monodisperse and spherical shaped, as can be seen in the micrograph of fig. 1b.

The particles are not broken into shells and fragments as would be obtained in the case of insufficient predrying and cristal water extraction. This is reported by Kanapilly et al. (1970) and confirmed by our studies. However the density of the large particle fraction with values between 1 g/cm^3 and 3 g/cm^3 is reduced drastically compared to the bulk density.

Similarly the density of the condensation particles is determined to be less than 1 g/cm^3 for mean geometric diameters between 0.08 μm and 0.15 μm. These monodisperse samples of the polydisperse condensation particle fraction are taken from certain positions of the centrifuge foil.

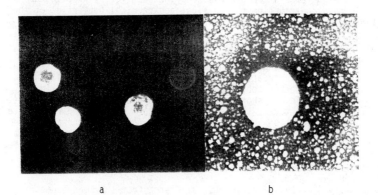

a   b

Figure 1. SEM micrographs from 2 μm cobalt oxide particles degraded from Co(NO$_3$)$_2$ (a) and from CoCl$_2$ (b) with condensation particles.

The actual density of the aerosol particles is about half or less than that of the bulk material indicating porous particle bodies. As demonstrated in both micrographs of fig. 1 the particles are not perfect spheres but show a rough surface area. No pores can be observed by the SEM. So due to its resolution power of about 20 nm the pores must be smaller.

The specific surface area of a sample of monodisperse cobalt oxide particles with a geometric mean diameter of 1.5 μm degraded from cobalt nitrate at 800°C is determined using a surface area meter according to the method of Brunauer-Emmett-Teller. The specific surface area is 12 ± 3 m²/g which is ten times larger than the surface area of spheres with 1.5 μm diameter. This confirms the moderate porosity of the particles. According to the measured specific surface area a mean pore size of about 20 nm can be calculated. As mentioned above pores of this size are hardly be detected by the SEM which agrees to the observations by the SEM. Since the structure of the surface area might be different from inner structures pores of other size cannot be excluded.

In order to determine the chemical composition of the particles generated at various temperatures either by thermal degradation or by evaporation and subsequent condensation the following analysing techniques are used: infrared (IR) spectroscopy, X-ray diffraction analysis by the Debye-Scherrer method, X-ray photoelectron spectroscopy for chemical analysis (ESCA), and Auger electron spectroscopy (AES). Reference powders are for comparison and calibration purposes.

Table 1. Chemical compositions of various cobalt oxide aerosols and reference powders.

Aerosolparticles			IR-Spectroscopy		X-ray Diffraction	
Degrading salt	Geom. Diameter (μm)	Degrad. Temp. (°C)	Co_3O_4	CoO	Co_3O_4	CoO
$Co(NO_3)_2$	1.2	800	yes	no	yes	yes
"	1.1	950	yes	no	yes	yes
"	1.1	1050	yes	yes	yes	yes
$CoCl_2$	1.6/<0.3	1100	yes	yes	yes	yes
"	1.6	"	yes	yes	yes	yes
"	<0.3	"	yes	yes		
Reference Powders						
Co_3O_4			yes	no	yes	no
Cobalt oxide 1300°C *			yes	yes		
Cobalt oxide 1450°C *			yes	yes		

* $Co(NO_3)_2$ and metallic cobalt powder decomposed in a tube furnace.

The IR spectra of Co_3O_4 and of CoO are clearly distinguishable as reported by Barnes et al. (1976). The former shows three narrow absorption minima and the latter a broad minimum. The samples treated at 800°C and 950°C show no CoO but only a spectrum for Co_3O_4 as indicated in table 1. The higher fired particles show both spectra. Because of the lack of a CoO reference sample cobalt nitrate and metallic cobalt powder is decomposed in the ambient air of a tube furnace at temperatures of 1300°C and 1450°C in order to obtain CoO. However the IR spectra again show both CoO and Co_3O_4 absorption minima in all cases indicating no pure CoO. Hence no aerosol particles of pure CoO can be produced in ambient air at high temperatures as long as there is no reducing atmosphere.

Similar results are obtained by the X-ray diffraction analysis as shown in table 1. Additionally this method indicates that the particles consist of a mixture of polycristalline and amorphous structures except of the condensation particles. These do not show any polycristalline structures. It is interesting that even those particles degraded at 800°C and 950°C show some amount of CoO which could not be observed by IR spectroscopy.

Using ESCA it is very difficult to distinguish between CoO and Co_3O_4 because the binding energies of the photoelectrons differ very little between the two chemical compositions. Calculating the fractions of cobalt and oxygen is another possibility of analysing the chemical composition which is in progress.

Using AES it is possible to analyse a depth profile of the cobalt and oxygen fraction in single particles. Analysing a few particles which are degraded from cobalt nitrate at a temperatue of 800°C or 950°C the cobalt to oxygen ratio is constant from the outer surface to 25% of the radius and is in between the ratios for CoO and Co_3O_4. This agrees with the analysis by X-ray diffraction. The constant ratio along the radius indicates a homogeneous particle composition.

Conclusion

In conclusion all cobalt oxide particles generated by either of the two methods at various temperatures consist of amorphous and polycristalline structures of both cobaltous oxide CoO and cobaltosic oxide Co_3O_4. The density of the particles degraded from cobalt nitrate is about half of the bulk density resulting in porous particle bodies with an increased surface area. The density of the small sized particles condensed from vaporized $CoCl_2$ is about 15% of the bulk density.

Acknowledgements

The authors are grateful for the support in chemical analyses to Prof. Dr. H.P. Böhm, Prof. Dr. G. Zundel, and Dr. W. Steurer of the Ludwig-Maximillian Universität, and to Dr. R. Henkelmann of the Technische Universität München.

References

Barnes, J.E., Kanapilly, G.M., Newton, G.J. (1976) Health Physics 30: 391.

Ferron, G.A., Kreyling, W.G., Haider, B. (1979) Proceedings of the 7th Annual Conference of the Association for Aerosol Research held in Düsseldorf, p.163 abstract in (1980) J. Aerosol Sci. 11: 248.

Kanapilly, G.M., Raabe, O.G., Newton, G.J. (1970) J. Aerosol Sci. 1: 313.

Kreyling, W.G., Ferron, G.A., Haider, B. (1984) J. Aerosol Sci. 15: (in press).

Kreyling, W.G., Ferron, G.A. (1984) J. Aerosol Sci. 15: (in press).

Published 1984 by Elsevier Science Publishing Co., Inc.

Aerosols, Liu, Pui, and Fissan, editors

AN ESTIMATION OF THE TAR PARTICULATE MATERIAL DEPOSITING IN THE RESPIRATORY
TRACTS OF HEALTHY MALE MIDDLE- AND LOW-TAR CIGARETTE SMOKERS

J.N. Pritchard and A. Black
Environmental and Medical Sciences Division,
AERE Harwell, Oxfordshire OX11 ORA, England.

EXTENDED ABSTRACT

Introduction

Some concern has been expressed over the relevance of machine smoking
cigarettes in the estimation of the tar inhaled by cigarette smokers.
Further, individual brands may be smoked in different ways. For example, in
low-tar cigarettes, the smoke is diluted with air drawn through perforations
in the filter and, in consequence, the cigarette may be more intensively
smoked. This has raised doubts as to whether actual tar yields to the smoker
are less from cigarettes with lower machine tar yields. Hence, it is important
that estimates are made of the tar deposition in smokers of different tar groups
of cigarette, and of the effects of changing grades.

Attempts to measure tar deposition in the past have relied on indirect
methods (e.g. see Hinds et al., 1983). In many cases, the data must be
related to mainstream deliveries estimated by machine smoking. At Harwell, a
radioactive label for the tar (particulate matter, water free - PMWF) has
been developed which is suitable for administration to volunteers. This has
the advantage that the amount of PMWF deposited may be measured directly in
vivo using whole-body counting techniques. Further, by use of a collimation
system, deposition may be categorised into different regions of the respiratory tract. These techniques have been utilised in a study of the regional
deposition and short-term clearance in smokers of middle-tar cigarettes who
switched to a low-tar brand for a period of six months.

Methods

The tracer developed for the PMWF is ^{123}I-labelled 1-iodohexadecane
(IHD), which has a boiling point falling in the middle of the range of compounds present in tar. It is introduced uniformly along the cigarette in a
solution of methylene chloride using a modified syringe drive. It is nontoxic, relatively inert and appears to undergo little pyrolysis during
cigarette combustion. It has been demonstrated that the IHD attaches itself
uniformly to the tar across the range of particle sizes encountered when
cigarettes are smoked under a wide variety of puff flow-profiles. Thus, the
quantity of radio-activity can be directly related to PMWF.

Deposition was measured using a single, 22 cm diameter uncollimated
NaI(Tl) detector in a geometry whose counting efficiency is insensitive to
the distribution of material in the body. Radioactivity present in the head,
thorax and abdomen was determined using a system of six 15 cm diameter
collimated NaI(Tl) detectors (Pritchard and Black, 1983). Material present in
the abdomen initially, has been rapidly cleared from airways external to the
lung by swallowing. This is combined with material still in the head to give

the deposit external to the lungs, i.e. in the upper respiratory tract (URT). Clearance from the lungs was subsequently followed for up to 2 days.

A pool of 23 healthy male middle-tar smokers were used in the study. They covered a wide range in terms of age, stature and cigarette consumption. All exhibited lung function within the normal range, as measured by forced expiratory manœuvres. The results from a group of 4 low-tar smokers are included for comparison. Cigarettes were issued ad libitum on a weekly basis to allow acclimatisation to the brand representing each tar group. On the majority of these occasions, the subject's smoking and breathing patterns were recorded and subsequently analysed using an Apple microcomputer. In order to minimise the effect of the experimental environment, a distraction in the form of 8 mm cine film was provided. After 1 month's acclimatisation, volunteers smoked a radiolabelled cigarette in as natural a manner as possible, subject to the constraints imposed by the radiological protection of the personnel involved. The enclosure in which the cigarettes were smoked was ventilated such that exposure to sidestream and exhaled smoke was negligible. Only respiratory pattern was monitored on these occasions. Middle-tar smokers then changed to the low-tar brand and radiolabelled low-tar cigarettes were smoked 1, 3 and 6 months after the switch.

At the beginning and end of their participation, subjects inhaled 2.5 μm polystyrene microspheres, labelled with ^{99m}Tc, in order to compare the tar deposition and clearance data with those of insoluble particles. The apparatus and procedures used have already been described (Walsh et al., 1976). In brief, the aerosol was inhaled by mouth breathing through a tube at a pre-determined flow-rate and breathing frequency, controlled by the subject with the aid of audio and visual signals.

Results

Clearance of both the IHD and the polystyrene microspheres could be expressed as two phase exponentials, with similar fast, but different slow phases. In Table 1, the mean values of half-times ($t_{\frac{1}{2}}$) for the fast and slow phases of clearance for the two materials are given, together with the proportion cleared by the slow phase. The larger error in the determination of the fast $t_{\frac{1}{2}}$ for polystyrene is due to its relatively short duration (typically 1 h). Since the slow phase $t_{\frac{1}{2}}$ for polystyrene is of the order of tens to hundreds of days (Bailey et al., 1982), it is not measurable over the duration of the study. These data are in good agreement with other subjects tested at this laboratory (Pritchard et al., 1980). That slow phase clearance is detectable for IHD indicates that the label is soluble in the lung. For inert particles such as polystyrene, it is commonly accepted that the slow phase of clearance represents pulmonary (P) clearance and the fast phase, mucociliary clearance of particles deposited in the tracheobronchial (TB) region of the lung. Since the fast phase $t_{\frac{1}{2}}$ for IHD is not significantly different to that of the polystyrene particles, we can infer that this represents mucociliary clearance of tar and, hence, TB deposition. Thus, by following lung clearance over two days, the lung deposit can be subdivided into P and TB regions for both tar and polystyrene particles. The regional deposition of tar (PMWF) thus derived, expressed as a fraction of the mean total deposit from a middle-tar cigarette, is summarised in Table 2.

Table 1. Short-term clearance of polystyrene particles and tar (Mean ± S.E.M.)

Experiment (No. of subjects)	Half-time of clearance (h) for:- a) Fast phase	b) Slow phase	Proportion of slow phase (%)
Polystyrene particles (52)	1.98+0.36	-	70.6+3.4
Middle-tar cigarettes (23)	1.98+0.12	17.8+1.1	38.1+2.4
Low-tar cigarettes (74)	1.92+0.09	17.5+0.5	35.1+1.1

Using a paired t-test, it can be shown that the total, TB and P depositions all drop significantly on switching from middle- to low-tar cigarettes, with a further significant drop between 3 and 6 months after the change (p <0.01). This may reflect the subjects becoming accustomed to the drop in smoke concentration and no longer smoking the cigarettes as intensively. None of the low-tar data differ significantly from the small group (4) of habitual low-tar smokers (p<0.01). The exception to these findings was URT deposition, which remained constant over the course of the study. Since a large proportion of this deposition will occur in the mouth and throat, its constancy could suggest a modification in the smoking pattern to achieve a similar degree of taste from the comparatively diluted smoke of the low-tar cigarette. Certainly, this could not be explained by the small difference in particle size (0.1 μm) observed between the two tar brands.

Table 2. Regional deposition of tar (PMWF) relative to total MT (Mean ± S.E.M.)

Region	Deposition (%) MT	LT1	LT2	LT3	LT
Total	100.0 ± 6.8	74.7 ± 5.5	74.7 ± 5.5	67.1 ± 5.1	57.8 ± 5.9
URT	13.9 ± 3.4	14.3 ± 3.4	8.0 ± 2.1	15.6 ± 3.4	8.4 ± 2.5
TB	51.1 ± 3.4	38.4 ± 3.4	42.2 ± 3.8	33.8 ± 3.0	32.9 ± 5.1
P	35.0 ± 4.2	21.5 ± 2.1	24.5 ± 2.5	18.1 ± 2.1	15.2 ± 2.1
Lung (P+TB)	86.1 ± 6.8	60.3 ± 5.5	66.7 ± 6.3	51.5 ± 4.2	48.9 ± 7.2

MT Habitual middle-tar   LT1 Low-tar month 1   LT3 Low-tar month 6
LT Habitual low-tar   LT2 Low-tar month 3

The U.K. Government publishes tables of the tar yields, excluding water
and nicotine, of cigarettes smoked under standard conditions. The ratio of
these values for the low- and middle-tar brands used (8-9 and 16-17 mg
respectively) is approximately 50%. That the subjects deposit a higher proportion than this (58-75%), even in the lung, confirms that the low-tar
cigarette is smoked somewhat more intensively than the middle-tar cigarette.
The ranking of the products in the published tables, however, remains correct.

Hinds et al. have estimated the proportion of inhaled smoke that is
deposited to range from 33 to 75% in only 6 male smokers (a factor of approximately 2.5). When combined with an observed range in deliveries of about a
factor of 20 (Green, 1978), we would expect the deposition also to range by
this factor. However, we only observe an average 4.8 fold range in deposition
over the different experiments. This suggests that the pattern of smoke
inhalation exerts a controlling influence on deposition, reducing the overall
range of tar exposure. Analysis of subjects' smoking and inhalation patterns,
not yet complete, may serve to elucidate these findings.

Summary

Smokers who switch from middle- to low-tar cigarettes will deposit less
tar in their respiratory tract. Compensation for the drop in smoke concentration on switching manifests itself in constant upper, but significantly
reduced lower, respiratory tract deposits. Due to the complexity of the
relationship between yield and deposit, the role of tar tables is to rank
brands rather than predict absolute levels of deposition.

References

Bailey, M.R. et al. (1982) Int. Environ. Safety, February, 18.

Comer, A.K. and Creighton, D.E. (1978) "The effect of experimental
conditions on smoking behaviour" in 'Smoking behaviour, physiological and
psychological influences' (Ed. R.E. Thornton), Churchill Livingstone, London.

Green, S.J. (1978) "Ranking cigarette brands on smoke deliveries" Ibid.

Hinds, W. et al. (1983) Am. Ind. Hyg. Assoc. J., 44, 113-118.

Pritchard, J.N. et al. (1980) "A comparison of the regional deposition of
monodisperse polystyrene aerosols in the respiratory tracts of healthy male
smokers and non-smokers" in 'Aerosols in science medicine and technology'
(Edited by W. Stöber and D. Hochrainer), Gesellschaft für Aerosolforschung,
Schmallenberg, W. Germany.

Pritchard, J.N. and Black, A. (1983) U.K.A.E.A. Report AERE-R 10736 (Revised)
Harwell, Oxfordshire, U.K.

Walsh, M. et al. (1976) U.K.A.E.A. Report AERE-R 8221, Harwell,
Oxfordshire, U.K.

Published 1984 by Elsevier Science Publishing Co., Inc.

Aerosols, Liu, Pui, and Fissan, editors

DEPOSITION OF AEROSOL PARTICLES IN COMPLEX FLOW FIELDS IN A CIRCULAR TUBE

Muhilan Pandian, Leonard K. Peters, and Asit K. Ray
Department of Chemical Engineering
University of Kentucky
Lexington, Kentucky 40506
U.S.A.

EXTENDED ABSTRACT

Introduction

The interior walls of the human lung are vulnerable regions for the collection and accumulation of inhaled aerosol particles. The deposition processes involved along with the lung morphology and its complex flow conditions are important in calculating the regional deposition of aerosol particles. Research thus far on the deposition of aerosol particles in the various generations of the human lung has assumed either flat or parabolic flow profiles. The mathematical schemes attempted do not permit an easy introduction of complex velocity profiles. In this study, a method that simultaneously combines inertial impaction, gravitational settling, and diffusion is used to calculate the fractional deposition of aerosol particles in simple geometries with complex flow fields.

Theoretical

The starting point of the method is Langevin's equation for a fine particle of relaxation time τ in a fluid with velocity $\vec{v}$,

$$\frac{d\vec{u}}{dt} = \frac{1}{\tau} (\vec{u} - \vec{v}) + \vec{u}_g ,$$

where $\vec{u}$ is the velocity of the particle, and $\vec{u}_g$ is the particle settling velocity. The particle trajectory is obtained by numerically solving the above equation using the stiff methous of Gear. This is done for different starting positions at the entrance to the domain of interest.

Deposition by Brownian diffusion is included by assuming that a particle diffuses about its trajectory, which is considered the most probable particle path. At any time t, the location of the particle with respect to its trajectory is given by

$$\frac{\partial w}{\partial t} = D_{Br} \frac{\partial^2 w}{\partial \vec{s}^2} ,$$

where $w(\vec{s},t)d\vec{s}$ is the probability that the particle with position $\vec{s}$ reaches some surface boundary in time t, and D_{Br} is the Brownian diffusion coefficient. The probability that the particle is collected can be obtained by applying the initial condition $w(\vec{s},0) = 0$ and the boundary condition $w(\vec{s}_1,t) = 1$, where $\vec{s}_1$ is the distance of the surface boundary from the particle trajectory. The results pertaining to the theory of Brownian motion

994

are applicable as long as the time intervals in the trajectory calculations are sufficiently large compared to the particle relaxation time. The fraction of an incoming uniformly distributed aerosol that is deposited within a certain distance downstream is then computed by a simple quadrature routine that sums up the effects of the various starting positions.

Results

The analysis has been compared for extreme conditions of the Peclet number (Pe = Uh/D_{Br}; where U = mean flow velocity; h = characteristic dimension of geometry studied, for example, the diameter in the case of a circular tube) with the available solutions from standard approaches. For particle diameters greater than about 5 μm, the effects of diffusion become negligible as expected. The numerical procedures handle particle diameters as small as 0.01 μm, although the equations become stiffer and thus more difficult to solve.

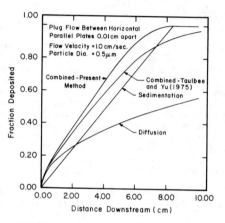

FIGURE 1. Comparison of fractional deposition in a horizontal parallel plate system.

The results obtained for deposition by the combined mechanisms for flow between parallel plates can be compared with those obtained by Taulbee and Yu (1975), and for flow in a circular tube with those obtained by Taulbee (1978). In Figure 1, the data obtained from the approach of Taulbee and Yu are compared with those of the present study for plug flow between horizontal parallel plates separated by 0.01 cm. The plot also includes values of fractional deposition computed for the individual mechanisms. Velocity profiles representative of those occurring in the first few generations of the lung are used to determine the deposition patterns in the lung. These flows are transient and quite complex exhibiting various circulations. The results help to identify the important processes at the various levels of the lung configuration.

Acknowledgements

This research was supported by the Kentucky Tobacco and Health Research Institute.

References

Taulbee, D. B. and Yu, C. P. (1975) "Simultaneous Diffusion and Sedimentation of Aerosols in Channel Flows", J. Aerosol Sci. 6:433.

Taulbee, D. B. (1978) "Simultaneous Diffusion and Sedimentation of Aerosol Particles from Poiseuille Flow in a Circular Tube", J. Aerosol Sci. 9:17.

Published 1984 by Elsevier Science Publishing Co., Inc.

Aerosols, Liu, Pui, and Fissan, editors

996

DEPOSITION AND RETENTION OF LARGE PARTICLES IN RODENTS*

H. C. Yeh, M. B. Snipes and R. D. Brodbeck
Inhalation Toxicology Research Institute
Lovelace Biomedical and Environmental Research Institute
P. O. Box 5890
Albuquerque, NM 87185
U. S. A.

EXTENDED ABSTRACT

Introduction

There is a general lack of data on the deposition and retention of large particles (> 8 μm) in the respiratory tracts of people. Especially lacking are data on long-term retention of these particles. Large particles like this are of special concern because they are frequently encountered in occupational settings with materials presumed to have low toxicity. Data from our laboratory using intratracheal instillation techniques indicated that large particles have long retention half-times in the rat lung (Snipes and Clem, 1981). To extend these observations, we have initiated studies with inhalation exposure of rats to particles ranging from 3 to 15 μm in diameter. The initial problems that had to be overcome related to difficulties in generating and delivering large particles to animal noses because of sedimentation and inertial losses of large particles during generation and transport. This paper describes the design, construction and evaluation of a large particle exposure system for rodents.

Materials and Methods

To minimize aerosol losses during its transport from the aerosol generator to the animal exposure ports, the dry powder aerosol generator and the animal exposure chamber were integrated into a single unit. The system consists of two parts, a fluidized bed generator and an exposure chamber (Raabe et al., 1973). Figure 1 is a cross-sectional diagram of the large particle exposure chamber for the small animals. The aerosol from the fluidized bed generator first passes up a central column "A" where a ^{85}Kr ionization tube is seated to bring aerosol closer to Boltzmann charge equilibrium. The aerosol then flows out the top of a convexed male radius nose piece into the exposure chamber. Clean air is injected through ports in the base plate and exists under the radius cap of column "A," sheathing the inner surface of the exposure chamber (space between columns B and C) to reduce aerosol losses to the inner wall. The top of column "C" is threaded to accommodate a concaved female radius cap that reduces turbulance while directing the aerosol flow down between columns "B" and "C." Column "C" has 40 ports arranged in four rows of ten ports each to allow the aerosol to reach the animals' breathing zone or aerosol sampling devices. Column "D" has the same configuration of ports as "C," but also has an exhaust vent

---

*Research performed under U. S. Department of Energy Contract Number DE-ACO4-76EV01013.

above each port from which the aerosol is exhausted through filters, assuring that each animal breathes a fresh separate supply of the same aerosol. The exposure unit is housed in a Lucite glove box, which is maintained at a slightly negative pressure relative to the exposure room. The glove box is equipped with two pass boxes to facilitate handling of animals and aerosol sampling devices.

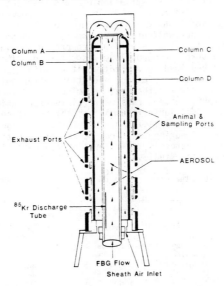

Figure 1. Schematic diagram of the large particle exposure unit.

A 5-cm I.D. brass fluidized bed aerosol generator developed at this laboratory was used to produce the test aerosols (Carpenter and Yerkes, 1980). The operating characteristics of the exposure unit were tested using nearly monodisperse polystyrene latex (PSL) particles and [46]Sc-labeled PSL microspheres (Medical Products Division/3M, New Brighton, MN 55112) were used for deposition and retention studies. The particles were supplied as dry black powders with a density of 1.23 ± 0.05 g/cm^3. The system was evaluated with 3.0, 9.0, and 15 μm aerosols. The powders were sonicated for 30 min in 2.0 ml of ethanol containing 0.25 cm^3 of clean stainless steel bed material to break up the clumps. After sonication, the suspension was poured onto 100 cm^3 of stainless steel bed material in the generator. An additional 100 cm^3 of stainless steel bed material was added and air-dried at a flow rate of 4.0 lpm. At this flow rate, the bed will not be fluidized.[2]

Results and Discussion

The background aerosol generated from the cleaned stainless steel bed material was found to be 127 ± 72 ng/L for the flow conditions used in this study. Filter samples were taken simultaneously at two different exposure port positions to evaluate the uniformity of the mass distribution within the chamber. An aerosol mass variation of less than 10% was achieved. The sonication pretreatment of the microsphere powder was found to be critical for deagglomeration to achieve single particles. Electron microscopic observations of aerosols produced from sonicated powders indicated that more than 99% of the aerosol was comprised of singlets, and no change in the particle shape or morphology was observed. The PSL aerosol mass output from a single bed loading was not constant with time. There was a large initial peak in the first 10 min of operation, after which the mass concentration decreased rapidly for the next few minutes and then leveled off for a time. Mass outputs between runs were fairly reproducible under the same operating conditions. Table 1 summarizes the operating characteristics of the large particle exposure system. Average mass concentrations up to 37 mg/m³ during the first 60 min of operation are achievable. For radioactive aerosols, this level of concentration is more than enough for most inhalation studies.

Table 1
Summary of the Operating Characteristics of the Large-Particle Exposure System

Nominal PSL Particle Size[a]	FBG Flow Rate, lpm	PSL/Bed Material	Sizing				Average Aerosol Mass Concentration (µg/L)[e]
			EM[b]		Impactor		
			CMD,[c] σ_g		AMAD,[d] σ_g		
3 ± 1 µm	42.5	130 mg/200 cm³	3.13 µm, 1.17		NA		17.0
		90 mg/200 cm³	3.34 µm, 1.12		3.40 µm, 1.15		1.11g
9 ± 1 µm	75.0	156 mg/200 cm³	NA		NA		4.71
		43 mg/200 cm³	9.53 µm, 1.08		9.22 µm, 1.21		0.50g
15 ± 3 µm	75.0	125 mg/200 cm³	NA		NA		37.3
		68 mg/200 cm³	15.98 µm, 1.06		NA		0.14f,g

[a]Given by manufacturer; mean ± range.
[b]EM = electron microscopy with Zeiss analyzer or optical comparator.
[c]CMD = count median diameter, σ_g = geometric standard deviation.
[d]AMAD = activity median aerodynamic diameter.
[e]Average over 60-min sampling.
[f]Average over 90-min sampling.
[g]46Sc-labeled PSL particles.
NA = not available, either not taken or beyond the size range of instrument.

References

Carpenter, R. L. and K. L. Yerkes (1980) "Relationship Between Fluid Bed Aerosol Generator Operation and the Aerosol Produced", Am. Ind. Hyg. Assoc. J. 41:888-894.

Raabe, O. G., J. E. Bennick, M. E. Light, C. H. Hobbs, R. L. Thomas, and M. I. Tillery (1973) "An Improved Apparatus for Acute Inhalation Exposure of Rodents to Radioactive Aerosols", Toxicol. Appl. Pharmacol. 26:264-273.

Snipes, M. B. and M. F. Clem (1981) "Retention of Microspheres in the Rat Lung After Intratracheal Instillation", Environ. Res. 24:33-41.

Published 1984 by Elsevier Science Publishing Co., Inc.

Aerosols, Liu, Pui, and Fissan, editors

COARSE MODE AEROSOL BEHAVIOR IN MAN: THEORY AND EXPERIMENT

F.J. Miller,[1] M.A. Grady,[2] and T.B. Martonen[2]
[1]Health Effects Research Laboratory
Environmental Protection Agency
Research Triangle Park, NC 27711
[2] Northrop Services, Inc.
Research Triangle Park, NC 27709
U.S.A.

EXTENDED ABSTRACT

Introduction

Because the potential of atmospheric aerosols to cause health effects is a function of the mass median aerodynamic diameter (MMAD) of the aerosol, hazard evaluation requires size-selective sampling of ambient contaminants. It is an issue of importance to determine the size of particles penetrating oro- and nasopharyngeal airways to reach tracheobronchial (TB) and pulmonary (P) regions and pose a potential threat to human health. Swift and Proctor (1982) used the data of Yu et al. (1981) and Chan and Lippmann (1980) in analyses that addressed deposition with oronasal breathing. In the current study, published data for extrathoracic (ET) and TB deposition were fit to logistic regression models prior to examining the possible influence of oronasal breathing on TB deposition.

Statistical Analyis Methods

Since between 40 and 60 percent of the inhaled air passes nasally during oronasal breathing (Niinimaa, 1981), the current analyses assumed a 50/50 split between the nasal and oral pathways. Using the data of Fig. 1 in Yu et al. (1981) and the recent data of Heyder (1984), particle losses in the ET region during nasal inspiration were examined using a logistic regression model of the form

$$N_I = [1 + e^{\alpha + \beta \log (\rho d^2 Q)}]^{-1} \qquad (1)$$

where ρ is the particle mass density (g), d is the particle diameter (μm), and Q is the airflow rate (cm^3 sec^{-1}). The term $\rho d^2 Q$ essentially defines an impaction parameter and has been found to be useful in characterizing the deposition of particles in the head and TB airways where inertial impaction is the dominant deposition mechanism. Nonlinear regression estimation techniques were used to fit Eq. (1). The use of a logistic model and nonlinear estimation have the advantage that no "breakpoint" for a series of linear regressions needs to be assumed, yet the predicted deposition efficiency will asymptotically approach a lower bound of zero and an upper bound of one. Inspiratory mouth mean deposition efficiency was estimated in a similar manner using the data of Fig. 3 in Yu et al. (1981) and the data of Heyder (1984). Because the data contained in Fig. 1 and 3 of Yu et al. (1981) were digitized, linear regressions were fit; the coefficients obtained were essentially the same as proposed by those authors. The percentage of variation in the data explained by the model (R^2) was obtained during these reanalyses (see Table 1).

Table 1. Deposition in the ET and TB Regions Using Linear and Nonlinear Regression Models[a] and a Comparison of These Models for Deposition with Oronasal Breathing

D_{ae} (μm)	ET Deposition Nose Lin	ET Deposition Nose Nlin	ET Deposition Mouth Lin	ET Deposition Mouth Nlin	TB Deposition[c] (Logistic Model) H	TB Deposition[c] (Logistic Model) L	TB Deposition[c] (Logistic Model) C&L	TB Deposition[c] (Logistic Model) ALL	Oronasal Breathing[d] ET Dep. Lin	Oronasal Breathing[d] ET Dep. Nlin	Oronasal Breathing[d] TB Dep. Lin	Oronasal Breathing[d] TB Dep. Nlin
1	0.18	0.18	0.00	0.01	0.00	0.13	0.10	0.02	0.09	0.09	0.01	0.01
2	0.42	0.42	0.00	0.04	0.01	0.27	0.22	0.08	0.21	0.23	0.06	0.06
3	0.56	0.59	0.12	0.09	0.06	0.38	0.33	0.19	0.34	0.34	0.13	0.13
4	0.66	0.70	0.20	0.17	0.19	0.48	0.42	0.32	0.43	0.44	0.18	0.18
5	0.74	0.78	0.27	0.26	0.37	0.55	0.50	0.45	0.50	0.52	0.23	0.22
6	0.80	0.83	0.32	0.35	0.56	0.61	0.56	0.56	0.56	0.59	0.25	0.23
7	0.85	0.86	0.36	0.43	0.70	0.66	0.61	0.65	0.61	0.65	0.26	0.23
8	0.90	0.89	0.40	0.51	0.81	0.69	0.66	0.72	0.65	0.70	0.25	0.22
9	0.94	0.90	0.43	0.58	0.87	0.73	0.69	0.78	0.69	0.74	0.24	0.20
10	0.98	0.92	0.46	0.64	0.91	0.75	0.72	0.82	0.72	0.78	0.23	0.18
11	1.00[b]	0.93	0.49	0.69	0.94	0.78	0.75	0.85	0.74	0.81	0.21	0.16
12	1.00	0.94	0.51	0.73	0.96	0.79	0.77	0.88	0.76	0.84	0.20	0.14
13	1.00	0.95	0.54	0.77	0.97	0.81	0.79	0.90	0.77	0.86	0.18	0.13
14	1.00	0.95	0.56	0.80	0.98	0.83	0.80	0.91	0.78	0.88	0.16	0.11
15	1.00	0.96	0.58	0.82	0.98	0.84	0.82	0.92	0.79	0.89	0.14	0.10
α	-0.959	7.16	-1.117	12.74	7.29	1.93	2.23	4.12				
β	0.397	-1.97	0.324	-2.73	-9.65	-3.04	-3.18	-5.64				
R^2	0.68	0.91	0.46	0.81	>0.99	0.95	0.97	0.96				

[a] The data for the linear (Lin) models are derived using Eqs. 7b and 9a in Yu et al. (1981) of the form $Y = \alpha + \beta \log(\rho d^2 Q)$. The nonlinear (Nlin) models are based on an equation of the form $Y = [1 + e^{\alpha + \beta \log X}]^{-1}$ where $X = \rho d^2 Q$ for ET deposition and $X = D_{ae}$ for TB deposition.

[b] For this and larger particle sizes, the equation predicts deposition greater than one.

[c] H=Heyder (1984), L=Lippmann (1977), C&L=Chan and Lippmann (1980), ALL=combined data.

[d] A total minute ventilation of 45 l/min was used with a 50/50 split in nasal and oral airflow.

Regression analyses for TB deposition data, as a percent of particles entering the trachea, utilized a logistic model based upon log D_{ae}, where D_{ae} is the aerodynamic diameter of the particle. By definition D_{ae} is the diameter of a spherical particle of unit density (cgs system) of the same terminal settling velocity as the particle under consideration (arbitrary shape, density, etc.). For a spherical particle, $D_{ae} = \sqrt{\rho}\, d$. To enable comparisons between controlled and spontaneous breathing patterns, separate regression analyses were performed on the TB data of Lippmann (1977), Chan and Lippmann (1980), and Heyder (1984). An additional analysis involved combining all of the data of these investigators. Unlike the ET deposition data, the available TB data are not evenly distributed across the range of particle sizes of interest so that weighted nonlinear regression was used. Weights that were a function of the reciprocal of the frequency of observations in a size interval allowed size intervals to contribute more evenly to the overall regression without masking the dispersion present in the data within an interval.

Results and Discussion

A summary of the predicted deposition for various regression analyses for the ET and TB regions is given in Table 1. Parameter estimates, along with the percentage of variation (R^2) in the data explained by the model, are provided for the various nonlinear estimation analyses.

Deposition of particles in the ET region for nasal inspiration is essentially the same for the linear and logistic models for particles up to about 8 μm D_{ae}. For larger particles, the linear model proposed by Yu et al. (1981) overestimates deposition. Discrepancies exist between the two models beginning at about 6 μm D_{ae} for ET deposition with oral breathing. As particle D_{ae} increases from 9 to 15 μm, the difference in the absolute deposition probabilities between the linear and nonlinear models increases monotonically from a low of 0.15 to a high of 0.24. The logistic model yields significantly higher deposition probabilities and explains 81% of the variation in the data compared to 46% for the linear model.

When separate logistic models were fit to TB deposition data from investigators using spontaneous breathing (Lippmann, 1977; Chan and Lippmann, 1980) compared to a more controlled breathing pattern (Heyder, 1984), it was evident that reproducible differences do exist between the breathing patterns employed. For small particles, spontaneous breathing yields a greater TB deposition of particles, while for larger particles the reverse holds. When the digitized data were plotted, intersubject variability for TB deposition with spontaneous breathing was greater than the mean difference between the two breathing patterns, thereby largely eliminating concerns about which breathing pattern is more appropriate for estimating deposition in man following exposure to ambient airborne contaminants.

Oronasal breathing was simulated for a minute ventilation of 45 l/min and a 50/50 split in airflow between the nose and mouth. With the high filtering efficiency of the ET region for large particles, the TB deposition predicted from the logistic regression models agrees with the recent TB deposition data of Heyder (1984) involving nasal and oral breathing with a 750 $cm^3\ s^{-1}$ mean flow rate and 4s breathing cycle period. Upon combining the deposition data for the two pathways, oronasal breathing would result in TB deposition efficiencies of 21, 17, 14, and 9% for particles of 8, 9, 10, and 12 μm D_{ae}

1002

respectively. Thus, the data presented by Heyder (1984) are in agreement with the deposition probabilities obtained using the logistic model for oronasal breathing (see last column of Table 1). Such analyses show that the TB deposition of large particles should not be ignored in the evaluation of potential health hazards.

References

Swift, D.L. and Proctor, D.F. (1982) "Human respiratory deposition of particles during oronasal breathing", Atmos. Environ. 16:2279.

Yu, C.P., Diu, C.K., and Soong, T.T. (1981) "Statistical analysis of aerosol deposition in nose and mouth", Am. Ind. Hyg. Assoc. J. 42:726.

Chan, T.L. and Lippmann, M. (1980) "Experimental measurements and empirical modelling of the regional deposition of inhaled particles in humans", Am. Ind. Hyg. Assoc. J. 41:399.

Niinimaa, V., Cole, P., Mintz, S., and Shephard, R.J. (1981) "Oronasal distribution of respiratory airflow", Respir. Physiol. 43:69.

Heyder, J. (1984) "Studies of particle deposition and clearance in humans", Symposium on "Problems of Inhalatory Toxicity Studies" Berlin, April 2-4.

Lippmann, M. (1977) "Regional deposition of particles in the human respiratory tract", In: Handbook of Physiology – Reaction to Environmental Agents (Edited by D.H.K. Lee et al.), pp. 213-232. The American Physiological Society, Bethesda, MD.

Disclaimer: This report has been reviewed by the Health Effects Research Laboratory, U.S. Environmental Protection Agency, and approved for publication. Mention of trade names or commercial products does not constitute endorsement or recommendation for use.

Published 1984 by Elsevier Science Publishing Co., Inc.

Aerosols, Liu, Pui, and Fissan, editors

MEASUREMENTS OF HYGROSCOPIC GROWTH RATES OF MEDICAL AEROSOLS[1,2,3]

T.B. Martonen
Northrop Services, Inc. -
    Environmental Sciences
Research Triangle Park, NC 27709

J.E. Lowe
Pulmonary Division, Dept. of Medicine
University of California
Irvine, CA 92717

EXTENDED ABSTRACT

Introduction

Aerodynamic characteristics of a hygroscopic particle may continually change during a breathing cycle, say of 1.74 s inspiration, 0.2 s pause, and 2.06 s expiration as used by the Task Group on Lung Dynamics (1966), because of water vapor uptake. One must account for the *rate of change* in size and density of a hygroscopic particle to ascertain its probable site of deposition within the human respiratory tract (Martonen, 1982a).

The temperature (T) and relative humidity (RH) profiles within the unanaesthesized human respiratory tract have not been systematically mapped; limited data are available for anaesthesized subjects (Déry *et al.* 1967). Indeed, there is even a paucity of *in situ* data taken at any given fixed location (Ingelstedt, 1956). A review of several T and RH studies has been presented elsewhere (Martonen and Miller, In press). The available information suggests that it is reasonable to assume a 90% RH in the trachea for mouth breathing, with a monotonic progression in distal airways to an equilibrium 99.5% RH in the deep lung. The latter value may be assumed in the trachea for nose breathing. In both instances a constant temperature of 37°C may be assumed in tracheobronchial (TB) airways.

We have measured the growth rates of NaCl and two bronchodilator aerosols using a laboratory system designed to simulate T, RH, and fluid dynamics of upper human airways.

Methods

The laboratory system is illustrated in Figure 1. Salient components of the system are discussed briefly below.

[1]This report has been reviewed by the Health Effects Research Laboratory of the U.S. Environmental Protection Agency and approved for publication. Mention of trade names or commercial products does not constitute endorsement or recommendation for use.

[2]Requests for reprints should be mailed to T.B. Martonen at the address above (P.O. Box 12313).

[3]The work reported here was funded by U.S. Environmental Protection Agency contract 68-02-2566 and National Institutes of Health grant NHLBI-RO1-HL-19704. We thank David S. Mukai for technical assistance.

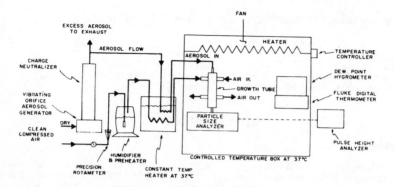

Figure 1. A laboratory system to generate and measure the hygroscopic
growth rates of monodisperse aerosols.

Aerosols were produced by either a Berglund-Liu Vibrating Orifice
Monodisperse Aerosol Generator (Model 3050, TSI, Inc., St. Paul, MN) or a
May Spinning Top Homogeneous Aerosol Generator (BGI, Inc., Waltham, MA).
Aerosols were sampled with a Climet 208 optical particle counter (Climet
Instruments Co., Redlands, CA) before and after passing through a growth
chamber in which the atmosphere (T and RH) of the upper TB airways was
simulated. Climet responses were pulse-height analyzed with a multichannel
analyzer (Model 8100, Canberra Industries, Inc., Meriden, CT).

A dry aerosol jet entered at the center of the growth chamber and was
mixed with clean, saturated compressed air. Particle growth occurred during
the time required to traverse the distance between the entrance and the
Climet's sensing volume. Growth was regulated by changing either the
saturated air flowrate or chamber length to control particle residence time.
T and RH were monitored below the aerosol-air mixing site and at a chamber
exhaust port.

Results and Discussion

The multichannel analyzer CRT showed a sharp peak when a monodisperse
aerosol was sampled. Figure 2 is a photograph of a CRT display for a
spinning top-produced bronchodilator (1% Isuprel®) aerosol. This particular
aerosol had a mass median aerodynamic diameter of 3.5 µm and geometric
standard deviation of 1.03.

Experimental results are compared with other laboratory data in
Figure 3. Theoretical curves of Crider *et al.* (1956) for NaCl are shown as
solid curvilinear lines. The experimental NaCl data indicate that growth is
slower than predicted by theory. This could be ascribed to the delay in
growth induced by the action of the aerosol jet entering the chamber
atmosphere, i.e., particles are not immediately introduced to the high RH
environment. Physiologically, this may simulate the effect of the laryngeal
jet in man.

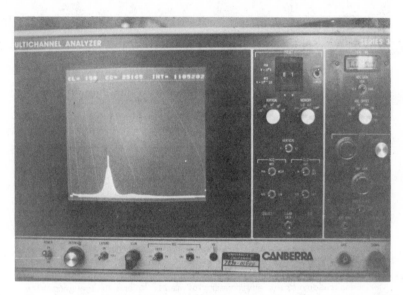

Figure 2.   The CRT display of a multichannel analyzer for a monodispere
            aerosol sampled with an optical particle counter.

The 0.5% and 1% Isuprel® bronchodilator drugs both have isoproterenol
hydrochloride as the active agent. They differ primarily in that the 0.5%
solution contains glycerin. The new data comparing 0.5% and 1% Isuprel®
aerosols substantiate earlier observations (Bell and Ho, 1981) that the
addition of a humectant such as glycerin can inhibit the *rate* of hygroscopic
growth of a given particle.

Bell and Ho (1981) stated that: "No experiments were done to verify
that the flow in the growth tube was indeed perfectly uniform plug flow
throughout its length." In these tests Climet samples were taken at the
center of the chamber exit and at 90°-spaced locations along the perimeter.
The multichannel analyzer CRT display of particle size distribution did not
vary with sampling position suggesting that corresponding sampled particle
residence times were the same, thereby supporting the assumption of Bell and
Ho (1981). This implies that the data of Figure 3 do indeed reflect change
in particle size as a function of time in a prescribed T and RH environment,
and supports the use of corresponding empirical growth rate formulas in
mathematical models of hygroscopic aerosol behavior in man (Martonen *et al.*,
1982b).

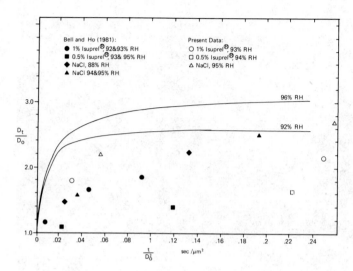

Figure 3. Hygroscopic aerosol growth in a chamber simulating the upper respiratory tract environment of man. D_o and D_t are geometric particle diameters orginally and at time t.

References

Bell, K. A., and Ho, A. T. (1981) J. Aerosol Sci. 12:247.

Crider, W. L., Milburn, R. H., and Morton, S. D. (1956) J. Meteorol. 13:540.

Déry, R., Pelletier, J., Jacques, A., Clavet, M., and Houde, J. J. (1967) Can. Anaes. Soc. J. 14:287.

Ingelstedt, S. (1956) Acta Oto-Laryng., Suppl. 131.

Martonen, T. B. (1982a) Bull. Math. Bio. 44:425.

Martonen, T. B., Bell, K. A., Phalen, R. F., Wilson, A. F., and Ho, A. T. (1982b) Inhaled Particles V (Edited by W. H. Walton), p. 93, Pergamon Press, Oxford, England.

Martonen, T. B., and Miller, F. J. (In press) J. Aerosol Sci.

Task Group on Lung Dynamics. (1966) Health Phys. 12:173.

Published 1984 by Elsevier Science Publishing Co., Inc.
Aerosols, Liu, Pui, and Fissan, editors

An Aerosol Dispersion Test: Comparison of Results From
Smokers and Non-Smokers

M. McCawley and M. Lippmann
Dept. of Environmental Medicine
New York University Medical Center
New York, N.Y. 10016
USA

Introduction

Dispersion of gases in the lungs has been extensively studied in
relation to closing volume and compliance. The theory is that regional
pulmonary pressure differences cause regional ventilation differences.
The regional ventilation differences are exhibited in a sequential
filling of different regions of the lung as shown by the distribution of
inhaled boluses of Xenon-133, (Jeanne and Tashkin, 1983). The pulmonary
pressure differences are a result of variations in resistance and
compliance among the lung units (Robertson, et al., 1969). Typically,
the lung is considered to consist of basal and apical regions.
Preferential ventilation of one area or the other depends on the
inhalation rate and at what level above residual volume the breath begins.

One problem in the detection of early changes in lung function is
that the small airways (<2mm) are predominantly involved. These small
airways in a normal lung contribute only 10 to 30% of the total
resistance of the tracheo-bronchial region (Hogg, et al., 1968). Since
only a fraction of the small airways are likely to be involved in
disease, the total contribution of this diseased portion of the airways
will be even less. Obstruction in the small airways, however, greatly
affects the distributon of ventilation, as has been shown by the
frequency dependence of compliance measurements (Woolcock, et al., 1969).

It has been pointed out by Altshuler, et al. (1959) that
submicrometer size aerosols may be used as tracers of pulmonary
ventilation and this technique may possibly be a more sensitive measure
of pulmonary function than more commonly used means. Other studies also
report differences in the exhaled aerosol concentration profile between
normals and bronchitics were seen by Muir (1970) and between smokers and
non-smokers by Love and Muir (1976).

Aerosols with particle diameters around 0.5 µm should be good tracers
of convective air motion. Unlike a tracer gas, relatively little
diffusion or sedimentation of particles of this size will occur during
the time period required for a breath. Therefore, measurement of the
dispersion of a bolus of aerosol was selected as a method of measuring
the change in the distributon of ventilation due to an early or minimal
lung disease.

Methods

Following the original design of Altshuler, et al. (1957), the aerosol is inhaled through an in-line aerosol photometer with an output to a strip chart recorder. The air flow travels, prior to this, through a pneumotachograph attached to a flow transducer and flow integration device. This transforms the flow to volume which can be displayed on a volt meter. Subjects listening to a metronome and watching the volume display can thereby control their own breathing rate.

Prior to the pneumotachograph a solenoid valve allows either a stream of particle-free air or an aerosol stream to be breathed. A trigger device connected to the flow integrator can be pre-set to alternately switch the solenoid on and off at the desired volumes thus creating a pulse of aerosol at any point during the inhalation.

Results

Table 1 shows the results of a group of non-smoking asymptomatic adults compared to a group of asymptomatic smokers. The index of dispersion of the aerosol bolus is the exhaled peak concentration normalized by dividing by the inhaled peak concentration.

Table 1. Comparison of index of dispersion of an aerosol bolus between smokers and non-smokers.

		Non-Smokers			
		Number of Subjects	Age (Yrs.)	Height (cm)	Index of Dispersion (%)
Men	Mean	12	31	181	22.04
	Std. Dev.		5.6	4.4	3.87
Women	Mean	8	29	164	25.70
	Std. Dev.		6.2	6.2	3.66
Total	Mean	20	30	175	23.39
	Std. Dev.		5.7	9.7	4.11

Smokers

		Number of Subjects	Age (Yrs.)	Height (cm)	Pack Years	Index of Dispersion (%)
Men	Mean	12	35	181	19.9	13.73*
	Std. Dev.		9.3	6.5	15.4	5.72
Women	Mean	8	35	161	18.0	14.48*
	Std. Dev.		5.0	5.6	10.9	4.64
Total	Mean	20	35	173	19.2	14.01*
	Std. Dev.		7.8	11.7	13.6	5.23

*P value < 0.0005 for comparison between smoking and non-smoking group.

Discussion

The smoking and non-smoking groups were both asymptomatic and clinically healthy. They were closely matched by the numbers in each sex and lung size as indicated by height. The smokers were approximately five years older, but in each group there was no significant difference in the index of dispersion by age.

The mean dispersion values (Table 1) were lower and more variable for the smokers. Actually, there was little overlap in the index of dispersion between the smokers and non-smokers. Only two non-smokers were below 19.0 while only two smokers had an index above this value. There was also no difference in sensitivity to smoking between the males and females in the smoking group.

Conclusions

The index of dispersion appears to be a sensitive means of differentiating between smokers and non-smokers. It thus offers the potential for early diagnosis of changes in the small airways since small airways changes would be expected in smokers with the number of pack-years experienced in this study.

1010

References

Altshuler, B., L. Yarmus, E.D. Palmes and N. Nelson. (1957) "Aerosol Deposition in the Human Respiratory Tract." A.M.A. Arch. of Environ. Health. 15:293.

Altshuler, B., E.D. Palmes, L. Yarmus, and N. Nelson. (1959) "Intrapulmonary Mixing of Gases Studied with Aerosols." J. Appl. Physiol. 14(3): 321.

Hogg, J.C., P.T. Macklem and W.M. Thurlbeck. (1968) "Site and Nature of Airway Obstruction in Chronic Obstructive Lung Disease." N. Engl. J. Med. 278: 13355.

Jenne, J.W. and D.P. Tashkin. (1983) "Bronchodilators as a Bronchial Challenge: A Critical Assessment." Provocative Challenge Procedures: Bronchial, Oral, Nasal and Exercise. (Edited by Sheldon B. Spector), CRC Press, Boca Raton, FL.

Love, R.G. and D.C.F. Muir. (1978) "Aerosol Deposition and Airway Obstruction." Am. Rev. Resp. Dis. 114: 891.

Muir, D.C.F. (1970) "The Effect of Airways Obstruction on the Single Breath Aerosol Curve." Airway Dynamics. (Edited by A. Bouhuys), Charles C. Thomas, Springfield, IL.

Robertson, P.C., N.R. Anthonisen and D. Ross. (1969) "Effect of Inspiratory Flow Rate on Regional Distribution of Inspired Gas." J. Appl. Physiol. 26(4): 438.

Woolcock, A.J., N.J. Vincent and P.T. Macklem. (1969) "Frequency Dependence of Compliance as a Test for Obstruction in the Small Airways." J. Clin. Invest. 48: 1097.

Published 1984 by Elsevier Science Publishing Co., Inc.

Aerosols, Liu, Pui, and Fissan, editors

STUDIES OF PARTICLE DEPOSITION AND RETENTION IN THE
HUMAN RESPIRATORY TRACT

M. R. Bailey and S. P. Stewart
National Radiological Protection Board
Chilton, Didcot, Oxon. OX11 0RQ UK

EXTENDED ABSTRACT

It is proposed to carry out a series of experimental studies of
particle deposition and retention in the human respiratory tract designed to
provide information which will enable improvements to be made to lung
clearance models, especially those used to assess the consequences of
exposure to airborne radioactive particles. Any results available at the
time of the conference will be reported.

The nasal passage

In the lung model used by the International Commission on Radiological
Protection (ICRP) to determine Limits on Intakes of Radionuclides by Workers
(ICRP Lung Model) it is assumed that part of the material deposited in the
nasal passage (N-P) is absorbed into the blood, the amount depending on the
chemical form of the material. The rest of the deposit is assumed to be
cleared to the pharynx and swallowed with a half-time of 0.4 day (ICRP,
1979). Recent experiments, however, indicate that particles deposited in
the posterior, ciliated N-P would be cleared and swallowed within a few
minutes, while particles deposited on the hairs and lining of the anterior
N-P are mainly removed by extrinsic means such as nose-blowing (Lippmann,
1970; Fry and Black, 1973; Bailey, 1984). Only particles deposited in the
anterior N-P would therefore be available for recovery by a nose-blow sample
taken at the end of a working shift. Absorption of material into the blood
is probably much more effective in the posterior N-P than in the anterior
N-P, because of differences in the lining epithelium. It is likely that for
radionuclides such as cerium-144 and plutonium-239 inhaled in soluble form a
significant fraction of the deposit in the posterior N-P would be retained
in the epithelium, which might then receive a higher radiation dose than any
other tissue (Bailey and James, 1979). It is important therefore, both for
biological monitoring and for the assessment of radiation doses resulting
from inhalation of radionuclides, to know how the nasal deposit is
distributed between the anterior N-P and the posterior N-P (Bailey, 1984a).
There are at present insufficient data available to permit this distribution
to be determined with confidence for a given particle size and breathing
pattern, although attempts have been made to calculate regional deposition
within the nasal passage from the dimensions of the airways and the
mechanics of particle deposition (Landahl 1950; Scott et al, 1978). It is
therefore proposed to carry out studies of the deposition and retention in
the human nasal passage of inhaled monodisperse γ-labelled particles
covering a wide range of sizes (0.1 - 100 μm). The kinetics of particle
clearance from the nasal passage as a function of particle size and
breathing pattern will be determined, and the clearance pattern will be
related to regional deposition within the nasal passage. The contribution

to clearance of spontaneous nose-blowing and wiping, and the effectiveness
of deliberate nose-blow sampling will be investigated.

The tracheobronchial region

In the ICRP Lung Model it is assumed that particles deposited
throughout the trachea and bronchial tree, the tracheobronchial region
(T-B), are cleared to the pharynx with a half-time of 0.2 days. The T-B
region is not, however, homogeneous: the length and diameter of the airways
decrease from the trachea to the terminal bronchioles, and the nature of the
epithelium and the fluid covering it also change (Lippmann et al, 1980).
Thus systematic differences in clearance rates are to be expected in
different generations in addition to random variations in each airway.
There is general agreement that mucus velocities, and hence particle
clearance rates decrease distally, but not about the rate at which this
occurs: estimates of the mean retention time in the terminal bronchioles
range from a few hours to a few weeks (Bailey, 1984). It has also been
suggested that there may be areas of static mucus throughout the T-B region,
and that clearance may be less effective around bifurcations than elsewhere.
The possibility that some particles deposited in the T-B may not be cleared
within a day affects not only lung modelling but also the interpretation of
deposition experiments, which generally rely on the assumption that material
cleared in the initial rapid phase represents the deposit in the tracheobronchial tree, while the remaining material represents that deposited in
the pulmonary parenchyma.

It is planned that volunteers will inhale, under closely controlled
conditions, an aerosol bolus of γ-labelled polystyrene particles, and
retention of the particles deposited will be followed for two weeks. In
this way, clearance of particles deposited at different depths in the
respiratory tract: major airways, distal airways, and alveoli, will be
investigated, and clearance of particles deposited preferentially at major
bifurcations will be compared with that of particles deposited uniformly
throughout airways.

The pulmonary region

Pulmonary retention of 1 μm and 4 μm diameter monodisperse γ-labelled
fused aluminosilicate particles has been followed for 12-18 months after
inhalation by 12 healthy non-smoking male volunteers (Bailey and Fry, 1983;
Bailey et al, 1984). Approximately 7% of the initial deposit of the 1 μm
particles and 40% of that of the 4 μm particles were associated with a
distinct rapid clearance phase. These figures correspond closely to the
calculated T-B deposits (Bailey et al, 1982), indicating insignificant rapid
pulmonary clearance. This is consistent with recent observations made
elsewhere (Bailey, 1984), but not with the assumption made in the ICRP Lung
Model that 40% of the initial pulmonary deposit (IPD) of a relatively
insoluble material is cleared with a half-time of 1 day. Retention of the
remaining material R(t) generally followed a two-component exponential
function, the phases having half-lives of the order of tens of days and
several hundred days respectively. At 350 days after inhalation R(t) was

46±11% IPD ($\bar{x}$±SD) and 54±11% for the 1 μm and 4 μm particles respectively. Retention of 1 μm particles was correlated with that of 4 μm particles and with the subject's physiological lung function as measured by the ratio of Forced Expiratory Volume in one second to Forced Vital Capacity. Dissolution rates based on urinary excretion of the γ-labels were estimated to be 7×10^{-4} d^{-1} and 2×10^{-4} d^{-1} for the 1 μm and 4 μm particles respectively. Mean pulmonary retention over this period, after correction for dissolution, followed a two-component exponential function with about 20% clearing with $t_{\frac{1}{2}}$ about 30 d, and the rest with $t_{\frac{1}{2}}$ about 700 d, in reasonable agreement with the 500 day half-time chosen for the slow phase of pulmonary clearance in the ICRP Lung Model.

A similar study is to be carried out on current cigarette smokers since a substantial fraction of the population smokes and evidence has emerged indicating that chronic cigarette smoking impairs pulmonary clearance of particles. Cohen et al (1979) and Bohning et al (1982), using magnetite and polystyrene particles respectively, found pulmonary clearance to be slower in smokers than in non-smokers. In neither study, however, were particles of uniform size employed, nor was any attempt made to determine the contributions to clearance made by the transport of particles from the lung to the gastro-intestinal tract and by translocation of material from the lung to the blood, and the results cannot therefore be used to predict quantitatively how smoking would affect the clearance of other materials.

Acknowledgement

This work was partially supported by the CEC under contract BIO-D-489-UK.

References

Bailey, M. R. (1984) "Human data on clearance", presented at the Workshop on Lung Modelling for Inhalation of Radioactive Materials, Oxford, March 26-28 1984, Commission of the European Communities (in press).

Bailey, M.R. (1984a) "Assessment of the dose to the nasopharyngeal region from inhaled radionuclides", presented at the Workshop on Lung Modelling for Inhalation of Radioactive Materials, Oxford, March 26-28 1984, Commission of the European Communities (in press).

Bailey, M.R. and Fry, F.A. (1983) "Pulmonary retention of insoluble particles in man", Current Concepts in Lung Dosimetry" (Edited by D. R. Fisher) USDOE.

Bailey, M.R., Fry, F. A. and James, A. C. (1982) "The long-term clearance kinetics of insoluble particles from the human lung" Ann. occup. Hyg. 26:273.

Bailey, M. R., Fry, F.A. and James, A.C. (1984) "Long-term retention of particles in the human respiratory tract", submitted for publication.

Bohning, D.E. Atkins, H. L. and Cohn, S. H. (1982) "Long-term particle clearance in man: normal and impaired", Ann. occup. Hyg. 26:259.

Cohen, D., Arai, S. F. and Brain, J.D. (1979) "Smoking impairs long-term dust clearance from the human lung", Nature 215:30.

Fry, F. A. and Black, A. (1973) "Regional deposition and clearance of particles in the human nose", Aerosol Sci. 4:113.

ICRP (1979) ICRP Publication 30 Part 1. Ann. ICRP 2: No 3/4.

Landahl, H. D. (1950) "On the removal of airborne droplets by the human respiratory tract II. The nasal passages", Bull. Math. Biophys. 12:161.

Lippmann, M. (1970) "Deposition and clearance of inhaled particles in the human nose", Ann. Otol. Rhinol. Laryngol. 79:519.

Lippmann, M., Yeates, D. B. and Albert, R.E. (1980) "Deposition, retention and clearance of inhaled particles", Brit. J. Industr. Med. 37: 337.

Scott, W. R., Taulbee, B. and Yu, C.P. (1978) "Theoretical study of nasal deposition", Bull. Math. Biol. 40:581.

Published 1984 by Elsevier Science Publishing Co., Inc.

Aerosols, Liu, Pui, and Fissan, editors

DYNAMIC SURFACE TENSION LOWERING BY IMPACTION OF ULTRASONICALLY
PRODUCED PHOSPHOLIPID AEROSOLS AT THE AIR-WATER INTERFACE

John F. Wojciak, Robert H. Notter, and Günter Oberdörster
Depts. of Chemical Engineering, Radiation Biology and Biophysics,
and Pediatrics, University of Rochester
Rochester, New York   14642 U. S. A.

Introduction

Surfactant aerosols may have therapeutic applications in the treatment of
lung surfactant deficient states such as the neonatal Respiratory Distress
Syndrome (Notter and Morrow, 1975). The characterization of the surface
activity of deposited aerosols is essential to evaluating their suitability
for surfactant replacement therapy. Interfacial films of phospholipid
components of lung surfactant are known to lower surface tension much more
effectively under dynamic compression than at equilibrium conditions (Tabak
et al., 1977). The dynamic compression state for such films is known to
arise under continuous compression-expansion cycling in a surface balance.
The present experiments indicate that a dynamic compression state can also be
generated in films formed during impaction of phospholipid aerosols on an
aqueous subphase of constant area.

Methods

Aerosols were produced by ultrasonic nebulization of a 1.5 mg/ml
dispersion of dipalmitoyl phosphatidylcholine (DPPC) and egg
phosphatidylglycerol (PG) (7:3 molar ratio) in 0.15M NaCl. The lipid-saline
aerosol had a mass median aerodynamic diameter of 4.1 μm and a geometric
standard deviation of 2.3. Deposition on a distilled water subphase was
accomplished by directing the aerosol stream at a water surface in a
specially constructed chamber. The chamber permitted exposure of part of the
subphase surface to the aerosol stream, while surface tension was monitored
outside the impaction zone by means of a platinum slide dipped into the
interface. Aerosol was directed at the surface at a flowrate of 7 1/min
through a circular tube of diameter 1 cm, with exiting jet perpendicular to
and located 1 cm above the water surface. All experiments were performed in
an environmental chamber at 30°C.

Results and Discussion

Continuous deposition of the phospholipid aerosol reduced the surface
tension of the aqueous subphase from 71 dyn/cm to 0.5-4 dyn/cm in a period of
10-11 minutes. This minimum surface tension was stable below 4 dyn/cm until
aerosol deposition was halted (Figure 1a), which resulted in relaxation of
the surface tension toward the minimum equilibrium value of about 22 dyn/cm
(Figure 1b). A similar phenomenon has been reported (in abstract form) for
the surface tension of a water droplet suspended in a flowing phospholipid
aerosol stream (Sweeney and Lee, 1982).
Surface tension reduction below the value defined by the equilibrium
spreading pressure of the surface active component(s) is indicative of a
dynamic compression state in the interfacial film. However, to cause this
state, a dynamic compression force is necessary, such as film compression in

1016

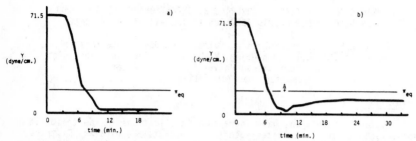

Figure 1: Surface Tension as a Function of Aerosol Deposition Time
a) Continuous Deposition; b) Deposition stopped at 'A'.

a surface balance (Tabak et al., 1977). Although there are some exceptions, this force must be continuous to maintain dynamic values of the surface tension, and its removal leads to relaxation of surface tension toward equilibrium values (Tabak et al., 1977, Notter et al., 1980). This kind of behavior was also found in the present aerosol experiments, as shown in Figure 1.

In a mechanistic sense, two processes might be responsible for the observed dynamic compression state in the deposited film. These are forces from the impinging airflow, the momentum of the depositing aerosol particles, or a combination of both. To differentiate between these mechanisms, additional experiments studied the effects of the airflow alone, as well as the impaction of a saline aerosol (without phospholipids) on previously spread equilibrium films and on the relaxation of dynamically compressed films. The airflow alone was found to have no effect in changing the surface tension of equilibrium films, or in reversing the course of surface tension relaxation in dynamically compressed films. This implied that forces associated with aerosol droplet deposition were responsible for the observed dynamic compression state in deposited phospholipid films. However, demonstrating this in practice was complicated by the presence of two competing effects when saline aerosols were directed against a surfactant film. Specifically, surfactant-free saline aerosols exert compressive forces when directed against the film (which can cause lower dynamic surface tension), but they also cause local depletion of surfactant (which causes higher surface tension). This was reflected in experimental results where deposition of saline aerosol droplets on a previously spread equilibrium film caused an initial dynamic surface tension decrease of 3 dyn/cm over several seconds, which was then followed by a monotonic rise in surface tension beyond the equilibrium value, as shown in Figure 2. This effect was not seen with phospholipid aerosol deposition in Figure 1a, for the aerosol itself continuously replenished interfacial surfactant.

These results strongly suggest that aerosol particle deposition was responsible for the observed dynamic compression state in the phospholipid film. However, they do not define the relative effects of the major aerosol deposition mechanisms (impaction, sedimentation, diffusion). The relative importance of these deposition mechanisms is determined primarily by aerosol particle size and airstream velocity. Because the experiments reported here involved significant bulk aerosol flow, inertial impaction forces would be

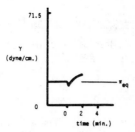

Figure 2: Response of Subphase Surface Tension to Saline Aerosol
Deposition: Surface Previously Covered with Equilibrium Film

expected to dominate. However, it is conceivable that tangential compression
forces (parallel to the surface) could be generated from hydrodynamic flows
in the subphase associated with particle deposition by any mechanism. To
determine specific deposition processes responsible for dynamic film
compression, a comparison was made of the energy necessary to cause dynamic
film states and the energy available from various deposition mechanisms.

The amount of energy necessary to dynamically compress a phospholipid
film from its equilibrium surface tension to near zero surface tension can be
approximated from compression isotherms obtained in surface balance studies.
The surface tension - area isotherm for dynamic compression of a DPPC:PG film
is nearly identical to the quasi-static isotherm until the equilibrium
spreading pressure is reached, which for DPPC:PG is approximately 48 dyn/cm
(equivalent to a surface tension of 22 dyn/cm) (Notter et al., 1980).
Thermodynamic equilibrium would then require that the surface tension remain
constant as the film was compressed further. However, for DPPC:PG films
under dynamic compression, the surface tension continues to fall below
22 dyn/cm, and reaches a minimum value < 1 dyn/cm at interfacial
concentrations of 36-40 $Å^2$/molecule (dynamic monolayer collapse). The area
between the dynamic and quasi-static isotherms gives a measure of the minimum
surface work required for dynamic compression above equilibrium values. The
lower surface concentration limit of integration for this calculation is the
area per molecule at the equilibrium spreading pressure (50 $Å^2$/molecule).
An apropriate upper limit of integration (most compressed film) is
36 $Å^2$/molecule, because each phospholipid molecule has two hydrocarbon
residues, each with limiting area of 18 $Å^2$. With these assumptions, the
calculated minimum work or energy is 10^{-14} ergs/molecule. This corresponds
to a minimum surface work of 3.7×10^2 ergs necessary for dynamic compression
to near zero surface tension at experimental surface concentrations.

Table 1 compares this energy with that available from deposition by
sedimentation and inertial impaction. For particles of the size in these
experiments (~4.0 μm), deposition by sedimentation is much larger than
deposition by diffusion (Hinds, 1982), and the kinetic energy carried by the
diffusive flux is small compared to the other deposition mechanisms.
Therefore, diffusion was not considered in this calculation. The kinetic
energies in Table 1 were calculated from the mass of deposited aerosol (from
tracer measurements) and the calculated velocity of the aerosol particles at
deposition (i.e. bulk air velocity for impaction, and terminal settling
velocity for sedimentation).

Table 1: Kinetic Energy of Depositing Aerosol Particles

Deposition Mechanism	Velocity	Mass of Aerosol Deposited	Kinetic Energy Delivered to Surface Upon Deposition
Impaction	150 cm/sec	0.5 grams	5.5×10^3 ergs
Sedimentation	5×10^{-2} cm/sec	0.5 grams	6.3×10^{-4} ergs

Energy needed to compress film = 3.7×10^2 ergs

The results in Table 1 show that the kinetic energy available at the surface from sedimentation is 6 orders of magnitude less than the minimum energy required for dynamic film compression. However, inertial impaction generates a kinetic energy well in excess of the required value. Similar energy will be generated by both saline and phospholipid-saline aerosols, but the former also gives rise to competing effects of surfactant depletion at the interface.

Conclusions

1) Aerosol impaction at a surface can cause dynamic compression of phospholipid films.
   a) In the presence of surfactant in the aerosol, the dynamic compression state is maintained as long as deposition is continuous.
   b) In the absence of surfactant in the aerosol, a competing effect of surface depletion of surfactant molecules becomes dominant and surface tension rises after an initial decrease.
2) The kinetic energy required to generate dynamic compression in deposited phospholipid films can be rationalized quantitatively with that available from an impacting stream of aerosol droplets.

References

Hinds, W.C. (1982): Aerosol Technology; Properties, Behavior and Measurement of Airborne Particles, Wiley-Interscience, New York.

Notter, R.H. and Morrow, P.E. (1975): Pulmonary surfactant: a surface chemistry viewpoint, Annals of Biomedical Engineering 3:119-159.

Notter, R.H., Tabak, S., Holcomb, S. and Mavis, R.D. (1980): Postcollapse dynamic surface pressure relaxation in binary surface films containing dipalmitoyl phosphatidylcholine, J. Colloid Interface Sci. 73:370-377.

Sweeney, R.J. and Lee, J.S. (1982): Surface tension reduction by aerosol surfactant deposition, Federation Proceedings Abstracts 41(4):998.

Tabak, S., Notter, R.H., Ultman, J. and Dinh, S. (1977): Relaxation effects in the surface pressure behavior of dipalmitoyl lecithin, J. Colloid Interface Sci. 60:117-125.

Published 1984 by Elsevier Science Publishing Co., Inc.
Aerosols, Liu, Pui, and Fissan, editors

DEPOSITION IN THE ALVEOLAR PART OF THE HUMAN LUNG

M. Svartengren, K. Philipson, L. Linnman and P. Camner
Department of Environmental Hygiene
Karolinska Institute
S-104 01 Stockholm, Sweden

Section of Inhalation Toxicology
Department of Toxicology
National Institute of Environmental Medicine
S-104 01 Stockholm, Sweden

Introduction

The health hazard of airborne particles depends on where in the airways the particles are deposited. The majority of particles deposited on respiratory epithelia with cilia are eliminated within 24 hrs with the mucociliary transport or by coughing (Camner, 1980; Pavia et al., 1980; Lippmann et al., 1980; Raabe, 1982). Insoluble particles deposited in the alveolar part of the lung clear much slower; the majority with a half-time of about one to several years (Bohning et al., 1982; Bailey et al., 1982; Philipson et al., submitted for publication). Camner and Philipson (1978) found that 24-hr lung retention of inhaled 4 µm (aerodynamic diameter 6 µm) Teflon particles varied considerably within the group of 10 healthy men but variation of the 24-hr retention was small for each individual when the exposure was repeated. This indicates that morphology of the airways is important for deposition of particles of this size. Svartengren et al. (1984) found Forced Vital Capacity (FVC) and especially Forced Expiratory Volume in 1 sec (FEV_1) to be significantly correlated with the 24-hr lung retention. These results suggest that deposition of particles might be highly dependent on airway resistance.

Subjects and design

Eight nonsmoking students, all without history of pulmonary disease, were exposed twice to Teflon particles at an interval of 14 days. For one exposure bronchoconstriction was induced before the inhalation of test particles and for the other exposure after inhalation of test particles when radioactivety in the lungs had been measured for about half an hour. The exposure procedures are described schematically in fig 1.

Methods

Bronchoconstriction was induced by inhalation of methacholinebromide for 30 sec with normal breaths. Peak expiratory flow (PEF) decreased by about 20%. This corresponds roughly to a doubling in airway resistance (R_{aw}). R_{aw} was measured 5 min after with a whole body pletysmograph (TEXAS CPI 2000), using the "panting technique". The Teflon particles were produced and labelled with ^{111}In (half-life 2.83 days) by a spinning disc technique, (Camner and Philipson, 1978; Camner, 1971; Philipson, 1977). The subjects inhaled the test particles with maximally deep inhalations at 0.5 l/sec during 3-8 min. After inhalation of the test aerosol the subjects rinsed their mouths with water. Radioactivity was measured using two 5 x 2 inch NaI crystals fitted with collimators.

Fig 1

Schematic description of the procedures for the two exposures to ^{111}In-labelled Teflon particles.

Exposure to the test aerosol <u>before</u> the induction of the bronchoconstriction

| a | b | c | d | a | e | f |

Exposure to the test aerosol <u>after</u> the induction of the bronchoconstriction

| a | d | a | b | e | f |

a = Measurement of R_{aw}
c = Measurement of the radioactivity in the lung for about 20 min
e = Measurement of radioactivity in the lung for up to 180 min

b = Inhalation of the test particles
d = Induction of bronchoconstriction
f = Measurements of radioactivity in the lung at 22-26 hrs after inhalation

Results

Table 1 presents the measurements of R_{aw} and the dose of methacholine-bromide for the two exposures to ^{111}In-labelled Teflon particles. Table 2 presents the percentage 24-hr lung retention. When the test aerosol was inhaled before the induction of the bronchoconstriction, R_{aw} and the 24-hr lung retention correlated significantly, r = 0.72 (p < 0.05). When the test particles were inhaled after the induction of the bronchoconstriction there was no significant correlation between R_{aw} and retention, (r = 0.12).

Table 1

Airway resistance (R_{aw}) at the exposures to ^{111}In-labelled Teflon particles.

	Bronchoconstriction induced <u>before</u> inhalation of Teflon particles		Bronchoconstriction induced <u>after</u> inhalation of Teflon particles	
	$R_{(aw)}$, cm H_2O/1.sec.		$R_{(aw)}$, cm H_2O/1.sec.	
	Before induction of bronchoconstriction	After induction of bronchoconstriction	Before induction of bronchoconstriction	After induction of bronchoconstriction
Mean	1.2	2.8	1.1	3.3
SD	0.4	1.1	0.4	1.3

Table 2

The percentage 24-hr lung retention of 4 μm Teflon particles labelled with
^{111}In.

	Bronchoconstriction induced after inhalation of Teflon particles	Bronchoconstriction induced before inhalation of Teflon particles
Mean	52	19
SD	10	5

Calculation of deposition of the Teflon particles

We have used the lung model and equations for deposition due to impaction
and sedimentation, described by Task Group on Lung Dynamics (1966). The procedure to obtain deposition in various airway generations was that used by Landahl (1950). We calculated with 3.9 um particles with a density of 2.13, inhaled at 0.5 l/sec with a tidal volume of 4 litres. To theoretically simulate
a bronchoconstriction the diameters of all generations were changed to the
same proportion (0.4-1.3). The airway resistance in the lung model was calculated using Poiseuille's law. 0.5 cm H$_2$O/1 sec was added to the calculated
values on account of airway resistance in mouth, throat and larynx. Fig 2
shows the experimental data on R$_{aw}$ and the 24-hr lung retention together with
the theoretical values in the non-ciliated or alveolar part of the lung. The
experimental data agree fairly well with the theoretical data.

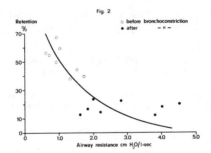

Fig. 2

Summary

Eight healthy subjects inhaled 4 μm Teflon particles labelled with ^{111}In
before and after a bronchoconstriction was induced. When bronchoconstriction
was induced after inhalation of Teflon particles R$_{aw}$ and 24-hr lung retention
correlated significantly. Retention was markedly lower when bronchoconstriction was induced before inhalation of the Teflon particles than when it was
induced after. The ranges were 13-25% and 38-68%, respectively. The experimental data agreed fairly well with theoretical data from a lung model where the
diameters of the airways were varied. The results indicate that a magnitude
of bronchoconstriktion occurring in real life can protect the alveolar part
of the lung by reducing the amount of inhaled particles which deposit there.

References

Bailey, M.R., Fry, F.A., James, A.C. (1982) "The long-term clearance kinetics of insoluble particles from the human lung", Ann. Occup. Hyg. 26:273.

Bohning, D.E., Atkins, H.L., Cohn, S.H. (1982) "Long-term particle clearance in man: normal and impaired", Ann. Occup. Hyg. 26:259.

Camner, P. (1971) "The production and use of test aerosols for studies of human tracheobronchial clearance", Environ. Physiol. Biochem. 1:137.

Camner, P., Philipson, K. (1978) "Human alveolar deposition of 4 um Teflon particles", Arch. Environ. Health 33:181.

Camner, P. (1980) "Clearance of particles from the human tracheobronchial tree", Clin. Sci. 59:79.

Landahl, H.D. (1950) "On the removal of air borne droplets by the human respiratory tract: 1. The lung", Bull. Math. Biophys. 12:43.

Lippmann, M., Yeates, D.B., Albert, R.E. (1980) "Deposition, retention, and clearance of inhaled particles", Br. J. Ind. Med. 37:337.

Pavia, D., Bateman, J.R.M., Clarke, S.W. (1980) "Deposition and clearance of inhaled particles", Clin. Respir. Physiol. 16:335.

Philipson, K. (1977) "Monodisperse labelled aerosols for studies of lung clearance", In abstracts of Uppsala. Dissertations from the Faculty of Science 433, Uppsala.

Philipson, K., Falk, R., Camner, P. (1984) "Long-term lung clearance in humans studied with Teflon particles labelled with ^{51}Cr. Submitted for publication.

Raabe, O.G. (1982) "Deposition and clearance of inhaled aerosols", Mechanisms in respiratory toxicology (Edited by H. Witschi and P. Nettesheim), Florida, CRC Press Inc. Boca Raton.

Svartengren, M., Hassler, E., Philipson, K., Camner, P. (1984) "Spirometric data and penetration of particles to the alveoli", Br. J. Ind. Med. (in press).

Task Group on Lung Dynamics (1966) "Deposition and retention models for internal dosimetry of the human respiratory tract", Health Phys. 12:173.

Published 1984 by Elsevier Science Publishing Co., Inc.

Aerosols, Liu, Pui, and Fissan, editors

SIMULTANEOUS DIFFUSIONAL AND GRAVITATIONAL DEPOSITION OF AEROSOL PARTICLES

J. Gebhart, J. Heyder and W. Stahlhofen

Gesellschaft für Strahlen- und Umweltforschung mbH.,
D-6000 Frankfurt/M., Paul-Ehrlich-Str. 20, F. R. Germany

Aerosol particles in the size range between 0.1 and about 1 um suspended in stationary air or in a viscous air stream are subjected to diffusional and gravitational transport. In the present study a battery consisting of a bundle of cylindrical tubes and a granular bed filter consisting of uniform glass spheres served as aerosol filters. The interaction of diffusional and gravitational particle transport was studied by measuring with optical instruments the removal of particles from resting as well as from moving monodisperse aerosols inside these filters.

NOMENCLATURE:

DE : particle deposition in a filter syst.
RC : fraction of particles escaping deposition; RC = 1 - DE
d_D : particle deposition due to diffusion
r_D : fraction of particles escaping diffusional deposition; $r_D = 1 - d_D$
d_G : particle deposition due to gravitational sedimentation
r_G : fraction of particles escaping sedimentation; $r_G = 1 - d_G$
α_D : deposition coefficient due to diffusion; $r_D = \exp -\{\alpha_D\}$
α_G : deposition coefficient due to sedimentation; $r_G = \exp -\{\alpha_G\}$
D : diffusion coefficient of aerosol particle
v : terminal settling velocity of aerosol particle
t_p : residence time in the filter system
R : radius of a glass bead in a filter-bed
a : radius of cylindrical tube

CYLINDRICAL TUBES: The aerosol battery comprised a bundle of 27 cylindrical tubes of 2 mm diameter and 980 mm length. Monodisperse aerosols in the diameter range 0.082 to 0.505 um were generated by nebulization of commercially available polystyrene latices. Changes in particle number concentration during passage of the aerosols through the tubes were determined with a Laser aerosol spectrometer.
Values of particle deposition are listed in Tab. 1. When the battery was vertically positioned, deposition was found to be in close agreement with values predicted by Ingham (1975) for pure diffusion. Deposition in a horizontally positioned battery agreed well with the values predicted by Taulbee (1978) for simultaneous diffusional and gravitational particle transport. Using the values derived by Pich (1972) for pure gravitational transport the experimental data can be described by (Heyder et al., 1984):

$$DE = d_D + d_G - d_D d_G / (d_D + d_G) \tag{1}$$

If the two mechanisms, however, are superimposed as independent mechanisms according to:

$$DE = 1 - r_D r_G = d_D + d_G - d_D d_G \qquad (2)$$

particle deposition is overestimated.

Particle deposition from stationary air in a horizontal cylindrical tube has been investigated by Goldberg et al. (1978). Their results obtained for a ratio $D / va = 0.1$ are summarized in Fig.1. Again the independent superimposition of the two mechanisms overestimates particle deposition whereas the values derived from the rigorous solution of the differential equation are met well by (Gebhart et al., 1984):

$$RC = 1 - DE = \exp -\{\alpha_D + \alpha_G - 0.75 \, \alpha_D \alpha_G / \alpha_D + \alpha_G\} \qquad (3)$$

In the range DE< 0.4 the single as well as the exponential combination of diffusion and sedimentation can be used to approximate the resulting effect. Higher deposition values, however, can be better expressed in terms of deposition coefficients only (equ.3).

GRANULAR BED FILTERS: The granular bed was a stainless steel cylinder of 156 mm diameter and 310 mm height containing glass beads of 2.69 or 5.91 mm dia.. Monodisperse droplets in the diameter range 0.09 to 3 um were generated by condensation of di-2-ethylhexyl sebaçate on silver chloride nuclei. Driven by a piston pump, 1500 cm^3 of an aerosol passed at a flow rate of 500 cm^{3}s^{-1} through a flow meter and a laser photometer into and out of the granular bed (steady state conditions).
Values of particle deposition are given in Fig. 2. Comparing pure diffusional and pure gravitational transport onto surfaces of different beads deposition for pulsatile flow can be described by two dimensionless groups of variables according to (Heyder et al. 1984):

$$d_D = 22 \, (Dt_p)^{1/2}/2R \qquad resp. \quad d_G = 1.05 \, (vt_p/2R)^{2/3} \qquad (4)$$

A detailed analysis shows that deposition is overestimated by equation (2) but well described by equation (1) or (3).
In the case of particle deposition from stationary air the granular bed was filled with 1000 cm^3 of a monodisperse aerosol and emptied again after a certain time period t_p. In the diffusion domain the experimental findings can be described by:

$$d_D = 1 - \exp - \{7.62 \, (Dt_p)^{1/2}/2R\} = 1 - \exp -\{\alpha_D\} \qquad (5)$$

whereas for pure sedimentation the results follow the equation (Gebhart et al., 1984):

$$d_G = 1 - \exp -\{1.88vt_p/2R\} = 1 - \exp -\{\alpha_G\} \qquad (6)$$

The simultaneous action of diffusion and sedimentation within stationary air is demonstrated in Fig. 3. Again the experimental data can be well approximated by equ. (1) or equ. (3) whereas the independent superimposition of the two mechanisms overestimates particle deposition.

REFERENCES:

Gebhart, J., Heyder, J. and Stahlhofen, W. (1984): submitted to J. Aerosol Sci.
Goldberg, I.S., Lam, K.Y., Bernstein, B. and Hutchens, J.O. (1978): J. Aerosol
Sci. 9 : 209 - 218
Heyder, J., Gebhart, J. and Stahlhofen, W. (1984): submitted to Aerosol Sci.
and Technology
Ingham, D.B. (1975): J. Aerosol Sci 6 : 125 - 132
Pich, J. (1972) : J. Aerosol Sci 3 : 351 - 361
Taulbee, D.B. (1978): J. Aerosol Sci. 9 : 17 - 23

Part. dia, μm	vertical		horizontal				
	d_D		d_G	DE			
	Ingham	Exper.	Pich	Taulbee	Exper.	equ. (1)	equ. (2)
0.082	0.25	0.24	0.02	0.25	0.24	0.251	0.265
0.102	0.21	0.21	0.02	0.21	0.21	0.211	0.226
0.163	0.13	0.14	0.04	0.13	0.14	0.139	0.165
0.22	0.10	0.10	0.07	0.115	0.12	0.128	0.163
0.353	0.06	0.06	0.15	0.16	0.16	0.167	0.201
0.465	0.05	0.05	0.22	0.24	0.24	0.23	0.26

Tab. 1: Particle deposition in the aerosol battery under steady flow.

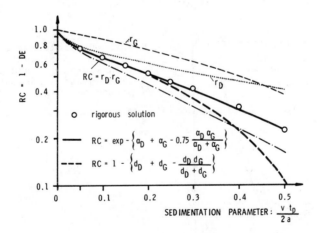

Figure 1. Fraction of particles escaping deposition from a
stagnant aerosol in a horizontal cylindrical tube
after Goldberg et al. (1978).

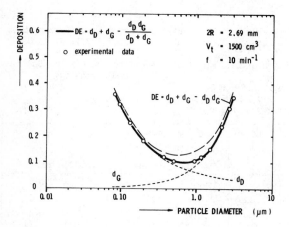

Fig. 2. Particle deposition from an aerosol with pulsatile flow in a granular bed of 2.69 mm glass beads.

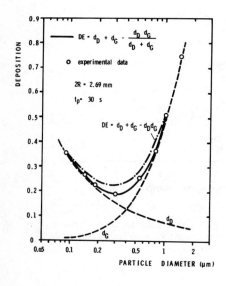

Fig. 3. Particle deposition from a stagnant aerosol in a granular bed of 2.69 mm glass beads over a period of 30 s.

Published 1984 by Elsevier Science Publishing Co., Inc.

Aerosols, Liu, Pui, and Fissan, editors

THE AEROSOL GENERATION, DILUTION AND EXPOSURE SYSTEM FOR THE CCMC'S DIESEL AND GASOLINE ENGINE EMISSION STUDY

David M. Bernstein, Hermann Pfeifer and John Brightwell
Battelle-Geneva Research Centres
7 route de Drize, 1227 Carouge-Geneva, Switzerland

[EXTENDED ABSTRACT]

The increasing use of light diesel-powered vehicles and the greater concentration of particulate matter emitted by diesel compared with gasoline fueled engines, has prompted much research into the consequences of such emission. A number of studies have been performed to investigate the potential health effects of diesel emissions in rodents. These studies, how-ever, have not included comparative investigations of gasoline engine emissions.

As part of the Health Effects Research Programme of the Committee of Common Market Automobile Constructors (CCMC) a chronic inhalation study is being performed on rodents exposed to automobile emissions. Using three engines running on the EPA's US-72 FTP cycle, rats and hamsters are exposed to four types of emission: gasoline, gasoline-catalyst, diesel and filtered diesel (gaseous components).

Experimental Design

Fischer 344 rats and Syrian Golden hamsters are being exposed for 16 hours per day, five days per week for up to two years. The exposures are carried out overnight and run using an automatic monitoring and control system.

The exhaust gases are diluted with clean air to obtain the exposure concentrations. The high dose concentrations are produced by a dilution with clean air at the tail pipe. The medium and low dose concentrations are obtained by subsequent dilution of the high dose concentrations. All exposure concentrations are defined on the basis of dilution ratios which remain constant with time.

Engine Test Facilities

The engine test facilities are located in a separate building, adjacent to the animal facilities, and consist of separate engine test cells for exhaust gas generation, a central control room and a workshop.

Each of the three engine test benches is equipped with an eddy current brake, inertia flywheel, disc brake, and an engine. The diesel emissions are generated by one VW Rabbit 1.5 liter engine and the gasoline emissions by two Renault R18 1.6 liter engines. The gasoline engines are micro-processor-controlled fuel injection US models and are identical except that one engine has no catalytic converter.

The engines are equipped with gearboxes (gasoline: automatic and diesel: manual) and are controlled by a stepping motor throttle or injection pump setting device. In order to reduce the size of the inertia flywheel (the simulated automobile is of the 2500 pound class) a multiplication gear (4:1) is used between the differential gear output and the flywheel.

This configuration permits dynamic simulation of a complete automobile. The engines are driven according to the US-72 (FTP) driving cycle (∿ 23 minutes) which is repeated continuously throughout the 16-hour exposure period. The cycle pattern, gear changing information (manual gear box), and the braking start and stop commands are stored in a microcomputer.

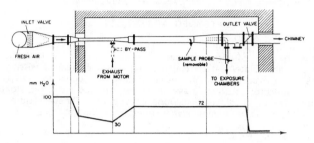

Figure 1: Primary dilution system

Safety and Performance

The computerized engine system automatically controls the speed over time and detects cycle deviation, and monitors general engine safety as well as exhaust gas composition. An extensive system of sensors and alarms in both the engine and animal facilities evaluates any deviation in the system and, if necessary, automatically commutes the atmosphere of the exposure chambers from engine emission to fresh air. Project personnel are simultaneously called in by a long distance paging system to ascertain and repair the cause of the alarm. With this system, after approximately one year of exposure nearly 250,000 km "distance" has been covered by each engine on the driving cycle (∿ 7,900 hours of operation), with a ratio between scheduled and actual running time of 97.5%.

Emission transfer and primary dilution system

The transfer of emissions from the engines to the animal exposure chambers (Hazleton 2000) is accomplished by a multistage process, driven by a slight overpressure following a primary dilution at the engine tailpipe exit, and a negative pressure supplied by the extraction system of the inhalation chamber.

The primary dilutions of between 11:1 and 30:1, depending upon motor type, are accomplished by the dilution tunnel shown in Figure 1. The system is designed to immediately reduce coagulation and growth of particles and produce the concentrations for high dose exposure. Filtered conditioned air is supplied to the inlet of a venturi. At the point of restriction there is an increase in velocity and hence an increase in kinetic energy. From an energy balance (Bernouilli's theorem) there must be a corresponding reduction in pressure. The exhaust from the engines is introduced into the air flow at the point of the greatest kinetic energy. The subsequent expansion produces very good mixing. This is followed by a straight tube of over 10 diameters in length before the take off for the exposure chambers.

The sample for the chambers is removed isokinetically with a slight increase in diameter of the dilution tunnel at this point in order to maintain constant velocity.

The take-off tube and transfer line to the exposure chambers have interior diameters of 72.5 mm. The emissions are transferred over a length of approximately 25 m at a mean linear velocity of ∿ 5.4 m/sec with a total mean residence time in the transfer line of ∿ 4.6 sec.

Secondary Dilution at Exposure Chambers

The emission transfer line enters the animal rooms through a fast acting pneumatic ball valve as shown in Figure 2.

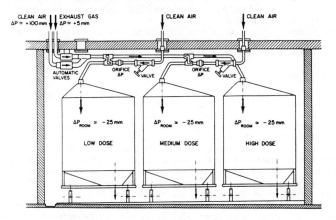

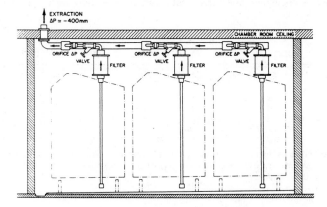

Figure 2: Secondary dilution system and animal exposure chambers
Top: Clean air and exhaust gas supply system
Bottom: Chamber air extraction system

The high dose concentration is delivered directly from the transfer line to the high dose chamber. All the chambers have a common clean air supply and extraction. The chambers are effectively isolated one from another by the installation of high pressure drop orifice plates in the supply and extraction of each chamber. With a +100 mm H_2O clean air supply there is a pressure drop of 25 to 50 mm across each supply orifice depending upon dilution air flow rate.

The common extraction has a pressure of -400 mm H_2O with orifice pressure drops of $\sim$ 150 mm for each chamber. With this highly pressure stable system secondary emission dilution is accomplished by aspiration from the high dose line into parallel lower dose clean air lines feeding the medium and low dose chambers using controlled pressure differential between the two lines. The system is balanced by adjusting the negative pressure in the exposure chambers.

During non-exposure periods or alarm conditions the emission ball valve automatically closes and clean air is supplied to the animals through a second pneumatic valve which opens.

Emission Transfer Characterization

Of primary interest in the design of this facility was the ability to deliver to the animals a representative sample of the materials emitted by the engine. With the engines operating on the US-72 cycle the efficiency of transfer from the primary dilution tunnel to the exposure chambers of the EPA regulated gaseous components CO, CO_2, total hydrocarbons, NO_x and NO was found to be 99% for all components. With the diesel over 70% of the aerosol component was delivered to the animals. Analyses of over 40 different aldehydes, ketones, phenols, polycyclic aromatic hydrocarbons as well as SO_2 and sulfate indicated that the diesel particles that were delivered to the animals were representative of those emittted by the engines.

Each of the 22 exposure chambers is sampled over one complete driving cycle during every night of exposure. The mean high dose concentrations in the rat exposure chambers over the first year of operation were found to be:

High dose chamber	CO (ppm)	CO_2 (%)	HC (ppm)	NO_x (ppm)	NO (ppm)
Gasoline	219 + 42	0.56 + 0.04	60.4 + 7.0	49.7 + 9.8	44.5 + 13.7
Gasoline-catalyst	20 + 9	0.62 + 0.02	1.3 + 1.2	7.5 + 3.3	6.7 + 3.9
Diesel	36 + 9	0.46 + 0.03	18.9 + 4.1	7.3 + 1.6	5.8 + 2.0
Diesel-Filter	36 + 9	0.47 + 0.03	18.8 + 4.1	7.4 + 1.7	6.0 + 2.0

Similar concentrations were measured in the hamster exposure chambers.

The mean concentration of particles in the high dose diesel chamber was found to be 7.1 $\pm$ 0.9 mg/m^3.

Published 1984 by Elsevier Science Publishing Co., Inc.

Aerosols, Liu, Pui, and Fissan, editors

COMPARISON OF EXPERIMENTAL AND CALCULATED DATA FOR THE TOTAL AND LOCAL DEPOSITION IN THE HUMAN LUNG

G. A. Ferron, W. G. Kreyling, B. Haider
Gesellschaft für Strahlen- und
Umweltforschung mbH München
Institut für Strahlenschutz
Ingolstädter Landstr. 1
D-8042 Neuherberg
F. R. G.

S. Hornik
Bundesgesundheitsamt

Institut für Strahlenhygiene
Ingolstädter Landstr. 1
D-8042 Neuherberg
F. R. G.

EXTENDED ABSTRACT

Introduction

Besides the total particle deposition in the human lung this contribution compares regional depositions calculated by different lung deposition models and with experimental data. An lung deposition model for inhaled aerosol particles is mostly composed of a lung structure model, a set of deposition equations for a single airway, and a mathematical computation method for the regional and total lung deposition. The used models are combined from four different lung structure models, two sets of deposition equations, and three mathematical methods.

Lung structure models

The used lung structure models has been developed by Weibel (1963) and modified by Olsen (1970), Hansen & Ampaya (1975) and Yeh & Schum (1980). The original models have different lung volumes. Because the volume influences directly the local deposition probabilities the computations have been carried out with a reduced lung volume of 3000 ccm, as published by Yu & Diu (1982). See this pubication for the references of the lung structure models. All four models have a structure with or nearly with a regular dichotomy. The sizes of the individual generations may differ up to 30% from the corresponding generation in the volume corrected model of Weibel. These models describe the human bronchial and pulmonary region with the generation numbered, resp., from 0 to 16 and from 17. Alveoli are included in the models depending on the mathematical method for the total and regional deposition calculation. These models have been extended by a mouth and a pharynx with a total volume of 50 ccm as derived by Rudolf (1984) from the experimental data. The sizes of the mouth and pharynx can be optimized for extrathoracic deposition by diffusion.

Deposition equations

The deposition equations determine the deposition probability of aerosol particles in an airway by sedimentation, diffusion and impaction. First a set of equations proposed by Landahl (1950) is used. These equations approximate the deposition by sedimentation and diffusion for a slighty disturbed laminar airflow in an airway. The impaction equation is based on experimental data. A second set published by Taulbee et al. (1978) uses equations for sedimentation and diffusion in a tube from a laminar airflow, and an impaction equation based on experimental data. Calculated deposition data of both sets correspond well for sedimentation and impaction. Higher deposition values are found by

Landahl for small diffusion parameters. Taulbee et al. calculate impaction only during inhalation.

For the extrathoracic deposition the experimental data are taken into account. No theory is used here.

Computation method

Three methods for the calculation of the total and regional particle deposition have been used. Generally the penetration of an aerosol through an airway generation is determined by multiplying the local penetrations for sedimentation, diffusion and impaction as derived from the corresponding deposition probabilities. The local deposition is derived from the local penetration corrected for the inspired aerosol volume and preceding aerosol losses. Gerrity et al. (1979) neglect the existance of the alveoli and add their volume to the last airway generation, by increasing its length but not its diameter. Crawford (1980) adds the alveoli as an additional last generation. As an alternative the alveoli can be introduced at a lower generation number. This makes sense, since the last four generations in Weibel's model are fully alveolated and the alveoli replace the tube walls. The deposition in the alveoli is determined assuming that no air mixing occurs and that the aerosol fills up an alveolus entirely or not at all.

Results and Discussion

The comparison of experimental and calculated data is carried out for three inhalation conditions with constant airflows of 250, 250 and 750 ccm/s and respiration times of 4, 8 and 4 s, resp. This corresponds to tidal volumes of 500, 1000 and 1500 ccm, resp. Experimental data on total and regional deposition have been published by Stahlhofen et'al. (1980) and have been recently summarized by Heyder (1984). They determine the deposition in the total lung (TL) and the regional depositions in the extrathoracic (E), the bronchial (B) and pulmonary region (P) in three healthy subjects. The functional residual capacity of the three subjects is about 3000 ccm.

Table 1 and fig. 1 present subsets of the results for particles with sizes of 0.08, 0.5 and 4.0 um. The data are selected in order to show the influence of the breating parameters and the components of the deposition models.

The calculated total deposition values are mostly in good agreement to the experimental data. Generally the data for 0.08 um particles are larger and the ones for 0.5 um particles smaller than the experimental data. Comparable results have been found by Yu and Diu (1982) by testing the lung structure models with their mathematical method. The model with alveoli at generation 19 fits less with the experimental data.

All models are equal suitable for calculating the regional particle deposition. However the calculated bronchial depositions are in general larger than the experimental values and beyond the experimental error of 2% (personal communication, Stahlhofen). The pulmonary deposition is mostly smaller than the experimental values. Possible explainations are wrong sets of deposition equations particularly for diffusion, or a wrong definition for the transition of the bronchial to the pulmonary region, as suggested by Hansen and Ampaya (1975).

In conclusion most of the lung deposition models for inhaled aerosol particles approximate the experimental deposition data well. However the calculated bronchial deposition is too large for all models.

1033

References

Crawford, D. J. (1980) "A Generalized Age-Dependent Lung Model with Applications to Radiation Standard" ORNL/TM-7454.

Gerrity, T. R., P. S. Lee, F. J. Hass, A. Marinelli, P. Werner, R. V. Lourenco (1979) "Calculated Deposition of Inhaled Particles in the Airway Generations of Normal Subjects" J. Appl. Physiol. 47: 867.

Heyder, J. (1984) "Studies of particle deposition and clearance in humans" Symposium of problems of Inhalatory Toxicity Studies, Federal Health Office, Institute for Drugs, Berlin, April 2-4.

Landahl, H. D. (1950) "On the Removal of Airborne Droplets by the Human Respiratory Tract. 1. The Lung" Bull. Math. Biophys. 12: 43.

Rudolf, G. (1984) "Ein Mathematisches Model zur Deposition von Aerosol teilchen im Atemtrakt des Menschens" Dissertation Johann-Wolfgang-Goethe Universität, Frankfurt am Main.

Stahlhofen, W., J. Gebhart, J. Heyder (1980) " Experimental Determination of the regional Deposition of Aerosol Particles in the Human Respiratory Tract" Am. Ind. Hyg. Assoc. J. 41: 385.

Taulbee, D. B., C. P. Yu, J. Heyder (1978) " Aerosol Transport in the Human Lung from Analysis of Single Breaths" J. Appl. Physiol. 44: 803.

Weibel, E. R. (1963) "Morphometry of the Human Lung" Springer Verlag, Berlin.

Yu, C. P., C. K. Diu (1982) "A Comparative Study of Aerosol Deposition in Different Lung Models" Am. Ind. Hyg. Assoc. J. 43: 54.

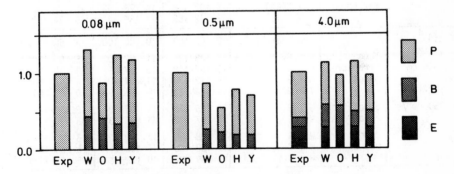

Figure 1. Comparison of different lung structure models normalized to experimental data. Tidal volume is 500 ccm and airflow is 250 ccm/s. Exp = experimenta data, W = Weibel, O = Olson, H = Hansen & Ampaya, Y = Yeh & Schum.

Table 1. Experimental and calculated deposition data for the human bronchial (B) and pulmonary (P) region, and total lung (TL) for three inhalation conditions and three particle sizes. The data are divided by the corresponding experimental total deposition data and given as percentages. The total lung deposition includes the extrathoracic deposition.

inhalation conditions	model tests	particle size / lung region	0.08 um			0.5 um			4.0 um		
			B	P	TL	B	P	TL	B	P	TL
tidal volume: 500 cm³, airflow: 250 cm³/s		exp. data rel.	0	100	100	0	100	100	13	59	100%
	lung structure model	W T G	44	89	131	25	60	86	29	55	113%
		O T G	42	46	89	23	30	54	28	39	96%
		H T G	35	88	124	19	59	78	21	65	114%
		Y T G	35	83	118	19	51	71	22	46	96%
	deposition equations	W T G	44	89	131	25	60	86	29	55	113%
		W L G	58	72	131	50	73	123	53	47	128%
	comp. method	W T G	44	89	131	25	60	86	29	55	113%
		W T C(24)	44	87	132	25	60	86	29	55	113%
		W T C(19)	38	125	164	22	94	116	29	56	113%
tidal volume: 1000 cm³, airflow: 250 cm³/s		exp. data rel.	0	100	100	0	100	100	9	81	100%
	lung structure model	W T G	24	97	121	12	73	89	25	72	106%
		O T G	24	82	107	12	62	74	24	65	98%
		H T G	22	99	121	12	76	88	18	80	107%
		Y T G	21	97	119	10	67	77	18	70	98%
	deposition equations	W T G	24	97	121	12	73	89	25	72	106%
		W L G	31	75	107	24	82	106	45	58	112%
	comp. method	W T G	24	97	121	12	73	89	25	72	106%
		W T C(24)	26	102	128	14	78	92	25	72	106%
		W T C(19)	23	142	165	11	130	141	25	74	108%
tidal volume: 1500 cm³, airflow: 750 cm³/s		exp. data rel.	0	100	100	0	100	100	14	58	100%
	lung structure model	W T G	16	108	125	11	92	102	15	60	102%
		O T G	16	102	118	10	87	97	11	58	97%
		H T G	15	112	128	11	94	105	14	59	100%
		Y T G	14	109	123	8	85	93	9	58	94%
	deposition equations	W T G	16	108	125	11	92	102	15	60	102%
		W L G	25	87	112	31	108	139	53	35	115%
	comp. method	W T G	16	108	125	11	92	102	15	60	102%
		W T C(24)	17	120	138	12	106	117	16	59	102%
		W T C(19)	15	189	204	10	172	182	15	69	111%

Lung structure models: W = Weibel, O = Olson, H = Hansen & Ampaya, Y = Yeh & Schum.
Deposition equations: T = Taulbee et al., L = Landahl.
Computation method: G = Gerrity et al., C(24) = Crawford (alveoli at generation 24), C(19) = Crawford (alveoli at generation 19).

Published 1984 by Elsevier Science Publishing Co., Inc.

Aerosols, Liu, Pui, and Fissan, editors

RELATIONSHIPS BETWEEN AMBIENT PARTICULATE MATTER, INHALED MASS FRACTION AND DEPOSITED FRACTION IN THE HUMAN RESPIRATORY TRACT

Jürgen Müller

Umweltbundesamt, Pilotstation Frankfurt

45 Feldbergstr., 6000 Frankfurt (M)

EXTENDED ABSTRACT

Introduction

Airborne particulate matter (PM) has a complex structure which locally, diurnally and seasonally demonstrates significant variations. Up to date general laws which describe the behaviour quantitatively are right only within certain limits (1,2). Therefore, since a long time several different sampling devices are in use which all pretend to measure the "relevant quantitative number" of PM concentration. Harmonizations on the national or international plane in order to express the PM-matrix in a scalar unit are difficult to attain. Nevertheless a senseful compromise has to be settled in order to attain international comparability between the measured data.

PM measuring has to be related to the health aspects as well as the airphysical and instrumental requirements:

1. The measured value should be comparable to the PM concentration inhaled by human beings.

2. The measured value should be relevant to damages of organic (animals and plants) and inorganic surface objects (buildings etc.).

3. The concentration of PM in ambient air should be measured as close to reality as possible.

Appropriate instruments designated in accordance with these requirements should have the following properties:

a) The instruments should be easily constructable, cheap, of low costs, adequate to the environment and applicable in most climatic regions

b) The instruments should not gravely be influenced by variations
of wind speed.

c) The collected samples (loaded filter) should be applicable for
subsequent chemical and physical analysis.

In a study for the European Community (3) several particulate samplers collocated together with an impactor were investigated in parallel
measurements. PM was collected at two stations, one located in an urban
residential area, the other one at a site under influence of heavy industry.

In a parallel study at the Warren Spring Lab., Stevenage, U.K. the
instruments were investigated in the wind tunnel (4).

Results of Measurements

The mass size distributions of PM measured at the two stations with
an 8-stage Andersen impactor are represented in Fig. 1. At the Frankfurt station 2/3 of the PM-mass was found in the fine mode (diameter
$< 2 \mu m$) of the particle spectrum. The main sources in this area are combustion processes of traffic and domestic heating which predominantly
contribute to fine particles.

At the Wetzlar station heavy industry sources (cement and iron
smelter) emitted a large amount of coarse particles which were added to
the urban emissions of traffic and domestic heating. In contrary to Frankfurt about one half of ambient PM were coarse particles ($> 2 \mu m$).

The collected mass of the impactor compared with the results of the
total PM-samplers in average was 15 % less. At Warren Spring in the
field tunnel experiments for the total PM-sampler at a wind speed of
$3 \, m/s$ a 50 %-cut off diameter of $22 \mu m$ was measured. For the Andersen-Impactor under the same conditions a 50 %-cut off of $9,7 \mu m$ was
determined. In order to get an estimate for the realistic ambient air mass
distribution the impactor results were extrapolated by 20 %.

In Fig. 2 the resulting relative mass size distribution at Frankfurt is
plotted. This curve with a coarse mode maximum in the range between 10
and 12 μm diameter is representative for urban air at sites not influenced
by nearby heavy industry, sand storm or sea salt sources. Under these
PM circumstances which were also found in American urban air (5)
most people in the world have to live.

Caused by increasing sedimentation the mass contribution of coarse
mode particles bigger than 12 um diameter rapidly falls off. Under normal

conditions beyond 25 μm diameter not more than 3 % of total particle mass may be detected.

By investigations of several groups during the last decade the human particulate inhalation in dependence of the particle size was revealed (6,7). The results of these measurements were summarized in an ISO/ Paper (8). The ISO-function found with articial particles for thoracic and extrathoracic inhalation can be applied to the measured urban air conditions. The result of this transformation is plotted in the second curve of Fig. 2 which represents the inhaled mass fraction of urban air. The inhaled air mass fraction consists roughly of 85 % of the total ambient particle spectrum.

However, in conjunction with the inhalated particle mass the deposited mass in the respiratory tract has also to be considered. For these measurements with artificial aerosols (6, 7, 9), up to now, no ISO-standardisation exists. The deposition in function of the particle size by the lower curve in Fig. 2 is applied to the urban particle spectrum. From the figure can be seen that almost all inhaled coarse particles are deposited, mostly extrathoracically. The fine particles inhaled thoracically in an amount of two third of it's mass are exhaled again.

Activities in order to define a reference method for PM sampling have to take into account the results summarized in Fig. 2.

References

1. Jaenicke, R. and C. Junge, Beitr. Phys. Atmos. 40, 129 (1967)

2. Whitby, K.T., W.E. Clark, V.A. Marple, G.M. Sverdrup, G.J. Sem, K. Willeke, B.Y.H. Lin and D.Y.H. Pui, Atm. Envir. 9 pp 463 - 482 (1975)

3. Müller, J.: "Field measurements of suspended particulate matter (SPM) sampled with different instruments", Study for the Commission of the European Communities (DG XI), Brussels (1984)

4. Barrett C.F., M.O. Ralph, S.L. Upton: "Wind tunnel measurements of the inlet efficiency of four samplers of suspended particulate matter", Study for the Commission of the Commission of the European Communities (DG XI), Brussels (1983)

5. Lundgren, D.A. and H. J. Paulus, JAPCA 25, pp 1227-1231 (1975)

6. Stahlhofen, W., J. Gebhart, J. Heyder, Am. Ind. Hyg. Ass. Journ.
   41, pp 385-398a (1980)

7. Lippmann, M.; c/o Handbook of Physiology: Section 9, Bethesda,
   MD, American Physiological Society, pp 213 - 232 (1977)

8. ISO-Technical Report 7708: "Air quality-Particle size fraction
   definitions for health-related sampling"

9. "Inhalation of Aerosols: Particle Deposition and Retention",
   edited by Klaus Willeke, Ann Arbor Science Publishers, Inc.
   Ann Arbor, MI

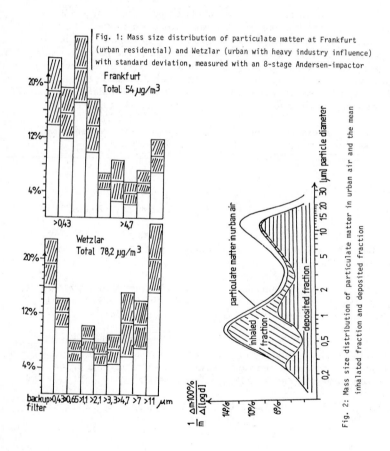

Fig. 1: Mass size distribution of particulate matter at Frankfurt (urban residential) and Wetzlar (urban with heavy industry influence) with standard deviation, measured with an 8-stage Andersen-impactor

Fig. 2: Mass size distribution of particulate matter in urban air and the mean inhalated fraction and deposited fraction

Published 1984 by Elsevier Science Publishing Co., Inc.
Aerosols, Liu, Pui, and Fissan, editors

EFFECT OF RESPIRATORY PATTERN ON DEPOSITION OF
ULTRAFINE AEROSOLS IN THE HUMAN RESPIRATORY TRACT

P. J. Anderson, F. J. Wilson, Jr., F. C. Hiller, R. C. Bone, and I. Lau
Pulmonary Division
University of Arkansas for Medical Sciences
University of Arkansas Graduate Institute of Technology
Little Rock, Arkansas 72205
U.S.A.

EXTENDED ABSTRACT

Human respiratory tract deposition of ultrafine particles (<0.1 μm diameter) is important because (1) pollutant aerosols such as engine exhaust and photochemical smog contain particles in this size range, (2) deposition of ultrafine particles is thought to be mainly in the peripheral gas exchange regions where efficient clearance mechanisms are lacking, (3) particles in this size range remain suspended in air for long periods of time, and (4) the large surface area to mass ratio for ultrafine particles makes them potential carriers for large quantities of adsorbed toxic gases. Theoretical models predict high respiratory tract deposition of these particles and that deposition increases with decreasing size (Task Group on Lung Dynamics, 1966). Brownian diffusion is the deposition mechanism that is most important in particles of diameters less than 0.5 μm, and as size decreases, diffusibility increases. Theoretical models also predict that deposition is increased by increasing tidal volume and inspiratory pause time (or breath-holding time), but there are few experimental data concerning the effect of respiratory pattern on deposition of ultrafine particles. The purpose of this study was to determine the effect of variations in tidal volume and inspiratory pause time on the fractional deposition of nonhygroscopic inhaled particles of 0.024 to 0.24 μm diameter in the normal human respiratory tract.

The ultrafine aerosol used in this study is produced from bis (2, 3 ethylhexyl) sebacate (BES) used as a 0.01% solution in 95% ethanol and contains adequate numbers of particles ranging from 0.02 to 0.2 μm in diameter. BES is a nontoxic, nonhygroscopic oil of very low volatility which has been widely used in aerosol inhalation experiments. Aerosol generation is carried out using a three-orifice Collison nebulizer with a condensation generator. Aerosol analysis is performed by means of an electrical aerosol analyzer (Model 3030, TSI, Inc.) (Liu, 1974). As programmed by the manufacturer, it produces counts of particles within 11 size range increments from 0.003 to 1.0 μm; for data reduction purposes, the size of the particles in each increment is taken to be the geometric midpoint of the increment. In our hands, this device has been accurate over a size range of 0.02 to 0.04 μm. This BES aerosol has a count median diameter of 0.037 μm and a geometric standard deviation of 1.7.

Deposition studies are performed using an aerosol inhalation system consisting of a 240 liter chamber containing 3 collection bags and a micro-

processor-controlled valve system (Hiller, 1982). The subject can inhale
either aerosol from the chamber or clean air from the clean air bag. The
exhaled air is collected either in the waste bag or in the sample
collection bag. Inhaled and exhaled volumes are measured by a differential pressure pneumotachograph. The signal from the pneumotachograph
is displayed on one channel of a dual beam oscilloscope and a prerecorded
respiratory pattern is displayed on the other channel of the oscilloscope.

Five different respiratory patterns were chosen, all with constant
flow rates of 1 L/second and minute ventilation of 12 liters. Three
patterns have tidal volumes of 1 liter with inspiratory pause times of
0, 1 and 2.5 seconds. The remaining 2 patterns have tidal volumes of
0.75 and 2 liters, both with a 1 second inspiratory pause.

Five adult males ranging in age from 23-34 years were studied. All
were non-smokers and had normal flow rates as measured by spirometry and
normal lung volumes. Each subject performed 3 separate deposition studies
for each of the five respiratory patterns. Each study was initiated from
functional residual capacity. Deposition was calculated as the difference
in concentration between inhaled and exhaled aerosol of five size fractions
corrected for system deposition and dead space constants.

Mean deposition fractions ($\pm$ S.E.) for all subjects on each of the
five respiratory patterns are given according to particle size in Table 1.

TABLE 1

MEAN DEPOSITION FRACTION ($\pm$S.E.) FOR VARIOUS RESPIRATORY PATTERNS

	0.02 μm	0.04 μm	0.08 μm	0.13 μm	0.24 μm
0 s Insp. pause 1 L tidal vol.	.49 (.06)	.39 (.04)	.32 (.05)	.23 (.04)	.12 (.04)
1 s Insp. pause 1 L tidal vol.	.56 (.03)	.48 (.04)	.39 (.05)	.29 (.06)	.29 (.06)
2.5 Insp. pause 1 L tidal vol.	.68 (.04)	.60 (.03)	.52 (.04)	.41 (.05)	.32 (.07)
1 s Insp. pause .75 L tidal vol.	.50 (.06)	.44 (.03)	.36 (.04)	.27 (.04)	.21 (.04)
1 s Insp. pause 2 L tidal vol.	.63 (.03)	.58 (.02)	.46 (.04)	.40 (.04)	.36 (.04)

The measurements in this study demonstrate the validity of theoretical
predictions in showing that (1) deposition increases as particle size falls
below 0.1 μm, (2) deposition increases with increasing tidal volume, and
(3) deposition increases with increasing inspiratory pause time. These data
vary slightly from theoretical models in that measured total deposition values
appear somewhat lower than predicted for particles from 0.02 to 0.08 μm in
diameter.

References

Hiller FC, Mazumder MK, Wilson JD, McLeod PC, Bone RC. Human respiratory tract deposition using multimodal aerosols. J Aerosol Sci 13:337-343, 1982.

Liu BYH, Whitby KJ, Pui DYH. A portable electrical aerosol analyzer for size distribution measurement of submicron aerosols. J Air Poll Cont Assoc 24: 1067-1072, 1974.

Task Group on Lung Dynamics. Deposition and retention models for internal dosimetry of the human respiratory tract. Health Physics 12:173-207, 1966.

Published 1984 by Elsevier Science Publishing Co., Inc.

Aerosols, Liu, Pui, and Fissan, editors

1042

EXTRAPOLATION MODEL FOR AEROSOL DEPOSITION IN MAMMALIAN LUNGS
AND COMPARISON WITH EXPERIMENTAL RESULTS

F. Daschil and W. Hofmann

Division of Biophysics, University of Salzburg
Erzabt Klotz-Strasse 11, A-5020 Salzburg/Austria

Most conclusions concerning biological effects of inhaled particles in
the human lung are largely based on experimental data obtained from different
animal species. Two strategies can be employed to extrapolate this information to man: either development of single animal-specific lung models or
application of a scaling procedure assuming the same anatomical lung structure
for all animals and man. The latter approach has been chosen mainly for
radiation protection purposes, particularly for the evaluation of relevant
parameters for dose calculations (stochastic sensitivity analysis) and the
assessment of risk where additional uncertainty of the biological response
has to be considered.

Based on the widely used symmetrical Weibel morphology for the human
lung, appropriate scaling factors have been determined by utilizing experimental information available as well as theoretical considerations. All
variations of anatomical and respiration parameters were expressed in terms
of one single quantity, namely the total body weight.

Predictions of regional or total aerosol deposition by the extrapolation model have been compared to experimental deposition data published in
the open literature for a variety of animal species and experimental conditions, e.g. aerosol characteristics. Generally this comparison showed a
solid agreement between theoretically predicted and experimentally measured
deposition data. It should be noted that systematic deviations between
certain groups of animals have been observed and that the quality of the
model for regional deposition depends on aerosol parameters and respiration
characteristics. On the other hand, this experimental feedback allows us
to refine the theoretical basis of our extrapolation model. Particularly
the problem how to parameterize morphologically similar subgroups or the
question whether this single parameter reduction should be substituted by
a two-parameter representation, need further detailed investigations.

Published 1984 by Elsevier Science Publishing Co., Inc.
Aerosols, Liu, Pui, and Fissan, editors

In VITRO DISSOLUTION STUDIES OF ULTRAFINE MIXED-MATRIX METAL AEROSOLS*

M. D. Hoover, M. T. Carter and W. C. Griffith
Lovelace Inhalation Toxicology Research Institute
P. O. Box 5890
Albuquerque, New Mexico 87185
U. S. A.

EXTENDED ABSTRACT

Introduction

Inhalation exposures of humans after accidents in the fission or fusion nuclear fuel cycles are likely to involve aerosols that are a complex mixture of materials. These may include neutron activation products, fission products, actinide radionuclides, and any structural materials involved in the accident. Because accidents usually involve high temperatures, potential aerosols are expected to be in the form of branched chain ultrafine aggregate particles. These may be uniform mixtures of metals, or they may consist of a core of high boiling point metals coated with a layer of other metals that condense at lower temperatures. We conducted our study to determine if the details of particle composition influence particle dissolution and are, therefore, important for determining the biological behavior of mixed metal particles.

Methods

Cobalt and iron were chosen because they are major neutron activation product radionuclides. They account for the largest fraction of the projected inhalation dose to people from a high- temperature fusion-related accident involving a design with a stainless steel first wall (Piet et al., 1982). The isotopes cobalt-57 ($t_{1/2}$ = 270 d) and iron-59 ($t_{1/2}$ = 45 d) were used as radioactive labels for in vitro dissolution studies. The physical properties of cobalt and iron are similar (melting point = 1500°C; boiling point = 2800°C). Any differences in dissolution rates should be related to differences in the chemical properties and crystalline forms of the materials and the amount of surface area of the materials available to interact with the solvent.

Ultrafine aerosols were generated using a metal chelate method (Kanapilly et al., 1978 and Hoover and Lee, 1982) in a special tube furnace system (Figure 1). Aerosols of oxides of the single metals were generated by vaporizing the chelated metal at 200°C in zone 1, degrading the chelate at 1000°C in zone 2, and allowing the free metal atoms to condense outside the furnace to form ultrafine particles. Mixed metal oxide particles were formed by vaporizing a chelate containing equal amounts of iron and cobalt. Coated particles were formed by creating an ultrafine aerosol of the core material in zones 1 and 2, vaporizing the coating in zone 3, and allowing

*Research performed under U. S. Department of Energy Contract Number DE-AC04-76EV01013.

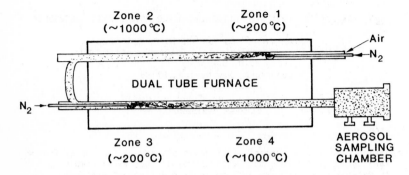

Figure 1. Schematic representation of the ultrafine aerosol generation
system used to produce ultrafine, mixed matrix aerosols for in
vitro dissolution studies.

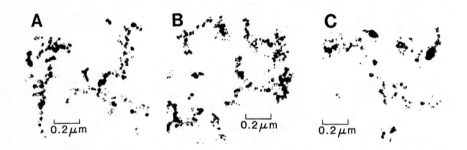

Figure 2. Transmission electron photomicrographs of ultrafine, branched
chain aggregate aerosol of (a) cobalt, (b) iron and cobalt, and
(c) cobalt coated on iron.

Table 1
Summary of the Particle Size, Specific Surface Area and Fraction Dissolved in 30 Days
for the Ultrafine Metal Oxide Aerosols Studied

Composition of Metal Oxide Particles	Projected Area Diameter of Primary Particles (μm ± SD)	Specific Surface Area (m²/g ± SD)[a]	Fraction Iron Dissolved in 30 Days (fraction ± SD)[a]	Fraction Cobalt Dissolved in 30 Days (fraction ± SD)[a]	Fraction Iron Dissolved Per Specific Surface Area (fraction/(m²/g) ± SD)	Fraction Cobalt Dissolved Per Specific Surface Area (fraction/(m²/g) ± SD)
Iron	0.033 ± 0.009	28.8 ± 3.0	0.005 ± 0.001	--	0.0002 ± 0.00004	--
Cobalt	0.028 ± 0.008	16.2 ± 1.1	--	0.02 ± 0.01	--	0.001 ± .0006
Iron and Cobalt	0.020 ± 0.009	14.4 ± 2.6	0.020 ± 0.010	0.15 ± 0.02	0.001 ± 0.0007	0.01 ± .002
Iron on Cobalt	0.029 ± 0.014	21.1 ± 1.4	0.020 ± 0.001	0.04 ± 0.01	0.001 ± 0.00008	0.002 ± .0005
Cobalt on Iron	0.035 ± 0.011	8.1 ± 1.0	0.007 ± 0.001	0.08 ± 0.02	0.001 ± 0.0002	0.01 ± .003

[a]Standard deviations based on triplicate samples.

the chelate of the coating to be degraded in zone 4 so that the core particles provided the nuclei for condensation of the coating material.

Triplicate samples of the particles containing approximately 50 µCi iron-59 or cobalt-57 were collected on Millipore type AE filters for in vitro dissolution in a serum ultrafiltrate solvent with diethylenetriamine-pentaacetic acid (Kanapilly et al., 1974). Sample changes were made at 1, 3, 5, 7, 24, 26, 28, 30, 48, 54, 72, and 78 hours and 4, 7, 9, 11, 14, 21, and 28 days after the start of the study. A Quantisorb surface area measurement device (Quantachrome Corp., Greenvale, NY) was used to measure the specific surface area of samples of the various particles.

Results

Figure 2 shows transmission electron photomicrographs of representative coated and uncoated particles. Table 1 summarizes the projected area diameters of the primary particles, the specific surface area measured by Quantasorb, the fraction of the materials dissolved in 30 days, and the fraction dissolved per specific surface area for the five aerosols studied.

Discussion

The projected area diameters of the primary ultrafine particles were statistically the same for the five aerosols studied, although the trend appeared to be for the coated particles to be larger than the uncoated particles (see Table 1). The addition of a coating material (e.g. iron onto cobalt or cobalt onto iron) resulted in an increase in the standard deviation of the primary particle sizes, indicating a nonuniform condensation process. It would be useful if some method could be developed to quantitate the fraction of the surface area of the core particles that is coated. Surface area studies are continuing.

In light of the similarity of the primary particle sizes, the specific surface areas of the five aerosols would have been expected to be similar. They range over a factor of nearly three, with the coated particles generally having lower specific surface area than the uncoated particles. This could result from a smoothing of available surface during the condensation of the coating.

The dissolution rates of the core materials were statistically similar to those of the corresponding single materials. This was to be expected because formation conditions of the core were independent of a later addition of coating. The presence of a coating did not appear to retard the dissolution of the core, apparently confirming that the core was not uniformly coated and shielded from the solvent. The dissolution of the iron or cobalt coating, however, was enhanced by a factor of 4 relative to that of the corresponding single materials. This is explained in part by the theoretically higher surface-to-volume ratio of a coating. If the addition of a uniform coating doubles the volume of the particle, the particle diameter increases by 26% and the surface-to-volume ratio increases by 58%. A second mechanism for the increased dissolution may be the imperfect nature of the crystal structure established at the interface between the core and coating.

1046

The apparent influence of a disruption of crystal structure can be seen in the fourfold and eightfold increases in the dissolution of the cobalt and the iron, respectively, in the mixed matrix particles as compared to the single composition particles. The study is continuing, with investigation by x-ray diffraction of the crystal structure of the five aerosols studied. The results to date indicate that details of particle composition can influence dissolution behavior by a factor of four or more. This could strongly influence the biological behavior of moderately soluble materials and should be studied further.

Acknowledgments

Research performed under U. S. Department of Energy Contract Number DE-AC04-76EV01013. M. T. C. participated in this work as a summer student researcher at the Inhalation Toxicology Research Institute under the auspices of the Associated Western Universities.

References

Piet, S., M. Kazimi, and L. Lidsky (June 1982) "Potential Consequences of Tokamak Fusion Reactor Accidents: The Materials Impact", DOE UC-20C,D,E, Massachusetts Institute of Technology.

Kanapilly, G. M., K. W. Tu, T. B. Larson, G. R. Fogel, and R. J. Luna (1978) "Controlled Production of Ultrafine Metallic Aerosols by Vaporization of an Organic Chelate of the Metal", J. Colloid Interface Sci. 65: 533-547.

Hoover, M. D. and J. Lee (December 1982) "Aerosols from Fusion Energy Systems", in Inhalation Toxicology Research Institute Annual Report 1981-1982, LMF-102, Albuquerque, NM, pp. 71-73.

Kanapilly, G. M., O. G. Raabe, and H. A. Boyd (1974) "A Method for Determining the Dissolution Characteristics of Accidentally Released Radioactive Aerosols", Proceedings of 3rd International Congress, IRPA, USAEC, Oak Ridge, TN.

Published 1984 by Elsevier Science Publishing Co., Inc.
Aerosols, Liu, Pui, and Fissan, editors

RESPIRATORY TRACT CLEARANCE OF INHALED PARTICLES
IN LABORATORY ANIMALS*

R. O. McClellan, M. B. Snipes and B. B. Boecker
Lovelace Inhalation Toxicology Research Institute
P. O. Box 5890
Albuquerque, NM 87185
U.S.A.

EXTENDED ABSTRACT

Introduction

A wide variety of particles may be inhaled by people and cause disease
in the respiratory tract or other tissues. For particles of a few materials,
epidemiological data may exist to provide evidence, generally of a
qualitative nature, of the relationship between exposure and an excess
incidence of disease. For these materials, investigations using laboratory
animals may provide insight into the mechanisms responsible for the induction
of disease. There may be no epidemiological data available for other
materials that may be inhaled in a particulate form, perhaps because the
material has only recently been synthesized. In such instances, studies in
people may not be feasible because of ethical considerations. Thus, the
only alternative may be to conduct investigations using laboratory animals,
tissues, cells or macromolecules to aid in assessing the health risks to
people for exposure to a material of unknown human toxicity. A common need
in studies with people and intact laboratory animals is to establish the
relationship between the exposure (duration, concentration and particle
characteristics such as size) and the spatial and temporal dimensions of
dose to tissues and cells. When studying effects in macromolecules, cells,
tissues, whole animals or people, there is also a need to understand the
relationship between dose and the observed effect. Thus, dose to tissue
becomes the common denominator for interpreting a wide range of
investigations with particulate materials.

For many years it has been recognized that despite qualitative
similarities, there are marked quantitative differences in the extent of
particle deposition and clearance in different species. The basis for these
differences has been open to speculation and included the possibility that
when polydisperse aerosols are inhaled, the size distribution of particles
depositing in the major compartments of the respiratory tract differs
between the species and influences the clearance rates. To investigate this
question, the clearance of inhaled particles was investigated in mice, rats,
guinea pigs and dogs. A single type of aerosol, fused aluminosilicate
particles with a radiolabel, was used either as a monodisperse or
polydisperse size distribution. These results are compared to similar data
from human studies conducted in other laboratories to identify species
similarities and differences in the clearance of inhaled particles.

---

*Research performed under U. S. Department of Energy Contract Number
DE-ACO4-76EVO1013.

1048

Methods

The exposure aerosols were fused aluminosilicate particles labeled with [134]Cs that had been generated in a dynamic system where the particles were fused at 1100 C. Some exposures were conducted with polydisperse aerosols. Others were conducted with monodisperse size fractions that had been separated using a spiral duct aerosol centrifuge. The methods of aerosol preparation and animal exposures have been previously described (Newton et al., 1980). The size distribution of the exposure aerosols was determined using a Mercer cascade impactor.

The animals used were CD-1 mice, Fischer-344 rats, 49 Hartley guinea pigs and Beagle dogs. Groups of mice, rats and guinea pigs were exposed to 0.7, 1.5 or 2.8 μm monodisperse aerosols or a polydisperse aerosol. Guinea pigs were exposed to a 2.0 μm monodisperse aerosol. All of the animals were exposed nose only to minimize external contamination. Exposure durations ranged from 25 to 75 minutes. Animals were assigned to a sacrifice schedule; the ones scheduled for sacrifice at the latest times after inhalation exposure had excreta collected for [134]Cs analysis at intervals using metabolism cages. At sacrifice components of respiratory tract and other tissues and organs were collected for radioanalysis to determine the distribution of [134]Cs in the body.

Tissue and excreta content of [134]Cs as a function of time post-exposure were evaluated with a simulation language approach that considered the two dominant pathways of [134]Cs removal; physical translocation of particles and dissolution of activity from the particles. Our intent was to (a) define a model to account for particle-associated and solubilized [134]Cs, (b) express the data as percent of the initial body burden in various compartments, (c) establish pathways by which the [134]Cs could be transferred between compartments, and (d) empirically fit the simulation to the observations to achieve a materials balance.

The model is presented in Figure 1. Definitions for model parameters (fraction per day) are as follows, with t representing time in days:

MNP(t) = mechanical transfer constant for particles from the
nasopharynx to the gastrointestinal tract;
MTB(t) = mechanical transfer constant for particles from the
tracheobronchial region to the gastrointestinal tract;
MP(t) = mechanical transfer constant for particles from the pulmonary
region to the gastrointestinal tract;
ML(t) = mechanical transfer constant for particles from the pulmonary
region to lung-associated lymph nodes (LALNs);
S(t) = solubilization or dissolution rate for the particles in the
major regions of the respiratory tract, LALNs, or
gastrointestinal tract.

Results and Discussion

The pulmonary retention of 1.5 μm particles in mice, rats & dogs and 2.0 μm particles in guinea pigs is shown in Figure 2. For comparison, human data developed by Bailey et al. (1982) using similar particles is also shown.

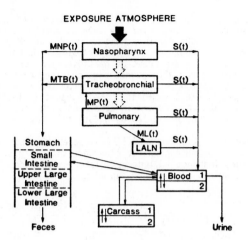

FIGURE 1. General kinetic model to simulate organ distribution and retention patterns for [134]Cs in mice, rats, guinea pigs, and dogs after inhalation exposure to [134]Cs-labeled aluminosilicate particles.

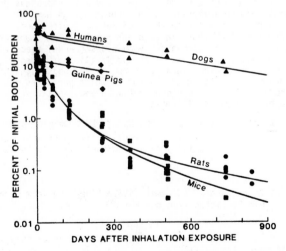

FIGURE 2. Pulmonary retention of labeled fused aluminosilicate particles inhaled by dogs, guinea pigs, rats, mice, and humans.

The results of the simulation modeling indicated that the clearance related to dissolution was similar for the several species studied. The long-term solubility constant for 0.7-, 1.5-, and 2.8-μm AMAD particles was 0.0015, 0.0008, and 0.0002 per day, respectively. Mechanical clearance of deposited particles was independent of particle size for dogs, rats, and mice. In the follow-up study with guinea pigs only one particle size was used, based on the assumption that the results from the other species would apply to guinea pigs. Table 1 summarizes the mechanical clearance factors of the four species. The results are in agreement with predictions that long term clearance would be dominated by dissolution (Mercer, 1967).

TABLE I Mechanical transport of ^{134}Cs-labeled aluminosilicate particles from the pulmonary region for clearance via the gastrointestinal tract (MP(t)) or for clearance to lung-associated lymph nodes (ML(t)) after inhalation by dogs, guinea pigs, rats, and mice.

| Species | Fraction transported per day: | | |
	MP(t), early	MP(t), late	ML(t)
Dogs	0.005 exp (-0.03t)	0.0001	0.0002
Guinea pigs	0.005 exp (-0.03t)	0.0001	0.00003
Rats	0.02 exp (-0.007t)	0.001	0.0007 exp (-0.5t)
Mice	0.02 exp (-0.006t)	0.0015	0.0007 exp (-0.5t)

Results of our studies indicate that deposition and retention of aerosols are similar in dogs and humans; rats and mice represent poor models for studying long-term retention in humans. Guinea pigs have deposition and early retention patterns similar to those of rats and mice, but their long-term patterns are more like those in dogs and humans. The translocation of particles from lung to lung associated lymph nodes (LALNs) seems similar among dogs, guinea pigs, and humans. The similarities of deposition and retention patterns for aerosols inhaled by humans and dogs suggest the dog is a useful model for predicting the extent to which particles deposited in the pulmonary region of humans would be retained and the extent to which the particles would be accumulated in LALNs.

The results suggest that the clearance patterns of mice and rats are sufficiently different than for man that these differences must be considered is extrapolating inhalation toxicity data from these species to man.

References

Bailey, M. R., F. A. Fry and A. C. James (1982) "The Long Term Clearance Kinetics of Insoluble Particles from the Human Lung," Ann. Occup. Hyg. 26 273-290

Mercer, T. T. (1967) "On the Role of Particle Size in the Dissolution of Lung Burdens," Health Phys. 13: 1211-1221.

Newton, G. J., G. M. Kanapilly, B. B. Boecker and O. G. Raabe (1980) "Radioactive Labeling of Aerosols: Generation Methods and Characteristics," Generation of Aerosols (Edited by K. Willeke), Ann Arbor Science Publications, Ann Arbor, MI.

Published 1984 by Elsevier Science Publishing Co., Inc.

Aerosols, Liu, Pui, and Fissan, editors

MONTE CARLO MODELLING OF INHALED PARTICLE DEPOSITION IN THE HUMAN LUNG[*]

L. Koblinger[1] and W. Hofmann[2]

[1]Central Research Institute for Physics, P.O. Box 49, H-1525 Budapest, Hungary
[2]University of Salzburg, Erzabt-Klotz-Strasse 11, A-5020 Salzburg, Austria

EXTENDED ABSTRACT

Introduction

Current models of inhaled particle deposition in the human respiratory tract make use of stylized forms which are approximations of the real anatomical structure of the lung (Weibel, 1963; Horsfield et al., 1971; Yeh and Schum, 1980). Successive generations of air passages are represented by straight cylindrical tubes of specified diameters and lenghts, branching at fixed angles into daughter tubes. Significant inter- as well as intra-subject variabilities of linear airway dimensions and branching patterns have been observed in morphometric studies, leading to experimentally observed random variations of aerosol deposition (Heyder et al., 1978, 1980; Tarroni et al., 1980). Current lung models do not, however, reflect this variability of the structural components of the lung. Thus it is the goal of this paper to describe a stochastic lung model which incorporates airway parameter variability as well as asymmetrical branching and which can be used for aerosol deposition and radiation dose calculations (Hofmann et al., 1983).

Random selection scheme of airway parameters

Instead of tracing all aerosol particles of an inhalation pulse through all airways of a generation at the same time, we use a Monte Carlo method, i.e. a repeated step-by-step random selection of pathways from statistical distributions of airway parameters and their correlations. If the statistics of the human airway structure are known, then a random selection procedure for aerosol deposition calculations can be constructed in the following way (for inspiration):

Step 1: Select inhaled aerosol diameter (either monodisperse or from a probability distribution) and initial velocity (dependent on tidal volume and respiratory frequency). Set time to t=0 and generation number k to 1 (trachea).

Step 2: Select a diameter (d_k) from the probability distribution of diameters in generation k.

Step 3: Select the tube length (l_k) from the probability distribution of lenghts in generation k, taking into account the correlation between diameters and lengths of tubes in generation k.

Step 4: Select the diameters of two daughter tubes $(d_{k+1}^{(1)}$ and $d_{k+1}^{(2)})$ from the distribution of cross section ratios of the parent tube to both daughter tubes and of the minor to the major tube.

---

[*]This research was carried out within the frame of the scientific corporation between the Hungarian and the Austrian Academy of Sciences.

Step 5: Select the lengths $(l_{k+1}^{(1)}$ and $l_{k+1}^{(2)})$ from the diameters and the correlation between diameters and lengths in generation k+1.

Step 6: Select branching angles of the two daughters from the distributions for the major and the minor daughter. Select gravity angles.

Step 7: Select the particle path at the bifurcation from diameters, branching angles and laws of flow.

Step 8: Calculate the deposition probability for different physical mechanisms in the selected tube, calculate the time elapsed since leaving the last bifurcation, modify velocity of aerosol particle.

Step 9: Decide from the diameter and the generation number of the daughter tube whether the particle has already reached the alveolar region. If so, terminate geometry selection. If not, set k=k+1 and continue from step 4.

The turning point of the particle flight from inhalation to pause and exhalation is defined by reaching zero velocity. This physical termination of the particle movement stops the cycle steps 4 to 9. For expiration the same path will be followed backward.

As can be seen from the above scheme the physical data (deposition) are calculated analytically while the airway geometry is selected randomly. During the flight of the particle time is recorded continuously and the selected dimensions are corrected for the change of the linear dimensions (particularly diameters in the higher generations) during the breathing cycle.

Statistical analysis of airway geometry

Our statistical analysis of the human airway structure is based on morphometric data measured at the Lovelace Inhalation Toxicology Research Institute, Albuquerque, NM, USA* (Raabe et al., 1976). Average values and distributions of airway diameters and lengths, criteria for the termination of pathways (reaching the alveolar region after different numbers of bifurcations), as well as correlations between cross sections of succeeding generations and those between diameters and lengths of tubes from the same generations were investigated. Since special emphasis has been placed upon correlations among the various airway parameters, some of these results will be discussed briefly:

An important question for the random sampling process is whether there is any correlation between the cross sections of the parents and both daughter tubes. Figure 1 illustrates the cross section ratios averaged for generations 6 through 17 in all lobes, which are used in step 4 of the selection procedure. The average value of 0.82 is in very good agreement with both the theoretically expected value of 0.794 for symmetrical branching (D'Arcy Thompson, 1942) and the previously published experimental value of 0.88 for real i.e. asymmetrical branching by Horsfield and Cumming, 1967. The latter also found a minimum ratio of 0.4, i.e. there were cases where the total cross section of both daughters exceeded twice that of the parent tube cross section. We also found such cases, but the 37 observed cases out of the total 3704 bifurcations are negligible or can simply be measurement- or registration errors.

---

*The authors are indebted to Dr. Hsu-Chi Yeh, Lovelace ITRI for supplying us with the magnetic tapes containing the morphometric data.

The asymmetry of branching can be characterized by the ratio of the diameters of the major and minor daughters of the same parents. Horsfield and Cumming (1967) found that in 27.6% of the cases the ratio was larger than 0.9, in 69% larger than 0.7, and the smallest ratio they found was 0.37. In our much larger data base the diameter ratio exceeds 0.9 in 36% of the cases and 0.7 in 82%. The histogram in Figure 2 shows the quite plausible results that high asymmetries, i.e. small diameter ratios, are more frequent with relatively large diameter parents. The information presented in this figure is used again in step 4 for the random selection of the daughter diameters.

Conclusions

This morphometric data base together with the described random selection procedure allows us to perform Monte Carlo deposition calculations for inhaled aerosols. Despite the vast amount of data stored in these files, still not all questions can be answered. Also the problem of enhanced deposition at bronchial airway branching sites (particularly at carinal ridges) needs still further investigation. Thus the Monte Carlo code is written in modular form to allow updating of deposition equations and morphometric data.

References

D'Arcy Thompson, W. (1942) Growth and Form, University Press, Cambridge.

Heyder, J., Gebhart, J., Roth, C., Stahlhofen, W., Stuck, B., Tarroni, G., De Zaiacomo, T., Formignani, M., Melandri, C. (1978) "Intercomparison of Lung Deposition Data for Aerosol Particles", J. Aerosol Sci. 9:147.

Heyder, J., Gebhart, W., Stahlhofen, W. (1980) "Inhalation of Aerosols: Particle Deposition and Retention", Generation of Aerosols and Facilities for Exposure Experiments (edited by K. Willeke), Ann Arbor Science Publ. Inc., Ann Arbor.

Hofmann, W., Koblinger, L., Daschil, F., Fehér, I., Baláshazy, I. (1984) "Stochastic Lung Dose Modelling", Proc. XI Regional IRPA Congress, Vienna (in press).

Horsfield, K., Cumming, G. (1967) "Angles of Branching and Diameters of Branches in the Human Bronchial Tree", Bull. Math. Biophys. 29:245.

Horsfield, K., Dart, G., Olson, D.E., Filley, G.F., Cumming, G. (1971) "Models of the Human Bronchial Tree", J. Appl. Physiol. 31:207.

Raabe, O.G., Yeh, H.C., Schum, G.M., Phalen, R.F. (1976) Tracheobronchial Geometry: Human, Dog, Rat, Hamster, Lovelace Foundation report LF-53, USA.

Tarroni, G., Melandri, C., Prodi, V., De Zaiacomo, T., Formignani, M., Bassi, P. (1980) "An Indication on the Biological Variability of Aerosol Total Deposition in Humans", Am. Ind. Hyg. Assoc. J. 41:826.

Weibel, E.R. (1963) Morphometry of the Human Lung, Springer Verlag, Berlin.

Yeh, H.C., Schum, G.M. (1980) "Models of Human Lung Airways and Their Application to Inhaled Particle Deposition", Bull. Math. Biol. 42:461.

1054

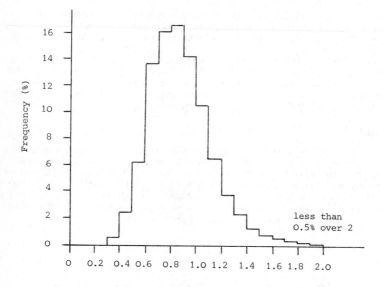

Figure 1. Ratio of the parent tube cross section
to the cross section of both daughters

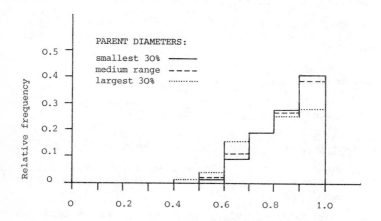

Figure 2. Ratio of the diameters of the
minor and the major daughters

Published 1984 by Elsevier Science Publishing Co., Inc.
Aerosols, Liu, Pui, and Fissan, editors

Evaluation of the Concentration and Size Distribution of Aerosols
Greater Than 5 Micrometers in the Hazelton Animal Exposure Chamber

F. Pisano and M. McCawley
West Virginia University
Morgantown, WV   26506   U.S.A.

Introduction

Uniformity of concentration and size distribution has been
characterized for particle sizes < 1 µm in existing medium-sized aerosol
exposure chambers (Willeke, 1980). Particles deposited in the lungs,
however, may be as large as 25 µm (ISO, 1983). Therefore, in order to
closely simulate the atmosphere of an underground coal mine, where the
distribution is bimodal, with modes at 5 and 15 µm (Burkhart, et al.,
1983), particles much greater than 1 µm must be characterized for uniform
mixing and size distribution within these chambers. The purpose of our
study was to characterize particle sizes in the range of <5 µm, 5-10 µm
each, and 10-20 µm in the Hazelton animal inhalation exposure chamber.

Theoretical

Polydisperse aerosols of coal dust with a mass median aerodynamic
diameter of >5 µm were used for exposure studies involving Syrian Golden
Hamsters in Hazelton Animal Exposure Chambers. The concentration and
size distribution of the aerosol was evaluated by light-scattering using
a Continuous Aerosol Monitor (CAM) and Cascade Impactors. A comparison
was also made of the results obtained using both methods. Since it was
assumed that there was a large difference between the upper and lower
portions of the chamber, the animal exposure procedure was organized to
assure uniform dosing.

This research involved the first reported use of the Marple Personal
Impactor for exposure chamber evaluation. The small size of the personal
impactor allowed as many as twelve different sites to be monitored
simultaneously within the chamber for the duration of the exposure.   It
was also possible to attach a personal impactor behind the sensor in the
CAM for direct comparison of the two methods.

The aerosol delivery system using respirable coal mine dust and
calibration of the monitoring equipment was tested. Specifically, dust
(a high volume airborne sample, collected at a facility mining the
Pittsburgh Seam near Morgantown, West Virginia) was classified with an
aerodynamic p rticle size classifier. The approximate cut was intended
to be between 5-10 micrometers, median size. Approximately one liter of
this dust was resuspended by the TSI Model 9310 fluidized bed aerosol

1056

generator and introduced into the Hazelton HIC-1000 exposure chamber.
Four sampling runs were made on separate days to determine the day-to-day
variability of the dust size distribution. This data is presented as a
frequency distribution in Figure (1). Dust concentrations in the chamber
without cages and pans, were calculated using side-by-side impactors in
four runs. These data are given in Table 1.

Table 1 along with Figure 1 show that varying concentrations in
various locations did not effect the size distribution appreciably.
Although the mass median aerodynamic size distribution ranged from 10 to
11 micrometers in four runs, this was within the acceptable bounds for
the classifier. Therefore, the use of the TSI model 9310 fluidized bed
aerosol generator, and the aerodynamic particle size classifier assured
little variability in particle size distribution from run to run.

Further tests were done on the size distribution between layers in
the chamber with animal cages and pans in place. Twelve personal
impactors were placed six to a level on two levels. These were the only
levels intended to be used for animal exposure. The results (Figure 2)
showed little difference in size distribution throughout the chamber.

Conclusions

The aerosol generation system can deliver a uniform concentration and
size distribution throughout the chamber. Therefore, a monitor needs to
be put in only one location in order to characterize the animal exposures
in the entire chamber.

References

1. Burkhart, J., M. McCawley, and R. Wheeler (1983) "Particle Size
Distribution in Underground coal Mines" presented at American Industrial
Hygiene Conference, Philadelphia, PA.

2. International Standards Organization (1983) "Air Quality-Particle
Size Fraction Definitions for Health Related Sampling" Technical Report
ISO/TR 7708-1983.

3. Willeke, K. (1980) Generation of Aerosols and Facilities for
Exposure Experiments. Ann Arbor Science, Ann Arbor, MI.

Table 1. Variability in mean mass concentration.

Run #	Mean mass conc. (mg/m³)	(S.D.)	(N)
1	10.97	(.90)	(3)
2	8.07	(.24)	(3)
3	7.13	(.23)	(3)
4	6.60	(.52)	(2)

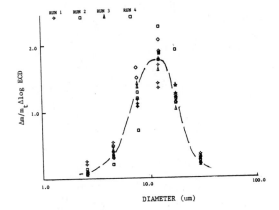

Figure 1. Variability in size distribution with change in concentration. Side-by-side impactors with no pans or cages in chambers.

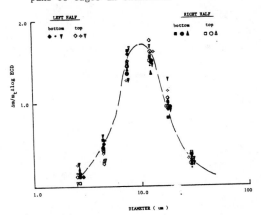

Figure 2. Variability in size distribution from simultaneous impactor samples. Samples taken with two levels of trays and pans in place.

AUTHOR INDEX

SUBJECT INDEX